EVOLUTION OF THE EYE AND VISUAL SYSTEM

Vision and Visual Dysfunction

General Editor Professor John Cronly-Dillon
Dept of Optometry and Vision Sciences, UMIST, Manchester, UK

Volume 1. Visual optics and instrumentation
Edited by W. N. Charman

Volume 2. Evolution of the eye and visual system
Edited by J. R. Cronly-Dillon and R. L. Gregory

Volume 3. Neuroanatomy of the visual pathways and their development
Edited by B. Dreher and S. R. Robinson

Volume 4. The neural basis of visual function
Edited by A. G. Leventhal

Volume 5. Limits of vision
Edited by J. J. Kulikowski, V. Walsh and I. J. Murray

Volume 6. The perception of colour
Edited by P. Gouras

Volume 7. Inherited and acquired colour vision deficiencies
Edited by D. H. Foster

Volume 8. Eye movements
Edited by R. H. S. Carpenter

Volume 9. Binocular vision
Edited by D. Regan

Volume 10. Spatial vision
Edited by D. Regan

Volume 11. Development and plasticity of the visual system
Edited by J. R. Cronly-Dillon

Volume 12. Visual agnosias
By O-J. Grüsser and T. Landis

Volume 13. Vision and visual dyslexia
Edited by J. F. Stein

Volume 14. Pattern recognition by man and machine
Edited by R. J. Watt

Volume 15. The man-machine interface
Edited by J. A. J. Roufs

Volume 16. The susceptible visual apparatus
Edited by J. Marshall

Volume 17. Index

VISION AND VISUAL DYSFUNCTION
VOLUME 2

Evolution of the Eye and Visual System

Edited by

John R. Cronly-Dillon

Dept of Optometry and Vision Sciences
UMIST, Manchester, UK

and

Richard L. Gregory

Dept of Psychology
University of Bristol, UK

CRC Press, Inc.
Boca Raton   Ann Arbor   Boston

© The Macmillan Press Ltd 1991

All rights reserved. No reproduction, copy or transmission of this publication
may be made without written permission.

Any person who does any unauthorized act in relation to this publication
may be liable to criminal prosecution and civil claims for damages.

First published 1991

Published in the USA, its dependencies, and Canada by
CRC Press, Inc.
2000 Corporate Blvd., N.W.
Boca Raton, FL 33431, USA

Typeset in Monophoto Ehrhardt by August Filmsetting, Haydock, St Helens, UK
Printed and bound in Great Britain

Library of Congress Cataloging-in-Publication Data
Vision and visual dysfunction/edited by John Cronly-Dillon.
   p. cm.
   Includes index.
   ISBN 0-8493-7500-2 (set)
   1. Vision.   2. Vision disorders.   I. Cronly-Dillon, J.
   [DNLM: 1. Vision.   2. Vision Disorders.   WW 100 V831]
QP474.V44
612.8'4—dc20
DNLM/DLC
Library of Congress   90–1881
                                                                            CIP

Evolution of the eye and visual system / edited by R. Gregory and J.
   Cronly-Dillon.
      p. cm.— (Vision and visual dysfunction; v.2)
   Includes index.
   ISBN 0-8493-7502-9
   1. Eye–Evolution.   I. Gregory, R. L. (Richard Langton)
   II. Cronly-Dillon. J.   III. Series.
   [DNLM: 1. Evolution.   2. Eye.   3. Perception.   WW 100 V831 v. 2]
QP474.V44 vol. 2
[QP475]
612.8'4 s—dc20
[591.1'823]
DNLM/DLC
for Library of Congress   90–1883
                                                                       CIP

Contents

Preface   ix
The contributors   xi

PART I: GENERAL CHARACTERISTICS

1. Evolution of visual behaviour   3
J. N. Lythgoe
Introduction   3
Evolution of eye structure   3
Physical and physiological constraints   5
Co-evolution   12

2. Origin of invertebrate and vertebrate eyes   15
J. R. Cronly-Dillon
Introduction   15
Morphological characteristics of animal photoreceptor organs   19
Vertebrates evolved from a nemertine-like ancestor   29
Origin of pineal and parietal eyes   31
Cephalic organ   35
Origin of the vertebrate lens   44
Evolution of decussation of the vertebrate visual pathway   46
Concluding remarks   48

3. Origins of eyes – with speculation on scanning eyes   52
R. L. Gregory
Introduction   52
Eye and brain: a hen-and-egg problem   53
Simple and compound eyes   55
Scanning eyes   55

PART II: EVOLUTIONARY DEVELOPMENTS

4. The evolution of vertebrate visual pigments and photoreceptors   63
J. K. Bowmaker
Introduction   63
Visual pigments   64
Photoreceptors   65
Evolution of photoreceptors and visual pigments   66
Conclusions   78

5. The vertebrate dioptric apparatus — 82
W. N. Charman

Introduction — 82
Basic optics — 82
Physical constraints on optical design — 86
Limits on optical image quality at the retina — 92
Overall retinal image quality — 99
Retinal aspects — 100
Depth-of-focus, ametropia and accommodation — 104
Conclusion — 112

6. Optics of the eyes of the animal kingdom — 118
M. F. Land

Introduction — 118
Camera eyes with corneal lenses — 119
Aquatic lens eyes — 121
Multi-element lenses — 122
Concave mirror optics — 123
Biological mirrors — 126
Compound eyes — 127
Apposition optics — 127
Superposition optics — 128
Conclusions — 133

7. Considering the evolution of vertebrate neural retina — 136
I. Thompson

Introduction — 136
Photoreceptors — 137
Between the plexiform layers — 139
Retinal ganglion cells — 141
General observations — 147

8. Functions of subcortical visual systems in vertebrates and the evolution of higher visual mechanisms — 152
D. J. Ingle

Introduction — 152
Visuomotor systems in amphibia — 153
Modulation of tectal functions by satellite visual structures — 157
Comparisons of motion sensitivity between fish and mammals — 160
Shape recognition in vertebrates: advantages of a visual cortex — 162

9. Neural control of pursuit eye movements — 165
R. Eckmiller

Introduction — 165
Sensory inputs for smooth pursuit system — 167
Possible role of cerebral cortex in smooth pursuit control — 174
Neural activity in the cerebellum and brain stem — 180
Missing links and model considerations — 189
Conclusions — 191

10. Patterns of function and evolution in the arthropod optic lobe — 203
D. Osorio

Introduction — 203
Organization of visual processing — 204
Evolution — 212

11. Evolution of visual processing — 229
G. A. Horridge

Introduction — 229
The laws of optics govern eye evolution — 230
The evolution of vision — 237
Insect vision — 240
The evolutionary approach — 241
The template model — 243
Other models — 251
Two-dimensional vision — 253
Semi-vision performance — 254
Larger templates — 260
Whole-eye templates — 263
Beyond semivision — 267
Conclusion — 268

12. Evolution of binocular vision — 271
J. D. Pettigrew

Introduction — 271
Hallmarks of binocular vision — 271
Cats, owls, monkeys and machines — 276
Evolution of avian stereopsis — 280
Stereopsis in other vertebrate groups? — 281

13. Evolution of colour vision — 284
C. Neumeyer

Introduction — 284
Ability to discriminate wavelengths — 285
Dimensionality of colour vision — 289
Phylogeny of colour vision — 295
Comparative colour perception — 297
The question "Why?" — 298

14. Uses and evolutionary origins of primate colour vision — 306
J. D. Mollon

Introduction — 306
Advantages of colour vision — 306
Trichromacy and its evolution — 310
Polymorphism of colour vision in platyrrhine primates — 314

PART III: PHYLOGENETIC EVOLUTION OF THE EYE AND VISUAL SYSTEM

15. Photosensory systems in eukaryotic algae — 323
J. D. Dodge

Introduction — 323
Description of the eyespots and associated structures — 323
Function of the photosensory apparatus — 334
Evolution of the photosensory apparatus — 336

16. Evolution of the cellular organization of the arthropod compound eye and optic lobe — 341
I. A. Meinertzhagen

Introduction — 341
The compound eye — 342
Evolution of the ommatidium — 344
The optic lobe — 353
Conclusions — 360

17. Photoreception and vision in molluscs
J. B. Messenger — 364
- Introduction — 364
- Classification — 364
- Extra-ocular photoreception — 364
- Non-cephalic eyes — 368
- Cephalic eyes — 371
- Visual processing — 383
- Summary — 391

18. Evolution of vision in fishes
S. M. Bunt — 398
- Introduction — 398
- Constraints on evolution — 398
- Deuterostomes — 399
- Agnatha — 401
- The transition from the agnathan to the gnathostome — 405
- The aquatic visual environment — 407
- Conclusions — 417

19. Central visual pathways in reptiles and birds: evolution of the visual system
T. Shimizu and H. J. Karten — 421
- Introduction — 421
- Phylogeny: reptiles, birds and mammals — 421
- Basic Sauropsida plan (reptiles and birds) — 423
- Reptilian plan — 426
- Avian plan — 429
- Evolution of the visual system — 435
- Conclusion — 436

20. Evolution of mammalian visual pathways
G. H. Henry and T. R. Vidyasagar — 442
- Introduction — 442
- Mammalian characteristics in the primary visual pathway — 442
- Comparative view of features of the visual pathway — 444
- Conclusions — 459

Appendix. Mammals
G. L. Walls — 467
- Introduction — 467
- Monotremes and marsupials — 467
- Placentals — 472
- Tentative schema of the evolution of the visual cells in vertebrates — 480

Index — 483

Preface

Three related questions have dominated our enquiry into the nature of the living world. Briefly these are concerned with:

1. The origin and nature of life itself.
2. The origin and diversity of living forms.
3. The problem of growth and development.

Historically, concern over such questions led to two doctrines: namely those reflecting the *creationist* and *evolutionary* schools of thought. The former proposed that the diverse forms of life were created as they are, by a transcendental God. *A fortiori* this led to the preformationist-type view of development in which the zygote was considered to be simply a miniature individual that grows. Indeed the creationist view of development was neatly summarized almost 2000 years ago by the Roman philosopher Seneca, who declared that 'In the seed are enclosed all the parts of the body of the man that shall be formed'.

By contrast, evolutionists considered speciation as being concerned with the transmutation of living forms (a view originally championed in 1721 by the French philosopher Montesquieu). Basically the position was that the various types of animals and plants have their origin in other pre-existing types, where the distinguishable differences are due to modifications of these pre-existing forms. However, it was only when the idea of natural selection was introduced to explain the origin and diversity of species that Darwin and Wallace's great evolutionary idea established itself as the cornerstone of all modern thinking in biology. The view they propounded was that major phyletic changes occur primarily through the accumulation of small modifications by natural selection of inheritable characteristics. As to the problem of growth and development: it now seems almost inevitable (given the resemblance between the transformations undergone by a zygote through the various stages of development, and the emergence of new species in the process of speciation) that sooner or later evolutionists would be lead into suggesting a relationship between the stages of ontogeny and those occurring in phylogeny.

Recapitulationist theories, such as those of Haeckel and Agassiz (who opposed Darwin), argued in favour of a view that stages of ontogeny repeat the *adult* forms of animals lower down the scale of organization; a notion that preserves certain features of preformationism. Indeed for the recapitulationist, evolutionary change proceeds by adding (preformed) stages to the end of ancestral ontogeny. Recapitulationism also postulated that development is accelerated as ancestral features are pushed back to earlier stages of descendant embryos. By contrast other evolutionary theories, such as that proposed by Von Baer (which was favoured by Charles Darwin), accepted the idea of the transmutation of species. Von Baer proposed that increasing differentiation calls only upon a conservative principle of heredity to preserve stubbornly the early stage of ontogeny, while evolution proceeds by altering later stages. Thomas Hunt Morgan wrote 'that the resemblance between certain stages in the embryos of higher animals and corresponding stages in the embryos of lower animals is most plausibly explained by the assumption that they have descended from the same ancestors'.

Tracing the lineage of the vertebrate eye has proved to be one of the most challenging and intractable problems encountered by evolutionary biologists. Traditionally, those seeking to establish a homologous sequence attempted to do so by comparing structures or organs in different species which appear to serve broadly similar functions. While the application of this approach has produced many possible indications as to the eye's origin, the problem remains unresolved. Indeed Darwin himself singled out the eye as one of the structures that presented some difficulty to his theory, even though he retained the firm conviction that it was only a matter of time before a proper sequence of homologies was established, and the problem solved. Part of the difficulty was the lack of fossil evidence. Because soft tissues perish and disintegrate, early evolutionary biologists were forced to turn to seemingly related living species and attempt to guess at appropriate homologous sequences, by comparing and observing similarities between gross anatomical and morphological features. Another approach was to exploit the suggested relationship between ontogeny and phylogeny to provide an index of the possible lineage of some of the soft tissues, including the eye.

In modern times, the elucidation of the structure of DNA, and developments in cell and molecular biology, have enabled biologists to establish a firm logical framework on which to rest the suggested relationship between the three central questions of biology. In addition, modern techniques for distinguishing the fine similarities and differences between tissues, cells, subcellular organization and molecular structure (and architecture) of cells greatly extended the range of homologies which evolutionary biologists could look for.

For Darwin, and others, the approach of trying to account for the details of structure and function through similarities and differences in adaptation, although insufficient by itself, was a useful adjunct in tracing ancestral relationships. This indeed was the approach used by Gordon Walls in his monumental treatise *The Vertebrate Eye and its Adaptive Radiation*. Since that seminal work was written, a great deal has been learned of the structure and functional organization of the eye and visual system in a variety of animal species, which makes a re-evaluation of the evidence timely.

The aim of this volume is to try and present, in an evolutionary context, a perspective of the vast accumulation of data on the structure and function of the visual system in different animal species that has accrued over the last 30 years. Inevitably there are some gaps, some intended, some unavoidable, and others simply due to oversight. Our intention is to correct for these omissions in the 'update volumes' to follow in this series. Meanwhile, the chapters in this book have been grouped under three sections:

Part 1 deals with general characteristics, beginning with an account of the evolution of visual behaviour by J. N. Lythgoe. It also includes two largely speculative chapters which attempt to paint a broad picture of the origin of animal eyes.

Part 2 is concerned primarily with the structural and functional adaptations in each of the subsystems of the visual system, in an evolutionary context. Each of the chapters therefore examines the evolution of a particular visual structure and/or function, e.g., photoreceptors and photopigments, dioptric system, visual processing, binocular vision, colour vision, etc.

Part 3 discusses the evolution of the visual system (both structure and function) in eukaryotic algae and in specific groups of animals, e.g., arthropods, molluscs, fishes, reptiles, birds and mammals. Also included is an extract on mammalian vision reprinted from Gordon Walls' classic 1942 monograph, *The Vertebrate Eye and its Adaptive Radiation*.

Finally the editors wish to thank the many colleagues who advised or assisted in the preparation of this volume; and especially to the following for their advice, and trouble they have taken in refereeing the contributions: T. R. Vidyasagar, L. Margulis, V. Walsh, J. J. Kulikowski, G. H. Hardie, M. Land, D. H. Paul, K. Ruddock and M. Tyrer.

J. R. Cronly-Dillon
R. L. Gregory
Manchester and Bristol
1991

The Contributors

J. K. Bowmaker
Dept of Visual Science
Institute of Ophthalmology
University of London
Judd St
London WC1H 9QS
UK

S. M. Bunt
Dept of Anatomy & Physiology
University of Dundee
Dundee DD1 4HN
UK

W. N. Charman
Dept of Optometry & Vision Sciences
UMIST
PO Box 88
Manchester M60 1QD
UK

J. R. Cronly-Dillon
Dept of Optometry & Vision Sciences
UMIST
PO Box 88
Manchester M60 1QD
UK

J. D. Dodge
Dept of Biology
Royal Holloway and
Bedford New College
University of London
Egham Hill
Egham
Surrey TW20 0LB
UK

R. E. Eckmiller
Dept of Biophysics
University of Düsseldorf
Universitätstr. 1
4000 Düsseldorf
Germany

R. L. Gregory
Dept of Psychology
University of Bristol
8–10 Berkeley Square
Bristol BS8 1HH
UK

G. H. Henry
John Curtin School of Medicine
Australian National University
GPO Box 334
Canberra
ACT 2601
Australia

G. A. Horridge
Centre for Visual Sciences
Australian National University
GPO Box 475
Canberra
ACT 2601
Australia

D. Ingle
Dept of Psychology
Boston College
Chestnut Hill
MA 02167
USA

H. Karten
Dept of Neuroscience
University of California School of Medicine at
San Diego
La Jolla
CA 92093
USA

M. F. Land
School of Biological Sciences
University of Sussex
Falmer
Brighton
UK

J. N. Lythgoe
Dept of Zoology
School of Biological Sciences
University of Bristol
Woodlands Road
Bristol BS8 1UG
UK

I. A. Meinertzhagen
Life Science Centre
Dalhousie University
Halifax
Nova Scotia B3H 4J1
Canada

J. B. Messenger
Dept of Animal and Plant Sciences
University of Sheffield
Western Bank
Sheffield S10 2TN
UK

J. D. Mollon
Dept of Experimental Psychology
University of Cambridge
Downing St
Cambridge CB 3EB
UK

C. Neumeyer
Institut für Zoologie
Johannes Gutenberg-Universität
Saarstr. 21
6500 Mainz
Germany

D. Osorio
Centre for Visual Sciences
Australian National University
Canberra
ACT 2601
Australia

J. D. Pettigrew
Dept of Physiology & Pharmacology
University of Queensland
Brisbane
Queensland 4072
Australia

T. Shimizu
Dept of Neuroscience
University of California School of Medicine at San Diego
La Jolla
CA 92093
USA

I. Thompson
University Laboratory of Physiology
University of Oxford
Parks Road
Oxford OX1 3PT
UK

T. R. Vidyasagar
John Curtin School of Medical Research
Australian National University
GPO Box 334
Canberra
ACT 2601
Australia

PART I

GENERAL CHARACTERISTICS

1 Evolution of Visual Behaviour

J. N. Lythgoe

Introduction

The visual capacity of all animals is limited by the physical nature of light. Using ourselves as an example, the difficulty in playing fast ball games like tennis in the evening is a consequence of the photon nature of light, and our inability to see very small objects is a consequence of its wave nature (Snyder, 1977). At a somewhat less fundamental level, the optical qualities of an animal's environment can be crucial to its visual performance and many examples are known where the optical environment has influenced the evolution of visual systems and their associated behaviours.

At the extreme, animals that live in total darkness have to rely on senses that use other modalities such as sound and electrical fields (Lythgoe, 1977; Coombs *et al.*, 1988; Hopkins, 1988; Kalmijn, 1988). For such animals the investment in eyes is no longer worth the return and in the course of evolution their visual sense has dwindled to nothing or almost nothing. For most other animals the optical environment and the visual tasks they have to carry out have caused various specializations in the visual apparatus. The physical environment does not evolve in the Darwinian sense. However, organisms living in it interact with each other and the evolution of one is linked to the evolution of the others.

Evolution of Eye Structure

It is reasonable to think that all eyes have evolved from the simple light-sensitive eye spot found today in many protozoa. If such an eye has an opaque backing of pigment, it will respond to light from one hemisphere only, and two eye spots with different orientations should give an animal information about the general direction of very large areas of light and shade. It is debatable whether such very simple photoreceptors qualify as eyes. However, the next stage in complexity, the simple eye-cup, is certainly an eye and may be the basic type from which all other eyes have evolved (Land, 1981, 1990). Simple eye-cups are found in many different phyla including flatworms, annelid worms, molluscs, crustacea, echinoderms, and the chordates which are on the same evolutionary branch as ourselves. Eye-cup eyes have up to 100 photoreceptors lining the base of the cup, and the resolution of the eye is limited by the angle subtended at each receptor by the rim of the cup. This is typically between 10° and 90°. Such eyes enable an animal to navigate with respect to a light source, but is unlikely to be much use for image-forming vision (Land, 1990).

An essential step in the evolution of eyes is the creation of an image-forming mechanism. To over-simplify, there are four strategies that have been used to improve on the image formed by simple eye-cup eyes (Fig. 1.1): (a) deepen the eye cup to form a pinhole eye; (b) team many eye cups together to form a compound eye; (c) use a lens to project a good-quality image of the visual scene onto an array of photoreceptors; (d) line the concave back of the eye with reflecting cells to project an image onto a photoreceptor array within the eye. Vertebrates have only camera-like lens eyes, but the protochordates, which are more lowly members of the same phylum (Chordata) have simple eye-cup eyes. The variety of eye types can be quite large in other phyla such as the Arthropoda which have all the main types of eye.

Pinhole eyes seem to be something of an evolutionary cul-de-sac. They are found in a few molluscs including the abalone *Haliotis*, which spends its adult life permanently fixed to a rock and may not have great need of an advanced visual system. However, a relatively enormous pinhole eye is found in the primitive cephalopod *Nautilus*. *Nautilus* is a 'living fossil' with a chambered spiral shell which bears a close resemblance to the fossil ammonites. The *Nautilus* eye is very large, almost 1 cm across, and appears to be quite sophisticated in everything except its optics. For example, its pupil aperture can change with light intensity and the eye itself is kept horizontal, even though the animal itself rocks back and forth as it swims (Hartline and Lange, 1984). The problem with pinhole eyes is that a sharp image requires a small pupil aperture, which in *Nautilus* inevitably results in an image two orders of magnitude less bright than the image cast by the lens eye of a

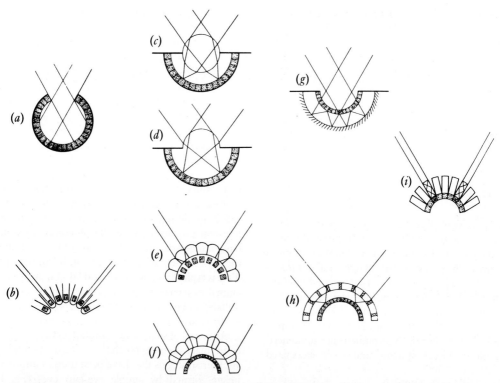

Fig. 1.1 *Nine different methods of image formation. (a) and (b) rely on shadowing: common in polychaete worms, flatworms, gastropod molluscs, crustaceans and chordates. (c)—(f) rely on refraction; present in cephalopod molluscs and vertebrates. (g) and (h) rely on reflection; lamellibranch molluscs, some crustacea including shrimps and prawns (Malacostrica). (i) relies on total internal reflection and is found in some Malacostraca (Land, 1981).*

fish. Increasing the pupil aperture results in a loss of sharpness of the image (Land, 1990). The optics of the *Nautilus* eye do not seem to be ideal, but it has survived unchanged for 400 million years, and it cannot have suffered too badly from poor optical design.

Compound eyes are usually found in arthropods, although they also occur in sabellid tubeworms and in at least three genera of bivalve molluscs. The characteristic faceted appearance of compound eyes is due to their construction from an array of optic units which usually consist of a lens and eight light-sensitive rhabdomeric receptor cells. In most cases except the dipteran flies (Strausfeld and Nassel, 1981), each set of rhabdomeres functions as a single unit, and all spatial information within the ommatidium is lost. There are two main types of compound eye: apposition and superposition. In the former, each ommatidium is optically isolated; in the latter, a single image is projected onto the whole array of photoreceptors. Apposition eyes are generally associated with bright light conditions and superposition eyes with vision in twilight (Land, 1981).

Single-lens eyes are most developed in vertebrates and in octopus, cuttlefish and squids, but are found in several other groups of invertebrates including spiders. In vertebrates and cephalopods there is an internal lens eye, and in spiders the cornea itself is the only lens. Lenses based on the external curvature of the cornea will only work in air because the refractive index of air, and the material of the eye is too similar to form an effective lens. However, many compound eye lenses owe their focusing power to layers of different refractive index within the lens itself (Exner, 1891; Nilsson, 1982), and many spherical lenses owe their aberration-free optics to the same principle (Land, 1981).

Reflector eyes are found among simple and compound eyes. In reflector compound eyes, the facets are four-sided and the sides act as mirrors forming an image in the superposition manner (Vogt, 1980; Land, 1981). This type of eye is found chiefly in shrimps and prawns. Quite different is the rare type of reflector eye found in the scallop *Pecten* (Land, 1965). Here the light is reflected from the back of the eye to form an image on a concentric layer of receptors deeper within the eye. In some ways this type of eye resembles the situation in vertebrate camera eyes, like those of the cat, which have a reflecting tapetum lining the back of the eye, from which the light that has traversed the photoreceptors without being absorbed is directed back again for a second chance of absorption.

All the examples quoted above depend upon the projection of an image onto an array of photoreceptors. The planktonic crustacean *Copilia* and its relatives are some of the very few exceptions (Gregory, Chapter 3). *Copilia* uses a vibrating lens to scan across the visual scene and project elements of the image onto a small group of receptor cells.

Physical and Physiological Constraints

The Nature of Light

Wavelength Nature of Light
Visual acuity is limited by the wave nature of light in two important ways; interference of light passing through small apertures limits the amount of detail in the retinal image, and the wavelength of light imposes a lower limit on the diameter of photoreceptors, thus limiting the amount of information they can read from the retinal image (Snyder, 1977; Land, 1981).

The retinal image of a point of light is not a simple spot, but is surrounded by concentric dark and light rings which decrease sharply in intensity. The bands are caused by light that has passed through one point of the aperture, interfering with light that has passed through another point. The radius of the first dark ring is given by the equation

$$r = 1.22 \, \lambda f / A \qquad (1.1)$$

where f is the focal length of the eye and A is the diameter of the aperture. It is conventionally considered that two images can be distinguished only if they are separated by more than r, which depends upon the ratio f/A and the wavelength of light (λ).

It does not matter how fine is the detail in the retinal image if the photoreceptors are spaced too far apart to resolve it. If the spacing between adjacent receptors is greater than r then diffraction is not a limiting factor. In fact the lower limit of receptor diameter, which is probably set, at least in part, by the waveguiding properties of very narrow cylinders, appears to be in the range 1–1.5 μm. For humans $f = 16.7$ mm and $A = 2$ mm, and r is about 5 μm, or about twice the spacing of the retinal cones. In fact most animals have f-numbers (A/f) of less than 4, which means that diffraction is not usually the limiting factor for resolution, although many eyes do approach that limit. Although the diffraction limit to image quality is set by the proportions, rather than the absolute size of the eye, the smallest practicable diameter of the photoreceptors remains constant. Thus it is always possible to get more information from a large retinal image than a small one, and in general large eyes give the larger retinal images.

Many photoreceptor cells like the outer segments of vertebrate rods and cones act as light guides, directing the light along their length where the visual pigment molecules are arrayed to catch it. When the diameter of a light-conducting fibre approaches the wavelength of light, interference effects within the fibre become important. Only a proportion of the light travelling down these very narrow fibres passes within the fibre; the rest is guided down the outside, where it may be absorbed by neighbouring photoreceptors or by the surrounding media (Kirschfeld and Snyder, 1975). Few, if any, photoreceptors are known that are less than about 1000 nm (1 μm) in diameter; they would not improve visual acuity and would be inherently insensitive.

Photon Nature of Light
Light travels through space as a wave, but its energy is transferred to matter as discrete parcels of energy called photons. Flies may be able to see the arrival of individual photons (Laughlin, 1981), although we humans cannot. However, there are commercial and military image intensifiers which allow us to 'visualize' individual photons and it is clear that humans, and other animals as well, are limited in dim light by the number of photons that are available to form an image.

The image on the retina is partitioned into picture elements, which cannot be more numerous than the number of photoreceptors in the retinal mosaic. A basic visual task is to distinguish one element of the image from its neighbours and to mark changes in the same element from one instant in time to the next. The time and place of arrival of individual photons can be predicted only as a statistical probability, and whether or not two image elements can be distinguished becomes a statistical question. As with any other statistical question, the larger the sample, the finer the differences that can be detected.

The obvious way to increase the photon count from each element is to arrange for a bright retinal image. For ordinary scenes which involve extended sources of light, the brightness of the image depends on the angle subtended by the pupil aperture at the retina (f/A). It follows that the image brightness of a small eye is the same as that of a larger one of the same proportions. However, when there

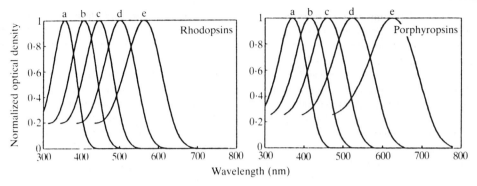

Fig. 1.2 *Absorbance spectra of five rhodopsins compared with their porphyropsin analogues. At longer wavelengths the porphyropsins become broader and sensitive to longer wavelengths than their rhodopsin analogues (Lythgoe and Partridge, 1989).*

are only point sources of light, then the image brightness is proportional to the area of the aperture.

The number of photons counted from each element of the image can be increased by enlarging the area of each element, by increasing the time over which the counts are made, by increasing the spectral bandwidth over which they are collected and by increasing the effective optical density of the receptor pigment.

Increasing the integration area can be achieved by enlarging the cross-sectional area of the photoreceptors, or by summing the information from many photoreceptors either neurally or by grouping receptors into larger optical units as in some deep-sea fish (Locket, 1971). The penalty for solutions of this kind is that the image is divided into fewer elements and the ability to see fine detail is lost.

Increasing the time over which photons are counted (the retinal integration time) carries the penalty of reduced temporal resolution and hence a reduced ability to see fast-moving objects (Lythgoe, 1979, for a review).

Increasing the spectral bandwidth of light that is collected may be of limited effectiveness because visual pigments have broad absorption spectra (Fig. 1.2). Increasing the packing density of rhodopsin molecules within the photoreceptor does not seem to be an option, but increasing the pathlength through the pigment-containing photoreceptor will increase absorption at all wavelengths, so that eventually all wavelengths will suffer 100% absorption (Knowles and Dartnall, 1977). In deep-sea fish and some nocturnal species living in less deep water, the retinal rods are arranged in banks so that light that is not absorbed by the first bank of rods is likely to be absorbed by the second, third or fourth. An intriguing possibility of this multi-bank system is that the light reaching the first bank of rods will have a different spectral distribution from that traversing the next bank, and so on. This would give the fish the opportunity for colour vision, even though it has only one visual pigment (Denton, 1990).

All optical filters in the light path will reduce the number of photons that are absorbed by the visual pigment and thus reduce overall sensitivity. Many diurnal vertebrates have filters that block shortwave light in the lens, the cornea or the retina itself, although the reason for their presence is not certain (Muntz, 1972). They absorb UV light that may damage the retina, especially in long-lived animals. Shortwave light is more strongly scattered by the atmosphere and optical media and will thus contribute 'noise' to the signal. Another suggestion is that the chromatic aberration in the lens makes it more difficult to bring all wavelengths to the same point of focus, with a consequent degradation of the image.

Some animals, particularly birds and turtles, have sets of different-coloured oil droplets that filter the light reaching individual photoreceptors (Lythgoe, 1979; Goldsmith *et al.*, 1984, Partridge, 1989). There is a convincing case that these are involved in the mechanism of colour vision. Coloured oil droplets must reduce sensitivity, and it may be no accident that most birds are diurnal, and the few nocturnal birds like owls and night-hawks have very few coloured oil droplets in the retina. Also, cone oil droplets in birds correlate better with their ecology than with their phylogeny (Partridge, 1989).

Spectral Distribution of Visual Information

Many animals can see further into the ultraviolet than we can and others can see further into the infrared. Indeed, some fish are able to see further into both the ultraviolet and infrared. The long-wavelength limit to vision is likely to be set by the longest-wavelength visual pigment which appears to be a porphyropsin with a maximum absorption (λ_{max}) of around 630 nm (Lythgoe, 1988). The absorption of this pigment is at 10% of its maximum value at 740 nm, and useful vision is likely to extend at least as far as this into the near-infrared. Ultraviolet vision has been well established in insects, fish and birds by electrophysiology and microspectrophotometry (Bowmaker, Chapter 4 and Neumeyer, Chapter 14). The lower limit to shortwave vision is likely to be around 280 nm, where absorption by aromatic amino acids effectively prevents vision further into the ultraviolet. The shortwave-absorbing screening pigments found at various locations in the optical path in many diurnal animals may be to protect the eye from the damaging effects of ultraviolet light and limit ultraviolet vision where they occur (Muntz, 1972; Miller, 1979).

Information-carrying Wavelengths

Many animals have more than one type of photoreceptor that differ in spectral sensitivity. This may mean that the animal has true colour vision (see below), but it may also mean that particular photoreceptors are used for particular visual tasks that do not involve the perception of colour. Goldsmith has outlined several of these waveband-specific mechanisms. It is not always obvious why particular wavebands are used for particular tasks. For example, humans, in common with almost all other mammals except some deep-diving whales, dolphins and seals, have a rhodopsin in the rods with a λ_{max} near 500 nm (Lythgoe, 1979). It is natural to suppose that this rhodopsin gives the greatest possible sensitivity in dim light. This supposition is reinforced by the discovery that deep-sea fish have rhodopsins with a λ_{max} nearer 485 nm, which neatly matches the spectral distribution of available light in the deep oceans (Denton and Warren, 1957; Munz, 1958; Crescitelli *et al.*, 1985; Partridge *et al.*, 1989), whereas shallow-living coastal fishes have rod rhodopsins of λ_{max} near 500 nm. However, it can be calculated that the rhodopsins of terrestrial animals, and those of animals living in shallow marine and fresh water (Lythgoe, 1979), absorb at shorter wavelengths than would be expected if sensitivity was the only important factor (Fig. 1.3).

Animals do not 'care' about abstract notions of sensitivity; what matters is whether objects in the environment that are important to them can be detected and recognized. In dim light this usually means that the respective photon catches from an object (o) and its background (b) are sufficiently different to be distinguished:

$$c = (o - b)/(o + b) \quad (1.2)$$

For humans c needs to be greater than about 0.01 if the object is to be visible.

There now seems to be a general agreement that 'noise' is an important limiting factor in visual discrimination. This noise may come from the photon nature of light, when it is often called 'shot noise'; it may have an environmental origin, for example fog or temperature differences in the atmosphere, or it may have physiological origins within the photoreceptors or higher up the neural pathway (Allo *et al.*, 1988). Shot noise, which comes from the uncertain time and arrival of individual photons in dim light, has been described by Land (1981), and Laughlin (1981). It is different from noise of environmental and physiological origin, because it is an integral part of the signal itself. Environmental and physiological noise is additional to the signal, although it may be difficult to distinguish from the signal.

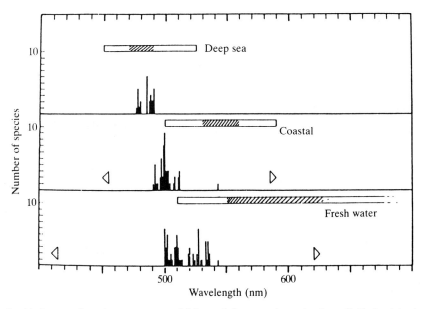

Fig. 1.3 *The relationship between the rod outer segments of fishes and the spectral wavebands available for vision in different aquatic environments. The shaded areas of the horizontal bars represent absorption maxima of visual pigments which would give greatest sensitivity to fishes living in those waters (Lythgoe, 1980).*

The reduction in visibility comes when a veil (v) of scattered light is interposed between the eye and the visual scene and the brightness of object and background become ($o+v$) and ($b+v$). On substituting into Equation 1.2, the contrast C is reduced:

$$C = (o-b)/v(o+b) \qquad (1.3)$$

The rate at which visual contrasts are reduced as an object recedes from the eye is described by the same equation underwater and on land:

$$C_z = C_0 e^{-cz} \qquad (1.4)$$

where C_0 is the contrast at zero range and C_z is the contrast at distance z. c is the narrow-band attenuation coefficient, and under water is heavily dependent upon wavelength (Duntley, 1963; Lythgoe, 1968). The reduction of visual contrasts by veiling scattered light under water is by far the most important factor that determines the range at which objects can be seen when there is reasonably good ambient light. In the clearest ocean water one large fish might be able to see another about 30 m away, but in turbid inshore water visibility might be reduced to 1 m or less.

Two important originators of physiological noise are the photoreceptors themselves (reviewed by Goldsmith, 1990) and synapses in the neural pathways (Laughlin, 1987). Physiological noise, like environmental noise, has the effect of reducing perceived contrasts and thus reduces visual discrimination. The effects of spurious signals are likely to be worse when the light is dim, and it is possible that they are one of the more important factors that limit visual performance. It has been noted several times that in most animals, vertebrate and invertebrate, that live on land or in water of moderate depth, tend to have scotopic pigments with λ_{max} in the 480–510 nm waveband. For example, all terrestrial mammals have rod rhodopsins absorbing near 500 nm. The reason for this is not clear, since they absorb at too short a wavelength to confer maximum sensitivity. Goldsmith considers the hypothesis that they are inherently more stable to chance isomerizations and he points out that we are obviously missing something important (Goldsmith, 1990).

Colour Vision

Colour is the sensation that the brain uses to report the shape of spectral radiance curves. Two or more receptor types must serve the same areas of the retinal image and the proportional difference in the number of photons that each receives is computed and coded as colour. There is good evidence from fish that the spectral quality of the environmental light has influenced their colour vision (Loew and Lythgoe, 1978; Levine and McNichol, 1979; Lythgoe and Partridge, 1989; Partridge, 1990). Fish live in an environment that is strongly coloured by the selective absorption of water and the coloured material that is dissolved and suspended in it to varying degrees.

Pure water is blue, owing partly to the selective absorption by the water molecule itself, and partly to Rayleigh scatter. In addition to the inherent blue colour, natural water contains two other important colouring materials: green chlorophyll and the yellow-brown products of organic decay that is variously called *Gelbstoff*, yellow substance or dissolved organic matter (DOM) (Jerlov, 1976). Within wide limits, water from any particular area has its characteristic colour. The waters of the open oceans, the Mediterranean and the Red Sea are clear and blue except where rivers bring nutrient-rich water into them. The coastal waters of temperate regions are green, owing partly to the presence of yellow dissolved organic matter. The green colour of chlorophyll often dominates in fresh water, although most fresh water is maximally transparent at longer wavelengths than most temperate coastal water. Fresh water that has leached through forest and swamp is often deeply stained by brown organic material which completely disguises the natural blue of the water itself. Such waters transmit most strongly in the red and the near infrared.

As daylight penetrates down through the water it suffers an exponential decay:

$$I_d = I_0 e^{-kd} \qquad (1.5)$$

where I_0 is the light at one depth and I_d is the light at the deeper depth d. k is the diffuse (broad-beam) attenuation coefficient and is strongly dependent on the wavelength. The result of this exponential decay is that light at the surface is fairly evenly distributed over the spectrum, but the waveband of available light becomes narrower as the depth increases. The light intensity also decreases and fish living in optically deep water have to cope with daylight that is both dim and restricted to a narrow waveband.

There have been reasonably extensive studies of the rod and cone visual pigments in coastal-marine and freshwater fishes (Loew and Lythgoe, 1978; Levine and McNichol, 1979; Lythgoe and Partridge, 1989). Fishes that live in very shallow water in either environment have visual pigments that are similar to those in land animals (Fig. 1.4). In rather deeper water, where the selective spectral absorption is significant, marine and freshwater fishes have different sets of cone pigments (Fig. 1.5). Fishes caught at between about 6 m and 60 m in the English Channel often have two rhodopsin cone pigments absorbing maximally at about 460 nm and 530 nm. Fishes living in various types of fresh water at an optically equivalent depth also tend to have two cone pigments, but these are porphyropsins absorbing maximally at about 540 and 600 nm (Fig. 1.6).

Animals that migrate from one optical environment to another may show corresponding shifts in their visual pig-

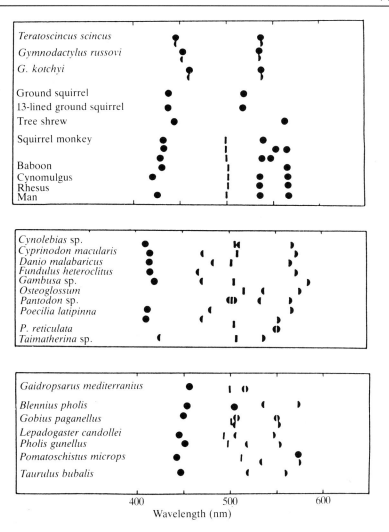

Fig. 1.4 *Comparison between the absorption maxima of visual pigments of terrestrial animals and those living in very shallow fresh water and very shallow coastal water. Note that they are all located in approximately the same region of the spectrum. Filled circles: single cones; half circles: one member of twin or paired cones; bars: rods (Lythgoe and Partridge, 1989).*

ments. Visual pigments consist of a protein (opsin) which has seven transmembrane segments to which is attached a prosthetic (chromophore) group. In vertebrates this is either retinal, forming a rhodopsin, or dehydroretinal, forming a porphyropsin. The effect is small at short wavelengths, but is as much as 60 nm at long wavelengths (Fig. 1.2). Rhodopsins are usually present in animals living in the sea or on land. Porphyropsins, which shift the visual pigment absorption to longer wavelengths, are more common in freshwater animals. The detailed amino acid sequence in the opsin also affects the spectral absorption of the molecule (see Goldsmith, 1990, for a review), although the detailed relationship between amino acid sequence and λ_{max} has yet to be worked out.

Some fish, like salmon and trout and the freshwater eel, that migrate between fresh water and the ocean show chromophore substitution (Beatty, 1984). The fully aquatic tadpole larvae of frogs have porphyropsins, while the more terrestrial adults have rhodopsins. Opsin changes have been observed in the eel and the northern Atlantic cod-like fish, the pollack. Eels spend their young adult life in fresh water and have typical freshwater porphyropsins. When they have grown sufficiently large, they begin their breeding migration, which takes them down the rivers, through the coastal waters of the continental shelf and then into the deep-blue ocean water of the Sargasso Sea. In anticipation of their time in coastal water, the porphyropsin in the rods change to a rhodopsin, giving a shift

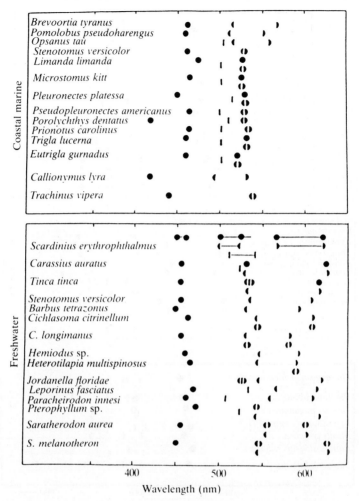

Fig. 1.5 *The visual pigments of various species of fish that live in moderate depths in coastal and fresh water. Long-wave-sensitive visual pigments are more common in freshwater fishes than marine fishes. Symbols as in Fig. 1.4 (Lythgoe and Partridge, 1989).*

in λ_{max} from 522 to 500 nm. In anticipation of their life in the deep sea, the opsin in the rods changes from the common 500-nm λ_{max} pigment to a 487-nm pigment typical of many deep-sea fishes. The pollack lives in the green coastal waters of the north-eastern Atlantic. The juveniles live near the surface and feed on plankters; the adults move deeper and feed on fish. The change in feeding habits coincides with a change in the λ_{max} of the blue-sensitive cone from 420 to 460 nm. This change is very likely to be due to a change in opsin (Shand et al., 1988). The reason for the change in the blue-sensitive cones but no other photoreceptors is not clear, but the presence of shortwave-sensitive visual pigments in many very shallow-living fishes may be associated with feeding on small plankters and insects in very shallow water (Bowmaker and Kunz, 1987; Lythgoe and Partridge, 1989).

The amino acid sequence of primate cone pigments indicates that the blue-sensitive rhodopsin is very ancient (Mollon, Chapter 14); the longwave-sensitive rhodopsins are less ancient, and, more recently on an evolutionary time scale, the longwave rhodopsin separated into red- and green-sensitive rhodopsins. Mathematical models (Lythgoe and Partridge, 1989), based on the assumption that a common visual task is to distinguish different types of green and brown plant material, indicate that for dichromats it is always important to have a blue-sensitive cone pigment teamed with one sensitive to longer wavelengths. It is always important that the blue cone is most sensitive in the 420–450-nm range. The spectral location of the longwave-sensitive cone is less critical, but appears to depend on whether it is the chromaticity of foliage that needs to be distinguished when a pigment in the 510–520-nm spectral region is best, or the chromaticity of brown

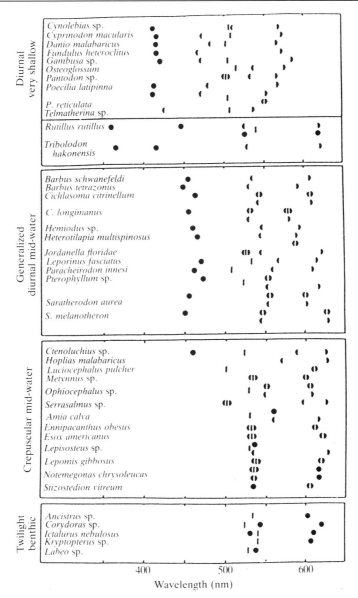

Fig. 1.6 *The relationship between visual pigments and the depth at which the fish lives in fresh water. Note that the spectral range of visual pigments is reduced as the depth increases and the spectral bandwidth of incident light available for vision becomes narrower. Symbols as in Fig. 1.4 (Lythgoe and Partridge, 1989).*

city of brown forest litter, when a longer-wavelength visual pigment is better.

It is difficult to relate studies based on chromaticity differences between real natural objects, and physiological studies of wavelength discrimination along the spectrum (see Jacobs, 1981, for a review). Consider, for example, the often-quoted example of fruit ripening from green to red. There is a decrease in chlorophyll and an increase in red pigment such as anthocyanins in apples and carotenoids in tomatoes. There is not a general slide in maximum reflectance along the spectrum from green through orange to red, but rather a simultaneous reduction in green pigment and an increase in red pigment. It is often said that it is important for the spectral sensitivity of the photoreceptors to overlap, because if they did not there would be bands of the spectrum which would appear dark. In nature this is less of a disadvantage because the spectral reflection curves of most natural objects are broad.

Co-evolution

It can be argued that almost all visual adaptations are due partly to the optical environment, and partly to inter-relationships between species. Herbivorous mammals, which are often at risk from predators, have eyes placed at the sides of the head, giving a very good all-round field of view. Predators, on the other hand, are less at risk from other predators, but need enhanced information in the forward direction. Their field of view is chiefly directed forward, but covers a relatively narrow angle (Walls, 1942).

Flowers often use animals as pollination vectors and they have an enormous range of shapes and colours that advertise to different species (Baker and Hurd, 1968; Kevan and Baker, 1983; Goldsmith, 1990). Insect-pollinated flowers have bright colours and bold patterns. The patterns have ultraviolet components that are not visible to us, but are visible to many insects, including bees. Flowers normally pollinated by birds often have their nectar accessible to birds with long bills but inaccessible to insects. Bird-pollinated flowers are often red or orange, colours that are well discriminated by birds. Flowers pollinated by bats are often white, or dark and strong-smelling; petals are reduced and there is a wide brush of stamens that deposit their pollen on the fur of the bat.

There is often a visual interplay between the sexes which may each develop their own colours, patterns and behaviours. For example, in honeybees the workers have relatively small eyes, though they are presumably large enough to find the correct flowers. Drones have bigger eyes, presumably because the demands made by the nuptial flight in pursuit of the queen requires better vision. Sex differences in visual function appear to be uncommon in vertebrates. The very strongly sexually dimorphic deep-sea angler fish are examples where the male has large, forward-looking tubular eyes which he presumably uses to find the female (Marshall, 1971). Once the male finds the female he attaches to her and continues to live a parasitic or semi-paratistic life with her. The female continues to feed by using a bioluminescent lure to attract small fish. Her eyes are smaller and appear to be less specialized for forward vision.

Behavioural experiments with vertebrates leave no doubt that particular colours and shapes are important in sexual and agonistic displays (see for example Neumeyer, Chapter 13). However, there are few convincing reports that particular displays have led directly to the evolution of particular retinal mechanisms. Perhaps the need to find food and avoid predators is essential to remain alive, and many food items and predators are camouflaged to avoid attention. Colours and patterns that are intended to invite attention like courtship signals and warning colours are usually easily visible using existing retinal mechanisms, and it is the way the signals are presented and the meaning the brain attaches to them that are significant.

The light from different species of fireflies sharing the same air space are similar in spectral radiance, but are distinguished by characteristic train of light pulses. In the New World fireflies of the genus *Photuris* are predators on other firefly species. Their technique is to mimic the flash sequence of the females of the prey species and attack the males when they come near (Lloyd, 1983). Several insects, including hoverflies, are inherently attracted to yellow, while naive bumblebees are attracted by any bright, saturated colour (Lunan, 1990). It is only later that they learn to associate particular colours with sources of nectar.

In vertebrates, particular responses to particular visual images may be innate or unlearned or acquired later in life. Naive chicks avoid bold barred patterns of yellow and black, although red and black, or black, or yellow alone seem to have little significance for them (Roper and Cooke, 1989). In his classic study of animal behaviour, Tinbergen (1951) tells how male sticklebacks behave aggressively to anything that is red. These examples are of instinctive behavioural response that appear to be 'hard wired' into the brain. However, behaviour patterns such as the development of strong social preferences can be established later in life and can cause the establishment of search images for cryptic items of food or the avoidance of particular predators (see Suboski, 1989, for a review). Hatchlings of several species of birds show strong social preferences, both filial and sexual, for the first animals they encounter in the few hours after hatching (Lorenz, 1935; ten Cate, 1989). Such behaviour involves appearance, as well as other characteristics such as song.

Where an animal is able to make a choice between several possible mates, it is often the biggest and the brightest that is preferred. A good example is found among wild-type native guppies *Poecilia reticulata* (Endler, 1988). The females are a drab brown, but the males, although smaller, have a variety of red, blue, white, yellow, black and iridescent spots in various sizes and combinations. The upper lobe of the tail fin is also extended. Males pester the females to mate, but it is she who actually chooses, mostly on the principle that gaudiest is best. In habitats where there are no predators, the males become more and more brightly coloured and their fins become longer. Where there are predators, the more bizarre males tend to get eaten. In general, the worse the predation, the less conspicuous are the males.

The visual pigments of guppies are known in some detail (Archer and Lythgoe, 1990). All the pigments appear to be rhodopsins; the rods contain a 501-nm-λ pigment and the λ_{max} of the cones are 408, 464, 533 and 572 nm. There is also a class of cone containing a mixture of the 533 and 572 pigment giving a λ_{max} in the region of

543 nm. All individuals have the 408, 464 and 501 pigments, but the population studied had individuals that contained one or both of the longwave pigments and the mixed intermediate pigment. In some ways this situation resembles the polymorphism present in New World primate cones (Bowmaker, Chapter 4), but in guppies the intermediate pigment is a mixture. There is no difference between the sexes and it is therefore difficult to argue that colour vision mechanisms in these fish is principally influenced by the need of the female to see the mating colours of the male.

The evolution of visual behaviour depends in part upon the optics of the eye, and in part upon the brain it serves. All the information that the brain has to use is provided by the eye and the brain cannot put back information that was not gathered by the eyes in the first place. We can be certain that the information capacity of an eye is limited by the wave and photon nature of light. Compound eyes can be as sensitive in dim light and have perception of movement as good as, or better than, camera eyes, but size for size the camera-type eyes like those of vertebrates and cephalopods appear to give the best vision for detail. Small animals have relatively larger eyes than big animals, but the eyes and brains are still small (Kirschfeld, 1976). It is likely that the capacity of a brain is limited by its size, while a large brain is needed to process all the information that a large eye can provide. In mammals the evolution of complex visual behaviour appears to be linked to the number of sensory areas in the middle level of the cortical processing (Kaas, 1989). Visually primitive mammals such as opossums may not have many more than 3 visual areas, whereas cats have at least 12 and New World monkeys at least 15. These areas are larger as well as more numerous in visually advanced animals, and that means the brains must also be bigger. In the evolution of visual behaviour there is clearly an advantage in being bigger, because only large animals are able to carry the greater bulk of brain and eyes that is needed for advanced visual systems.

References

Allo, A.-C., Donner, K., Hyden, C., Larsen, L. O. and Reuter, T. (1988). Low retinal noise in animals with low body temperature allows high visual sensitivity. *Nature Lond.*, **334**, 348–350.
Archer, S. N. and Lythgoe, J. N. (1990). The visual pigment basis for cone polymorphism in the Guppy, *Poecilia reticulata*. *Vision Res.*, **30**, 225–233.
Baker, H. G. and Hurd, P. D. (1968). Intrafloral ecology. *Annu. Rev. Entomol.*, **13**, 385–414.
Beatty, D. D. (1984). Visual pigments and the labile scotopic visual system of fish. *Vision Res.*, **24**, 1563–1573.
Bowmaker, J. K. and Kunz, Y. W. (1987). Ultraviolet receptors, tetrachromatic colour vision and retinal mosaics in the Brown Trout (*Salmo trutta*); age dependent changes. *Vision Res.*, **27**, 2101–2108.
Coombs, S., Janssen, J. and Webb, J. C. (1988). In *Sensory Biology of Aquatic Animals*. eds. Atema, J., Fary, R. R., Popper, A. N. and Tavolga, W. N. pp. 553–594. New York: Springer.
Crescitelli, F., McFall-Ngai, M. and Horwitz, J. (1985). The visual pigment sensitivity hypothesis: further evidence from fishes of varying habitats. *J. Comp. Physiol. A*, **157**, 323–333.
Denton, E. J. (1990). Light and vision at depths greater than 200 metres. In *Sensory Biology of Aquatic Animals*. eds. Atema, J., Fary, R. R., Popper, A. N. and Tavolga, W. N. pp. 127–148. New York: Springer.
Denton, E. J. and Warren, F. J. (1957). The photosensitive pigments in the retinae of deep-sea fish. *J. Mar. Biol. Ass. UK*, **36**, 651–662.
Duntley, S. Q. (1963). Light in the sea. *J. Opt. Soc. Am.*, **53**, 214–233.
Endler, J. A. (1988). Frequency-dependent predation, crypsis and aposomatic coloration. *Phil. Trans. R. Soc. Lond. B*, **319**, 505–523.
Exner, S. (1891). *Die Physiologie der facettirten Augen von Krebsen und Insecten*. Leipzig-Wien: Deuticke.
Goldsmith, T. H. (1990). Optimization, constraint, and evolution of eyes. *Q. Rev. Biol.*, **65**, 281–322.
Goldsmith, T. H., Collins, J. S. and Licht, S. (1984). The cone oil droplets of avian retina. *Vision Res.*, **24**, 1661–1671.
Hartline, P. H. and Lange, G. D. (1984). Visual systems of cephalopods. In *Comparative Physiology of Sensory Systems*. ed. Bolis, L. pp. 335–355. Cambridge: Cambridge University Press.
Hopkins, C. D. (1988). Social communication in the aquatic environment. In *Sensory Biology of Aquatic Animals*. eds. Atema, J., Fary, R. R., Popper, A. N. and Tavolga, W. N. pp. 233–267. New York: Springer.
Jacobs, G. H. (1981). *Comparative Color Vision*. New York: Academic.
Jerlov, N. G. (1976). *Marine Optics*. Amsterdam: Elsevier.
Kalmijn, J. (1988). Hydrodynamic and acoustic field detection. In *Sensory Biology of Aquatic Animals*. eds. Atema, J., Fary, R. R., Popper, A. N. and Tavolga, W. N. pp. 83–130. New York: Springer.
Kass, J. H. (1989). The evolution of complex sensory systems in mammals. *J. Exp. Biol.*, **146**, 165–176.
Kevan, P. G. and Baker, J. G. (1983). Insects as flower visitors and pollinators. *Annu. Rev. Entomol.*, **28**, 407–453.
Kirschfeld, K. (1976). The resolution of lens and compound eyes. In *Neural Principles of Vision*. eds. Zettler, F. and Weiler, R. pp. 354–369. Berlin: Springer.
Kirschfeld, K. and Snyder, A. W. (1975). Waveguide mode effect, birefringence and dichroism in fly photoreceptors. In *Photoreceptor Optics*. eds. Snyder, A. W. and Menzel, R. pp. 56–77. Berlin: Springer.
Knowles, A. and Dartnall, H. J. A. (1977). The photobiology of vision. In *The Eye*, Vol. 28, ed. Davson, H. pp. 540–545. New York: Academic.
Land, M. F. (1965). Image-formation by a concave reflector in the eye of the scallop, *Pecten maximus*. *J. Physiol.*, **179**, 138–153.
Land, M. F. (1981). Optics and vision in invertebrates. In *Handbook of Sensory Physiology*, Vol. VII/6B. ed. Autrum, H. pp. 471–594. New York: Springer.
Land, M. F. (1990). Optics of the eyes of marine animals. In *Light and Life in the Sea*. eds. Herring, P. J., Campbell, A. K., Whitfield, M. and Maddock, L. pp. 149–166. Cambridge: Cambridge University Press.
Laughlin, S. B. (1987). Form and function in retinal coding. *Trends Neurosci.*, **10**, 478–483.
Laughlin, S. B. (1981). Neural principles in the visual system. In *Handbook of Sensory Physiology*, Vol. VII/6B. ed. Autrum, H. pp. 133–280. New York: Springer.
Levine, J. S. and MacNichol, E. F. (1979). Visual pigments in teleost fishes: effects of habitat, microhabitat and behavior on visual system evolution. *Sensory Processes*, **3**, 95–130.
Lloyd, J. E. (1983). Bioluminescence and communication in insects.

Annu. Rev. Entomol., **28**, 131–160.

Locket, N. A. (1971). Adaptations to the deep-sea environment. In *Handbook of Sensory Physiology*, Vol. VII/5. ed. Crescitelli, F. pp. 67–192. Berlin: Springer.

Loew, E. R. and Lythgoe, J. N. (1978). The ecology of cone pigments in teleost fish. *Vision Res.*, **18**, 715–722.

Lorenz, K. (1935). Der Kumpan in der Umwelt des Vogels. *J. Ornithol.*, **83**, 137–213, 289–413.

Lunan, K. (1990). Colour saturation triggers innate reactions to flower signals: flower dummy experiments with Bumblebees. *J. Comp. Physiol. A*, **166**, 827–834.

Lythgoe, J. N. (1968). Visual pigments and visual range underwater. *Vision Res.*, **8**, 997–1012.

Lythgoe, J. N. (1977). Fishes: vision in dim light and surrogate senses. In *Sensory Ecology*. ed. Ali, M. A. pp. 155–168. New York: Plenum.

Lythgoe, J. N. (1979). *The Ecology of Vision*. Oxford: Clarendon.

Lythgoe, J. N. (1988). Light and vision in the aquatic environment. In *Sensory Biology of Aquatic Animals*. eds. Atema, J., Fary, R. R., Popper, A. N. and Tavolga, W. N. pp. 57–82. New York: Springer.

Lythgoe, J. N. (1980). Ecological adaptations. In *Vision in Fishes*. ed. Ali, M. A. pp. 431–445. New York: Plenum.

Lythgoe, J. N. and Partridge, J. C. (1989). Visual pigments and the aquisition of visual information. *J. Exp. Biol.*, **146**, 1–20.

Marshall, N. B. (1971). *Exploration in the Life of Fishes*. Cambridge, Mass.: Harvard University Press.

Miller, W. H. (1979). Ocular optical filtering. In *Handbook of Sensory Physiology*, Vol. VII/6A. ed. Autrum, H., pp. 69–143. New York: Springer.

Muntz, W. R. A. (1972). Inert absorbing and reflecting pigments. In *Handbook of Sensory Physiology*, Vol. VII/1. ed. Dartnall, H. J. A. pp. 529–565. Berlin: Springer.

Munz, F. W. (1958). Photosensitive pigments from the retinae of certain deep-sea fish. *J. Physiol. Lond.*, **140**, 220–235.

Nilsson, D. E. (1982). The transparent compound eye of *Hyperia* (Crustacea): examination with a new method of analysis of refractive index gradients. *J. Comp. Physiol. A*, **147**, 339–349.

Partridge, J. C. (1989). The visual ecology of avian oil droplets. *J. Comp. Physiol. A*, **165**, 415–426.

Partridge, J. C. (1990). The colour sensitivity and vision in fishes. In *Light and Life in the Sea*. eds. Herring, P. J., Campbell, A. K., Whitfield, M. and Maddock, L. pp. 167–184. NY: Cambridge University Press.

Partridge, J. C., Archer, S. N., Lythgoe, J. N. and van Groningen-Luyben, W. (1989). Interspecific variation in the visual pigments of deep-sea fishes. *J. Comp. Physiol. A*, **164**, 513–529.

Roper, T. J. and Cooke, S. E. (1989). Responses of chicks to brightly coloured insect prey. *Behaviour*, **110**, 276–293.

Shand, J., Partridge, J. C., Archer, S. N., Potts, G. W. and Lythgoe, J. N. (1988). Spectral absorbance changes in the violet/blue sensitive cones of the juvenile pollack. *Pollachius pollachius. J. Comp. Physiol. A*, **163**, 699–703.

Snyder, A. W. (1977). Acuity of compound eyes, physical limitations and design. *J. Comp. Physiol.*, **116**, 161–182.

Strausfeld, N. J. and Nassel, D. R. (1981). Neuroarchitectures serving compound eyes of crustacea and insects. In *Handbook of Sensory Physiology*, Vol. VII/6B. ed. Autrum, H. pp. 1–132. New York: Springer.

Suboski, M. D. (1989). Recognition learning in birds. In *Perspectives in Ethology*. Eds. Bateson, P. P. G. and Klopfer, P. H. pp. 137–172. New York: Plenum.

ten Cate, C. (1989). Behavioral development: towards understanding processes. In *Perspectives in ethology*. eds. Bateson, P. P. G. and Klopfer, P. H. pp. 243–270. New York: Plenum.

Tinbergen, N. (1951). *The Study of Instinct*. Oxford: Clarendon.

Vogt, K. (1980). Die Spiegeloptik des flusskebsauges: the optical system of crayfish eye. *J. Comp. Physiol.*, **135**, 1–19.

Walls, G. L. (1942). *The Vertebrate Eye and its Adaptive Radiation*. Cranfield, Michigan: Cranfield Institute of Science.

2 Origin of Invertebrate and Vertebrate Eyes

J. R. Cronly-Dillon

To suppose that the eye, with its inimitable contrivances for adjusting the focus to different distances, for admitting different amounts of light, and for the correction of spherical and chromatic aberration, could have formed by natural selection, seems, I freely confess, absurd in the highest degree. Yet reason tells me that if numerous gradations from a perfect and complex eye to one very imperfect and simple, each grade being useful to its possessor, can be shown to exist; if further, the eye does vary ever so slightly, and the variations be inherited, which is certainly the case; and if any variation or modification in the organ be ever useful to an animal under changing conditions of life, then the difficulty of believing that a perfect and complex eye could be formed by natural selection, though insuperable by our imagination, can hardly be considered real.

Charles Darwin, The Origin of Species

Introduction

The origin of the vertebrate eye has long been a mystery. This was clearly acknowledged by Darwin in a chapter he wrote in *The Origin of Species* (1859), entitled 'Problems for the theory' an extract from which is reproduced above. This passage is interesting for another reason. It expresses Darwin's view that major phyletic changes are brought about by a series of microevolutionary steps in which the transformation of organisms (or of an organ) can be traced through small progressive changes that lead from one morphological form to another. Consequently, palaeontologists seeking to establish a homologous sequence generally approach the problem by comparing structures or organs in different species that appear to serve similar functions. However, attempts to trace the origin of the vertebrate eye by application of Darwin's simple 'microevolutionary rule' have proved to be extraordinarily difficult. Even today the apparent gaps in the fossil record leave us with a picture that is far from clear, so much so that one is tempted to think that the events that transformed a simple photosensory precursor into the sophisticated visual organ found in vertebrates may have occurred over a remarkably short evolutionary time scale. From this, it may reasonably be argued that the apparently unheralded appearance of a complex structure such as the vertebrate eye may perhaps have derived through a process of punctuational or macro evolution, in the sense advocated by Stanley (1981). For example, a rapid evolutionary transformation of the eye's precursor could have occurred, say, as the result of mutations that affected the path of its development in the embryo (for an erudite account of the relationship between 'ontogeny and phylogeny' see Gould, 1977).

In the embryo, the first obvious sign of the developing eye is the appearance of the optic vesicles at the cephalic end of the neural tube. This observation, among other things, led Ray Lankester (1890) to suggest a cerebral origin for the vertebrate eye. Later, his idea was taken up and extended by Studnicka (1912, 1918) and subsequently adapted by Polyak (1957) and Sarnat and Netsky (1981) to account for the derivation of the visual centres, such as the optic tectum etc., as well as of the eye itself (Fig. 2.1). In their search for the eye's origin, Lankester and his successors went as far back as *amphioxus*, a protochordate. In the present review the aim is to extend this enquiry to some earlier antecedents and to consider structures that may have served as an ancestral precursor, not only of vertebrate, but also of some invertebrate eyes. In pursuing this course I shall explore a fascinating but curiously neg-

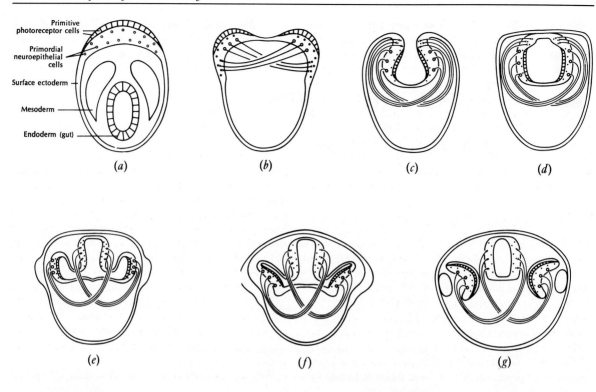

Fig. 2.1 *Sarnat and Netsky's adaptation of Polyak's (1957) theory of the origin of the vertebrate visual system. (a) Part of the surface ectoderm of an early ancestor differentiated into neuroectoderm where some cells had crude photosensory properties. (b) With the evolution of bilateral symmetry, an incipient medial invagination concentrated the photoreceptive epithelium into paired placodes. (c) Invagination of neural plate. (d) The closed neural tube is formed. (e) Cerebral tissue proliferated in the dorsal region, allowing the photoreceptor cells to retain proximity to sources of light at the skin. Epidermis overlying the optic vesicles thickens to form the lens placode. (f), (g) The optic vesicle invaginated and the retina was formed. The optic nerves continued to decussate ventral to the neural tube. That portion of the wall of the neural tube receiving the decussated optic nerve fibres developed into the optic tectum (from the caudal part) and the rostral portion evolved into the dorsal thalamus. Extensive connections between the optic tectum and the thalamus, particularly between the ventral nucleus of the lateral geniculate body, found in almost all vertebrates (After Sarnat and Netsky, 1981).*

lected proposal by Willmer (1970) in which he suggests the possibility that both vertebrate and other image-forming eyes (such as those found in molluscs, as well as the compound eyes of arthropods) may have derived from a structure in a nemertine-like ancestor, i.e. the cephalic organ, whose function was not primarily visual. The particular attraction of Willmer's proposal is that it not only points to this organ as a possible common precursor of both vertebrate and many invertebrate eyes, but also suggests a solution to a number of thorny problems raised by Schwalbe (1874), Gordon Walls (1942) and others, concerning the origin of rods and cones as distinct cell types. Willmer's analysis centres on two central notions, as follows: (a) that the eye may have originated as a structure whose function was not primarily visual, but rather one that may also have been concerned with sensing local changes in salinity; (b) that photoreceptors, both of vertebrates and invertebrates, are of epithelial origin, and that epithelial tissue consists of at least two types of cells, which differ in polarity with respect to the regulation of ion and water movement. Willmer suggests that the difference between vertebrate and invertebrate photoreceptors stems from the duality in cell type in epithelial tissues. The notion is further extended to account for the origin of vertebrate rods and cones as distinct cell types.

A key element in Willmer's (1970) exposition is the role of vitamin A, on the one hand in photoreception and on the other in the regulation of ionic equilibria. The significance of this with respect to the lineage of the vertebrate eye is that the organ from which the eye was derived may previously have served several different functions by, for example, combining photoreception, sensitivity to salinity, and possibly touch, with secretory activity. These sensory functions may then have become structurally differentiated from one another in the course of phylogeny to give rise to a variety of vertebrate and invertebrate organs. Such a line of reasoning led Willmer to consider, on the basis of morphological and functional criteria, that different portions of the cephalic organ of nemertines may have given rise to the eyes of vertebrates and of some invertebrates, e.g. the cephalopod chambered eye and arthropod (compound) eyes, as well as other structures that have secretory functions where the activity is regulated by light (such as the hypothalamus and the pituitary complex, etc.).

Cerebral Origin of the Vertebrate Eye: the Classical Picture

During normal vertebrate development the surface ectoderm invaginates to form the neural tube and the eyes develop as paired outgrowths of the embryonic diencephalic portion of this tube. Studnicka (1912, 1918) remarked that if one traces the inner layer of the optic cup around the latter and through the optic stalk into the central nervous system one emerges into the ependymal layer of the brain wall. He also noted that the eye of the larval lamprey was precociously functional and able to convey crude directional information while still merely an optic vesicle. The lens at that stage exists in the form of a flat cushion and is incapable of performing a dioptric function, and Studnicka suggested that before it took up its dioptric role in the eye it must have existed phylogenetically as something else.

Previously Balfour (1881) had pointed out that while the retina derived from the brain it had originated in the integument as patches of photosensory epithelium located on the area of ectoderm that became involuted during formation of the neural tube (Figs 2.1 and 2.2). This provided a possible explanation of the inversion of the vertebrate retina. Lancaster and his associates reasoned that the eye must have originated in some light-sensitive cells buried in the nervous system of some protochordate such as the lancelet *amphioxus*, which was transparent enough so that light could travel through the body to reach the light-sensitive cells in the brain. Then, as the body became opaque, this cluster of light-sensitive cells within the neural tube became displaced closer to the surface until eventually it was separated from the latter by a layer of ectoderm which retained its transparency, and which ultimately became differentiated to become the cornea and lens.

Following this line of reasoning, Studnicka (1898) drew

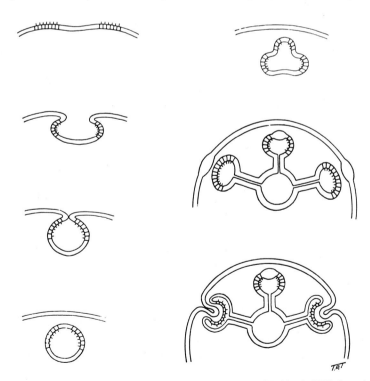

Fig. 2.2 *Phylogenetic development of vertebrate eyes (Schimkewitsch's 1921 version of Balfour's 1881 theory). Left: photosensitive ciliated ectoderm on the dorsal aspect carried inwards by the invagination of surface ectoderm during formation of the neural tube (and in the process, enclosing photosensitive epithelium as ependyma). Right: formation of two lateral eyes and one median eye.*

attention to the fact that if the retina is thought of as a photosensory portion of the brain wall, growing outwards from the latter, to keep it near the skin in an ancestor whose body was becoming larger and more opaque as evolution proceeded, then the sclerotic and uveal coats were homologous to the meningeal envelopes of the central nervous system, the dura mater and the pia arachnoid. In particular, he noted that the vascularity and pigmentation of the choroid were strongly pia-like in their characteristics, and that, particularly in lampreys, there were striking histological similarities between choroid and pia. The evidence showing that the photoreceptors of vertebrate eyes are ependymal in origin, accords with the view that they are homologous to the light-sensitive infundibular organ of *amphioxus* (Boeke, 1908), although it should be remembered that direct evidence connecting the two is still lacking. Both Studnicka's theory and those of his predecessors concerning the origin of the retina are supported by the photosensitivity of the lining of the diencephalon of many species, where in some instances it acts as a photic organ that controls the annual sexual cycle. A description of the photoreceptive processes of cells in *amphioxus*, which line the cerebral vesicle (third ventricle), has been given by Eakin.

While the ideas elaborated by Studnicka have much to commend them, we are reminded by Gordon Walls (1942) that in many primitive vertebrates the nervous system begins its development as a solid column of tissue and that it is canalized only secondarily. Also, if the lineage of the vertebrate retina is obscure, that of the lens has been shrouded in mystery. Several investigators (Sharpe, 1885; Schimkewitsch, 1921), have speculated that the lens may have originated as an independent organ which was 'captured' by the progressive outpouching of the photosensory portion of the neural tube (the optic vesicle). However, such an explanation is unlikely as, among other, the optic cup in some animals can induce lens formation in an indifferent piece of ectoderm, e.g. a piece grafted from the ventral surface of the embryo.

Similarities between Visual and Tactile Sensory Coding

It is worth noting some of the consequences that relate to the coding of visual information that may derive from the integumentary origin of the eye. Von Bekesy (1959, 1960) demonstrated that many of the auditory effects and illusions that arise from the spatio-temporal pattern of vibration of the cochlea (a structure of integumentary origin) can be mimicked experimentally in human subjects by vibrating the skin. The ear and the skin therefore appear to share certain common features in the neural processing of sensory information. Might this also be the case between the skin and the eye (which is also of integumentary origin)?

Current wisdom has it that tactile sensations are mediated by slowly and rapidly adapting mechanoreceptors. The former, which respond continuously to an enduring stimulus, are more suited for detecting and locating 'static' tactile stimuli. The latter respond transiently to the onset and perhaps the offset of the stimulus, and are more suited for coding spatio-temporal information. Thus receptors in the skin already appear to possess characteristics that appear to foreshadow the sustained (X) and transient (Y) information-coding channels later found in the vertebrate retina (Enroth-Cugell and Robson, 1966), the former coding 'form' (and, where appropriate, 'colour'), the latter serving 'flicker' and 'movement'. The cephalic organ of nemertines and its ganglion (which Willmer proposes is the homologue of the vertebrate retina) also have an integumentary origin, and it seems not unlikely that the neural processing in the cephalic organ may also reflect this X–Y dichotomy.

The Diverse Physiological Roles of Photoreception in Animals

It is quite clear to any observer of animal behaviour that the general reactions of many organisms are regulated by light (see Lythgoe, Chapter 1 and Dodge, Chapter 15). Perhaps one of the most primitive of all the responses to light is the cycle of diurnal rhythms that keep pace with, and track, the alternation of day and night. Sensitivity to light, linked to a 'biological clock', can also be used to measure day length and therefore provide information of time of year or season. Some protists change their form in response to light, and many species are activated by light (e.g. flatworms; Loeb, 1893). In many species light affects reproductive activity. Therefore, in seeking to trace the antecedents of the vertebrate eye we should bear in mind that animals utilize photoreception for a variety of purposes, for example:

1. As a detector of UV radiation (that may harm the animal). For example, UV radiation may stimulate the animal to retreat from the harmful stimulus; or it may activate the movement of melanophores, which effectively 'pulls a screen over the animal' (Young, 1935).
2. As a visual sense organ for the detection of form and movement, etc.
3. As a detector of day length, which serves to regulate the reproductive cycle.

Correspondingly, a variety of separate photosensory systems (including the lateral eyes) appear to have been developed at various times in phylogeny, for the specific purpose of providing the animal with these different sorts of information. Whence, we may ask, did these different photosensory structures originate, i.e. did they arise and differentiate from a single ancestral photosensory organ or did they evolve independently from a variety of photosensory structures? Before addressing this question we need briefly to consider the overall form and structure of some invertebrates and prevertebrate eyes and photoreceptor systems.

Morphological Characteristics of Animal Photoreceptor Organs

Eye Spots and Light-Sensitive Cells

In the most primitive unicellular organisms, such as the amoeba, there is a diffuse reaction to light whereby the whole cell, in the absence of any apparent specialization, responds by a simple sol-gel alteration of its cytoplasm.

We may conjecture that perhaps the earliest stage in the evolution of an eye occurred when a small area or region of the organism differentiated to become specially sensitive to light (Fig. 2.3(b),(c)).

Following the appearance of primitive multicellular organisms, one finds a general dermal photosenstive reaction in the absence of any specialized photoreceptors. In addition, certain ectodermal cells acquired a special sensitivity to light; some developed initially in the surface epithelium while others occasionally migrated below the surface layer (Figs. 2.3(a) and 2.4(a), (b)).

Primitive light-sensitive cells usually assumed a specialized form which may be differentiated into two main types. One type, found in worms and molluscs, is apolar (Fig. 2.3(a)). The other, also found in worms, has a polar

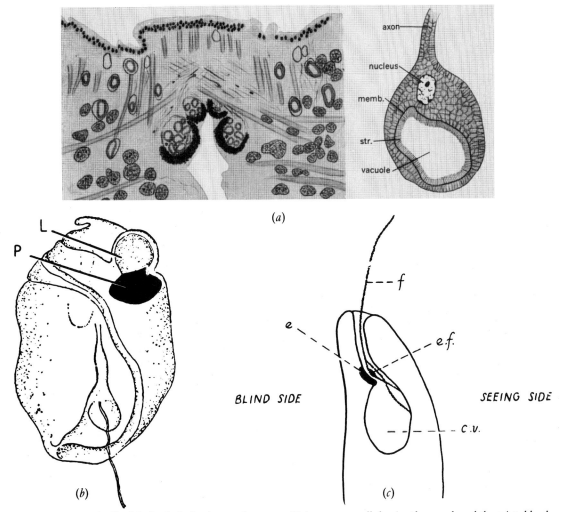

Fig. 2.3 (a) *Apolar visual cells of the leech. Left: pigmented eye cups. Right: receptor cell showing the vacuole and the striated border (after Scriban and Autrum, 1932). (b) Localized eye spot of* Pouchetia *showing the pigmented zone P, and the lens, L. (After Schutt, 1896; Duke-Elder, 1958). (c) Eye spot of* Euglena viridis *associated with a flagellum, f, the flagellum with the enlargement, ef, which is the photosensitive area; cv, contractile vacuole; e, pigment shield (After Wager, 1900; Duke-Elder, 1958).*

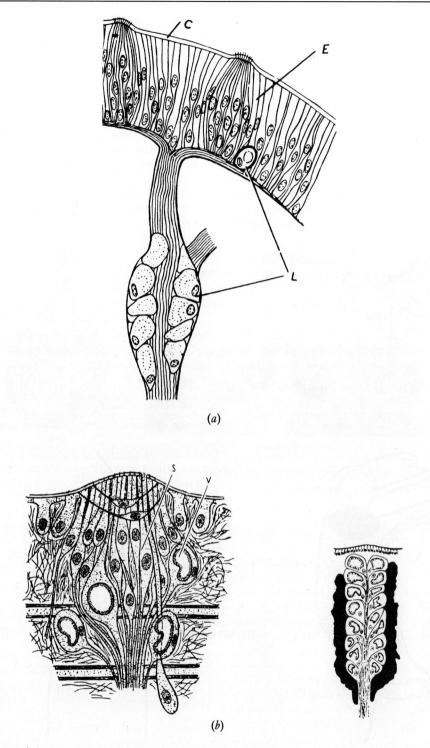

Fig. 2.4 *(a) Single light-sensitive cells in the earthworm* Lumbricus terrestris. *L, photoreceptor cells lying in the basal region of the epidermis and also in the enlargements of the nerve in close relation to the epidermis (After Hess, 1919). (b) Left: collection of undifferentiated sensory cells, S, among which are large light-sensitive cells, V. (After Butschli, 1921). Right: each eye consists of a cluster of apolar cells provided with optic organelles (After Hesse, 1908).*

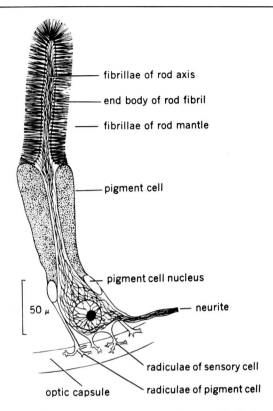

Fig. 2.5 *Diagrammatic representation of a group of retinal cells of* Limax *(Smith, 1906)*.

(or a bipolar) configuration in which a distal end is specialized to receive the stimulus of light and a proximal end to conduct away the excitation via a conducting nerve fibre (Fig. 2.5).

The Simple Eye or Ocellus

In general characteristics, these photoreceptor systems consist either of a single light-sensitive cell or a group of such cells, which generally act without functional interaction (Figs 2.6, 2.7(a), 2.8, 2.9 and 2.10). Usually these 'eyes' are associated with pigment which forms an absorbent screen within and around the visual cells.

Unicellular Eyes

These may be epithelial or subepithelial, and consist of either one of two types of light-sensitive cell, i.e. a bipolar form with a specialized sensory terminal or an apolar form characterized by an intracellular organelle (see Fig. 2.3(a)).

The Multicellular Eye

The aggregation of a large number of light-sensitive epithelial cells generated a crude receptor mosaic with a potential for providing directional information. The aggregation of photoreceptor cells was a prelude for the development of a flat, cupulate, vesicular and, eventually, image-forming eye with a lens system (as seen in the polychaete worm *Vanadis* and, in its most developed form, in cephalopods) (see Figs. 2.9, 2.13).

The subepithelial eye

Here a number of light-sensitive apolar cells migrate from the surface epithelium to form a subepithelial mass in association with nerve fibres (see Fig. 2.4(b)). In the leech, subepithelial eyes form by clumping together a number of these cells within a dense pigmentary mantle.

Invaginated and chambered invertebrate eyes

Directional sensitivity was enhanced by the epithelial invaginated eye which is found in various elaborated forms, from the flat (Fig. 2.9(a)) to the invertebrate chambered eye (through the cupulate eye pinhole eye to the vesicular forms and the invertebrate lens image-forming eye) (Figs 2.8–2.11, 2.13, 2.14).

CEPHALOPOD EYES In cephalopods (and in polychaete worms) we find the invertebrate chambered eye attaining its highest form of development (Figs. 2.13, 2.14). The eye has an elaborate intrinsic and extrinsic musculature, is able to accommodate for near and far objects and has active pupillary control. The retina consists of a simply organized epithelium containing visual cells and supporting cells. From the outer end of each visual cell an axon proceeds to the brain. In this respect the cephalopod retina is similar to that in other invertebrate chambered eyes. However, the organization of the cephalopod retina differs from that of other simple invertebrate chambered eyes in that at the outer end of its visual cells (where the axon arises) the cells have one or more short branched dendritic processes that end among adjacent visual cell bases (Fig. 2.13(b)), or, in some cases ascend some distance into the nuclear layer (Young, 1962a). This allows the possibility for functional interaction between receptor elements. A feature of the cephalopod eye of particular interest is that there is communication between the anterior chamber and the sea (Fig. 2.13(a)). We will return to the latter point when discussing the possibility that the cephalopod eye (as well as that of vertebrates) may have originated from a secretory structure (i.e. the cephalic organ) in a primitive nemertine ancestor whose receptor function included both photoreception and the detection of salinity changes in its aquatic environment.

Aggregate and Composite Eyes (Simple Eye)

The aggregate eye

This is an accumulation of ocelli that are clustered together in a manner that superficially resembles a compound eye (see Fig. 2.12(a)). However, each ocellus is ana-

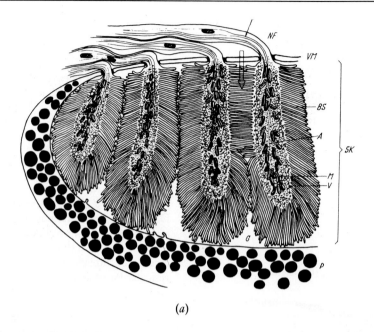

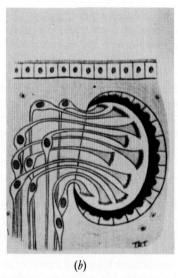
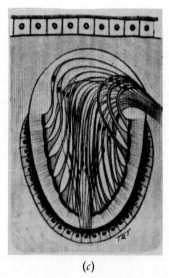

Fig. 2.6 *(a) Diagram of electron microscope section of the eye cup of planarian. BS, microvilli. NF, nerve fibres. V, vesicles. M, mitochondria (After Rohlich and Torok, 1961). (b) Eye of the turbellarian worm* Planeria gonocephala. *Eye of the nemertine worm* Drepanophorus *(After Duke-Elder, 1958).*

tomically separate from the other and there is no interaction between them at the level of the retina.

The composite ocellus
This is formed by the fusion of two or more ocelli, each with its own retina and pigment cup (see Fig. 2.12(b)).

The Compound Eye
The simple ommatidial or compound eye appears as an evolutionary development from an unknown invertebrate simple eye (see Fig. 2.12(c)). However, unlike the aggregate eye, to which it bears a resemblance, the sensory elements in a compound eye are structurally and functionally associated in groups. This has been brought about by the division of the individual sensory cells of a simple

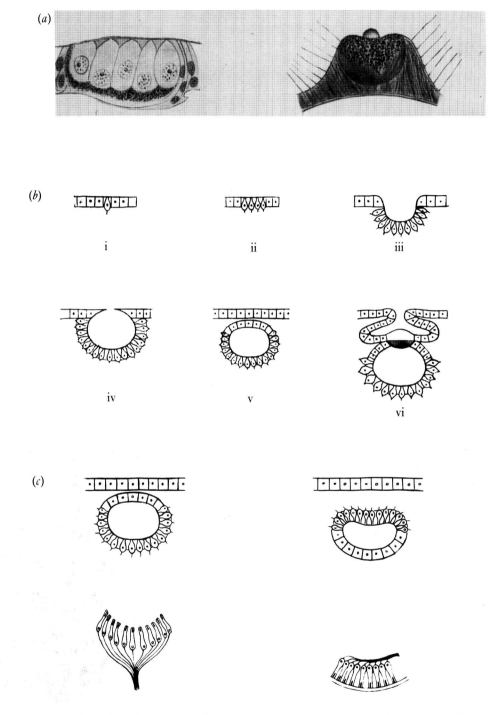

Fig. 2.7 (a) Flat eyes. Left: ocellus of the aquatic annelid worm, Stylaria lacustris (After Hesse, 1908). Right: ocellus of a hydromedusan (Lizzia); the sensory cells are capped by lens-like thickening of the cuticle (Hertwig, 1878, and Jourdan, 1889). (b) Development of the simple (non-inverted) invaginated eye of invertebrates (according to Duke-Elder, 1958). (i) Single epithelial light-sensitive cell. (ii) Collection of visual cells forming a flat eye. (iii) Cupulate eye. (iv) Formation of a pinhole eye. (v) Vesicular eye. (vi) Image-forming eye of cephalopods. (c) Top: comparison of development of non-inverted (left) and inverted (right) retinae of invertebrate vesicular eyes. Bottom: arrangement of visual cells in non-inverted and inverted retinae (After Duke-Elder, 1958).

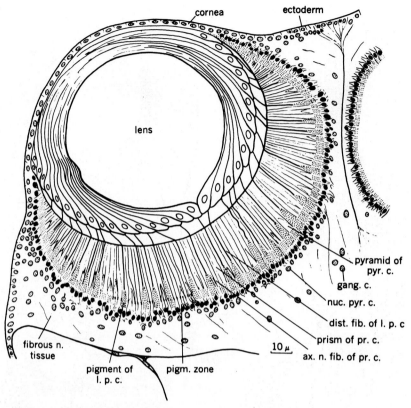

Fig. 2.8 *The large complex eye of the coelenterate* Charybdea *has a biconvex cellular lens, a stratified grouping of pigment cells and packed retinal sensory cells of two kinds and a vitreous mass (from Berger, 1900).*

eye to form a coordinated colony. The development of ocelli and of each ommatidium of a compound eye display many similarities, as well as important differences, which suggests that, ontogenetically and phylogenetically, aggregate and compound eyes may have arisen as parallel developments from some unknown primitive ancestor.

Photosensitive Structures in Protochordates
These lowly animals, of which there are three classes (i.e. the hemichordates, the tunicates and the lancelets) possess a rudimentary nerve cord, a notochord and gill clefts. Apart from the pelagic tunicates, these creatures are either sessile or have a burrowing habit. In *Balanoglossus* (a hemichordate) the nervous system develops as a longitudinal groove of ectoderm, which develops into a tube but shows no indication of optic vesicle formation. However, the larvae (but not the adult forms) of some species contain simple eyespots (simple depressions of cells surrounded by pigment) in the apical plate. In tunicates, typified by the ascidian, the eye arises in the larval form as an outpouching of the cerebral vesicle to form a single sensory organ consisting of a sac containing a statocyst and a very rudimentary eye. The retina is composed of a few sensory cells derived from the inner wall of the neural tube. It is capped with pigment and above it lies a primitive cellular lens which faces the brain (Duke-Elder, 1958). The lancelets, of which *amphioxus* is a well-studied example, possess a dorsal tubular nerve cord, a notochord and gill slits but lack a differentiated brain or eyes. However, the body is divided into several myotomes. Although it possesses no definitive eyes, *amphioxus* is strongly photo-negative. *Amphioxus* appears to possess two types of photosensitive organ (a third, i.e. the cells of Joseph, were once thought by Johannes Muller (1842) to be photosensory, but this has now been discounted). One acknowledged photosensitive mechanism, the organs of Hesse (1899), consists of individual photosensitive cells scattered on the ventral and lateral aspects of the nerve cord towards its posterior end (Fig. 2.15(a)). These photoreceptors are single, large ganglion cells, each provided with a brush-like ciliated margin and an axon, and each capped by a crescent-shaped pigment cell. Despite its many interesting features, it is unlikely, among other things, because of its location, to have been an antecedent of vertebrate lateral eyes. A second photosensory mechanism is located towards the cephalic end of the animal. Here, a small median area of

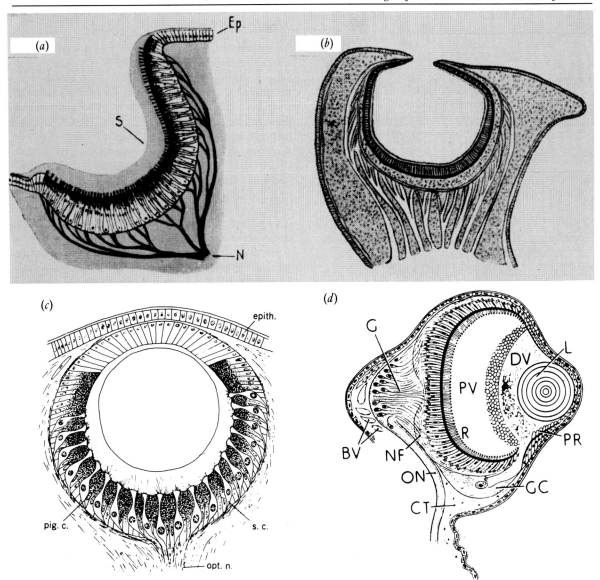

Fig. 2.9 *(a) Cupulate eye: the ocellus of the limpet* Patella. *Ep, epithelium. S, secretory substance covering the visual cells. N, nerve (After Hesse). (b) Pinhole eye: the ocellus of the mollusc. (c) Vesicular eye:* Helix *(After Hesse, 1908). s.c., photoreceptor sense cell. (d) Image-forming eye of the polychaete worm* Vanadis. *DV, distal vitreous. PV proximal vitreous. L, lens. R, retina. G, ganglion cells (After Hesse, 1908).*

ependymal cells lining the central canal of the nerve cord is differentiated to form an infundibular organ, which is light-sensitive (Fig. 2.15(b)). It would be interesting to study this organ in greater detail to determine the possible relationship to the cephalic organ of nemertines.

Inverse Retinas in Invertebrates

Some of the differences between the vertebrate eye and the lateral eyes of invertebrates have already been noted. The invertebrate eye develops directly through an invagination of the surface ectoderm, and the visual cells are connected to the nervous system secondarily. By contrast, the eye of vertebrates arises directly from the neural ectoderm and therefore is related only secondarily to the surface ectoderm by virtue of the fact that the neural ectoderm itself ultimately derives as an infolding from the surface layer (which produces the inverted retina). Despite these differences, some invertebrate eyes also have an inverted or inverse retina, i.e a retina where the visual cells are

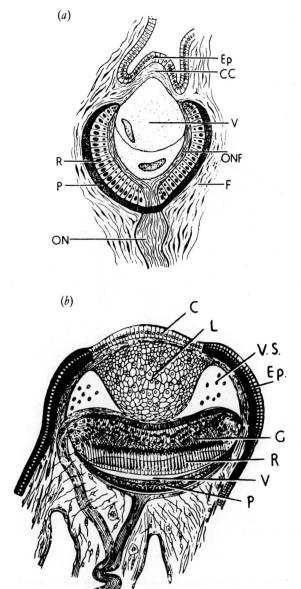

Fig. 2.10 *Inverted retinae.* (*a*) *Dorsal eye of* Onchidium, *showing the inverted retina pierced by the fibres of the optic nerve. Ep, epithelium. R, visual cells. CC, connective tissue. V, large vitreous cells (Semper, 1883, from Duke-Elder, 1958).* (*b*) *Eye of the scallop* Pecten *(After Hesse, 1902).*

oriented so that the photoreceptor end is directed away from the incident light. Their suggested mode of development from surface ectoderm is illustrated in Fig. 2.7(c) and examples of such eyes are shown in Figs. 2.10 and 2.11.

The Eye of *Pecten*

A particularly noteworthy example of an invertebrate eye with an inverted retina features in the common scallop *Pecten maximus* (Fig. 2.11). This bivalve is remarkable in that it possesses an image-forming eye with a lens and a tapetum which forms the image by reflection within a distal layer of visual cells (Land, 1965). The retina contains two layers of photoreceptor cells (Dakin, 1928). In studying the electrical responses of this eye, Hartline (1938) found that the distal layer of the retina mediated strong 'off' responses, while impulses would arise from the proximal layer only at the onset of illumination. Careful study of the retina of *Pecten* with the electron microscope has revealed that the visual cells of the distal layer are composed of concentric lamellae and that each lamella is continuous with the stalk and basal body of a cilium (Miller, 1960). In this respect, the visual cells of the distal layer are similar to the photoreceptor outer segments of the vertebrate retina, which are also derived from cilia (Wolken, 1963). The photoreceptors of the proximal layer, on the other hand, derive from microvilli, and in this respect resemble the photoreceptor cells of octopus and insect compound eyes.

Photoreceptor Systems that Derive from Microvilli and from Cilia Show Opposite Polarities in Their Response to Light

Although the development of image-forming eyes in molluscs and vertebrates is thought to have occurred independently, the vitamin A system is used in the photoreceptor process. In fact, it would appear from the apparently universal distribution of the carotenoid-protein system in animal photoreception that the system was probably inherited from some primitive common ancestor. Recent DNA hybridization experiments, utilizing a DNA strand of the bovine rhodopsin gene as a hybridization probe, have examined the degree of homology between different visual pigment genes (see chapter 5 by Goldsmith and chapter 6 by Piantanida in Volume 6 of this series). The results obtained from such experiments suggest that a primordial gene, through repeated gene duplication and mutation etc., gave rise to all vertebrate and most invertebrate visual pigments (Goldsmith, Chapter 5, in Volume 6). In contrast, it is worth while noting that the photoreceptor pigment in *Euglena* (as well as in *Phycomyces*) is thought to be a flavin complex (based on the action spectrum), even though (in *Euglena*) it is reported (Miller, 1960) that the eye spot also contains a carotenoid which probably acts as a light filter. Likewise, the functional processes of lateral inhibition and the ON and OFF systems that partake in the coding of primary visual information seem also to enjoy a similar widespread distribution

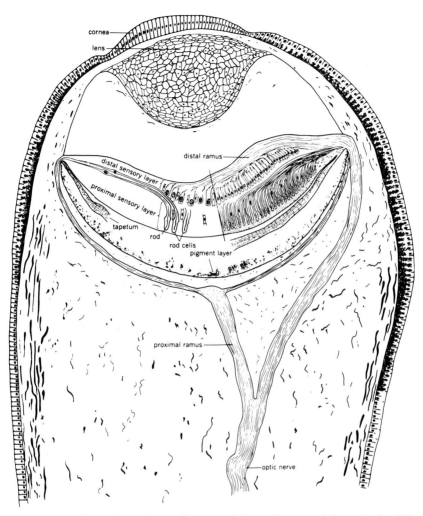

Fig. 2.11 *Schematic representation of* Pecten's *mantle eye, illustrating the inverted retina and the proximal and distal layers of photoreceptors. The proximal retinal sensory cells develop from microvilli, while those of distal retina derive from cilia. (After Dakin, 1928; Bullock and Horridge, 1960).*

throughout the vertebrate and invertebrate taxa (Hartline, 1938; Ratliff and Miller, 1961). One possibility is that some of these functional characteristics may be intimately linked to the nature of the photoreceptor process itself and that this may conceivably impose certain functional restrictions on the manner in which photoreceptor information is initially coded. Indeed as the *Pecten* retina does not appear to have provision for functional interaction between its sensory elements, the existence of two different response patterns to the ONSET and OFFSET of light recorded from the proximal (ON) retina and the distal (OFF) retina reflect differences in the polarity of the generator potential produced by light shining on the receptors in each of these retinae (Land, 1984). The OFF response pattern recorded from the distal retina (which contains the ciliated visual cells) is consistent with a generator potential in which light hyperpolarizes the cell. In this respect the ciliated receptor cells of the distal retina of *Pecten* resemble vertebrate photoreceptors where light also produces a hyperpolarization of the photoreceptor membrane (Baylor and Fuortes 1970; George and Hagins, 1983; Biernbaum and Bowndes, 1985), although in *Pecten* it is the potassium rather than the sodium channels that are affected. Indeed in vertebrate photoreceptors the hyperpolarization is brought about by a light-induced closure of sodium channels and a decrease in Na^+ conductance, whereas in the ciliated photoreceptors of *Pecten*, the hyperpolarization is due to a light-induced opening of potassium channels and an increase in K^+ conductance (Cornwell and Gorman, 1983). This difference in the way

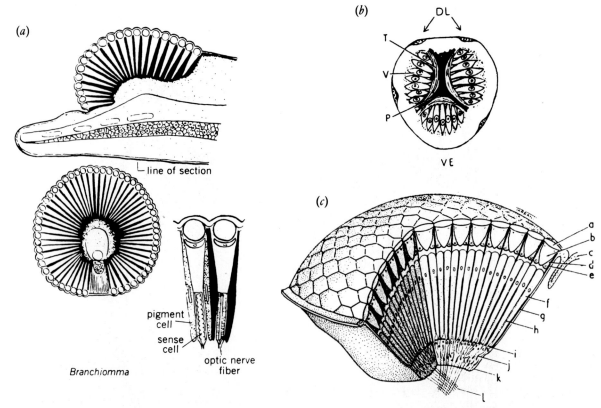

Fig. 2.12 (a) *Aggregate eye of the polychaete worm* Branchiomma vesiculoseum *(Hesse after Milne and Milne, 1959).* (b) *Composite ocellus of* Cypris: *the unpaired median eye represents fusion of three ocelli. DL, dorso-lateral ocelli. VE, ventral ocellus. P, pigment mantle. V, visual cells. T, tapetum. (After Claus: Duke-Elder, 1958).* (c) *Schematic diagram of the compound eye of an insect. a, corneal facet. b, crystalline cone. c, surface epithelium. d, matrix cells of the cornea. e, iris pigment cell. f, cell of the retinule. g, retinal pigment cell. h, rhabdome. i, fenestrated basement membrane. j, nerves from retinular cells. k, lamina ganglionaris. l, outer chiasma (After Hesse; Duke-Elder, 1958).*

in which hyperpolarization is generated in vertebrate and invertebrate ciliated photoreceptors has been interpreted as indicating that there is no evolutionary connection between them (see Messenger, Chapter 18). However it may be premature to conclude this until we can ascertain whether or not, cyclic nucleotides (such as cGMP), Ca^{2+}, and transducin-like proteins, which have a critical role in vertebrate photoreception (Lamb, 1986) are (or are not) implicated in the light induced responses of the *Pecten* retina. Whatever the answer, it is remarkable that there appears to be a correlation between hyperpolarization (induced by light) and ciliated forms of photoreceptors.

By contrast, the ON response of photoreceptor cells of the proximal retina in *Pecten* (which are derived from cells with microvilli) is brought about by *depolarization* of the receptor membrane by the action of light on the photoreceptor pigment. This was confirmed by MacReynold and Gorman (1970) who demonstrated that the photoreceptors of the distal and proximal layers of *Pecten*

irradiens respond to light with different polarities. This suggests a dichotomy in the response to light of ciliated and microvillous-type photoreceptors. i.e. in photoreceptors derived from cilia, light acts to hyperpolarize the membrane, whereas in those derived from microvilli, light depolarizes the membrane (Land, 1984). Thus the action of light on the photoreceptor pigment generates different ionic movements across the plasma membrane in these two types of cell.

In the case of the *Pecten* eye it is perhaps dangerous to assume that the output from both photoreceptor layers is concerned with vision. Undoubtedly, the distal OFF retina (which is at the focus of the reflection image from the tapetum) is responsible for the behavioural response to visual shadows and its function is therefore clearly visual. On the other hand there is no clear evidence that the output of the proximal ON retina affects the animal's visual behaviour and it is possible that the latter may simply be a photic organ that is involved in the regulation of the

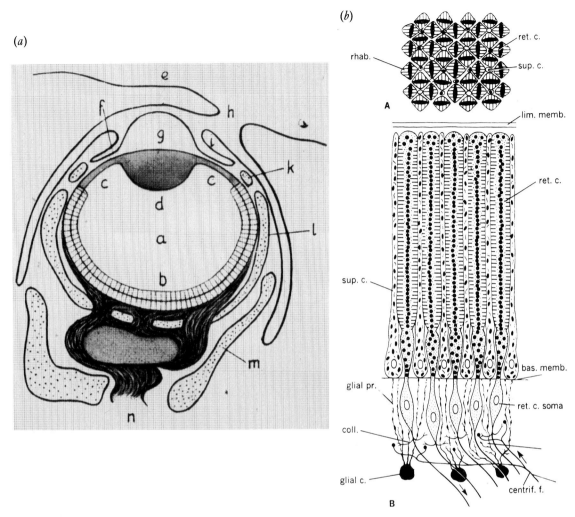

Fig. 2.13 (a) Cephalopod eye. a, invaginated epithelium forms the optic vesicle. b, retina. f and c, anterior and posterior parts of ciliary body. g and d, anterior and posterior surfaces of the lens. e, surface epithelium forms the cornea. h, hole or perforation (e.g. in Oegopsidae) so that the cavity is flushed by sea water (a possible homologue of the opening of the ampulla of the cephalic organ of a nemertine). i, iris. k, l, m, cartilage. n, optic nerve (After Duke-Elder, 1958). (b) Arrangement of elements of the retina of octopus. A, tangential section showing rhabdomeres. B, radial section of retina, illustrating lateral interaction between adjacent retinal elements (After J. Z. Young).

animal's reproductive cycle. In cephalopods, secretory structures such as the optic gland, white body, etc. that are in close association with the posterior parts of the eye are concerned with regulation of the animal's breeding cycle.

Vertebrates Evolved from a Nemertine-like Ancestor

There are now compelling reasons to suppose that vertebrates evolved from an ancestral nemertine in which the proboscis was everted through the mouth (Fig. 2.16). Indeed, Willmer (1970) notes that the controversial Silurian fossil *Jamoytius* shows what appears to be a notochord, and scales which apparently reflect an underlying myotomal segmentation, a dorso-ventrally flattened head with large lateral vesicles that could be interpreted as eyes, and no signs of any limbs, fins, etc. (Fig. 2.17(b)) (White, 1946; Ritchie, 1960). The fossil *Anaspid* fishes (Goodrich, 1909) were probably not much more advanced than these creatures, at least in some respects (Fig. 2.13(a)). Whatever the shortcomings of this suggestion, Willmer has pointed out how vertebrates fit in with a general scheme in

30 *Evolution of the Eye and Visual System*

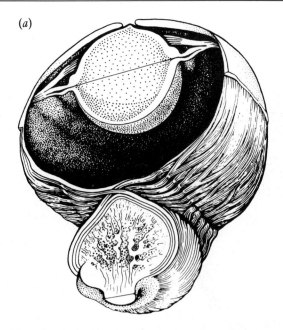

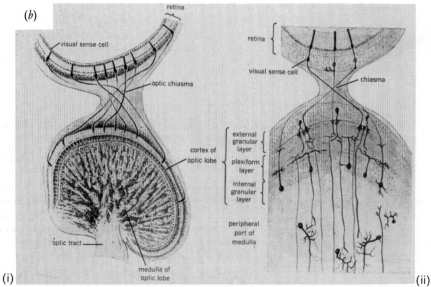

Fig. 2.14 *(a) Eye and optic lobe in octopus showing optic nerves running along the wall of the bulb to the bulb (After J. Z. Young). (b) General scene of the cephalopod retina low- (i) and high- (ii) magnification views, showing the relation of the visual cells with the optic lobe and the layers of the latter's cortex (After Ramon y Cajal, 1917).*

which a nemertine-like ancestor acquired a swimming habit and developed eyes to guide it.

However, it is pertinent to examine the sense organs of a nemertine such as *Lineus* and explore how they may have developed towards the photosensory organs of lower vertebrates. Another equally valid approach is to consider the photosensory organs that occur in primitive vertebrates and examine the various groups of invertebrates in search of possible precursors. Both approaches are based on the principle that evolution works by modification of existing mechanisms rather than by the sudden introduction of a new one.

In primitive rhabdocoeles (tubellarians) the animals react directly by movement to changed salt concentration in the region of the cephalic pits. The cephalic organ in creatures like *Lineus* and *Cerebratulus* and other littoral

Origin of Invertebrate and Vertebrate Eyes 31

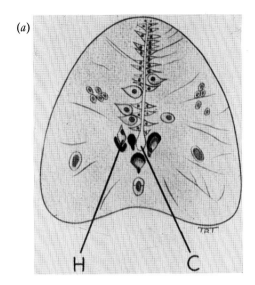

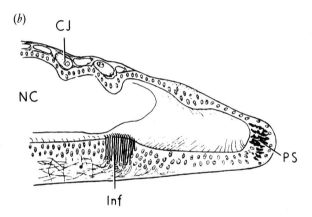

Fig. 2.15 *(a) Neural visual cells of* Amphioxus*: section of the spinal cord in the region of the fifth segment. C, central canal. H, large visual cells of Hesse with associated pigment cells (Hesse). (b) Sagittal section of the anterior portion of* Amphioxus*. Inf, infundibulum. CJ, cells of Joseph. NC, neural canal. PS, anterior pigment spot (Boeke, 1908).*

ing the following: (a) the vertebrate eye; (b) the cephalopod eye and the latter's associated optic gland; also (c) the curiously large eyes of polychaete worms *Aciopa* and *Vanadis*; and, even, (d) by evolving along rather different lines, the lateral eyes of the arthropods. In addition, Willmer has argued that the whole complex of the cephalic organ and its neighbouring tissues is remarkably suggestive of the sort of complex that must have been present in organisms which gave rise to the primitive vertebrates of which the existing representatives (with the exception of the hagfishes) all have well-formed eyes, a hypothalamus and a pituitary complex containing both anterior and posterior portions whose secretory activity is regulated by light. The proposal is that in the course of evolution the functions represented by these different structures became compartmentalized and separated from the ancestral cephalic organ. Indeed, among the cyclostomes, the eye of the hagfish *Myxine* is not directionally sensitive to light and is covered by skin and muscle. Histologically, the eye of *Myxine* is similar to certain sections of the cephalic organ of nemertines (Fig. 2.18(b)). *Myxine* has well-developed neurosecretory cells in its central nervous system. This might indicate that the cephalic organ rudiment, or placode, had incorporated its neurosecretory part into the nervous system, but that the eye-like part had been kept separated and had not developed much further, or that it had become degenerate.

Origin of Pineal and Parietal Eyes

In advancing his theory on the origin of the vertebrate eye, Balfour (1881) supposed that a portion of surface ectoderm, that was photosensory in some primitive ancestor, differentiated into neural ectoderm and was carried inside the animal by the evolution of the neural tube. The *foveolae opticae*, which expand to form the optic vesicles, were considered as representing an ancestral stage in which the eyes were essentially a pair of photosensory epithelial pits in the skin (Schimkewitsch's version of Balfour's theory, quoted by Walls, 1942). This theory was also adapted to explain the appearance of a third, dorsally placed 'pineal eye' that occurs in some lower vertebrates (Fig. 2.2). This eye may lie within the integument, beneath a depigmented area of the skin, or in a depression in the roof of the skull. A median or pineal eye was present in placoderms, in all major groups of bony fish of the Devonian period, in ancestral amphibians and in reptiles. By the Triassic period, many fishes, amphibians and reptiles lacked the pineal eye (Romer, 1970). Among living vertebrates, it is still found in some fishes, anurans, amphibians and lizards.

Embryologically, the pineal eye develops from the distal

nemertines is extremely well developed (and it is from such nemertine-like organisms that the vertebrate stock probably arose). The physiological role of the cephalic organ in nemertines is somewhat unclear, although it is known to react with secretory activity in response both to changes in salinity of its aquatic environment and also to the level of illumination. There is also some indication that it can respond to the direction of illumination. On the basis largely of morphological criteria, Willmer (1970) has suggested that if the nemertine or late rhabdocoeloid stock has branches in several directions then the cephalic organ could be the precursor of several later structures, includ-

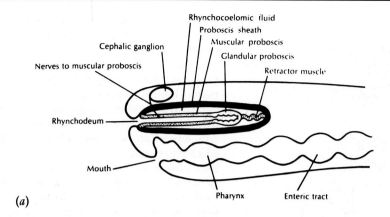
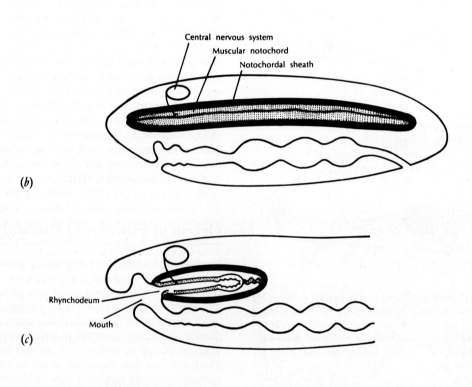
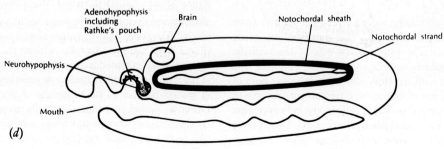

Fig. 2.16 *Diagram of the conversion of two types of nemertine ancestors (a,c) into notochords of protochordates (b) and of vertebrates (d) (After Sarnat and Netsky, 1981).*

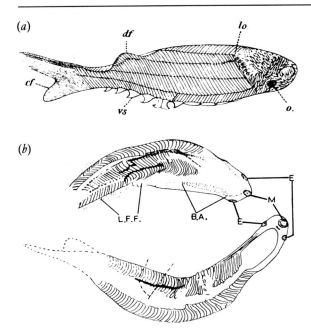

Fig. 2.17 *(a) Fossil anapsid* Birkenia elegans. *cf, caudal fin; df, dorsal fin. lo, lateral openings. o, orbit. vs, ventral scales (Goodrich, 1909, after Willmer). (b) Two specimens of the Silurian fossil* Jamyotius kerwoodi. *E, eye. L.F.F., lateral fin fold. M, mouth. B.A., branchial apertures (After Ritchie, 1960; Willmer, 1960).*

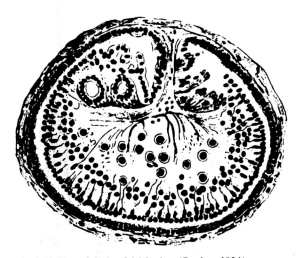

Fig. 2.18 *Eye of the hagfish* Myxine *(Ducker, 1924).*

end of the pineal or parapineal evagination of the epithalamic ependyma. It becomes separated from this primary evagination and continues to develop in a similar fashion to the paired lateral eyes, forming retina, lens and cornea. Nerve fibres from the retina of the parietal eye grow into the habenula (Ortman, 1960; Kappers, 1965). The

Origin of Invertebrate and Vertebrate Eyes 33

physiological role of the pineal that is associated with its response to light is suggested by the observations of the ammocoete larvae of *Lampetra planera* (Young, 1935). The latter displays a daily rhythm of contraction (in darkness) and expansion (in light) of the melanophores of the body. This rhythm is abolished by removal of the pineal. Similarly, in larval amphibia, darkness induces pallor and light brings about the expansion of the melanophores. In these animals pinealectomy abolishes the effect, although other factors besides the pineal and the epiphysis are involved (Bargnara, 1963). In the adult lamprey it may, in addition, be necessary to remove the lateral eyes before the rhythm is abolished. These results suggest that the pineal eyes were more primitively organized, and, among other things, their function was to mediate this darkening reaction. However, with the greater development of the lateral eyes the pineal eye has gone into a secondary position or into relative disuse.

In mammals the pineal body itself is probably not directly sensitive to light, but nevertheless stores serotonin and contains high concentrations of melatonin (a hormone that contracts melanophores), and other indole compounds, when the animal is kept in the dark. The enzyme, hydroxyindole-*o*-methyl-transferase, which is present in pineal tissue (and needed in the synthesis of melatonin), is inhibited by light. Consequently, less melatonin is formed during exposure to light (Wurtman *et al.*, 1963). The pineal eye, or some part of the epiphyseal complex, seems also to be able to fulfil similar functions. Light stops the production of melatonin, the hormone which in vertebrates contracts the melanophores (Wurtman *et al.*, 1963a), and UV inhibits the neural responses to light of other wavelengths (Dodt, 1964). Thus it seems quite possible that the pineal eye acts primarily to protect the organism from UV light. When the light gets too strong, it activates a mechanism for pulling the blinds over the rest of the body in the form of expanding melanophores, assisted by the pituitary which liberates a melanophore-expanding principle.

Biological Clock Function

Willmer (1970) remarked that the production of melatonin, or of a similar transmitter in the dark, and its cessation in the light, could provide a means for measuring the length of day, of initiating diurnal rhythms, and thus of controlling other vital processes that are affected by night and day and by seasonal changes. Eakin and Westfall (1960) reported that a PAS-positive material accumulates in the parietal eye in the dark-adapted lizard (Wurtman *et al.*, 1963b); they also noted that the pineal gland stores serotonin. Both of these observations are consistent with the action of the pineal organ as a clock. Thus, if melatonin were made only in darkness then a store of melatonin

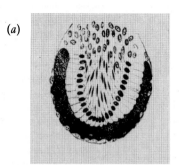

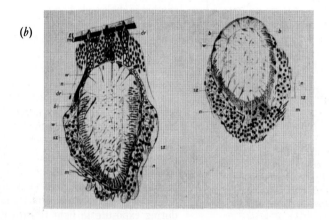

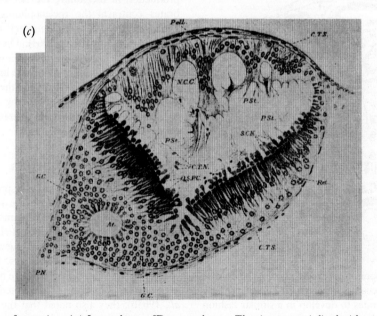

Fig. 2.19 *(a)–(b) Eyes of nemertines. (a) Inverted type of* Drepanophorus. *The pigment cup is lined with sensory cells leading to nerve cells. (b)* Geonemertes *(After Schroder, 1918, quoted by Willmer, 1970). (c) Section through the pineal eye of the lamprey,* Geotria australis. *(After Dendy, 1907; Willmer, 1970).*

could accumulate during the night and would start to run down during the day as its synthesis was inhibited by light.

Relation to the Eye Spots of Nemertines

Many nemertines, such as *Lineus*, have pigmented eye spots, similar to those of turbellarians, distributed over the anterior end of the body. Willmer drew attention to the fact that in the terrestrial nemertine *Geonomertes palaensis* (Schroder, 1918) the structure of the eye spot appears to be directly comparable to the pineal eye of the lamprey, *Geotria australis* (Fig. 2.19; Dendy, 1907). In the lamprey, the retina of the pineal eye is a direct one with sensory cells interspersed between supporting cells and it is not difficult to envisage its derivation from such eye spots as those of *Genomertes*. In lizards the pineal eye has a more elaborate structure (Fig. 2.20; Eakin and Westphall, 1959) and the sensory cells have an ultrastructure similar to that of cones of the mammalian retina (Eakin *et al.*, 1961), with the outer segment apparently formed from flagella (Fig. 2.21). However, unlike the vertebrate eye it is not endowed with much of a retinal ganglia (apart from its sensory and supporting cells) and in this respect resembles the eye spots of e.g. *Genomertes*. (Fig. 2.19(b)). This accords with the suggestion that some eye spot rudiment was internalized with the invagination of the neural ectoderm, and that the microvilli-derived photoreceptor cells, characteristic of primitive eye spots, were transformed (see pp 26–28) into photosensitive ependymal-like cells as an adaptation to a changed requirement for regulating and stabilizing the ionic conditions within the cavity formed by the closure of the neural tube.

The following summarizes some of the principal morphological and functional features that led Willmer (1970) to figure that the pineal eye (but not the lateral eyes) of vertebrates may have derived from the eye spots of nemertines:

1. The eye spots of nemertines are not provided with much in the way of retinal ganglia and can be removed without grossly interfering with directional light sensitivity.

2. In nemertines the eye spots are connected to the brain by nerves to the dorsal ganglia or commissure. In vertebrates the pineal eye is also connected to the brain via a dorsal nerve, whereas the optic nerves approach the brain along its ventral aspect.

3. The function of eye spots in nemertines may be similar to that in planarians where the eye spots have been shown (Merker, 1934) to be sensitive to ultraviolet light. A worm coming into shallow water or meeting terrestrial conditions might benefit from being able to detect and avoid UV light. The pineal eye, or some part of the hypophyseal complex, also appears to fulfil similar functions. Thus it is possible that the eye spots of a nemertine became converted into the pineal and became concerned with the detection of the total amount of light for protective purposes.

While the foregoing points to a possible connection between primitive eye spots and vertebrate pineal or parietal eyes, it would seem to exclude eye spots as the direct precursors of the vertebrate lateral eyes. Alternately, a modified version of Balfour theory, in which the placode of some peripheral structure such as the cephalic organ fuses and is carried inside by the evolution of the neural tube could reasonably account for the vertebrate eye and its inverted retina.

Cephalic Organ

The following description of the structure of the cephalic organ in nemertines (Fig. 2.22) follows the account given by Ling (1969) as reported by Willmer (1970):

In Lineus *the cephalic organ consists of a ciliated groove opening up into a wide ampulla that is lined with exceptionally long flagella between which are located numerous very long microvilli. This appears to be a sensory surface, in which the sensory cells have their flagella and much of their cytoplasm on the outside of an 'external limiting membrane' somewhat in the manner of receptors of the vertebrate eye. In addition, the physiological appearance and structure of the cephalic organ in response to changes in salinity and light*

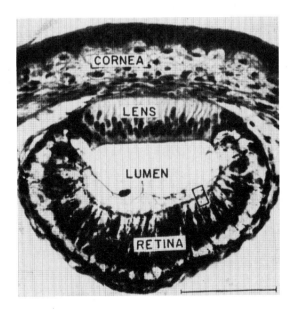

Fig. 2.20 *Section through the pineal eye of a lizard (Eakin and Westphall, 1959).*

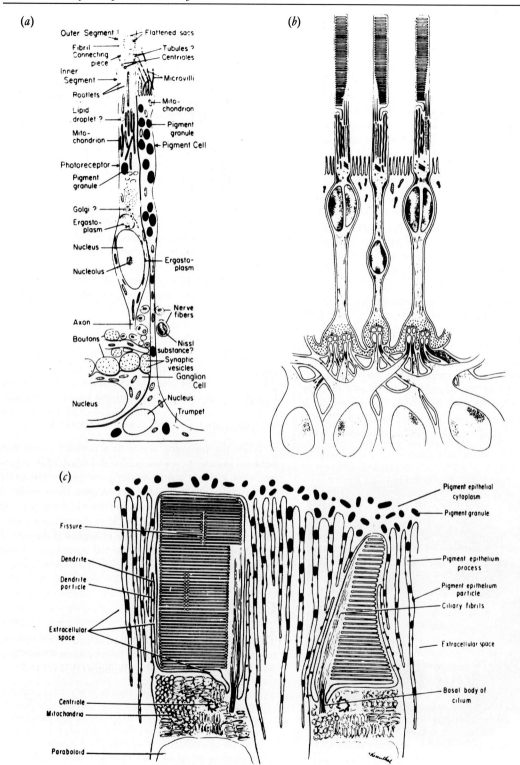

Fig. 2.21 *(a) Receptor and supporting cells in the pineal eye of a lizard (Eakin and Westphall, 1960). (b) Receptors in the guinea pig (Sjostrand, 1961). (c) Receptors and pigment epithelium of the mudpuppy (Brown et al., 1963).*

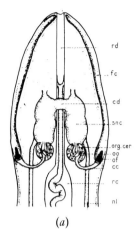 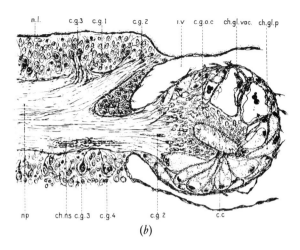

Fig. 2.22 Cephalic organ. (a) Anterior end of Lineus showing the position and general form of the cephalic organ. cc, cephalic canal. cd, dorsal commissure. fc, ciliated groove. nl, lateral nerve. of, opening of canal. oo, canal in cephalic organ. org.cer, cephalic organ. rc, rhynchocoel. rd, rhynchodaeum. snc, cephalic ganglion. (b) Longitudinal section of the cephalic organ. c.c., cephalic canal. c.g. 1–4, types of ganglion cells. c.g.o.c., ganglion cells in the cephalic organs. ch.gl.vac., vesicular tissue. ch.gl.p, secretory cells. i.v., blood vessel. np, neuropil. n.l., cephalic ganglion. ch. ns, neuro-secretory fibres (Lechenault, 1963, after Willmer).

levels reveal that it also has a secretory function.

The centre of the floor of the ampulla is raised into a conical hump, from the middle of which a narrow ciliated tube penetrates more deeply into the animal, bends three times at right angles, and finally terminates in a diverticulum surrounded by a group of highly vacuolated cells. The cilia in this tube are arranged partly radially and partly (along one side) longitudinally. The two parts are separated by a palisade of very long and large flagella. At the first two bends of this tube two groups of secretory cells can pour out their secretion into the lumen. When secretion does not occur the cells become engorged with their own products. The secretion stains with the same staining reaction (PAS, paraldehyde, fuchsin, etc.) as those of neurosecretory substances elsewhere, e.g. in the preoptic nucleus of the hypothalamus in vertebrates.

Around this flagellated or ciliated tube are several groups of nuclei corresponding to different classes of cells. The cells which separate the group bearing longitudinally oriented flagella from those with radial flagella, are structurally quite distinct and have very swollen flagella. The first group of nuclei around the tube belong to the cells forming the tube. Outside it there is a second group of cells. These send processes both toward the flagellate cells, where they appear to end between the cell bodies in a process bearing a simple cilium, and also to join the nerve fibres which run towards the main cerebral ganglia. Three sorts of nuclei are clearly distinguishable among these 'bipolar' cells (Ling, 1969). A second group of nuclei belongs to 'ganglion' cells that are similar to those of the cerebral ganglion themselves and while the detailed connexions of these cells is obscure, they nevertheless send fibres which pass into the main nerve going to the brain. This nerve may correspond to the optic nerve of vertebrates. Thus in the cephalic organ three main groups of cells (flagellated cells, bipolar cells and ganglion cells) are established as separate entities, and the main nerve is surrounded by neuroglia cells.'

Possibility that the Vertebrate Eye Evolved from the Cephalic Organ of Nemertines

Willmer suggests that an eye of the vertebrate type may have arisen by modification of the cephalic organ in nemertoid creatures as soon as they became actively swimming and required more directional control of their movements.

First, he noted that the existence of a large ampulla with closely packed cells already connected to the nervous system, suggests the possibility of directional sensitivity. The lining cells have both well-developed flagella and stereocilia (or very large microvilli), which on the one hand could have given rise (a) to the rods and cones; and on the other (b) to pigment epithelium whose long microvilli (e.g. in amphibia) surround the photoreceptors. The tube leading from the ampulla contains cells that are immediately connected to at least two groups of nerve cells in the organ before the issuing nerve enters the cerebral ganglia. There is therefore already a potential retinal ganglion with three distinct groups of integrating neurones (Willmer, 1970). Cytologically, the three types of nuclei present in the second group of cells bear a strong resemblance to the three types in the layer of bipolar cells in the vertebrate retina (Fig. 2.23).

Ontogenetically the vertebrate eye develops as a vesicle which evaginates from the side of the neural tube. The

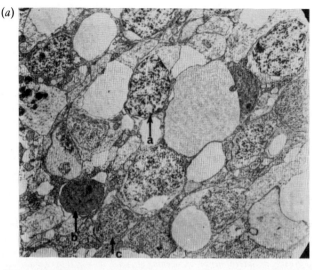
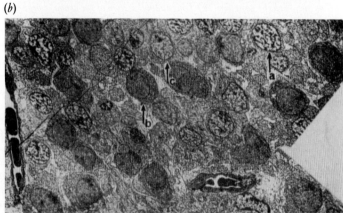

Fig. 2.23 *Among the cells that make up the cephalic organ, several are distinguishable on the basis of the cytological appearance of their nuclei. Among the different 'bipolar' cells of the cephalic organ, three sorts of nuclei are clearly distinguishable. Another distinct group of nuclei belongs to 'ganglion' cells, which send fibres which pass into the main nerve going to the brain. A comparison of the nuclei of bipolar cells of the nemertine cephalic organ with those of bipolar cells in the primate retina shows a remarkable resemblance. (a) Electron micrograph of the nuclei of 'bipolar' cells in the cephalic organ of Lineus (Photo by Ling). (b) Electron micrograph of nuclei of 'bipolar' cells in the primate retina (Villegas, 1960). Compare the appearance of nuclei a, b, c in the top and bottom pictures (After Willmer, 1970).*

cephalic organ of nemertines develops from a separate placode. During phylogeny this vesicle could have become fused with the invaginating neural tube and eventually become transferred to it. This would explain both the origin and position of the optic vesicles in the lateral wall of the neural tube, and also the inverted retina. Such a transference and attachment to another placode does in fact occur in some species of nemertines (Salensky, 1896). Since the cephalic organ is essentially a tubular structure derived by invagination of the surface, the inner part, where flagellate cells preponderate, could develop into the retina.

Possible Relationship Between the Cephalic Organ and the Vertebrate Lens

In advancing his theory about the origin of the vertebrate eye, Willmer considered the cephalic organ may also have given rise to the lens. The suggestion is that the outer part of the ampulla, which would be the last to invaginate (and would thereby retain its anterior position) could have given rise to the lens either (a) from the reorganization of its cells, perhaps in an attempt to produce another cephalic organ placode or ampulla; or (b) by production of mucoid secretion in the cavity, or (c) by a combination of both.

The concept that the retina and lens both derive initially from the same invaginated tubular structure provides a ready explanation for the anomalous regeneration of the amphibian lens from antero-dorsal retinal cells (Stone, 1967; Twitty, 1956).

The Diverse Physiological Roles of Vitamin A: Photoreception and/or Ion Transport

It is pertinent to reflect on the fact that a major problem that had to be solved in the evolution of cells and organisms was the need for them to sense and adjust to the changes in salinity in their aqueous environment. To achieve this it was necessary to evolve active membrane processes that enabled cells to regulate ionic conditions within the cell interior. Equally, for many aquatic species, both light and daylength, and the salt concentration of their environment, were implicated in the regulation of reproductive activity. This is still true of many invertebrate and some vertebrate species (notably the euryhaline fishes, marine turtles, etc). Since in some species, day length and salt sensitivity are sometimes both involved in the regulation of breeding it is perhaps not surprising to find that in some the two became linked, and sensitivity to light became coupled to ion regulation.

Secondly, the vast majority of visual mechanisms in the animal kingdom depend on the pigment rhodopsin or at any rate vitamin A_1 or A_2 (retinol), and a protein opsin. The importance of vitamin A is not only its role in photoreception but also its use in the regulation of membrane permeability, e.g. of erythrocytes and lysosomes (Dingle and Lucy, 1965), and of retinal in egg membranes of fish (Plack et al., 1959) etc. A particularly interesting example where vitamin A is involved in ion transport is to be found in the marine purple bacterium *Halobacterium habolium* which lives in an aquatic environment where the salt concentration is exceptionally high (namely the Dead Sea).

In *H. habolium*, a form of bacteriorhodopsin in the bacterial membrane is used primarily as part of a proton pump. However, there is also evidence that *H. habolium* also uses its bacteriorhodopsin for sensory transduction (Spudich and Bogomolni, 1984). Here then is an example of a photopigment being used for a dual purpose. Retinal (vitamin A aldehyde) extracted from *H. habolium* will complex with bovine opsin, although bacteriorhodopsin differs from vertebrate rhodopsin in a number of ways, i.e.: (a) there is little or no homology between the amino acid sequence (primary) structure of bacteriorhodopsin and animal visual pigments (Hargrave et al., 1983), and (b) retinal in bacteriorhodopsin cycles between all-*trans* and 13-*cis* instead of 11-*cis* as in vertebrate, e.g. cattle, rhodopsin (Stockenius and Bogomolni, 1984). Despite these differences there are some important conformational features which bacteriorhodopsin shares with visual pigments, i.e. both classes of molecule make seven traverses of the membrane (Engelman and Zaccai, 1980). This has led to the suggestion that the secondary and tertiary structure may perhaps have been the only common foci of natural selection and that the lack of sequence homology may simply be a manifestation of the fact that the divergence of visual opsins and bacteriorhodopsin took place such a long time ago (see Goldsmith, Volume 6, Chapter 5).

Willmer has suggested that the ampulla and tube of the cephalic organ may also contain retinal-like substances. While this has yet to be confirmed, similar changes of cell permeability or ionic balance could be produced in these structures, either directly, by changing the ionic distribution, or indirectly, by the action of light or darkness on their flagella; i.e., light might act on a presumptive retinal-protein complex associated with flagellated epithelial cells. Because of the central role of vitamin A in regulating membrane permeability, and in other cases photoreception, the change-over, from an organ responding to NaCl or to salinity changes, to an organ detecting changes in light intensity could be effected rather easily if this change had some biological advantage.

The Origin of Vertebrate and Invertebrate Photoreceptors

Comparative ultrastructural studies of photoreceptors by Wolken (1958, 1959a) and others have led, on the basis of certain morphological features, to the general classification of photoreceptors into two broad categories. One group, characteristic of vertebrates and typified by rods and cones, derive from flagellated (ependymal-like) cells. The second, characteristic of the majority of photoreceptors in invertebrates, derive from microvilli and consist of an array of microvilli or tubules called a rhabdomere. It is normally supposed that these two categories of photoreceptor evolved independently from one another; however, Willmer (1970) has outlined a scheme that allows vertebrate and invertebrate photoreceptors to evolve from the same ancestral structure.

Basing his argument on the possibility that some invertebrate phyla, as well as vertebrates, arose from creatures related to nemertoid stock, Willmer suggests that the cephalic organ, as well as being a precursor of the vertebrate eye, might equally be a precursor of invertebrate simple eyes. For example, he suggests that the cephalopod eye could have been derived by more or less direct development of the ampulla, giving rise to the retina (with its direct sensory elements), while the lens could have developed from mucoid secretions from cells in the mouth of the tube. The cephalopod eye develops embryologically by direct invagination of the surface, and such development would be consistent with the above idea. A feature of the cephalopod eye that is of particular interest is that the anterior chamber of its eye is in direct communication with the sea (Fig. 2.13(b)), suggesting that it may have retained some of the characteristics of an organ for detecting changes in salinity of its aquatic environment. This would be consistent with its presumed derivation from the ampulla of nemertines.

Unlike vertebrates, the photoreceptor cells of the squid are modified microvilli (Wolken, 1958; Zonana, 1961). Cilia do not seem to be present. This would seem to discount development from the ampulla because the latter have very well developed cilia (or flagella). However, some of the cells of the ampulla have very well developed microvilli, while the deeper and more tubular part of the organ, from which the vertebrate eye is more likely to have originated, has mostly cilia and fewer microvilli. Furthermore, in the squid the eye is spatially separated from the optic lobe, whereas in vertebrates the integrating cells lie juxtaposed with photoreceptors in the retina. This spatial arrangement is consistent with the idea that the photoreceptors of vertebrates and squid derived from different locations of the cephalic organ and that in each case, the nerve cells in the cephalic ganglion proper became the integrating cells. Equally, the idea that both vertebrate and some invertebrate eyes share a common lineage with the cephalic organ may account partly for the structure of photoreceptors in the curious two-layered retina of *Pecten*. In this bivalve, the photoreceptors located in the proximal retinal layer derive from microvilli, and are therefore like those of most invertebrate simple eyes, while those in the other, distal layer are vertebrate-like and based on cillia.

Phylogenetic Relationship Between Vertebrate Rods and Cones

In the development of the vertebrate eye, an interesting feature is the dichotomy between rods and cones. The initial distinction between the two types of cell, based on morphological (and later functional) criteria was originally proposed by Schultze (1866). This simple distinction has been greatly complicated by the discovery of intermediate forms, e.g. cones with rod-like properties in nocturnal geckoes (Crescitelli, 1965), or rod systems with cone-like properties such as the green rod system in frogs, as well as many other intermediate forms (see Bowmaker, chapter 4 and Thompson, chapter 7, and Goldsmith, chapter 5 in Volume 6). Such observations led Gordon Walls (1942) to propose that in some phyletic lines rods and cones could be transformed into one another by a process, whose mechanism was unknown, which he called transmutation, i.e. that some cone systems somehow became transmuted into rods and vice versa. However, despite the complications introduced by the appearance of intermediate forms, the classical distinction between rods and cones would appear to represent a fundamental structural and functional polarity between two different classes of cell. In addition to the morphological and functional criteria commonly used to distinguish between a whole variety of rod and cone types, it has now been established that a further distinction can be made on the basis of the molecular structure of rod and cone opsins (Underwood, 1968, 1970) although the genes for both vertebrate rhodopsin and cone visual pigments appear to have evolved, through gene duplication and mutation, from a common primordial visual pigment gene (see Goldsmith, chapter 5 and Piantanida, chapter 6 in Volume 6).

Both vertebrate and invertebrate retinae derive ultimately from epithelium, as indeed does the cephalic organ of nemertines. Willmer notes that during the early evolution of multicellular organisms (for example in the embryo of the sponge) two main groups of cells become evident in epithelial tissues, and these display opposite polarities with respect to fluid and ion translocation (Fig. 2.24(a)). One group, broadly referred to as mechanocytes, usually have flagella or cilia and contain several nucleoli. The other, designated amoebocytes, include amoeboid cells and those with microvilli. Among invertebrate eyes this fundamental duality/polarity of cell types (which reflects the retina's epithelial origin) is manifested prominently in the eye of *Pecten irradiens* (Fig. 2.11) and also in the distal eye of the coelenterate *Charybdea* (Fig. 2.8).

Amoebocytes are generally capable of pinocytosis, and their morphological features generally include a single nucleolus, lobosed pseudopodia, flattened surfaces, or indented surfaces as seen in the distal tubules of the kidney. Cells at this end of the scale have a propensity to store carbohydrates or mucopolysaccharides within their cytoplasm. These characteristics are also seen in invertebrate photoreceptors (which develop from microvilli) and in some cells of the pigment epithelium of vertebrate retina, which places these cells at the amoebocytic end of the series. Similar features may be found in the typical cone which has a single nucleolus, stores carbohydrate in a paraboloid as well as lipids and carotenoids, and has a broad (or lobose) cone pedicle. This contrasts with the typical rod, which generally has more than one nucleolus, no stored lipid, usually no paraboloid, and a synaptic terminal that is a simple synaptic spherule. On the basis of such criteria Willmer suggested that rods correspond more to the flagellate end of the series and cones to the microvillus end. The fact that the cone outer segment develops from a cilium would seem to be inconsistent with this argument until one recalls that in *Triturus* the retina (which contains both rods and cones) is capable of regenerating from the pigment epithelium, where many cells display prominent villi. Rods and cones probably lie at some intermediate position in the series, where each displays some features that are characteristic of cells that

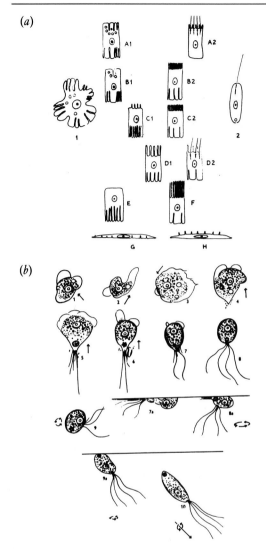

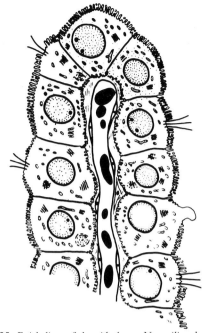

Fig. 2.25 *Epithelium of choroid plexus. Note ciliated and non-ciliated cells. The surfaces of some cells have both cilia and microvilli, others only microvilli (Millen and Rogers, 1956).*

Fig. 2.24 *(a) Simplified diagrams of the main gradations (a–e) of cell surfaces in various epithelia in relation to the amoebocytic cell on the left (1) and to the flagellate cell on the right (2). In the epithelial cells the upper surface is the distal one, and the lower surface is the basal or proximal one adjacent to the tissue fluids. In E the infolded membranes lie next to the aqueous humour from which they are separated by a basement membrane. A, Trachia (a cell with microvilli only has also been described): A1, goblet cell; A2, ciliated cell. B, intestine: B1, goblet cell; B2, brush-border cell. C, kidney: CV1, distal tubule cell: C2, proximal tubule cell. D, choroid plexus: D1, 'polypoid cell; D2, ciliated cell. E, cell of ciliary process (adjacent to vitreous body). F, cell from epididymis with stereocilia. G, endothelial cell. H, cell from peritoneal surface (After Willmer, 1970). (b) Diagrammatic representation of the change in form of* Naegleria gruberi *when the organism is placed in distilled water. 1–3, amoeboid form. 4, polarized form. 5, 6, filiform pseudopodia present at posterior pole. 7–10, development of active flagella. (After Willmer, 1956, 1970).*

lie on either side of the divide, namely cones tending towards the microvillus end and rods tending more towards the flagellate end of the series. This organization is illustrated by the epithelium of the choroid plexus (Fig. 2.25) which is made up of ciliated and non-ciliated cells yet all the cells display villi to a greater or lesser degree. According to Willmer's classification, those cells that only have microvilli are clearly amoebocytes, while those that have villi and cilia display a mix of amoebocytic and mechanocytic characteristics. The contention is not only that the classical distinction between rods and cones is related to the fundamental duality of cell types in epithelial tissues but also that photoreceptors, i.e. cone-like rods, rod-like cones, and indeed the great variety of cone and rod types, reflect the intermediate forms in epithelial tissues (Fig. 2.25). In this respect it is interesting to note that isolated cilia were observed by Atler (1965) in inner retinal neurones and in retinal pigment epithelium which also display microvilli. In addition it is suggested that the spectrum of different cell types found in epithelial tissues (and in different eyes) may be dictated by the ionic conditions prevailing in those eyes and by the specific requirements within those tissues for maintaining the ionic equilibrium.

Organization of the Vertebrate Pigment Epithelium

The fundamental polarity represented by the different epithelial cell types in the translocation of fluid and in the maintenance of ionic equilibria is well illustrated by the retinal pigment epithelium. Beyond the retina and over the ciliary processes, the pigment epithelium continues as a pigmented layer until it reaches the iris. There it folds back on itself to form a double-layered epithelium in which both layers are pigmented. When it again reaches the ciliary processes it continues to cover the underlying pigment epithelium, but it loses its pigment until it arrives at the margin of the retina. Over the ciliary processes the cells of the inner and outer layers are very different in their microstructure and there is a secretion of aqueous humour across the two layers of the epithelium. The outer pigmented layer secretes into the erstwhile cavity of the neural tube (i.e. the cavity that became occluded during the process of invagination of the optic cup). The unpigmented layer secretes from this potential cavity of the bulb. Thus the inner and outer layers are working in opposite directions with respect to the original cavity, but with a net transfer of fluid into the eye (Willmer, 1970).

In nemertines, the function of the epithelium of the cephalic organ appears to be related to salt balance. Consequently, if the vertebrate optic vesicle is indeed phylogenetically linked with the original placode of the cephalic organ, then the lining cells of the vesicle may well have a role in the maintenance of ionic equilibria. Willmer's (1970) hypothesis focuses on the fact that fluid leaves the cavity of the optic vesicle when invagination of the vesicle begins. During this process, the inner and outer parts of the epithelium come close to each other until the layer of cells, which will form the rods and cones of the retina, lies in intimate contact with the pigment layer. All that is left of the original cerebrospinal fluid between the two layers lies in the minute spaces between the rods and cones and the processes of the pigment cells. The composition of this fluid is presumably determined by the interplay of the pigment cells and the presumptive retinal cells acting on the original cerebrospinal fluid, on materials available from the choroidal blood vessels and on those from retinal vessels on the other.

In the adult vertebrate retina vitamin A is transported, via retinol binding protein, from the blood to the pigment epithelium and from there along the villi of the pigment epithelium to the receptors themselves. Vitamin A is also likely to be present in the developing optic vesicle where its role may be predominantly to regulate ionic equilibria. The outer segments are among the last features to differentiate in presumptive photoreceptor cells in the developing retina. Therefore visual cells may in the early stages have well been concerned with ionic equilibria. Indeed, when tissue cultures are made from suspensions of retinal cells they aggregate into clusters and form hollow vesicles, tubules and rosettes. At other times, solid clusters of larger and more cytoplasmic units are formed as 'lentoids' and these two groups may coexist side by side (Tansley, 1933; Pomerat and Littlejohn, 1956; Moscona, 1957). These formations are reminiscent of the behaviour in culture of the proximal and distal tubules of the avian kidney and are strongly suggestive of oriented cells of two types with respect to the passage of fluid through them.

Commenting on this dual classification of epithelial cells in terms of their surface morphology, Willmer remarks that this is to some degree arbitrary. He notes that in metazoa the epithelia separate external or luminal cavities from the tissue fluid and blood. In for example a blastula, the animal–vegetal gradient might mean that cells strongly organized for the ejection of water and for the conservation of ions would concentrate at one end, while ion water movement of the cells at the other end would display the opposite polarity. Cells of intermediate type might act less strongly in one direction or the other, so that at some axial location along the blastula (not necessarily equatorially), there would be cells where the movement of water and ions in either direction was balanced. In almost all blastulae the cells grade in morphology more or less sharply from animal pole to vegetal pole.

Epithelial Origin of the Integrative Properties of Nerve Cells

Among the photoreceptor responses found in animal eyes, there appear to be two fundamental types: those where the cell is depolarized by the action of light and others where the cell is hyperpolarized. The former is common among photoreceptors that develop from microvilli. The latter is characteristic in those derived from cilia. While this simple dichotomy is perhaps obscured by the existence of intermediate forms there is nevertheless a close correlation between photoreceptor response types and the fundamental duality of response and morphological types found in epithelial cells, with respect to the polarity of ion translocation.

Photoreceptors and neurones derive from a common epithelial origin. Indeed an animal photoreceptor is simply a nerve cell (formerly an epithelial cell) that has become specialized to detect and respond to light. All neurones, other than primary receptor cells, carry out a sensory function at their postsynaptic membrane sites. That is, receptors sited in postsynaptic membrane bind specific chemical transmitter molecules released by presynaptic terminals; these may be either depolarizing or hyperpolarizing in their action, depending on the specific nature of the transmitter and receptors in postsynaptic membrane. Thus the synaptic membranes of neurones, like the light-sensitive membranes of photoreceptors, display an epithelial-like duality in response type with regard to the polarity of stimulus-induced translocation of ions across their synaptic membranes. As light-sensitive and postsynaptic membranes both perform a 'sensory' function, it is perhaps not unreasonable to enquire if the similarity between photoreceptors and neurones also extends to the operation of their transduction machinery at these 'sensory surfaces'. In this regard it is interesting that recent work has pointed to similarities between opsins, the beta-adrenergic receptor (Dixon et al., 1986) and the muscarinic receptor (Kubo et al., 1986). This indeed suggests that such receptor proteins are part of a larger group of related proteins that function by activating second messenger cascades (see also chapter 5 by Goldsmith in Volume 6). One further point worth noting relates to the possible epithelial origin of the integrative properties of nerve cells, which depend so critically on the interplay between depolarizing and hyperpolarizing membrane potentials. In neurones such potentials are usually generated in postsynaptic membrane at excitatory and inhibitory synaptic sites, at different locations over the cell surface. In some epithelial cells, such as are found in choroid plexus (Fig. 2.25), the cell features microvilli in some parts of its surface and cilia in others. According to the scheme we have been following, this suggests that different areas of the plasma membrane surface may be translocating ions in opposite directions, and this indeed may have foreshadowed the evolution of excitatory and inhibitory synaptic contacts.

What Determines the Type and Variety of Photoreceptors Found in the Eyes of Different Species?

Willmer (1970) notes that if the retina is considered as being initially like any other epithelium, i.e. as containing basically two types of cell, then exactly how these cells have to function in the photoreceptive layer of a given species must depend on the equilibrium which they must maintain with their immediate surroundings and also with the environment of the various conducting and integrating cells in the rest of the retina. This equilibrium with the surroundings must depend not only on the cells of the rod and cone layer, but also on the activity of cells of the pigment layer and on the extent to which the composition of the blood in the neighbourhood of these cell groups is controlled and suitable without further adjustment. In one species, rods may have to be very different from cones to maintain an ionic equilibrium which might otherwise be ill-regulated. In another species the internal environment may be already so well regulated that rods and cones have only to act as fine adjustments 'pushing and pulling' about the equilibrium point. To do this, both may have to act more or less 'amoebocytically' if the concentration tends to be high, or as 'flagellated' cells (with the opposite polarity with respect to water and ion transport) if they tend to be on the low side. Both units might therefore appear cone-like, or both might look like rods. Against this background one may perhaps view the selection pressures which lead to the phylogenetic development of new functional adaptations (such as the acquisition of new cone systems for colour vision etc.), as operating by changing ionic conditions within the eye, shifting the equilibrium and thereby refashioning the characteristics and proportions of different photoreceptors within the retina. How might this transmutation of cell types be brought about?

One possibility arises from the finding that within epithelial tissues, a variety of cells display intermediate forms of the two fundamental types by combining various proportions of amoebocytic and flagellate characteristics. Secondly, under appropriate physiological circumstances these two features may be interconvertible. An illustration of how an amoeboid cell may transform into a flagellate one, and vice versa, in response to changes in the moisture and salinity of its environment is afforded by the protozoan *Naegleria*, whose sexual phase has flagellated gametes (Fig. 2.24(b)).

During its normal life history, *Naegleria* can change reversibly from being a creeping phagocytic and amoeboid organism to become a free-swimming flagellate. *Naegleria gruberi* is normally amoeboid and lives in the soil. During this phase of its life cycle it is without a definite polarity and feeds on bacteria. If these organisms are transferred to distilled water, they immediately become polarized in that pseudopodia continue to form anteriorly while a tail region develops at the other end. The latter after a while develops one or more flagella beating from the posterior end of the cell; meanwhile, the anterior end has become round or oval in shape, and has lost, or almost lost, the capacity to put forward pseudopodia (Fig. 2.24(b)). The formation of the flagellum and basal body in *Naegleria* in response to, for example, a change in moisture and salinity of its environment, takes place entirely *de novo*. There is no trace of the flagellum in the amoeboid phase (Schuster, 1963; Dingle and Fulton, 1966). Observations on the behaviour of *Naegleria*, as well as of sponge embryos, indicate that the pinocytosing and mucoid cells, and those with microvillous surfaces, are more closely related to the original amoeboid cells (Fig. 2.24 (a)). As to the mechanism which brings about the transformation it suggests that, at the genetic level, *Naegleria* already has all the necessary information and synthesizing apparatus to produce either amoeboid or flagellate activity. If different forms of proteins are required in the two forms, then there must be regulation of some sort to minimize production of the unwanted proteins. Since the behaviour of *Naegleria* is determined by the environment, it could be suggested that changes in the cell surface are perhaps among the effective agents for calling for different proteins and so redirecting, by some means, the activity of the nucleus to meet that demand. If this is so, then it is conceivable that similar changes of cell permeability or ionic balance could be produced in them, either directly by changing the ionic distribution or indirectly by the action of light or darkness on their flagella, i.e. light might act on the retinal-protein system to produce effects on the cells similar to those caused by a hypotonic solution. In the alga *Chlamydomonas reinhardi*, gamete formation may be induced by adding distilled water to the vegetative phase living on agar slopes. Gamete formation involves production of flagella, and requires or is hastened by the action of light. Light, flagella and ionic concentration thus appear to be involved in gamete formation (Lewin, 1953; Sagar and Granick, 1954). Hence a connection between flagella, light sensitivity and salt balance is perhaps not so remarkable.

For the gecko retina, it has been suggested firstly (Goldsmith, Volume 6, Chapter 5) that its curious cone-like rods may have derived from a diurnal ancestor which lost its gene for rod opsin, and that the loss of this pigment gene may have been a contributory factor to transmutation, and secondly that the transformation of cone opsin, displacing it towards the rod end of the scale, occurred as an adaptive response to nocturnal life. If light and the character of the visual pigment affect ion movement, as is the case for bacteriorhodopsin, then the loss of a pigment gene, coupled with a nocturnal habit, may alter the ionic equilibrium within the retina in favour of rod-like receptor cells.

Origin of the Vertebrate Lens

The vertebrate lens is essentially a structure derived from ectoderm, and its formation is generally induced directly or indirectly as the result of the approach and contact of the optic vesicle with the under-surface of this epithelium (Huxley and de Beer, 1934; Spemann, 1938). In some frog species, only the normal overlying ectoderm is competent, while in others any ectoderm is competent. In yet others, the lens develops in its normal position with or without the underlying optic vesicle (Karp and Berrill, 1981; Karkinen-Jaaskelainen, 1978). Unrelated tissues can also induce lens differentiation (Jacobson, 1966; Karp and Berrill, 1981).

The suggestion is that while the retina and the lens formed from the outer part of the cephalic organ, the latter would be the last to invaginate, and could remain in its original position to give rise to the lens from a reorganization of its cells. To explain this reorganization, Willmer proposed a similar mechanism to that suggested previously to account for the development of rods and cones as distinct cell types.

A key stage in normal lens development is the inductive influence exercised by the optic vesicle when it contacts the surface ectoderm. Except in some simple systems such as the slime mould, the mechanism of embryonic induction is generally unclear, although it is still thought to depend on some hormone-like influence. Despite the concerted efforts of many able investigators, no one has yet succeeded in isolating or demonstrating a (diffusible) inductive agent for the lens. Willmer suggests that similar changes in a cell's organization or behaviour may be determined by the aggregate of local changes in a cell's environment produced by the activities of neighbouring cells – specifically, that local differences in concentration, either up or down, in essential metabolites or specific ion concentrations, may be all that is necessary to remodel the activities of a cell. This indeed could account for the fact that, given the appropriate conditions, a lens may form in a piece of ectodermal tissue irrespective of its site of origin (Jackson, 1966).

During development of the eye, the optic vesicle enlarges and fills with fluid, but on touching the ectoderm it begins to invaginate and loses fluid from the inside. At this point the ectoderm cells overlying the vesicle invaginate coincidentally with the invaginating cup, and eventually form a small separate vesicle which is closed off from the outside world (Fig. 2.26). The lens vesicle at this stage, while it is filled with fluid, is reminiscent in some ways of a little blastula. The cells of the lens then become visibly separated into two groups. The outer, more superficial cells remain as a low cuboidal epithelium; the inner cells elongate and begin to form the special lens proteins (crystallins). When this occurs, the contained fluid disappears and the lens continues to grow as a solid body by the progressive addition of cells to both its inner and outer layers through cell divisions that occur mostly equatorially.

In *Xenopus*, the lens can regenerate from the inner layer of cells of the corneal epithelium under the influence of the eyecup. Willmer's hypothesis about how cells in the surface ectoderm reorganize to form the cellular lens derives in part from the cytology of normal corneal epithelium from which such regeneration can take place. An examination of normal epithelium shows an even (and random) distribution between cells that feature one nucleolus (an amoebocytic characteristic) and those having two or more nucleoli (commonly found in cells at the flagellate end of the series). However, in the particular localized area of surface ectoderm from which the lens will actually form, there is a great preponderance of cells with only one nucleolus. Close examination also suggested that most of the large cells that form the inner layer (i.e. those that form the lens fibres) possess only a single nucleolus (Freeman, 1963). The change in the number and distribution of these two types in the corneal epithelium before lens formation actually begins emphasizes possible changes in the local environment and a means of getting two types of cells into

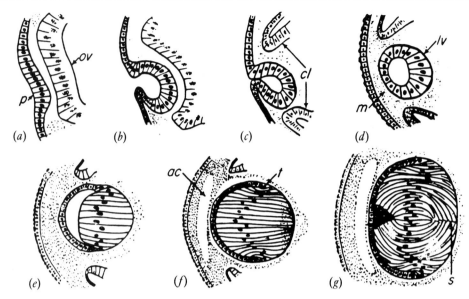

Fig. 2.26 *Development of the vertebrate lens: changes in the epithelium over the optic vesicle that lead to the formation of a lens. Note the blastula-like arrangement of the cells in (d), (c) and (e). Anterior chamber: cl, lips of optic cup. lv, lens vescile. m, mesenchyme. ov, optic vesicle. p, primordium of lens. s, suture. t, transitional zone (Adapted from Walls, 1942).*

the vesicle in proportions that lead to successful lens formation.

Possibility that the Cephalic Organ of Nemertines May also Be Linked Phylogenetically to the Cephalopod Optic Gland and the Vertebrate Hypothalamus

In addition to the strictly neural tissue connected with the eye, cephalopods have other structures, i.e. the white body, the optic gland, the subpedunculate body and the parolfactory bodies, which lie in close association with the posterior parts of the eye. Willmer notes that the cytology of some of these bodies (Boycott and Young, 1956) has sufficient similarity to the glandular and vesicular tissues of the cephalic organ of nemertines to suggest that the two could be related.

The optic gland, in particular, has secretory material in its cells and a nuclear structure very like that of the secretory cells of the cephalic organ. In cephalopods the optic gland is concerned with the sexual development of the animal, and removal of the organ causes the gonads to degenerate (Wells and Wells, 1959; Wells, 1964). In this context, but also in that of the possible evolution of the cephalic organ in the vertebrate direction, it may be asked what happens to the glandular part of the cephalic organ if the rest of the organ becomes converted into the eye.

For the cephalopods the answer may lie in the optic gland and associated structures. It has been suggested that in vertebrates, in ontogeny, and subsequently in phylogeny, the cephalic organ anlage may have fused with the anlage of the neural tube so that it now appears as an outgrowth of the latter, the optic vesicle. It would not be surprising therefore to find that in the process the glandular cells may have separated to some extent from the neuroretina and become incorporated into the main nervous system, where they reappeared as the cells of the preoptic and other nuclei of the hypothalamus, where similar neurosecretory cells and vesicular cells are found. These cells in the hypothalamus have the capacity to respond to changes in salt concentrations much as do the cells of the cephalic organ (Andersson and Jewel, 1957). Also, Oztan and Gorbman (1960) have shown that in larval lampreys the neurosecretory cells of the hypothalamus also become depleted of their secretion in response to light.

In the nemertoid stage, the salt concentration of the environment could well have been important in regulating breeding activity. If the stimuli of light and of low salt both produced similar effects, in environments in which the salt was relatively constant, some animals might have tuned their sexual activities to the strength or duration of illumination. This would provide a rationale for the well-known relationship between light and breeding in vertebrates which is mediated by the hypothalamic region. Equally, the well-developed cephalic organs of the nemertines are integrated with the cephalic ganglia by means of a large nerve which essentially could have become the optic nerve. However, in the nemertines the nerve connecting the cephalic organ to the cephalic ganglion does not cross the midline, and in this respect differs significantly from the decussation of visual afferents in vertebrates.

Evolution of Decussation of the Vertebrate Visual Pathway

In the embryo, each of the lateral eyes in vertebrates is formed from an optic vesicle that emerges from the same side of the neural tube. Yet in all vertebrates optic axons of retinal ganglion cells cross the midline under the floor of the forebrain to connect to visual centres on the opposite side of the brain. In man and higher animals, there is hemi-decussation at the chiasm, and the optic tract contains an almost equal number of fibres from each eye. In other mammals fibres from the contralateral eye predominate in the optic tract and, while it is often stated that in submammalian vertebrates there is complete decussation of the optic nerve, many lower vertebrates possess a small ipsilateral projection to the visual centres. Hence the presence of a small uncrossed component of the optic nerve is therefore phylogenetically old (for details see Sarnat and Netsky, 1981).

In arthropods and cephalopods, fibres within the optic nerve are likewise arranged in a retinotopic manner and project topographically in the visual centres of the brain. However, it is only in vertebrates that optic fibres cross the midline to connect directly with centres on the opposite side of the brain. To explain this visual decussation Ramon y Cajal (quoted by Sarnat and Netsky, 1981) pro-

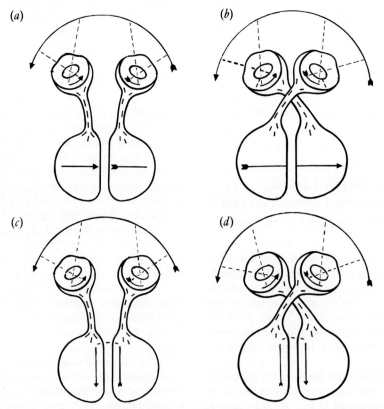

Fig. 2.27 *Evolution of visual decussation. (a) Ramon y Cajal noted that with the evolution of an image-forming chambered eye the image of a horizontally moving object is reversed in each eye. He then argued that, for a non-decussating visual pathway, the central representation of such a moving object along the medio-lateral axis of each optic lobe would be discontinuous and incoherent because the direction of movement in each lobe would be the opposite of that in the other. (b) He therefore suggested that decussation of the visual pathway evolved to compensate for this, so as to enable the central representation of the direction and continuity of the movement to be preserved in both optic lobes. In advancing his hypothesis, Cajal assumed that the horizontal axis of the visual field was represented in each case along the medio-lateral axis of the optic lobe. For cephalopods, arthropods and vertebrates, this is now known to be incorrect. (c,d) Instead, the horizontal visual axis is represented along the rostro-caudal axis of the optic lobe to which each eye is connected. This allows the direction and continuity of the represented movement to be preserved in both optic lobes, irrespective of whether the visual pathway decussates or not, or whether it is a compound or a chambered eye.*

posed that because the lens in the chambered eye produces a reversed image of the visual object on the retina it was necessary for the optic fibres to decussate to compensate for this image reversal (Fig. 2.27(b)). Cajal argued that if the optic nerves from such a chambered eye failed to decussate and projected ipsilaterally (i.e. to the same side of the brain as the eye) then the continuity of parts of images appearing, for example, in the visual fields of both eyes would be retinotopically projected to the optic tectum in a fragmented fashion that would disrupt the coherence of the retinal image (Fig. 2.27(a)).

This argument applies only if the horizontal visual axis is represented medio-laterally in the visual centres of the brain. However, it is now know that, for example, in lower vertebrates (e.g. fish and amphibia) the horizontal visual axis for each eye is represented rostro-caudally in the optic tectum and not mediolaterally as was originally supposed. Given this information the 'map' of the visual field, on say the tectum, it becomes immaterial whether or not the optic fibres decussate or remain on the same side; the coherence of the visual field representation remains unaffected (Fig. 2.27(d) and (c)). Many cephalopods such as squid and octopus possess chambered image-forming eyes, but their optic fibres enter the optic lobe on the same side, i.e. no optic fibres cross the midline to the opposite side. Indeed, the central projection of their optic fibres is similar to that in arthropods, which have a compound eye. In cephalopods (as well as in arthropods, crustacea, etc.), visual activity in each optic lobe is co-ordinated with the other via the optic commissure linking the two optic lobes (see Pettigrew, chapter 12, for an account of the evolution of binocular vision).

Development of Visual Decussation in Vertebrates Possibly Associated With the Acquisition of a Swimming Habit that Was Visually Guided

Sarnat and Netsky (1981) describe the early phylogenetic stages of vertebrate vision as providing for crude differentiation of intensity of light in different parts of visual field. This allowed spatial localization in the absence of a well-defined image. In vertebrates the actual perception and processing of complex images could not occur until the optic cup, the lens and the other structures of the chambered eye evolved for the focusing of images and until the visual centres in the brain had achieved the degree of complexity necessary to process such complex information. Concomitant with the development of the chambered eye, retinal ganglion cells sent axons across the midline to innervate motor neurones in the contralateral neuroectoderm of the bilaterally symmetric animal. These crossing axons were the primordial optic nerve. Any

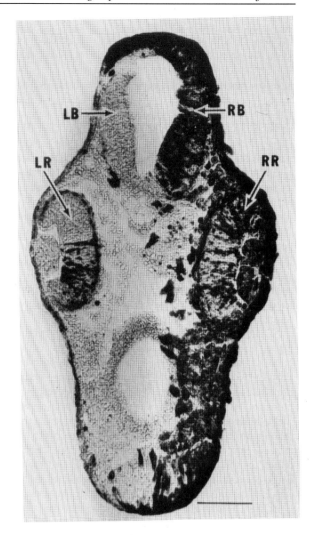

Fig. 2.28 Cells originating from one of the first two blastomeres of Xenopus embryo were labelled with horseradish peroxidase. Generally, the labelled cells were confined to the same side of the brain as the labelled blastomere except for cells that have moved from the opposite side into the ventral diencephalon and ventral part of the retina. Reciprocal movement of cells from each side of the prospective forebrain into the prospective retina on the opposite side starts before the neural tube closes and results in the formation of an incipient optic chiasma which may provide the pathway for optic axons to grow from the retina to the opposite side of the brain. The figure depicts a coronal section through Xenopus at stage 37, showing the localization of the dark reaction product for horseradish peroxidase, a label which was injected into the right blastomere at the two-cell stage. The label is seen in the right side of the brain (RB), the right retina (RR), and the ventral half of the left retina (LR), while the dorsal half of the left retina is unlabelled. Note the absence of label from the left side of the brain (LB), and the reduced labelling of the ventral half of the right retina (After M. Jacobson; Jacobson and Hirose, 1978).

change in illumination, either a flash of brightness or a shadow, signalled the possible presence of a predator or other danger and the decussation provided a reflex pathway for coiling away from the threatened side.

In this respect the optic nerves may be similar to the decussating interneurones of the tactile system (see Coghill, 1929, 1930a,b). Indeed, the coiling reflex in response to non-specific cutaneous stimulation is the most primitive spinal reflex. In the coiling reflex, activity from the sensory neurone is conducted to motor neurones on the opposite side of the spinal cord by an interneurone with a decussating axon. The decussation allows the animal to coil away from the noxious stimulus. Also, the alternating contraction of myotomes on the two sides of the body required for swimming, and for ambulation, in simple tetrapods was necessarily mediated by decussating interneurones (Sarnat and Netsky, 1981). Thus the development of the visually guided undulating movement associated with a swimming habit in some early predecessor seems likely to have provided the selection pressure which led to the acquisition of a decussating visual pathway.

Sarnat and Netsky (1981) note that it is unlikely that the earliest optic fibres directly innervated muscles on the opposite side of the body, as originally suggested by Polyak (1957), because motor neurones do not generally decussate elsewhere in the nervous system. However, they point out that in a few cases, such as the trochlear nerve, the neurones of the nucleus secondarily migrate across the midline during embryonic development. Significantly in the case of the optic system Jacobson and Hirose (1978) found, in the developing *Xenopus* tadpole, that some embryonic cells in the floor of the forebrain (located in the region of the future optic chiasma) migrate across the midline (Fig. 2.28). As a result they suggested that the latter may form a pathway for optic axons to grow from the retina to the opposite side of the brain.

Concluding Remarks

This chapter has attempted to outline and explore some of the more promising ideas that have been advanced to explain the phylogeny of the vertebrate eye. Particular attention was given to Willmer's (1970) analysis of the cephalic organ of nemertines, which provides a tantalizing picture of how a variety of vertebrate and invertebrate organs (including the vertebrate eye) may have evolved through transformation of a single structure in a pre-vertebrate animal. Meanwhile there is a need to strengthen the suggested homology between the vertebrate eye and the nemertine cephalic organ. For a start, it would help if it were demonstrated that the cephalic organ was related to the infundibular organ in *amphioxus*. However attractive the theory, the evidence presented by Willmer is only suggestive and has not fully established the lineage between the vertebrate eye and its presumed predecessor, the cephalic organ of nemertines – at least not in a sense that would have satisfied Darwin. A full description along Darwinian lines would have necessitated defining a phylogenetic sequence of homologous structures, each of which differs only in some minor way from its predecessor, leading eventually to the vertebrate eye. Alternatively, if the emergence of the vertebrate eye was arrived at, not along a strictly Darwinian series of microevolutionary steps, but instead arose through a punctuational mode of evolutionary advance, then it may be fruitless to expect a clear set of homologies based solely on a classical-style comparison of morphological and functional characteristics.

Despite the clarification which the ideas and associated studies described in this chapter have brought about in suggesting the ancestry of some of the eye's components, many questions relating to the eye's origin remain unanswered. For example, our notions of the origin of the vertebrate lens are somewhat hazy, although of all the suggested alternatives, those of Willmer and of Studnicka seem to be the most attractive. Thus one hundred years or more after Charles Darwin we find ourselves facing many of the same problems in defining the phylogeny of the vertebrate eye. While this in itself may be disconcerting, there is hope that help may come through the application of techniques derived from molecular biology that have already been applied succesfully in other evolutionary studies (see Goldsmith, chapter 5, in Volume 6). Also, recent advances in methods for characterizing the cell surface lend hope that it may soon be possible to detect subtle similarities in cell surface characteristics, or to resolve fine differences between cells and tissues in different species. Such methods allow the identification and differentiation of specific cell markers that are characteristic of certain tissues (or even of cell types). This would allow us the possibility of establishing homologies between structures based on certain shared patterns of cell surface molecules. Only then may we perhaps find a proper sequence of homologies of molecular structure (e.g. of cell surface antigens) that would finally resolve the mystery of the eye's origin.

References

Andersson, B. and Jewel, P. A. (1957). The effect of long periods of continuous hydration on the neurosecretory material in the hypothalamus of the dog. *J. Endocrinol.*, **15**, 332.

Assheton, R. (1896). Notes on the ciliation of the ectoderm of the amphibian embryo. *Q. J. Microsc. Sci.*, **38**, 465.

Atler, P. A. (1965). Isolated cilia in inner retinal neurones and in retinal pigment epithelium. *J. Ultrastruct. Res.*, 123, 730.

Bagnara, J. T. (1963). The pineal body lightening reaction of larval amphibians. *Gen. Comp. Endocrinol.*, 3, 465.

Balfour, J. (1881). *A Treatise on Comparative Embryology*. London.

Baylor, D. A. and Fuortes, M. G. F. (1970). Electrical responses of single cones in the retina of the turtle. *J. Physiol.*, 207, 77–92.

Bekesy, G. von (1959a). Neural funnelling along the skin and between inner and outer hair cells of the cochlea. *J. Acoust. Soc. Am.*, 31, 1236–49.

Bekesy, G. von (1960). Neural inhibitory units of the eye and skin: qualitative demonstration of contrast phenomena. *J. Opt. Soc. Am.*, 50, 1060–1070.

Berger, E. W. (1900). Physiology and histology of the cubosmesudae. *Mem. Biol. Lab. Johns Hopkins Univ.*, 4(4), 1–84.

Biernbaum, M. S. and Bownde, M. D. (1985). Light induced changes in GTP and ATP in frog rod photoreceptors. Comparison with recovery of dark current and light sensitivity during dark adaptation. *J. Gen. Physiol.*, 85, 107–121.

Boeke, J. (1908). Das infundibularorgan im Gehirne des Amphioxus. *Anat. Anz.*, 32, 473–488.

Boycott, B. B. and Young, J. Z. (1956). The subpendunculate body and nerve and other organs associated with the optic tract of cephalopods. In *Bertil Hanstrom, Zoological Papers in Honour of his Sixty-fifth Birthday*. Ed. Wiungstrand, K. G. pp. 76–102. Lund: Zoological Institute.

Brown, P. K., Gibbons, I. R. and Wald, G. (1963). The visual cells and visual pigment of the mud puppy, *Necturus. J. Cell. Biol.*, 19, 79.

Butschli (1921). *Vorlesungen uber vergleichender Anatomie*. pp. 817, 826, 872. Berlin.

Cajal, S. R. y (1909–1911). *Histologie du Système Nerveux de l'Homme et de les Vertèbres*. Madrid: Consejo Superior de Investigaciones Cientificas.

Cajal, S. R. y (1917). Contribucion al conocimiento de la retina y centros opticos de los Cefalopodos. *Trab. Lab. Invest. Biol. Univ. Madr.*, 15, 1–82.

Cameron, G. (1953). Secretory activity of the choroid plexus in tissue culture. *Anat. Rec.*, 117, 115.

Coghill, G. E. (1929). *Anatomy and the Problem of Behaviour*. Cambridge: Cambridge University Press.

Coghill, G. E. (1930a). Individuation versus integration in the development of behaviour. *J. Gen. Psychol.*, 3, 431–435.

Coghill, G. E. (1930b). Correlated anatomical and physiological studies of the growth of the nervous system of amphibia. IX. The mechanism of association of *Ambystoma punctatum. J. Comp. Neurol.*, 51, 311–375.

Coghill, G. E. (1931). Corollaries of the anatomical and physiological study of *Ambystoma* from the age of the earliest movement to swimming. *J. Comp. Neurol.*, 53, 147–168.

Cornwell, M. C. and Gorman, A. L. F. (1983). The cation selectivity and voltage dependence of the light-activated potassium conductance in scallop distal photoreceptor. *J. Physiol.*, 340, 287–305.

Crescitelli, F. (1965). The spectral sensitivity and visual pigment content of the retina of *Gekko gekko*. CIBA Found. Symp. Colour Vision. *Physiol. Exp. Psychol.*, p. 301.

Dakin, W. J. (1921). The eye of *Peripatus. Q. J. Microsc. Sci.*, 65, 163.

Dakin, W. J. (1928). The eyes of *Pecten, Spondylus, Amussium* and allied lamellibranchs with a short discussion on their evolution. *Proc. R. Soc. Lond. B*, 103, 355–365.

Darwin, C. (1859). *The Origin of Species by Means of Natural Selection*. London: John Murray.

Dendy, A. (1907). On the parietal sense organs and associated structures in the New Zealand lamprey (*Geotria australis*). *Q. J. Microsc. Sci.*, 51, 1.

Dingle, A. D. and Lucy, J. A. (1965). Vitamin A, carotenoids and cell function. *Biol. Rev. Camb. Physiol. Soc.*, 40, 422.

Dingle, A. D. and Fulton, C. (1966). Development of the flagellar apparatus of *Naegleria. J. Cell. Biol.*, 31, 43.

Dixon, R. A. F., Kobilka, B. K., Strader, D. J. et al. (1986). Cloning of the gene and cDNA for mammalian beta-adrenergic receptor and homology with rhodopsin. *Nature*, 321, 75–79.

Dodt, E. (1964). Physiologie des Pinealorgans Anurer Amphibien. *Vision Res.*, 4, 23.

Dowling, J. E. and Boycott, B. B. (1966). Organization of the primate retina: electron microscopy. *Proc. R. Soc. Lond. B*, 166, 80.

Ducker, M. (1924). Uber die Augen der Zyklostomen. *Jena Z. Med. Naturw.*, 60, 471.

Duke-Elder, S. (1958). *The Eye in Evolution*. Vol. 1 of *System of Ophthalmology*. London: Henry Kimpton.

Dunn, R. F. (1966). Studies on the retina of the gecko. *Coleonyx variegatus*. 1: The visual cell classification. *J. Ultrastruct. Res.*, 16, 651.

Eakin, R. M. and Westphall, J. A. (1959). Fine structure of the reptilian third eye. *J. Biophys. Biochem. Cytol.*, 6, 133.

Eakin, R. M. and Westphall, J. A. (1960). Further observations on the fine structure of the parietal eye of lizards. *J. Biophys. Biochem. Cytol.*, 8, 483.

Eakin, R. M., Quay, W. B. and Westphall, J. A. (1961). Cytochemical and cytological studies of the parietal eye of the lizard, *Sceloporus occidentalis*. *Z. Zellforsch. Mikrosk. Anat.*, 53, 449.

Engelman, D. M. and Zaccai, G. (1980). Bacteriorhodopsin is an inside-out protein. *Proc. Natl. Acad. Sci. USA*, 77, 5894–5898.

Enroth-Cugell, K. and Robson, J. (1966). The contrast sensitivity of retinal ganglion cells of the cat. *J. Physiol.*, 187, 517–552.

Freeman, G. (1963). Lens regeneration from the cornea in *Xenopus laevis. J. Exp. Zool.*, 154, 39.

George, J. S. and Hagins, W. A. (1983). Control of Ca^{++} on rod outer segment disks by light and cyclic GMP. *Nature*, 303, 344–348.

Goodrich, E. S. (1909). Cyclostomes and fishes. In *A Treatise on Zoology*. Ed. Lankester, E. R. Vol. 9. London: Black.

Gotcharoff, M. (1953). Le phototropisme chez *Lineus ruber* et *Lineus sanguineux* au cours de la régénération des yeux. *Ann. Sci. Nat. Zool. Biol. Animale*, 15, 369.

Gould, S. J. (1977). *Ontogeny and Phylogeny*. Cambridge, Mass.: Harvard University Press.

Haeckel, E. (1866). *Generelle Morphologie der Organismen: Allgemeine Grundzuge der organischen Formen-Wissenschaft, mechanisch begrundet durch die von Charles Darwin reformierte Descendez-Theorie*. Berlin: George Reimer.

Hargrave, P. A., McDowell, J. H., Curtis, D. R. et al. (1983). The structure of bovine rhodopsin. *Biophys. Struct. Mech.*, 9, 235–244.

Hartline, J. K. (1938). The discharge of impulses in the optic nerve of *Pecten* in response to illumination of the eye. *J. Cellular Comp. Physiol.*, 11, 465.

Henderson, R. (1977). The purple membrane from *Halobacterium halobium*. *Ann. Rev. Biophys. Bioeng.*, 6, 87–109.

Hertwig (1878). *Das Nervensystem und die Sinnesorgane der Medusen*. Jena.

Hertwig (1893). *Die Zelle und die Gewebe*. Jena.

Hess, W. N. (1920). Notes on the Biology of some common Lampyridae. *Biol. Bull.*, 38, 291–308.

Hesse, R. (1902). Untersuchungen uber die organe der Lichtempfindung bei niederen Thieren. VIII Weitere Thatsachen. *Allg. Z. Wiss. Zool.*, 72, 565–656.

Hesse, R. (1908). Die Sehen der neideren Thieren. *Jena Z. Naturwiss.*

Holtfreter, J. (1947). Changes of structure and the kinetics of differentiating embryonic cells. *J. Morphol.*, 80, 57.

Huxley, J. S. and de Beer, G. R. (1934). *The Elements of Experimental Embryology*. Cambridge: Cambridge University Press.

Jacobson, A. G. (1966). Inductive processes in embryonic development. *Science*, 152, 25–35.

Jacobson, M. and Hirose, G. (1978). Origin of the retina from both sides of the embryonic brain: A contribution to the problems of crossing at the optic chiasma. *Science*, **202**, 637–639.

Jourdan (1889). *Les Sens Chez les Animaux Inferieure*. Paris.

Kappers, J. A. (1965). Survey of the innervation of the epiphysis cerebri and the accessory pineal organs of vertebrates. *Progr. Brain Res.*, **10**, 87.

Karkinen-Jaaskelainen, M. (1978). Permissive and directive interactions in lens induction. *J. Embryol. Exp. Morphol.*, **44**, 167–179.

Karp, G. and Berrill, N. J. (1981). *Development*. 2nd edn. New York: McGraw-Hill.

Kubo, T., Fukada, K., Mikami, A. *et al.* (1986). Cloning, sequencing and expression of complementary DNA encoding the muscarinic actylcholine receptor. *Nature*, **323**, 411–416.

Lamb, T. D. (1986). Transduction in vertebrate photoreceptors: the roles of cyclic GMP and calcium. *TINS*, 224–228.

Land, M. F. (1965). Image formation by a concave reflector in the eye of the scallop, *Pecten maximus*. *J. Physiol.*, **179**, 138–153.

Land, M. F. (1984). Molluscs. In *Photoreceptors and Vision in Invertebrates*. Ed. Ali, M. A. pp. 699–725. New York: Plenum.

Lankester, E. (1880). *Darwinism & Parthogenesis*. London.

Lechenault, H. (1963). Sur l'existence de cellules neurosecrétrices chez les Hoplonemertes. Caractéristiques histochimiques de la neurosécrétion chez les Nemertes. *Compt. Rend.*, **256**, 3201.

Lewin, R. A. (1953). Studies on the flagella of algae. (ii) Formation of flagella by *Chlamydomonas* in light and darkness. *Ann. NY Acad. Sci.*, **56**, 1091.

Ling, E. A. (1969). The structure and function of the cephalic organ of a nemertine, *Lineus ruber*. *Tissue Cell*, **1**, 503.

Loeb, J. (1893). Ueber kunstliche Umwandlung positiv heliotropischer Thiere in negativ heliotropischer und umgekehrt. *Pflügers Arch. ges. Physiol.*, **54**, 81.

Loeb, J. (1906). *The Dynamics of Living Matter*. London.

McReynolds, J. S. and Gorman, A. L. S. (1970). Photoreceptor potentials of opposite polarity in the eye of the scallop *Pecten irradiens*. *J. Gen. Physiol.*, **56**, 376–399.

Merker, E. von (1934). Die Sichtbarkeit, ultravioletten Lichtes. *Biol. Rev. Camb. Phil. Soc.*, **9**, 49.

Millen, J. W. and Rogers, G. E. (1956). An electron microscope study of the choroid plexus of the rabbit. *J. Biophys. Biochem. Cytol.*, **2**, 407.

Miller, W. H. (1960). In *The Cell*, Vol. 4. Eds Brachet, J. and Mirsky, A. E. New York: Academic.

Milne, L. J. and Milne, M. J. (1956). Invertebrate photoreceptors. In *Radiation Biology*. Ed. Hollaender, A. New York: McGraw-Hill.

Milne, L. J. and Milne, M. J. (1959). Photosensitivity in invertebrates. In *Handbook of Physiology*, Sect. 1, *Neurophysiology*, **1**, 621–645.

Moscona, A. (1957). Formation of lentoids by dissociated retinal cells of the chick embryo. *Science* **125**, 598.

Noell, W. K. (1960). The impairment of visual cell structure by iodoacetate. *J. Cell. Comp. Physiol.*, **40**, 25.

Ortman, R. (1960). Parietal eye and nerve in *Anolis carolinensis*. *Anat. Rec.*, **137**, 386.

Oztan, N. and Gorbman, A. (1960). Responsiveness of the neurosecretory system of larval lampreys (*Petromyzon marinus*) to light. *Nature*, **186**, 167.

Plack, P. A., Kon, S. K. and Thompson (1959). Vitamin A1 aldehyde in the eggs of the herring (*Clupea harenga* L.) and other marine teleosts. *Biochem. J.*, **71**, 467.

Polyak, S. (1957). *The Vertebrate Visual System*. Chicago: University of Chicago Press.

Pomerat, C. M. and Littlejohn, J. (1956). Observations on tissue culture of the human eye. *Southern Med. J.*, **49**, 230.

Ratliff, F. and Miller, W. H. (1961). In *Nervous Inhibition*. Ed. Florey, E. London: Pergamon Press.

Ritchie A. (1960). A new interpretation of *Jamoytius kerwoodi*

(White). *Nature*, **188**, 647.

Rohlich, P. and Torok, L. J. (1961). Electronmikroskopische Untersuchungen des Auges von Planarien. *Z. Zellforsch. Mikrosk. Anat.*, **54**, 361.

Romer, A. S. (1970). *The Vertebrate Body*, 4th edn. Philadelphia: Saunders.

Sagar, R. and Granick, S. (1954). Nutritional control of sexuality in *Chlamydomonas reinhardi*. *J. Gen. Physiol.*, **37**, 729.

Salensky, W. (1896). Bau und Metamorphose des Pilidium. *Z. Wiss. Zool.*, **43**, 481.

Sarnat, H. B. and Netsky, M. G. (1981). *Evolution of the Nervous System*. Oxford University Press.

Schimkewitsch (1921). *Lehrbuch der vergleichender Anatomie der Wirbeltiere*. Stuttgart.

Schroder, O. (1918). Beitrage zur Kenntnis von *Geonemertes palaensis* (Semper). *Abhandl. Senckenberg. Naturforsch. Ges.*, **35**, 155.

Schultze, M. (1886). Zur Anatomie und Physiologie der Retina. *Archiv. Mikr. Anat.*, **2**, 175–186.

Schuster, F. (1963). An electron-microscope study of the amoeboflagellate, *Naegeria gruberi* (Schardinger). The amoeboid and flagellate stages. *J. Protozool.*, **10**, 297.

Schwalbe, W. (1874). *Graefe-Saemisch Handbuch der gesamte Augenheilkunde*, **1**, 398.

Scriban, I. A. and Autrum, H. (1932). Hirudinae. In *Kukenthal's Handb. Zool.*, **2**(8), 119–352.

Smith, G. (1906). The eyes of certain pulmonate gasteropods with special reference to the neurofibrillae in *Limax maximus*. *Bull. Mus. Comp. Zool. Harv.*, **48**, 231–283.

Sjostrand, F. S. (1961). Electron microscopy of the retina. In *The Structure of the Eye*. Ed. Smelser, G. K. New York: Academic Press.

Spemann, H. (1938). *Embryonic Development and Induction*. New Haven: Yale University Press.

Spudich, J. L. and Bogomolni, R. A. (1984). Mechanism of colour discrimination by a bacterial sensory rhodopsin. *Nature*, **312**, 509–513.

Stanley, S. M. (1981). *The New Evolutionary Timetable*. New York: Basic Books.

Stockenius, W. and Bogomolni, R. A. (1984). Bacteriorhodopsin and related pigments of halobacteria. *Ann. Rev. Biochem.*, **52**, 587–616.

Stone, L. S. (1967). An investigation recording all salamanders which can and cannot regenerate a lens from the dorsal iris. *J. Exp. Zool.*, **164**, 87.

Studnicka, F. K. (1898). Untersuchungen uber Bau der Sehnerven der Wirbelthiere. *Jena Z. Naturwiss.*, **31**, 1–28.

Studnicka, F. K. (1912). Uber die Entwicklung und die Bedeutung der Seitenaugen von Ammocoetes. *Anat. Anz.*, **41**, 561–578.

Studnicka, F. K. (1918). Das Schema der Wirbeltieraugen. *Zool. Jahrb. Abt. Anat.*, **40**, 1.

Tansley, K. (1933). The formation of rosettes in the rat retina. *Br. J. Ophthalmol.*, **17**, 321.

Tokuyasu, K. and Yamada, E. (1959). The fine structure of the retina studied with the electron microscope. IV: Morphogenesis of outer segments of retinal rods. *J. Biophys. Biochem. Cytol.*, **6**, 225.

Twitty, V. (1956). Eye. In *Analysis of Development*. Eds. Willier, Weiss and Hamburger. Philadelphia: Saunders.

Underwood, G. (1968). Some suggestions concerning vertebrate visual cells. *Vision Res.*, **8**, 483–488.

Underwood, G. (1970). The eye. In *Biology of the Reptilia*, Vol. 2, *Morphology*. Eds. Gans, C. and Parsons, T. S. pp. 1–7. New York: Academic Press.

Van der Kamer, J. C. (1965). Histological structure and cytology of the pineal complex in fishes, amphibians and reptiles. *Progr. Brain Res.*, **10**, 30.

Villegas, G. M. (1960). Electronmicroscope study of the vertebrate retina. *J. Gen. Physiol.*, **43**, Suppl. 2, 15.

Wager, J. (1900). On the eye-spot and flagellum in *Euglena viridis*. *J.*

Linn. Soc. (Zool.). Lond., 27, 463.

Walls G. L. (1942). *The Vertebrate Eye and Its Adaptive Radiation.* Bloomfield, Mich.: Cranbrook.

Wells, M. J. (1964). Hormonal control of sexual maturity in cephalopods. *Bull. Natl Inst. Sci. India*, 27, 63.

Wells, M. J. and Wells, J. (1959). Hormonal control of sexual maturity in *Octopus. J. Exp. Biol.*, 36, 1.

White, E. I. (1946). *Jamoytius kerwoodi*, a new chordate from the Silurian of Lanarkshire. *Geol. Mag.*, 83, 89.

Willmer, E. N. (1956). Factors which influence the acquisition of flagella by the amoeba *Naegleria gruberi. J. Exp. Biol.*, 33, 583.

Willmer, E. N. (1970). *Cytology and Evolution.* New York: Academic Press.

Wolken, J. J. (1958). Studies on photoreceptor structures. *Ann. NY Acad. Sci.*, 75, 161.

Wolken, J. J. (1963). Structure and molecular organization of retinal photoreceptors. *J. Opt. Soc. Am.*, 53, 1–19.

Wurtman, R. J., Axelrod, J. and Phillips, L. S. (1963a). Melatonin synthesis in the pineal gland: control by light. *Science*, 142, 277.

Wurtman R. J., Axelrod, J. and Fischer, J. E. (1963b). Melatonin synthesis in the pineal gland: Effect of light mediated by the sympathetic nervous system. *Science*, 143, 1328.

Young, J. Z. (1935). The photoreceptors of lampreys. II. The functions of the pineal complex. *J. Exp. Biol.*, 12, 254.

Young, J. Z. (1962a). The retina of cephalopoda and its degeneration after optic nerve section. *Phil. Trans. R. Soc. Lond. B*, 245, 1–18.

Young, J. Z. (1962b). The optic lobes of *Octopus vulgaris. Phil. Trans. R. Soc. Lond. B*, 245, 19–58.

Young, R. W. and Droz, B. (1968). The renewal of protein in retinal rods and cones. *J. Cell Biol.*, 39, 169.

Zonana, H. W. (1961). Fine structure of the squid retina. *Bull. Johns Hopkins Hosp.*, 109, 185–205.

3 Origins of Eyes – with Speculations on Scanning Eyes

Richard L. Gregory

Introduction

Eyes are remarkable for their ingenuity and precision as optical instruments – even though they are made from inherently unsatisfactory materials – and for their diversity and variety in nature. They are dramatic evidence of brilliant creative design, so it is not surprising that eyes have always been seen as a principal challenge to theories of evolution – giving Charles Darwin 'a cold shudder'.

Among the early conceivers of evolution of species was Erasmus Darwin, Charles Darwin's grandfather, who in his poem *Temple of Nature* (1803) wrote (in the section, 'Production of Life' V):

> ORGANIC LIFE beneath the shoreless waves
> Was born and nurs'd in Ocean's pearly caves;
> First forms minute, unseen by spheric glass,
> Move on the mud, or pierce the watery mass;
> These, as successive generations bloom,
> New powers acquire, and larger limbs assume;
> Whence countless groups of vegetation spring,
> And breathing realms of fin, and feet, and wing.
>
> Thus the tall Oak, the giant of the wood,
> Which bears Britannia's thunders on the flood;
> The Whale, unmeasured monster of the main,
> The lordly Lion, monarch of the plain,
> The Eagle soaring in the realm of air,
> Whose eye undazzled drinks the solar glare.

Evolution of plant and animal forms was considered as early as the Latin poet Lucretius. But to Charles Darwin and, later in conception though simultaneously announced, to Alfred Russel Wallace, lies the credit for seeing essentially statistical processes capable of generating the incredibly successful inventions of life. So although blind, Darwinian evolution is a powerful intelligence, and it is not misleadingly anthropomorphic to speak of biological 'design' (cf. Gregory, 1987).

From Charles Darwin's Note Book of 1837 it is clear that he was by then a convinced evolutionist; but in 1831, when he started his great voyage on the *Beagle*, he was not convinced. According to his son Francis Darwin, writing in 1906: 'On his departure in 1831, Henslow gave him vol. 1 of Lyell's *Principles of Geology*, then just published, with the warning that he was not to believe what he read. But believe it he did, and it is certain (as Huxley has forcibly pointed out) that the doctrine of uniformitarianism when applied to Biology leads of necessity to Evolution. If the extermination of a species is no more catastrophic than the natural death of an individual, why should the birth of a species be any more miraculous than the birth of an individual? It is quite clear that this thought was vividly present to Darwin when he was writing out his early thoughs in the 1837 Note Book: "If *species* generate other *species*, their race is not utterly cut off." '

By 1844, 15 years before *The Origin of Species*, Darwin had got his theory well worked out and had considered what might refute it. In the 1844 Essay (in the Section '*Difficulties in the acquirement by Selection of complex corporeal structure*') Darwin saw the eye as a challenge to his theory of evolution by selection of the survival of the fittest. (The term 'survival of the fittest' was coined by Herbert Spencer in 1852.) Darwin wrote (1844):

> *In the case of the eye, as with the more complicated instincts, no doubt one's first impulse is to utterly reject every such theory. But if the eye from its most complicated form can be shown to graduate into an exceedingly simple state,—if selection can produce the smallest change, and if such a series exists, then it is clear (for in this work we have nothing to do with the first origin of organs in their simplest forms) that it might possibly have acquired by gradual selection of slight, but in each cases, useful deviations. . . . In the case of the eye, we have a multitude of different forms, more or less simple, not graduating into each other, but separated by sudden gaps or intervals; but we must recollect how incomparably greater would the multitude of visual structures be if we had the eyes of every fossil which ever existed. . . . Notwithstanding the large series of existing forms, it is most difficult even to conjecture by what intermediate stages very many simple organs could possibly have graduated into complex ones: but it should*

be borne in mind, that a part having originally a wholly different function, may on the theory of gradual selection be slowly worked into quite another use; the gradations of forms, from which naturalists believe in the hypothetical metamorphosis of part of the ear into the swimming bladder of fishes, and in the insects of legs into jaws, show the manner in which this is possible.... In conjecturing by what stages any complicated organ in a species may have arrived at its present state, although we may look at analogous organs in other existing species, we should do this merely to aid and guide our imaginations; for to know the real stages we must look only through one line of species, to one ancient stock, from which the species in question has descended. In considering the eye of a quadruped, for instance, though we may look at the eye of a molluscous animal or of an insect, as a proof how simple an organ will serve some of the ends of vision; and at the eye of a fish as a nearer guide of the manner of simplification; we must remember that it is a mere chance (assuming for the moment the truth of our theory) if any existing organic being has preserved any one organ, in exactly the same condition, as it existed in the ancient species at remote geological periods.*

Darwin's account of the origin and evolutionary stages or 'gradations' of eyes, as given in *The Origin of Species* (sixth edition, p. 135) is:

The simplest organ which can be called an eye consists of an optic nerve, surrounded by pigment cells and covered by translucent skin, but without any lens or other refractive body. We may, however, according to M. Jourdain, descend even a step lower and find aggregates of pigment-cells, apparently serving as organs of vision, without any nerves, and resting merely on sarcodic tissue. Eyes of the above simple nature are not capable of distinct vision, and serve only to distinguish light from darkness. In certain star fishes, small depressions in the layer of pigment which surrounds the nerve are filled... with transparent gelatinous matter, projecting with a convex surface, like the cornea of higher animals. Jourdain suggests that this serves not to form an image, but only to concentrate the luminous rays and render their perception more easy. In this concentration of the rays we gain the first and by far the most important step towards the formation of a true, picture-forming eye; for we have only to place the naked extremity of the optic nerve, which in some of the lower animals lies deeply buried in the body, and in some near the surface, at the right distance from the concentrating apparatus, and an image will be formed on it.

In the great class of the Articulata, we may start from an optic nerve simply coated with pigment, the latter sometimes forming a sort of pupil, but destitute of a lens or other optical contrivance. With insects it is now known that the numerous facets on the cornea of their great compound eyes form true lenses, and that the cones include curiously modified filaments....

When we reflect on these facts, here given much too briefly, with respect to the wide, diversified, and graduated range of structure in the eyes of the lower animals; and when we bear in mind how small the number of living forms must be in comparison with those which have become extinct, the difficulty ceases to be very great in believing that natural selection may have converted the simple apparatus of an optic nerve, coated with pigment and invested by transparent membrane, into an optical instrument as perfect as is possessed by any member of the Articulate Class.

Darwin goes on to compare the evolutionary-developed eye with man-invented telescopes:

It is scarcely possible to avoid comparing the eye with a telescope. We know that this instrument has been perfected by the long-continued efforts of the highest human intellects; and we naturally infer that the eye has been formed by a somewhat analogous process. But may not this inference be presumptuous? Have we any right to assume that the Creator works by intellectual powers like those of man?

How was this received? In the month of publication, November 1859, Charles's grandfather Erasmus Darwin describes in a letter to Charles the reaction of a friend, Henry Holland, on first seeing *The Origin*: 'He is in a dreadful state of indecision, and keeps stating that he is not tied down to either view, and that he has always left an escape by the way he has spoken of varieties. I happened to speak of the eye before he had read that part, and it took away his breath—utterly impossible—structure—function, &c., &c., &c., but when he had read it he hummed and hawed, and perhaps it was partly conceivable, and then he fell back on the bones of the ear, which were beyond all probability or conceivability.'

Charles Darwin wrote to Asa Grey, probably in February 1860: 'The eye to this day gives me a cold shudder, but when I think of the fine known gradations, my reason tells me I ought to conquer the cold shudder.' Writing to the geologist Sir Charles Lyell on 15 February 1860, Charles Darwin says: 'Henslow is staying here [at Down house]; I have had some talk with him... and [he] will go a very little way with us, but brings up no real argument against going further. He also shudders at the eye!'

On 3 April 1860 he wrote again to Asa Grey: 'I remember the time when the thought of the eye made me cold all over, but I have got over this stage of the complaint, and now small trifling particulars of structure often make me very uncomfortable. The sight of a feather in a peacock's tail, whenever I gaze at it, makes me sick!'

Eye and Brain: A Hen-and-Egg Problem

How could eyes have developed without visually competent brains to make them useful? How could visually competent brains have developed without signals from eyes? We may see this as a hen-and-egg problem (Gregory, 1968). The origins of the proximal senses of taste and touch do not have this problem, for they monitor, rather directly, immediately life-important features without requiring subtle interpretations for their signals to be useful. Taste and smell monitor nutrients, and check for toxicity – often with gradients leading to sources of food and giving early warnings of pollutants. Touch and related senses monitor immediately important biological features, such as hardness and wetness and temperature, and body damage to self and sometimes to foe. Touch is vital for mating. As these are immediately useful, with a minimum of interpretation, these proximal senses do not have this hen-and-egg problem of senses needing subtle brains, and brains needing senses, for their early development. Yet once developed, the 'distance' senses of vision and of hearing have the great advantages of probing distance – and so providing early warning – without giving their owner's presence away. By giving early warning they allow time for planning – making possible the development of predictive creative intelligence. It may be appropriate to think of perceptions as *hypotheses* of sources of stimuli – objects – requiring not only hard-wired processing of sensed stimuli, but also cognitive processes for generating predictive hypotheses of the *causes*, or *sources*, of stimuli (Gregory, 1980). Thus, perception takes off from the stimulus–response behaviour of simple creatures, to conceptual planning and thinking in vertebrates and in the more advanced cephalopods and crustacea. This broad sweep from stimulus-driven reflexes and tropisms to knowledge-based perceptual hypotheses, including characteristics beyond what can be sensed, to conceptual thinking free from the senses, is what needs to be described and explained. The anatomical development of the brain reflects this kind of development. Thus Sarnat and Netsky (1974) write, in *Evolution of the Nervous System*, which I cannot do better than quote in full (p. 29):

> Eyes and olfactory receptors developed early in the evolution of vertebrates. These structures are already differentiated in the most primitive of living vertebrates, the cyclostomes. Tactile perception and taste give information about the distant or remote environment. The importance of distant information is evidenced by the evolution of the phylogenetic series of vertebrates, contrasted with the failure of creatures lacking distance receptors to evolve further, exemplified by amphioxus.
>
> The anatomical organization of the nervous system established in the hypothetical ancestral vertebrate was repeated and expanded in all subsequent vertebrates. That basic pattern involved specialization of the hindbrain for receiving information about the immediate environment, and of the midbrain and forebrain for receiving information about the distant environment. Sensory impulses related to touch, temperature, taste, and balance thus entered the medulla for quick reflexive responses by motor nuclei. Information from distance receptors, however, entered the midbrain from the eyes, or the forebrain from the olfactory epithelium. Because the distance from the object perceived by sight or smell was greater, more time was available before motor responses were required, so that a longer delay in conducting impulses to medullary motor centres was not a disadvantage. Remote information also required more interpretation before responses were made, and the forebrain therefore became more associative while the medulla remained reflexive. With the further evolution of the forebrain, all sensory information eventually was relayed rostrally for interpretation and correlation, but the primitive medullary reflexes persist, even in man.

Although it is true that subtle brain-processing of optical images enormously increases their usefulness, it is not quite true to say they have *no* use without neural facilities of 'interpreting' or 'computing' to give meaning in terms of the world of objects. Sensing distant movement, for example, is useful even though the source of the movement is not recognized – for movement is often associated with danger, and so it signals probable danger. (It is noteworthy that many radar systems reject stationary echoes allowing only moving 'targets' to be visible.) Early eyes would be useful even though they signalled only movement. But although this is a simple task compared with cognitively developed object perception it does require some neural processing (Reichardt and Poggio, 1976, 1979). So there is a hen-and-egg problem here, for this first use of optical stimuli, for warning of unspecified danger, though the danger is not itself optical.

It is generally thought that light receptors developed in regions of skin. Darwin thought that any nerve could become sensitive to light. The notion is that these regions became cups, increasing shadow contrast. When the eye cups became deep – considerably increasing shadow contrast – they closed except for a small hole, as today in *Nautilus*. At this point there is a sudden reversal of movement – as the eye becomes an image-forming pinhole *camera obscura*. We may conjecture that the curious crossings of the touch and motor representations of many, including mammalian, brains can be attributed to this optical reversal of pinhole and single-lens 'simple' eyes. The presumption is that this greatly simplifies and shortens the association pathways between retinal signals and somatosensory and motor control maps of the brain. (It does not of course matter that retinal images are inverted, laterally and vertically, except for relations to other brain mapping, for the retinal image is not an object of perception: it is not seen. It is just one of the very many cross-sections of the visual pathway; though the retina is special, being the transducer from light to neural activity.) This optical reversal does not occur in compound eyes, and arthropods do not have at all the same neural cross-overs.

This generally accepted development of light-sensitive detectors from touch receptors may suggest (Gregory, 1967, 1968) a solution to the hen-and-egg problem – if in the beginning moving shadows activated already formed touch movement-processing. The notion is: the first vision was a take-over from touch – which itself developed without the hen-and-egg problem, for the most primitive touch sense would have biological use, as it monitors threats very directly, and so be valuable for survival without subtle signal analysis.

Simple and Compound Eyes

We may now develop this speculation, to account for the origin of the two essentially different kinds of eyes; simple and compound. Simple eyes have one lens serving a mosaic of receptors, with many (for the human eye 10^6) parallel neural channels. Compound eyes have many lenses, each with its own receptor (or functionally unitary cluster, of around 7 receptors) with a single optic nerve for each ommatidium channel. Perhaps suggestively, there are also two very different kinds of touch. There is *passive* touch depending on many receptors and many parallel neural channels, and there is *active*, or 'haptic', touch depending on exploring a shape with a moving probe. Thus, we may sense a shape from passive contact on the palm of the hand, or we may trace a shape with a moving finger. Passive touch requires parallel signalling from many channels; active or haptic touch can work with but a single channel. Can we suppose that if the first eyes took over touch, they took over the two very different *kinds* of touch – for simple eyes the many parallel channels of passive touch, for compound eyes the single channel of a moving touch probe? Haptic touch may be given by several fingers, rather than one, preferably forming a line of moving probes. So we should expect eyes with a few in-line moving receptors.

Scanning Eyes

Understanding how eyes work depends on concepts which have been developed through inventions of technology. Thus, appreciating the optical imaging of eyes required understanding image formation by man-made lenses; it is remarkably late in the history of science, being unknown to the Greeks, and not known much before Johannes Kepler (1571–1630), who described image formation by the lens of simple eyes. Before this knowledge eyes literally could not be 'seen' – and the same remains true now for much of perception and brain function. This need for technological understanding has surfaced more recently with what we may call 'scanning' eyes – working essentially like television cameras.

'Scanning' is a technical term meaning creating a two- (or three-) dimensional spatial representation from a moving one-dimensional probe. It is spreading out a time series into a spatial pattern, as in the raster-driven electron beam of television. Scanning movements of the eye would not be necessary when there are very many channels, or ommatidia (Kuiper, 1966), as in bees and dragon flies. These are stationary; but they may have evolved from eyes having a small number of scanning optical channels.

On this account the development of compound eyes converges through evolutionary development to the initial solution of simple eyes: simultaneous transmission of information down many parallel channels. As we know from engineering systems, this is a much better design for biological eyes, because of the low channel capacity of neuronal channels, compared with electronic 'eyes' such as TV cameras whose components work thousands and even millions of times faster, making temporal scanning a good option for TV though not for biological eyes.

If compound eyes started by temporal scanning, we should expect the first compound eyes to have only a few channels with large-amplitude, more or less continuous movements. This is easily seen and is well known in *Daphnia*, with its 22 ommatidia. But shouldn't we expect the start of compound eyes to be just *one* moving optic probe? It would be interesting to find a *single-channel* scanning eye in nature.

The distinguished naturalist Selig Exner described in 1891 what seemed to be a uniquely interesting eye, of the copepod *Copilia quadrata*. Exner made important discoveries on the optics of compound eyes, including a new principle of focusing light by graded refractive index in ommatidia; but at first he failed to realize that *Copilia*'s eyes were indeed eyes, as they were so unusual. He described *Copilia* as a beautiful, highly transparent pinhead sized creature, having a pair of lens-like structures deep within the body which 'showed the most lively movements'. When I came across this brief account in *The Science of Mind and Brain* by J. S. Wilkie (1953) it made me wonder whether this could be a *scanning* eye – which though obvious to us with our familiarity with television, might well have been opaque to Exner at that time, even though he was a brilliant naturalist and an expert in optics.

The principle of converting one dimension into two dimensions, with a time series of signals, was invented by F. C. Bakewell in about 1850, for his Copying Telegraph (Bakewell, 1853); but it was not well known or its significance at all generally appreciated until considerably after 1884, when Paul Nipkow invented the scanning disc, which became the heart of Baird's mechanical television of the 1930s. We may assume that the principle of sending spatial information down a single channel by scanning would not have been known to Exner at that time, so it would not be surprising that he failed at first to recognize it as an eye and never understood it.

Here is a translation of part of Exner's 1891 paper (Wilkie, 1953):

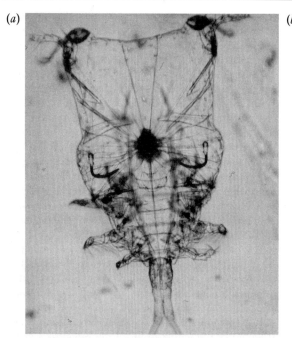 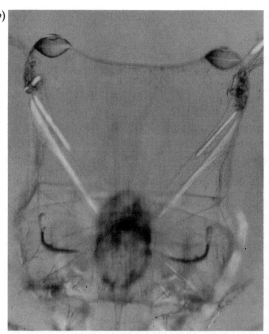

Fig. 3.1 *(a) A living female specimen of* Copilia quadrata. *(b) A more detailed view of the optical apparatus of* C. quadrata. *Each eye has two lenses: a large anterior lens and a second, smaller lens deep in the body with an attached photo receptor and a single optic nerve fibre to the central brain. The second lens and photoreceptor are in continual movement across the image plane of the first lens. This seems to be a scanning eye, a 'mechanical television camera'.*

Copilia, which I have had the opportunity of examining living and dead ... is a copepod of a few millimetres in length flattened from above downwards, and seen from above about the shape of an isosceles triangle. The narrow base of this triangle is formed by the front edges of the animal, and at either end of this edge is a strikingly beautiful lens ... Grenacher has observed, and I can confirm, that the lens is composed of two substances: one cuticular, which itself has the form of a concavo-convex lens; and one posterior to this which is a powerful biconvex lens. The lenses are the most anterior parts of the whole animal, and behind them is not, as one would have expected, a retina, but transparent parts of the body. Only far behind, about half the length of the body away, does one discover a structure which at first one does not at all recognize as related to the lens. It is a crystalline body having the form of a cone, rounded in front, with high refractive power, superposed upon a yellow rod. ... This rod is the only pigmented part of the animal's body. The crystalline cone is anchored anteriorly by suspensory ligaments, which extend to the region of the lens. Laterally, a nerve enters the yellow rod, and this is the optic nerve. Also a striate muscle is attached to the rod.

The yellow rod showed the most lively movements, which were remarkably constant. The rods, of the two eyes, were drawn towards the median plane or moved from it together, and, as far as could be seen without measurements, they remained at the same distance from the lenses. I found by micrometer that in the living animal the distance between the posterior pole of the lens and the convexity of the crystalline cone was 0.87 mm. ... I cut a narrow slice from the anterior extremity of the animal, and was able to arrange this in the water at such an angle that the hinder surface of the lens was turned towards the objective of the microscope. In this way one sees a surprisingly beautiful image thrown by the lens. Its distance from the posterior pole of the lens I found to be 0.93 mm.

As *Copilia* seemed to have been forgotten since Exner's description of 1891 I determined to look for her, hoping that with our knowledge of electronic scanning we might, as it were, see this unusual eye with new eyes. I set up an expedition in 1962 to look for *Copilia* in the Bay of Naples, where she had been seen by Exner. The expedition was funded by the US Air Force, from their European office in Brussels, from whom I held a grant for working on perceptual problems of astronauts. Professor J. Z. (John) Young, then Head of the Department of Anatomy at University College London – who had extensive experience of work at Naples on octopus – kindly arranged for laboratory space at the Zoologica Stazioni, with collection of specimens, from beyond the Bay, by laboratory staff. I was joined by Neville Moray (Oxford zoologist and psychologist) and Helen Ross (a graduate student working with me on perceptual problems of astronauts.) Our knowledge and experience of this kind of work was extremely limited.

Receiving several litres of teaming plankton each day,

we came near to despair of ever identifying *Copilia* from Exner's description. Then, after about 10 days, we saw unmistakably a living specimen – with indeed a pair of lenses deep within the transparent body, in lively motion. They moved horizontally, in exact opposition (by a single muscle), in a rapid, saw-tooth, scan-like motion.

The first strange thing about *Copilia quadrata* is that, though a Copepod, she is not of 'oar-foot' shape. She is squared in front, with her two huge anterior lenses like car headlamps; hence the very appropriate '*quadrata*'. '*Copilia*' is also appropriate, for she is evocatively beautiful, and all her charms are visible, as she is perhaps uniquely transparent. Indeed she is exceedingly hard to see and is easily lost even within the confines of a glass jar or a Petri dish. The female *Copilia* is 5–6 mm in length and about 1 mm in width. The huge anterior lenses are fixed, while the moving inner lens of each eye is attached to a 'rod' photoreceptor, which is curved inwards, looking like a hockey stick. The receptor is orange in colour, this being the only pigment of this extraordinarily transparent creature, in which all the inner structure is clearly visible with a low-power microscope. We became convinced that this is a single-channel eye, which apparently works by (horizontal) scanning. For there is only a single optic nerve for each eye, and the movements are 'saw-tooth' in velocity. They move fast inwards and slow outwards. A saw-tooth motion should be expected for scanning, the scanned information being in the slow direction. A symmetrical sweep is unlikely, as precision would be needed to avoid overlap errors, and the signals would be alternately reversed in sequence, requiring sophisticated interpretation. So saw-tooth movements are signs or symptoms of scanning.

Measuring the separation of the anterior and posterior lenses, it was clear that there is no convergence, as for stereo vision; the 'scans' move outward. The elements of each eye are very similar to a normal ommatidium; the difference is the enormous separation between the optical elements. The almost continuous movement of the posterior lens and its attached receptor has a frequency from about 0.5 to 5 'scans' per second. We later found that there can be long resting periods, generally with a violent burst of 'scanning' just before moving off. *Copilia* appears to be dead in these resting periods, as nothing moves inside, since she has no heart.

On a later expedition to Naples (August 1972), with Anthony Downing (an ex-student and colleague), Philip Clark (technician), Freja Gregory and my schoolgirl daughter Romilly, we extended the observations somewhat. Using infrared illumination, to which we believe *Copilia* is blind, we found that this marked burst of 'scanning' also occurs in darkness just before swimming (often more like leaping) off. Tony Downing (Downing, 1972) established, by placing a grating in front of the eye and looking at the lens's image from behind, with a high-power microscope, that the inner lens lies at plane of the in-focus image of the anterior lens. This would be appropriate for scanning; but we have no more direct evidence that this is a truly scanning single-channel eye. We found that the male *Copilia*, which is much larger and very different in appearance, with rudimentary eyes and like a woodlouse in shape, can emit bright flashing light from specialized light organs, and that this occurs only when there is some ambient ultraviolet light (Gregory, 1986). We guess this is to attract the female, the necessary ambient ultraviolet being available on the surface in moonlight. Vertical migration (necessary for finding a mate, by reducing the three dimensions under water to the two dimensions at the surface) may be provided by different colour response of the small median eye. However, vertical migration in Copepods is, in spite of Sir Alistair Hardy's beautiful experiments, mysterious (Hardy, 1956).

Scanning is appropriate for television, as electronics provides high-frequency (many MHz), large-bandwidth channels; but neural channels are extremely limited in frequency (maximum about 100 Hz), with low bandwidths, though there may be large numbers in parallel, which is costly and difficult for electronics. So a scanning biological eye is surprising. However, it may have been a necessary step for the evolution of multi-channel, many-ommatidia compound eyes. If so, we should expect scanning eyes with more than one, though not with large numbers of ommatidia. We should expect this because the channel capacity of a single neural channel is so low that visual development requires more channels.

There are strong candidates for scanning with more than one optical channel. Most familiar is the eye of *Daphnia* (well known as goldfish food, though much more interesting than the goldfish!), which is in continuous, large-amplitude, oscillatory rotary movement. Without such movement a small number of ommatidia would not provide useful acuity. There are gaps in the visual field between the individual ommatidia fields (Downing, 1972), so this is a strong candidate for a multi-channel scanning eye.

The large marine copepod *Labidocera* was described in a major paper by Parker (1891). The male's eye was described as having two retinas, which may be rotated on their lenses through an angle of about 45°: '... by contraction of the posterior muscle, the retina may be drawn upward and backward over the surface of the lens, till its axis, instead of pointing dorsally, is directed forward and upward at an angle of about forty-five degrees with its original position. The retina is not usually held for any length of time in this position, but is soon returned by the contraction of the anterior muscle to its normal place. The backward motion of the retina is accomplished with such rapidity that the animal has the appearance of winking.'

These observations are essentially confirmed and extended by Land (1988). He finds that the movements occur in, '... bouts lasting from a few seconds to a minute,' often with many minutes between bouts. This is the same as for *Copilia*.

Contrary to Parker's early observations, Michael Land (Land, 1988) finds that the backward motion in *Labidocera* is slower, the forward movements probably being by elastic recoil. Scanning must be slow with the limited information rate of a *neural* channel. Land found these scans to take between half and one second. A moving light did not affect the rhythm of the scanning, which we found also for *Copilia*. But unlike *Copilia*, a moving light source pulled the eye with it, over the full range of the eye's movement, which is as much as 25° either side of the central position. The 'two retinas' of this eye are remarkably interesting. One is a line of five 'slab-like' rhabdoms lying in the focal plane of the lens (Land, 1984), and three other receptors lying much closer to the lens, and so receiving out-of-focus images. Land finds the eyes of the female to be somewhat different, though with a 'hint' of the linear arrangement of the male. The female eyes are separate; the male eye cups are joined. So the male has ten in-line, evidently scanning receptors. Land suggests that the lines of in-focus receptors are used for scanning, while the other receptors are used for tracking moving lights or objects – presumably especially for seeking mates. The up–down following movement induces a related movement of the tail (and so of the swimming position) while the line of 'scanning' receptors does not evoke tail or eye movements. That the orienting receptors are out of focus should not make them useless for this purpose. Land suggests that the extra receptors serve to keep the creature stable in spite of waves, so the scanning is more effective and less ambiguous. *Daphnia* also has a stabilizing system, though this is for rotation, as its eye moves (scans) rotationally.

Land (1984) describes a ribbon retina in some heteropods (e.g. *Pterotrachea*, *Oxygyrus*), having 'a long ribbon, only 3–6 cells wide, but many hundreds long.' Fresh specimens of a small heteropod with a ribbon of 3 receptors just over 400 receptors long were 'in almost constant scanning motion.' This is also described by Land and colleagues (Land *et al.*, 1990) as in almost constant scanning motion, through an arc of just over 90°. In this paper, the mantis shrimp *Odontodactylus scyllarus* is described as having spontaneous eye movements independent for each eye, at a maximum rate of three per second. Land *et al.* (1990) report: 'Most movements are made in a direction

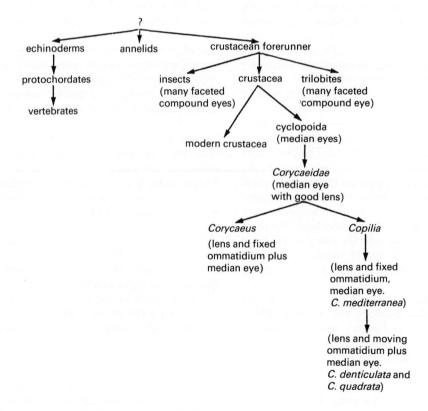

Fig. 3.2 *Evolutionary sequence for the scanning eye of* Copilia *suggested by Neville Moray. (Redrawn from Moray, 1972).*

approximately at right angles to the orientation of the specialized band.... The slow speed of the eye-movements is compatible with scanning, that is, the uptake of visual information during the movement rather than its exclusion as in conventional saccades.' The point here is that the very rapid eye jumps, for example of human saccades, serve just the opposite of scanning. They are so rapid that they prevent information uptake during such eye movements. However, human following eye movements are not saccadic but are smooth, allowing continuous uptake while objects are tracked.

The necessity for slow eye movements for scanning is discussed by Moray (1972), with special reference to *Copilia*, with a suggested evolutionary sequence as shown in Fig. 3.2.

We are left with many questions. In particular, is some ancient creature, somewhat like the rare and beautiful *Copilia*, the first step towards the massive complex compound eyes of insects? Are simple and compound eyes derived, as we have speculated, respectively from passive contact and active probing touch? Questions such as these are surely worthy of continuing attempts to answer, with more sophisticated speculations and more data for testing these ideas and discovering new facts of life.

Philosophy, pure science and technology combine in studies of evolutionary developments of eyes – to cast light on the nature and origins of our perception and our intelligence.

References

Bakewell, F. C. (1853). *Electric Science; Its History, Phenomena and Applications*. London: Ingram Cooke.

Downing, A. C. (1972). Optical scanning in the lateral eyes of the copepod *Copilia*. *Perception*, 1, 247–262.

Exner, S. (1891). *Die physiologie der facettirten Augen von Krebsen und Insecten*. Leipzig and Vienna: Deuticke.

Giesbrecht, W. (1892). *Pelagischen Copepodem des Golfes von Neapel*. Berlin: Friedlander.

Gregory, R. L. (1967). The origins of eyes and brains. *Nature Lond.*, 213, 369.

Gregory, R. L. (1968). The evolution of eyes and brains—a hen-and-egg problem. In *The Neuropsychology of Spatially Oriented Behaviour*. ed. Freedman, S. J. New York: Academic. Reprinted in Gregory, R. L. (1974). *Concepts and Mechanisms of Perception*, pp. 602–613. London: Duckworth.

Gregory, R. L. (1980). Perceptions as hypotheses. *Phil. Trans. R. Soc. Lond. B*, 290, 181–197.

Gregory, R. L. (1986). See Naples and live: the scanning eye of Copilia. In *Odd Perceptions*. London: Methuen and Routledge.

Gregory, R. L., Ross, H. E. and Moray, N. (1964). The curious eye of Copilia. *Nature Lond.*, 201, 1166. Reprinted in Gregory, R. L. (1974) *Concepts and Mechanisms of Perception*, pp. 390–584. London: Duckworth.

Hardy, A. C. (1956). *The Open Sea: The World of Plankton*. London: Collins.

Kuiper, J. W. (1966). On the image formation in a single ommatidium of the compound eye in Diptera. *Int. Symp., Functional Organization of the Compound Eye*. ed. Bernhard, C. G. London: Pergamon.

Land, M. F. (1984). Molluscs. In *Photoreception and Vision in Invertebrates*. ed. Ali, M. A. NY: Plenum.

Land, M. F. (1988). The functions of eye and body movements in *Labidocera* and other copepods. *J. Exp. Biol.*, 140, 381–391.

Land, M. F., Marshall, D., Brownless, D. and Cronin, T. W. (1990). The eye movements of the mantis shrimp *Odontodactylus scyllarus* (Crustacea: Stomatopoda). *J. Comp. Physiol. A.*, 167, 155–166.

Moray, N. (1972). Visual mechanisms in the copepod *Copilia*. *Perception*, I, 193–207.

Parker, G. H. (1891). The compound eyes in crustaceans. *Bull. Mus. Comp. Zool. Harvard*, 21, 44–140.

Reichardt, W. and Poggio, T. (1976). Visual control of orientation behaviour in the fly. *Q. J. Biophys.*, 9(3), 311–375.

Reichardt, W. and Poggio, T. (1979). Figure-ground discrimination by relative movement in the visual system of the house fly. Part 1: Experimental results. *Biol. Cybern.*, 35, 81–100.

Sarnat, H. B. and Netsky, M. G. (1974). *Evolution of the Nervous System*. Oxford: Oxford University Press.

Wilkie, J. S. (1953). *The Science of Mind and Brain*. London: Hutchinson.

PART II

EVOLUTIONARY DEVELOPMENTS

4 The Evolution of Vertebrate Visual Pigments and Photoreceptors

J. K. Bowmaker

Introduction

The visual system of vertebrates and many higher invertebrates has evolved to provide a detailed picture of the surrounding environment. The perception of the relative positions of objects, their contours, surface texture, colour and brightness, and their movement within the visual field is based ultimately on information extracted from the two-dimensional image of the environment focused on the photoreceptive layer of the retina. For this to occur, light must first be absorbed and this is achieved through specialized photosensitive molecules, visual pigments, which are contained within the photoreceptor cells.

Visual pigments belong to a very large family of structurally similar transmembrane proteins that act as receptor molecules in a wide range of different cell types: all function through the activation of a G-protein that binds guanosine triphosphate (GTP). The family includes not only all the visual pigments, but also acetylcholine muscarinic receptors (of which there may be at least five pharmacologically distinct subtypes), noradrenergic receptors (again at least five subtypes), serotonin or 5-hydroxytryptamine (5-HT) receptors (at least three subtypes), dopaminergic receptors and probably many others (North, 1989).

From an evolutionary viewpoint, there was presumably an ancestral membrane protein whose basic structure enabled it to function as a receptor molecule. Its structure can be inferred from the known structures of visual pigment proteins and other G-protein receptor molecules. These consist of a single polypeptide chain, normally made up of between 400 and 500 amino acids that form seven hydrophobic regions spanning the membrane linked by extra-membrane hydrophilic loops (Fig. 4.1). The transmembrane segments, comprising about 50% of the molecule, are thought to be α-helices, whereas the remaining hydrophilic regions, constituting about 25% of the molecule on each side of the membrane, are presumed to be straight chains (Hargrave et al., 1984). The seven helices are thought to form a bundle or palisade within the membrane creating a cavity on the extracellular side which acts as a ligand-binding pocket.

The molecules have other common features (Fig. 4.1). There are two glycosylation sites near the amino terminus on the extracellular side and a series of phosphorylation sites near the carboxyl terminus on the cytoplasmic side. Also, regions of the third cytoplasmic loop linking helices 5 and 6 are thought to be involved in interacting with the membrane-bound G-protein. In addition, there appear to be two highly conserved cysteine residues in the extracellular loops linking helices 2 and 3 and 4 and 5, which form a disulphide bridge thought to be essential for the correct folding and formation of the molecule (see Saibil, 1990). Indeed, a rare form of colour blindness in man, a type of blue cone monochromacy, has been attributed to a single-point mutation which leads to one of the cysteines in the opsins of the long- and middle-wave visual pigments being replaced by arginine (Nathans et al., 1989). Since the disulphide bridge cannot form in these opsins, the long- and middle-wave visual pigments are not functionally expressed, leaving only the short-wave, blue-sensitive cone pigment.

The postulated ancestral receptor molecule has evolved by numerous amino acid substitutions into a great family of receptors with a wide range of functions. In the case of visual pigments the ligand (retinal) is bound into the molecular pocket and must be activated by light, whereas in other membrane receptors the binding of the ligand itself (for example, a neurotransmitter or hormone) activates the receptor molecule. The 'excited' molecule then activates a G-protein (transducin in phororeceptors) that either directly affects ionic membrane channels or modifies a second messenger that in its turn regulates ionic channels. The

(a)

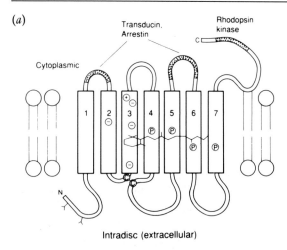

(b)

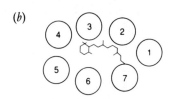

Fig. 4.1 (a) Schematic view of the proposed transmembrane arrangement of rhodopsin. There are two glycosylation sites near the amino terminus (N) and a series of phosphorylation sites near the carboxyl terminus (C). 11-cis retinal lies in the centre of the ring of helices (see b) and is linked to a lysine on helix 7. Features that tend to be common to the G-protein binding family of receptors are the prolines (P) in helices 4 to 7 and the disulphide bridge shown between two cysteines (C–C) at the intradisc (extracellular) surface. Sites of interaction with rhodopsin kinase, transducin (G-protein) and arrestin (involved in inhibition of activated rhodopsin) on the cytoplasmic loops are shaded. (b) Proposed packing of the helices within the membrane, viewed from the cytoplasmic surface and the location of retinal between them. (From Saibil, 1990).

second messenger may be any of a number of enzymes such as adenylate cyclase in the case of β-adrenergic receptors, but a cyclic guanosine monophosphate (GMP) phosphodiesterase in the case of visual pigments (for recent reviews, see North, 1989; Saibil, 1990).

Visual Pigments

All visual pigments are therefore built to the same pattern and consist of a large protein moiety, opsin, that binds a prosthetic group derived from vitamin A. The evolution of visual pigments within the vertebrates has led to a wide range of pigments, with maximum sensitivities (λ_{max}) extending from the ultraviolet, with λ_{max} as short as 350 nm, to the far red, with λ_{max} as long as 630 nm (for a review, see Bowmaker, volume 6, chapter 7 of this series). This spectral range is limited by two factors. First, because of the filtering effects of the atmosphere, most (about 80%) of the spectral radiation at the earth's surface is restricted to wavelengths between about 300 and 1100 nm, with the energy distribution having a maximum at about 480 nm and the quantum distribution maximal at about 555 nm. Secondly, the quantal energy of photons above about 850 nm is too low to lead to photoisomerization of organic molecules, whereas photons below about 300 nm have sufficiently high energies to be destructive to proteins (see Knowles and Dartnall, 1977). Thus an animal's visual system is constrained to function within a maximum spectral range from about 300 to 850 nm. Many animals have much smaller spectral ranges than this, for example our own, which is limited from about 380 nm to 780 nm. However, some diurnal freshwater teleosts and reptiles probably take full advantage of the entire available spectrum.

The wavelength of maximum sensitivity of a particular visual pigment is determined primarily by the structure of its opsin and secondarily by the prosthetic group bound to the opsin. The prosthetic group may be either retinal (found in all vertebrates and many invertebrates), 3-dehydroretinal (found in some freshwater fish, amphibians and reptiles as well as some crustaceans) or 3-hydroxyretinal (found in some insects). The amino acid sequences of several opsins have now been determined including the rod and three cone opsins of man (Nathans et al., 1986a), the rod opsins of cattle, sheep and chicken, a presumed cone opsin of a blind cave fish, as well as octopus rhodopsin and three opsins from Drosophila (Applebury and Hargrave, 1986; Ovchinnikov et al., 1988; Takao et al., 1988; Yokoyama and Yokoyama, 1990). Opsin molecules vary in length from 348 amino acids in human and bovine rhodopsin to 455 in octopus rhodopsin. The differences in the numbers of amino acids are due to variations in the lengths of some of the loops and of the N and C terminal regions.

Although the opsins are homologous and are clearly derived from a common ancestor, the percentage homology varies greatly. The vertebrate rod opsins show about 85% or greater homology, but only about 40% homology with the human cone opsins. The human 'red' and 'green' cone opsins have a 96% homology, but only about a 40% homology with the human 'blue' cone opsin. The invertebrate opsins so far sequenced are much more divergent. Their homology with vertebrate opsins is only about 15 to 30% and they have a similar low homology with each other.

Within opsin, retinal is bound via a protonated Schiff's base linkage to a lysine in the seventh membrane helix (position 296 in the vertebrate rod opsin, but 312 in the human 'red' and 'green' cone opsins) and it is the amino acid sequences in the transmembrane helices surrounding retinal that determine the λ_{max} of the molecule. The absorption of light by retinal is a consequence of the conjugated chain of single and double bonds within the molecule and of its three-dimensional configuration; the more linear the conjugated chain, the longer the λ_{max}. Retinal itself absorbs maximally at about 380 nm, whereas the protonated Schiff's base has a λ_{max} at about 440 nm. Charged amino groups within the opsin surrounding retinal will affect its conformational structure and thus spectrally displace the peak absorbance of the molecule. In this way, the amino acid sequences of opsin can 'tune' the visual pigment to absorb maximally at a specific spectral location. The proposed three-dimensional structure of visual pigments suggests that the Schiff's base region of retinal will be affected by polar groups in helices 1, 2 and 7, whereas the cyclohexane region will be affected by its surrounding helices 3, 4 and 5 (Fig. 4.1) (Honig et al., 1979; Kosower, 1988; Nathans, 1990).

Since vitamin A exists in two forms, A_1 and A_2, a given opsin can form two different visual pigments, either a rhodopsin when combined with retinal, the aldehyde of vitamin A_1, or a porphyropsin when combined with 3-dehydroretinal, the aldehyde of vitamin A_2. These two pigments will be spectrally distinct, since the additional double bond in the cyclohexane ring of 3-dehydroretinal causes a bathochromic shift in the absorbance of the molecule. Thus a rhodopsin with λ_{max} at 500 nm ($P500_1$) will have a 'paired' porphyropsin containing an identical opsin, but with λ_{max} at about 527 nm ($P527_2$). The spectral difference between 'paired' pigments is wavelength-dependent, with the difference increasing with longer λ_{max}, so that a long-wave rhodopsin, $P565_1$, has a 'paired' porphyropsin, $P620_2$. At short wavelengths in the blue and ultraviolet spectral regions, the difference is reduced to as little as 5 nm (Whitmore and Bowmaker, 1989).

Photoreceptors

In vertebrates, visual pigments are contained in two distinct classes of photoreceptor, rods and cones, with both classes being found in most species. Both types of photoreceptor are modified ciliary cells and have common features. The cells are composed of two segments, an outer segment that is embryonically a modified cilium and consists of a great quantity of cell membrane containing visual pigment, and an inner segment that contains the basic cellular organelles such as the nucleus and mitochondria (Fig. 4.2). The mitochondria are concentrated in the ellipsoid region close to the base of the outer segment, where they are involved in regulating the metabolism of the outer

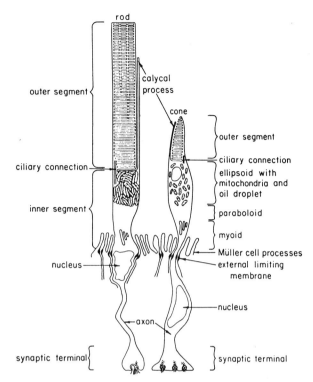

Fig. 4.2 Schematic drawing of a 'typical' rod and cone. Note the isolated discs within the rod outer segment and the oil droplet in the ellipsoid region of the cone. Oil droplets are not present in the cones of teleosts and mammals. (From Fein and Szuts, 1982; after Nilsson, 1964).

segment. The two segments are linked by a ciliary connection that contains the typical nine pairs of ciliary microtubules, but does not have the central pair common to motile cilia. The inner segment terminates in a synaptic foot that connects the photoreceptor with secondary retinal interneurones, bipolar and horizontal cells (for recent detailed reviews, see Dowling, 1987; Rodieck, 1988).

Nevertheless, rods and cones show markedly different morphological features (Fig. 4.2). The outer segments of rods are typically elongated cylinders, ranging in diameter from about 1 to 10 µm and in length from about 10 to 200 µm. Normally, the outer segment contains a stack of membrane discs that are formed from the plasmalemma at the proximal end of the segment close to the ciliary connection. Visual pigment molecules are incorporated into the membrane as the discs form and comprise about 80% of the membrane protein. The discs pinch off from the plasmalemma and lie free within the cytoplasm of the outer segment. As new discs form they cause the older discs to migrate towards the distal end of the segment. This process would result in the continuing elongation of the outer segment, but the distal region embedded in the pigment epithelium is continually removed phagocytically by processes of the pigment epithelial cells (Bok, 1985). In man, the discs are completely replaced in about 10 days, whereas in poikilotherms the process can take up to 30 days.

The outer segments of cones are much more variable in shape than those of rods: they tend to be small and conical, only 10 to 30 µm long with a diameter at the base of about 2 to 5 µm, though in man the foveal cone outer segments are more 'rod-shaped'. The discs are formed in a fashion similar to that in rods, but they are not regularly pinched off and the intra-disc spaces remain in contact with the extracellular space. Nevertheless, the cell membrane is continually forming and the distal region is removed by the pigment epithelium.

In many animal groups, double (and in some cases triple) cones are found in which the two halves may be more or less closely connected. The morphology of double cones is highly variable, especially within the teleosts: the two halves may be identical and closely fused and connected by gap junctions or they may be significantly different in shape and joined only loosely with no apparent membrane connections. In these cases the larger of the two halves is the principal member and the smaller is the accessory. The visual pigment complement of the two members may also be identical or different and a complex terminology has been introduced to identify the many different types of cone, such as identical and non-identical twins (Levine and MacNichol, 1979).

The inner segments of cones also differ from those of rods. The synaptic terminal is more complex and in many animal groups inclusions are found in the ellipsoid region. In some 'primitive' fish, amphibians, birds, reptiles and monotreme mammals, oil droplets are found which may be either optically clear or brightly coloured owing to the presence of high concentrations of carotenoid. Coloured oil droplets are most highly developed in diurnal birds and reptiles and must play a role in the colour vision of these species. In the cones of a number of species of surface-living tropical fish, the position of the oil droplet is occupied by an ellipsosome consisting of greatly modified mitochondria. These structures may also be coloured owing to high concentrations of mitochondrial enzymes such as cytochrome C (for a review see Bowmaker, volume 6, Chapter 7 of this series).

Rods and cones are also functionally different. The visual system operates over a range of light intensity of almost 10 log units, from almost total darkness to bright sunlight. At low light levels the rods mediate scotopic vision and have evolved to detect single photons, saturating at high light levels. The transducing enzyme cascade system amplifies the signal of a single photoisomerization sufficiently for the receptor potential to significantly alter the release of neurotransmitter at the receptor cell synapse.

However, a steady light of about 100 photons absorbed per second saturates the rod and the extended dynamic range of the scotopic system (about 5 log units) is achieved through the pooling of individual rod signals by the retinal neural network (Rodieck, 1988; Lamb and Pugh, 1990).

In contrast, cones, mediating photopic vision, are slightly less sensitive than rods, operate at higher levels of illumination and do not saturate. Individual cones are about five times (0.7 log unit) less sensitive than rods and cannot signal the absorption of individual photons. This is a characteristic of their transduction process and does not reflect a difference in the quantum efficiency of the visual pigment. However, because they are able to respond when the rate of photoisomerizations is equal to the rate of pigment regeneration, they are able to continue functioning even under constant bright illumination.

In general, animals possess only a single class of rod containing a visual pigment with λ_{max} close to 500 nm in the case of rhodopsins, but about 530 nm if a porphyropsin. Such a scotopic system can give information about brightness, but cannot confer wavelength discrimination. However, most animals possess more than one class of cone, each class containing a visual pigment with a different spectral sensitivity. The animal is then able to compare the rate of quantum catch of one class of cone with the rate of quantum catch of a second, spectrally distinct class and is therefore able to obtain both brightness and wavelength (colour) information.

Evolution of Photoreceptors and Visual Pigments

Two questions are often asked with respect to the evolution of the vertebrate photoreceptor system. Firstly, which is the ancestral (more primitive?) photoreceptor, the rod or cone, and secondly, what was the ancestral visual pigment? There are no simple answers to either question, though an answer to the first might suggest an answer to the second.

Rods are more conserved throughout the vertebrates, in terms of both structure and visual pigments, whereas cones are highly diverse in structure and can be divided into many spectrally distinct classes (possibly as many as five) within a given species. Rods could then be considered the simpler, and therefore ancestral, photoreceptor. However, in other respects rods could be considered to be more specialized than cones: they have a more complex morphology with isolated discs and are capable of signalling the detection of a single photon. A possible way to approach the question of the ancestral vertebrate photoreceptor is to look at modern species that are thought to be direct descendants of the earliest vertebrates.

The present-day cyclostomes, lampreys and hagfish,

are considered to have arisen directly from ancient ostracoderms, extinct jawless, fish-like primitive vertebrates of the Silurian and Devonian periods. There has been much debate as to whether the photoreceptors of lampreys are rods or cones (see Crescitelli, 1972), though it is clear that they have two classes of photoreceptor: one with long outer segments and the other with short outer segments. Both types of outer segment are morphologically cone-like in having many discs that are infoldings of the outer plasma membrane, but recent studies (Govardovskii and Lychakov, 1984; Ishikawa *et al.*, 1987; Negishi *et al.*, 1987) clearly demonstrate that the two classes are functionally different.

Microspectrophotometry (Govardovskii and Lychakov, 1984) has revealed that the shorter outer segments have λ_{max} close to 517 nm whereas the longer contain a pigment with λ_{max} at about 555 nm. Further, the spectral sensitivity of the dark-adapted eye is maximal at about 510 to 520 nm, but with short-wave background illumination the maximum sensitivity is displaced to about 555 nm. These data imply strongly that the lamprey has a duplex retina with both cone-like and rod-like photoreceptors and exhibits a Purkinje shift between scotopic and photopic vision similar to that seen in most vertebrates. However, the rod-like photoreceptors also have cone-like features: the pigment reacts with hydroxylamine in a similar manner to cone pigments (Hisatomi *et al.*, 1988), while the cells apparently do not saturate at high light intensities and are also involved in photopic vision, perhaps even subserving a dichromatic colour vision system (Govardovskii and Lychakov, 1984). Immunocytochemical reactions also support the idea that the shorter of the two outer segment types is rod-like and the longer cone-like (Ishikawa *et al.*, 1987; Negishi *et al.*, 1987).

Lampreys appear to possess both a rhodopsin (λ_{max} about 500 nm) and a porphyropsin (λ_{max} about 520 nm) (see Crescitelli, 1972) and in migratory species may possess the porphyropsin when migrating upstream and the rhodopsin when migrating downstream. However, Govardovskii and Lychakov (1984) suggest that on the basis of the shape of the absorbance spectra, the pigments of both classes of receptor are rhodopsins. This is somewhat surprising, since most rods containing rhodopsin have λ_{max} close to 500 nm. The lampreys studied by Govardovskii and Lychakov were from fish migrating upstream, so that it is quite possible that the photoreceptors contained mixtures of rhodopsin and porphyropsin, shifting the λ_{max} to longer wavelengths.

Porphyropsins are generally found only in freshwater species and the discovery of a porphyropsin in the lamprey led Wald (1957) to conclude, since he believed vertebrates (fish) evolved in fresh water, that porphyropsins were the ancestral vertebrate visual pigments. Such a view of the environment of ancestral vertebrates is probably no longer tenable and it is thought that the early ostracoderms lived in a marine or estuarine environment (McFarland *et al.*, 1985). If this was the case then it is more probable that the ancestral rod pigment was a rhodopsin with λ_{max} close to 500 nm.

The spectral sensitivity of an ancestral cone pigment is perhaps more debatable. The lamprey's 'cone' with λ_{max} at 555 nm may also contain a mixture of a rhodopsin and a porphyropsin and, if so, then the rhodopsin would have a λ_{max} at a somewhat shorter wavelength at about 530 nm. This would agree with the cone pigments of other shallow-water marine fish, both elasmobranchs and teleosts, that have long-wave cones with maximum sensitivities around 520 to 540 nm (for a review, see Bowmaker, 1990a). Elasmobranchs were once thought to have pure rod retinas, but this is probably only true for skate, *Raja* sp., and electrophysiological studies in the lemon shark, *Negaprion brevirostris*, reveal a photopic system maximally sensitive at about 540 nm (Cohen, 1990). However, if the 555-nm pigment in the lamprey is a pure rhodopsin then it may reflect the longer-wave cone rhodopsins found in some freshwater teleosts, amphibians, reptiles, birds and mammals. It is interesting to note that the quantal distribution of light at the earth's surface is also maximal at about 555 nm.

A conclusion that can be drawn from these studies of lampreys is that the ancestral vertebrate visual system was based on relatively unspecialized photoreceptors that were neither rods nor cones (though perhaps more cone-like) and that at least two spectral classes of photoreceptor were present at a very early stage. There is no reason why an ancestral vertebrate should not have been carrying more than a single opsin gene, and colour vision (requiring two spectral classes of photoreceptor), as distinct from photopic vision, may have been present in the earliest of vertebrates.

As vertebrates evolved, giving rise to the great radiation of fish, amphibians, reptiles, birds and mammals, the visual pigments and photoreceptors became adapted to the very wide range of photic environments to which the animals were exposed and to the visual tasks required of them. It is perhaps simpler to review this evolutionary radiation by looking at rod and cone systems separately within each major vertebrate group.

Fish

Teleost fish are ideal for surveying a wide range of visual pigments since they occupy relatively clearly defined photic environments with markedly different spectral ranges. Water acts as a monochromator and selectively absorbs both short- and long-wave light. With increasing depth not only is the light attenuated, but the spectral bandwidth is reduced first at long wavelengths and then at short wavelengths. In clear oceanic water the intensity of the downwelling light is reduced by about 2 log units within the first 100 m and is restricted to a narrow spectral band centred around 470 nm (for a review, see Loew and McFarland, 1990). By about 1000 m the intensity of the light is reduced to a level below the threshold for scotopic vision.

The classical example of the adaptive fit of photoreceptors and visual pigments to environmental light is that of the visual system of deep-sea fish. Clarke (1936) proposed that since the underwater light at depth in the ocean was both greatly attenuated in brightness and restricted to the blue region of the spectrum, deep-sea fish should possess pure rod retinas containing a rhodopsin with λ_{max} at about 470 nm. The peak sensitivity of the rods would match the wavelength of maximum transmission of clear water, thus giving the fish maximum sensitivity. In addition, many deep-sea creatures are bioluminescent and the maximum emission of their bioluminescence is also at about 480 nm (Herring, 1983).

Clarke's proposal has indeed proved to be the case. Most deep-sea fish, both teleosts and elasmobranchs, have pure rod retinas and possess rhodopsins with λ_{max} at about 470 to 480 nm (Denton and Warren, 1957; Denton and Shaw, 1963; Partridge et al., 1988). In order to increase sensitivity, not only are cones absent from the retina, but the rod outer segments are often elongated (up to 200 µm), or the rod outer segment layer may be composed of up to six rows of shorter outer segments (for a detailed review, see Locket, 1977).

However, there are some interesting exceptions to the 'deep-sea rhodopsin' rule. Some species have not only blue-emitting photophores, but also light organs emitting deep-red light with a maximum emission at about 700 nm (Widder et al., 1984). Such deep-red light will be effectively invisible to species with 480 nm rhodopsins, but species with red photophores have been found to possess two classes of rod containing different visual pigments, one with λ_{max} at about 515 to 520 nm and the other at about 540 to 550 nm (Bowmaker et al., 1988; Partridge et al., 1989).

The pigments appear to be 'pigment-pairs', the longer a porphyropsin and the shorter a rhodopsin, but their location in separate rods is distinctly different from most rhodopsin–porphyropsin systems in which the pigments are mixed in the same photoreceptor (see below). These deep-sea species have not evolved an opsin for a deep-sea rhodopsin, but an opsin that is able to form pigments that will give the fish sensitivity not only to the blue downwelling light but also to both the dominant blue bioluminescence and its own red bioluminescence. In addition, because the pigments are located in separate photoreceptors, the fish have the possibility of being able to distinguish the two photophores on the basis of wavelength, that is they have the potential for a colour vision system based not on cones but on rods. If this is the case then it is interesting that the fish have developed two spectral classes of photoreceptor by utilizing a rhodopsin–porphyropsin system and not by evolving a second opsin. In contrast, some species of deep-sea fish do have two classes of rod that contain spectrally distinct 'deep-sea' rhodopsins with λ_{max} only a few nm apart (Partridge et al., 1989).

Fish living at shallower depths will not be exposed to such a restricted photic environment and can perhaps be thought of as living under lighting conditions similar to those experienced by their primitive vertebrate ancestors. The downwelling light in shallow seas in the regions of continental shelves is sufficiently bright for photopic vision and has a relatively broad spectral range extending into the green and yellow regions of the spectrum. Fish living in this environment have duplex retinas with rods and, in the majority of cases, two classes of cone. The rods contain a rhodopsin with λ_{max} at about 500 nm, whereas the two cone populations have rhodopsins with λ_{max} at about 520 to 540 nm and about 440 to 460 nm. The longer-wave cone population is normally composed of double cones which, in the case of the cod, *Gadus morrhua*, are identical twin doubles with a $P517_1$, whereas the single short-wave cones contain a $P446_1$ (Bowmaker, 1990a). Such fish thus have a complete but basic photoreceptor system: a rod population for scotopic vision and two cone populations for a dichromatic photopic system. The evolution of the short-wave cone pigment presumably occurred at an early stage in the separation of photopic and scotopic vision. With a photopic system having an ancestral pigment at long wavelengths, it would not be surprising for a second pigment to evolve at short wavelengths.

As discussed previously, marine species normally have a visual system based on rhodopsins, whereas porphyropsins are a general feature of freshwater species and are found not only in freshwater fish but also in amphibians, especially in their aquatic phases, and freshwater turtles. Porphyropsins have not been reported in birds or mammals. The presence of porphyropsins in species inhabiting freshwater can be related to the spectral properties of the ambient light in the water. A characteristic of many bodies of fresh water is that they contain quantities of suspended particles such as silt and also contain varying amounts of dissolved organic materials derived from water drainage off the land. These substances scatter and absorb light, primarily at shorter wavelengths, so that light is attenuated much more rapidly than in clear water and the maximum transmission of the water is displaced from that of about 470 nm in clear water to longer wavelengths in the green to red regions of the spectrum. As a consequence, the visual systems of many animals occupying these environments have evolved to be maximally sensitive at longer wavelengths by taking advantage of the bathochromic shifts of porphyropsins: the spectral sensitivity of rods can be displaced from about 500 nm to about 530 nm and that of long-wave-sensitive cones from about 565 nm to about 625 nm. Nevertheless, many species retain rhodopsins and both rods and cones are found to contain mixtures of the two pigments (for reviews, see: Bridges, 1972; Lythgoe, 1972; Beatty, 1984).

The change of prosthetic group from retinal to 3-dehydroretinal is clearly seen in species that migrate between marine and freshwater environments. In general these species, such as some salmon and lampreys, show clear anatomical and physiological changes prior to their migration demonstrating that the changes are genetically programmed and are not environmentally dictated. The European and North American eels, *Anguilla* sp., are classical examples: they spend most of their lives in freshwater river systems as yellow eels, but on reaching maturity they undergo a metamorphosis becoming silver eels, and migrate to the Atlantic ocean, eventually breeding at depths of up to 400 m in the Sargasso Sea.

Before migration, in addition to undergoing great anatomical changes, the eel's visual system also changes dramatically. In fresh water they possess a duplex retina containing both cones and rods and their visual pigments are dominated by porphyropsins. The rods contain a pigment pair, $P523_2$–$P501_1$, with a λ_{max} close to 520 nm, but during the seaward migration the ratio of the mixture changes and the λ_{max} shifts towards 500 nm. In addition, the metamorphosis leads to a loss of cones and to the inactivation of the 'freshwater' rod opsin gene and to the expression of a further opsin gene that codes for a 'deep-sea' pigment (Carlisle and Denton, 1959; Beatty, 1975). The new rhodopsin is a $P482_1$ and its expression displaces the λ_{max} of the rods to below 500 nm.

The expression of a new opsin during development was once thought to be unique to the eel, but Munz and McFarland (1973) reported an opsin change in the rods of a cardinal fish, *Apogon brachygrammus*: the larval stage has a rhodopsin with λ_{max} at 482 nm, whereas the adults have a rhodopsin with λ_{max} at 494 nm. Recently Shand *et al.* (1988) have demonstrated a somewhat similar phenomenon in a coastal marine fish, the pollack, *Pollachius pollachius*. This species, closely related to the cod, has a duplex retina with rods (λ_{max} = 498 nm) and two classes of cones. One class consists of identical paired cones containing a $P521_1$, whereas the other, a class of single cones maximally sensitive at short wavelengths, changes its visual pigment during development. The juvenile stages of the fish live in shallow water, feeding on plankton, and the single cones contain a rhodopsin with λ_{max} at about 410 to 420 nm. However, as the fish matures it migrates into deeper water, begins feeding on crustaceans, and the spectral sensitivity of the single cones shifts to longer wavelengths around 450 to 460 nm. The new, longer-wave visual pigment is still a rhodopsin and the assumption is that during development the opsin expressed in the single cones changes. At depth, there will be less short-wave light available so that the shifted spectral sensitivity of the fish will more closely match the new photic environment, though the change in feeding habits may also be an important factor.

A somewhat surprising feature of a number of species of freshwater fish is the presence of a population of single cones maximally sensitive in the near ultraviolet with λ_{max} between about 350 and 380 nm. These were first identified in cyprinids (Avery *et al.*, 1983; Hárosi and Hashimoto, 1983), but have since been reported in yearling brown trout, *Salmo trutta* (Bowmaker and Kunz, 1987) and guppies, *Poecilia reticulata* (Archer and Lythgoe, 1990), and may also occur in larval perch, *Perca flavescens*, and bluegill, *Lepomis macrochirus* (Loew and McFarland, 1990). We have also recently identified these cones in the goldfish, *Carassius auratus* (Bowmaker *et al.*, 1991) (Fig. 4.3). The finding of ultraviolet sensitivity in freshwater fish was unexpected, since it has been generally assumed that ultraviolet light was very rapidly absorbed and scattered by the first few centimetres of water. However, this proves not to be the case (Loew and McFarland, 1990) and, at least in the trout, the ultraviolet cones are common and presumably play a significant role in vision (Douglas *et al.*, 1989).

The evolution of ultraviolet-sensitive photoreceptors in vertebrates is not unique to freshwater teleosts: they probably also occur in reptiles and birds (see below), though there is no evidence for their occurrence in marine fish and amphibians. Ultraviolet vision has often been considered to be 'specialized' and to serve specific functions different from the 'normal' visual system, though this assumption may be a simple consequence of our blindness to this spectral region. There can be no *a priori* reason why opsins capable of forming ultraviolet-sensitive visual pigments should not have evolved at an early stage in the evolution of vertebrates along with all the other opsins: visual information is available in the ultraviolet and it would be surprising if it were not utilized. Nevertheless, when the opsins of these pigments have been sequenced, it will be interesting to discover whether they evolved from a single vertebrate ancestral ultraviolet-sensitive pigment or if they evolved independently in the teleosts and in the birds and reptiles. The lack of homology between the invertebrate rhodopsins and the vertebrate pigments must make it probable that the ultraviolet-sensitive pigments in insects evolved completely separately from those in vertebrates.

The presence of ultraviolet-sensitive cones in freshwater fish gives them the potential for tetrachromatic colour vision. In general, in a rhodopsin-based system the cones have maximum sensitivities at about 355, 450, 500 and 565 nm, but at about 360, 460, 530 and 620 nm with a to vary their spectral sensitivity and colour vision, especially at long wavelengths. Within the cyprinids, two groups of species are found, those that retain a porphyropsin-dominated visual system throughout the year and those that vary their rhodopsin–porphyropsin ratio seasonally. Seasonal variations were first reported for the rod pigments in the rudd, *Scardinius erythrophthalmus* (Dartnall *et al.*, 1961), in which the ratio of the two pigments is biased in favour of rhodopsins in the summer, but in favour of porphyropsins in the winter. This has since been shown to occur in parallel within the cones (Loew and Dartnall, 1976; Whitmore and Bowmaker, 1989), so that in the summer the fish are less sensitive at long wavelengths. The replacement of the chromophore can be manipulated experimentally, apparently being controlled by day length and temperature (for reviews, see Bridges, 1972; Beatty, 1984).

It would be expected that the seasonal change in long-wave sensitivity could be easily related to either a change in the photic environment of the fish or in its behaviour patterns. However, although these do change, there is no obvious direct correlation and no totally convincing explanation for the spectral shifts has been forthcoming (for a review, see Knowles and Dartnall, 1977). It is interesting that the goldfish, which does not show any natural seasonal variation from its 100% porphyropsin complement, can be driven artificially to an almost complete rhodopsin system if slowly adapted to long day lengths (16L : 8D) and high temperatures (28°C) (Tsin and Beatty, 1978).

The complexity of the visual pigment system in freshwater fish, especially in the cyprinids, raises some interesting evolutionary and genetical problems. Has the seasonal variation in the ratio of rhodopsin to porphyropsin evolved independently in land-locked freshwater fish or is it a remnant of the system seen in migratory species? Certainly this is probably the case with land-locked species of salmonids, since migratory species of salmon show seasonal variations as they mature in fresh water before migrating to the ocean (Bridges, 1972), but the situation in cyprinids is far from clear.

In cyrinids, closely related species do not show the same seasonal variations. Roach, *Rutilus rutilus*, and rudd are known to interbreed and produce viable offspring (Wheeler, 1969), but roach do not show seasonal variations in the levels of porphyropsin whereas rudd do. What is the situation in hybrid offspring? The hybrid problem is compounded because blue-sensitive cones in cyprinids appear to be polymorphic: some species have a population of single cones maximally sensitive at about 450 nm whereas other species have a population with λ_{max} at about 410 nm. Roach have the longer-wave-sensitive cones (Downing *et al.*, 1986) whereas rudd probably have the shorter (Whitmore and Bowmaker, 1989). However, in an earlier study of rudd cone pigments in which Loew and Dartnall

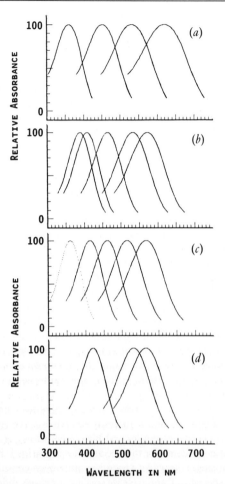

Fig. 4.3 *Absorbance spectra template curves of cone visual pigments from two teleosts, a bird and a mammal. (a) Goldfish,* Carassius auratus, *with porphyropsins with λ_{max} at 360, 542, 532 and 620 nm. Note the relatively broad absorbance spectra and the long-wave peak of the longest-wave pigment. The $P532_2$ and the $P620_2$ are located in double cones. (b) Guppy,* Poecilia reticulata, *with rhodopsins with λ_{max} at 389, 408, 464, 533 and 572 nm. The $P533_1$ and $P572_1$ are located in double cones. (c) Pigeon,* Columba livia, *with rhodopsins with λ_{max} at 413, 460, 515 and 567 nm. An ultraviolet-sensitive pigment has not been directly measured in the pigeon, but is probably present and shown as the dotted curve with λ_{max} at 360 nm. The $P567_1$ is found in both members of double cones. (d) Human with rhodopsins with λ_{max} at 420, 531 and 564 nm. These species with multiple cone pigments probably represent the culmination of the evolution of cone pigments (and colour vision) within these three vertebrate classes. λ_{max} data from Bowmaker et al. (1991) (goldfish), Archer and Lythgoe (1990) (guppy), Bowmaker (1989) (pigeon) and Bowmaker (1990b) (human).*

porphyropsin-based system (Fig. 4.3). Because the rhodopsins and porphyropsins can form pigment pairs within each cell type, the fish have the added benefit of being able

(1976) demonstrated seasonal variation of cone pigments, they also reported a blue-sensitive cone population with λ_{max} at about 450 nm. A possible explanation for this discrepancy of cone populations in rudd is that Loew and Dartnall's rudd were from a hybrid population that carried the 'roach' blue-sensitive cone, but possessed the 'rudd' seasonal variation. An alternative is simply that rudd are polymorphic and that isolated populations occur with either the 410- or 450-nm cone types.

The effect of hybridization on visual pigments was demonstrated by McFarland and Munz (1965) in the rod pigments of two closely related salmonids, the brook char, *Salvelinus fontinalis*, and the lake char, *S. namaycush*. The two species contain spectrally different rhodopsin/porphyropsin mixtures with the rhodopsins being a $P503_1$ and a $P512_1$ respectively. Fertile hybrid fish were produced by artificial fertilization and the first-generation (F_1) hybrids were all found to contain a mixture of the two rhodopsins whereas in the F_2 generation some individuals contained the $P503_1$, some a mixture and some the $P512_1$, in the ratio $1:2:1$. This is the expected outcome if the opsins are inherited as a single genetic factor and there is codominance of the two genes.

Another example of a polymorphism of visual pigments in fish has been reported in the guppy (Archer and Lythgoe, 1990). These fish live in the surface layers of West Indian streams, but, in contrast to most temperate freshwater species, have visual pigments which are probably pure rhodopsins (Schwanzara, 1967). Being surface-living, guppies possess a number of cone types, probably three classes of single cone containing a $P389_1$, a $P408_1$ and a $P464_1$ and double cones containing pigments maximally sensitive in the green–red region of the spectrum (Fig. 4.3). The two members of the double cones appear to be able to have one of three spectral sensitivities with λ_{max} at about 533, 548 and 572 nm. The shape of the absorbance spectra of the 533-nm and 572-nm cones are typical of rhodopsins, but the spectrum of the 548-nm cones is too broad for a rhodopsin with λ_{max} at 548 nm and Archer and Lythgoe (1990) suggest that these cones contain a mixture of the $P533_1$ and the $P572_1$.

Individual guppies may have one, two or all three of these cone types, though in a given population of fish, individuals tend to favour the same complement. The variations that occur naturally are probably related to the photic environment and behavioural patterns of each isolated fish population and it would seem likely that the proportions of the different cone spectral sensitivities could be varied artificially in the laboratory by maintaining fish under different lighting conditions. If this is the case then it is an interesting example of environmental control over gene expression: whether one or the other opsin gene or both genes are expressed in any single cone would be dependent on the photic environment.

In addition to the complexity of their double cones, guppies have a total complement of at least five cone opsins and, if they are all involved in colour vision (a pentachromatic system) then perhaps this species could be considered to represent the culmination of the evolution of colour vision in teleosts (Fig. 4.3). If it is assumed that an ancestral fish possessed only a green-sensitive (λ_{max} about 530 nm) and a blue-sensitive (λ_{max} about 450 nm) cone rhodopsin then it may be that the longer-wave pigment was ancestral to all cone pigments maximally sensitive in the red–green region of the spectrum (from about 500 to 565 nm), whereas the shorter-wave pigment gave rise to the cone pigments maximally sensitive in the blue–violet–ultraviolet spectral region (from about 350 to 500 nm). The exact status of cone pigments with λ_{max} around 500 nm will be of great interest, especially in their relation to rod opsins.

Amphibians

The amphibians appear to show a transitional stage between the visual systems of some 'primitive' fish and the highly developed systems of diurnal reptiles and birds (see below). The evolution of amphibians (and of tetrapods generally) is not clear, but it is probable that they evolved from rhibistrian crossopterygian fish sometime in the middle Devonian period, about 350 to 400 million years ago (McFarland *et al.*, 1985). The rhibistrians were lobed-finned fish related to the coelacanths. The modern coelacanth, *Latimeria chalumnae*, a deep-sea fish presumably living in an environment very different from that of its ancestors, has long, slender rods and some cones with colourless oil droplets (see Crescitelli, 1972), with the rods containing a typical deep-sea rhodopsin (Dartnall, 1972). Modern amphibians also have duplex retinas, but with two classes of rod and at least two classes of cone; double cones and single cones with a clear oil droplet. Amphibians that remain completely aquatic generally possess porphyropsins, whereas those that become more terrestrial after metamorphosis generally possess rhodopsins.

The majority of the rods ('red' rods) contain either a rhodopsin with λ_{max} at about 500 nm or a porphyropsin with λ_{max} at about 530 nm (or a mixture of the two), but in addition there is a small population of rods ('green' rods) that are blue-sensitive with λ_{max} at about 430 to 440 nm (Liebman and Entine, 1968; Bowmaker, 1977a). The visual pigment of the 'green' rods behaves rather like a cone visual pigment in that it is sensitive to hydroxylamine and has cone-like photoproducts (Bowmaker, 1977a). Behavioural experiments also suggest that the 'green' rods may be involved in colour vision (Muntz, 1966).

In the adult leopard frog, *Rana pipiens*, the principal member of the double cones contains a $P575_1$, whereas the accessory member contains a $P502_1$, spectrally similar to the rhodopsin of the 'red' rods (Liebman and Entine, 1968). The single cones also contain the $P575_I$. More recently, Hárosi (1982) has identified a minority population of small single cones lacking an oil droplet that contain a pigment spectrally similar to that of the 'green' rods. He also found a red–blue double cone in which both members lacked an oil droplet.

Amphibians thus have a complex of photoreceptors and visual pigments that give them the potential for wavelenth discrimination at both mesopic and photopic levels and with the possibility of trichromatic colour vision. However, electrophysiological and behavioural investigations are not consistent and the status of colour vision in anurans is not well established (see Jacobs, 1981). Presumably, the visual pigment complement of modern amphibians has evolved separately from that of the teleosts and an analysis of the opsin structures, especially that of the 'green' rod pigment, should indicate the relationships and evolutionary status of these pigments.

Reptiles and Birds

A complexity of photoreceptors similar to that found in freshwater teleosts is also found in diurnal reptiles and birds, perhaps culminating in an avian pentachromatic colour vision system. Because of the close evolutionary relationship of birds and reptiles (excluding the crocodiles), their visual systems can be considered together. The most striking feature of the retinas of diurnal forms is the presence of brightly coloured oil droplets that are located in the ellipsoid region of the cone inner segments and therefore interpose between the light incident on the photoreceptor and the visual pigment. The role of these droplets has long fascinated those interested in comparative aspects of colour vision, but their exact functions have not as yet been fully elucidated.

The cone population of turtles and passerine birds consists of a single class of double cones and at least four classes of single cone (for recent reviews, see Goldsmith *et al.*, 1984; Lipetz, 1985; Bowmaker, volume 6, chapter 7 of this series). The double cones, with a large principal member and a smaller accessory member, normally have a yellow- or orange-coloured droplet in the principal member, but may or may not have a droplet is the accessory member, whereas the single cones can be classified (at least to a first approximation) by the colour or spectral transmission of their droplets: R (red), Y (yellow), C (clear) and T (transparent). The R, Y and C types often contain sufficiently high concentrations of carotenoid to act as cut-off filters, transmitting only the longer wavelengths. These coloured droplets have presumably evolved from the colourless droplets found in the cones of lower vertebrates as in some 'primitive' fish (such as the sturgeons, *Acipenser* spp., and the coelacanth) and amphibians (see Crescitelli, 1972).

Given a set of coloured filters, it is theoretically possible to design a colour vision system based on a single visual pigment located in different classes of cone, each containing a filter of a different colour. This suggestion was first put forward by Klause in 1863 (see Crescitelli, 1972) and was not finally countered until microspectrophotometry demonstrated the presence of multiple cone pigments in turtles (Liebman and Granda, 1971) and birds (Bowmaker and Knowles, 1977, Bowmaker, 1977b). Four cone pigments have now been clearly identified in a number of species with, in the case of rhodopsins, maximum sensitivities in the red at about 565 nm, in the green at between about 500 and 520 nm, in the blue at about 450 and 460 nm and in the violet at between about 400 and 420 nm (Fig. 4.3) (Bowmaker, volume 6, chapter 7 of this series). Freshwater turtles have a porphyropsin system, with the maximum sensitivities of the pigments displaced to longer wavelengths. The maximum sensitivities of the cone pigments in reptiles and birds are similar to those found in many shallow-living fish and this correspondence of λ_{max} may reflect the results of convergent evolution from an ancestral two-pigment system common to both teleosts and reptiles.

The visual pigments and oil droplets in birds and reptiles are combined in a very logical fashion. Thus, in single cones, the long-wave visual pigment is normally combined with the R-type droplet, the green-sensitive pigment with the Y-type droplet, the blue-sensitive pigment with the C-type droplet and the violet-sensitive pigment with the T-type droplet (though since these droplets transmit the full spectrum, there is no reason why transparent droplets should not be combined with any visual pigment). Where the droplets act as cut-off filters, they not only reduce the absolute sensitivity of the cone, but also narrow the effective spectral sensitivity of the cell by cutting off the short wavelengths and so displacing the peak sensitivity to longer wavelengths. It has been suggested that narrowing the spectral sensitivity of cones and removing the 'confusing' overlap of visual pigments at short wavelengths could improve wavelength discrimination, though behavioural experiments with birds and turtles do not reveal great improvements in wavelength discrimination over species without oil droplets (see for example, Wright, 1979; Barlow, 1982; Arnold and Neumeyer, 1987; Govardovskii and Vorobyev, 1989).

A number of birds (e.g. Goldsmith, 1980) and the turtle, *Pseudemys scripta elegans* (Arnold and Neumeyer, 1987), have been shown to be sensitive to ultraviolet light and, although an ultraviolet-sensitive cone has not been directly identified, behavioural and electrophysiological evidence (Chen and Goldsmith, 1986) implies strongly that such a cone must be present. Since all the cones in these species have oil droplets, an ultraviolet-sensitive cell must have a transparent droplet. Thus some birds (and possibly some turtles) may have five-cone visual pigments similar to those of the guppy and which could form the basis of a pentachromatic colour vision system (Fig. 4.3). However, the cone population containing transparent droplets in most birds comprises only about 5 to 10% of the total cone population, so that acuity in the ultraviolet will be very poor, but may nevertheless serve a useful function in colour vision similar to that of the short-wave sensitive cones in man that also account for only about 5 to 10% of the total cone population.

From a theoretical viewpoint, there is no real limit to the number of spectrally distinct classes of cone that may be involved in colour vision. The minimum requirement is of course two classes with overlapping spectral sensitivities, though the maximum identified in vertebrates is five and as many as eight spectral classes of ommatidia have been identified in a mantis shrimp (Cronin and Marshall, 1989). However, by applying the sampling theorem from communication theory, Barlow (1982) has suggested, as a consequence of the broad absorbance spectra of visual pigments, that in an animal with a wavelength discrimination range similar to that of man (from about 430 to 650 nm), trichromacy is perhaps the most satisfactory system for colour analysis. Additional overlapping cone sensitivities within the same spectral range would not be efficient in extracting any further wavelength information.

The corollary of Barlow's hypothesis is that a polychromatic colour vision system with more than three channels would be useful to an animal only if either the absorbance spectra of the channels were narrower than visual pigment absorbance spectra or if the wavelength discrimination range was greater than in man (Bowmaker, 1983). Both of these criteria are satisfied in many birds and reptiles where the spectral sensitivities of the cone channels are narrowed by the cutoff characteristics of the coloured oil droplets and the visual range is extended into the ultraviolet by the transmission characteristics of the lens and cornea. Similarly, in fish such as the guppy, which appear to have five cone visual pigments (Archer and Lythgoe, 1990), the absorbance spectra of at least some of the cone channels are narrowed by coloured filters, ellipsosomes that contain high concentrations of cytochrome C (MacNichol et al., 1978; Avery and Bowmaker, 1982).

Geckos

The geckos represent an interesting 'oddity' in the evolution of visual pigments and photoreceptors. The Gekkonidae include diurnal, crepuscular and nocturnal species and show a range of photoreceptor types from rather cone-like cells in diurnal species to more classical rods in nocturnal species. The cone-like cells have long, slender outer segments and colourless or pale yellow oil droplets, whereas the rod-like receptors have typical rod-shaped outer segments, but may have oil droplets and may show infoldings of the plasma membrane. Double cells, either cone-like or rod-like, may also be present.

The photoreceptors and visual pigments of the large nocturnal Tokay gecko, *Gekko gekko*, have been studied in detail (for a review, see Crescitelli, 1977). The retina contains both single and double rods with no oil droplets and only two visual pigments have been identified, a $P521_1$ and a $P467_1$. The majority of single rods contain the longer-wave pigment, but the double rods can have any combination of the two pigments (Crescitelli et al., 1977). The electroretinogram exhibits two b-waves, the faster apparently generated by the longer-wave photoreceptors. The two pigments are also biochemically distinct, the longer having characteristics normally associated with cone pigments and the shorter behaving more like a typical rod pigment. The $P521_1$ is destroyed in the dark by hydroxylamine and borohydride and exhibits an ionochromic shift in response to changes in chloride and nitrate concentrations, whereas the $P467_1$ is stable in the dark and shows no ionochromic effect. In addition, when bleached, the $P521_1$, in contrast to typical rod pigments, does not show a long-lived metarhodopsin III photoproduct (Crescitelli, 1977).

The implication of these findings is that the longer-wave rods are functionally more cone-like than the shorter-wave rods and there is the possibility that geckos may have a dichromatic colour vision system. Unfortunately, there appear to be no published electrophysiological or behavioural data to establish whether geckos exhibit a Purkinje shift or are able to discriminate wavelength. The structure of the photoreceptors of geckos and other reptiles led Walls (1942) to propose a transmutation theory: modern geckos possess photoreceptors that are in the process of transmutation to rods from the cones of their diurnal ancestors. The behaviour of the visual pigments has been considered to support this idea and it will be interesting to see whether the opsins of the two pigments are more homologous with cone or with rod opsins.

Mammals

The mammals show a great diversity in their visual systems, since they have radiated into almost every environmental niche. Mammals range from primarily nocturnal species, such as rats and mice, Myomorpha, with almost pure rod retinas, to the highly diurnal squirrels, Sciuromorpha, with well-defined dichromatic colour vision and to the primates, where the Old World monkeys, Catarrhini, are the only mammals with well-developed trichromatic colour vision.

Mammals evolved from early therapsid reptiles of the Permian and, by the Jurassic, five distinct orders of mammals can be recognized. However, modern mammals show a dichotomy in their evolution during the Cretaceous: the Jurassic Theria giving rise to the great Cenozoic radiation of the marsupial, Metatheria, and placental, Eutheria, mammals, whereas the monotremes, Protheria, probably evolving from a separate multituberculate Jurassic order (McFarland et al., 1985).

The early Jurassic and Cretaceous mammals were small and most probably nocturnal, the multituberculates being herbivores and the therians insectivores. It may not be unreasonable to assume that the visual system of these insectivorous early mammals was similar to the ancestral vertebrate system suggested above: a dominant scotopic system based on rods with a typical rhodopsin maximally sensitive at about 500 nm and a limited photopic system based on two spectral classes of cone maximally sensitive to the longer wavelengths around 530 to 550 nm and to the shorter wavelengths around 440 to 460 nm.

Such a system is retained in a wide range of modern mammals that are both nocturnally and diurnally active and in which the duplex retina has about 10 to 20% cones. Within the carnivores, the domestic cat, *Felis domesticus*, and dog, *Canis familiaris*, have been shown to be behaviourally dichromatic. The cat has cones maximally sensitive at about 555 and 450 nm (Hammond, 1978; Loop et al., 1987) and the dog has cones with maxima at about 555 and 429 nm (Neitz et al., 1989). A similar system has been found in an ungulate, the domestic pig, *Sus scrofa*, that has cones with maxima at about 556 and 439 nm (Table 4.1) (Neitz and Jacobs, 1989).

In more truly nocturnal mammals, such as the nocturnal rodents, a more reduced photopic mechanism appears to be present. In rats, *Rattus norvegicus*, where cones are rare, only a single spectral class of cone has been identified with a peak sensitivity at about 510 nm (Neitz and Jacobs, 1986), only slightly displaced from the scotopic sensitivity close to 500 nm. Also, in the gerbil, *Meriones unguiculatus*, only a single cone class appears to be present, but in this case the photopic maximum sensitivity is at about 493 nm, slightly shorter than the scotopic sensitivity that has a maximum at about 500 nm (Jacobs and Neitz, 1989). If these animals have only a single class of cone, then they would be completely colour blind and probably represent a specialization to the nocturnal habit. The evolutionary steps leading to these cone pigments absorbing near 500 nm are not immediately clear. Have they evolved from a longer-wave cone opsin or from a rod opsin?

Among the diurnal rodents, the squirrels, *Sciuridae*, have a well-developed photopic system and have been thought of as possessing a pure cone retina. However, it is clear from behavioural and electroretinographic studies that both ground squirrels and tree squirrels have scotopic and photopic sensitivity, though the cone-to-rod ratio may be as high as 10:1 (Green and Dowling, 1975). The rods of all rodents that have been studied have a rhodopsin

Table 4.1 *Cone visual pigments of non-primate mammals*[1]

Class	Genus	λ_{max} (nm)	
Insectivora	tree shrew *Tupaia*	440	555
Rodentia	rat *Rattus*	510	
	gerbil *Meriones*	493	
	ground squirrel *Spermophilus*	437	517
	grey squirrel *Sciurus*	?	540
Carnivora	cat *Felis*	450	555
	dog *Canis*	430	555
Artiodactyla	pig *Sus*	439	555

[1] Most mammals have rod visual pigments with λ_{max} close to 500 nm (see Lythgoe, 1972). The exceptions are the aquatic mammals: many cetaceans have 'deep-sea' rhodopsins with λ_{max} at about 480 to 490 nm (McFarland, 1971).

with λ_{max} close to 500 nm. In the ground squirrel, *Spermophilus (Citellus) lateralis*, two classes of cone are present with λ_{max} at about 517 and 437 nm (Kraft, 1988) and in the grey tree squirrel, *Sciurus carolinensis*, a long-wave cone with λ_{max} at about 540 nm has been reported (Table 4.1) (Loew, 1975). However, squirrels have very yellow lenses (indeed they may be amber in colour) that greatly attenuate light at shorter wavelengths, so that the effective spectral maxima of the cone mechanisms in the ground squirrel are at about 525 and 460 nm (Jacobs et al., 1985; Kraft, 1988).

Prosimians

A group that has been of interest in the evolution of the mammals, and more specifically of the primates, are the tree shrews, Tupaiidae. These small, diurnal animals are arboreal omnivores, and have sometimes been classified as primitive primates and included in the prosimians. The visual system of *Tupai* has been studied in some detail: the retina is duplex, but is cone-dominated with only about 10% of the photoreceptors being rods (Müller and Peichl, 1989). Two spectral classes of cone are present (Table 4.1) with λ_{max} at about 556 and 440 nm (Jacobs and Neitz, 1986; Petry and Hárosi, 1990; Bowmaker et al., 1990), though, as in most mammals, the short-wave cones are rare, comprising only about 5 to 10% of the cone population (Müller and Peichl, 1989). The rods have λ_{max} at about 498 nm (Petry and Hárosi, 1990; Bowmaker et al., 1990).

The true prosimians, the lemurs, Lemuridae, lorises, Lorisoidae, and bush babies, Tarsoidea, all tend to be nocturnal. In general, they have rod-dominated retinas, and indeed the bush baby, *Galago crassicaudatus*, was thought to have a pure rod retina (Dartnall et al., 1965). However, Dodt (1967) demonstrated a Purkinje shift in the bush baby that suggests a cone population with λ_{max} at about 552 nm (Fig. 4.4), though a recent microspectrophotometric study (Petry and Hárosi, 1990) identified only rods with λ_{max} near 501 nm. Nevertheless, this cannot exclude the possibility of small cone populations being present, since the random-sampling technique of microspectrophotometry can easily fail to identify minority photoreceptor types.

In the lemurs there is more positive evidence of cones and a number of species are diurnal. In the ringed-tailed lemur, *Lemur catta*, behavioural and electroretinographic experiments (Blakeslee and Jacobs, 1985) demonstrate clearly that colour vision is present. However, it is not clear whether their colour vision is dichromatic or trichromatic and the most parsimonious conclusion may be that they have only two classes of cone, but achieve trichromacy by inclusion of their rods. Blakeslee and Jacobs

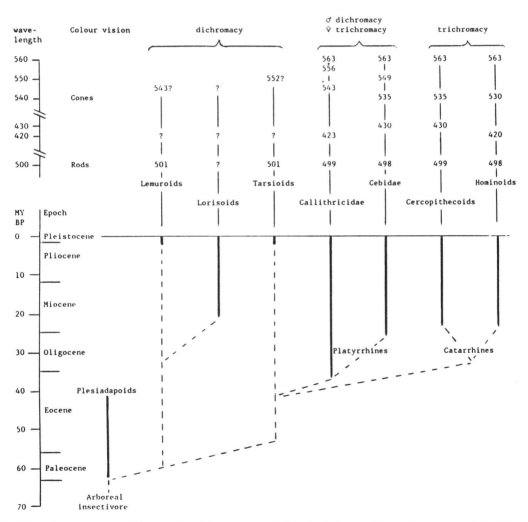

Fig. 4.4 *Schematic representation of the evolution of the primates and their visual pigments. The prosimians are probably dichromatic with two cone pigments, one maximally sensitive in the red–green spectral region at about 543 to 555 nm and the other in the blue at about 430 to 450 nm. Trichromacy appeared with the evolution of the simians about 40 million years ago and is only available to both males and females in the catarrhines. For details and sources of the λ_{max} of the visual pigments, see the text. The evolutionary tree is greatly simplified from Szalay and Delson (1979).*

(1985) classify the lemur's colour vision as severely protanomalous in human terms, which implies that their long-wave cone has a λ_{max} at relatively short wavelengths. In a limited microspectrophotometric study of the black lemur, *Lemur macaco* (Bowmaker, 1991), we have been able to identify only a single population of long-wave cones with an estimated λ_{max} at about 540 to 545 nm (Fig. 4.4). The rods have λ_{max} at about 501 nm.

All the evidence from the non-primate and prosimian mammals strongly suggests that at best they have dichromatic colour vision based on a long-wave pigment with maximum sensitivity at about 555 nm or less and a short-wave pigment with maximum sensitivity between about 430 and 450 nm. Their rods have λ_{max} close to 500 nm. A more complex photopic system appears to have evolved only recently within the mammals and it is only the anthropoids (simians) that possess forms of trichromatic colour vision.

The evolutionary transition from the prosimians to the anthropoids is not clearly defined, but must have occurred during the Eocene, about 40 to 50 million years ago, and the major division of the higher primates into the New World platyrrhines and the Old World catarrhines was well established by the beginning of the Oligocene, some 35 million years ago (Fig. 4.4). It may also be the case that the two groups of monkey evolved from different lineages of prosimians (McFarland *et al.*, 1985).

New World Monkeys

The New World monkeys are divided into two families, the more primitive Callithricidae, including the marmosets and tamarins, and the Cebidae, including the capuchins and squirrel monkeys. Both families are normally considered to be more primitive than the Old World monkeys and this may be reflected in their photopic visual system. Platyrrhines are basically dichromatic, but a polymorphism of the long-wave pigment permits some two-thirds of females to have the benefit of trichromacy (Mollon *et al.*, 1984; Jacobs and Neitz, 1985, 1987a; Travis *et al.*, 1988). Both the Callithricidae and the Cebidae have three cone pigments available in the red–green region of the spectrum, in addition to a short-wave pigment. In individual males only one of the three long-wave pigments is expressed and the animals are dichromatic, but in females either one or a combination of any two of the long-wave pigments may be present (Fig. 4.4).

The most likely explanation for this diversity of pigment distribution is that these species have only a single gene locus on the X-chromosome that determines the long-wave pigment, but that three alleles of the locus are available, coding for slightly different opsins that are expressed as cone pigments with slightly different peak sensitivities. Females homozygous for the locus will be dichromats whereas those heterozygous for the locus will be trichromats, since they can express two long-wave pigments. X-chromosome inactivation is necessary to ensure the segregation of the two pigments into separate classes of cone.

In the Callithricidae, the common marmoset, *Callithrix jacchus*, and the saddle-backed tamarin, *Saguinus fusicollis*, have long-wave pigments with λ_{max} at about 543, 556 and 563 nm (Jacobs *et al.*, 1987; Travis *et al.*, 1988), whereas in the Cebidae the squirrel monkey, *Saimiri sciureus*, has pigments with maxima at about 536, 549 and 563 nm (Mollon *et al.*, 1984, Jacobs and Neitz, 1987a). These pigment locations also appear to occur in the tufted capuchin, *Cebus apella*, and the dusky titi, *Callicebus moloch*: Jacobs and Neitz (1987b) report pigments with λ_{max} at about 549 and 562 nm and we have recorded a P535 in a male *Cebus* (Bowmaker *et al.*, 1983).

We have recently studied, both behaviourally and microspectrophotometrically, the inheritance of cone visual pigments within family groups of marmosets (Tovée, 1990). In one family, the parents and four offspring, two sets of twins, were studied. The mother was presumed to be trichromatic, since microspectrophotometry showed that she possessed both the P556 and P563 pigments, whereas the father possessed only the P556 and was shown behaviourally to be dichromatic. In one pair of twins, a male and a female, the male was behaviourally dichromatic and possessed the P556, whereas the female was shown behaviourally to be trichromatic and microspectrophotometry demonstrated the presence of the P556 and P563. The inheritance of the pigments in these twins appears to follow the single gene hypothesis: the son inheriting the P556 from his mother, and the daughter inheriting the paternal P556 and the maternal P563. It is interesting to note that the female possessing long- and middle-wave cones separated by only 7 nm clearly showed evidence of wavelength-opponent processes in the red–green spectral region, with a distinct notch in the spectral sensitivity function at about 580–590 nm.

The short-wave cones of the platyrrhines also appear to vary. Those of the squirrel monkey have maximum absorbance at about 430 nm (Mollon *et al.*, 1984), whereas those of the marmoset absorb maximally at about 423 nm (Travis *et al.*, 1988). These are both displaced to shorter wavelengths than that of the tree shrew, which peaks close to 440 nm (Petry and Hárosi, 1990; Bowmaker *et al.*, 1991).

Old World Monkeys and Man

Until recently our knowledge of the colour vision and cone pigments of catarrhine monkeys was based almost solely on macaques (e.g. Bowmaker *et al.*, 1978; Hárosi, 1987; Baylor *et al.*, 1987). We now have microspectrophotometric data from seven other species of Old World monkey, all from Africa (Bowmaker et al., 1991). These include the baboon, *Papio papio*, patus monkey, *Erythrocebus patus*, and five species of *Cercopithecus*. These monkeys come from different geographical regions and from a wide range of ecological niches. They differ considerably in their body weight and colouring, some are exclusively arboreal and others predominantly terrestrial. Nevertheless, all have long-wave and middle-wave cone pigments with absorbance maxima close to 565 and 535 nm and short-wave cones with λ_{max} at about 430 nm.

The spectral location of the middle-wave and long-wave cone pigments in man appear to be well established with maxima at about 530 and 565 nm (Dartnall et al., 1983; Shnapf et al., 1987). Our best estimate from microspectrophotometry puts the λ_{max} at 531 and 563 nm (Bowmaker, 1990b). These values are very similar to those of catarrhine monkeys, though the middle-wave cone pigment appears to be somewhat shorter. The short-wave cone pigment in man has not been so clearly specified and microspectrophotometry suggests an absorbance maximum close to 420 nm, distinctly shorter than that in Old World monkeys.

The genes encoding the opsins of the four visual pigments in humans (and presumably in Old World monkeys) are located on three chromosomes. Those for the short-wave pigment and the rod pigment are located on chromosomes 7 and 3 respectively, whereas the genes for the middle- and long-wave pigments are found in tandem on the long arm of the X-chromosome (Nathans et al., 1986a). The presence of separate genes on the X-chromosome coding for the two pigments in the red–green spectral region ensures that trichromacy will normally be present in both males and females and the inheritance of colour vision deficiencies that are a consequence of genetic variations of these two genes will be sex-linked (Nathans et al., 1986b).

The opsins for the long- and middle-wave pigments are 96% homologous, the two genes having presumably evolved recently by gene duplication occurring though unequal crossing over during cell division. This event must have occurred during the evolution of the catarrhines from their prosimian ancestors in the Eocene about 40 to 50 million years ago. The rod opsin and the short-wave opsin are both about 40% homologous with each other and with the long- and middle-wave pigments, suggesting that they may have diverged from each other at about the same time. It has been suggested from the amino acid sequence differences between human and bovine rod opsins that opsins may diverge at a rate of about 1% per 10 million years (Nathans and Hogness, 1984), suggesting that the three classes of pigment diverged more than 500 million years ago (Nathans et al., 1986a). This would fit nicely with the concept put forward earlier that these pigments were present in the early ostracoderms that appear in the fossil record in the late Cambrian to early Ordovician periods about 500 million years ago.

The homology and tandem arrangement of the middle- and long-wave opsin genes on the X-chromosome that occurs in man and Old World primates should allow frequent opportunity for the formation of hybrid genes that code for pigments with intermediate λ_{max} (Nathans et al., 1986b). However, the apparently fixed positions of the Old World cone pigments implies that either there is a functional or ecological constraint that maintains the peak sensitivities at about 535 and 563 nm, or else it must be more difficult than was previously thought for the structure of opsins to be modified so as to displace the absorbance spectra of visual pigments. The divergence of a single mammalian visual pigment maximally sensitive in the red–green spectral region into two spectrally distinct classes of cone has been correlated with the need of early primates to detect ripe (red) fruit against a highly variable background of green foliage (e.g. Mollon, 1989, Chapter 14) and similar arguments have been applied to the spectral sensitivities of the cones of diurnal birds (e.g. Bowmaker and Knowles, 1977).

In contrast to the catarrhines, the platyrrhines have at least five spectral locations for the middle- and long-wave pigments, at about 536, 543, 549, 556 and 563 nm with the two extreme locations being similar to those in the catarrhines. The only spectral location common to all species is the longest, i.e. close to 563 nm. This might be thought to reflect the spectral location of the ancestral pigment, but such a long-wave pigment has not been found among the prosimians and indeed has not been identified in any other mammal. The longer-wave pigments of the lemur (P543) and tree shrew (P555) are located at positions spectrally similar to two of those of the Callithricidae and many mammals have a long-wave cone maximally sensitive at about 555 nm. It would thus seem more likely that the longer-wave pigment of the ancestral simian had λ_{max} at an intermediate wavelength between 535 and 563 nm, perhaps at 555 nm (Table 4.1, Fig. 4.4).

A further feature of these longer-wave primate cone pigments is that they appear to 'cluster' at intervals of about 6 to 7 nm. This is reminiscent of the clustering of the λ_{max} of extracted pigments from rods noted by Dartnall and Lythgoe (1965) and also appears to occur within the rhodopsins of deep-sea fish (Partridge et al., 1989). Unfortunately, insufficient data are available from short-wave pigments to establish whether clustering also occurs at these wavelengths, though the short-wave cones of man and marmosets have λ_{max} around 420–424 nm, whereas those of Old World monkeys and the squirrel monkey have λ_{max} at about 430 nm. If 'clustering' does occur, it must reflect constraints on the structure of opsin: substitutions of non-polar and polar amino-acids close to the relevant regions of the chromophore cannot yield an infinite range of visual pigments with different spectral maxima. This would appear to be the case for cone pigments responsible for anomalous trichromacy in man, where severely anomalous observers may have anomalous pigments that are displaced by only about 6 to 7 nm from their 'normal' pigment (Pokorny *et al.*, 1973; Nagy *et al.*, 1985; Neitz *et al.*, 1989).

Conclusions

The evolution of visual pigments and photoreceptors within the vertebrates must have begun at a very early stage. The earliest photoreceptors may well have been rather unspecialized cells that would not readily be classified as either rods or cones. Some of the 'most primitive' of modern vertebrates, the cyclostomes, probably have a dichromatic colour vision system and it would seem likely that the differentiation of photoreceptors into a single class of rods underlying scotopic vision and at least two spectrally distinct classes of cone subserving colour vision at photopic levels was present even in the free-swimming ostracoderms, perhaps as early as the Ordovician period, some 500 million years ago.

However, the evolution of the more polychromatic forms of photopic vision has probably occurred independently within the major vertebrate classes. Diurnal surface-living teleosts and diurnal reptiles and birds probably represent the culmination of the evolution of the photopic system in vertebrates: some species may well have five spectrally distinct classes of cone subserving their colour vision. The spectral location of the cone visual pigments in these animal classes are very similar and presumably illustrates the result of convergent evolution. Likewise, the coloured oil droplets in the cones of some reptiles and birds and the pigmented ellipsosomes in the cones of some teleosts may also have evolved independently to serve similar functions.

The evolution of cone visual pigments and polychromacy within mammals appears to have occurred relatively recently. The early mammal-like reptiles from the Permian period were probably dichromatic and it is only with the appearance of the primates within the last 50 million years that trichromacy has developed. Here again, although the evolution of more than two spectral classes of cone has occurred independently from those in fish and birds, the spectral location of the cone pigments are similar to those of many other vertebrates (and indeed are almost identical to those of some species of cichlid fish (Levine and MacNicol, 1979).

With the development of techniques to isolate and sequence the genes coding for visual pigments (e.g. Nat-

hans *et al.*, 1986) and the ability to express the genes in mammalian cell lines (Nathans, 1990), it must be only a matter of time before we have a better and more complete understanding of the evolution and relationships of the great variety of visual pigments. The limited amount of data so far available has already proved useful (Goldsmith, Volume 6, Chapter 5) and, because of the relative ease of determining the absorbance spectra of the visual pigments of a given species, opsins offer a powerful tool with which to study the evolution of species, especially within a restricted environment, but also in a much wider context.

References

Applebury, M. L. and Hargrave, P. A. (1986). Molecular biology of visual pigments. *Vision Res.*, **26**, 1881–1885.

Archer, S. N. and Lythgoe, J. N. (1990). The visual pigment basis for cone polymorphism in the guppy, *Poecilia reticulata. Vision Res.*, **30**, 225–233.

Arnold, K. and Neumeyer, C. (1987). Wavelength discrimination in the turtle, *Pseudemys scripta elegans. Vision Res.*, **27**, 1501–1511.

Avery, J. A. and Bowmaker, J. K. (1982). Visual pigments of the four-eyed fish, *Anableps anableps. Nature Lond.*, **298**, 62–63.

Avery, J. A., Bowmaker, J. K., Djamgoz, M. B. A. and Downing, J. E. G. (1983). Ultraviolet sensitive receptors in a freshwater fish. *J. Physiol. Lond.*, **334**, 23P.

Barlow, H. B. (1982). What causes trichromacy? A theoretical analysis using comb-filtered spectra. *Vision Res.*, **22**, 635–643.

Baylor, D. A., Nunn, B. J. and Schnapf, J. L. (1987). Spectral sensitivity of cones of the monkey *Macaca fascicularis. J. Physiol. Lond.*, **390**, 145–160.

Beatty, D. D. (1975). Visual pigments of the American eel, *Anguilla rostrata. Vision Res.*, **15**, 771–776.

Beatty, D. D. (1984). Visual pigments and the labile scotopic visual system of fish. *Vision Res.*, **24**, 1563–1573.

Blakeslee, B. and Jacobs, G. H. (1985). Color vision in the ringed-tailed lemur (*Lemur catta*). *Brain, Behav. Evol.*, **26**, 154–166.

Bok, D. (1985). Retinal photoreceptor-pigment epithelium interactions. *Invest. Ophthalmol. Vis. Sci.*, **26**, 1659–1693.

Bowmaker, J. K. (1977a). Long-lived photoproducts of the green-rod pigment of the frog. *Vision Res.*, **17**, 17–23.

Bowmaker, J. K. (1977b). The visual pigments, oil droplets and spectral sensitivity of the pigeon. *Vision Res.*, **17**, 1129–1138.

Bowmaker, J. K. (1983). Trichromatic colour vision: why only three receptor channels? *Trends Neurosci.*, **6**, 41–43.

Bowmaker, J. K. (1989). Avian colour vision and the environment. In *Proceedings 19th International Ornithological Congress.* ed. Ouellet, H. pp. 1284–1294. Ottawa: University of Ottawa Press.

Bowmaker, J. K. (1990a). Visual pigments of fish. In *The Visual System of Fish.* eds. Douglas, R. H. and Djamgoz, M. B. A. pp. 81–107. London: Chapman & Hall.

Bowmaker, J. K. (1990b). Cone visual pigments in monkeys and humans. In *Advances in Photoreception: Proceedings Symposium on Frontiers of Visual Science.* pp. 19–30. Washington: National Academy Press.

Bowmaker, J. K. (1991). Visual pigments and colour vision in primates. In *Advances in Understanding Visual Processes: Convergence of Neurophysiological and Psychophysical Evidence.* eds. Valberg, A. and Lee, B. B. Oxford: Plenum.

Bowmaker, J. K., Astell, S., Hurst, D. M. and Mollon, J. D. (1991).

Photosensitive and photostable pigments in the retinae of Old World monkeys. *J. Exp Biol.*, in press.

Bowmaker, J. K., Dartnall, H. J. A. and Herring, P. J. (1988). Longwave sensitive visual pigments in some deep-sea fishes: segregation of 'paired' rhodopsins and porphyropsins. *J. Comp. Physiol. A*, **163**, 685–698.

Bowmaker, J. K., Dartnall, H. J. A., Lythgoe, J. N. and Mollon, J. D. (1978). The visual pigments of rods and cones in the Rhesus monkey, *Macaca mulatta*. *J. Physiol.*, **274**, 329–348.

Bowmaker, J. K. and Knowles, A. (1977). The visual pigments and oil droplets of the chicken retina. *Vision Res.*, **17**, 755–764.

Bowmaker, J. K. and Kunz, Y. W. (1987). Ultraviolet receptors, tetrachromatic colour vision and retinal mosaics in the brown trout (*Salmo trutta*): age-dependent changes. *Vision Res.*, **27**, 2102–2108.

Bowmaker, J. K., Mollon, J. D. and Jacobs, G. H. (1983). Microspectrophotometric results for Old and New World primates. In *Colour Vision*. eds. Mollon, J. D. and Sharpe, L. T. London: Academic.

Bowmaker, J. K., Thorpe, A. and Douglas, R. H. (1991). Ultraviolet-sensitive cones in the goldfish. *Vision Res.*, **31**, 349–352.

Bridges, C. D. B. (1972). The rhodopsin–porphyropsin visual system. In *Handbook of Sensory Physiology*, Vol. VII/1. ed. Dartnall, H. J. A. pp. 417–480. Berlin: Springer.

Carlisle, D. B. and Denton, E. J. (1959). On the metamorphosis of the visual pigments of *Anguilla anguilla*. *J. Mar. Biol. Ass. UK*, **38**, 97–102.

Chen, D. M. and Goldsmith, T. H. (1986). Four spectral classes of cones in the retinas of birds. *J. Comp. Physiol. A*, **159**, 473–479.

Clarke, G. L. (1936). On the depth at which fish can see. *Ecology*, **12**, 452–456.

Cohen, J. L. (1990). Vision in elasmobranchs. In *The Visual System of Fish*, eds Douglas, R. H. and Djamgoz, M. B. A. pp. 465–490. London: Chapman & Hall.

Crescitelli, C. (1956). The nature of the gecko visual pigment. *J. Gen. Physiol.*, **40**, 217–231.

Crescitelli, F. (1972). The visual cells and visual pigments of the vertebrate eye. In *Handbook of Sensory Physiology*, vol. VII/1. ed. Dartnall, H. J. A. pp. 245–363. Berlin: Springer.

Crescitelli, F. (1977). The visual pigments of geckos and other vertebrates: an essay in comparative biology. In *Handbook of Sensory Physiology*, vol. VII/5. ed. Crescitelli, F. pp. 391–449. Berlin: Springer.

Crescitelli, F., Dartnall, H. J. A. and Loew, E. R. (1977). The gecko visual pigments: a microspectrophotometric study. *J. Physiol.*, **268**, 559–573.

Cronin, T. W. and Marshall, N. J. (1989). A retina with at least ten spectral types of photoreceptors in a mantis shrimp. *Nature Lond.*, **339**, 137–140.

Dartnall, H. J. A. (1972). Visual pigment of the coelacanth. *Nature Lond.*, **239**, 341–342.

Dartnall, H. J. A., Arden, G. B., Ikeda, G. B., Luck, C. P., Rosenberg, M. E., Pedler, C. M. H. and Tansley, K. (1965). Anatomical, electrophysiological and pigmentary aspects of vision in the bush baby: an interpretative study. *Vision Res.*, **5**, 399–424.

Dartnall, H. J. A., Bowmaker, J. K. and Mollon, J. D. (1983). Human visual pigments: microspectrophotometric results from the eyes of seven persons. *Proc. R. Soc. Lond. B*, **220**, 115–130.

Dartnall, H. J. A., Lander, M. R. and Munz, F. W. (1961). Periodic changes in the visual pigments of fish. In *Progress in Photobiology*. eds. Christensen, B. and Buchman, B. pp. 203–213. Amsterdam: Elsevier.

Dartnall, H. J. A. and Lythgoe, J. N. (1965). The spectral clustering of visual pigments. *Vision Res.*, **5**, 81–100.

Denton, E. J. and Shaw, T. I. (1963). The visual pigments of some deep-sea elasmobranchs. *J. Mar. Biol. Ass. UK*, **43**, 65–70.

Denton, E. J. and Warren, F. J. (1957). The photosensitive pigments in the retinae of deep-sea fish. *J. Mar. Biol. Ass. UK*, **36**, 651–652.

Dodt, E. (1967). Purkinje-shift in the rod eye of the bush baby, *Galago crassicaudatus*. *Vision Res.*, **7**, 509–517.

Douglas, R. H., Bowmaker, J. K. and Kunz, Y. W. (1989). Ultraviolet vision in fish. In *Seeing Contour and Colour*. eds. Kulikowski, J. J., Dickinson, C. M. and Murray, I. J. pp. 601–606. Oxford: Pergamon.

Dowling, J. E. (1987). *The Retina. An Approachable Part of the Brain.* Cambridge, MA: Harvard University Press.

Downing, J. E. G., Djamgoz, M. B. A. and Bowmaker, J. K. (1986). Photoreceptors of cyprinid fish: morphological and spectral characteristics. *J. Comp. Physiol. A*, **159**, 859–868.

Fein, A. and Szuts, E. Z. (1982). *Photoreceptors: Their Role in Vision*. Cambridge, UK: Cambridge University Press.

Goldsmith, T. H. (1980). Humming birds see near U.V. light. *Science Wash. DC.*, **207**, 786–788.

Goldsmith, T. H., Collins, J. S. and Licht, S. (1984). The cone oil droplets of avian retinas. *Vision Res.*, **24**, 1661–1671.

Govardovskii, V. I. and Lychakov, D. V. (1984). Visual cells and visual pigments of the lamprey, *Lampetra fluviatilis*. *J. Comp. Physiol. A*, **154**, 279–286.

Govardovskii, V. I. and Vorobyev, M. V. (1989). The role of oil droplets in colour vision. *Sensory Systems*, **2**, 150–158.

Green, D. G. and Dowling, J. E. (1975). Electrophysiological evidence for rod-like photoreceptors in the gray squirrel, ground squirrel and prairie dog retinas. *J. Comp. Neurol.*, **159**, 461–472.

Hammond, P. (1978). The neural basis of colour discrimination in the domestic cat. *Vision Res.*, **18**, 233–235.

Hargrave, P. A., McDowell, J. H., Feldmann, R. J., Atkinson, P. H., Rao, J. K. M. and Argos, P. (1984). Rhodopsin's protein and carbohydrate structure: selected aspects. *Vision Res.*, **24**, 1487–1499.

Hárosi, F. I. (1982). Recent results from single-cell microspectrophotometry: cone pigments from frog, fish and monkey. *Color Res. Appl.*, **7**, 135–141.

Hárosi, F. I. (1987). Cynomologus and Rhesus monkey visual pigments. *J. Gen. Physiol.*, **89**, 717–743.

Hárosi, F. I. and Hashimoto, Y. (1983). Ultraviolet visual pigment in a vertebrate: a tetrachromatic cone system in the Dace. *Science Wash. DC*, **222**, 1021–1023.

Herring, P. J. (1983). The spectral characteristics of luminous marine organisms. *Proc. R. Soc. Lond. B*, **220**, 183–217.

Hisatomi, O., Iwasa, T. and Tokunaga, F. (1988). The visual pigment of lamprey. In *Molecular Physiology of Retinal Proteins*. ed. Hara, T. pp. 371–372. Osaka: Yamada Science Foundation.

Honig, B., Dinur, U., Nakanishi, K., Balough-Nair, V., Gawinovicz, M. A., Arnaboldi, M. and Motto, M. G. (1979). An external point charge model for wavelength regulation in visual pigments. *J. Am. Chem. Soc.*, **101**, 7084–7086.

Ishikawa, M., Takao, M., Washioka, H., Tokunaga, F., Watanabe, H. and Tonosaki, A. (1987). Demonstration of rod and cone photoreceptors in the lamprey retina by freeze replication and immunoflourescence. *Cell Tissue Res.*, **249**, 241–246.

Jacobs, G. H. (1981). *Comparative Colour Vision*. New York: Acadamic.

Jacobs, G. H. and Neitz, J. (1985). Color vision in squirrel monkeys: sex-related differences suggest a mode of inheritance. *Vision Res.*, **25**, 141–143.

Jacobs, G. H. and Neitz, J. (1986). Spectral mechanisms of color vision in the tree-shrew (*Tupaia belangeri*). *Vision Res.*, **26**, 291–298.

Jacobs, G. H. and Neitz, J. (1987a). Inheritance of colour vision in a New World monkey (*Saimiri sciureus*). *Proc. Natl Acad. Sci. USA*, **84**, 2545–2549.

Jacobs, G. H. and Neitz, J. (1987b). Polymorphism of the middle wavelength cone in two species of South American monkey: *Cebus apella* and *Callicebus moloch*. *Vision Res.*, **27**, 1263–1268.

Jacobs, G. H. and Neitz, J. (1989). Cone monochromacy and a reversed Purkinje shift in the gerbil. *Experientia*, **45**, 317–403.

Jacobs, G. H., Neitz, J. and Crognale, M. (1985). Spectral sensitivity of ground squirrel cones measured with ERG flicker photometry. *J. Comp. Physiol. A*, **156**, 503–509.

Jacobs, G. H., Neitz, J. and Crognale, M. (1987). Color vision polymorphism and its photopigment basis in a callitrichid monkey (*Saguinus fusicollis*). *Vision Res.*, **27**, 2089–2100.

Knowles, A. and Dartnall, H. J. A. (1977). The photobiology of vision. In *The Eye*, vol. 2B. ed. Davson, H. London: Academic.

Kosower, E. M. (1988). Assignment of groups responsible for the 'opsin shift' and absorptions of rhodopsin and red, green and blue iodopsins (cone pigments). *Proc. Natl Acad. Sci. USA*, **85**, 1076–1080.

Kraft, T. W. (1988). Photocurrents of cone photoreceptors of the golden-mantled ground squirrel. *J. Physiol.*, **404**, 199–213.

Lamb, T. D. and Pugh, E. N. (1990). Physiology of transduction and adaptation in rod and cone photoreceptors. *Trends Neurosci.*, **2**, 3–13.

Levine, J. S. and MacNichol, E. F. (1979). Visual pigments in teleost fishes: effects of habitat, microhabitat and behaviour on visual system evolution. *Sensory Processes*, **3**, 95–130.

Liebman, P. A. and Entine, G. (1968). Visual pigments of frogs and tadpoles. *Vision Res.*, **8**, 761–775.

Liebman, P. A. and Granda, A. M. (1971). Microspectrophotometric measurements of visual pigments of two species of turtle, *Pseudemys scripta* and *Chelonia mydas*. *Vision Res.*, **11**, 105–114.

Lipetz, L. E. (1985). Some neural circuits of the turtle retina. In *The Visual System*. eds. Fein, A. and Levine, S. pp. 107–132. New York: Liss.

Locket, N. A. (1977). Adaptations to the deep-sea environment. In *Handbook of Sensory Physiology*, vol. VII/5. ed. Crescitelli, F. pp. 67–192. Berlin: Springer.

Loew, E. R. (1975). The visual pigments of the gray squirrel, *Sciurus carolinensis leucotis*. *J. Physiol.*, **251**, 48–49P.

Loew, E. R. and Dartnall, H. J. A. (1976). Vitamin A1/A2-based visual pigment mixtures in cones of the rudd. *Vision Res.*, **16**, 891–896.

Loew, E. R. and McFarland, W. N. (1990). The underwater visual environment. In *The Visual System of Fish*. eds. Douglas, R. H. and Djamgoz, M. B. A. pp. 1–43. London: Chapman & Hall.

Loop, M. S., Millican, C. L. and Thomas, S. R. (1987). Photopic spectral sensitivity of the cat. *J. Physiol.*, **382**, 537–553.

Luckett, W. P. (1990). *Comparative Biology and Evolutionary Relationships of Tree Shrews*. New York: Plenum.

Lythgoe, J. N. (1972). List of vertebrate visual pigments. In *Handbook of Sensory Physiology*. Vol. VII/1. ed. Dartnall, H. J. A. pp. 604–624. Berling: Springer.

Lythgoe, J. N. (1979). *The Ecology of Vision*. Oxford: Oxford University Press.

MacNichol, E. F., Kunz, Y. W., Levine, J. S., Hárosi, F. I. and Collins, B. A. (1978). Ellipsosomes: organelles containing a cytochrome-like pigment in the retinal cone of certain fishes. *Science Wash. DC*, **200**, 549–551.

McFarland, W. N. (1971). Cetacean visual pigments. *Vision Res.*, **11**, 1065–1076.

McFarland, W. N. and Munz, F. (1965). Codominance of visual pigments in hybrid fishes. *Science Wash. DC*, **150**, 1055–1057.

McFarland, W. N., Pough, F. H., Cade, T. J. and Heiser, J. B. (1985). *Vertebrate Life*, 2nd edn. New York: Macmillan.

Mollon, J. D. (1989). 'Tho' she kneel'd in that place where they grew...'. *J. Exp. Biol.*, **146**, 21–38. (See also Chapter 14).

Mollon, J. D., Bowmaker, J. K. and Jacobs, G. H. (1984). Variations of colour vision in a New World primate can be explained by a polymorphism of retinal photopigments. *Proc. R. Soc. Lond. B*, **222**, 373–399.

Müller, B. and Peichl, L. (1989). Topography of cones and rods in the tree shrew retina. *J. Comp. Neurol.*, **282**, 581–594.

Muntz, W. R. A. (1966). The photopositive response of the frog (*Rana pipiens*) under photopic and scotopic conditions. *J. Exp. Biol.*, **45**, 101–111.

Munz, F. W. and McFarland, W. N. (1973). The significance of spectral position in the rhodopsins of tropical marine fishes. *Vision Res.*, **13**, 1829–1874.

Nagy, A. L., Purl, K. F. and Houston, J. S. (1985). Cone mechanisms underlying the colour discrimination of deutan color deficients. *Vision Res.*, **25**, 661–669.

Nathans, J. (1990). Determinations of visual pigment absorbance: role of charged amino acids in the putative transmembrane segments. *Biochemistry*, **29**, 937–942.

Nathans, J., Davenport, C. M., Maumenee, I. H., Lewis, R. A., Hejtmancik, J. F., Litt, M., Lovrien, E., Weleber, R., Bachynski, B., Zwas, F., Klingaman, R. and Fishman, G. (1989). Molecular genetics of human blue cone monochromacy. *Science Wash. DC*, **245**, 831–838.

Nathans, J. and Hogness, D. S. (1984). Isolation and nucleotide sequence of the gene encoding human rhodopsin. *Proc. Natl Acad. Sci. USA*, **81**, 4851–4855.

Nathans, J., Thomas, D. and Hogness, D. S. (1986a). Molecular genetics of human color vision: the genes encoding blue, green and red pigments. *Science Wash. DC*, **232**, 193–203.

Nathans, J., Piantanida, T. P., Eddy, R. L., Shows, T. B. and Hogness, D. S. (1986b). Molecular genetics of inherited variations in human color vision. *Science Wash. DC*, **232**, 203–210.

Negishi, K., Teranishi, T., Kuo, C.-H. and Miki, N. (1987). Two types of lamprey retina photoreceptors immunoreactive to rod- or cone-specific antibodies. *Vision Res.*, **27**, 1237–1241.

Neitz, J., Geist, T. and Jacobs, G. H. (1989). Colour vision in the dog. *Vis. Neurosci.*, **3**, 119–125.

Neitz, J. and Jacobs, G. H. (1986). Reexamination of spectral mechanisms in the rat (*Rattus norvegicus*). *J. Comp. Psychol.*, **100**, 21–29.

Neitz, J. and Jacobs, G. H. (1989). Spectral sensitivity of cones in an ungulate. *Vis. Neurosci.*, **2**, 97–100.

Neitz, J., Neitz, M. and Jacobs, G. H. (1989). Analysis of fusion gene and encoded photopigment of colour-blind humans. *Nature Lond.*, **342**, 679–682.

Neitz, J. and Jacobs, G. H. (1990). Polymorphism in normal human colour vision and its mechanism. *Vision Res.*, **30**, 621–636.

Nilsson, S. E. G. (1964). An electron microscopic classification of the retinal receptors of the leopard frog (*Rana pipiens*). *J. Ultrastruct. Res.*, **10**, 390–416.

North, R. A. (1989). Neurotransmitters and their receptors: from the clone to the clinic. *Sem. Neurosci.*, **1**, 81–90.

Ovchinnikov, Yu. A., Abdulaev, N. G., Zolotalev, A. S., Atomomov, I. D., Bespalov, I. A., Dergachev, A. E. and Tsuda, M. (1988). Octopus rhodopsin. Amino acid sequence deduced from cDNA. *FEBS Lett.*, **232**, 69–72.

Partridge, J. C., Archer, S. N. and Lythgoe, J. N. (1988). Visual pigments in the individual rods of deep-sea fish. *J. Comp. Physiol. A*, **162**, 543–550.

Partridge, J. C., Shand, J., Archer, S. N., Lythgoe, J. N. and van Groningen-Luyben, W. A. H. M. (1989). Interspecific variation in the visual pigments of deep-sea fishes. *J. Comp. Physiol. A*, **164**, 513–529.

Petry, H. M. and Hárosi, F. I. (1990). Visual pigments of the tree shrew (*Tupaia belangeri*) and greater galago (*Galago crassicaudatus*): a microspectrophotometric investigation. *Vision Res.*, **30**, 839–851.

Pokorny, J., Smith, V. C. and Katz, I. (1973). Derivation of the photopigment absorption spectra in anomalous trichromats. *J. Opt. Soc. Am.*, **63**, 232–237.

Rodieck, R. W. (1988). The primate retina. In *Comp. Primate Biol.*, **4**, 203–278.

Saibil, H. (1990). Cell and molecular biology of photoreceptors. *Sem. Neurosci.*, **2**, 15–23.

Schnapf, J. L., Kraft, T. W. and Baylor, D. A. (1987). Spectral sensitivity of human cone photoreceptors. *Nature Lond.*, **325**, 439–441.

Schwanzara, S. A. (1967). The visual pigments of freshwater fishes. *Vision Res.*, **7**, 121–148.

Shand, J., Partridge, J. C., Archer, S. N., Potts, G. W. and Lythgoe, J. N. (1988). Spectral absorbance changes in the violet/blue sensitive cones of the juvenile pollack, *Pollachius pollachius*. *J. Comp. Physiol. A*, **163**, 699–703.

Szalay, F. S. and Delson, E. (1979). *Evolutionary History of the Primates*. New York: Academic.

Takao, M., Yasui, A. and Tokunaga, F. (1988). Isolation and sequence determination of the chicken rhodopsin gene. *Vision Res.*, **28**, 471–480.

Tovée, M. J. (1990). *A polymorphism of the Middle- to Long-wave Cone Photopigments in the Common Marmoset* (Callithrix jacchus jacchus): *a Behavioural and Microspectrophotometric Study*. Ph.D. Thesis, University of Cambridge.

Travis, D. S., Bowmaker, J. K. and Mollon, J. D. (1988). Polymorphism of visual pigments in a callitrichid monkey. *Vision Res.* **28**, 481–490.

Tsin, A. T. C. and Beatty, D. D. (1978). Goldfish rhodopsin: P499$_1$. *Vision Res.*, **18**, 1453–1455.

Wald, G. (1957). The metamorphosis of visual systems in the sea lamprey. *J. Gen Physiol.*, **40**, 901–914.

Walls, G. L. (1942). *The Vertebrate Eye and its Adaptive Radiation*. Michigan: Cranbrook Inst. of Science Bull., **19**.

Wheeler, A. (1969). *The Fishes of the British Isles and N.W. Europe*. London: Macmillan.

Whitmore, A. V. and Bowmaker, J. K. (1989). Seasonal variation in cone sensitivity and short-wave absorbing visual pigments in the rudd *Scardinius erythrophthalmus*. *J. Comp. Physiol. A*, **166**, 103–115.

Widder, E. A., Latz, M. F., Herring, P. J. and Case, J. F. (1984). Far-red bioluminescence from two deep-sea species. *Science Wash. DC*, **225**, 512–514.

Wright, A. A. (1979). Color-vision psychophysics: a comparison of pigeon and human. In *Neural Mechanisms of Behavior in the Pigeon*. eds. Granda, A. M. and Maxwell, J. H. pp. 89–127. New York: Plenum.

Yokoyama, R. and Yokoyama, S. (1990). Isolation, DNA sequence and evolution of a colour visual pigment gene of the blind cave fish *Astyanax fasciatus*. *Vision Res.*, **30**, 807–816.

5 The Vertebrate Dioptric Apparatus

W. N. Charman

Introduction

Although a wide range of optical designs is found among invertebrate eyes, including eyes based on refraction, reflection, and mixed refractive and reflective elements (e.g. Land, 1980, 1981, 1988; Nilsson, 1988, 1990), the fundamental optics of vertebrate eyes follow an essentially standard pattern (Fig. 5.1). This basically comprises a transparent cornea and lens, which refract light to form an optical image on the retina lining the interior surface of a quasi-spherical eyeball. Nevertheless, many subtle variations on this simple theme occur, deriving from the differing environments, life styles, dimensions, neural capabilities and relative importance of vision for the animals concerned.

The evolutionary origins of the vertebrate eye remain obscure (see Cronly-Dillon, Chapter 2). As Duke-Elder (1958, p 238) notes, in the most primitive vertebrates known to man – the long-extinct agnathous fishes whose fossil remains date from the Silurian era over 400 million years ago – well-developed lateral eyes are already present. Duke-Elder concludes:

> *It would seem, therefore, that the vertebrate eye evolved not as a late off-shoot from the simple eye of the invertebrates after the latter had reached an advanced stage; it probably emerged at a very early stage, further back than geological evidence can take us, and developed along parallel but diverging lines. The apposite remark of the great German anatomist, Froriep (1906), that the vertebrate eye sprang into existence fully-formed, like Athene from the forehead of Zeus, expressed the frustration of the scientists of half a century ago; today we are little wiser.*

This remark retains much of its validity. Certainly the sole living representative of the ancient order of coelacanths, *Latimeria* – surprisingly rediscovered a generation ago – possesses eyes which follow the characteristic vertebrate design and are optically fairly typical of many deep-sea fish (Millot and Carasso, 1955). As, then, the various vertebrate groups developed (e.g. Olson, 1977) it appears that all their eyes shared certain major design features, including fundamental weaknesses such as the inverted retina through which the light must pass before reaching the visual pigments.

In this chapter an attempt will be made to outline the parameters that affect the detailed design of the optics of the vertebrate eye and to illustrate the way in which the eyes of different species are adapted to their needs. A great deal of relevant material is contained in early volumes on vertebrate vision (Franz, 1934; Walls, 1942; Rochon-Duvignaud, 1943; Polyak, 1957; Duke-Elder, 1958; Tansley, 1965) and several excellent extended reviews of the optical aspects of both vertebrate and invertebrate vision have appeared more recently (Weale, 1974; Hughes, 1977a; Land, 1981).

Basic Optics

It will be helpful at this stage to describe the essential optical elements of the eye in a little more detail. This may conveniently be done in terms of a schematic eye model, such as those shown in Fig. 5.2. Schematic eyes have been elaborated at various levels of complexity for humans (e.g. Charman, 1983, and in volume 1, chapter 1 of this series) and for many other vertebrate species (see Hughes, 1977a, 1979a, and Martin, 1983, for reviews). These models are based on anatomical sections, together with representative measurements of individual surface curvatures and refractive indices. The paraxial optical qualities of the real eyes are summarized in terms of the cardinal points of Gaussian thick-lens theory, i.e. the positions of the pairs of focal, principal and nodal points (e.g. Born and Wolf, 1975, pp 150–166; Freeman, 1990, pp 127–134). Since extreme accuracy in measuring individual parameters is required if the overall ametropia of the eye is to be accurately modelled (Hughes, 1977a), it is usual to adjust the values of the parameters slightly to make the ametropia of the schematic eye match that of the real eye, as measured

The Vertebrate Dioptric Apparatus 83

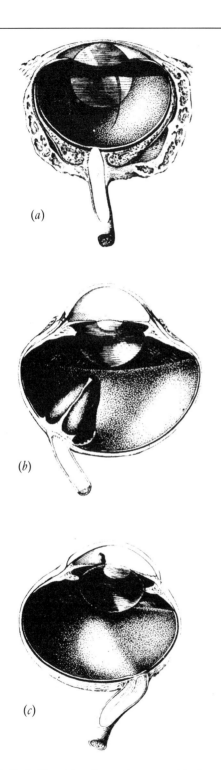

Fig. 5.1 *Early illustrations (not to the same scale) of several vertebrate eyes (After Soemmerring, 1818). Each eye is of similar basic form, although the relative dimensions and other details differ. (a) Cod. (b) Ostrich. (c) Horse.*

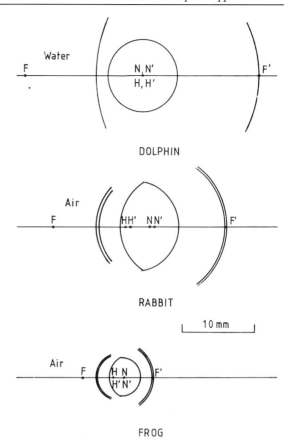

Fig. 5.2 *Examples of schematic eyes and cardinal points for different species: dolphin (Rivamonte, 1976), rabbit (Hughes, 1972) and frog (Du Pont and de Groot, 1976). In each case the positions of the first and second focal, principal and nodal points are shown by F, F', H, H' and N, N' respectively.*

by suitable objective procedures. It is normally assumed that for an emmetropic eye the second focal point must lie in the relevant entry plane of the receptor outer segments.

In fact there has in the past been considerable controversy over the degree of ametropia exhibited in the eyes of various species, owing in part to the need to employ objective techniques such as retinoscopy to establish the refractive error (see, e.g., Hughes, 1977a, pp 665–675 and Hughes, 1979b, for review). Glickstein and Millodot (1970) argued powerfully that there was likely to be an inherent artefact in retinoscopy, in that the reflex (and the reflexes in other objective techniques: see, e.g., Howland in volume 1, chapter 18 of this series) probably arose mainly from the interface between the retina and the vitreous, rather than from the physiologically relevant region of the receptor outer segments. Then, since all retinas tend to have approximately the same thickness, there would be a systematic tendency for retinoscopy to erroneously sug-

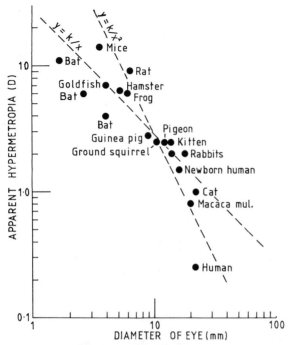

Fig. 5.3 *Logarithmic plot of apparent hypermetropia as measured by retinoscopy for various species against the ocular diameter. According to Glickstein and Millodot (1970), as the power of an eye is given by* $F = n'/f'$, *the error in measured power,* ΔF, *associated with an error in the focal length,* $\Delta f'$, *is given by* $\Delta F = -(n'f'^2)\Delta f'$. *If, then, the retinoscopic reflex originates at a plane anterior to the retinal receptor outer segments, there will be an error in the retinoscopic refraction. Glickstein and Millodot suggested that since the retinas of all species have a nearly constant thickness,* $\Delta f'$ *would be constant across species and hence that a plot of log* ΔF *against log* f' *would have a slope of* -2. *An alternative suggestion (Charman and Jennings, 1976; Nuboer and van Genderen-Takken, 1978; Nuboer et al., 1979), that if the colour of the retinoscopic reflex is predominantly red, the longitudinal chromatic aberration of the eye will lead to apparent hypermetropia, predicts that the hypermetropia is inversely proportional to the eye length, to give a slope of* -1. *The two dashed lines with slopes of* -2 *and* -1 *have been drawn to pass through the central data points. Based on data from Millodot and Glickstein (1970) and Hughes (1977a), together with bat data from Suthers and Wallis (1970).*

gest greater amounts of hypermetropia in small eyes. A plot of retinoscopically determined apparent hypermetropia against eye size for a variety of vertebrates agrees reasonably well with these predictions (Fig. 5.3).

Such factors as the interaction of longitudinal chromatic aberration with the colour of the reflex (Charman and Jennings, 1976; Millodot and Sivak, 1978; Nuboer and Van Genderen-Takken, 1978; Nuboer et al., 1979), the presence of a tapetum, or air-water refractive effects in retinoscopy of acquatic animals (Hueter and Gruber, 1980; Spielman and Gruber, 1983) may modify Glickstein and Millodot's conclusions somewhat in particular species, and the retina-vitreous interface may not be the reflecting surface in fish (Sivak, 1974). Nevertheless, it seems reasonable to conclude that most mammals and probably many other vertebrates display functional mean emmetropia, although the exact refraction may vary with pupil size owing to the effects of spherical aberration (e.g. Hughes, 1977b). This does not mean that some spread in refractive error may not be found between individual animals but that, as in man, the distribution of errors in any species peaks strongly around emmetropia, a phenomenon which is probably related to a natural 'emmetropization' process depending in part upon environmental visual experience (see, e.g., Young and Leary in volume 1, chapter 2 of this series for review). Certainly electrophysiological studies, which do not depend upon a retinal reflex, appear to show emmetropia in fish (Meyer and Schwassman, 1970), pigeons (Millodot and Blough, 1971), frogs (Millodot, 1971; Moser and Krueger, 1972), young rats (Meyer and Salinsky, 1977) and rabbits (Pak, 1984): only opossums (Picanco-Diniz et al., 1983) and old rats (Meyer and Salinsky, 1977) apparently have modest amounts of myopia (around $-2D$). Most schematic eyes therefore now show near-emmetropia on axis.

Returning to the details of the schematic eye, the dioptric power, F, of any refractive surface depends upon its radius of curvature, r, and the refractive indices n and n' of the media it separates, i.e. $F = (n'-n)/r$. For an animal in air (refractive index unity), the image of an external object is formed on the retina by the combined refractive action of the cornea (index about 1.38) and lens (index around 1.42), the indices of the aqueous and vitreous humours (about 1.34) being close to that of water (1.33): Hughes (1977a, pp 652-653) gives more exact index values for a range of species. The bulk of the aerial power is normally provided by the cornea, although the exact balance varies somewhat between species: in man the lens provides only about a quarter of the overall ocular power of some 60 dioptres. Exceptions to this basic pattern are provided by those animals having a transparent 'spectacle' formed by fusion of the lids, notably lizards and snakes (e.g. Walls, 1942, pp 449-461). This structure effectively takes over the refractive function of the cornea in some species (Citron and Pinto, 1973; Sivak, 1977, see Fig. 5.4) as well as providing protection against abrasion to the eye in the animal's environment or, perhaps, that of their ancestors. A variant on the spectacle is provided by transparent portions in the mobile lower lid of some species of lizard (Duke-Elder, 1958, pp 366-367).

In aquatic animals, the similarity of the refractive indices of water and the aqueous humour, combined with the

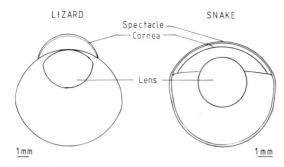

Fig. 5.4 *Comparison between the geometries of a lizard eye (Iguana iguana: Citron and Pinto, 1973) and a snake eye with a spectacle. The spherical lens of the snake eye makes a relatively greater contribution to the overall power of the eye (After Sivak, 1977).*

relatively small thickness of the cornea and its roughly concentric surfaces, result in the refractive power of the cornea being small as compared with that of the crystalline lens. Thus almost all of the optical power in the eye of an aquatic animal derives from the lens, which in fish is therefore usually spherical in order to maximize its power for a given diameter (e.g. Charman and Tucker, 1973). It has often been suggested that in some diving birds the lenticular power may be supplemented by the effect of a transparent, highly refracting nicitating membrane, which is drawn across the cornea when the bird is underwater (Duke-Elder, 1958, p 643; Tansley, 1965, p 105; Meyer, 1977, p 566): however, this now seems unlikely in view of the similarity of the index of the membrane of that of the cornea (1.37) and its apparent lack of refractive power (Sivak et al., 1978; Sivak, 1980a). While 'spectacles' exist in the eyes of many fish (e.g. in some cyclostomes and teleosts), these appear to have a largely protective role and contribute little to the refractive power of the eye.

The lens in all species is typically inhomogeneous, with a substantially higher index in its central (nuclear) regions than in its outer layers (cortex): this index gradient probably plays an important role in minimizing optical aberrations (see below). Some progress has been made in mapping the form of the index gradients of the lenses in man (Chan et al., 1988; Pierscionek et al., 1988), rat (Campbell, 1984), rabbit (Nakao et al., 1968), cat (Jagger, 1990) and goldfish (Axelrod et al., 1988), but the practical measurements are difficult and considerable uncertainties remain. If objects at various distances are to be clearly focused on the retina, some form of accommodation or focusing mechanism is necessary: this differs widely in form and efficiency between species, as will be discussed in more detail below.

The pupil of the eye, which constitutes its aperture stop, has an important influence on the optical performance of the eye, particularly on its light-gathering power,

retinal image quality and depth-of-focus. In vertebrates it is usually mobile and is often non-circular (e.g. Walls, 1942, pp 217–228).

In modelling the gross optical characteristics of any terrestrial eye it may for many purposes be adequate to use a reduced eye model in which the actual optical surfaces are replaced by a single surface (Fig. 5.5) separating air and the eye medium, which is often assumed to be water. The paraxial characteristics of this system, i.e. its imaging characteristics close to the optical axis, can then be summarized in terms of the positions of the two focal points (F and F'), together with the coincident pairs of principal (H) and nodal points (N). Hughes (1979a) tabulates the positions of these points for a selection of species. In such reduced eyes, it is usual to assume that the aperture stop, exit and entrance pupils all coincide at the cornea, although some authors (e.g. Thibos, 1987) assume that the true pupil (i.e. aperture stop) lies at an anatomically more realistic position within the eye medium.

Of course, it is important to remember that paraxial schematic and reduced eye models like those shown in Figs 5.2 and 5.5 do not adequately describe wide-angle, off-axis imagery. Such imagery is particularly important for the many species in which, unlike man, there is no fovea lying close to the optical axis. In the case of eyes

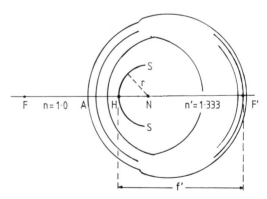

Fig. 5.5 *A reduced eye (based on Hughes, 1979a). The single spherical refracting surface SS, of radius of curvature r, separating air n = 1.0, and water, n' = 1.333, replaces the refractive effect of the cornea, lens and eye media of the real eye. F and F' are the first and second focal points respectively of the reduced eye. The two principal points coincide at H, the pole of the single refracting surface, and the two nodal points at N, the centre of curvature of that surface. The axial length HF' of the reduced eye corresponds to its second focal length, f'. Most authors assume that the true pupil (aperture stop) lies at the refracting surface, so that the centres of the exit and entrance pupils also coincide at the same position, H. A minority of authors (e.g. Thibos, 1987) place the true pupil at a more anatomically plausible position within the reduced eye; the exit pupil then coincides with the aperture stop and the entrance pupil lies slightly anterior to the other pupils.*

which possess a single fovea lying away from the optical axis there is a substantial angle between the visual and optical axes, so that good optical quality is required at substantial optical field angles: similar demands for good off-axis optical image quality may arise in species for which there is some form of visual streak or double fovea (Hughes, 1977a). Attempts have been made to produce eye models which adequately predict the wide-field performance of the human eye (e.g. Pomarantzeff *et al.*, 1971; Lotmar, 1971; Drasdo and Fowler, 1974; Drasdo and Peaston, 1980; Kooijman, 1983; Fitzke, 1987) and rat eye (Campbell and Hughes, 1981) but as yet not enough is known about the gradients of refractive index inside the lenses of most species to make general wide-angle models possible. However, Holden *et al.* (1987) have shown that a simple model based on measured ocular dimensions can give useful information on retinal magnification at the edge of the visual field in many species: their results indicate that while axial and peripheral magnification factors are essentially equal in birds, the peripheral magnification is only some 64% of the axial magnification in primates.

Although an approximate theoretical estimate of the extent of the uniocular field may be made on the basis of some form of schematic eye and a knowledge of the extent of the retina (e.g. Vakkur and Bishop, 1963), practical measurements are more reliable. Typically, a small light source is progressively moved around the outer circumference of the eye and its image is viewed through the pupil by transillumination through the sclera. Earlier work has been reviewed by Duke-Elder (1958, pp 669–670) and a variant of the same basic method has recently been used to map the projection of the visual field onto the pigeon retina (Hayes *et al.*, 1987). As an alternative, the reflex from the retina can be mapped with a reversible ophthalmoscope (e.g. Hughes, 1972, 1979c; Martin and Young, 1983; Martin, 1984, 1986; McFadden and Reymond, 1985), this method being suitable for the determination of both monocular and binocular fields of the living animal.

In comparing the eyes of different species, fuller understanding of the ocular dioptrics demands consideration of both the physical and neural constraints on their detailed design.

Physical Constraints on Optical Design

It is clear that any eye must make use of the electromagnetic radiation that is available in the animal's environment. This radiation will have a certain spectrum, which must be at least partly overlapped by the transmittance characteristics of the eye media and the absorption spectrum of any visual pigments. It will also have a spatial and directional distribution, so that an image of the environment can, in principle, be obtained. The wavelengths present in the radiation will, through the effects of diffraction, set limits to the maximal attainable optical resolution for any pupil size. The quantity of available radiation (and its variability) will necessarily influence the optical design of any eye: the lower the available flux the greater the need for effective light-gathering power in the optics is likely to be. Lastly, it might be desirable to preserve the polarization of the radiation to the retinal level, primarily because its distribution across the visual field might be of navigational value.

Within these basic physical constraints, it becomes possible to consider the likely optical performance of particular eye designs, having regard to the further possible limitations set by optical aberrations. It should, of course, always be borne in mind that the design of the optics will also be affected by constraints set by the size of the animal concerned and by the quality of the neural array that the optical image serves.

Spectrum of Available Radiation and Ocular Transmittance

Radiation may originate from objects in the environment of an animal in one or two basic ways. It may be emitted, either as a result of blackbody radiation associated with the temperature of the objects or by some bioluminescent process: the temperature of most terrestrial objects, including animals, is such that any emitted blackbody radiation is primarily in the infrared (IR) at wavelengths of around 10 µm. Alternatively, light from some external source may be scattered and reflected by objects, allowing them to be detected. Under natural conditions, the obvious source is the sun, although moonlight and starlight are potential alternatives: the latter have very similar spectra to the sun, although the available fluxes are much lower (see, e.g., Henderson, 1977; Lythgoe, 1979, ch. 1; Munz and McFarland, 1977).

Use of infrared radiation for sensory detection is viable and is employed by a variety of snakes to aid location of warm-blooded prey preparatory to striking (Bullock and Cowles, 1952; Bullock and Diecke, 1956; Barrett *et al.*, 1970; Newman and Hartline, 1982). Nevertheless, since potential angular optical resolution is proportional to wavelength (see below), it is obvious that, for the same eye size, there are severe resolution penalties in using wavelengths of the order of 10 µm, as compared with optical wavelengths of about 0.5 µm. In addition, the energy of a photon is inversely proportional to wavelength, so that absorption of an IR photon is less likely to be able to initiate a photochemical event in the retina. Lastly, there would appear to be potential problems associated with the fact that the animal itself is at a temperature similar to that

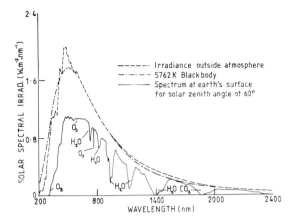

Fig. 5.6 *Solar spectrum outside the earth's atmosphere (dashed curve) in comparison with a 5762-K blackbody (chain-dotted curve). The full curve gives the spectrum at the earth's surface when the solar zenith angle is 60°; the atmospheric gases responsible for the major absorption bands are indicated (after Mechirikunnel et al., 1983).*

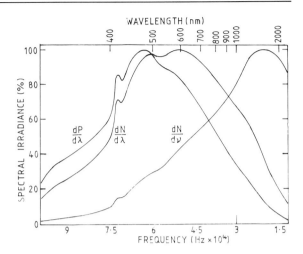

Fig. 5.7 *The shape of the solar spectrum depends on how it is plotted. Although it is conventional to plot* $dP/d\lambda$, *the power per unit area per unit wavelength interval, the spectrum changes shape if plotted as the number of photons per unit wavelength or frequency interval (* $dN/d\lambda$ *and* $dN/d\nu$ *respectively).*

of its environment so that it generates noise in the waveband to which its infrared detectors need to be sensitive. In practice, the thermally sensitive pit organs that are found in snakes and function rather like crude pinhole cameras appear to give the animals an infrared acuity of a few degrees (Newman and Hartline, 1982).

The solar spectrum outside the earth's atmosphere is shown in Fig. 5.6 (Mecherikunnel *et al.*, 1983). It can be seen that it is a reasonable approximation to a 5762-K blackbody. When plotted in terms of spectral irradiance as a function of wavelength, the spectrum peaks in the 'visible' region, between about 400 and 700 nm. However, it has been argued that it would be more sensible to plot in terms of numbers of photons per unit wavelength or frequency interval, rather than energy, since absorption of a photon may potentially initiate a photochemical event in the retina (Dartnall, 1975; Lythgoe, 1979). As the energy carried by a photon is inversely proportional to wavelength ($E = hc/\lambda$, where E is the photon energy, h is Planck's constant, c is the velocity of light and λ is the wavelength), this has the effect of shifting the spectral peak to rather longer wavelengths (Fig. 5.7).

The spectrum of the radiation directly reaching the eye of any animal will differ from that outside the atmosphere owing to spectrally selective light losses through absorption and scattering in the atmosphere for terrestrial species and in both atmosphere and water for aquatic species. In the atmosphere, strong absorption occurs in the ultraviolet at wavelengths less than about 290 nm owing to the presence of ozone and in the near-infrared owing to carbon dioxide and water vapour (Fig. 5.6). Of course, over the evolutionary time-scale there may have been changes in

atmospheric absorption and, indeed, there is currently concern over the apparent depletion of ozone caused by human activities, which may increase the ultraviolet flux reaching the earth's surface. It is important to note that the spectrum reaching the ground varies with the solar zenith angle and, of course, atmospheric conditions: in particular, when the sun is lower in the sky, leading to longer air paths, the energy peak in the available direct radiation shifts to longer wavelengths (e.g. Mecherikunnel *et al.*, 1983).

In an aquatic environment, water transmittance varies widely, depending upon the purity and the amount of suspended matter. Scattering from such matter is of particular importance in that it not only affects the quantity of light available but also its spectral and directional distributions and polarization (e.g. Lythgoe, 1972, 1979, 1984). Even very clear water can, at depth, constitute a very effective spectral filter, passing only a relatively restricted range of wavelengths (Fig. 5.8; Tyler and Smith, 1970) and hence affecting the possibilities for colour vision. The progressive loss in available light flux with increasing depth of water demands corresponding increases in the sensitivity of the visual system of the animals living at each depth.

The different states of polarization brought about by scattering in different regions of the sky or water environment may be of navigational value to some vertebrates (Waterman, 1975), particularly birds (Kreithen and Keeton, 1974; Delius *et al.*, 1976; Able, 1982) and, perhaps, fish (Waterman and Forward, 1970): its use by invertebrates, such as bees, is well known (e.g. Waterman,

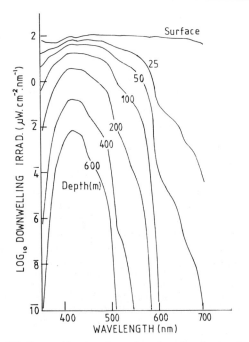

Fig. 5.8 *Downwelling irradiance through various depths of very clear lake water. The spectral bandwidth of transmitted sunlight and also its total amount reduce steadily as the depth increases. (After Tyler and Smith, 1970.)*

1981). However, Adler and Taylor (1973) have suggested that the apparent ability of amphibia to make use of the electric field vector of linearly polarized light (Taylor and Adler, 1973) is due to the pineal rather than the visual system.

Bioluminescence is of interest as a light source, in that it is produced by a wide variety of organisms. Unlike an animal's thermal radiation, its wavelengths lie within the visible region. Among vertebrates, luminescence is probably most important in deep-sea environments, where it not only may provide the major source of general illumination (Hemmings and Lythgoe, 1964; Jerlov, 1968) but also, through its spatial, temporal and spectral characteristics, can play a major role in communication and feeding. For example, there is evidence that the visual pigments of the deep-sea fish *Aristostomias scintillans* are specially adapted for perceiving the red and green bioluminescence emitted by the species (O'Day and Fernandez, 1974). It is interesting to note that some deep-sea vertebrate and invertebrate prey species appear to emit a luminous cloud in response to attack, in much the same way as military aircraft eject infrared decoys to confuse heat-seeking missiles (Locket, 1977, p 79).

The available external radiation can stimulate the retinal receptors only if it penetrates the eye. Although the spectral transmittance of the various ocular components is fairly well documented for man (see, e.g., Ruddock, 1972 and Charman, 1991a, for reviews), information for other species is rather sparse (see, e.g. Muntz, 1972, for review), apart from primates and rabbits (Geeraets and Berry, 1968), and some fish (e.g. Witkovsky, 1968; Muntz, 1975; Van den Berg and Mooij, 1982; Bassi *et al.*, 1984; Douglas, 1989). The cornea, whose transparency in comparison with the opaque sclera appears to depend upon the quasi-regular arrangement of its constituent collagen fibres (Maurice, 1957, 1969, 1970; Hart and Farrell, 1969; Feuk, 1970; Benedek, 1971), absorbs below about 300 nm. However, the short-wave cutoff of the complete eye is usually provided by the lens, with many species showing progressively greater short-wave absorption with increasing age (lens yellowing), this being associated with increases in lens thickness. Examples of such transmittance changes with age are shown in Fig. 5.9 for the case of the brown trout: they may be of significance in relation to the existence of ultraviolet-sensitive cones in the retinas of young trout and their subsequent, almost complete disappearance by the age of 2 years (Bowmaker and Kunz, 1987; Douglas, 1989; Douglas *et al.*, 1989).

However, it is of interest that the spectral sensitivity of the retina may in some cases extend beyond the spectral band that is transmitted by the eye's optics. Humans whose lenses have been removed owing to cataract show extended spectral sensitivity at shorter wavelengths (Gaydon, 1938; Wald, 1949; Ruddock, 1972). In the case of the human lens, at least, it appears that two processes contribute to the age-dependent increase in short-wavelength absorption: the pathlength through the lens

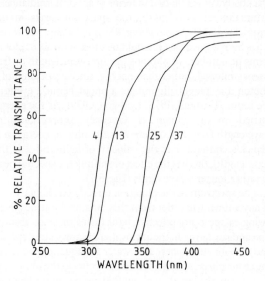

Fig. 5.9 *Relative spectral transmittance of the lens of the brown trout for the ages (months) indicated. The short-wave transmittance cutoff moves to longer wavelengths as the age increases. (After Douglas, 1989.)*

nucleus remains constant but there is increased pigment deposition, while the pigment concentration remains constant in the lens cortex but the pathlength increases (Mellerio, 1987). In general, since the eye media inevitably scatter and absorb some light throughout their volume, light losses are potentially greater in larger eyes, which involve longer lightpaths.

In addition to these general transmittance characteristics, additional localized regions of absorbing pigment are found in many species. The corneas of many fish contain areas of coloration, usually orange or green, which may be changeable under different environmental conditions owing to pigment migration (e.g. Moreland and Lythgoe, 1968; Orlov and Gamburtzeva, 1976; Kondrashev *et al.*, 1986). The exact function of these filters is still uncertain. It has been suggested that they may protect the retina from high light levels in shallow waters, reduce the effects of light scatter, or improve retinal image quality by reducing chromatic aberration. A related role is probably played by corneal iridescence in fish, which is presumably caused by lamination in the corneal substantia propria. This iridescence may serve to reduce intraocular flare by reflecting away the bright rays of the sun that strike the cornea at a grazing angle (Lythgoe, 1971, 1975): its nature depends upon the level of dark adaptation (Shand and Lythgoe, 1987).

At the retinal level, the central fovea of primates is screened by macular pigment (e.g. Ruddock, 1972; Weale, 1974), which absorbs heavily at wavelengths less than about 540 nm and may therefore also be helpful in reducing the adverse effects of chromatic aberration on image quality (Reading and Weale, 1974). More controversial is the role of the coloured oil droplets which are found at the junction of the receptor inner and outer segments of many invertebrates, including all birds and some reptiles, amphibians, mammals and fish (e.g. Walls, 1942; Muntz, 1972). The droplets are in some cases colourless but may otherwise be red, orange or yellow (e.g. Granda and Dvorak, 1977; Goldsmith *et al.*, 1984): their distribution across the retina of different species is often non-uniform. Although a role in colour vision has been strongly favoured (see, e.g., Lythgoe, 1979, pp 58–60 for review), it may be that they enhance the contrast of objects against certain backgrounds, such as the sky, or that they simply reduce glare and the effects of chromatic aberration (Tansley, 1965, pp 46–47). However, Weale (1974) dismisses the suggestion of a role in reducing chromatic aberration as inconceivable. Young and Martin (1984) have argued, on the basis of electromagnetic theory appropriate to the small, 2–4 μm diameter of the droplets, that they may act as microlenses which enhance light capture in the receptor outer segments as well as, perhaps, providing the basis for a potential mechanism to detect light polarization for navigation purposes (see also Baylor and Fettiplace, 1975; Ives *et al.*, 1983). However, if the light-gathering effect was of great importance it is difficult to see why all receptors should not receive the benefit conferred by the oil drops. In any case, it may be significant that oil drops in nocturnal species always appear to be colourless, implying that the spectral absorption of the drops in diurnal species must have some purposive function which has to be sacrificed in the photon-poor nocturnal environment.

Limitations Set by Photon Flux

If we accept that the visual process starts with the absorption of individual photons, then we would expect that the spatial and temporal distribution of such photons on the retina would set limits to the potential acuity (e.g. Barlow, 1964; Pirenne, 1967; Land, 1981). Assuming that the relative radiance distribution of the object scene and other relevant factors do not change, the number of photons arriving on any unit area of retina in any particular interval of time will depend on the photon flux, the individual photons being distributed randomly in space and time. Thus, although the mean number of photons on any such area may be constant, the number in each time interval will follow Poissonian statistics.

The consequences of such a situation are illustrated in Fig. 5.10 (after Barlow, 1964), which shows the counting statistics for a hypothetical, 'perfect' retinal detector recording a mean number of 2, 6 or 18 photons in each counting interval. In each case rather than always registering the mean number, other numbers may be recorded, with frequencies following the Poisson distributions shown in the figure. The distributions overlap, so that a count of, e.g., 4 could correspond to the mean value of 2 or the mean value of 6. Although the means of the neighbouring frequency distributions differ by the same factor, 3 (i.e. the contrasts are the same), it is obvious that the overlap between the distributions is greater when the

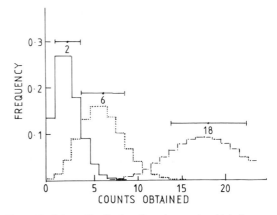

Fig. 5.10 *Poisson distributions for a detector in which the mean numbers of counts are 2, 6 and 18. In each case the horizontal bar gives the standard deviation. (After Barlow 1964.)*

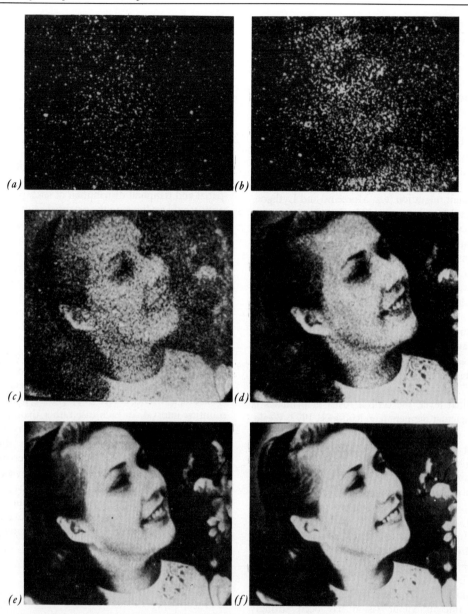

Fig. 5.11 *Effect of increasing the number of quanta in an image. As the number of photons contributing to the image increases, so greater amounts of detail can be discriminated. Number of photons are as follows: (a) 3×10^3; (b) 1.2×10^4; (c) 9.3×10^4; (d) 7.6×10^5; (e) 3.6×10^6; (f) 2.8×10^7. (After Rose, 1953.)*

neighbouring means are 2 and 6 than when they are 6 and 18. Hence supposing, for example, that we have two such detectors side by side on the retina and compared their counts to determine whether they were receiving differing mean numbers of photons, if the mean numbers were 2 and 6 we would quite often record the same number of photons in both detectors, so that we could not distinguish between the two outputs, whereas with mean numbers of 6 and 18 distinction between the levels would be much more reliable. This argument about spatial contrast discrimination can equally be applied to temporal discrimination. Although real retinal receptors detect only a fraction of the available photons, and system noise slightly complicates the argument (e.g. Barlow, 1964), the principles discussed above remain broadly valid. Thus, as photon flux increases, giving improved counting statistics, the potential acuity (or contrast discrimination) will improve. It follows that it is easier to achieve good spatial vision at high light levels. A pictorial illustration of the advantages of increasing the photon numbers is given in Fig. 5.11 (after Rose, 1953).

The quantum statistics can be improved in ways other than by an increase in photon flux. We might, for example, increase the integration or exposure time over which the number of photons were counted: this is essentially the procedure used in astronomical photography. However, in the case of vision it carries the obvious penalty of reducing temporal discrimination and this might be a severe disadvantage to some animals. Increasing the area of individual retinal receptors, or pooling the outputs of several receptors, increases the probability of a photon being collected by one functional unit and at the expense of reduced spatial resolution.

A better way of improving the photon statistics would be to increase the light-gathering power of the eye. For a vertebrate eye of any given overall axial length this is approximately equivalent to increasing the pupil diameter, D. In the simple reduced eye model of Fig. 5.5, where the pupil is assumed to coincide with the refracting surface and the axial length equals the second focal length, f', the retinal illuminance is, in fact, proportional to $(D/f')^2$. Clearly, then, a scaled up, larger eye with the same value of (D/f') will still yield the same retinal illuminance. For an eye with any given focal length, f', the retinal illuminance can be increased only by an increase in pupil diameter.

The demand for higher values of (D/f') or, more formally, for higher values of numerical aperture, $n' \sin u'$, where n' is the vitreous index and u' is the semiangle of the image-forming cone of rays, leads to the classic form of 'nocturnal' eye seen in many small animals where only modest acuity is called for. While the eye is small, the cornea is relatively large and the lens is nearly spherical, to give a high numerical aperture while keeping spherical aberration within reasonable limits (Fig. 5.12).

However, for the same numerical aperture an overall increase in the scale of the eye does increase the light collected from unit solid angle of object space. Hence the potential absolute threshold and increment sensitivity may be improved if the right strategy is used to sample the image space (Barlow, 1964; Hughes, 1977a, pp 654–655). For example, if the diameter of a retinal detector is p, its area is proportional to p^2. The number of photons collected by an individual detector unit is proportional to the product of retinal illuminance and the area of the detector, i.e. to $(Dp/f')^2$. If then we scale up D, f' and p equally, the retinal illuminance will remain constant but the number of photons collected by each detector will increase. Large receptor units in large eyes will therefore tend to give good sensitivity. Alternatively, if we keep p constant while scaling up both D and f' then the number of photons collected by each detector will remain constant. However, the angular subtense of the diameter of our detector at the nodal point of the reduced eye is pn'/f'. Hence, if the eye size increases but the retinal detectors remain constant in size, the quantum catch of individual receptors will not improve but the retinal grain will effectively become finer in angular terms, potentially allowing better acuity. A large eye is therefore an attractive design for animals which are active in a low-light environment but which nevertheless desire good acuity. However, large eyes demand more body space and this introduces new problems. A compromise, adopted by both terrestrial and deep-sea species, is to sacrifice visual field (see, e.g., Martin, 1984, for the case of the tawny owl) while retaining the benefits of an eye of relatively long focal length and high numerical aperture. Although the resultant eyes are usually termed 'tubular' (Fig. 5.13), they may lack cylindrical symmetry in some deep-sea fish (Locket, 1977, pp 98–104), which may also have an accessory retina or other adaptations to give some sensitivity to light and movement in parts of the visual field not served by the

OPOSSUM　　HOUSE MOUSE　　LYNX

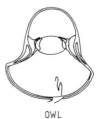  

OWL　　BUSH BABY　　DEEP-SEA FISH

Fig. 5.12 *Nocturnal eyes (not to the same scale) in which the numerical aperture has been maximized to yield the highest possible retinal illuminance (After Walls, 1942, p. 173). Schematic eyes have been described for the opossum (Oswaldo-Cruz et al., 1979) and mouse (Remtulla and Hallett, 1985). Maffei et al. (1990) have recently measured the visual acuity of the lynx, using visual evoked potentials, and find it very similar to that of the cat, at around 5–6 c/deg.*

Fig. 5.13 *Tubular eyes, which have a high numerical aperture and a long focal length to improve acuity but sacrifice visual field. Based on Walls (1942, p. 213) with the deep-sea fish eye modified after Locket (1977): the acessory retina is on the left-hand side of the deep-sea eye illustrated. The bush baby* Galago crassicausatus *has similar spatial resolution to the cat (Langston et al., 1986), although its eye length is substantially smaller (15 mm compared with 22 mm).*

main retina (Locket, 1977, pp 104–120). There are problems in moving any tubular eye within its socket, and eye movements in species like owls that possess such eyes therefore tend to be very limited (e.g. Steinbach and Money, 1973; Steinbach and Angus, 1974).

Modifications to these simple ideas that are necessary in view of the wave nature of light will be discussed further below.

Of course, it must be remembered that it is not merely the number of photons that reach the retina that is of importance but also how many are actually absorbed to lead to visual detection. In part this will depend upon the characteristics of the absorption of the visual pigment and the receptor design, particularly outer segment length, since a longer outer segment enhances the chances of photon absorption (see, e.g. Locket, 1977, p 91). However, if we confine ourselves to the purely optical aspects, we may note that since it is likely that some of the photons that strike the retina will pass through without being absorbed, it is possible to enhance the chances of photon capture by adding to the retina a reflecting layer, or tapetum lucidum. This redirects the wasted light to pass a second time through the receptors: that not absorbed on the second pass will emerge from the eye to be seen by an observer as 'eye glow' from the animal concerned. Tapeta are found in many mammals, fish and crocodiles and in some birds (Nicol and Arnott, 1974) that are active under low light conditions. Three types have been distinguished, varying in the nature and efficiency of their reflecting materials: the reflection may be either specular or diffuse (see, e.g. Walls, 1942, pp 228–246; Nicol, 1981). The guanine tapetum, which depends on constructive interference from a quarter-wavelength stack of alternating layers of high and low refractive index, is particularly efficient (Land, 1972). Unfortunately, since the original incident images of points are formed by convergent pencils of light, there is a divergent spread of light in the reflected beam, even for a specular tapetum. Thus there is some lateral displacement of the photons from the original image point on their second pass through the retina, leading to image blur and a loss in potential acuity. For this reason, in some species the tapetum is occlusible with pigment granules, which migrate to cover and obscure the reflecting surfaces under brighter conditions (e.g. Denton and Nicol, 1964). Very similar tapetal structures are found in many invertebrates (Ribbi, 1980; Land, 1981).

Limits on Optical Image Quality at the Retina

Diffraction

Even if an optical system such as a vertebrate eye is 'perfect' from the geometrical optical viewpoint, in that all the rays of light from a single object point converge to a single image point, the wave nature of light results in the in-focus, nominally point image having a finite lateral extent.

The form of the point image formed by such a 'perfect' or 'diffraction-limited' optical system depends on the shape and dimensions of the aperture stop or pupil, as well as the wavelength, λ, of the light. Formally, the spatial distribution of light amplitude in the in-focus point image is given by the Fourier transform of the spatial form of the aperture stop of the system, with wavelength as a scaling constant. The corresponding intensity distribution, the point-spread function (PSF), is given by the square of the amplitude distribution. In the case of a system with a circular pupil and monochromatic light, the point image is an Airy diffraction pattern (Airy, 1835) having a central bright core surrounded by a series of progressively weaker diffraction rings (e.g. Born and Wolf, 1975, pp 395–398; Freeman, 1990, pp 436–438). The radius of the central bright core is equal to $0.61 \lambda/(n' \sin u')$, where n' is the refractive index in the image space and u' is the semiangle of the cone of rays forming the point image ($n' \sin u'$ is the numerical aperture, as discussed earlier). This leads to the well-known Rayleigh criterion for the resolution of two similar self-luminous object points, that they should be separated by a distance corresponding to $0.61 \lambda/(n' \sin u')$ in the image space. Using our reduced eye model (Fig. 5.5), the Rayleigh criterion gives a two-point angular resolution equal to $1.22 \lambda/D$ radians, where D is the pupil diameter. If, rather than expressing resolution capability in terms of two-point resolution, we specify the highest spatial frequency of sinusoidal grating that can be imaged by the system, we find that its period is given by λ/D radians, i.e. the cutoff spatial frequency is $\pi D/180\lambda$ c deg^{-1}.

Instead of considering only limiting resolution, it is often helpful to know how an optical system images spatially extended objects. Under the conditions that normally apply in vision, the illumination is incoherent: there are no systematic phase relationships between the light emanating from different object points and it is therefore valid to add the intensities due to those object points to determine the final image. It is also reasonable to make the futher assumption for aberrated systems that the aberrations can be treated as constant over the local field area of interest. Selwyn (1948) showed that, under these constraints, a grating object with a sinusoidal or cosinusoidal radiance profile was unique in that its image was always unchanged in basic spatial form, apart from magnification, and merely differed in modulation (contrast) and spatial phase from the original object grating.

Thus one useful description of imaging performance for extended objects is the optical transfer function (OTF), which is complex and comprises the modulation transfer function (MTF), which gives the degradation in

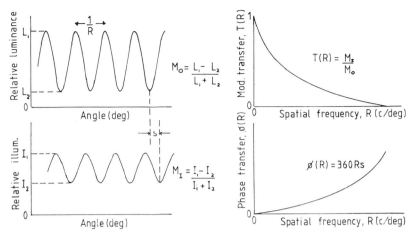

Fig. 5.14 *Modulation and phase transfer. Any optical system will normally degrade the modulation, M_O, of an object grating of spatial frequency R (period 1/R) and will also, if the corresponding line-spread function is asymmetric, change its spatial phase: s is the lateral shift of the image in comparison with paraxial expectations. Plots of the modulation transfer $T(R) = M_I/M_O$ and phase transfer, $\phi(R) = 360Rs$ degrees, as a function of R comprise the modulation transfer function (MTF) and phase transfer function (PTF) respectively.*

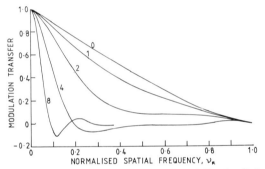

Fig. 5.15 *Modulation transfer functions for a diffraction-limited optical system with a circular pupil, working in monochromatic light. It suffers from the errors of focus indicated, where defocus is expressed in quarter-wavelengths (Rayleigh units) of wavefront aberration. The normalized cutoff frequency, $v_R = 1.0$, corresponds to $(\pi D \times 10^6)/180\lambda$ c deg^{-1}, where D mm is the pupil diameter and λ nm the wavelength of the light. One Rayleigh unit corresponds to $(2 \times 10^{-3}\lambda)/D^2$ dioptres.*

modulation in the image of a sinusoidal grating as a function of spatial frequency, and the phase transfer function (PTF), which gives the corresponding changes in spatial phase (Fig. 5.14; see Charman, 1983). Systems like a diffraction-limited lens with a circular pupil which yield symmetrical point-spread functions always have a PTF equal to zero.

Fig. 5.15 shows the MTF for a diffraction-limited eye with a circular pupil, both in focus and for various amounts of defocus. The spatial frequency has been expressed in normalized terms, so that a value of unity for the normalized spatial frequency corresponds to the cutoff frequency of $\pi D/180\lambda$ discussed above. Note that whereas modulation transfer falls steadily to zero at the cutoff in the in-focus case, this is not necessarily the case if the eye is out of focus.

Of course, in practice, not all vertebrate eyes have circular pupils. The most obvious result of this lack of circular symmetry is a corresponding lack of symmetry in the PSF and an orientation dependence in the optical resolution and OTF. An optically perfect (diffraction-limited) eye with a vertical slit pupil 0.5 mm wide and 5 mm long, working with light of wavelength 500 nm, would, for example, have an optical cutoff frequency of about 17 c deg^{-1} for vertical gratings and 170 c deg^{-1} for horizontal gratings. Thus, clearly, although a slit-shaped pupil may have advantages over a circular pupil in its capability of closing more competely under very bright environmental conditions, this is at the expense of orientation dependence in its diffraction-limited optical resolution. The numerous other asymmetric shapes found among vertebrate pupils, from the ovals of the ungulates through the bizarre shapes of some amphibia (Fig. 5.16) to the multiple apertures of some sharks under bright conditions, must all similarly involve asymmetric retinal PSFs and orientation-dependent optical resolution.

In principle, the image of any extended object can be calculated by regarding it as the summation of an array of points, each of which is imaged as a PSF. The final image is then given by the summation of all these point images. Formally this process is known as taking the convolution of the PSF with the radiance distribution of the object. Alternatively, the object may be broken into its spatial frequency components by taking the Fourier transform of its radiance distribution: in the imaging process each of these is changed in modulation and phase according to the

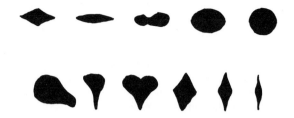

Fig. 5.16 *Shapes of a selection of contracted amphibian entrance pupils: all the pupils are circular when dilated (after Walls, 1942).*

OTF of the system. These degraded gratings are then superimposed to form the image by taking the inverse Fourier transform (e.g. O'Neill, 1963; Born and Wolf, 1975). Calculations of this type have been presented for humans (Gubisch, 1967) and various other animals (e.g. Wässle, 1971; Krueger and Moser, 1972).

Aberrations

Any system of spherical surfaces having centres of curvature lying on a common optical axis suffers from aberrations or faults which may result in an image which is more degraded than that affected by diffraction alone (see, e.g., Born and Wolf, 1975, ch. 5; Freeman, 1990, ch. 7).

On the optical axis (and at all other points in the image field) the image is blurred by longitudinal chromatic aberration and spherical aberration. This blur can be reduced by pupil constriction. Off the axis, further blur may be caused by coma and oblique astigmatism: the aberrations field curvature and distortion do not blur an image point but result in it being in the wrong longitudinal and lateral position respectively. The off-axis aberrations tend to increase with field angle.

Although the magnitudes of some aberrations are modified if the surfaces are aspheric, as in many vertebrate eyes, their form remains similar. The various tilts, decentrations and other asymmetries that occur in real eyes, and the frequent lack of a true optical axis, further modify the aberrations and the imaging effects may be additionally complicated by a lack of rotational symmetry in the retinal surface. In spite of these modifications in comparison with man-made systems, it is still fruitful to discuss ocular imaging defects in terms of the classical forms of aberration.

It is evident that, in general, image quality is likely to deteriorate with increasing distance from the optical axis, owing to the additional off-axis aberrations. This is less of a problem in those groups, such as primates, which have a well-developed fovea lying close to the optical axis, the decrease in optical performance in the periphery being matched by the fall-off in neural performance (see, e.g., Charman in volume 5, chapter 7 of this series for man). It is intrinsically more of a problem for those species having a visual streak or twin foveas, for which reasonable image quality must presumably be maintained over an extended field.

However, as Guidarelli (1972) points out, the optical design of the vertebrate eye approximates reasonably closely to that of a homocentric system, particularly in some fish. In such a system, the refracting and detector surfaces are all concentric, with a common centre of curvature at the centre of the aperture stop: the chief ray from any object point, passing through the centre of the aperture stop, is always an optical axis, so that no off-axis aberrations occur. Note, too, that the nodal points will also lie at the centre of the aperture stop, so that the lateral magnification on the spherical detector surface will be constant at all points in the field. In fact, this is approximately true for many bird and fish eyes but not for primates (Holden *et al.*, 1987).

Although the vertebrate eye is not exactly homocentric, its characteristics are such that the off-axis aberrations are relatively small. In particular, the curved surface of the retina can be matched to the field curvature and astigmatic image shells of the optical system, while distortion can presumably be adapted to at the central perceptual level.

Chromatic Aberration

Chromatic aberration arises from the dispersive nature of the ocular media, which have a higher refractive index at the blue end of the spectrum than at the red, conferring correspondingly greater ocular power at shorter wavelengths. Figure 5.17 shows the two forms of chromatic aberration. Longitudinal (axial) chromatic aberration results in a shift in the longitudinal position of a point image with wavelength, whereas lateral (transverse) chromatic aberration results in the long-wavelength images being larger than their short-wavelength counterparts, the image of each white-light object point being drawn out into a short radial spectrum. We can easily estimate the approximate magnitude of the longitudinal chromatic aberration using our reduced eye model. The power, F, of the eye is given by

$$F = n'/f' = (n'-1)/r$$

where f' is the second focal length, n' the refractive index of the eye medium, and r the radius of curvature of the refracting surface. Differentiating with respect to wavelength we find:

$$\frac{dF}{d\lambda} = \frac{1}{r}\frac{dn'}{d\lambda} = \frac{F}{n'-1}\frac{dn'}{d\lambda}$$

If we are interested in the power change, ΔF, correspond-

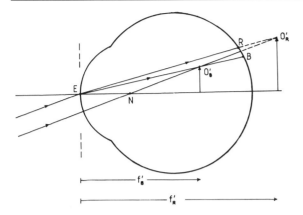

Fig. 5.17 *Ocular longitudinal and transverse chromatic aberration. In the reduced eye model illustrated, a ray from a distant off-axis object point passes undeviated through the eye's nodal point N. Red and blue principal or chief rays passing through the centre of the aperture stop E are refracted to generate red and blue images O'_R and O'_B, which it is assumed lie respectively behind and in front of the retina. The axial distance between these two images represents the longitudinal aberration (usually expressed in dioptric terms as $(1/f'_B - 1/f'_R)$. The chief rays strike the retina at the points R and B, which thus represent the centres of the corresponding red and blue blur patches. The angle subtended by RB at the nodal point N is the corresponding angular TCA.*

ing to the wavelength interval, $\Delta\lambda$, between the blue and red, we can approximate:

$$\Delta F = F_{yellow}(n'_{blue} - n'_{red})/(n'_{yellow} - 1) = F_{yellow}/V$$

where $V = (n'_{yellow} - 1)/(n'_{blue} - n'_{red})$ is the constringence (Abbe number) for the medium. Taking this to be the same as the value for water, we find that we would expect the chromatic aberration, expressed in dioptres, to amount to about one-sixtieth of the power of the eye.

Hughes (1979a) has tabulated values of longitudinal chromatic aberration obtained essentially on this basis for several species. Sivak and Mandelman (1982) measured the dispersion of the different ocular media in a variety of species and found that while the aqueous and vitreous dispersions differed little from that of water, those of the cornea and lens were substantially higher. Therefore, they recalculated the chromatic aberration using their data, to find somewhat greater amounts of aberration, particularly for human and cat, although they point out that their human values agree well with experimental data (Mandelman and Sivak, 1983). Millodot and Sivak (1978) have made direct retinoscopic measurements of ocular chromatic aberration for gerbils, rats, rabbits (see also Nuboer and van Genderen-Takken, 1978; Nuboer *et al.*, 1979) and leopard frogs and find amounts of aberration broadly compatible with these predictions. More recent measurements of the chromatic aberration of the rat (Chadhuri *et al.*, 1983), mouse (Remtulla and Hallett, 1985) and an African cichlid fish (Fernald and Wright, 1985a) are also in agreement. The chromatic aberration of the isolated vertebrate lens has been explored in more detail by Kreuzer and Sivak (1985), who again find that it is rather higher than expected, presumably because of the gradient index structure of this refractive component.

The transverse chromatic aberration expressed in terms of visual angle for a reduced eye can be approximated by

$$u(n'_{blue} - n'_{red})/n'_{yellow}$$

or about

$$u(n'_{yellow} - 1)/(V \times n'_{yellow})$$

so that it becomes increasingly significant as the field angle, u, increases: of course, it should be remembered that a reduced eye model can give only a rough guide to the likely magnitude of the aberration (see also Howarth, 1984) and that the aberration reduces as an eye approaches closer to homocentricity. Since each point is blurred into a short radial spectrum, grating structures in which the bars are oriented radially are not degraded, whereas those with a tangential orientation are (Thibos, 1987). Direct experimental measurement across the visual field has been attempted only in the case of humans (Ogboso and Bedell, 1987) but Cheney and Thibos (1987) and Thibos *et al.* (1987) have demonstrated the orientation dependence in retinal imagery that is introduced by the aberration.

Clearly, the effects of chromatic aberration will be much reduced if the spectrum of available light is limited in bandwidth, as applies in some underwater environments, of if the prereceptoral media act as a spectral filter to reduce the available bandwidth, as in the case of human macular pigment (Reading and Weale, 1974). Absence of short-wave receptors in the central fovea of the human is also beneficial in reducing the effects of image blur that is due to chromatic aberration and hence maximizing the potential acuity. In fact, there is a strong tendency for the media of all diurnal species to act as yellow filters, whereas this filtering effect is absent in nocturnal species (Walls, 1942, pp 191–205; Tansley, 1965, pp 44 and 62; Lythgoe, 1979, pp 118–124). Such filters may not only be beneficial in reducing the effects of chromatic aberration: they may also serve to minimize the photochemical damage that blue and ultraviolet radiation can cause to the retina in bright environments (Ham *et al.*, 1976; Sliney, 1977).

Spherical Aberration

Spherical aberration is of particular importance in that it occurs on the optical axis of any eye, as well as in the rest of the image field. Thus even for eyes in which there is a single, well-developed fovea lying close to the optical axis, spherical aberration must be adequately controlled if a good-quality optical image is to be available. In general, increases in the numerical aperture of any system (i.e. effectively the value of D/f') tend to increase the amount of spherical aberration. Thus, spherical aberration is likely to be of particular significance for eyes in which the numerical aperture is large and yet a good-quality retinal image is still required (in the human eye a large natural pupil is found only under scotopic conditions when optical image quality is less important, owing to the limited resolution of the neural network).

In the form most frequently met with in eyes, the effect of spherical aberration is that rays passing through the more marginal zones of the pupil are refracted more strongly than those that pass through the pupil centre (under-corrected or positive spherical aberration; see Fig. 5.18(a)).

In the vertebrate eye, spherical aberration can be controlled by using aspheric surfaces for the cornea and lens, so that the surfaces become flatter for the more marginal rays, thus reducing their optical power (Fig. 5.18(b)), by the more central regions of the crystalline lens having a higher index, thereby giving it more power for near-paraxial rays (Fig. 5.18(c); Le Grand, 1967), or by reducing the pupil diameter (Fig. 5.18(d)), although this reduces the light flux collected. In terrestrial eyes, controls of all these types are found, but for many fish, in which the lens is spherical and provides almost all of the optical power, and the pupil is relatively immobile, the index gradients are solely responsible for reducing the spherical aberration. For human eyes, El Hage and Berny (1973) have argued that the aberrations of the cornea and lens are of opposite sign (positive for the cornea and negative for the lens) and cancel each other out, whereas Millodot and Sivak's (1979) measurements suggest that each component is separately corrected: in any case, the whole eye still has significant spherical aberration at larger pupil diameters, so that best optical performance is obtained with smaller pupils (see below).

Direct measurements of isolated lenses from many species have been carried out by Sivak and Kreuzer (1983), using multiple laser beams to determine the ray intersection patterns. They find that, among goldfish, yellow perch and rock bass, only the rock bass has negligible spherical aberration, the other fish showing residual under-corrected aberration, although less than that for a glass bead of homogeneous index. Sroczynski (1975, 1977, 1979) had earlier found species and age-related differences in the amount of spherical aberration in the fish lens, as do Seltner et al. (1989). Among mammals, Sivak and Kreuzer (1983) found that the cow, pig, lamb and rabbit showed little spherical aberration, while the dog and rat (see also Chadhuri et al., 1983) showed negative spherical aberration, as also do the lenses of the ground squirrel (Sivak et al., 1983) and mouse (Remtulla and Hallett, 1985). Cat (see also Jagger, 1990) and human lenses showed a mixture of positive and negative aberrations. Duck and frog lenses displayed negative aberration. Thus it would appear that exact compensation of spherical aberration is by no means the rule. However, it must be remembered that the significance of spherical aberration for those animals with mobile pupils depends, as in the human case, upon the pupil

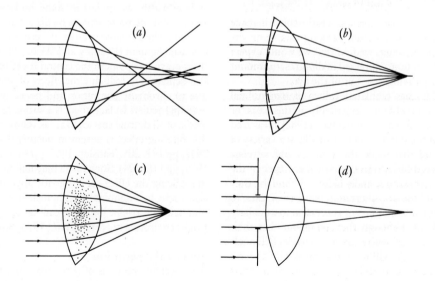

Fig. 5.18 *(a) Under-corrected spherical aberration. The marginal rays are refracted more strongly than the paraxial rays nearer the axis, leading to a blurred image point. (b) Control of spherical aberration by an aspheric anterior lens surface. (c) Control of spherical aberration by a gradient index lens: stippling represents regions of higher index. (d) Control by stopping down the pupil.*

diameter and is much reduced with small pupils. Moreover, for those animals possessing accommodation, the aberration is likely to vary with the state of accommodation: there is some evidence that in humans spherical aberration changes from positive to negative as the level of accommodation is increased, being minimal at an intermediate viewing distance (Ivanoff, 1956).

The case of the spherical fish lens immersed in a medium of constant refractive index (the eye humours, approximating closely to water) is of particular theoretical interest. More than a century ago, Matthiessen (1880) observed empirically that for such lenses there was an almost constant ratio (2.55:1) between the distance separating the lens centre from the focal point, and the radius of the lens. Fletcher *et al.* (1954) were able to show that, for a spherical lens to satisfy Matthiessens' ratio and also be aplanatic, or free from spherical aberration and coma, it had to have a spherically symmetric gradient of refractive index, the index decreasing along any radius from a central value of 1.51 to that of the eye humours at its outer surface. The non-zero amounts of spherical aberration and departures from Matthiessens' ratio seen in some eyes presumably reflect departures from the ideal index gradient.

Other Aberrations
Although the oblique astigmatism of the human eye has been well documented (see, e.g., Charman 1983, 1991a for reviews) little information is available for other species. Hughes (1977b) finds little variation in the refraction across the field of the rat eye and Hughes and Vaney (1978) have shown that the rabbit eye, with its extended visual streak, is within 1 D of emmetropia across its full horizontal visual field and has less than 1 D of astigmatism at all points. The pigeon eye also appears to be free of astigmatism (Erichsen, 1979; Fitzke *et al.*, 1985a). Campbell and Hughes's (1981) calculations with a gradient index eye model for the rat suggest that coma is probably well controlled.

Scattered Light
Stray light can occur within any eye as a result of reflections from the optical surfaces and retina, together with scattering from optical inhomogeneities. Although a tapetum confers the advantages discussed earlier in increasing retinal sensitivity, it may also increase the amount of stray light, particularly if the tapetal reflection is diffuse, when the interior of the eye may act as a crude integrating sphere. The circumlental gap or aphakic aperture that occurs in many fish, due to the pupil diameter exceeding that of the lens, may also be a source of non-image forming light. All stray light reaching the retina serves potentially to diminish the contrast of the retinal image.

The scattered light of the human eye has been extensively studied, mainly by the use of psychophysical techniques (see Charman, 1991a, for review). Vos *et al.* (1976) summarize many of these studies in terms of an equivalent veiling luminance L_{eq} (measured in $cd\,m^{-2}$) that would produce the same masking effect as a glare source giving an illuminance E at the eye, as a function of the angular distance, α (in degrees), between the glare source and the fixation point. The following approximate relationship is found:

$$L_{eq} = \frac{29E}{(\alpha+0.13)^{2.8}} \qquad (0.15° < \alpha < 8°)$$

Scattering increases throughout life by a factor of at least 2–3 times (Allen and Vos, 1967; Hemenger, 1984; Ijspeert *et al.*, 1990).

DeMott and Boynton (1958) made direct measurements of entoptic stray light in steer, sheep, cat and pig eyes by cutting a small hole through the sclera and retina in the back of the eye and measuring the flux through a small aperture in front of a photomultiplier as the light beam incident on the eye was rotated about the pupil centre. They found surprisingly similar results for their different species under a range of different wavelengths, pupil diameters and other conditions (Fig. 5.19): the results also compared well with both psychophysical and objective studies on the human eye. The illuminance found by DeMott and Boynton is well described as being inversely proportional to the $\frac{5}{2}$ power of the distance from the image centre, very similar to the human psychophysical result given above and to later measurements on the

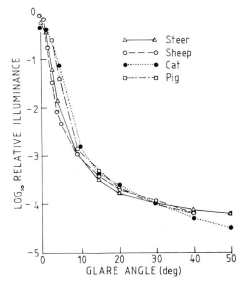

Fig. 5.19 *The angular distribution of retinal stray light in the eyes of the four species indicated (After DeMott and Boynton, 1958).*

cat by Robson and Enroth-Cugell (1978). Few studies have been made of scattered light in other species, although Jagger (1988) comments that scattered light is the major culprit for the poor image quality present in the eye of the cane toad. In birds, the shadowing effect of the pecten may help to minimize the adverse effects of stray light scattered from the retinal image of the solar disc (Barlow and Ostwald, 1972).

As will be discussed below, some of the adverse, contrast-reducing effects of intraocular stray light may be lessened by the directional characteristics of the retinal receptors.

Effect of the Shape of the Retinal Surface

There seems little doubt that in several species the effects of optical aberrations are modified by the development of either optical components or retinas that lack rotational symmetry about the optical axis. For example, although there is little astigmatism in the pigeon eye and its lateral fields are emmetropic (Glickstein and Millodot, 1970), its field in front of the bill appears to be myopic (Catania, 1964; Millodot and Blough, 1971; Nye, 1973; Erichsen, 1979). Fitzke et al. (1985b) have argued that this is an example of an adaptation that allows a ground-feeding bird to keep the ground in focus when standing erect, and that it is possible to express the refractive error in terms of the equation

$$R = -\frac{\sin A}{H} = -\frac{1}{V}$$

where R is the refractive state of the eye in dioptres, A is the angle of elevation below the horizon, H is the pupil height (the distance in metres from the centre of the pupil to the ground when the bird is standing erect) and V is the viewing distance to the ground at angle A (Fig. 5.20(a)). Hodos and Erichsen (1990) have found that the refraction in the lower visual fields of the quail, chicken and crane eyes closely follows these theoretical predictions (Fig. 5.20(b)). Although asymmetric refractive components could, perhaps, be responsible for this behaviour, it seems much more plausible to attribute it to the asymmetric shape of the retina and eyeball. Anatomical sections of the pigeon eye appear to show such asymmetry (Marshall et al., 1973). Recent experiments in which chicks raised in an environment of restricted height developed appropriate upper-field myopia (Miles and Wallman, 1990) suggest strongly that partial-field myopia of this type in birds develops after hatching as a response to the visual environment. Thus lower-field myopia is normally an adaptive matching of refractive error to the needs of ground-foraging birds. In principle the supposed 'ramp' retinas of the horse and stingray (Walls, 1942, p 255; Duke-Elder, 1958, pp 642–643; but see Sivak and Allen, 1975; Sivak,

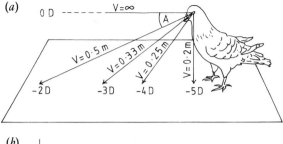

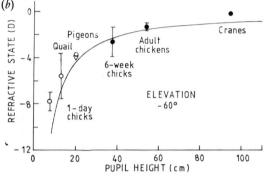

Fig. 5.20 (a) Relationship between the dioptric distance of the ground in various viewing directions and the height, H, of the eye of a ground-foraging bird above the ground (After Hodos and Erichsen, 1990). In the case shown the vertical height H is 0.2 m. (b) Mean refractive state, R, of the lower visual field (A = 60°) plotted as a function of the distance, H, from the pupil centre to the ground for several species of bird. The smooth curve is that predicted from R = −(sin A)/H (After Hodos and Erichsen, 1990).

1976) and, perhaps, of the rabbit (De Graauw and van Hof, 1980) could have similar origins, and future studies may reveal many other examples.

Retinal Illuminance

One additional feature of off-axis performance that deserves comment is the retinal illuminance. Ignoring possible absorption effects, this is a function of both the area of the entrance pupil and the area of the retina on which the light from unit solid angle of visual field falls. As is well known, in the case of the human eye the entrance pupil becomes elliptical when viewed peripherally, its area at 90° to the visual axis dropping to about 20% of its axial value (e.g. Jay, 1962). However, calculation shows that the area of the retinal image corresponding to unit solid angle of field also falls with increasing field angle (Drasdo and Fowler, 1974). Rather remarkably, the changes in these two factors appear to almost exactly compensate for each other, so that the retinal illuminance is maintained at an approximately constant value for field angles up to some 70–80° (Fitzke, 1981; Bedell and Katz, 1982; Charman,

1983, 1989; Kooijman, 1983). Pflibsen *et al.* (1988) have shown that this conclusion is not substantially altered by pathlength-dependent absorption within the lens, and Kooijman and Witmer (1986) have confirmed experimentally that there is, indeed, approximately constant retinal illuminance across the retina of human and rabbit eyes, even though the entrance pupil area varies markedly with field angle.

However, it seems unlikely that this can be true for all vertebrate eyes. In a truly homocentric eye, in which the optical surfaces and retina are concentric about the centre of the aperture stop, the retinal representation of unit solid angle of visual field is independent of the field angle. However, it is obvious that in many species with approximately homocentric eyes, for example fish in which the retina is approximately concentric with the lens, the area of the entrance pupil must vary roughly as cos θ, where θ is the field angle. Thus it appears probable that in such eyes the retinal illuminance for a ganzfeld must fall gradually as the field angle is increased, with possible consequences for the relative sensitivities of the central and peripheral areas of retina.

Overall Retinal Image Quality

Although the general layout of the eye has been studied in many species, few investigators have attempted to explore the quality of the optical image on the retina other than in the case of man.

The human retinal image close to the optical axis has been assessed by a variety of psychophysical and objective techniques (see, e.g., Charman, 1983, 1991a, 1991b for reviews). As would be expected, for small pupils the performance is almost diffraction-limited, but as the pupil diameter increases aberrations start to play an increasingly dominant role. The result is that absolute performance is optimal at pupil diameters of 2–3 mm (Fig. 5.21).

The earliest measures of the non-human retinal image were made by DeMott and his colleagues (DeMott and Boynton, 1958; DeMott, 1959). They examined the retinal image directly in enucleated eyes of steers and a cat, by cutting a small window in the sclera near the posterior pole, a method also used by Röhler (1962). The cat's retinal image was further investigated by several other early authors (Westheimer, 1962; Wässle, 1971; Bonds *et al.*, 1972). In Wässle's typical study, the MTF was estimated by examining ophthalmoscopically the contrast in the retinal images of square-wave gratings of different frequencies. A 3-mm artificial pupil was used, but unfortunately the aberration characteristics of the eye were altered by the use of a contact lens (the cornea of the cat eye is aspheric). The results showed an MTF that approached zero at about 12 c/deg. Interestingly, Wässle went on to calculate the corresponding forms of the retinal images of a range of discs, borders, bars, points and other stimuli. Bonds (1974), using an ophthalmoscopic method to measure the double-pass line-spread function, deduced broadly similar MTFs for contact-lens wearing cats, with performance getting worse as the pupil diameter was increased. The artificial pupils used were circular rather than mimicking the slit-shaped natural pupil of the cat, although a 1.5×7 mm slit pupil apparently gave results similar to those from a 3 mm diameter round pupil. Image quality declined somewhat in the periphery. Robson and Enroth-Cugell (1978) refined such measurements further by using a fibre-optic probe to measure the single-pass MTF; this method is free from any ambiguities in the nature of the retinal reflection that are present with ophthalmoscopic techniques and probably gives the most reliable results (Fig. 5.22). In comparing Figs 5.21 and 5.22, note that at similar pupil sizes the performance of the cat eye is worse than that of the human eye (the axial length of the cat eye, about 22 mm, is only slightly smaller than that of man, about 24 mm).

Krueger and Moser (1972) estimated the MTF of the frog *Rana esculata* by photographic recording and subsequent analysis of the double-pass line-spread function, using an ophthalmoscopic technique. Although this method is unlikely to yield very accurate results, they found that modulation transfer approached zero at about 6 c/deg, well below that found for man. Krueger and Moser went on to calculate the corresponding retinal images of a range of discs, bars, double bars and annuli, their results clearly demonstrating the substantial contrast reductions and other degradations occurring in the retinal images. Du Pont and de Groot (1976) found even worse optical performance for the frog than did Krueger and Moser. More recently, Jagger (1988) has used a fibre-optic

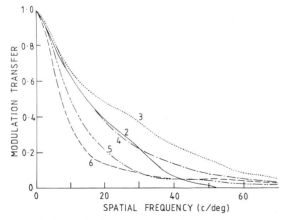

Fig. 5.21 *MTFs for the human eye for the pupil diameters (mm) indicated (After Artal, 1990).*

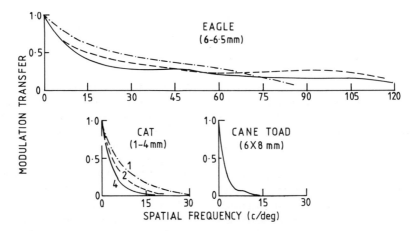

Fig. 5.22 *Ocular MTFs for the three vertebrate species and the pupil sizes (mm) indicated: serpent eagle, with active pupil and accommodation (after Shlaer, 1972); cat, with paralysed accommodation, an artificial pupil and a contact lens refractive correction (After Robson and Enroth-Cugell, 1978); cane toad (After Jagger, 1988).*

probe to measure the single-pass MTF for the cane toad, an eye that also has a high numerical aperture. He found that scatter severely degrades retinal image quality, the effective cut-off frequency for the MTF being about 10 c/deg (Fig. 5.22), well below the diffraction-limited value of 200 c/deg for the dilated pupil used. However, image quality is well maintained across the visual field and, since the retinal receptors are visible ophthalmoscopically (albeit using a slightly smaller pupil diameter: Jagger, 1985, 1988), it is evident that the optical quality is adequate to meet the animal's neural needs.

Among birds, Shlaer (1972) has directly measured the retinal image quality of an African serpent eagle, using the double-pass ophthalmoscopic apparatus of Campbell and Gubisch (1966). The length of this bird's eyeball is very similar to that of the human eye. The bird was alert and only lightly restrained. Its pupil diameter was about 6–6.5 mm. Since no drugs were used, Shlaer selected the narrowest external line-spread functions in his recordings as indicating the optimal image quality that was likely to be achieved. The resultant MTFs are illustrated in Fig. 5.22. The eagle MTF is much superior to the human MTF (Fig. 5.21), and indeed is not too far below diffraction-limited performance, implying good aberration control in its large-pupil eyes.

Although much more work is required before any comprehensive picture of comparative retinal image quality is available, it is clear that the extent to which diffraction-limited optical performance is approached varies widely for different species at the same pupil diameter, implying widely different standards of aberration control. To understand why such variations are acceptable we need to consider the interaction of the optical image with the retina.

Retinal Aspects

The characteristics of the retinal receptors can interact with the optical design of the vertebrate eye in two important ways. First, since the long, quasi-cylindrical receptors differ in refractive index from the surrounding medium and their diameters are comparable with the wavelength of light, they act as optical waveguides, preferentially trapping light which is travelling nearly parallel to their lengths. Second, the limited spatial resolution of the neural elements may allow the tolerances on optical quality to be relaxed.

Retinal Fibre Optics

Brücke (1844) was the first to note that if the long receptor outer segments had a refractive index which was greater than that of the surrounding medium, light might be trapped within the receptor by multiple total internal reflections. The resultant greater pathlength would then yield greater absorption. Both Brücke and Helmholtz (1856, vol. 1, p 229) further noted that each receptor would have a limiting numerical aperture within which light could be trapped and that it would therefore be advantageous if the receptors were all oriented to point towards the centre of the exit pupil of the eye.

Such speculations remained a curiosity until Stiles and Crawford (1933) demonstrated psychophysically that, in the human eye, light which entered the centre of the pupil, and hence had normal incidence at the retina, was a more effective stimulus than light which passed through the pupil periphery. This observation caused Wright and Nelson (1936) to revive the suggestion that fibre-optic effects might occur within the retina, although it was not until the work of Tansley and Johnson (1956) on the grass snake that total internal reflection effects within retinal cones were directly observed.

Toraldo di Francia (1949) introduced an important new concept when he pointed out that the diameter of the receptors was of the same order as the wavelength of light and hence that an explanation on the basis of geometrical optics alone was inadequate. The receptors would act as dielectric waveguides, having specific directional or modal patterns, depending upon their dimensions, the refractive indices, and the wavelength and polarization of the light. Jean and O'Brien (1949) were the first of a number of workers to demonstrate, using microwaves and appropriately scaled models of the receptors, that waveguide modes could account for many features of the Stiles–Crawford effect, the light being funnelled down the inner segments into the outer segments. Waveguide modes were soon demonstrated to occur in the retinas of a variety of species, including rats, rabbits, monkeys, humans and goldfish (e.g. Enoch, 1963, 1967; Tobey *et al.*, 1975) and appropriate theoretical treatment has been provided by such workers as Snyder and Pask (1973). It is interesting to note that Hannover had, in fact, already apparently observed waveguide modes in the large rod photoreceptors of the frog in the mid-nineteenth century (see Enoch and Tobey, 1973). The strong tendency of the primate photoreceptors to be oriented towards the centre of the exit pupil was confirmed by Laties and Enoch (1968). Subsequent theoretical and experimental work has been thoroughly reviewed by Enoch and Tobey (1981) and Enoch and Lakshminarayanan (in volume 1, chapter 12 of this series). While psychophysical methods have been used to study the Stiles–Crawford effect in man, it has also been demonstrated electrophysiologically in turtles (Baylor and Fettiplace, 1975), although in the latter species absorption by the oil droplets present in the cones modifies the angular variation in sensitivity.

A rough idea of the acceptance angle of a retinal receptor can be gained by applying geometrical optics to the case of an ideal cylindrical photoreceptor having refractive index n_1, surrounded by a medium of index n_2. The geometrical approximation becomes increasingly valid as the receptor diameter increases. If light is incident from a medium of refractive index n_3 it will be trapped by total

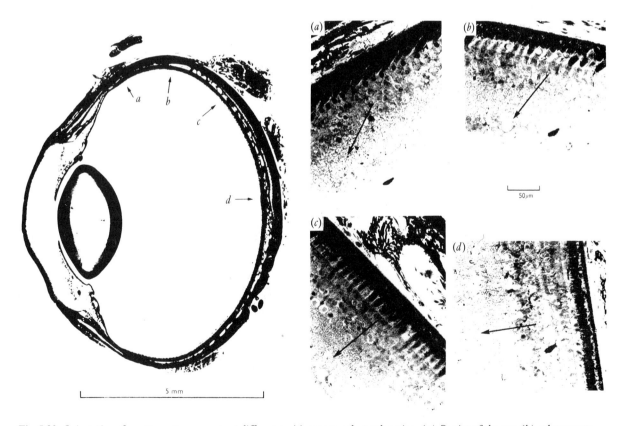

Fig. 5.23 *Orientation of receptor outer segments at different positions across the turtle retina. (a) Section of the eye; (b) enlargements at higher magnification of the areas of retina indicated. The arrows show the direction of the centre of the exit pupil in each case: note that the receptor orientation always matches this direction (After Baylor and Fettiplace, 1975).*

internal reflection if its incidence angle on the entry face of the receptor is less than θ_L, where the numerical aperture of the receptor, $n_3 \sin \theta_L$ is given by $(n_1^2 - n_2^2)^{\frac{1}{2}}$ (see, e.g., Enoch and Lakshminarayanan, in volume 1, chapter 12). Using the refractive index measurements of Barer (1957) and Sidman (1957) it is possible to estimate θ_L for cones as typically being around 8°. Outside the cone with this semi-angle, light cannot be trapped within the receptor.

It can be seen that these directional characteristics provide a valuable means of suppressing the contrast-reducing effects of stray light which has been reflected and scattered within the eye to strike the retina obliquely: such light is ineffective at stimulating the receptors, since its incidence angle exceeds θ_L. Similarly, if image-forming light from the pupil is to stimulate the retina, the receptors should point towards the centre of the exit pupil (e.g. Laties and Enoch, 1968; Baylor and Fettiplace, 1975; see Fig. 5.23). Further, if the full benefits of a large pupil are to be reaped, the acceptance angle $2\theta_L$ of the receptors obviously must exceed the subtense of the exit pupil as viewed from the retina. Bossomaier et al. (1989) have recently shown that although this condition is at first sight not fulfilled in the case of the eye of the diurnal garter snake, which has a very high value of (D/f') and an all-cone retina, the inner segments of the retinal cones contain numerous small, refractive 'microdroplets', about 0.1 μm in diameter, which effectively increase the local refractive index and hence the acceptance angle of the receptors, so that all the pupillary light is trapped.

Fibre-optic effects obviously have the further benefit that once light is trapped within a receptor it cannot escape to stimulate neighbouring receptors. There is no 'cross talk' between neighbouring receptors and hence the quality of the optical image at the entry face of the receptors is preserved along their length. Although these simple arguments need some modification if waveguide effects are taken into account, they do emphasize the importance of these directional effects in relation to stray light and the maintenance of overall image quality.

The Receptor Mosaic

Over a century ago Helmholtz (1856, vol. II, p 35) emphasized the importance of the spatial organization of the retinal receptors when he suggested that, for the retinal image of a grating to be resolved, it was necessary that the retinal cones stimulated by neighbouring bright bars of a grating should be separated by one unstimulated cone. This simple, one-dimensional concept has gradually been elaborated within the framework of sampling theory. If the retinal receptors are packed within a two-dimensional hexagonal array (more strictly described as a triangular array, Williams, 1988), as is essentially the case with the human retina (Hirsch and Hylton, 1984), with neighbouring receptors having a centre-to-centre spacing, s, the highest or critical spatial frequency, f_c, of grating that can be unambiguously resolved is given by:

$$f_c = \pi f' / [180 sn'(3)^{\frac{1}{2}}] \quad \text{c deg}^{-1}$$

where f' is, as before, the second focal length of the eye and n' is the refractive index of the vitreous (Snyder and Miller, 1977; Miller and Bernard, 1983). Gratings of frequencies higher than this 'Nyquist limit' are undersampled and may appear as spurious gratings of lower spatial frequency, a phenomenon known as 'aliasing' (see, e.g. Hughes, 1986; Williams, 1986; Charman, volume 5, chapter 1 of this series). The possibility of aliasing being observed is reduced by the fall in ocular MTF with spatial frequency but, in the case of humans, it can be observed under some conditions of natural viewing (Williams and Collier, 1983; Williams et al., 1983; see Charman, volume 5, chapter 1 of this series for review).

Taking into account the Nyquist limit of the retinal mosaic and the possibility of aliasing, the view has often been expressed that the cutoff frequency of the optics should always approximately 'match' that of the retina. It is assumed that if the optical quality is inadequate the potential resolution of the retina is wasted and, if too good, distortions in the perceived image may arise owing to aliasing. However, there may be real advantages in allowing the cut-off frequency of the optics to exceed that of the retina (Hughes, 1986; Snyder et al., 1986). For example, if, as discussed earlier, the optical modulation transfer tends to fall steadily with spatial frequency, it might be better to have a relatively high optical cutoff in order to ensure reasonably high values of modulation transfer at spatial frequencies up to the retinal Nyquist limit. The case for such a strategy may be strengthened by the fact that the spatial frequency spectrum of many natural scenes varies approximately as the inverse of the spatial frequency, so that it is desirable to keep reasonable modulation transfer at higher spatial frequencies in order to ensure that they are represented at adequate contrast in the retinal image. It may be noted that, since in general the spatial distribution of receptors may vary asymmetrically across the retina, notably in those species with twin foveas or visual streaks, there is no way in which an optical system with approximate rotationally symmetry could have an optical cutoff that matched the neural cutoff at all points across the retina: in order to provide adequate image quality to the retinal areas having highest neural resolution it is inevitable that notionally too high a quality of optical image will be available to those areas of poorer neural capability (see, e.g., Reymond, 1985, 1987, for the cases of an eagle and a falcon).

It can be argued that further factors diminish the possibility of aliasing artefacts. Averaging over the receptor aperture (Miller and Bernard, 1983) and disorder in the receptor array (Yellott, 1982; Hirsch and Miller, 1987) may be beneficial, although Bossomaier et al. (1985) argue that disorder is tolerable only if there is marked convergence of receptors to ganglion cells. Small eye movements (e.g. Eizenman et al., 1985) might convey similar advantages to 'dynamic scanning' in man-made fibre bundles (e.g. Kapany, 1967), where rapid image movement across the fibres reduces both noise and regular aliasing.

There is, in fact, little doubt that optical resolution can sometimes exceed neural resolution: in some species, it is possible to see the individual retinal receptors of the living animal through an ophthalmoscope or similar device, using the eye's optics as part of the imaging chain. Thus Land and Snyder (1985) were able to observe ophthalmoscopically the quasi-uniform cone mosaic of the garter snake and to show that optical modulation transfer remained quite high (35%) at the cutoff frequency (2.1 c deg^{-1}) of the retinal mosaic. Jagger (1985) has shown similar effects in the eye of the cane toad. However, it should be stressed that in both these animals the optical performance is well below the limit set by diffraction and a much closer match between optical and neural quality may occur in species with higher acuity, such as eagles and falcons (Reymond, 1985, 1987). In the latter cases there is a close similarity between the optical and anatomical resolutions, and behavioural acuity for the deep convexiclivate fovea: however, this implies that the optical resolution must be markedly superior to the anatomical resolution away from the foveal areas. Undersampling by retinal receptors is probably responsible for the relatively low level of acuity in young primates (Jacobs and Blakemore, 1988).

It is clear that a complicating factor in comparisons of optical and neural capabilities is that both may be evolving and that part of any mismatch may arise from this cause. Weale (1976) presents an interesting argument based on sampling theory (Toraldo di Francia, 1955) to suggest that the photosensitive surface evolved first and the optics then followed. The results of many recent studies on the development of refraction suggest that the retina is part of a feedback loop in which the growth of the optical components of the eye is controlled by the need to provide a sharply focused optical image of adequate clarity on the retina (e.g. Schaeffel and Howland, 1988; see Young and Leary, volume 1, chapter 2 of this series for review). Thus the characteristics of the optics may be largely controlled by the demands of the detector array: improvements in the optical image occur only up to the level at which further improvement cannot be detected by the retina. In this sense, 'matching' is the inevitable product of normal growth processes, the neural capabilities being set by the biological needs of the animal with the optical characteristics being adjusted accordingly (Snyder et al., 1986).

Other Retinal Factors

In addition to the general retinal factors already discussed, it is of interest to note several unusual retinal adaptations which allow animals to make optimal use of the available photon flux. Photomechanical changes, in which either pigment or the photoreceptors themselves move in the retina in order to ensure that, depending upon the light level, rods or cones have the optimal chance of capturing the photons, are widespread in fish, amphibia and birds (Walls, 1942, pp 145–153). Miller and Snyder (1977) have discussed the tiered retinas of the pig and cat-eyed snake (*Leptodiera*). In the snake, the light passes first through a tier or layer of rods, with the cones lying somewhat posterior: the order is reversed in the pig. Miller and Snyder show that this arrangement works so that much of the radiation not absorbed by the first layer of receptors is available to the second layer: each type of receptor simultaneously occupies most of the light-capture area of the retina. Multiple bank retinas in deep-sea fish are usually thought to be an adaptation to increase retinal sensitivity at low light levels (e.g. Locket, 1977, pp 140–151) although, puzzlingly, it appears that only the inner layer of receptors in the tiered retina of the conger eel contribute to visual sensitivity (Shapley and Gordon, 1980).

One retinal feature of considerable, and still controversial, optical interest is the foveal pit, where the cells of the inner retina are displaced laterally and the cones are thinner and more closely packed: such foveal pits are, perhaps, most fully developed in birds, but occur also among primates, lizards (e.g. Makaretz and Levine, 1980) and some teleost fish. The most obvious advantage of the clearing of the retinal tissue is that it might reduce intra-retinal scatter in the area yielding most acute vision. However, although this might explain the shallow concaviclivate fovea, it is difficult to see how it could account for the very deep convexiclivate fovea (Fig. 5.24(a,b)).

Walls (1937, 1942, p 180) suggested that, owing to the difference in refractive index between the vitreous and retinal tissue, the sloping walls of such a foveal pit serve to magnify the image falling on the receptor layer of the retina (Fig. 5.24(a)). This suggestion was discounted by Pumphrey (1948), who suggested instead that a deep fovea distorts any image that does not lie on the foveal centre and that this might serve as a stimulus for correct fixation. The concept of the foveal pit as a magnifying device has been revived in terms of the hemispherical bottom of the pit acting as the negative lens component of a telephoto system, with the cornea and crystalline lens forming the positive component (Snyder and Miller, 1978). The longer focal length of the telephoto lens effectively means that the angular subtense of the foveal cones, referred to the outside world, is smaller than it would be in the absence of the pit. An alternative suggested role is that of a

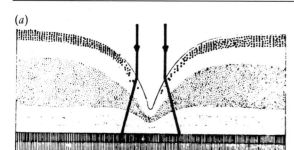

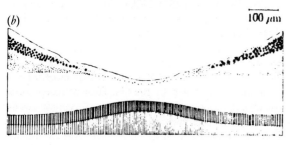

Fig. 5.24 *(a) Diagrammatic cross-section of the deep convexiclivate fovea found in reptiles, birds and some fish. The rays indicate the possible magnifying effect of the foveal pit. (b) The concaviclate fovea found in primates and the binocular field of some birds and reptiles (Based on Walls, 1942; Harkness and Bennett-Clark, 1978).*

focus detector, an image appearing undistorted if in correct focus but with either barrel or pincushion distortion for incorrect focus (Harkness and Bennet-Clark, 1978). At present any optical role is unproven, although Maraketz and Levine (1980) note that the low value of convergence of cones to ganglion cells in this retinal region makes it well suited as an anatomical substrate for any of the suggested optical mechanisms.

Depth-of-focus, Ametropia and Accommodation

So far we have implicitly assumed that the vertebrate eye is always focused to produce a sharp image at the relevant depth in the retina. In fact, any combination of optical focusing elements and a detector array always has some finite depth-of-focus, or range of focal positions over which the output of the detector array remains essentially constant. This depth-of-focus will depend not only on the characteristics of the optical system, in particular its numerical aperture (NA, proportional to D/f') but also upon the characteristics of the detector array, in that any specific value of optical defocus is less likely to be detected by an array with coarse resolution capabilities than by one with densely spaced detector elements.

The Diffraction-limited Eye

If we consider a diffraction-limited optical system combined with an 'ideal' detecting surface, in the sense that all the spatial information present in the image can be registered by the detector, it is usually somewhat arbitrarily assumed that the limits on defocus are set by the Rayleigh criterion, which states that the actual imaging wavefronts must not deviate by more than a quarter-wavelength from the ideal spherical form centred at the required image point. The corresponding effect on the MTF is shown in Fig. 5.15. It can be seen that the defocus-induced changes

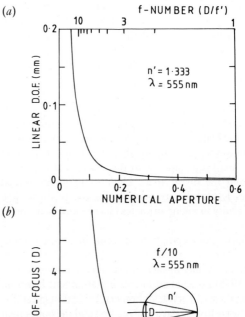

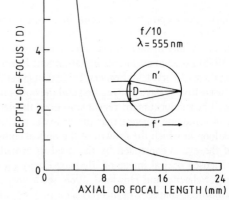

Fig. 5.25 *(a) Optical depths-of-focus in the image space of a diffraction-limited eye as a function of its numerical aperture, at a wavelength of 555 nm. The Rayleigh quarter-wavelength criterion has been used and a vitreous index of 1.333 has been assumed. The upper abscissa scale gives the approximate corresponding values of* D/f', *assuming a reduced eye model. (b) Dioptric depth-of-focus for eyes with a* D/f' *corresponding to f/10 (i.e. a numerical aperture of 0.067) as a function of the second focal or axial length* f'. *It has again been assumed that the wavelength is 555 nm and the vitreous index is 1.333.*

are relatively modest. The criterion leads to a longitudinal error of focus given approximately by $\pm \lambda n'/2 (\mathrm{NA})^2$. The value of this depth-of-focus as a function of numerical aperture is plotted in Fig. 5.25 (a), assuming a wavelength of 555 nm and a refractive index for the vitreous of 1.333. It can be seen that this sets a relatively tight tolerance on focus for eyes of high NA but that the depth-of-focus is quite substantial for lower NAs (the numerical aperture of the human eye is about 0.1 for a 3 mm pupil). Note that the permitted longitudinal error of focus for any particular numerical aperture is independent of the ocular dimensions. However, as noted by Glickstein and Millodot (1970), any given longitudinal shift in focus represents a bigger dioptric change for a small eye than for a large eye so that, in dioptric terms, a small, diffraction-limited eye can tolerate a bigger dioptric error of focus (Fig. 5.25).

Real eyes

Although experiments with the human eye using small artificial pupils show that sensitivity to defocus may approach the diffraction limit (see Charman, volume 1, chapter 1 of this series for review), this is unlikely to occur under natural viewing conditions. Not only do ocular aberrations begin to play a role but, more importantly, the limited resolution performance of the retina makes the overall optical–neural performance less sensitive to defocus. Green *et al.* (1980), following arguments employed much earlier by Emsley (1952) for the human eye, developed useful formulae for calculating the approximate dioptric depth-of-focus based on the concept of retinal blur circles.

Figure 5.26 illustrates their basic argument, using a reduced eye and a circular pupil. It is assumed that the eye is emmetropic, so that an object at infinity is focused clearly on the retina. If now the object is moved to a distance l in front of the eye, the image is formed at a distance l' behind the refracting surface. The diameter, d, of the corresponding retinal blur circle is then given by similar triangles as:

$$d/D = (l' - f')/l'$$

Introducing the power of the eye $F = n'/f'$, the object and image vergences $L = n/l$ and $L' = n'/l'$, and the basic refraction formula $L' = L + F$, we find:

$$d = DL/F \quad \text{or} \quad d = D\,\Delta F/F$$

where we have written ΔF as the dioptric error of focus from the in-focus, zero vergence position at infinity. Obviously, for any given dioptric error of focus, the linear diameter of the retinal blur circle varies directly with the pupil diameter D and inversely with the power of the eye, F. Alternatively, we may write the same expression in terms of the second focal length f' of the reduced eye as:

$$d = Df'\,\Delta F/n' \quad (5.1)$$

so that the blur circle diameter is proportional to the focal length (i.e. to the overall length) of the reduced eye. Since the posterior nodal distance of the eye is given by f'/n', the angular subtense of the blur circle diameter at the nodal point is:

$$d\,(\text{radians}) = D\,(\text{metres}) \times \Delta F\,(\text{dioptres}) \quad (5.2)$$

The problem now is to decide what is an 'acceptable' diameter of blur circle. Green *et al.* (1980) initially made

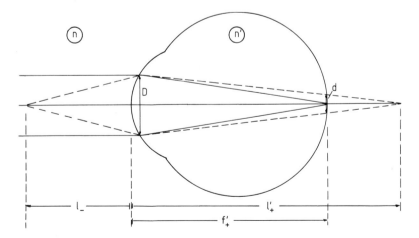

Fig. 5.26 *Reduced eye illustrating the blur circle resulting from defocus. The eye accurately focuses a distant object point (full rays), but a near object point at distance l (dashed rays) is focused at a distance (l' − f') behind the retina. If the eye pupil (coincident with the refracting surface) is circular with diameter D, the out-of-focus image point appears as a blur circle of diameter d on the retina.*

two simplifying assumptions: that D/f' was constant for all eyes and that the tolerable linear diameter of blur circle was also the same for all eyes (effectively this is equivalent to stating that the neural receptor array is the same in all species). Thus we have $D/f' = K_1$, and $d = K_2$, where K_1 and K_2 are constants. Substituting these in Equation (5.1) and reorganizing gives us:

$$\Delta F = \pm (n'K_2)/(K_1 f'^2) \quad (5.3)$$

Hence, on these assumptions, the permissible error of focus, or depth-of-focus, is inversely proportional to the square of the second focal length of the eye or, more generally, to the square of the ocular length.

However, as Green et al. (1980) point out, it is not true that all eyes have the same value of (D/f') or the same tolerance on linear diameter of blur circle. If we assume that in Equation 5.2 the tolerable blur circle diameter simply corresponds to the angular subtense of the target detail, ϕ degrees, that can just be resolved by the animal (e.g. the width of one black or white bar in an acuity target, so that if ϕ is $\frac{1}{60}$ of a degree then the acuity corresponds to 30 c deg^{-1}) we find, from Equation 5.2:

$$\Delta F \text{ (dioptres)} = \pm 17.5\, \phi \text{ (degrees)}/D \text{ (mm)} \quad (5.4)$$

This can obviously be alternatively expressed in terms of the visual acuity ω c/deg, using $\omega = 1/2\, \phi$, as:

$$\Delta F \text{ (dioptres)} = \pm 8.8/[D \text{ (mm)} \times \omega \text{ (c deg}^{-1})] \quad (5.5)$$

Using arguments based on the MTF of a defocused optical system, Green et al. (1980) refined Equation 5.5 slightly to yield:

$$\Delta F \text{ (dioptres)} = \pm 7/[D \text{ (mm)} \times \omega \text{ (c/deg)}] \quad (5.6)$$

Figure 5.27 reproduces plots from Green et al. (1980) showing the expected depths-of-focus for different animals based on the assumptions of Equations 5.3 and 5.5 and published values of ϕ and D ($\frac{1}{2}$ maximum or at 10 cd m^{-2}). While there are obviously various simplifications in the approach, it is of great value in emphasizing that small eyes with poor acuity would be expected to have very large dioptric depths-of-focus, whereas the necessity for exact focus becomes much more compelling when acuity is high.

Although the above approximations are based on geometrical optics, the importance of visual acuity in controlling effective depth-of-focus can be understood easily by considering the through-focus modulation transfer for a diffraction-limited eye at different spatial frequencies (Fig. 5.28). Whereas modulation transfer (and hence image contrast) at low spatial frequencies is relatively insensitive to changes in focus, that at high spatial frequencies falls rapidly with focal error. Hence, if the neural array can register only low spatial frequencies, depth-of-focus is likely to be large, whereas it will be smaller if the retina can respond to finer details. Some indication of the variations between the spatial bandwidths of the visual systems of different species can be gained by considering their contrast sensitivity functions (Fig. 5.29, after Petry et

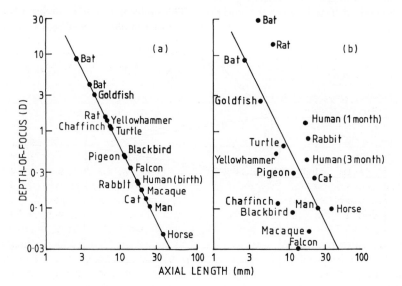

Fig. 5.27 *Logarithmic plots of predicted depths-of-focus (i.e. the maximum tolerable error of focus) against axial lengths for various species (After Green* et al., *1980). (a) Figures derived on the assumption that all eyes have the same value of* D/f' *and the same tolerable blur circle diameter on the retina. (b) More realistic estimates based on published values of pupil diameter and visual acuity for the individual species. The straight lines in each case have slopes of* -2 *(see text for explanation).*

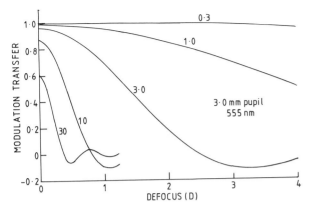

Fig. 5.28 *Through-focus modulation transfer for a diffraction-limited eye with a 3 mm pupil at a wavelength of 555 nm at the spatial frequencies (c/deg) indicated. Note that image modulation (contrast) at low spatial frequencies is much less sensitive to defocus than that at high spatial frequencies.*

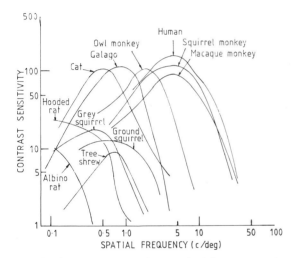

Fig. 5.29 *Contrast sensitivity functions for different mammals (based on compilation by Petry* et al., *1984). Precise focus is likely to be more important for those animals which are capable of using high spatial frequency information.*

al., 1980). Although the latter include both optical and neural factors, and all species do not have the same pupil sizes, the obvious differences between species do serve to emphasize that depth-of-focus is likely to show substantial variation.

Evidently, both exact ametropia and any necessity for accommodation are likely to be less important for those animals possessing eyes with large depths-of-focus. It would be expected that even though emmetropia might remain desirable as a mean condition for such animals, considerable variation in refraction about that mean could be tolerated. At present there is no published information to indicate whether the spread in refractive error in, say, bats and rats is greater than among humans.

Accommodation

We have seen that some animals with good acuity have small depths-of-focus and hence, by implication, might be expected to need some form of focusing or accommodation mechanism. However, the aim of that accommodation system will be to ensure that the retina lies always within the depth-of-focus for the range of object distances of interest to the particular animal. A large animal with a small depth-of-focus, like the horse, may have little requirement for accommodation, since even with its head lowered the grass is still some 0.3 m away, requiring only 3 D of accommodation. On the other hand, small birds might, for example, be interested in feeding their young at distances of a few centimetres and could therefore benefit from the ability to accommodate by up to perhaps 40 D. Thus there is not necessarily a simple correlation between amplitude of accommodation and depth-of-focus alone. Also, we might guess that the speed at which the accommodation changes, as well as the potential amplitude of that change, might vary with the requirements of the animal. Particular problems in accommodation might be anticipated for those species, like cormorants, which require good vision both in air and under water: neutralization of the power of the cornea demands either that the lens can vary in power over a very wide range or that some other optical adaptation is called for.

In fact, there is surprising diversity in the accommodation mechanisms used by different vertebrates (see, e.g., Walls, 1942, pp 247–288; Duke-Elder, 1958, pp 640–655; Tansley, 1965, ch. 7; Sivak, 1980b). Both static (steady-state) and dynamic aspects of accommodation performance are of interest. These have been widely investigated in man (see, e.g., Ciuffreda, volume 1, chapter 11 of this series for review) but studies of other species are in their infancy, largely because of difficulty in reliably eliciting a normal, voluntary accommodation response. Most workers have used drugs or electrical stimulation, giving little more than an indication of the maximal amplitude that might be achieved, but remote measurement techniques like photorefraction (Howland and Howland, 1974; Howland, volume 1, chapter 18 of this series) now offer greater potential for non-invasive investigations. Smith and Harwerth (1984) have succeeded in measuring the accommodative amplitudes in young rhesus monkeys by behavioural methods (at 17–18 D the amplitudes were rather higher than for equivalent-aged humans) and it may be possible to extend such measurements to other species.

In man, accommodation forms part of the 'near triad'. Under binocular conditions of observation, the frontally placed eyes must converge to maintain single vision of a near object, as well as accommodating to give sharp retinal images. Pupil constriction also occurs (accommodative miosis). The input from the convergence system undoubtedly forms an important part of the overall near-control system (see, e.g., Schor and Ciuffreda, 1983, for review), although the exact role of the near-pupil response remains somewhat enigmatic. However, while similar accommodation characteristics might be expected in other species with frontally placed eyes and mobile pupils, exact analogues of the human accommodation systems are unlikely to occur in, for example, the many species with more laterally placed eyes. Nevertheless, it is helpful to review briefly the major findings in man to indicate the type of response characteristics that might be of interest.

The main feature of the human static response is that the retinal image is rarely in exact focus. Instead, over-accommodation is typical in distance vision and under-accommodation in near vision (Fig. 5.30(a)). The slope of the quasi-linear portion of the accommodation response/stimulus curve is found to vary with observing conditions such as target illuminance and spatial frequency content, and pupil size (Charman, 1986). All these are related to depth-of-focus (and acuity) and effectively the accuracy of the response becomes worse as the acuity degrades and depth-of-focus increases: the accommodation control system does not attempt to maintain exact focus but merely keeps the image within the depth-of-focus over the available range of accommodation. Thus it might be expected that similar effects would occur in other species, with the slope of the response/stimulus curve being lower in those animals which, while possessing active accommodation, have poorer acuity. As yet virtually no experimental evidence is available. However, it is true that the slope of the curve in human infants steepens towards the 'ideal' value of unity as their acuity improves (Green et al., 1980). Note that the objective amplitude of accommodation, which is necessarily what is measured in non-human species, is always smaller than the subjective amplitude over which satisfactory vision is maintained, owing to the depth-of-focus of the eye.

The typical human dynamic response to a change in target distance is shown in Fig. 5.30(b): there is a latency of about $\frac{1}{3}$ s followed by a response time of about $\frac{1}{2}$ s. Dynamic changes in target distance can be followed at frequencies up to a few Hertz. As is well known, accommodation in man falls with increasing age, a characteristic shared at least with other primates (Bito et al., 1982), chickens (Troilo and Wallman, 1985) and fish (Fernald and Wright, 1985b), although the origins of the loss differ in different classes of vertebrate. In darkness or in the absence of a stimulus, human accommodation reverts to a slightly myopic tonic or 'resting' state (Fig. 5.28(a)). It is possible that similar effects occur in other species: Marshall et al. (1973) have noted that pigeons accommodate by about 2 D when placed in the dark.

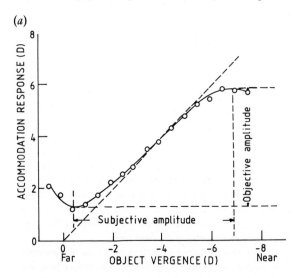

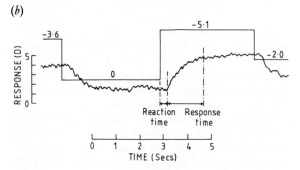

Fig. 5.30 (a) Typical static accommodation response–stimulus curve for man (the slightly myopic subject was aged 38 years). Note under-accommodation for near targets and over-accommodation for far targets. The subjective amplitude of accommodation exceeds the objective amplitude because the target still appears clear to a subject when it is at the limits of his depth-of-focus. (b) Dynamic responses to abrupt changes in target distance: the corresponding target vergences (dioptres) are also shown. Each time the target distance is changed there is a short reaction time or latency before the accommodation response commences.

Mechanisms for Accommodation

Dynamic

The major possible dynamic mechanisms were clearly recognized by Young (1801) during his pioneering work to demonstrate that the lens is the seat of human accommodation. They include:

1. Change in corneal curvature.
2. Movement of the lens.

3. Shape change in the lens.
4. Change in the axial length of the eye.

CORNEAL CURVATURE Since corneal power is proportional to corneal curvature (inversely to radius of curvature), a terrestrial animal can potentially increase the refractive power of its eye by steepening the curvature of its cornea, or decrease it by reducing the curvature. In birds Crampton's muscle is thought to be capable of producing such curvature changes (see, e.g., Walls, 1942, pp 279–280; Meyer, 1977, p 554). Although the earlier literature is somewhat contradictory, recent results support the concept of corneal accommodation in chicks (Troilo and Wallman, 1987). The corneal component amounts to up to about a third of the total amplitude of about 18 D, the rest being provided by lenticular change. Broadly similar results have been found for the chick by Schaeffel and Howland (1987), who also find that almost all of the pigeon's amplitude of accommodation of about 8 D is due to the cornea (see also Gundlach et al., 1945). (Surprisingly, Schaeffel et al., 1986, suggest that accommodation acts independently in the two eyes of chicks, which also possess some 17 D of accommodation immediately after hatching.)

It is probable that further work will confirm the importance of corneal change as a component of avian accommodation. One would expect that corneal change would involve marked changes in asphericity and hence in aberration, both on the optical axis and in the peripheral field: the latter might be of importance in relation to the lateral fovea of species with twin foveas.

LENTICULAR MOVEMENT According to Duke-Elder (1958, p 644) this mechanism is characteristic of more primitive vertebrates. It is not surprising that it is found among aquatic species. Under water, the cornea cannot contribute to ocular refraction since its power is almost neutralized, and the lens, being spherical and of high rigidity, is difficult to deform; therefore change in the position of the lens relative to the retina is the only viable focusing mechanism. However, in fish an optical problem arises when a spherical lens is moved by the *retractor lentis* muscle with respect to a roughly hemispherical retina: different portions of the retina are refocused by different dioptric amounts (Fig. 5.31). Thus the direction of movement varies between species depending upon the organization of their retinae and other factors. In the African cichlid fish *Haplochromis burtoni*, for example, the lens moves in the pupillary plane along the naso–temporal axis, movement being directed towards a region of retinal specialization characterized by a higher density of all cell types except rods. When the accommodation is relaxed, the temporal retina is focused for near vision. Active accommodation through lens movement adjusts the focus

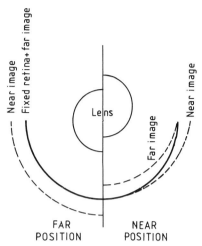

Fig. 5.31 *Schematic illustration of effect of lens movement on image surfaces in the fish eye. The image surfaces for distant and near objects are concentric with the rigid, spherical lens but the radius of curvature of the distance image surface is smaller than that for the near image. If the image shell for distant objects is made coincident with a spherical retinal surface (left-hand part of figure), forward movement of the lens (right-hand part of figure) will produce a sharp near image on only part of the retinal surface, since the fixed curvature of the retina cannot match that of both the distant and near image surfaces.*

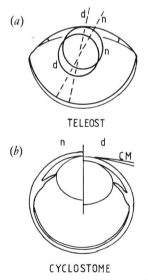

Fig. 5.32 *(a) Accommodation in teleosts.* n *is the position of the lens for relaxed near accommodation and* d *the position for distance vision when the lens has been moved backwards and sideways through the action of the retractor lentis muscle. (b) Accommodation in cyclostomes. In the relaxed state (*n*) the eye is accommodated for near objects. The cornealis muscle* CM *pulls the cornea backwards to flatten it and push the lens backwards towards the retina for accommodation on distant objects (*d*) (After Duke-Elder, 1958, pp. 644–646).*

of the temporal retina for far vision (Fernald and Wright, 1985b). In general there is a combination of lens movement both along the pupil plane and along the pupil axis (Fig. 5.32(a)), although substantial variation occurs between species (Munk, 1973; Somiya and Tamura, 1973; Sivak, 1975). Amplitudes up to about 20 D have been observed (e.g. Sivak, 1975).

In the unusual case of the Atlantic flying fish *Cypserlurus heterus*, the problem of obtaining near-emmetropia in both air and water has apparently been solved through the development of a cornea in the form of a low, three-sided pyramid, with each face bulging only slightly, in contrast to the hemispherical cornea of most fish (Baylor, 1967): thus in air the cornea should have negligible power, so that the eye remains emmetropic. However, the optics of this arrangement remain to be fully explored, since on the 'axis' of such an eye in air one might expect image tripling to occur. A related eye is found in the 'four-eyed' fish *Dialommus fuscus*, a Galapagos species which spends much of its time on rocks. Its cornea appears to consist of two relatively flat parts, joined by a pigmented band (Munk, 1969). Since a flat cornea has zero power in both air and water, no unusual accommodation demands are made on moving between the two media.

Variants of the lens movement mechanism occur among the cyclostomes, such as the lamprey, the 'spectacle' covering the cornea being drawn taut by the action of the cornealis muscle, to flatten the cornea against the lens and push the lens backwards towards the retina (Fig. 5.32 (b)). Active accommodation therefore brings about clearer distance vision (e.g. Walls, 1942, p 258). In the selachians it has been suggested that the lens moves in a forward direction to make the eye more myopic for near vision (Duke-Elder, 1958, p 647). In snakes and amphibia lens movement also occurs, although part of the amplitude of about 8 D observed in the toads *Bufo viridis* and *Bufo americanus* may be associated with changes in axial length (Douglas *et al.*, 1986; Mathis *et al.*, 1988). The amplitudes of amphibia that feed in air appear to be inadequate to ensure good underwater vision, although aquatic amphibian species like the newt are emmetropic underwater and become myopic in air (Tansley, 1965, p 68). The eye of snakes is thought to be degenerate with respect to other reptiles, the loss of scleral ossicles, annular pad and ciliary muscle having occurred during the evolutionary phase in which snakes became burrowing creatures and lost their legs. For distance vision, muscle fibres at the root of the iris contract to exert pressure on the vitreous, which in turn pushes the lens forward towards the cornea. The eyeball may also lengthen slightly, but there are only modest changes in the shape of the lens.

CHANGE IN LENS SHAPE This is the principal means of accommodation in the higher vertebrates, the reptiles, birds and mammals. However, whereas in most reptiles (excluding snakes) and birds this is brought about by direct squeezing of the lens by the ciliary body and iris, the eyeball being strengthened against deformation by a circle of bony plates (scleral ossicles) inserted into the sclera round the base of the cornea, in mammals the shape changes normally result from changes in the tension exerted by the zonular fibres on the elastic lens capsule and lens substance.

The lens is unusually soft and malleable in birds and reptiles with good accommodation, and develops a thick annular pad (ringwulst) to ensure good contact with the ciliary processes and to distribute the forces exerted (Duke-Elder, 1958, pp 649–652; Meyer, 1977, pp 551–556). In the cormorant and dipper (Goodge, 1960), which both actively seek for food underwater, the loss of corneal power is compensated for by a massive amplitude of accommodation of up to 50 D (Levy and Sivak, 1980). Sivak *et al.* (1985) found even larger potential amplitudes of accommodation in the goldeneye (67 D) and hooded merganser (up to 80 D), although non-diving ducks like the mallard and wood duck had amplitudes of only about 3 D and 6 D respectively. In the same study, which used direct electrical stimulation of the eye, the bulk of the full amplitude accommodation change occurred within about 0.3 s, although the very short latency of 0.03–0.08 s between the onset of the electrical stimulus and the beginning of the response may not, perhaps, be typical of natural conditions. Penguins, although having good vision under water, have much smaller amplitudes, largely because the accommodative demand is reduced by the cornea having relatively low power in air (Howland and Sivak, 1984; Sivak *et al.*, 1987). It does not appear that crocodiles can focus accurately under water (Fleishman *et al.*, 1988), although they may have a modest accommodative ability of about 8 D for aerial vision (Duke-Elder, 1958, p 158).

In mammals, the zonular tension that determines the shape of the lens is in turn controlled by changes in the ciliary body. When the ciliary ring is relaxed, the zonular fibres are under tension and the lens is flattened for distance vision: contraction of the ciliary muscle and ring relaxes the zonular tension, allowing the lens to take up the more convex form dictated by capsular forces and its own elasticity. As a mechanism, this is probably better suited as a fine-focus adjustment rather than as a means of obtaining the very large changes required for a transition from aerial to underwater vision. Equally, if the lens approaches quasi-spherical form, as it tends to in nocturnal mammals with eyes having large numerical apertures (high values of D/f'), changes in zonular tension become relatively ineffective at producing lens shape changes (Fisher, 1969).

Although this is the basic mechanism in the mammalian eye, other processes may also contribute to refractive change in some species. Thus in the cat it appears that both lens movement and shape change are responsible for accommodation (Vakkur and Bishop, 1963; Sunderland and O'Neill, 1976). Particular interest attaches to accommodation in the sea otter, a carnivorous species which is active both above and underwater. Although its corneal power in air amounts to some 59 D, it is apparently emmetropic in both air and water (Murphy et al., 1990), so that its amplitude of accommodation must at least match the corneal power: this may be due to an unusual, well-developed lenticular accommodation mechanism, involving the forward movement of the lens and the subsequent formation of an anterior lenticonus by contraction of the iris sphincter (Murphy et al., 1990). However, such a mechanism may involve some sacrifice in underwater acuity at low light levels, since if accommodation is to be maintained then the iris sphincter cannot dilate to compensate for the lack of light; alternatively, if dilation occurs, the efficiency of accommodation suffers (Schusterman and Barrett, 1973). Among other marine mammals, it appears likely that dolphins have little accommodation and are myopic in air and hypermetropic in water (Dral, 1975; Rivamonte, 1976), the ametropia being kept within reasonable bounds by the relative flatness of the cornea, which has a power on only about 20 D in air (Rivamonte, 1976). Seals and other pinnipeds apparently have eyes optimized for underwater vision (Schusterman and Balliet, 1971): maintenance of reasonable vision in air may be aided by the combination of a slit pupil with an astigmatic cornea (Walls, 1942, pp 444–448; Piggins, 1970).

It is of interest that, when feeding, some species may be able to use monocular information derived from the amount of accommodation exercised, to judge prey distance. Harkness (1977) has clearly demonstrated this in elegant experiments in which a chameleon was fitted with a range of converging and diverging spectacle lenses: the chameleon's tongue extension when striking at prey was systematically in error by an amount directly predictable from the apparent distance of the prey as viewed through the lenses.

CHANGES IN LENGTH OF THE EYE As has already been noted, changes in length may occur as an accompaniment to other mechanisms, but as yet no species has been found that uses them as the main process for focusing the eye.

Static systems

ASYMMETRIC OR 'RAMP' RETINAE As noted in the earlier discussion of optical factors, a retina that does not possess rotational symmetry about the optical axis allows different portions of the retina to be conjugate with objects at different distances. Thus ground-feeding birds or animals that maintain an approximately constant posture with respect to their environment can benefit from such asymmetry and avoid the need for continuous active accommodation (Fitzke et al., 1985b; Hodos and Erichsen, 1990). However, the existence of a 'ramp' retina in the horse has been disputed (Sivak and Allen, 1975), and some doubts exist too as to whether the spatially asymmetric retina of the stingray is useful as a static accommodation mechanism in view of the existence of an effective dynamic mechanism and other factors (Sivak, 1976).

STENOPAIC PUPILS It is clear from earlier discussion that, for a circular pupil, depth-of-focus is inversely proportional to pupil size, so that the necessity for accommodation is reduced under conditions where such a pupil is constricted. Duke-Elder (1958, p 641) suggests that this is the mechanism whereby sea snakes bridge the refractive difference between aquatic and aerial vision. However, to obtain the full optical benefit from a small pupil, it must be round: a slit-shaped pupil leads to out-of-focus images in which the blur patches are themselves slit-shaped, so that linear detail perpendicular to the axis of the slit is substantially blurred. Similarly, it is incorrect to assume that those species in which the pupil contracts to give multiple pinholes will have a large depth-of-focus. For example, as Murphy and Howland (1986) point out, the four pinhole apertures of the constricted gecko pupil will, if the eye is out of focus, simply give four laterally displaced images, the lateral displacement increasing with the magnitude of defocus. This is essentially the principle of the classic Scheiner double-pinhole used in a variety of clinical refraction instruments. Only when the object is conjugate with the retina will the multiple images merge into one. Thus Murphy and Howland reason that the role of the multiple pinhole pupil is, in some species, not a way of increasing depth-of-focus but rather a precise indicator of correct focus and hence potentially of object distance, under illumination conditions where a single contracted circular pupil would indeed have large depth-of-focus and hence fail to give such information. A knowledge of object distance might be used by the animal to strike accurately at prey, as in the case of the chameleon (Harkness, 1977): it would clearly be of interest to explore this suggestion experimentally.

THE CORRUGATED RETINA The larger bats (Megachiroptera) have eyes with choroidal papillae whose internal contours provide an undulating surface over which the photoreceptors are arranged. It has been suggested that this could be a static accommodation mechanism, with the vitread receptors near the internal peaks of the papillae receiving a clear distance image and those near the valleys a sharp near image (Walls, 1942; Duke-Elder, 1958); the dioptric difference involved is about 1.5 D (Neuweiler, 1962). Weale (1974, p. 31) derided this suggestion on the basis that it would result in appalling perceived image quality, since only a fraction of the area of any image would be sharply focused. Murphy *et al.* (1983) used dynamic photorefraction to show that the flying fox has active accommodation with an amplitude of at least 3.1 D, much greater than the maximum of the suggested 'static' mechanism, so that at best the papillations can only serve to increase the depth-of-focus.

DUPLICATED OPTICAL SYSTEMS   The eye of the four-eyed fish *Anableps* is well known as an example of this type of eye. It swims in such a way that the water-line cuts the middle of the cornea. There are two pupils, the upper serving aerial vision and focusing an image through the lens on the appropriate area of retina, the lower vision underwater. The optics of this arrangement in the case of *Anableps microlepsis* have been discussed by Schwassmann and Kruger (1965).

OTHER POSSIBLE STATIC MECHANISMS   Duke-Elder (1958, p 643) has suggested that the long photoreceptor elements found in deep-sea fish and nocturnal geckos might have significance in relaxing the requirements for precise focus. This seems improbable, in view of the low acuity of such species and optical waveguide considerations: it is, as Duke-Elder himself suggests, much more likely to be an adaptation to increase sensitivity. Similarly, suggestions that the accessory retinae of deep-sea fish might subserve distance vision seem equally untenable: they appear more likely to give simply a modest measure of light sensitivity and movement detection in part of the peripheral field. Kölmer (1924) has suggested that the twin foveae of the kingfisher might be a specialization to avoid the need for accommodation, the central fovea serving distance vision and the second, temporal, fovea, near the ora terminalis, serving binocular vision underwater.

Conclusion

It can be seen that the optical arrangements of vertebrate eyes show wide diversity within a single broad framework of design. The details of design of any particular species can be understood only in relation to the environmental needs, neural capacity, body size and other sensory capabilities of the animal concerned. With improved experimental methods for following optical and visual performance in the living animal, a better understanding of the subtleties of the interactions between these various factors is beginning to emerge.

References

Able, K. P. (1982). Skylight polarization patterns at dusk influence migratory orientation in birds. *Nature Lond.*, **299**, 550–551.
Adler, K. and Taylor, D. H. (1973). Extraocular perception of polarized light by orienting salamanders. *J. Comp. Physiol.*, **86**, 203–212.
Airy, G. (1835). On the diffraction of an object glass with circular aperture. *Trans. Camb. Phil. Soc.*, **5**, 283–291.
Allen, M. J. and Vos, J. J. (1967). Ocular scattered light and visual performance as a function of age. *Am. J. Optom.*, **44**, 717–727.
Artal, P. (1990). Calculations of two-dimensional foveal retinal images in real eyes. *J. Opt. Soc. Am. A*, **7**, 1374–1381.
Axelrod, R., Lerner, D. and Sand, P. J. (1988). Refractive index within the lens of a goldfish eye determined from the paths of thin laser beams. *Vision Res.*, **28**, 57–65.
Barer, R. (1957). Refractometry and interferometry of living cells. *J. Opt. Soc. Am.*, **47**, 545–556.
Barlow, H. B. (1964). The physical limits of visual discrimination. In: *Photophysiology*, Vol. 2. ed. Giese, A. C. pp. 163–202. New York: Academic.
Barlow, H. B. and Ostwald, T. J. (1972). Pecten of the pigeon's eye as an intra-ocular eye shade. *Nature New Biol.*, **236**, 88–90.
Barrett, R., Maderson, P. F. A. and Meszler, R. M. (1970). The pit organs of snakes. In: *Biology of the Reptilia*. eds. Gans, C. and Parsons, T. S. pp. 277–304. London: Academic.
Bassi, C. J., Williams, R. C. and Powers, M. K. (1984). Light transmittance by goldfish eyes of different sizes. *Vision Res.*, **24**, 1415–1419.
Baylor, D. A. and Fettiplace, R. (1975). Light path and photon capture in turtle photoreceptors. *J. Physiol. Lond.*, **248**, 433–464.
Baylor, E. R. (1967). Air and water vision of the Atlantic flying fish, *Cypselurus heterurus*. *Nature Lond.*, **214**, 307–309.
Bedell, H. E. and Katz, L. M. (1982). On the necessity of correcting peripheral target luminance for pupillary area. *Am. J. Optom. Physiol. Opt.*, **59**, 767–769.
Benedek, G. B. (1971). The theory of the transparency of the eye. *Applied Optics*, **10**, 459–473.
Bito, L. Z., DeRousseau, C. J., Kaufman, P. L. and Bito, J. W. (1982). Age-dependent loss of accommodative amplitude in rhesus monkeys: an animal model for presbyopia. *Invest. Ophthalmol. Vis. Sci.*, **23**, 23–31.
Bonds, A. B. (1974). Optical quality of the living cat eye. *J. Physiol. Lond.*, **243**, 777–795.
Bonds, A. B., Enroth-Cugell, C. and Pinto, L. H. (1972). Image quality of the cat eye measured during retinal ganglion cell experiments. *J. Physiol. Lond.*, **220**, 383–401.
Born, M. and Wolf, E. (1975). *Principles of Optics*. 5th edn. Oxford: Pergamon.
Bossomaier, T. R. J., Snyder, A. W. and Hughes, A. (1985). Irregularity and aliasing: a solution? *Vision Res.*, **25**, 145–147.
Bossomaier, T. R., Wong, R. O. and Snyder, A. W. (1989). Stiles-Crawford effect in garter snake. *Vision Res.*, **29**, 741–746.
Bowmaker, J. K. and Kunz, Y. W. (1987). Ultraviolet receptors, tetrachromatic colour vision and retinal mosaics in the brown trout (*Salmo trutta*): age-dependent changes. *Vision Res.*, **27**, 2101–2108.
Brücke, E. (1844). Ueber die Physiologische bedeutung der Stabformigen Korper und der Zwillingszapfen in den Auges der Wirbelthiere. *Arch. Anat. Physiol.*, **11**, 444–451.
Bullock, T. H. and Cowles, R. B. (1952). Physiology of an infra-red receptor. The facial pit of pit vipers. *Science Wash. DC*, **115**, 541–543.
Bullock, T. H. and Diecke, F. P. J. (1956). Properties of an infra-red detector. *J. Physiol. Lond.*, **134**, 47–87.
Campbell, F. W. and Gubisch, R. W. (1966). Optical quality of the human eye. *J. Physiol. Lond.*, **186**, 558–578.

Campbell, M. C. W. (1984). Measurement of refractive index in an intact crystalline lens. *Vision Res.*, **24**, 409–415.

Campbell, M. C. W. and Hughes, A. (1981). An analytic, gradient index schematic lens and eye for the rat which predicts aberrations for finite pupils. *Vision Res.*, **21**, 1129–1148.

Catania, A. C. (1964). On the visual acuity of the pigeon. *J. Exper. Analysis of Behavior*, **7**, 361–366.

Chadhuri, A., Hallett, P. E. and Parker, J. A. (1983). Aspheric curvatures, refractive indices and chromatic aberration for the rat eye. *Vision Res.*, **23**, 1351–1364.

Chan, D. Y. C., Ennis, J. P., Pierscionek, B. K. and Smith, G. (1988). Determination and modelling of the 3-D gradient refractive indices in crystalline lenses. *Applied Optics*, **27**, 926–931.

Charman, W. N. (1983). The retinal image in the human eye. *Progress in Retinal Res.*, **1**, 1–50.

Charman, W. N. (1986). Static accommodation and the minimum angle of resolution. *Am. J. Optom. Physiol. Opt.*, **63**, 915–921.

Charman, W. N. (1989). Light on the peripheral retina. *Ophthal. Physiol. Opt.*, **9**, 91–92.

Charman, W. N. and Jennings, J. A. M. (1976). Objective measurements of the longitudinal chromatic aberration of the human eye. *Vision Res.*, **16**, 999–1005.

Charman, W. N. and Tucker, J. (1973). The optical system of the goldfish eye. *Vision Res.*, **13**, 1–8.

Cheney, F. E. and Thibos, L. N. (1987). Orientation anisotropy for the detection of aliased patterns by peripheral vision is optically induced. *J. Opt. Soc. Am. A*. **4**, 92.

Citron, M. C. and Pinto, L. H. (1973). Retinal image: larger and more illuminous for a diurnal lizard. *Vision Res.*, **13**, 873–876.

Dartnall, H. J. A. (1975). Assessing the fitness of visual pigments for their photic environment. In: *Vision in Fishes*. ed. Ali, M. A. pp. 543–563. New York: Plenum.

De Graauw, J. G. and van Hof, N. W. (1980). Frontal myopia in the rabbit. *Behavioral Brain Res.*, **1**, 339–341.

Delius, J. D., Perchard, R. J. and Emmerton, J. (1976). Polarized light discrimination by pigeons and an electroretinographic correlated. *J. Comp. Physiol. Psychol.*, **90**, 65–69.

DeMott, D. W. (1959). Direct measures of the retinal image. *J. Opt. Soc. Am.*, **49**, 571–579.

DeMott, D. W. and Boynton, R. M. (1958). Retinal distribution of entopic stray light. *J. Opt. Soc. Am.*, **48**, 13–22.

Denton, E. J. and Nicol, J. A. C. (1964). The chorioidal tapeta of some cartilaginous fishes. (Chondrichthyes). *J. Mar. Biol. Ass. UK*, **44**, 219–258.

Douglas, R. H. (1989). The spectral transmission of the lens and cornea of the brown trout (*Salmo trutta*) and goldfish (*Carassius auratus*) – effect of age and implications for ultraviolet vision. *Vision Res.*, **29**, 861–869.

Douglas, R. H., Bowmaker, J. K. and Kunz, Y. W. (1989). Ultraviolet vision in fish. In: *Seeing Contour and Colour*. eds. Kulikowski, J. J., Dickinson, C. V. M. and Murray, I. J. pp. 601–616. Oxford: Pergamon.

Douglas, R. H., Collett, T. S., Wagner, H. J. (1986). Accommodation in anuran amphibia and its role in depth vision. *J. Comp. Physiol. A*, **158**, 133–143.

Dral, A. D. G. (1975). Vision in cetacea. *J. Zoo Animal Medicine*, **6**, 17–21.

Drasdo, N. and Fowler, C. W. (1974). Non-linear projection of the retinal image in a wide-angle schematic eye. *Br. J. Ophthalmol.*, **58**, 709–714.

Drasdo, N. and Peaston, W. C. (1980). Sampling systems for visual field assessment and computerised perimetry. *Br. J. Ophthalmol.*, **64**, 705–712.

Duke-Elder, S. (1958). *System of Ophthalmology*, vol. 1: *The Eye in Evolution*. London: Kimpton.

Du Pont, J. S. and de Groot, P. J. (1976). The quality of the optic system of the frog's eye (*Rana esculenta*). *Vision Res.*, **16**, 1179–1181.

Eizenman, M., Hallett, P. E. and Frecker, R. C. (1985). Power spectra for ocular drift and tremor. *Vision Res.*, **25**, 1635–1640.

El Hage, S. and Berny, F. (1973). Contribution of the crystalline lens to the spherical aberration of the eye. *J. Opt. Soc. Am.*, **63**, 205–211.

Emsley, H. H. (1952). *Visual Optics*, vol. 1. London: Hatton.

Enoch, J. M. (1963). Optical properties of retinal receptors. *J. Opt. Soc. Am.*, **53**, 71–85.

Enoch, J. M. (1967). The retina as a fibre optics bundle. In: *Fiber Optics*. ed. Kapany, N. S. pp. 372–396. New York: Academic.

Enoch, J. M. and Tobey, F. L. (1973). A special microscope microspectrophotometer: optical design and application to the determination of waveguide properties of frog rods. *J. Opt. Soc. Am.*, **63**, 1345–1356.

Enoch, J. M. and Tobey, F. L., eds. (1981). *Vertebrate Photoreceptor Optics*. Berlin: Springer.

Erichsen, J. T. (1979). *How Birds Look at Objects*. D.Phil. Thesis, Oxford University.

Fernald, R. D. and Wright, S. E. (1985a). Growth of the visual system in the African cichlid fish, *Haplochromis burtoni*: optics. *Vision Res.*, **25**, 155–161.

Fernald, R. D. and Wright, S. E. (1985b). Growth of the visual system in the African cichlid fish, *Haplochromis burtoni*: accommodation. *Vision Res.*, **25**, 163–170.

Feuk, T. (1970). On the transparency of the stroma in the mammalian cornea. *IEEE Trans. Bio-Medical Engng.*, BME-17, 186–190.

Fisher, R. F. (1969). The significance of the shape of the lens and capsular energy changes in accommodation. *J. Physiol. (Lond.)*, **201**, 21–47.

Fitzke, F. W. (1981). Optical properties of the eye (abstract). *Invest. Ophthalmol. Vis. Sci.*, **20** (suppl.), 144.

Fitzke, F. W. (1991). A representational schematic eye. In *Modelling the Eye with Gradient Index Optics*. ed. Hughes, P. Cambridge: Cambridge University Press.

Fitzke, F., Hayes, B. P., Hodos, W., Holden A. L. (1985a). Electrophysiological optometry using Scheiner's principle in the pigeon eye. *J. Physiol. (Lond.)*, **369**, 17–31.

Fitzke, F. W., Hayse, B. P., Hodos, W., Holden, A. L. and Low, J. C. (1985b). Refractive sectors in the visual field of the pigeon eye. *J. Physiol. (Lond.)*, **369**, 33–44.

Fleishman, L. J., Howland, H. C., Howland, M. J., Rand, A. S. and Davenport, M. L. (1988). Crocodiles don't focus underwater. *J. Comp. Physiol. A*, **163**, 441–443.

Fletcher, A., Murphy, T. and Young A. (1954). Solutions of two optical problems. *Proc. R. Soc. Lond. A*, **223**, 216–225.

Franz, V. (1934). Vergleichende Anatomie des Wirbeltierauges. In *Handbuch der vergliechenden Anatomie des Wirbeltiere*, vol. 2. eds. Bolk, L., Goppert, E., Kallius, E. and Lubosch, W. pp. 989–1292. Berlin: Urban and Schwarzenberg.

Freeman, M. H. (1990). *Optics*, 10th edn. London: Butterworths.

Froriep, D. O. (1906). *Handbuch des vergleichende und experimentale Entwicklungslehre des Wirbeltiere*. Jena, 2.

Gaydon, A. G. (1938). Colour sensations produced by ultra-violet light. *Proc. Phys. Soc. Lond.*, **50**, 714–720.

Geeraets, W. J. and Berry, E. R. (1968). Ocular spectral characteristics as related to hazards from lasers and other sources. *Am. J. Ophthalmol.*, **66**, 15–20.

Glickstein, M. and Millodot, M. (1970). Retinoscopy and eye size. *Science*, **168**, 605–606.

Goldsmith, T. H., Collins, J. S. and Licht, S. (1984). The cone oil droplets of avian retinas. *Vision Res.*, **24**, 1661–1671.

Goodge, W. R. (1960). Adaptation for amphibious vision in the dipper (*Cinclus mexicanus*). *J. Morphol.*, **107**, 79–91.

Granda, A. M. and Dvorak, C. A. (1977). Vision in turtles. In: *Handbook of Sensory Physiology*, vol. VII/5: *The Visual System in Vertebrates*. ed. Crescetelli, F. pp. 451–495. Berlin: Springer.

Green, D. G., Powers, M. K. and Banks, M. S. (1980). Depth of focus, eyesize and visual acuity. *Vision Res.*, **20**, 827–835.

Gubisch, R. W. (1967). Optical performance of the human eye. *J. Opt. Soc. Am.*, **57**, 407–415.

Guidarelli, S. (1972). Off-axis imaging in the human eye. *Atti d. Fond. G. Ronchi*, **37**, 449–460.

Gundlach, R. H., Chard, R. D. and Skahen, J. R. (1945). The mechanism of accommodation in pigeons. *J. Comp. Psychol.*, **38**, 28–42.

Ham, W. T., Mueller, H. A. and Sliney, D. H. (1976). Retinal sensitivity to damage from short wavelength light. *Nature Lond.*, **260**, 153–154.

Harkness, L. (1977). Chameleons use accommodation cues to judge distance. *Nature Lond.*, **267**, 346–351.

Harkness, L. and Bennet-Clark, H. C. (1978). The deep fovea as a focus indicator. *Nature Lond.*, **272**, 814–816.

Hart, R. W. and Farrell, R. A. (1969). Light scattering in the cornea. *J. Opt. Soc. Am.*, **59**, 766–774.

Hayes, B. P., Hodos, W., Holden, A. L. and Low, J. C. (1987). The projection of the visual field upon the retina of the pigeon. *Vision Res.*, **27**, 31–40.

Helmholtz, H. von (1856–66). *Handbuch der Physiologischen Optik*. ed. Gullstrand, A., Kries, J. and Nagel, V., 3rd edn (1909). Reprint: New York: Dover (1962) of translation by Southall, J. P. C., for Optical Society of America (1924).

Hemenger, R. P. (1984). Intraocular light scatter in normal visual loss with age. *Applied Optics*, **23**, 1972–1974.

Hemmings, C. C. and Lythgoe, J. N. (1964). Better visibility for divers in dark water. *Triton*, **9** (4), 28–31.

Henderson, S. T. (1977). *Daylight and its Spectrum*. 2nd edn. Bristol: Hilger.

Hirsch, J. and Hylton, R. (1984). Quality of primate photoreceptor lattice and limits of spatial vision. *Vision Res.*, **24**, 347–355.

Hirsch, J. and Miller, W. H. (1987). Does cone positional disorder limit resolution? *J. Opt. Soc. Am. A*, **4**, 1481–1492.

Hodos, W. and Erichsen, J. T. (1990). Lower-field myopia in birds: an adaptation that keeps the ground in focus. *Vision Res.*, **30**, 653–657.

Holden, A. L., Hayes, B. P. and Fitzke, F. W. (1987). Retinal magnification factor at the ora terminalis: a structural study of human and animal eyes. *Vision Res.*, **27**, 1229–1235.

Howarth, P. A. (1984). The lateral chromatic aberration of the eye. *Ophthal. Physiol. Opt.*, **4**, 223–226.

Howland, H. C. and Howland, B. (1974). Photorefraction: A technique for the study of refractive state at a distance. *J. Opt. Soc. Am.*, **64**, 240–249.

Howland, H. C. and Sivak, J. G. (1984). Penguin vision in air and water. *Vision Res.*, **24**, 1905–1909.

Hueter, R. T. and Gruber, S. H. (1980). Retinoscopy of aquatic eyes. *Vision Res.*, **20**, 197–200.

Hughes, A. (1972). A schematic eye for the rabbit. *Vision Res.*, **12**, 123–138.

Hughes, A. (1977a). The topography of vision in mammals of contrasting life style: comparative optics and retinal organisation. In: *Handbook of Sensory Physiology*, vol. VII/5: *The Visual System in Vertebrates*. ed. Crescetelli, F. pp. 613–756. Berlin: Springer.

Hughes, A. (1977b). The refractive state of the rat eye. *Vision Res.*, **17**, 927–939.

Hughes, A. (1979a). A useful table of reduced schematic eyes for vertebrates which includes computed longitudinal chromatic aberrations. *Vision Res.*, **19**, 1273–1275.

Hughes, A. (1979b). Artefact of retinoscopy in the rat and rabbit eye has its origin at the retina/vitreous interface rather than in longitudinal chromatic aberration. *Vision Res.*, **19**, 1293–1294.

Hughes, A. (1979c). A schematic eye for the rat. *Vision Res.*, **19**, 569–588.

Hughes, A. (1986). The schematic eye comes of age. In: *Visual Neuroscience*. eds. Pettigrew, J. D., Sanderson, K. J. and Levick, W. R. pp. 60–89. Cambridge: Cambridge University Press.

Hughes, A. and Vaney, D. L. (1978). The refractive state of the rabbit eye: variation with eccentricity and correction for oblique astigmatism. *Vision Res.*, **18**, 1351–1355.

Ijspeert, J. K., De Waard, P. W. T., Van den Berg, T. J. T. P. and DeJong, P. T. V. M. (1990). The intraocular straylight function in 129 healthy volunteers: dependence on angle, age and pigmentation. *Vision Res.*, **30**, 699–707.

Ivanoff, A. (1956). About the spherical aberration of the eye. *J. Opt. Soc. Am.*, **46**, 901–903.

Ives, J. T., Normann, R. A. and Barber, P. W. (1983). Light intensification by cone oil droplets: electromagnetic considerations. *J. Opt. Soc. Am.*, **73**, 1725–1731.

Jacobs, D. S. and Blakemore, C. (1988). Factors limiting the postnatal development of visual acuity in the monkey. *Vision Res.*, **28**, 947–958.

Jagger, W. S. (1985). Visibility of photoreceptors in the intact living cane toad eye. *Vision Res.*, **25**, 729–731.

Jagger, W. S. (1988). Optical quality of the eye of the cane toad *Bufo marinus*. *Vision Res.*, **28**, 105–114.

Jagger, W. S. (1990). The refractive structure and optical properties of the isolated crystalline lens of the cat. *Vision Res.*, **30**, 723–738.

Jay, B. S. (1962). The effective pupillary area at varying perimetric angles. *Vision Res.*, **2**, 418–428.

Jean, J. N. and O'Brien, B. (1949). Microwave test of a theory of the Stiles Crawford effect. *J. Opt. Soc. Am.*, **39**, 1057.

Jerlov, N. G. (1968). *Optical Oceanography*. London: Elsevier.

Kapany, N. S. (1967). *Fiber Optics*, pp. 88–99. New York: Academic.

Kölmer, W. (1924). Uber das Auge des Eisvogels (*Alcedo attis attis*). *Pfluger's Arch. Ges. Physiol.*, **204**, 266–274.

Kondrashev, S. L., Gamburtzeva, A. G., Gnjubkina, V. P., Orlov, O. J. and Pham Thi My (1986). Coloration of corneas in fish. A list of species. *Vision Res.*, **26**, 287–290.

Kooijman, A. C. (1983). Light distribution on the retina of a wide-angle theoretical eye. *J. Opt. Soc. Am.*, **73**, 1544–1550.

Kooijman, A. C. and Witmer, F. K. (1986). Ganzfield light distribution on the retina of human and rabbit eyes: calculations and *in vitro* measurements. *J. Opt. Soc. Am.*, **73**, 1544–1550.

Kreithen, M. I. and Keeton, W. T. (194). Detection of polarized light by the homing pigeon, *Columba livia*. *J. Comp. Physiol.*, **89**, 83–92.

Kreuzer, R. O. and Sivak, J. G. (1985). Chromatic aberration of the vertebrate lens. *Ophthal. Physiol. Opt.*, **5**, 33–41.

Krueger, H. and Moser, E. A. (1972). The influence of the modulation transfer function of the dioptric apparatus of the acuity and contrast of the retinal image in *Rana esculenta*. *Vision Res.*, **12**, 1281–1289.

Land, M. F. (1972). The physics and biology of animal reflectors. *Progr. Biophys. Molec. Biol.*, **24**, 75–106.

Land, M. F. (1980). Compound eyes: old and new mechanisms, *Nature Lond.*, **287**, 681–686.

Land, M. F. (1981). Optics and vision in invertebrates. In *Handbook of Sensory Physiology*, vol. VII/6B: *Comparative Physiology and Evolution of Vision in Invertebrates. B. Invertebrate Visual Centers and Behavior 1*. ed. Autrum, H. pp. 471–593. Berlin: Springer.

Land, M. F. (1988). Paradoxical superposition. *Nature Lond.*, **332**, 15–16.

Land, M. F. and Snyder, A. W. (1985). Cone mosaic observed directly through natural pupil of live vertebrate. *Vision Res.*, **25**, 1519–1523.

Langston, A., Casagrande, V. A. and Fox, R. (1986). Spatial resolution of the galago. *Vision Res.*, **26**, 791–796.

Laties, A. M. and Enoch, J. M. (1968). Photoreceptor orientation in the primate eye. *Nature Lond.*, **218**, 172–173.

Le Grand, Y. (1967). *Form and Space Vision*. Transl. Millodot, M. and Heath, G. C. Bloomington: Indiana University Press.

Levy, B. and Sivak, J. G. (1980). Mechanisms of accommodation in the bird eye. *J. Comp. Physiol.*, **137**, 267–272.

Locket, N. A. (1977). Adaptations to the deep-sea environment. In: *Handbook of Sensory Physiology*, Vol. VII/5: *The Visual System in Vertebrates*. ed. Crescetelli, F. pp. 67–192. Berlin: Springer.

Lotmar, W. (1971). Theoretical eye model with aspherics. *J. Opt. Soc. Am.*, **61**, 1522–1529.

Lythgoe, J.N. (1971). Irridescent corneas in fish. *Nature Lond.*, **233**, 205–207.

Lythgoe, J. N. (1972). The adaptation of visual pigments to the photic environment. In: *Handbook of Sensory Physiology*, vol. VII/1: *Photochemistry of Vision*. ed. Dartnall, H. J. pp. 566–603. Berlin: Springer.

Lythgoe, J. N. (1975). The structure and function of iridescent corneas in teleost fishes. *Proc. R. Soc. Lond. B*, **188**, 437–457.

Lythgoe, J. N. (1979). *The Ecology of Vision*. Oxford: Clarendon.

Lythgoe, J. N. (1984). Visual pigments and environmental light. *Vision Res.*, **24**, 1539–1550.

McFadden, S. A. and Reymond, L. (1985). A further look at the binocular visual field of the pigeon (*Columba livia*). *Vision Res.*, **25**, 1741–1746.

Maffei, L., Fiorentini, A. and Bisti, S. (1990). The visual acuity of the lynx. *Vision Res.*, **30**, 527–528.

Makaretz, M. and Levine, R. L. (1980). A light microscopic study of the bifoveate retina in the lizard *Anolis carolinensis*: general observations and convergence ratios. *Vision Res.*, **20**, 679–686.

Mandelman, T. and Sivak, J. G. (1983). Longitudinal chromatic aberration of the vertebrate eye. *Vision Res.*, **23**, 1555–1559.

Marshall, J., Mellerio, J. and Palmer, D. A. (1973). A schematic eye for the pigeon. *Vision Res.*, **13**, 2449–2453.

Martin, G. R. (1983). Schematic eye models in vertebrates. In: *Progress in Sensory Physiology*, vol. 4. eds. Autrum, H., Ottoson, D., Perl, E. R., Schmidt, R., Shimazu, H. and Willis, W. D. Berlin: Springer.

Martin, G. R. (1984). The visual fields of the tawny owl, *Strix aluco* L. *Vision Res.*, **24**, 1739–1751.

Martin, G. R. (1986). Total panoramic vision in the mallard duck, *Anas platyrhynchos*. *Vision Res.*, **26**, 1303–1305.

Martin, G. R. and Young, S. R. (1983). The retinal binocular field of the pigeon (*Columba livia*: English racing homer). *Vision Res.*, **23**, 911–915.

Mathis, U., Schaeffel, F. and Howland, H. C. (1988). Visual optics in toads (*Bufo americanus*). *J. Comp. Physiol. A*, **163**, 201–213.

Matthiessen, L. (1880). Untersuchungen über den Aplanatismus und die Periscopie der Krystallinsen in den Augen der Fische. *Pflüger's Arch. Ges. Physiol.*, **21**, 287–307.

Maurice, D. M. (1957). The structure and transparency of the cornea. *J. Physiol. (Lond.)*, **136**, 263–286.

Maurice, D. M. (1969). The cornea and sclera. In: *The Eye*. vol. 1. ed. Davson, H. New York: Academic.

Maurice, D. M. (1970). The transparency of the corneal stroma. *Vision Res.*, **10**, 107–108.

Mecherikunnel, A. T., Gatlin, J. A. and Richmond, J. C. (1983). Data on total and spectral solar irradiance. *Applied Optics*, **22**, 1354–1359.

Mellerio, J. (1987). Yellowing of the human lens: nuclear and cortical contributions. *Vision Res.*, **27**, 1581–1587.

Meyer, D. B. (1977). The avian eye and its adaptations. In: *Handbook of Sensory Physiology*, vol. VII/5: *The Visual System in Vertebrates*. ed. Crescetelli F. pp. 549–611. Berlin: Springer.

Meyer, G. E. and Salinsky, M. C. (1977). Refraction of the rat: estimation by pattern evoked visual cortical potentials. *Vision Res.*, **17**, 883–885.

Meyer, D. L. and Schwassman, H. O. (1970). Electrophysiological method for determination of refractive state in fish eyes. *Vision Res.*, **10**, 1301–1303.

Miles, F. A. and Wallman, J. (1990). Local ocular compensation for imposed local refractive error. *Vision Res.*, **30**, 339–349.

Miller, W. H., and Bernard, G. D. (1983). Averaging over the foveal receptor aperture curtails aliasing. *Vision Res.*, **23**, 1365–1369.

Miller, W. H. and Snyder, A. W. (1977). The tiered vertebrate retina. *Vision Res.*, **17**, 239–255.

Millodot, M. (1971). Measurement of the refractive state of the eye in frogs (*Rana pipiens*). *Rev. Can. Biol.*, **30**, 249–252.

Millodot, M. and Blough, P. M. (1971). The refractive state of the pigeon eye. *Vision Res.*, **11**, 1019–1022.

Millodot, M. and Sivak, J. (1978). Hypermetropia of small animals and chromatic aberration. *Vision Res.*, **18**, 125–126.

Millodot, M. and Sivak, J. G. (1979). Contribution of the cornea and lens to the spherical aberration of the eye. *Vision Res.*, **19**, 685–687.

Millot, J. and Carasso, N. (1955). Note préliminaire sur l'oeil de *Latimeria chalumnae* (Crossoptérygien coelacanthidé). *C. R. Acad. Sci. Paris*, **241**, 576–577.

Moreland, J. D. and Lythgoe, J. N. (1968). Yellow corneas in fish. *Vision Res.*, **8**, 1377–1380.

Moser, E. A. and Krueger, H. (1972). Retinoscopic and neurophysiological refractometry in *Rana temporia*. *Pflugers Arch.*, **335**, 235–242.

Munk, O. (1969). The eye of the 'four-eyed' fish *Dialommus fuscus*. *Vidensk. Meddr. dansk naturh. Foren.*, **132**, 7–24.

Munk, O. (1973). Early notions of dynamic accommodatory devices in teleosts. *Vidensk, Meddr. dansk naturh. Foren.*, **136**, 7–28.

Muntz, W. R. A. (1972). Inert absorbing and reflecting pigments. In: *Handbook of Sensory Physiology*, vol. VII/1: *Photochemistry of Vision*. ed. Dartnall, H. J. A. pp. 529–565. Berlin: Springer.

Muntz, W. R. A. (1975). The visual consequences of yellow filtering pigments in the eyes of fishes occupying different habitats. In: *Light as an Ecological Factor*, Vol. II. ed. Evans, G. C., Bainbridge, R. and Rackham, O. pp. 271–287. Oxford: Blackwell.

Munz, F. W. and McFarland, W. N. (1977). Evolutionary adaptations of fishes to the photic environment. In: *Handbook of Sensory Physiology*, vol. VII/5: *The Visual System in Vertebrates*. ed Crescetelli, F., pp. 193–274. Berlin: Springer.

Murphy, C. J. and Howland, H. C. (1986). On the Gekko pupil and Scheiner's disc. *Vision Res.*, **26**, 815–816.

Murphy, C. J., Bellhorn, R. W., Williams, T., Burns, M. S., Schaeffel, R. and Howland, H. C. (1990). Refractive state, ocular anatomy, and accommodative range of the sea otter (*Enhydra lutris*). *Vision Res.*, **30**, 23–32.

Murphy, C. J., Howland, H. C., Kwiecinski, G. G., Kern, T. and Kallen, F. (1983). Visual accommodation in the flying fox (*Pteropus giganteus*). *Vision Res.*, **23**, 617–620.

Nakao, S., Fujmoto, S., Nagata, R. and Iwata, K. (1968). Model of refractive index distribution in the rabbit crystalline lens. *J. Opt. Soc. Am.*, **58**, 1125–1130.

Neuweiler, G. (1962). Bau und leistung des Flughundauges (*Pteropus giganteus* gig. Brunn.). *Z. Vergl. Physiol.*, **46**, 13–56.

Newman, E. A. and Hartline, P. H. (1982). The infra-red 'vision' of snakes. *Sci. Am.*, **246**, March, 98–107.

Nicol, J. A. C. and Arnott, H. J. (1974). Tapeta lucida in the eyes of goatsuckers (Caprimulgaidae). *Proc. R. Soc. Lond. B.*, **187**, 349–352.

Nicol, J. A. C. (1981). Tapeta lucida of vertebrates. In: *Vertebrate Photoreceptor Optics*. ed. Enoch, J. M. and Tobey, F. L. pp. 401–431. Berlin: Springer.

Nilsson, D.-E. (1988). A new type of imaging optics in compound eyes. *Nature Lond.*, **332**, 76–78.

Nilsson, D.-E. (1990). From cornea to retinal image in invertebrate eyes. *Trends Neurosci.*, **13**, 55–64.

Nuboer, J. F. W. and van Genderen-Takken, H. (1978). The artifact of retinoscopy. *Vision Res.*, **18**, 1091–1096.

Nuboer, J. F. W., Bos, H., van Genderen-Takken, H., van den Hoeven, H. and van Steenbergen, J. C. (1979). Retinoscopy and chromatic aberration. *Experientia*, **35**, 1066–1067.

Nye, P. W. (1973). On the functional differences between frontal and lateral visual fields of the pigeon. *Vision Res.*, **13**, 559–574.

O'Day, W. T. and Fernandez, H. R. (1974). *Aristostomias scintillans* (Malacosteidae): a deep-sea fish with visual pigments apparently adapted to its own bioluminescence. *Vision Res.*, **14**, 545–550.

Ogboso, Y. U. and Bedell, H. E. (1987). Magnitude of lateral chromatic aberration across the retina of the human eye. *J. Opt. Soc. Am. A.*, **4**, 1666–1672.

Olson, E. C. (1977). The history of vertebrates. In: *Handbook of Sensory Physiology*, vol. VII/5: *The Visual System in Vertebrates*. ed. Crescetelli, F. pp. 1–45. Berlin: Springer.

O'Neill, E. L. (1963). *Introduction to Statistical Optics*. Reading, Mass: Addison-Wesley.

Orlov, O. J. and Gamburtzeva, A. G. (1976). Changeable coloration of the cornea in the fish *Hexagrammos octogrammus*. *Nature Lond.*, **263**, 405–406.

Oswaldo-Cruz, E., Hokoc, J. N. and Sousa, A. P. B. (1979). A schematic eye for the opossum. *Vision Res.*, **19**, 263–278.

Pak, M. A. (1984). Ocular refraction and visual contrast sensitivity of the rabbit, determined by the VECP. *Vision Res.*, **24**, 341–345.

Petry, H. M., Fox, R. and Casagrande, V. A. (1984). Spatial contrast sensitivity of the tree shrew. *Vision Res.*, **24**, 1037–1042.

Pflibsen, K. P., Pomarentzeff, O. and Ross, R. N. (1988). Retinal illuminance using a wide-angle model of the eye. *J. Opt. Soc. Am. A*, **5**, 146–150.

Picanco-Diniz, C. W., Silveira, L. C. L. and Oswaldo-Cruz, E. (1983). Electrophysiological determination of the refractive state of the eye of the opossum. *Vision Res.*, **23**, 867–872.

Pierscionek, B. K., Chan, D. Y. C., Ennis, J. P., Smith, G. and Augusteyn, R. C. (1988). Non-destructive method of constructing three-dimensional gradient index models for crystalline lenses: Theory and experiment. *Am. J. Optom. Physiol. Opt.*, **65**, 481–491.

Piggins, D. (1970). Refraction of the harp seal. *Pagophilus groenlandicus* (Erxleben, 1777). *Nature Lond.*, **227**, 78–79.

Pirenne, M. H. (1967). *Vision and the Eye*. London: Chapman & Hall.

Polyak, S. (1957). *The Vertebrate Visual System*. Chicago: University of Chicago Press.

Pomarentzeff, O., Fish, H., Govignon, J. and Schepens, C. L. (1971). Wide angle model of the human eye. *Ann. Ophthalmol.*, **3**, 815–819.

Pumphrey, R. J. (1948). The theory of the fovea. *J. exp. Biol.*, **25**, 299–312.

Reading, V. M. and Weale, R. A. (1974). Macular pigment and chromatic aberration. *J. Opt. Soc. Am.*, **64**, 231–234.

Remtulla, S. and Hallett, P. E. (1985). A schematic eye for the mouse, and comparisons with the rat. *Vision Res.*, **25**, 21–31.

Reymond, L. (1985). Spatial visual acuity of the eagle, *Aquila audax*: a behavioural, optical and anatomical investigation. *Vision Res.*, **25**, 1477–1491.

Reymond, L. (1987). Spatial visual acuity of the falcon, *Falco berigora*: a behavioural, optical and anatomical investigation. *Vision Res.*, **27**, 1859–1874.

Ribi, W. A. (1980). The phenomenon of eye glow. *Endeavour*, **5**, 2–7.

Rivamonte, L. A. (1976). Eye model to account for comparable aerial and underwater acuities of the bottlenose dolphin. *Netherlands J. Sea Res.*, **10**, 491–498.

Robson, J. G. and Enroth-Cugell, C. (1978). Light distribution in the cat's retinal image. *Vision Res.*, **18**, 159–173.

Rochon-Duvigneaud, A. (1943). *Les Yeux et la Vision des Vertébrés*. Paris: Masson.

Röhler, R. (1962). Die Abbildungseigenschaften der Augenmedien. *Vision Res.*, **2**, 391–429.

Rose, A. (1953). Quantum and noise limitations of the visual process. *J. Opt. Soc. Am.*, **43**, 715–716.

Ruddock, K. H. (1972). Light transmission through the ocular media and macular pigment and its significance for psychophysical investigation. In: *Handbook of Sensory Physiology*, vol. VIII/4:

Visual Psychophysics. ed. Jameson, D. and Hurvich, L. M. pp. 455–469. Berlin: Springer.

Schaeffel, F. and Howland, H. C. (1987). Corneal accommodation in chick and pigeon. *J. Comp. Physiol. A*, **160**, 375–384.

Schaeffel, F. and Howland, H. C. (1988). Mathematical model of emmetropization in the chicken. *J. Opt. Soc. Am.*, **A5**, 2080–2086.

Schaeffel, F., Howland, H. C. and Farkas, L. (1986). Natural accommodation in the growing chicken. *Vision Res.*, **26**, 1977–1993.

Schor, C. M. and Ciuffreda, K. J., eds. (1983). *Vergence Eye Movements: Basic and Applied Aspects*. London: Butterworths.

Schusterman, R. J. and Balliet, R. F. (1971). Aerial and underwater visual acuity in the California sea lion (*Zalophus californianus*) as a function of luminance. *Ann. NY Acad. Sci.*, **188**, 37–46.

Schusterman, R. J. and Barrett, B. (1973). Amphibious nature of visual acuity in the Asian 'clawless' otter. *Nature Lond.*, **244**, 518–519.

Schwassmann, H. O. and Kruger, L. (1965). Experimental analysis of the visual system of the four-eyed fish *Anableps microlepsis*. *Vision Res.*, **5**, 269–281.

Seltner, R. L., Weerheim, J. A. and Sivak, J. G. (1989). Role of the lens and vitreous humour in the refractive properties of the eyes of three strains of goldfish. *Vision Res.*, **29**, 681–685.

Selwyn, E. W. H. (1948). The photographic and visual resolving power of lenses. Part 1. Visual resolving power. *Photogr. J.*, **88B**, 6–12.

Shand, J. and Lythgoe, J. N. (1987). Light-induced changes in corneal iridescence in fish. *Vision Res.*, **27**, 303–305.

Shapley, R. and Gordon, J. (1980). The visual sensitivity of the retina of the conger eel. *Proc. R. Soc. Lond. B*, **209**, 317–330.

Shlaer, R. (1972). An eagle's eye: quality of the retinal image. *Science Wash. DC*, **176**, 920–922.

Sidman, R. (1957). The structure and concentration of solids in photoreceptor cells studied by refractometry and interference microscopy. *J. Biophys. Biochem. Cytol.*, **3**, 15–30.

Sivak, J. G. (1974). The refractive error of the fish eye. *Vision Res.*, **14**, 209–213.

Sivak, J. G. (1975). Accommodative lens movements in fishes: movement along the pupil axis vs movement along the pupil plane. *Vision Res.*, **15**, 825–828.

Sivak, J. G. (1976). The accommodative significance of the 'ramp' retina of the eye of the stingray. *Vision Res.*, **16**, 945–950.

Sivak, J. G. (1977). The role of the spectacle in the visual optics of the snake eye. *Vision Res.*, **17**, 293–298.

Sivak, J. G. (1980a). Avian mechanisms for vision in air and water. *Trends Neurosci.*, **3**, 314–317.

Sivak, J. G. (1980b). Accommodation in vertebrates: A contemporary survey. *Current Topics in Eye Research*, **3**, 281–330.

Sivak, J. G. and Allen, D. B. (1975). An evaluation of the ramp retina of the horse eye. *Vision Res.*, **15**, 1353–1356.

Sivak, J. G. and Kreuzer, R. O. (1983). Spherical aberration of the crystalline lens. *Vision Res.*, **23**, 59–70.

Sivak, J. G. and Mandelman, T. (1982). Chromatic dispersion of the ocular media. *Vision Res.*, **22**, 59–70.

Sivak, J. G., Bobier, W. R. and Levy, B. (1978). The refractive significance of the nictitating membrane of the bird eye. *J. Comp. Physiol.*, **125**, 335–339.

Sivak, J. G., Gur, M. and Dovrat, A. (1983). Spherical aberration of the lens of the ground squirrel. (*Spermophilis tridecemlineatus*). *Ophthal. Physiol. Opt.*, **3**, 261–265.

Sivak, J. G., Hildebrand, T. and Lebert, C. (1985). Magnitude and rate of accommodation in diving and nondiving birds. *Vision Res.*, **25**, 925–933.

Sivak, J., Howland, H. C. and McGill-Harelstad, P. (1987). Vision of the Humboldt penguin (*Spheniscus humboldti*) in air and water. *Proc. R. Soc. Lond. B*, **229**, 467–472.

Sliney, D. H. (1977). The ambient light environment and ocular hazards. In: *Retinitis Pigmentosa*. eds. Landers, M. B., Wolbarsht, M. L., Dowling, J. E. and Laties, A. M. New York: Plenum.

Smith, E. L. and Harwerth, R. S. (1984). Behavioural measurements of accommodative amplitude in rhesus monkeys. *Vision Res.*, 24, 1821–1827.

Snyder, A. W. and Miller, W. H. (1977). Photoreceptor diameter and spacing for highest resolving power. *J. Opt. Soc. Am.*, 67, 696–698.

Snyder, A. W. and Miller, W. H. (1978). Telephoto lens system of falconiform eyes. *Nature Lond.*, 275, 127–129.

Snyder, A. W. and Pask, C. (1973). The Stiles-Crawford effect – explanation and consequences. *Vision Res.*, 13, 115–1137.

Snyder, A. W., Bossomaier, T. R. J. and Hughes, A. (1986). Optical image quality and the cone mosaic. *Science Wash. DC*, 231, 499–501.

Soemmerring, D. W. (1818). *De Oculorum Hominis Animaliumque Sectione Horizontale Commentatio*. Gottingen: Vandenhoeck and Ruprecht.

Somiya, H. and Tamura, T. (1973). Studies on the visual accommodation of fish. *Jap. J. Icthyology*, 20, 193–206.

Spielman, S. L. and Grueber, S. H. (1983). Development of a contact lens for refracting aquatic animals. *Ophthal. Physiol. Opt.*, 3, 255–260.

Sroczynski, S. (1975). Die spharische Aberration der Augenlinse der Regenbogenforelle (*Salmo gairdneri* Rich.). *Zool. Jb. Physiol.*, 79, 204–212.

Sroczynski, S. (1977). Spherical aberration of crystalline lens in the roach. *Rutilus rutilus* L. *J. Comp. Physiol.*, 121, 135–144.

Sroczynski, S. (1979). Das optische System des Auges des Flussbarsches (*Perca fluviatilis* L.) *Zool. Jb. Physiol.*, 83, 224–252.

Steinbach, M. J. and Angus, R. G. (1974). Torsional eye movements of owls. *Vision Res.*, 14, 745–746.

Steinbach, M. J. and Money, K. E. (1973). Eye movements in the owls. *Vision Res.*, 13, 889–891.

Stiles, W. S. and Crawford, B. H. (1933). The luminous efficiency of rays entering the eye pupil at different points. *Proc. R. Soc. Lond. B*, 112, 428–450.

Sunderland, H. R. and O'Neill, W. D. (1976). Functional dependence of optical parameters on circumferential forces in the cat lens. *Vision Res.*, 26, 1151–1158.

Suthers, R. A. and Wallis, N. E. (1970). Optics of the eyes of echolocating bats. *Vision Res.*, 10, 1165–1173.

Tansley, K. H. (1965). *Vision in Vertebrates*. London: Chapman & Hall.

Tansley, K. and Johnson, B. K. (1956). The cones of the grass snake's eye. *Nature Lond.*, 178, 1285–1286.

Taylor, D. H. and Adler, K. (1973). Spatial orientation by salamanders using plane polarized light. *Science Wash. DC*, 181, 285–287.

Thibos, L. N. (1987). Calculation of the influence of lateral chromatic aberration on image quality across the visual field. *J. Opt. Soc. Am. A*, 4, 1673–1680.

Thibos, L. N., Walsh, D. J. and Cheney, F. E. (1987). Vision beyond the resolution limit: aliasing in the periphery. *Vision Res.*, 27, 2193–2197.

Tobey, F. L., Enoch, J. M. and Scandrett, J. H. (1975). Experimentally determined optical properties of goldfish cones and rods. *Invest. Ophthalmol.*, 154, 7–23.

Toraldo di Francia, G. (1949). Retina cones as dielectric antennas. *J. Opt. Soc. Am.*, 39, 324.

Toraldo di Francia, G. (1955). Resolving power and information. *J. Opt. Soc. Am.*, 45, 497–501.

Troilo, D. and Wallman, J. (1985). Mechanisms of accommodation in the chicken. *Soc. Neurosci. Abstr.*, 11, 1042.

Troilo, D. and Wallman, J. (1987). Changes in corneal curvature during accommodation in chicks. *Vision Res.*, 27, 241–247.

Tyler, J. E. and Smith, R. C. (1970). *Measurements of Spectral Irradiance Under Water*. New York: Gordon and Breach.

Vakkur, G. J. and Bishop, P. O. (1963). The schematic eye in the cat. *Vision Res.*, 3, 357–381.

Van den Berg, T. J. T. P. and Mooij, J. E. M. (1982). Eye media absorption in goldfish. *Vision Res.*, 22, 1229–1231.

Vos, J. J., Walraven, J. and Van Meeteren, A. (1976). Light profiles of the foveal image of a point source. *Vision Res.*, 16, 215–219.

Wald, G. (1949). The photochemistry of vision. *Doc. Ophthalmol.*, 3, 94–137.

Walls, G. L. (1937). Significance of the foveal depression. *Arch. Ophthalmol. NY*, 18, 912–919.

Walls, G. L. (1942). *The Vertebrate Eye and its Adaptive Radiation*. Bloomfield Hills, MI: Cranbrook.

Wässle, H. (1971). Optical quality of the cat eye. *Vision Res.*, 11, 995–1006.

Waterman, T. H. (1975). Natural polarized light and e-vector discrimination by vertebrates. In: *Light as an Ecological Factor*. eds. Evans, G. C., Bainbridge, R. and Rackham, O. Oxford: Blackwell. Oxford: Blackwell.

Waterman, T. H. (1981). Polarization sensitivity. In: *Handbook of Sensory Physiology*, vol. VII/6B: *Comparative Physiology and Evolution of Vision in Invertebrates. B. Invertebrate Visual Centers and Behavior I*. Ed. Autrum, H. Berlin: Springer.

Waterman, T. H. and Forward, R. B. (1970). Field evidence for polarized light sensitivity in the fish *Zenarchopterus*. *Nature Lond.*, 228, 85–87.

Weale, R. A. (1974). Natural history of optics. In: *The Eye*, vol. 6. eds. Davson, H. and Graham, L. T. pp. 1–110. London: Academic.

Weale, R. A. (1976). Ocular optics and evolution. *J. Opt. Soc. Am.*, 66, 1053–1054.

Westheimer, G. (1962). Linespread function of the living cat eye. *J. Opt. Soc. Am.*, 52, 1326.

Williams, D. R. (1986). Seeing through the photoreceptor mosaic. *Trends in Neuroscience*. May 1986, 193–198.

Williams, D. R. (1988). Topography of the foveal cone mosaic in the living human eye. *Vision Res.*, 28, 433–454.

Williams, D. R. and Collier, R. (1983). Consequences of spatial sampling by a human receptor mosaic. *Science*, 221, 385–387.

Williams, D. R., Collier, R. J. and Thompson, B. J. (1983). Spatial resolution of the short-wavelength mechanism. In: *Colour Vision*. ed. Mollon, J. D. and Sharpe, L. T. pp. 487–503. London: Academic.

Witkovsky, P. (1968). The effect of chromatic adaptation on color sensitivity of the carp. *Vision Res.*, 8, 823–837.

Wright, W. D. and Nelson, J. H. (1936). The relation between the apparent intensity of a beam of light and the angle at which it strikes the retina. *Proc. Phys. Soc.*, 48, 401–405.

Yellott, J. I. (1982). Spectral analysis of spatial sampling by photoreceptors: topological disorder prevents aliasing. *Vision Res.*, 22, 1205–1210.

Young, S. R. and Martin, G. R. (1984). Optics of retinal oil droplets: a model of light collection and polarization detection in the avian retina. *Vision Res.*, 24, 129–137.

Young, T. (1801). On the mechanism of the eye. *Phil. Trans. R. Soc. Lond. B*, 92, 23–88.

6 Optics of the Eyes of the Animal Kingdom

Michael F. Land

Introduction

Eye evolution seems to have proceeded in two stages. Eyes of some kind have been around for as long as the Metazoa have existed, judging from the presence of pigmented eye spots in coelenterates, platyhelminths and most of the other 'lower phyla'. It is now clear that rhodopsin itself, the molecule responsible for the absorption of light in eyes, is similar in amino acid composition throughout the Metazoa, and so one can presume a common molecular ancestor (Goldsmith, 1990). In contrast, eyes themselves do not seem to have a common origin. From details of anatomy and from known phylogeny, Salvini-Plawen and Mayr (1977) estimate that simple eye spots arose independently at least 40 and perhaps as many as 65 times. There is even a remarkable series of intracellular 'eyes' in the protozoan dinoflagellates (Warnowiidae) (Eakin, 1972; Greuet, 1984), although these are certainly not ancestors of metazoan eyes. However, eyes as we usually think of them – relatively large structures with an optical system and good resolution – are found in only 6 of the 33 metazoan phyla (the Cnidaria, Mollusca, Annelida, Onychophora, Arthropoda and Chordata) and are thus a relatively recent development dating from well into the Cambrian period, when the major phyla diverged. The six phyla that do have good eyes, capable of something more than just the selection of a congenial habitat, are also the most successful, accounting for 96% of known species (Barnes, 1987), and it is tempting to think that the attainment of optical 'lift-off' has contributed to this success.

The number of fundamentally different ways of producing an image in an eye is relatively small, somewhere between 8 and 10 (see Fig. 6.18). Some of them have been 'discovered' many times in evolution. The best example is perhaps the type of spherical lens found in fish (Figs. 6.3 and 6.4), which has evolved as many as 8 times in 4 different phyla. Convergence is thus common in eye evolution, and similarity of optical design is by no means a guarantee of common descent. Conversely, one kind of eye can evolve into another. Our own eyes, whose main optical structure is the cornea, evolved from aquatic eyes in which the lens provided all the power.

When we examine the different ways in which nature makes images, it is interesting to compare these with those available to human engineering. As in optical technology, all the known ways in which animal eyes form images are based on refracting lenses, on reflecting mirrors or, in the simplest cases, just on pinholes. However, when we look in detail at the way in which natural and manufactured lenses and mirrors are produced, the resemblances become less clear. A lens designer, trying to produce a lens with a minimum of defects, will usually use many curved surfaces and several different kinds of homogeneous glass; in contrast, the commonest well-corrected natural lens, the type found in the eyes of fish and in a modified form in our own eyes, is constructed from an optically inhomogeneous gradient of protein and water mixtures. Unpromising as this may sound to us, if the gradient is the correct one then the lens can be of excellent quality (Pumphrey, 1961). Similarly, natural mirrors are not made of metal – biological systems cannot reduce metals to their elemental state – but of multiple, quarter-wavelength-thick layers of materials with alternating high and low refractive indices. The scales of fish, for example, have multi-layer mirrors made from guanine ($n = 1.83$) and cytoplasm ($n = 1.34$) (Land, 1972). In fact, high-quality mirrors have been made this way for some time – but using vapour-deposited zinc sulphide and magnesium fluoride as the high- and low-index layers – and indeed, for a decade or so inhomogeneous lenses too have become a practical reality, so in some ways the gap between nature and technology has been closing. It would be wrong to think of the differences between natural and human optics just as matters of material composition; there are also important differences in overall design. The rather close resemblance between human eyes and cameras tends to obscure the fact that most other types of eyes are constructed on very different lines. For example, among the compound eyes of insects and crusta-

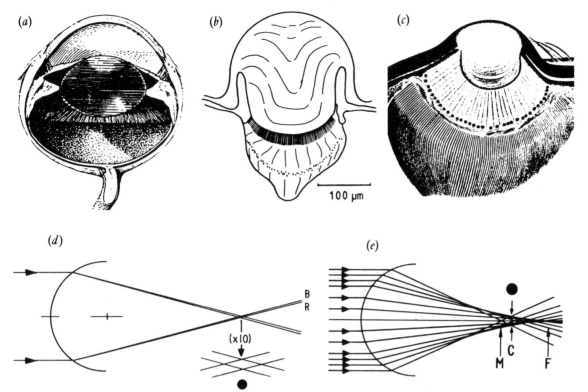

Fig. 6.1 *Eyes with optical systems based on refraction at a curved air–tissue interface: (a) lynx, (b) larva of a tiger beetle, (c) a jumping spider (various sources). Parts (d) and (e) show the relative magnitudes of chromatic and spherical aberration in eyes based on a single interface. Spherical aberration is potentially a serious problem at large apertures. B, R, blue and red rays; F is the Gaussian focus; M the focus for marginal rays and C the circle of least confusion.*

ceans there are multiple-array optical systems of various kinds, some based on lenses and others on mirrors, which so far have no counterparts in optical technology. There is much here that engineers may yet learn from the animal kingdom.

In this chapter I will try to give an up-to-date account of the various kinds of optical system found in the eyes of different animals. Although we know most about the optics of our own eyes and those other vertebrates, it is at the other end of the animal kingdom, among the marine molluscs and crustaceans, that some of the most intriguing optical systems are to be found. Some of theses have been discovered only in recent years, and indeed two whole classes of optical system based on mirrors rather than lenses were discovered within the lifetime of current textbooks. However, I will begin with some comments on the human eye, because it is easier to establish on familiar territory what some of the problems of 'eye design' are, before considering the different ways in which other animals have solved them. Those readers who would like a more systematic and complete account of the subject are referred to an earlier review (Land, 1981), and several recent accounts of vision in different invertebrate groups are given in Ali (1984).

Camera Eyes with Corneal Lenses

Terrestrial animals, including ourselves, have an air–tissue interface separating the contents of the eye from the surrounding atmosphere, and that interface is the main image-forming component in the eye (Fig. 6.1). In the human eye about two-thirds of the optical power is in the cornea, with the lens contributing the remainder. The main function of the lens in mammals is not so much the formation of the image as to act as an accommodative mechanism, varying the focal length of the system so as to alter the distance of best focus.

In all optical systems based on spherical interfaces the most serious potential problem is spherical aberration: peripheral rays tend to be over-focused, and uncorrected the image quality would be very poor (see Fig. 6.1(e)). Most other defects matter less; off-axis aberrations do not matter much in the human eye, where critical resolution is needed only in the centre, and obviously field curvature is no problem. Chromatic aberration is not corrected in our eyes, nor apparently in any other refracting eyes, but it is minimized by the fact that our visual pigments respond to fairly narrow (100 nm half-width) spectral bands, and we maintain image sharpness by using only a part of the spectrum for high acuity vision. Thus, spherical aberration remains one of the major design problems of this kind of eye, and evolution has come up with two different ways of dealing with it. The first is to make the refracting surface non-spherical: to flatten the outer regions relative to the centre to give an elliptical profile. This in fact occurs in the human eye, with the radius of curvature of the outer regions of the cornea about twice that of the central part (Weale, 1974). A second method, which we will meet again in connection with the spherical lenses of aquatic animals, is to incorporate into the optical system a lens with an appropriate refractive index gradient. This mechanism occurs also in the human eye; the lens is not homogeneous and its refractive index gradient tends to reduce the aberration (Millodot and Sivak, 1979). The two correcting mechanisms together result in an optical system that is near perfect on the axis; image quality is then set by the diffraction limit of about 1', when the pupil has its daylight diameter of 2 to 2.5 mm (Westheimer, 1972).

Curiously, the corneal type of eye is uncommon in the animal kingdom outside the land vertebrates. The only other large group to use it is the spiders (Figs. 6.1 and 6.2). Spider eyes vary greatly in structure, but all are small, and all have a single corneal lens (Land, 1985a). The largest are certainly those of the Australian net-casting spider *Dinopis*, where they attain a diameter of 1.4 mm (the human eye is about 24 mm in diameter). *Dinopis* eyes do not have particularly good resolution, but they achieve enormous sensitivity by having a very wide aperture ($F/0.58$) and large receptors (Blest and Land, 1977). *Dinopis* is a wholly nocturnal hunter which catches its prey by literally throwing a net at it, and its huge eyes allow it to see relatively large moving insects in the dark. At the other end of the scale are the jumping spiders. These two have fairly large eyes – for spiders – but here the emphasis is on resolution rather than sensitivity. In the best of these, *Portia fimbriata*, the minimum interreceptor angle is only 2.4' (Williams and McIntyre, 1980), which compares very favourably with the same angle in the very much larger eyes of primates (about 0.5' in man). Jumping spiders are diurnal, and need their excellent eyesight to stalk prey. The majority of web-building spiders have neither the resolution of jumping spider eyes nor the sensitivity of *Dinopis*. Unlike vertebrates, spider eyes lack a focusing lens – with a focal length of 1 mm or less the depth of focus is so enormous that they do not require one. However, they do have a lens behind the cornea which contributes to the power of the optical system and, interestingly, it has an inhomogeneous structure which has a correcting effect on the eye's spherical aberration, just as in our own eyes (Blest and Land, 1977).

There are corneal lens eyes in insects too, but mainly in the larvae (Fig. 6.1(b)) (Land, 1985b). These little eyes, which are quite like those of spiders, are discarded in the adult, and replaced with compound eyes. No one has yet come up with a good reason why this should happen. Simple eyes are in some important respects superior to

Fig. 6.2 *Spiders have 'camera-type' eyes as do land vertebrates. The jumping spider on the left has its large antero-median eyes specialized for acute vision in bright light, whereas the net-casting* Dinopis *(right) has its postero-median eyes specialized for seeing movement of prey in forests at night.* Dinopis *eye 1.4 mm in diameter; same magnification for both.*

compound eyes (as first noted by Mallock (1894), the main problem with compound eyes is the diffraction limit imposed by the small size of the lenses) and as we know that simple eyes can work perfectly well in the spiders, their eviction from most insects at metamorphosis is the more surprising.

Aquatic Lens Eyes

Aquatic animals do not have an optically usable cornea, because it has essentially the same medium on both faces, and thus the whole of the focusing power of the eye must be in the lens. This necessitates a lens with as short a focal length as can be attained, which is why most aquatic vertebrates and many invertebrates have spherical lenses. However, this raises problems because a spherical lens with the same focal length as that of a fish (about 2.5 radii) would have spherical aberration so bad that it would be virtually useless (Fig. 6.3). Furthermore, the refractive index would need to be about 1.66, and this is not attainable using normal biological materials like proteins. 1.56 is probably the upper limit for plausible refractive indices, which would give the lens a focal length of about 4 radii – considerably longer than that of a fish or squid.

The solution to this conundrum was worked out by Matthiessen in the 1880s, although Maxwell had already reached similar conclusions in 1854 'while contemplating his breakfast herring' (Pumphrey, 1961). These lenses are not optically homogeneous, but contain a gradient of refractive index, highest in the centre and falling at the periphery to a value not much exceeding that of the surrounding water. The beauty of this arrangement is that it kills two birds with one stone. The presence of the gradient means that rays of light are bent continuously within the lens (Fig. 6.3), rather than just at the front and rear surfaces, and this results in a shorter focal length than could be provided by a homogeneous lens with the same central refractive index. Furthermore, provided the gradient has the correct form, the lens will be free from spherical aberration, since the rays furthest from the axis

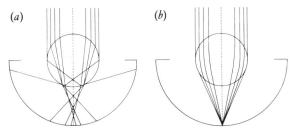

Fig. 6.3 *The fish-eye problem: a homogeneous lens (a) has appalling spherical aberration. This can be cured by having a gradient of refractive index, highest in the centre (b). The continuous ray bending can give a perfect image, as well as a shorter focal length (After Pumphrey, 1961).*

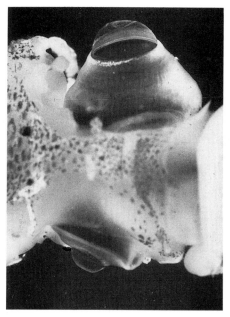 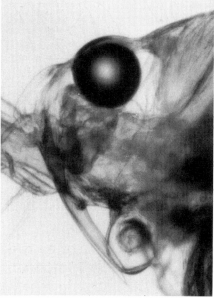

Fig. 6.4 *Two unusual spherical lens eyes of the 'Matthiessen' type. On the left is a squid (*Histioteuthis*) with one large and one small eye (large eye 8 mm across). The copepod* Labidocera *(right) is the only crustacean with this kind of eye. Lens diameter about 0.1 mm (From Land, 1981, 1984).*

will be bent less relative to the central ones than would be the case in a homogeneous lens. The exact form of the gradient was not known to Matthiessen – although he made an enlightened guess – and was not in fact worked out until many decades later (Fletcher *et al.*, 1954; Luneberg, 1944). A simple examination of a fish lens (cooked or raw) shows that there is a gradient: the outside is soft and easily squashed, but the centre is hard and crystalline. The texture reflects both the protein concentration and the refractive index.

This solution to the problem of spherical aberration has evolved independently many times (Fig. 6.4). Besides fish, eyes with essentially the same geometry are found in the cephalopod molluscs (octopus, squid, cuttlefish – but curiously not in *Nautilus*, which only has a pinhole; see Muntz and Raj, 1984), in the gastropod molluscs (conches and heteropods), once in the annelid worms (the Alciopidae) and once in the crustaceans (the copepod *Labidocera*) (Land, 1981, 1984). It is a simple matter to determine whether or not the lens is of this design, because the focal length–radius ratio is diagnostic. If it is around 2.5 ('Matthiessen's ratio') then the lens *must* be inhomogeneous. It seems that this is both the best and the most popular way of producing a well-corrected lens. It has the additional virtue that the spherical symmetry of the lens and eye ensures that the image is well resolved over a very wide field; there is no single optical axis as there is in the human eye.

The eyes of the copepod crustacean *Labidocera* and the heteropod snail *Oxygyrus* deserve special mention, not because of their optics but because they have linear scanning retinae (Land, 1982, 1984). In *Oxygyrus* the retina is only 3 receptors wide, but it is 410 receptors long, giving a field of view that is a long strip. The eye tilts through 90° every second or so, along an arc at right angles to the retinal strip, thus scanning the surrounding water for food particles. *Labidocera* is similar, but on a much smaller scale. Its retina contains only 10 elongated receptors resembling slab waveguides, and this receptor line scans through an arc of about 35°, as shown in Fig. 6.5. This remarkable behaviour was first described by G. H. Parker (1891).

Multi-element Lenses

In view of the excellence and ubiquity of spherical 'Matthiessen' lenses, I was very surprised indeed to find a lens in a marine animal that was constructed along totally different lines. The optical system of the single ventral eye of the copepod crustacean *Pontella* consists of a number of

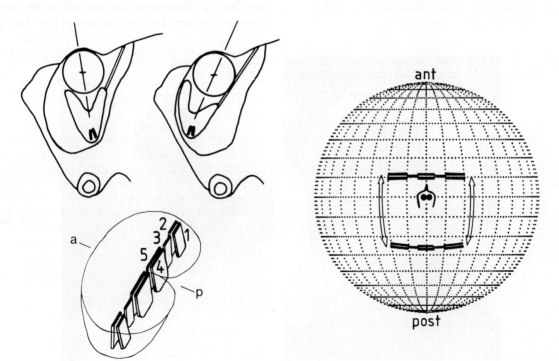

Fig. 6.5 *Scanning eye of* Labidocera: *top left: views of the eye cup in its extreme back and front positions; lower left: disposition of the ten receptors in the pair of eye cups (a, anterior; p, posterior); right: the hemisphere above the animal showing the field scanned by the retina (From Land, 1984; Parker, 1891).*

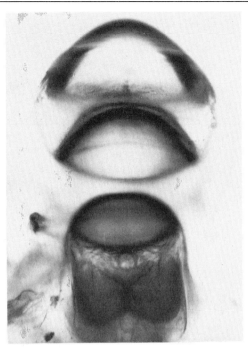 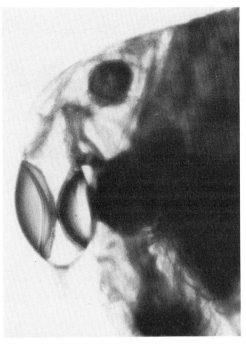

Fig. 6.6 *Triplet lens in* Pontella, *another copepod: left, male eye from below; right, from the side. Whole eye is 0.4 mm long. Females have only two lenses rather than three.*

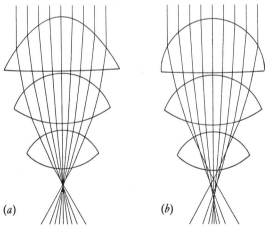

Fig. 6.7 *Ray paths in* Pontella *eye. (a) Ray-tracing through a model eye, assuming a homogeneous refractive index of 1.52. Replacing the actual parabolic first surface with a spherical one; (b) gives an image with serious spherical aberration.*

components, not unlike a camera lens, or more precisely a microscope objective (Fig 6.6). In the males the eyes have three elements, two outside the eye-cup in the rostrum, and one in the front of the eye-cup itself (Land, 1984). Curiously, the females have a doublet not a triplet, with only one component in the rostrum. The first lens of the triplet in the male has an interesting shape, its front surface being approximately parabolic. The reason for this became clear when I traced rays through it (Fig. 6.7). If this non-spherical surface is replaced by a spherical one of the same power, the final image is quite messy because of the spherical aberration of this and all the other surfaces. Resubstituting the parabolic surface gives a much better image in which the spherical aberration of the whole system is well corrected.

It is not clear what this eye does for the animal. It only contains six receptors in a curious concentric array, and the field of view of the eye is quite small. Nevertheless, the sexual dimorphism of the eyes and the colourful appearance of the animals themselves strongly suggests a role in mate finding. So far, however, the only known behaviour of these small planktonic animals is that they jump out of dishes! It is worth noting that *Labidocera* mentioned above as having spherical inhomogeneous lenses is in the same copepod family as *Pontella*. Thus in closely related genera we find the two alternative solutions to the problem of spherical aberration: a gradient of refractive index in one and a non-spherical surface in the other.

Concave Mirror Optics

Bivalve molluscs are perhaps not the kind of animal one would look to for optical surprises, or even much in the

124 *Evolution of the Eye and Visual System*

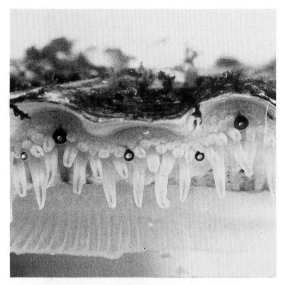 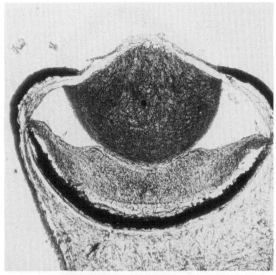

Fig. 6.8 *Left: eyes of the scallop* Pecten maximus. *Right: central section through an eye showing the 'lens', thick retina and the spherical mirror-like rear surface. Eye 1 mm across.*

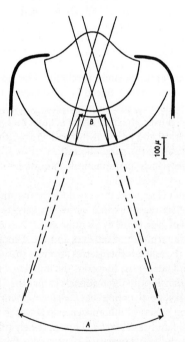 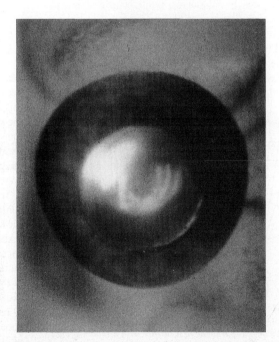

Fig. 6.9 *Left: optical system of the scallop eye; the lens alone forms an image at A which is refocused at B by the mirror. Right: image of the author's hand, in a scallop's eye (photographed through a microscope that the hand is holding).*

way of eyesight. However, in one case at least this judgment would be wrong. Scallops provide an almost unique example of a type of eye that *should* exist, but which had not been found until relatively recently. This is an eye that forms an image using a concave mirror, rather than a lens. Scallops (*Pecten* and related genera) have 50 to 100 eyes around the edge of the mantle of the two shells. They are quite large – 1 mm in diameter – and really quite 'eye-like' eyes (Fig. 6.8), but few are aware of their existence because they are on the inedible part of the animal, which is usually thrown away. A cursory look at a section through a scallop's eye (Fig. 6.8) shows an overall layout not unlike that of a fish eye. It has a single chamber, so is camera-like rather than compound; it has a lens of sorts, and behind this a retina filling the space between the lens and the back of the eye. When I first studied these eyes in 1964 there was some evidence that they could detect movement, implying a functional optical system, but the assumption was that the optics were conventional.

The first indication that this was not correct came from a more careful look at the section (Fig. 6.8). If we recall that the shortest focal length that a fish lens can have is about 2.5 radii, this implies that there must be a gap between the lens and the retina of at least 1.5 lens radii across which light can travel to a focus. The absence of such a space in *Pecten* is more than surprising – it means that the eye should not be able to see! The second indication came from the appearance of the eye when I looked into it (Fig. 6.9). I saw an inverted image of the room, including a distorted picture of myself looking through the microscope. After some thought it became clear that this image could not have been formed by the lens (if that were the case its apparent position would not be in the eye but near infinity), which left only one candidate, the spherical silvery mirror which lines the back of the eye. Here we have an eye that is the optical analogue of a Newtonian telescope (Land, 1965).

The reflected image lies on one of the two layers that make up the retina, and as early as 1938 the pioneer electrophysiologist H. K. Hartline had found that this layer gave responses to the offset rather than the onset of light (Hartline, 1938). This 'OFF-responding' layer lies immediately behind the lens, whereas the other, 'ON-responding' layer is in contact with the mirror, and receives no image (Fig. 6.9). The 'OFF' layer is undoubtedly the one that enables the scallop to see trouble coming (Land, 1965), and to shut its shells, but the function of the 'ON' layer remains a mystery.

Mirror eyes of this kind are rare, and the eyes of scallops are certainly the best of them from an optical point of view. A few other molluscs like the cockle *Cardium* have similar but much smaller eyes, and there are also examples in the Crustacea, in the nauplius eyes of some copepods and ostracods (Land, 1984). There are even examples in the flatworms and the rotifers (see Ali, 1984). However, these are all very small, and their image quality is unknown and probably poor. The only other *large* mirror eyes are in the big (10 mm) deep-sea ostracod *Gigantocypris* shown in Fig. 6.10. These were described by Sir Alister Hardy (1956) as follows:

> The paired eyes have huge metallic-looking reflectors behind them, making them appear like the headlamps of a large car; they look out through glass-like windows in the otherwise orange carapace and no doubt these concave mirrors behind serve instead of a lens in front.

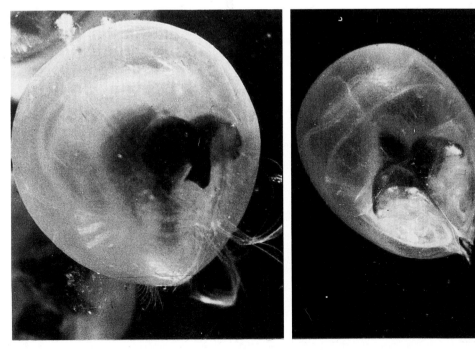

Fig. 6.10 Gigantocypris mulleri, *a deep-sea ostracod with parabolic reflecting eyes. It has a diameter of about 10 mm. (Photograph by Dr M. R. Longbottom).*

Hardy was undoubtedly right, but these eyes have a very odd structure, indicating enormous light-gathering power but poor resolution, which is consistent with life in the nearly lightless zone below 600 m (Land, 1984).

I suspect that the reason for the unpopularity of mirror eyes is that image contrast is inevitably poor, because focused light reaching the retina has already passed through it once unfocused (Fig. 6.9). Lens eyes do not suffer from this drawback. There is another kind of mirror eye that some crustaceans use, but this is a compound eye, and I will return to it later after I have outlined how compound eyes in general work. Before that, however, a brief comment is needed on the optics of biological mirrors.

Biological Mirrors

Fig. 6.11 is an electron micrograph of the mirror in the scallop's eye. It shows alternating layers of 'holes', which we know to be places where guanine crystals were present before the tissue was prepared, and darker spaces representing cell cytoplasm. A particularly interesting feature of the figure is the scale: there are about five layer-pairs μm^{-1}, and since the crystals and spaces are of similar thickness this means that each is about 0.1 μm thick. This will give them an *optical* thickness (thickness × refractive index) of about one-quarter of the wavelength of light (this structure reflects green light between 500 and 550 nm).

The properties of quarter-wavelength multi-layers are well known (Land, 1972). Although a single crystal of guanine reflects only a few per cent of the incident light, a stack of them, separated by water spaces of the same optical thickness, can reflect virtually all the light at the peak wavelength with as few as ten crystals in the stack (Fig. 6.11). With minor variations, this is the way all biological mirrors work. The materials may vary: in scallop mirrors and fish scales the principal component of the crystals is guanine, but in the silvery-blue patches on butterfly wings, for example, the layers are chitin and air. Other variants include cats' eyes, which use the zinc salt of

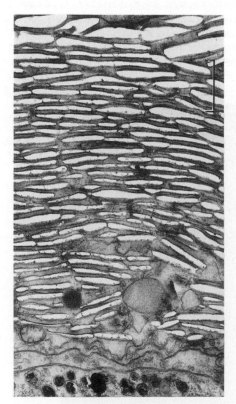
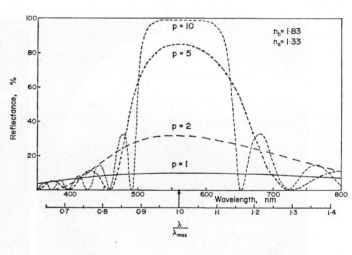

Fig. 6.11 *Left: electron micrograph of a section through the multi-layer mirror in a scallop eye; the 'holes' are the spaces left by guanine crystals, and the darker parts are cytoplasm. Scale bar, 1 μm; the light would normally come from the top of the page. Right: calculated reflectances of guanine–water quarter-wavelength multi-layers with different numbers of plates (p), and a wavelength of maximum reflectance of 560 nm. Notice that with only ten high refractive index layers the reflectance is close to 100% (From Land, 1972).*

cysteine, bush-babies which have riboflavin crystals in their tapeta, and the feathers of hummingbirds, which employ melanin 'pancakes' inflated with gas (Land, 1972). These interference mirrors are not only highly reflecting, they are also coloured because they are tuned, as a result of their thickness, to reflect best at a particular wavelength. In some cases, such as the scallop's eye, the reflecting tapeta of many vertebrates from sharks to cats, and the silvery sides of fish, it is the reflectance rather than the colour of the stacks that is exploited. But there are many instances, especially in insects where the colours are more important and used in sexual display. The brilliant blues of the *Morpho* butterflies are one example, and there are other examples where the 'colour' involved is in the ultraviolet (Ghiradella *et al.*, 1972).

Compound Eyes

Most people are aware that at least half the animal kingdom has a kind of eye that is fundamentally different from those I have already described, in that there are many optical systems in each eye, rather than a single system as in a camera. Since the turn of the century, following the publication of Sigmund Exner's famous treatise of 1891, compound eyes have been divided on optical grounds into two types, called apposition and superposition (Exner, 1891).

Apposition Optics

In apposition eyes each lens forms a separate image at the distal tip of a rod containing photosensitive pigment called the rhabdom. This rod, which is made up of contributions from about eight receptor cells, behaves as a lightguide, so that light entering its tip is effectively scrambled; if there were any detail in the pattern of the entering light, this is lost. This solves a problem that goes back to Leeuwenhoek, namely: How is it that each lens forms a small *inverted* image, when the geometry of the image as a whole is erect (see Figs. 6.12(a) and 6.13). In most apposition eyes this is simply not a problem because the rhabdom does not 'know' that the scrambled image it received is inverted. The overall image is made up of the contiguous fields of view of the rhabdom tips in the image plane of each lens (Fig. 6.12(a)), and these fields are typically limited by diffraction to a minimum of about 1 degree across (Mallock, 1894). This compares pretty unfavourably with less than 1' for the cones in our own eyes. Apposition eyes are found in most diurnal insects (bees and grasshoppers, for example) and some surface or shore-living crustaceans like the crabs. They also turn up in very unexpected places, such as the mantle eyes of clams of the family Arcacae and on the tentacles of the fan-worm *Branchiomma* (Land, 1981). Although we associate compound eyes with the arthropods, they are not confined to this phylum. For further discussion see Kunze (1979), Snyder (1979), Land (1985, 1989) and Nilsson (1989).

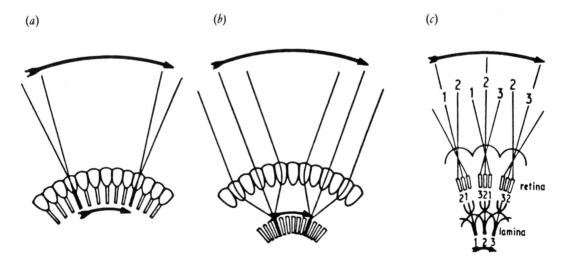

Fig. 6.12 *Compound eyes: in* apposition *eyes (a) each lens forms a small inverted image, although the total image is erect; in* superposition *eyes (b) the image at any point is formed by the superimposition of rays through many facets, giving an erect optical image deep in the eye. See Fig. 6.13. In* neural superposition eyes *several receptors in adjacent ommatidia image the same direction in space. Numbers indicate how each direction is brought together in the lamina. See text.*

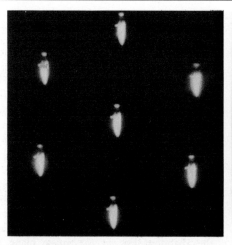 

Fig. 6.13 *Images in compound leyes. Left: photographs of the inverted images of a candle flame in the facets of the* apposition *eye of a robber fly (Asilidae). Right: single erect image of a famous nineteenth-century naturalist photographed in the* superposition *eye of firefly (Lampyridae). See Fig. 6.12. (The photographs were taken with the eyes held by a drop of fluid hanging from a microscope slide and the microscope focused on the image plane of the eyes' optics, with the eyes themselves facing an object some distance away.)*

Dipteran flies provide an exception to the rule that the inverted image is not further resolved within the ommatidium. In these insects the receptive regions of the 8 receptors (rhabdomeres) are not joined so as to form a single rhabdom, but are separated, and thus view different parts of each image (Fig. 6.12(c)). How, then, is the final erect image put together? The solution lies in the finding that the angle between the visual directions of the rhabdomeres in 1 ommatidium is the same as that between the ommatidial axes themselves, which means that the 6 eccentric rhabdomeres in 1 ommatidium all have fields of view that coincide with the central rhabdomeres in adjacent ommatidia (Kirschfeld, 1967). Beneath the retina all the 8 receptors whose rhabdomeres look in the same direction send their axons to the same synaptic 'cartridge' in the lamina, via an impressively complicated piece of neural rewiring (Braitenberg and Strausfeld, 1973). Thus at the level of the second-order neurones the overall image is the same as in an apposition eye. What dipteran flies gain from this is an effective increase in the photon signal from each direction, without the loss of resolution that would come from having wider receptors. Kirschfeld (1967) describes these as 'neural superposition' eyes, but from an optical point of view they are of the apposition type.

Superposition Optics

Exner (1891) found that in the cleaned eye of the male glow-worm it was not possible to see multiple images as in apposition eyes but only a single erect image. This image (Fig. 6.13) lay relatively deep in the eye, not immediately behind the optical systems as in apposition eyes. This obviously raised difficulties, because simple optical systems – single lenses and mirrors – produce inverted images, and an array of single lenses cannot be persuaded to produce a single erect image either. If one draws in the paths that rays of light would have to take in order to contribute to such an image (Fig. 6.12(b)), it becomes clear that what each optical element must do is to redirect rays so that they emerge from the back of the element at the same angle at which they entered it, but reflected across the element's axis. Exner asked what optical systems were able to do this, and concluded that the only device that would work was a two-lens combination: a simple inverting telescope with a magnification of -1 (a plane mirror will work too, as we shall see, but Exner knew that he was dealing with a refracting system in this case). Exner therefore proposed that the optical elements in the glow-worm eye, the so-called crystalline cones, actually consisted of two lenses somehow embedded in the same physical structure. We will see later how this is achieved; for the moment it is enough to note that Exner was right (at least about glow-worms), and that the telescopes themselves use inhomogeneous optics not dissimilar from those invoked by Matthiessen to explain fish lenses. The overall ray pattern in a superposition eye is shown in Fig. 6.12(b). It can be seen that parallel rays entering the eye over a large area contribute to the image at any point, and it is these 'superposed' contributions that give the eye its name. The principal merit of this kind of eye is its sensitivity, receptors receiving a hundred to a thousand times more light than in an apposition eye (for discussions see Kunze, 1979, and Nilsson, 1989). For this reason, superposition eyes are common in nocturnal insects (fireflies and other beetles,

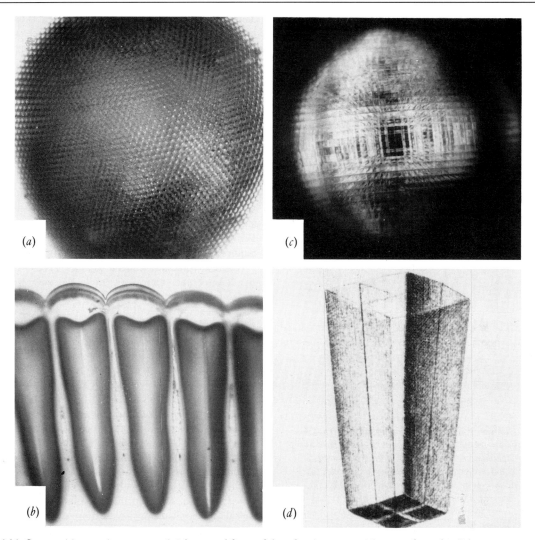

Fig. 6.14 *Superposition eyes in crustacea: (a) hexagonal facets of the refracting superposition eyes of a euphausiid* (Meganyctiphanes*); (b) bullet-shaped crystalline cones of this eye; (c) square-facetted reflecting superposition eye of a shrimp* (Palaemonetes*); (d) Grenacher's drawing of 1879 showing the square jelly-like optical structures in the shrimp eye. Both eye are about 1 mm in diameter.*

moths) and deep-water crustacea (krill, lobsters, shrimps).

There was a period of about 15 years between the early 1960s and 1975 when the eyes of the macruran decapod crustacea (shrimp, crayfish and lobsters) did not work. Not that this bothered the shrimps, but it did worry the scientists trying to understand them. The problem was that the two-lens telescope method of building a superposition eye, as advocated by Exner, requires optical elements that derive their power from a gradient of refractive index, and there appeared to be no such gradient in the equivalent structures in decapod crustaceans. Superposition image formation seemed to be ruled out, but there was not at the time anything sensible to put in its place.

The general structure of macruran eyes suggests a superposition plan – a deep-lying retina, and a clear zone across which rays can be brought to focus – but apparently without any appropriate optics. The answer to the puzzle was provided by Klaus Vogt in 1975. The optical elements in crayfish (Vogt, 1975) and shrimp (Land, 1976) are not lenses, but mirrors.

The optical structures in the eyes of crayfish and shrimp are not the hard, bullet-shaped objects found in the eyes of beetles, moths, and krill which have refracting superposition eyes; instead they are flat-faced, four-sided truncated pyramids of relatively low refractive index jelly (Fig. 6.14(b) and (d)). Vogt found that the faces of these

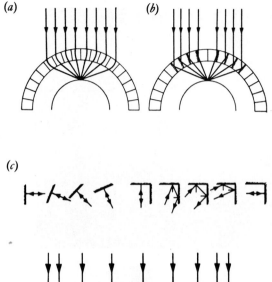

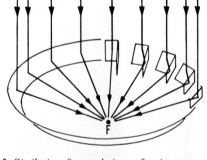

Fig. 6.15 *Similarity of raypaths in a refracting superposition eye (a) and a reflecting superposition eye (b). (c) shows how a reflecting superposition eye could work either with a series of 'saucer rim' mirrors as shown at the left, or an assembly of 'corner reflectors' as on the right. The inserts show that the raypaths seen from above are nearly identical. The corner reflector is better for a wide-angle eye because it has no single axis. F is the focal point.*

along straight rows of the square lattice are reflected from one face as in Fig. 6.15(b). Light rays not entering along these rows encounter one face of the element, are reflected to an adjacent face, and then out of the bottom (Fig. 6.15(c)). The question is, where will they finish up in the image plane? Obviously it would be a good thing if they were brought to a focus at the same point in the image as the light reflected once as in Fig. 6.15(b). In fact this is precisely what happens, because mirrors at right angles – or corner reflectors – have the property of reflecting light through two right angles, that is, in a plane parallel to their original direction. The effect of this can be seen in Fig. 6.15(c). The corner mirror behaves *as if* it were a single mirror in a plane at right angles to the incoming ray, and this in turn means that the simple diagram in Fig. 6.15(b) is valid for all rays, not just the ones that happen to coincide with the lines of mirrors.

There is a practical sequel to this story. It turns out that the 'lobster eye' optical design is the only wide-field image-forming system that can be used to produce an X-ray image (X-rays will reflect at grazing incidence, but do not reflect or refract usably at normal incidence). Such a telescope has been designed for deep space research, based on the principle described here (Angel, 1979).

I should re-emphasize that *most* eyes of the superposition type use the refracting telescope proposed by Exner (Exner, 1891; Kunze, 1979). It is only the macruran decapod crustacea that employ mirrors (Nilsson, 1989). Refracting superposition eyes occur in several insect orders, notably the Lepidoptera (moths) and some Coleoptera (beetles). In the crustacea they are found in the orders Mysidacea and Euphausiacea (krill, Fig. 6.14(a) and (c); Land *et al.*, 1979). It is probable that these represent four cases of independent evolution, although the mysids and euphausiids may be more closely related than current taxonomic systems suggest.

An interesting point concerns the way in which the tiny telescopes in refracting superposition eyes are constructed. Exner had himself concluded that there was not enough optical power in the end surfaces of the optical elements (or 'crystalline cones') for surface refraction to be the main ray-bending mechanism, and he came up with an alternative idea, the lens cylinder. A lens cylinder is a rod with a gradient of refractive index which is highest along the axis and falls to a minimum at the cylinder's surface. In this respect it is like a Matthiessen fish lens, except that the symmetry is radial, not spherical, and the form of the gradient is simpler. Exner believed that it should be parabolic, although later studies suggest that it should follow the very similar hyperbolic secant function (Fletcher *et al.*, 1954). In either event, the lens cylinder works by continuous ray bending (Fig. 6.16(a) and (b)). A ray incident at right angles to the end of the cylinder will be bent towards the higher refractive index regions near the axis, and if the

pyramids acted as plane mirrors, and indeed had a coating on them similar to the reflecting layer in the scallop eye (Vogt, 1980). A glance at Fig. 6.15(a) and (b) will show that a series of radially arranged plane mirrors produces almost the same raypaths as the radial inverting telescopes in the refracting superposition eyes proposed by Exner; light is reflected across the axis of each optical element in both cases. Clearly, superposition images can be produced by either kind of optical array.

One of the intriguing features of the decapod eyes with this kind of image formation is that they have square facets, rather than the hexagonal ones seen in almost all other compound eyes. Is there a good reason for this, or is it merely a phylogenetic oddity? There is indeed an optical explanation, and it has to do with the fact that most of the light entering the eye is reflected not from one face for each square element, but from two. (Only rays reaching the eye

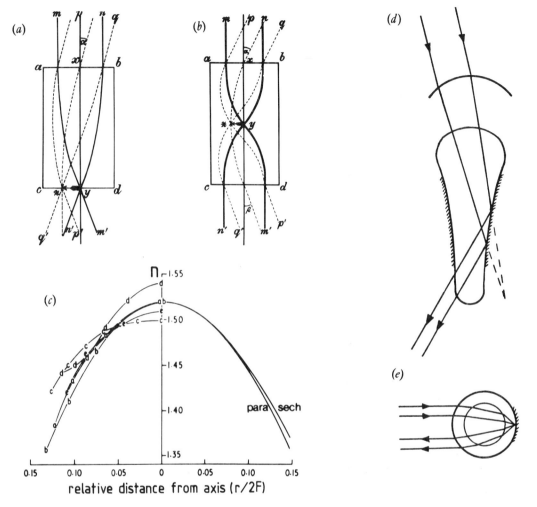

Fig. 6.16 (a) and (b) are Exner's diagrams showing the paths of rays through a focal (a) and an afocal 'lens cylinder' (From Exner, 1891). (c) shows the theoretical (right), and measured (left) gradient of refractive index in lens-cylinders (sech is the hyperbolic secant function and para the parabolic approximation). The gradient measured in five different species corresponds closely to both theoretical curves (From Land, 1980). (d) shows an optical element in a 'parabolic superposition' eye. Seen from the side, rays partially focused by the cornea are recollimated by the parabolic mirror-lining of the element. Seen from above (e), rays are focused on the mirror and refocused after reflection, by a homogeneous cylindrical lens within the structure (After Nilsson, 1988).

gradient is correct then all such rays will meet at a single focus on the axis – just as though the cylinder were a simple lens. Beyond this focus the rays will diverge, and finally, after a distance equal to that from the front face to the focus, they will be parallel again. In superposition eyes this is the situation. Parallel beams entering the crystalline cones emerge parallel, but redirected as indicated in Figs. 6.12(b) and 6.15(a).

In the recent past there has been a degree of healthy controversy over the relations between the different kinds of superposition eye, and this has centred around the question of whether functioning intermediates between the mirror and telescope versions could exist. Could there be a hybrid system that used both lenses and mirrors? Dan-Eric Nilsson has recently described a new type of superposition eye in a crab, which does indeed have features in common with the other two, as well as some of its own (Nilsson, 1988). He calls it 'parabolic' superposition, because the optical system is a combination of a short-focus lens and a convex parabolic mirror which intercepts the partially focused beam and recollimates it (Fig. 6.16(d)). A complication arises from the fact that the mirror, which forms the outer lining of each optical element, is cylindrical in cross-section, not parabolic, so that the light paths are quite different in the longitudinal and transverse planes. Another feature of this arrangement is

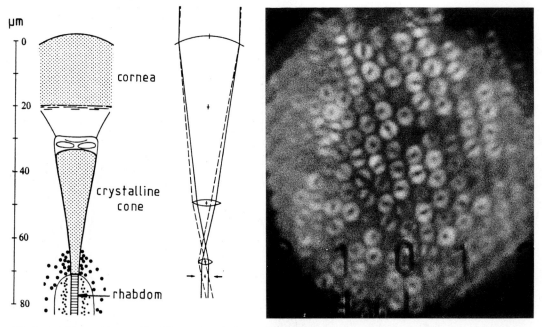

Fig. 6.17 *Optics of butterfly ommatidia. The structure is shown on the left and an equivalent lens diagram next to it. Notice that the cornea and crystalline cone behave together as a telescope with a magnification of about ten times (from Nilsson et al., 1988). A consequence of this arrangement is that the tip of the photopigment-containing rod (rhabdom) is imaged at the cornea, where it is possible to observe, magnified, the waveguide modes supported by the rhabdom (right). (In very fine fibres such as waveguides, light propagates as a series of transverse interference patterns called modes. The rhabdom of butterflies' eyes is wide enough to support two modes, the fundamental or LP_{01} mode and the second order or LP_{11} mode. The former, which was a bright centre, is excluded here with crossed polaroids leaving the second mode with its characteristic dark centre; see Van Hateren, 1989.)*

that axial rays are focused at the rear of the optical element, so that when the reflected rays are excluded – by screening pigment when the eye is light adapted – the eye effectively becomes an apposition eye. The significance of this optical system is that it forms a possible evolutionary link between apposition eyes and both the other types of superposition eye. The whole question of compound eye evolution has been discussed in detail recently by Nilsson (1989).

Lens cylinders are not confined to superposition eyes. It turns out that there are two ways of producing an *apposition* eye that involve lens cylinders rather than the more common corneal lenses. One simply employs a lens cylinder of one-half the length of those in superposition eyes. Such a structure has a focus at its inner surface (Fig. 6.16(a)) and behaves like a simple lens. *Limulus* employs such a mechanism (Exner, 1891; Land, 1979). The other mechanism is more subtle, and has been found recently in butterflies whose optical systems are probably derived evolutionarily from the superposition system of moths (Nilsson *et al.*, 1988). Here the lens cylinders emulate two lenses – as in moths (Figs.6.16(b) and 6.17) – but instead of the lenses having roughly equal power, the second has a much shorter focal length than the first, so that the structure as a whole behaves as an inverting telescope, but this time with a magnification of around ten times. The effect of this optical arrangement is to take a parallel beam entering the 20 μm facet, and squeeze it – still parallel – down the 2 μm receptor. The other effect is to change the basis for determining the acceptance angle of each receptor. In a conventional apposition eye this is determined (geometrically at least) by the subtense of the receptor tip at the nodal point of the lens. In the butterfly type of eye the receptor diameter is not important; it is the critical angle for total internal reflection that now matters, and this in turn is set by the receptor's refractive index (Fig. 6.17). If the complement of this angle is 20°, and the telescope magnification is ten times, then the overall hemi-acceptance angle will be 2°, which is a fairly typical value. In fact, if one takes into account waveguide mode effects in the butterfly's receptors (see Fig. 6.17) then the distinction between this afocal optical design and the more common focal type becomes rather less clear cut (Nilsson *et al.*, 1988). However, a point of biological interest is that the butterfly eye forms another link between the superposition and apposition types, providing a way across a gap that had seemed unbridgeable until recently.

As with the reflecting superposition design, so technology has caught up with lens cylinders too. Since about

1970 both glass and plastic lens cylinders with parabolic gradients have become available, and are used among other things as image-transmitting lightguides (Land, 1980).

Conclusions

It does seem that nearly every method of producing an image that exists has been tried somewhere into the animal kingdom. There are a few others – zone plates, Fresnel lenses, and zoom lenses coming to mind particulary – which may or may not exist in some corner of the animal kingdom. However, whether or not they come to light it is clear that the evolutionary process has both produced and – more surprisingly in some ways – sustained nearly as many ways of forming images as men have produced.

By way of a summary I include Fig. 6.18, which contains thumbnail sketches of the main types of eye known at the present time, and Table 6.1 which shows their distribution in the animal kingdom (non-biologists might need to consult a zoology textbook here such as Barnes's *Invertebrate Zoology* (1987)). In Fig. 6.18 I have added on the left two simple eye types that have not been specifically discussed in this essay; they represent the probable precursors of the other types. (a) is the kind of simple eye-cup without a lens or other optical arrangement which is common among the lower phyla – flatworms, annelids and some molluscs. (b) is a 'proto compound eye' in which

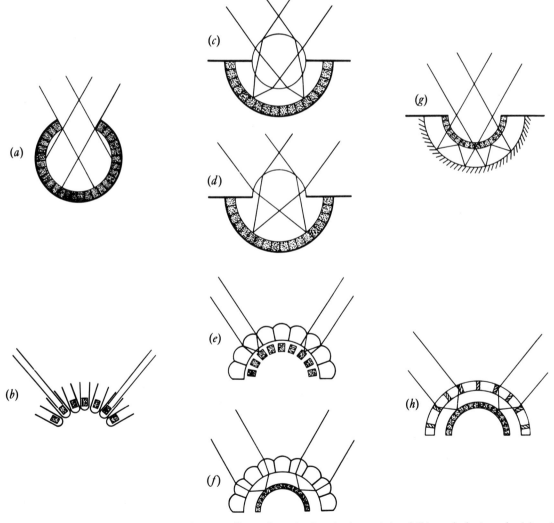

Fig. 6.18 *Summary of the eight principal mechanisms of image formation in animal eyes. (a) and (b) use shadowing only, (c) to (f) use refraction, and (g) and (h) reflection. Further explanation is given in Table 6.1*

Table 6.1 *Distribution of the types of optical system shown in Fig. 6.18.*

Type	Description	Distribution
(a)	Pigmented pit. No lens or mirror. Shadow provides some directionality. Precursor of (c), (d) and (g)	Most of the lower phyla: some coelenterates, platyhelminthes, protochordates, many annelids and molluscs. Pinhole eye in *Nautilus*
(b)	Multiple pigmented tubes. Precursor of the compound eyes (e), (f) and (h)	Sabellid tube-worms (Annelida). Probably early arthropods
(c)	Spherical lens eye	Fishes, cephalopod and some gastropod molluscs, alciopid annelids, copepod crustacean (*Labidocera*)
(d)	Corneal refraction	Terrestrial vertebrates. Most arachnids, some insect larvae, *Peripatus*, and myriapods
(e)	Apposition compound eye	Most diurnal insects and many diurnal and shallow-water crustacea. *Limulus*, trilobites, *Branchiomma* (Annelida) and *Arca* (Bivalvia)
(f)	Refracting superposition compound eye	Nocturnal insects (moths, lacewings, some beetles), some crustacea (mysids, euphausiids). Apposition version in butterflies
(g)	Simple mirror eye	*Pecten* (bivalve mollusc) and *Gigantocypris* (crustacean)
(h)	Reflecting superposition compound eye	Decapod crustacea only (shrimps, prawns, crayfish and lobsters)

single receptors lie at the bottom of pigmented tubes. Some sabellid tubeworms have eyes like this, as do a few starfish, but it has to be admitted that nothing like it is to be found in the arthropods – the phylum of compound eyes *par excellence*. The addition of a lens or mirror to the (a) type of eye-cup will produce the range of eyes we know as simple or single chambered (c, d and g), and similar additions to (b) produce the known range of compound-eyes (e, f and h).

Faced with this diversity, one inevitably asks whether some designs are better than others, and whether some are particularly good in certain habitats. It seems to be very hard to generalize. The basic vertebrate type of eye (c and d) is very versatile with respect to the range of light intensities over which it can be used, and this probably accounts for the fact that fish lookalikes have evolved so many times. The compound eyes are different in this respect, and two types – apposition and superposition – do seem to be needed to cover the full range of light intensities from daylight to night. However, ecological differences do not explain very much of the diversity. The reflecting and refracting superposition eyes coexist in closely related animals (shrimp and krill) with nearly identical lifestyles, and one has to assume that these are independently evolved but functionally equal solutions to the same problem. The spherical lens eyes and multi-element lens eyes in pontellid copepods present a similar pair of solutions. It does not seem to be necessary for a particular eye-type to be demonstrably better than its rivals for it to survive through long periods of evolutionary time.

Acknowledgement

This discussion is a modified and updated version of an article originally published in the *Proceedings of the Royal Institution of Great Britain* (1985, 57, 167–189), and is published with permission of the Royal Institution.

References

Ali, M. A., Ed. (1984). *Photoreception and Vision in Invertebrates*. New York: Plenum.

Angel, J. R. P. (1979). Lobster eyes as X-ray telescopes. *Astrophys. J.*, 233, 364–373.

Barnes, R. D. (1987). *Invertebrate Zoology*. Philadelphia: Saunders.

Blest, A. D. and Land, M. F. (1977). Physiological optics of *Dinopis subrufus* L. Koch: a fish-lens in a spider. *Proc. R. Soc. Lond. B*, 196, 197–222.

Braitenberg, V. and Strausfeld, N. J. (1973). Principles of the mosaic organization of the visual system's neuropil of *Musca domestica* L. In *Handbook of Sensory Physiology*, Vol. VII/3. Ed. Autrum, H. pp. 631–660. Berlin: Springer.

Eakin, R. M. (1972). Structure of invertebrate photoreceptors. In *Handbook of Sensory Physiology*, Vol. VII/1. Ed. Dartnall, J. A. pp. 625–684. Berlin: Springer.

Exner, S. (1891). *The Physiology of the Compound Eyes of Insects and Crustaceans*. Transl. R. C. Hardie (1989). Berlin: Springer.

Fletcher, A., Murphy, T. and Young, A. (1954). Solutions of two optical problems. *Proc. R. Soc. Lond. A*, 223, 216–225.

Ghiradella, H., Aneshansley, D., Eisner, T., Silberglied, R. E. and Hinton, H. E. (1972). Ultraviolet reflection of a male butterfly: interference color caused by thin-layer elaboration of wing scales. *Science Wash. DC*, 178, 1214–1217.

Goldsmith, T. H. (1990). Optimization, constraint, and history in the

evolution of eyes. *Quart. Rev. Biol.*, **65**, 281–322.
Greuet, C. (1984). Organisation et évolution des dispositifs visuels des protistes flagellés cas de l'ocelloïde des dinoflagellés *Warnowiidae*. In *La Vision chez des Invertébrés*. Eds Clément, P. and Ramousse, R. pp. 68–74. Paris: CNRS.
Hardy, A. C. (1956). *The Open Sea*. Part 1. London: Collins.
Hartline, H. K. (1938). The discharge of impulses in the optic nerve of *Pecten* in response to illumination of the eye. *J. Cell. Comp. Physiol.*, **11**, 465–477.
Kirschfeld, K. (1967). Die Projektion der optischen Umwelt auf der Raster der Rhabdomere im Komplexauge von *Musca*. *Exp. Brain Res.*, **3**, 248–270.
Kunze, P. (1979). Apposition and superposition eyes. In *Handbook of Sensory Physiology*, Vol. VII/6A. Ed. Autrum, H.-J. pp. 441–502. Berlin: Springer.
Land, M. F. (1965). Image formation by a concave reflector in the eye of the scallop, *Pecten maximus*. *J. Physiol. Lond.*, **179**, 138–153.
Land, M. F. (1972). The physics and biology of animal reflectors. *Progr. Biophys. Mol. Biol.*, **24**, 75–106.
Land, M. F. (1976). Superposition images are formed by reflection in the eyes of some oceanic decapod crustacea. *Nature Lond.*, **263**, 764–765.
Land, M. F. (1979). The optical mechanism of the eye of *Limulus*. *Nature Lond.*, **280**, 396–397.
Land, M. F. (1980). Compound eyes: old and new optical mechanisms. *Nature Lond.*, **287**, 681–686.
Land, M. F. (1981). Optics and vision in invertebrates. In *Handbook of Sensory Physiology*, Vol. VII/6B. Ed. Autrum, H.-J. pp. 471–592. Berlin: Springer.
Land, M. F. (1982). Scanning eye movements in a heteropod mollusc. *J. Exp. Biol.*, **96**, 427–430.
Land, M. F. (1984). Crustacea. In *Photoreception and Vision in Invertebrates*. Ed. Ali, M. A. pp. 401–438. New York: Plenum.
Land, M. F. (1985a). The morphology and optics of spider eyes. In *Neurobiology of Arachnids*. Ed. Barth, F. G. pp. 53–78. Berlin: Springer.
Land, M. F. (1985b). Optics of insect eyes. In *Comprehensive Insect Physiology, Biochemistry and Pharmacology*, Vol. 6. Eds Kerkut, G. A. and Gilbert, L. I. pp. 225–275. Oxford: Pergamon.
Land, M. F. (1989). Variations in the structure and design of compound eyes. In *Facets of Vision*. Eds Stavenga, D. G. and Hardie, R. C. pp. 90–111. Berlin: Springer.
Land, M. F., Burton, F. A. and Meyer-Rochow, V. B. (1979). The optical geometry of euphausiid eyes. *J. Comp. Physiol.*, **130**, 49–62.
Luneberg, R. K. (1944). *The Mathematical Theory of Optics*. PhD thesis, Brown University, Providence, RI. Republished (1964). Berkeley: University of California Press.
Mallock, A. (1894). Insect sight and the defining power of composite eyes. *Proc. R. Soc. Lond. B*, **55**, 85–90.
Millodot, M. and Sivak, J. (1979). Contribution of the cornea and lens to the spherical aberration of the eye. *Vision Res.*, **19**, 685–687.
Muntz, W. R. A. and Raj, U. (1984). On the visual system of *Nautilus pompilius*. *J. Exp. Biol.*, **109**, 253–263.
Nilsson, D.-E. (1988). A new type of imaging optics in compound eyes. *Nature Lond.*, **332**, 76–78.
Nilsson, D.-E. (1989). Optics and evolution of the compound eye. In *Facets of Vision*. Eds Stavenga, D. G. and Hardie, R. C. pp. 30–73. Berlin: Springer.
Nillson, D.-E., Land, M. F. and Howard, J. (1988). Optics of the butterfly eye. *J. Comp. Physiol. A*, **162**, 341–366.
Parker, G. H. (1891). The compound eye in crustaceans. *Bull. Mus. Comp. Zool. Harv.*, **21**, 44–140.
Pumphrey, R. J. (1961). Concerning vision. In *The Cell and the Organism*. Eds Ramsay, J.A. and Wigglesworth, V. B. pp. 193–208. Cambridge: Cambridge University Press.
Salvini-Plawen, L. V. and Mayr, E. (1977). On the evolution of photoreceptors and eyes. *Evol. Biol.*, **10**, 207–263.
Snyder, A. W. (1979). Physics of vision in compound eyes. In *Handbook of Sensory Physiology*, Vol. VII/6A. Ed. Autrum, H.-J. pp. 255–313. Berlin: Springer.
Van Hateren, J. H. (1989). Photoreceptor optics, theory and practice. In *Facets of Vision*. Eds Stavenga, D. G. and Hardie, R. C. pp. 74–89. Berlin: Springer.
Vogt, K. (1975). Zur Optik des Flusskrebsauges. *Z. Naturforsch.*, **30c**, 691.
Vogt, K. (1980). Die Spiegeloptik des Flusskrebsauges. The optical system of the crayfish eye. *J. Comp. Physiol.*, **135**, 1–19.
Weale, R. A. (1974). Natural history of optics. In *The Eye*, Vol. 6. Eds Davson, H. and Graham, L. T. pp. 1–110. New York: Academic.
Westheimer, G. (1972). Optical properties of vertebrate eyes. In *Handbook of Sensory Physiology*, Vol. VII/2. Eds Fuortes, M. G. M. pp. 449–482. Berlin: Springer.
Williams, D. S. and McIntyre, P. (1980). The principal eyes of a jumping spider have a telephoto component. *Nature Lond.*, **288**, 578–580.

7 Considering the Evolution of Vertebrate Neural Retina

Ian Thompson

Introduction

The basic design of all vertebrate retinas is essentially the same. Light passes through the neural retina and is transduced in the outer segments of photoreceptors which lie next to the pigment epithelium. The electrical signal is passed via bipolar cells to the retinal ganglion cells, which provide the brain with information about the visual image via the optic nerve. Horizontal interactions within the retina are provided by horizontal cells and by amacrine cells, which are located at different levels in the retina. The interplexiform cell appears to provide feedback across the retinal layers. Imposed on this basic uniformity, there is wide variation of detail.

Within each of six basic classes of retinal neurones, a wealth of subclasses has been revealed. Neurones in different subclasses may differ in their dendritic or axonal morphology, their synaptic connections, their neurotransmitters or their receptive field characteristics. Additionally, they may vary both in their absolute numbers and/or in their distribution across the retina. But what is the significance of this variety? The visual neurobiologist asks how retinal diversity constrains the processing of the visual image, from transduction to its transmission in the optic nerve. The developmental biologist can examine how the neural diversity is generated. This chapter will indicate how, given the underlying uniformity of retinal design, phylogenetic variation in vertebrate retina can be considered in the context of evolution.

In the absence of a fossil record, evolutionary studies of the retina must be based on comparisons between extant species. Traditionally, the most common approach has been to consider the adaptive radiation of the retina. Interestingly, Cajal (1893) in his pioneering comparative study of the vertebrate retina viewed the variations not so much phylogenetically as in the context of adaptations for visual analysis. Thus descriptions of rod and of cone pathways transcended different vertebrate orders and closely related species could have quite different neural elements.

Walls's monumental monograph (Walls, 1942) is an extensive description of the adaptive radiation of the vertebrate eye, including the retina. One of his concerns, the categorization into diurnal, nocturnal or arrhythmic eyes, echoes Cajal but Walls also views retinal variation from a phylogenetic perspective. His monograph concludes with a tentative schema of the evolution of vertebrate photoreceptors. As the understanding of the relation between retinal structure and function has increased (e.g. Rodieck, 1973; Dowling, 1987), the trend has been to relate comparative differences in retinal organization to behavioural differences (e.g. the terrain hypothesis of Hughes, 1977). This approach can be labelled adaptationist in that the retinal organization is considered to determine a behavioural phenotype that is optimally adapted to the environment (see Calow and Townsend, 1981). Adaptationist arguments can run into teleological difficulties, especially if there is no predictive element relating environmental niche to biological structure. Gould and Lewontin (1979) also point out that the adaptationist programme has obscured important biological constraints on the nature of possible structural variations. The adaptationist view is focused at the end-point of the evolutionary process, asking how a particular structure is adapted to the environment. Formally, all structures are possible. Gould and Lewontin are more interested in the biological rules that determine variations in structure. It is these mechanisms that provide the material for selective evolution and the significance of a particular structure may be misunderstood unless these biological constraints are known.

Considerations of adaptive value tend to be applied to species that are quite closely related and are used to identify selective pressures moulding small biological modifications. An important element of such considerations is to decide which characteristics are plesiomorphic, i.e. represent the shared, primitive ancestral condition, and which are apomorphic, i.e. derived characteristics possibly reflecting particular selection pressures. The identification of plesiomorphic and apomorphic homologies plays an important role in phylogenetic reconstructions, especially in cladistics (see Hennig, 1979; Eldredge and Cracraft, 1980). However, it is rare for differences in the organization of the nervous system, let alone the retina, to be employed in the determination of evolutionary sequences (see Johnson, 1986). But one interesting application has been that of Pettigrew et al. (1989) to the phylogenetic relation between megabats and primates.

A final context in which evolutionary considerations can be applied to retinal organization is in the relation between evolution and development. The modern synthesis of the ontogeny–phylogeny arguments of Haeckel and Von Baer is to ask in what way developmental mechanisms can be modified to generate the patterns of adult diversity (e.g. see Gould, 1977; Ebbesson, 1984; Maynard Smith et al., 1985; Oster et al., 1988). It is in this area that the different approaches come closest. The developmental biologist wants to understand how genetic mechanisms regulate orderly retinal development to generate specific retinal organizations. The neurobiologist needs to know the functional significance of the structural variations. For the evolutionary biologist, it is the functional variation that provides the behavioural phenotype on which selection acts.

In the following sections of the chapter, I first outline some of the variations observed in the vertebrate retina and then ask how evolutionary considerations have been brought to bear on this variation. It seems reasonable to assume that evolutionary pressures on the retina act via its role in visual performance. This role must be defined by the types and distributions of the retinal ganglion cells, as they constitute the only output to the central nervous system. In turn, their properties depend both on the nature of the input provided by the photoreceptors and on the intrinsic circuitry linking receptors and ganglion cells. I shall begin with a consideration of the photoreceptors and progress via the interneurones to the ganglion cell layer.

Photoreceptors

Since Schultze's (1866, 1873) duplicity theory of vision, the presence of rods in the retina has been associated with low-light, low-acuity vision and the presence of cones with photopic high-acuity, colour vision. Schultze arrived at his theory by comparing the relative distribution of rods and cones in the retinas of species varying in the degree of nocturnal behaviour. To say that rod vision evolved for low-light environments may be correct, but it is misleadingly simple. The duplicity theory has implications for the properties both of the receptors and also of subsequent retinal circuitry. In the immediately following sections, I consider the nature of the ancestral photoreceptor and phylogenetic variations in receptor types and distributions.

The Ancestral Receptor

Walls's 1942 monograph concludes with a figure in which he suggests a possible schema for the evolution of vertebrate photoreceptors. It has the familiar 'tree of life' form, in which the phenotypes of ancestral types are inferred from those of extant species. In the schema, the earliest vertebrates are thought to possess cones. In subsequent phylogenies, cones sometimes disappear and then re-emerge. What evidence is there for this schema and can anything be said about the nature of the ancestral photoreceptor?

One obvious problem is in the definitions of rods and cones. Schultze (1866) identified rods by the presence of a long cylindrical outer segment, whereas cones had shorter conical or tapered outer segments. These criteria are not always reliable (e.g. Duke-Elder, 1958; Pedler, 1965; Cohen, 1972; Jacobs, 1990), and one particularly interesting ambiguity is found in cyclostomes (hagfish and lampreys). These jawless fish are thought to represent an early divergent line of vertebrate evolution. As such, the lampreys are placed at the bottom of the 'tree' by Walls (1942). He also shows one group (Pteromyzonidae) with both rods and cones. But, although the lamprey does have long and short photoreceptors, ultrastructurally both have rod-like outer segments (Ohman, 1976; Holmberg, 1977; but see Negishi et al., 1987). Conversely, at the next branch Walls shows most of the elasmobranchs with all rod retinae. Current opinion is that only the ray has an all-rod retina; other elasmobranchs have duplex retinae, with one cone type that functions in bright light (see Cohen, 1990). Interestingly, the skate rods work in both dim and bright light and the retinal circuitry is as complex as in duplex retinae (Dowling and Ripps, 1970, 1972; Sakai et al., 1986). There are hints that the lamprey retina may be comparable (Holmberg, 1977). Duplicity within the rod population is also seen in amphibia, which have red and green rods (Walls, 1942).

It is apparent that simply comparing photoreceptor types in extant species produces no obvious conclusions about the nature of the primitive photoreceptor. Obviously, existing species are adapted to their current niche and may no longer resemble the primitive pattern. Evolution may have seen the emergence of duplex retinae several times. For instance, Walls (1942) suggested that the rod-like photoreceptors in snakes and gekkos evolved from the all-cone retinae of lizards. However, the difficulty in identifying photoreceptor types unambiguously complicates these interpretations: Cohen (1972) reports that disc membranes in gekko rods are open to the outside, a characteristic of cones. Similarly, the incidence of species with all-rod or all-cone retinae has changed since Walls (1942). It is now known that most elasmobranchs have duplex retinae and that lampreys have all-rod retinae. It would be very interesting to know if lizards do in fact have all-cone retinae. Diurnal ground squirrels were described by Walls as having all-cone retinae but subsequent studies have revealed a small percentage of rods (West and Dowling, 1975; Anderson and Fisher, 1976). In other mammals, while diurnal species have a higher percentage of cones and nocturnal species a higher percentage of rods, it is doubtful whether there are instances of all-rod or all-cone retinae (see Hughes, 1977; Samorajski et al., 1966; Carter-Dawson and La Vail, 1979; Szel and Rohlich, 1989). Was the nocturnal placental precursor actually all-rod, as suggested by Walls (1942)? It begins to appear as if the shared characteristic across vertebrate groups is a mixed photoreceptor layer in which the *proportion* of rods and cones varies. This pattern has very different implications for the evolution of rod- or cone-dominant retinae: changes in development mechanisms simply shift the balance between the two types.

Are there any other grounds for assessing the evolutionary relations of the photoreceptor types? Ontogenetic arguments are not clear-cut. At the turn of the century, it was thought that rods developed from cones as the immature rod outer segments are conical (Bernard, 1900; Cameron, 1905). Although this transformation does not occur, the developmental cell lineage of rods and cones is not clear. In mixed retinae, rods are born after cones, but do they share a common precursor cell? This appears unlikely in teleost fish, in which rods, but not cones, continue to be produced across the retina throughout life (Johns, 1982; Branchek and BreMiller, 1984). Individual rods, unlike cones, also continuously shed and replace discs (Young, 1971). Do these differences indicate precocious maturation (progenesis) in cones or a delayed and protracted development (hypermorphosis) in rods?

It is possible that a future analysis of the molecular homologies of photopigments (e.g. Nathans et al., 1986) will reveal the spectral sensitivity of the primitive pigment. But this knowledge may not be adequate to define the receptor as rod or cone. Although cone pigments do display a broader range of peak absorbances than rod pigments, there is no absolute relation between receptor type and photopigment type (Lythgoe, 1979; Bowmaker, 1990). Other molecular markers show little receptor specificity when different species are examined: antibodies labelling rod outer segments in mammals can label cones in birds and reptiles (see Szel and Rohlich, 1989). What does emerge from a consideration of photopigments and habitats (Munz and McFarland, 1977; Loew and Lythgoe, 1978; Lythgoe, 1979) is that the spectral sensitivity of the pigment is adapted to maximize contrast between object and background. For aquatic species, the appropriate wavelength can depend on a number of factors such as the water depth, the angle of viewing and, in shallow water, the brightness of an object seen horizontally. Thus for some environments, more than one pigment may have evolved to give optimal monochromatic (light–dark) contrast detection. From such a situation, it is not difficult to imagine how output from the different receptors might be compared in order to detect colour contrasts.

Searching for a primitive receptor type may not be particularly fruitful. Most vertebrate groups share a mixed retina, differing in the proportions of rods and cones. The significance of the earlier development of cones is unclear. It can be argued that cone outer segments have a simpler organization: they do not have the separated discs or high-gain biochemical cascade of the rods. But given the difficulty in deciding what are derived characteristics, one should be cautious. Perhaps we have been considering the wrong sensory organ. The pineal apparatus in some vertebrate species retains photoreceptors and an afferent nerve, known as the parietal eye in lizards. Photoreceptors in the parietal eye have morphological similarities to cones, even if they do synapse directly onto the ganglion cells (see Hamasaki and Elder, 1977; Engbretson and Anderson, 1990). Could the parietal eye reflect a plesiomorphic state, unadapted for fine visual analysis?

Conclusion

Speculation about the nature of the ancestral photoreceptor on the basis of extant species is not productive. Photoreceptors differ in many characteristics and it is not obvious which are the plesiomorphic features. The fact that modern studies have found that most species possess mixed rod-and-cone retinae throws into question Walls's idea that rods and cones had evolved more than once. Evolutionary variation is seen in the relative proportion of the two receptor classes.

Receptor Specializations

Oil droplets are found in the photoreceptors of many species. How did they evolve? The droplets may be coloured or colourless; they are found mainly in cones but occasionally in rods (Walls, 1942; Lythoge, 1979). Only amphibia, reptiles and birds possess brightly coloured oil droplets, with different colours present in different receptors. In extant species, there appears to be no instance in which colour vision depends on different colours of oil droplets screening a single photopigment. The combination of different oil drops and different cone pigments will increase the spectral selectivity of the photoreceptor at the expense of decreased sensitivity (Bowmaker, 1977; Lythgoe, 1979). This may be useful in discrimination between different types of foliage or, if the oil droplet is yellow, in reducing the effect of scattered light (Lythgoe, 1979; Walls, 1942). However, Walls considers coloured oil droplets to reflect a primitive evolutionary state, which appears strange if their function is to increase the selectivity of an existing multi-pigment receptor system. Perhaps, coloured oil droplets could have been an efficient way of generating spectral selectivity in receptors that had not yet evolved the mechanisms for expressing a single photopigment in one receptor?

Although the taxonomical distribution of colourless oil droplets is more widespread than that of coloured droplets, Walls considers that the loss of colour is a secondary adaptation to a low-light habitat. Walls (1934) has argued this for the reptiles. Diurnal lizards are described as having all-cone retinae and coloured oil droplets. Both snakes and gekkos, which have evolved from lizards through a fossorial or nocturnal phase, have only colourless droplets. Sturgeons (chondostreans) and lungfish also have colourless oil droplets in their photoreceptors. Walls interprets the lack of colour as a secondary adaptation and their presence as evidence that oil droplets appeared early in phylogeny. This interpretation raises a question as to why teleosts do not have oil droplets. Have they lost them, or is their appearance in sturgeons and lungfish a case of convergent evolution? A further unexplored question is whether colourless droplets have any adaptive value, and that colour is actually a secondary adaptation.

Receptor Distributions

Photoreceptors vary not only in their form but also in the pattern of their distribution across the retina. It is the packing of receptors that ultimately determines the sampling of the visual image. For instance, in duplex retinae how can the image be efficiently sampled by both rods and cones operating over different luminance ranges? One elegant solution is that of the anatomically tiered retina (Miller and Synder, 1977). In the pig and peripheral human retina, light first passes through the cone outer segments and then through the outer segments of the rods; in the cat-eyed snake, the tiering is reversed. In a variety of fish, amphibia and birds anatomical tiering is achieved by virtue of a photomechanical response in the receptors (Arey,1915; Walls, 1942). In photopic conditions, cones shorten and rods lengthen with the reverse occurring in scotopic illumination. The problem is avoided in animals that have a retina dominated by one receptor type, e.g. the rod-dominated retinae of rats and elasmobranchs in which spatial analysis by cones is rudimentary. It might appear that a similar solution had evolved in the case of the fovea of diurnal primates, which is a rod-free zone. However, Wikler and Rakic (1990) have demonstrated that both diurnal and nocturnal primates have a qualitatively similar rod distribution, which peaks dorsal to the foveal pit. The implication is not that rod analysis has been degraded in diurnal primates but that a cone-rich specialization has been added to the basic nocturnal primate pattern. Differential distributions of different receptor types are quite common (Walls, 1942; Rodieck, 1973; Lythgoe, 1979) and presumably reflect the dominant spectral composition of the relevant part of the visual field.

Regional differences in photoreceptor density have been described in many species and are commonly assumed to reflect variations in visual acuity. Of course, this makes assumptions both about diversity of receptor types and about post-receptoral convergence onto ganglion cells. Regional specializations in ganglion cell distribution are discussed in a subsequent section. In cone-rich retinae the receptors often show a mosaic packing. This is particularly dramatic in teleost fish with good visual acuity (e.g. Engstrom 1963; Wagner, 1978; Lythgoe, 1979). The evolutionary pressure for this organization is unclear, especially as such regular distributions of different cone types must have implications for the spatial sampling at different wavelengths, assuming that there is no compensatory chromatic aberration in the optics (see Williams, 1986). That these cone mosaics resemble insect ommatidia has long been apparent (Ryder, 1895). There are even ultraviolet sensitive receptors present in some species (Bowmaker and Kunz, 1987). One intriguing possibility is that these cone mosaics could be genetically controlled in a similar manner to the *Drosophila* ommatidia (e.g. Tomlinson, 1988; Banerjee and Zipursky,1990).

Conclusions

A number of solutions to receptor packing in duplex retinae have evolved. Receptors can be separated through the thickness of the retina (tiering) or across its surface (mosaics and separate areae). The solutions show little phylogenetic consistency and the underlying developmental mechanisms may give better insights into their evolution.

Between the Plexiform Layers

Although the division of the retina into outer nuclear, outer plexiform, inner nuclear, inner plexiform and ganglion cell layers is common to almost all vertebrates, the details of cell type and connections comprising these layers display considerable variation between species. This diversity was thoroughly documented by Cajal (1893), who viewed the retina not so much in the context of phylogeny but more as a reflection of the balance of rods and cones. In this section, I indicate the diversity in the processing of signals between photoreceptors and ganglion cells and consider whether any phylogenetic insights can be derived from the various strategies for processing rod and cone signals.

Although the basic lamination of most vertebrate retinae is very similar, the location of the ganglion cell layer vitread to the inner plexiform layer is not found in the cyclostomes (Walls, 1942; Holmberg, 1977; Fritzsch and Collin, 1990). Fritzsch and Collin demonstrate that, in the lamprey, the inner plexiform layer is the one nearest the vitreous and suggest that this may be the plesiomorphic pattern, rather than a degenerate one. Whether the argument can be extended to describe displaced ganglion cells (Cajal, 1893; Dogiel, 1895; Karten et al., 1977) as plesiomorphic is more debatable. Nevertheless, in spite of their peculiar retinal lamination, the cyclostomes share the cell types found in other vertebrate retinae.

Bipolar cells

Cajal (1893) concluded that two classes of bipolar cells could be distinguished in teleosts and mammals; rod bipolars or cone bipolars. He observed that the dendrites of the two types differed in laminar location in the outer plexiform layer, that the dendrites could be restricted or diffuse and that rod and cone pathways may remain separate at the level of ganglion cells.

Since Cajal, it has become clear that the distinction between rod and cone bipolars only applies to mammals. There is only one type of mammalian rod bipolar. It receives input from 15–40 rods, stains positively for protein kinase C and projects on the ON-sublamina of the inner plexiform layer where it synapses with amacrine cells (see Dowling, 1987; Sterling, 1990; Greferath et al., 1990). Mammalian cone bipolars are more variable. Sterling (1990) distinguishes 10 types of cone bipolar in the cat and at least 6 have been reported in the primate including the midget and diffuse bipolar classes (see Rodieck, 1988). Teleost bipolars either receive only cone input or mixed rod–cone input (Scholes, 1975; Wagner, 1990) and a shared rod–cone bipolar seems characteristic of non-mammalian vertebrates (Dowling, 1987). The organization of cone bipolars also reveals different strategies for preservation of chromatic information. The primate midget bipolar receives input from a single cone (Polyak, 1941; Rodieck, 1988), whereas one class of teleost cone bipolar has a wide dendritic field which contacts cones of a specific type and another class has a small dendritic field receiving mixed red–green and blue–green input (Scholes,1975).

Another important development has been the recognition of separate ON- and OFF-centre bipolars, a division common to all vertebrate bipolars (see Dowling, 1987; Sterling, 1990). ON- and OFF-bipolars can be distinguished physiologically and also by the location of their axons in the inner plexiform layer (Famiglietti and Kolb, 1976; Famiglietti et al., 1977; Nelson et al., 1978). In eutherian mammals virtually all bipolars terminate in either the ON- or OFF-sublaminae, although one class of cone bipolar does have a bistratified axonal arbour (Rodieck, 1988; Mariani, 1990). The echidna shares with non-mammalian species, including teleosts, bipolar cells whose axons arborize in more than one sublamina of the inner plexiform layer (Dowling, 1987; Negishi et al., 1988; Wagner, 1990; Young and Vaney, 1990). This circuitry implies much more mixing of the ON- and OFF-pathways in non-mammalian retina, although just how the mixing occurs depends on the details of synaptic circuitry in the inner plexiform layer.

The basic synaptic circuitry underlying the formation of ON- and OFF-centre pathways appears to be similar in different species. Photoreceptors synapse with bipolar cells and horizontal cells, making either conventional or ribbon synapses. Although the detailed morphology may vary with species and receptor type, the ON-bipolar is associated primarily with the ribbon synapse whereas the OFF-bipolar is associated with conventional synapses (see Dowling, 1987). Although centre–surround opponency is a common feature of bipolar cells, a variety of retinal circuitry may generate it. Horizontal cells can inhibit bipolar cells directly or feedback onto photoreceptors and it is possible that interactions in the *inner* plexiform layer could be involved, especially for chromatic opponency (see Dowling, 1987; Kaplan et al., 1989).

Horizontal Cells

Horizontal cells mediate lateral spread of visual responses in the outer plexiform layer and can contribute to the antagonistic surround observed in bipolar cells. There are two basic types of horizontal cell in vertebrate retinae: with and without axons (Boycott et al., 1978; Gallego, 1985, 1986). Axonless horizontal cells have a very similar morphology in all vertebrate species, although they have not been described in primate (see Boycott et al., 1987). Horizontal cells with axons have a more variable morphology, especially in the nature of their axonal arbour, and several subtypes have been defined. It is not difficult to view the different morphology of horizontal cells as variations on a single theme. What is more interesting is differences in their connections and roles in visual processing.

Electrical recordings from horizontal cells revealed two types of responses (see Dowling, 1987). Most horizontal cells hyperpolarize to illumination irrespective of the wavelength of the light; these are the luminosity, L-type cells. Some horizontal cells respond differently when wavelength is varied, hyperpolarizing to one wavelength and depolarizing to others; these are the chromaticity or C-type cells. C-type horizontal cells are found only in species with well-developed colour vision such as turtles and teleost fish (Svaetichin and MacNichol, 1958; Tomita,1965; Fuortes and Simon, 1974). The finding of C-type potentials in the retina of holostean fish was taken by Burkhardt et al., (1983) to support the idea that colour vision could have had a very early common origin from an aquatic, jawed vertebrate. This view implies that the C-type horizontal cell, with its complex connectivity (Stell et al., 1982; Toyoda et al., 1982), is a plesiomorphic feature associated with the appearance of colour vision. However, it should be noted that C-type horizontal cells are not necessary for good colour vision. All mammalian horizontal cells are L-type (but see Boycott et al., 1986, on the significance of imbalanced cone input in primate horizontal cells).

Interplexiform Cells

Retinal neurones that extend processes into both plexiform layers were described by Cajal (1893), but it is only recently that they have been recognized as a distinct group either as a sixth type of retinal neurone (see Dowling, 1987) or as a subtype of amacrine cell (e.g. Sterling, 1990). They have been found in a variety of vertebrates: teleosts, amphibia, elasmobranchs and mammals (see Dowling, 1987). Such cells provide a centrifugal pathway through the retina, from ganglion cells to receptors, but their exact function in different species is not certain. In teleost fish, the dopaminergic interplexiform cells probably modulate gap junctions between horizontal cells. They also receive input from the retinopetal fibres found in many non-mammalian species (Dowling, 1987).

Amacrine Cells

Generalizations about amacrine cells are difficult. Even their defining characteristic, as axonless neurones, has been questioned (Dacey, 1989). The safest generalization is that they constitute the most diverse group of retinal neurones. The teleost roach currently holds the record with 43 morphological types (Wagner and Wagner, 1988) but at least 26 types have been recognized by Golgi staining in the primate retina (Mariani, 1990). There is also considerable variety in the range of neurotransmitters and neuromodulators expressed in amacrine cells (see Rodieck, 1988; Lasater, 1990). How is this diversity reflected in the functional role of amacrine cells?

One common view, based on the organization of the inner plexiform layer, is that the complexity of ganglion cell receptive fields is correlated with the amount of amacrine cell input (e.g. Dowling, 1987; Sterling, 1990; Djamgoz and Yamada, 1990). In the inner plexiform layer, bipolar cells make only ribbon synapses onto two postsynaptic elements, either two amacrine cells or a ganglion cell and an amacrine cell. Amacrine cells make conventional synapses onto a variety of postsynaptic cells. In frogs and birds, the ratio of conventional amacrine synapses involving amacrine cells to bipolar ribbon synapses is much higher than in cat and primate (see Dowling, 1987). However, the absolute density of the bipolar ribbon synapse shows much less variation, which implies that variability in amacrine connectivity is responsible for differences in ganglion cell properties observed in the different species. More specific evidence is available for particular cell types. In many species, ON-OFF retinal ganglion cells are directionally selective and receive input from transient amacrine cells (see Dowling, 1987; Djamgoz and Yamada, 1990). Input from sustained amacrine cells may well generate the orientation selectivity displayed by some catfish ganglion cells (Naka, 1980). In the cat retina, alpha ganglion cells receive 85% of the synaptic input from amacrine cells, which may explain their transient visual responses (see Sterling, 1990). Cat beta cells have 30% of their synapses from amacrine cells, many of which arise from the AII amacrine cell that is specialized for processing of signals from the rods (see Sterling, 1990).

The variety of amacrine morphology, transmitters and connectivities poses a challenge for any phylogenetic synopsis. The obvious conclusion is that the evolution of amacrine cells permitted diversity of visual analysis in the retinal ganglion cell population. For instance, the emergence of transient visual responses would greatly facilitate motion detection. But as Masland (1988) points out, the cat and primate have relatively simple inner plexiform layers but a wide range of amacrine cell types. He suggests, on the basis of amacrine cell mosaics, that many classes of mammalian amacrine cells may not be directly involved in spatial vision.

Retinal Ganglion Cells

The literature on variation in vertebrate retinal ganglion cell populations is extensive, but many of the techniques employed have inherent limitations. The techniques that have been used most widely in comparative morphological studies have been the Golgi stain (e.g. Cajal,1893) and the Nissl stain (see Hughes, 1977; Stone, 1981). Nissl stains reveal cell bodies and have been widely used to plot ganglion cell isodensity distributions. With care it is possible to discriminate neurones and glia but the distinction between ganglion cells and displaced amacrine cells is not clear. Although different ganglion cell classes differ in the size of their cell bodies, this is not a reliable indicator. Thus estimates of visual acuity from ganglion cell density (e.g. Hughes, 1977) are likely to be overestimates in the absence of information concerning relative distribution of different classes. The reduced silver stain of Boycott and Wassle, which stains neurofibrils to give a Golgi-like staining of one class of retinal ganglion cell (see Peichl and Wassle, 1981; Peichl et al., 1987) has also proved very useful. The Golgi stain itself reveals neuronal types at random, so information about the density of ganglion cell classes is lost but exquisite dendritic detail is revealed (e.g. Cajal, 1893; Boycott and Dowling, 1969; Boycott and Wassle, 1974). Similar detail can be obtained by intracellular injection of tracer, which can also be combined with physiological recording or with retrograde tracers so that dendritic morphology can be compared with other properties of the ganglion cell. A major problem with all these studies is that there are no objective methods for classification on the basis of dendritic morphology. Visually, it is easy to detect differences between dendritic patterns, but defining similarities is more difficult.

Ganglion cell classification on the basis of physiological characteristics is also problematic (see Rodieck and Brening, 1983; Stone, 1983). Not only are there problems with qualitative parameters (one experimenter's 'sustained' may be another's 'transient'), but also it is difficult to identify quantitative parameters that can be used to define a class or an incidental property of that class (e.g. spatial frequency tuning alone does not discriminate ganglion cell classes). Finally, physiology gives no clear indication of the incidence of a given ganglion cell type and whether the visual image on the retina is reliably sampled by that ganglion cell population (see Wassle et al., 1983).

Given these problems, caution should be applied to the interpretations given below especially when comparisons are made between different orders of vertebrates. None the less, I shall proceed to examine variations in vertebrate ganglion cell numbers, in their retinal distribution and in their morphological and physiological diversity.

Ganglion Cell Numbers

There is tremendous variation between species in the total number of retinal ganglion cells. Can any phylogenetic conclusions be drawn from this variation? Not surprisingly, the lowest numbers of ganglion cells are found in those species with degenerate eyes, e.g. blind cavefish (Vonedia and Sligar, 1976), the burrowing snakes *Typhlops* and *Calabaria* (Halpern, 1973; Reperant et al., 1987) and the mole (Lund and Lund, 1965). Whether the lamprey eye, with 12 000–35 000 ganglion cells, can be considered degenerate is unclear (Holmberg, 1977; Vesselkin et al., 1989; Fritzsch and Collin, 1990). Most other studies, on non-degenerate retinae, have focused on mammals and have described a tremendous range, from a few thousand in some microchiropteran species (Pettigrew et al., 1988) to over 1 million in rhesus monkey and in man (Rakic and Riley, 1983; Provis et al., 1985). There are relatively few comparative studies on non-mammalian species. Estimates for teleost fish are complicated by the continual growth of the retina through life; values range from 200 000 to 800 000 in different species (Kock and Reuter, 1978; Johns and Easter, 1977; Ito and Murakami, 1984). A comparative paper on procellariiform seabirds (Hayes and Brooke, 1990) produced total ganglion cell estimates ranging from 600 000 in the common diving petrel to over 2 million in the Kerguelen petrel (Hayes and Brooke, 1990). Chick, pigeon and quail all have values in excess of 2 million (Bingelli and Paule, 1969; Rager and Rager, 1978; Ikushima et al., 1986).

The significance of total ganglion cell counts is limited in the absence of information on eye size and ganglion cell distributions. Nevertheless, it is interesting to speculate about the variations in numbers seen in different groups of vertebrates. For instance, does the range of values seen in mammals say anything about the evolution of retinal ganglion cell populations. For instance, is there a relation between the dominance of different photoreceptor types and ganglion cell numbers? Among rodents, the rod-dominated myomorph eyes (mouse, rats, etc.) have 50 000–150 000 ganglion cells, whereas the cone-dominated sciurid eyes have totals in the range of 500 000 to 1 million (Drager and Olsen, 1980; Tiao and Blakemore, 1975; Long and Fisher, 1983; Perry et al., 1983; Wakakuwa et al., 1987). Similarly, for amazonian hystricomorph rodents, ganglion cell number correlates better with degree of diurnal behaviour than with retinal area (Silveira et al., 1989a). Within the primates, man, the rhesus monkey and the dichromatic *Cebus* monkey all have more than 1 million ganglion cells, whereas the rod-dominated owl monkey appears to have many fewer ganglion cells (Webb and Kaas, 1976; Rakic and Riley, 1983; Povis, 1985; Silveira et al., 1989b; Curcio and Allen, 1990). It seems that, for closely related species, nocturnal animals have fewer ganglion cells than diurnal animals.

Fossil evidence suggests that the primitive mammal was a small-eyed nocturnal insectivore, presumably with a rod-dominated retina. The ancestral primate is also thought to be nocturnal, but with large eyes (Martin, 1989). It seems likely that good colour vision has evolved more than once in the mammalian line (the diurnal squirrels and the diurnal primates), and with it an increase in the numbers of ganglion cells that are necessary to convey the spectral information.

Obviously, it is impossible to determine the importance of vision for the ancestral mammal. It does not seem likely that its retina was degenerate but even degeneracy is not irreversible in evolutionary terms. Walls (1942) argues, on several grounds, that the ancestral form of the snakes was a fossorial lizard with atrophied eyes. Subsequent emergence of the snakes led to evolution of a duplex retina with relatively large numbers of ganglion cells e.g. 127 000 in the American garter snake and 60 000 in the European viper (Wong, 1989; Ward et al., 1987). Whether the microophthalmia of the burrowing snakes *Typhlops* and *Calabaria* (Halpern, 1973; Reperant et al., 1987) represents persistence of the ancestral form or a recurrence of degeneracy is unclear.

Conclusion

Few obvious consistencies emerge from a widespread phylogenetic comparison of total ganglion cell numbers. However, within closely related mammalian species, ganglion cell number does appear to correlate with the ratio of cones to rods. Does this imply a developmental linkage in which evolutionary selection for cone-vision also results in increased ganglion cell numbers?

Ganglion Cell Distributions

The packing of retinal ganglion cells is an important constraint on visual behaviour. A wide variety of ganglion cell distributions have been reported and have been correlated with species' behaviour. How much does this approach tell us about the evolution of such distributions?

There are numerous studies on the isodensity distributions of ganglion cells in vertebrate species. Most commonly, the studies have concentrated on the degree of foveal or visual streak specialization. The rationale is that regions of high ganglion cell density are associated with high visual acuity, and so mapping the isodensity curves onto the visual field indicates something of the behavioural specialization of the animal. Inevitably, many of the studies are of post-mortem material stained for Nissl material, which presents some problems in interpretation. Care needs to be taken in distinguishing between retinal ganglion cells, displaced amacrine cells and glia. Identification of ganglion cells is easier if retrograde axonal tracers are employed. Such tracers also facilitate identification of different classses of retinal ganglion cells that may be sampling the retinal image in parallel. Finally, in order to draw conclusions about absolute visual acuities, the retinal magnification factor has to be determined. Nevertheless, much interesting information has accumulated.

Inhomogeneity in ganglion cell distributions is most pronounced in species with foveal pits. Foveas are characterized by a high density of photoreceptors, usually cones, and a lateral displacement of the other retinal layers thus generating the pit. Various theories have been proposed to explain the advantages of the pit, ranging from the optical to the metabolic (see Hughes, 1977), but one consequence is that the density of ganglion cells surrounding the pit is very high (up to $38\,000\,mm^{-2}$ in man (Curcio and Allen, 1990) and $33\,000\,mm^{-2}$ in macaque (Perry and Cowey, 1985)). Even in the absence of a foveal pit, many species have distinct ganglion cell concentrations, which can take the form of either a roughly circular area centralis or an elongated visual streak. The form of the ganglion cell distribution does not always follow the photoreceptor density distributions. Photoreceptor distributions, especially those of rods, tend to be more homogeneous, so that changes in ganglion cell distribution must also reflect differences in receptor–ganglion cell convergence. However, the detail with which different parts of the visual world are analysed must be determined by the ganglion cells, which constitute the retinal output.

One of the most influential ideas about the functional significance of ganglion cell distributions has been the terrain hypothesis of Hughes (1977). Many early workers had noted that variations in ganglion cell topography had no simple phylogenetic basis, and suggested functional adaptation. Hughes looked at retinae from nearly 100 mammalian species and distinguished 5 classes of ganglion cell distribution, as follows:

1. low ganglion cell density with little centroperipheral density gradient (small nocturnal inhabitants of undergrowth);

2. high ganglion cell density with little centroperipheral density gradient (diurnal retinae of tree shrew or squirrel);

3. steep centroperipheral density gradient (man and other diurnal primates);

4. central or temporal area superposed on a visual streak (cat and many other carnivores);

5. visual streak unassociated with an area (rare in mammals: rabbit, if ganglion cell classes are ignored).

These patterns are not confined to the mammals, but reports of isodensity maps from other vertebrate groups are rarer. In a comparative study of reef teleosts, Collin and Pettigrew (1988a,b) found that fish had well-developed areae, which, in some species, were associated with pronounced horizontal streaks. Unlike mammalian retinae, the fish retinae sometimes possessed more than one area. Some birds also have more than one area, a pattern most pronounced in bifoveate raptors (Walls, 1942; Fite and Rosenfield-Wessels, 1975; Hayes and Brooke, 1990). Other species have a single area, a horizontal streak or both (Meyer, 1977; Ikushima et al., 1986; Wathey and Pettigrew, 1989; Hayes and Brooke, 1990).

The terrain hypothesis states: 'The visual streak is common to terrestrial species whose field of view is not completely obscured by nearby vegetation and not simply to those which occupy open country, as suggested by Luck (1965); the terrain surface is overt in all but the most overgrown woods, above the stream bed to a fish, or above the sea to a flying or floating bird' (Hughes, 1977). Hughes proceeds to consider the geometry of the terrain's projection on the eye. Briefly, the retinal projection of distant objects on the terrain is condensed round the horizontal meridian. (For an animal of eye height h, objects that are distant by more than $2.5h$ project to the lower 21° of the superior retina.) The terrain argument requires that the streak should coincide with the visual horizon with the head in its normal position. However, the reason why the streak is symmetrical about the horizontal meridian is not immediately obvious since, in open country, inferior retina views the sky. Whereas land or the surface of water has a planar projection, this is not true for the sky unless, as Hughes suggests, a relatively constant altitude of predators could define a 'terrain' in the superior visual field.

Hughes also considers the evolution of the circumscribed area centralis in mammals. Even in species with visual streaks, there is a region of particularly high ganglion cell density that is radially symmetric; in many mammals, especially the primates, the area centralis is more pronounced than the visual streak. Noting that the visual projection of the area centralis usually coincides with the intersection of the vertical and horizontal meridia, Hughes suggests that this is the region least affected by optical flow during forward motion of the animal, and so is best situated for fine spatial analysis. Stone's (1938) view is that the area centralis arises from a conjunction of the horizontal specialization of the streak and a vertical specialization associated with binocular depth vision. He sees it as the retinal fixation point for binocular vision in mammals. It coincides with the naso-temporal division of retinal ganglion cells into crossed and uncrossed populations. The zone of naso-temporal overlap should be densely sampled by ganglion cells to ensure good depth resolution for objects located straight ahead of the animal. Stone goes on to propose that the area centralis and the visual streak are basic and independent features of the mammalian retina and that differences between species are merely quantitative (although recently, Stone and Halasz (1989) have failed to find an area centralis in the elephant).

For Hughes (1977) and Stone (1983), high ganglion cell densities reflect high spatial resolution, which is associated with regions of visual importance, defined either by the retinal projection of the visual environment or by the constraints of binocular vision. The extension of these arguments to non-mammalian species has been less extensive. In their study of ten reef teleosts, Collin and Pettigrew (1988a,b) noted that those species with pronounced horizontal streaks lived in the open with a clear sand–water horizon and that those species that lacked a streak but possessed a symmetric area centralis lived among the coral in 'enclosed' environments. The work of Hayes and Brooke (1990) on the distribution of retinal ganglion cells in procellariiform seabirds (albatrosses, petrels, etc.) emphasizes that feeding behaviour may be just as important as habitat. All the species described spend most of the day and night flying over open sea (except at breeding time) yet there was considerable variation in ganglion cell density patterns. Hayes and Brooke suggest that a strong horizontal streak is associated with capture of prey located near the horizontal meridian of the visual field, whereas species with dorsal area centrales tend to seize prey under water after diving from either the air or the surface. Such associations are inevitably tentative, but given the remote habitats of procellariiform seabirds, more may be known about their retinae than their feeding behaviour, so that hypotheses about the former can be tested.

The phylogenetic relation of areae centrales in fish and birds to those in mammals is unclear. The two-axis model of Stone (1983) is not obviously applicable, as the axis defined by the naso-temporal divison does not exist, to a first approximation, in non-mammalian species. Similarly, the central fovea of bifoveate birds (Meyer, 1977) will not correspond either to a location of minimum optic flow (Hughes, 1977) or a centre of binocular fixation (Stone, 1983). However, the lateral fovea is associated with binocular vision (Meyer, 1977; Hodos et al., 1985; Hayes et al., 1987; Frost et al., 1990).

An ability to test hypotheses about the evolutionary significances of retinal ganglion cell distributions is obviously important, since it is clear that the majority of comparative descriptions adopt the adaptationist approach. Particular ganglion cell density patterns are adapted to particular visual analyses that are beneficial to life in a particular niche. A statement to the effect that concentrations of ganglion cells represent areas of visual importance obviously runs into the dangers of circularity. What is required from the adaptationist approach is a quantitative association between visual behaviour patterns and the manner in which the ganglion cells sample the visual image before theories that have good predictive value can be generated.

No grand phylogenetic conclusions can be drawn about adult ganglion cell distributions: visual streaks, areae centrales and foveas are apparent in a wide range of vertebrate species. Perhaps instead of asking how these patterns evolved we should ask how they develop or rather if there are common ontogenetic mechanisms that can explain the diversity of adult patterns. When the retinae of amphibia and mammals are examined early in development, the ganglion cells are much more uniformly packed than in the adult (e.g. Stone et al., 1982; Lia et al., 1987; Beazley et al., 1989; Wikler and Finlay, 1989). Subsequent heterogeneity can be generated by a variety of mechanisms: selective generation of ganglion cells, selective loss of ganglion cells or non-uniform expansion of the retina.

In all retinae examined so far, ganglion cells are added at the retinal margins. In amphibia there is an asymmetry in ganglion cell generation, which Beazley et al. (1989) suggest is responsible for the formation of a horizontal streak. Compared with amphibia and, especially, fish, the period of ganglion cell generation in mammals is short. In mammals, argue Wikler and Finlay (1989), a combination of relatively small variations in the initial distribution of ganglion cells followed by regional differences in retinal expansion, rather than cell death, determines the various adult patterns. They invoke heterochrony, variations in the rate or duration of development, as the evolutionary mechanism by which different mammalian retinal organizations emerge. Whether there is a fundamental disagreement between the heterochronic explanation and the 'common timetable' for retinal development espoused by Dreher and Robinson (1988) is not clear. Although Dreher and Robinson suggest that many events in retinal development bear a fixed relation to the period between conception and eye-opening, other aspects of development such as cell cycle time do not. What Wikler and Finlay provide is a basis for quantifying the mechanisms underlying the development of the retinal ganglion layer in different species and then asking whether these differences can be generated by relatively simple differences in timing. For instance, how can the different isodensity contours that are found associated with different cell classes in one retina (e.g. Stone, 1983; Bravo and Pettigrew, 1981) be generated?

Conclusion

Comparative studies of ganglion cell distributions tend to adopt the adaptationist view: that a particular retinal organization is optimized for the environment of the species showing that organization. Such arguments run into the danger of circularity. However, the studies have revealed that ganglion cell distributions do correlate better with behaviour and habitat than with phylogeny (closely related species can have different distributions and vice versa). The implication is that relatively small changes in the developmental sequence can generate these patterns of convergent evolution.

Ganglion Cell Classes

Projection Patterns

One basis for classification is the nature of the anatomical connections. As we have seen, ganglion cells can vary in the nature of their input, in terms of both cell type and laminar location in the inner plexiform layer. They can also vary in their outputs to central visual nuclei and to the sublaminae of those nuclei. The axons of retinal ganglion cells pass through the chiasm, either crossing the midline or remaining uncrossed, and then synapse in various parts of the diencephalon or in the tectum (for discussion of phylogeny and nomenclature of these areas see Sarnat and Netsky, 1981). The exact projection pattern for a given type of ganglion cell varies with species.

There seems to be general agreement that the basic vertebrae pattern is for ganglion cells to cross the midline and project to the optic tectum (or in the mammals the superior colliculus). Phylogenetic adaptations might be seen as the appearance of uncrossed retinal projections and of the retino–thalamic pathway. But such bald statements need qualification. Neither partial decussation nor retino–diencephalic projections are restricted to the mammals. Information is required about the relative numbers and types of ganglion cells comprising the different pathways.

Although the crossed retino–tectal projection is the numerically dominant pathway in most species, it is not a uniform one. Both anterograde and retrograde tracing studies reveal that it is comprised of several classes of retinal ganglion cell. The classes of cells projecting to the diencephalon may also differ from those going to the tectum. In rat and rabbit, virtually all cells project to the superior colliculus, but a distinct subset branches to the lateral geniculate (Provis and Watson, 1981; Linden and Perry, 1983; Drager and Hofbauer, 1984). In cat and monkey, the retino–geniculate projection is numerically dominant and few cells branch to the superior colliculus (e.g. Illing and Wassle, 1981; Perry and Cowey, 1984). However, it should be noted that the *absolute* numbers of retinal ganglion cells going to the superior colliculus of rats, cats and monkeys is approximately the same (100 000). The distributions between retino–diencephalic and retino–tectal classes is not confined to mammals. Bravo and Pettigrew (1981) have also described ganglion cells in owls that project only to the visual thalamus and not to the optic tectum.

The pattern of retinal decussation also varies both with species and ganglion cell class. While uncrossed retinal projections have been reported in fish (see Springer and Landreth, 1977; Prasada Rao and Sharma, 1982; Meek, 1990), transitorily in birds (O'Leary et al., 1983) and, after metamorphosis, in amphibia (e.g. Hoskins, 1989) they have been described in most detail for mammals. In all mammals, the vast majority of ganglion cells projecting ipsilaterally are confined to temporal retina, i.e. to that part of the retina viewing the binocular component of the contralateral visual field. What varies with species and cell class is the proportion of ganglion cells in temporal retina that project *contralaterally*. In this part of the rodent retina, the crossed projection is numerically dominant (Drager and Olsen, 1980) and is substantial for the cat's retinotectal pathway (Wassle and Illing, 1980). However, primates show a complete partial decussation, i.e. all temporal retina projects ipsilaterally and all nasal retina contralaterally (e.g. Cynader and Berman, 1972). It was the discovery that Microchiroptera had the plesiomorphic decussation pattern (that of the cat and rodent), whereas Megachiroptera had the primate pattern that led Pettigrew (1986) to challenge the monophyletic origin of the bats (see Pettigrew et al., 1989).

Form and Function

The diversity of retinal ganglion cell types was firmly established by Cajal (1893), on the basis of both dendritic field size and the lamination of dendrites in the inner plexiform layer. Subsequent morphological classifications also relied on the dendritic branching patterns seen in retinal whole-mounts after either Golgi staining or intracellular dye injection. The functional significance of this diversity emerged more slowly. The earliest recordings emphasized the centre-surround organization of receptive fields divided into ON-, OFF- and ON-OFF categories (Hartline, 1938; Barlow, 1953; Kuffler, 1953). As the variety of functional observations exploded, the grounds for classification of retinal ganglion cells on either functional or morphological criteria became unclear (see Rowe and Stone, 1977; Rodieck and Brening 1983; Stone, 1983). From an evolutionary perspective, one can ask two questions. Can the functional diversity be related to possible selection pressures (the adaptationist approach)? How can ontogenetic mechanisms be modified to generate such diversity?

One frequent phylogenetic generalization about ganglion cell diversity is that complexity of visual analysis in the retinal ganglion cell layer is inversely correlated with the importance of visual cortex (e.g. Mollon, 1990; Sterling, 1990). Thus sophisticated visual analysis reflects cortical mechanisms in primates and cats, whereas in frogs and rabbits it depends more on retinal processing. Is this generalization accurate, and if so what is the evolutionary route?

The lamination of the inner plexiform layer supports the view that more complex analysis is occurring in the non-mammalian retina. Cajal's (1983) drawings reveal much more complex patterns of dendritic lamination in non-mammal vertebrates. The mammalian inner plexiform has been divided into two zones, corresponding to ON- and OFF-channels (Famiglietti and Kolb, 1976; Nelson et al., 1978), although immunocytochemistry suggests more complex lamination patterns (see Koontz and Hendrickson, 1987). Morphological diversity among mammalian retinal ganglion cells is obviously greater than the ON-OFF dichotomy. This is particularly true for retino-collicular projection, in which the ganglion cells show considerable morphological diversity (Boycott and Wassle, 1974; Perry, 1979; Illing and Wassle, 1981; Kolb et al., 1981; Linden and Perry, 1983; Perry and Cowey, 1984; Leventhal et al., 1985). This morphological diversity is matched by physiological diversity; most retino-collicular cells belong to the W-cell class. This is a heterogeneous group that includes receptive field types that are more complex than the simple centre-surround organization described by earlier workers (see Stone, 1983; Levick, 1986). Properties like edge-detection and directionality have been described for certain mammalian W-cells and are very like those seen in non-mammalian vetebrates. Some of the early descriptions of retinal ganglion cells' receptive fields in frog and pigeon included categories such as dimming detectors, boundary detectors and direction detectors (Maturana et al., 1960; Lettvin et al., 1961; Maturana and Frenk, 1963). Some of these categories probably simply reflect the classic centre-surround organization, but subsequently more quantitative studies have confirmed the existence of complex receptive fields in the ganglion cells of amphibia and birds (e.g. Miles, 1972; Witpaard and ter Keurs, 1975; Holden, 1977a,b). The spatial and chromatic properties of teleost ganglion cell receptive fields have been well studied (see Guthrie, 1990). They include receptive fields that display both spatial and chromatic opponency (double-opponent cell: Daw, 1968), characteristics which, in the primate visual system, have not been found outside visual cortex (Livingstone and Hubel, 1984).

The implications of this brief survey of ganglion cell characteristics is that all vertebrate species display a wide range of receptive field types but that, in the mammals, diversity is particularly evident in the retino-collicular projection. Birds may well be similar. The retino-thalamic pathway in the owl is dominated by concentric ON- or OFF-centre units (Pettigrew,1979), although there are many directional ganglion cells in the avian retina, presumably projecting to the optic tectum. Thus the phylogenetic 'encephalization' of visual analysis referred to earlier reflects an increasing dominance of the retino-geniculate projection in which the retinal ganglion cells perform a relatively simple visual analysis. One can ask whether the evolution of a large retino-geniculate pathway reflects the selective amplification of an existing, but small, population of ganglion cells, or whether it reflects the emergence of new ganglion cell types. There are some grounds for favouring the latter possibility. Alpha retinal ganglion cells are common to many mammalian retino-geniculate projections (Peichl et al., 1987) but the equivalent of the carnivore beta-cells is missing in rodents (Perry, 1979). Similarly, there is no obvious homologue of the primate P-, or midget, retinal ganglion cells in subprimate species, certainly not the cat beta cell (see Shapley and Perry, 1986). Variations in the diversity of ganglion cells projecting to the geniculate are also suggested by the variability in geniculate lamination (see Kaas et al., 1972; Pettigrew et al., 1989).

The ontogenetic mechanisms that generate diversity of retinal ganglion cells are largely unknown, although it may be assumed that morphological diversity indicates functional diversity (e.g. Pomeranz and Chung, 1970; Saito, 1983; Amthor et al., 1989a,b). Morphological diversity among cat retinal ganglion cells is correlated with neuronal birthdates; small, beta, ganglion cells projecting to the geniculate are born first (Walsh et al., 1983). Perhaps heterochronic mechanisms bias the ganglion cell population in favour of a retino-thalamic projection. Such a suggestion raises a number of questions. We know very little about what determines class-specific targeting in ganglion cells, let alone how it may be altered by heterochrony.

Conclusion

Comparative studies lend support to the view that complexity of ganglion cell form and function is inversely correlated with the degree of encephalization of visual analysis. In mammals, and possibly birds, the evolutionary route appears to be via the addition of specific retino-geniculate ganglion cell types to the existing retino-collicular groups. Further understanding of the evolution of different cell types may come from understanding their development.

General Observations

It should be clear from the literature cited above that variations in retinal organization are those of detail rather than of substance. Moreover, the variations do not display any grand phylogenetic changes. The correlations that exist tend to be more closely related to the behaviour of a species rather than to its Linnaean pedigree. For instance, the 'terrain hypothesis' relates ganglion cell distributions to the environment and can be applied equally to fish, birds and mammals. Similarly, as Cajal observed, many of the variations in retinal pathways have a closer relation to the rod–cone balance than to phylogeny. These observations imply a high incidence of convergent evolution. Similar solutions have evolved separately in different lines, e.g. within mammals, cone-dominated retinae have appeared both in the sciurid rodents and in the primates. What are the evolutionary implications of this conclusion?

I think an important perspective is the ontogenetic one. We need to understand the developmental mechanisms that underlie the diversity of neurones and their connections in the retina. It is through genetic changes in these mechanisms that the species variation in the adult is derived. There are many questions about retinal development. We do not understand the developmental origins of the basic six retinal cell types, let alone the basis of their numerous variations. The importance of developmental interactions between the different cell types remains to be fully explored. Even in the adult, the different cell types are not independent: variation in one cell type is often associated with variation in another. It is highly probable that there is a developmental cascade in which interactions between cells define the adult pattern. We know little about such interactions and their implications for potential variations in retinal organization. Nor do we know much about the mechanisms that control the numbers of neurones. These could have considerable evolutionary implications, since one aspect of variation in adult retinae is in the relative proportions of different neuronal subclasses.

In emphasizing ontogeny, I am concentrating on the way in which evolution can change retinal organization and am ignoring the functional, adaptational consequences of those changes. This is partly because of the danger of circularity in the adaptationist approach. A species with a given behaviour pattern has a particular retinal organization; therefore that retinal organization is adapted to that behaviour pattern. The circle can be broken when we know sufficient about the functional implications of retinal organization to accurately predict unknown behaviour patterns from a known retinal organization or vice versa. But as our knowledge of the functional significance of retinal circuitry increases, such predictions will become easier and more accurate.

References

Amthor, F. R., Takahashi, E. S. and Oyster, C. W. (1989a). Morphologies of rabbit retinal ganglion cells with concentric receptive fields. *J. Comp. Neurol.*, 280, 72–96.

Amthor, F. R., Takahashi, E. S. and Oyster, C. W. (1989b). Morphologies of rabbit retinal ganglion cells with complex receptive fields. *J. Comp. Neurol.*, 280, 97–121.

Anderson, D. H. and Fisher, S. K. (1976). The photoreceptors of diurnal squirrels: outer segment structure, disk shedding, and protein renewal. *J. Ultrastruct. Res.*, 55, 119–141.

Arey, L. B. (1915). The occurrence and the significance of photomechanical changes in the vertebrate retina—an historical survey. *J. Comp. Neurol.*, 25, 535–554.

Banerjee, U. and Zipursky, S. L. (1990). The role of cell–cell interaction in the development of Drosophila visual system. *Neuron*, 4, 177–187.

Barlow, H. B. (1953). Summation and inhibition in the frog's retina. *J. Physiol.*, 119, 69–88.

Beazley, L. D., Dunlop, S. A., Harman, A. M. and Coleman, L.-A. (1989). Development of cell density gradients in the retinal ganglion cell layer of amphibians and marsupials: two solutions to one problem. In *Development of the Vertebrate Retina*. Eds Finlay, B. L. and Sengelaub, D. R. pp. 199–226. New York: Plenum.

Bernard, H. H. (1900). Studies in the retina: rods and cones in the frog and in some other amphibia. Part I. *Q. J. Microsc. Sci.*, 43, 23–27.

Bingelli, R. L. and Paule, W. J. (1969). The pigeon retina: quantitative aspects of the optic nerve and ganglion cell layer. *J. Comp. Neurol,.* 137, 1–18.

Bowmaker, J. K. (1977). The visual pigments and oil droplets of the chicken retina. *Vision Res.*, 17, 1129–1138.

Bowmaker, J. K. (1990). Visual pigments of fish. In *The Visual System of Fish*. Eds Douglas, R. H. and Djamgoz, M.B.A. pp. 81–107. London: Chapman & Hall.

Bowmaker, J. K. and Kunz, Y. W. (1987). Ultraviolet receptors, tetrachromatic colour vision and retinal mosaics in the brown trout (*Salmo trutta*): age-dependent changes. *Vision Res.*, 27, 2101–2108.

Boycott, B. B. and Dowling, J. E. (1969). Organization of the primate retina: light microscopy. *Phil. Trans. R. Soc. Lond. B*, 255, 109–184.

Boycott, B. B. and Wassle, H. (1974). The morphological types of ganglion cells of the domestic cat's retina. *J. Physiol.*, 240, 397–419.

Boycott, B. B., Hopkins, J. M. and Sperling, H. G. (1987). Cone connections of the horizontal cells of the rhesus monkey's retina. *Proc. R. Soc. Lond. B.*, 229, 345–379.

Branchek, T. and BreMiller, R. (1984). The development of photoreceptors in the zebrafish, *Brachydanio rerio*. I. Structure. *J. Comp. Neurol.*, 224, 107–115.

Bravo, H. and Pettigrew, J. D. (1981). The distribution of neurons projecting from the retina and visual cortex to the thalamus and tectum opticum of the barn owl, *Tyto alba*, and the burrowing owl, *Speotyto cunicularia*. *J. Comp. Neurol.* 199, 419–441.

Burkhardt, D. A., Gottesman, J., Levine, J. S. and MacNichol, E. F. (1983). Cellular mechanisms for color-coding in holostean retinas and the evolution of colour vision. *Vision Res.*, 23, 1031–1041.

Cajal, S. R. y (1893). La rétine des vertèbres. *La Cellule*, 9, 119–257. Transl. (1972) *The Structure of the Retina* by Thorpe, S. A. and Glickstein, M. Springfield: Thomas. Also transl. (1973) as *The Vertebrate Retina*. by Maguire, D. and Rodieck, R. W. pp. 775–790. San Francisco: Freeman.

Calow, P. and Townsend, C. R. (1981). Energetics, ecology and evolution. In *Physiological Ecology. An Evolutionary Approach to Resource Use*. Eds Townsend, C. R. and Calow, P. pp. 3–19. Oxford: Blackwell.

Cameron, J. (1905). The development of the retina in amphibia: an embryological and cytological study. *J. Anat. Physiol.*, 39, 135–153, 332–361, 471–488.

Carter-Dawson, L. D. and LaVail, M. M. (1979). Rods and cones in the mouse retina. II. Autoradiographic analysis of cell generation using tritiated thymidine. *J. Comp. Neurol.*, 188, 263–272.

Cohen, A. I. (1972). Rods and cones. In *Physiology of Photoreceptors: Handbook of Sensory Physiology, Vol. VII/2*. Ed. Fuortes, M.G. F. Pp. 63–110. Berlin, Heidelberg: Springer.

Cohen, J. L. (1990). Vision in elasmobranchs. In *The Visual System of Fish*. Eds Douglas, R.H. and Djamgoz, M.B.A. pp. 465–490. London: Chapman & Hall.

Collin, S. P. and Pettigrew, J. D. (1988a). Retinal topography in reef teleosts. I. Some species with well-developed areae but poorly-developed streaks. *Brain, Behav. Evol.*, 31, 269–282.

Collin, S. P. and Pettigrew, J. D. (1988b). Retinal topography in reef teleosts. II. Some species with prominent horizontal streaks and high-density areae. *Brain, Behav. Evol.*, 31, 283–295.

Curcio, C. A. and Allen, K. A. (1990). Topography of ganglion cells in human retina. *J. Comp. Neurol.*, 300, 5–25.

Cynader, M. and Berman, N. (1972). Receptive field organization of monkey superior colliculus. *J. Neurophysiol.*, 35, 187–201.

Dacey, D. M. (1989). The dopaminergic amacrine cell. *J. Comp. Neurol.*, 301, 461–489.

Daw, N. W. (1968). Colour-coded ganglion cells in the goldfish retina. Extension of their receptive fields by means of new stimuli. *J. Physiol.*, 197, 567–592.

Djamgoz, M. B. A. and Yamada, M. (1990). Electrophysiological characteristics of retinal neurones: synaptic interactions and functional outputs. In *The Visual System of Fish*. Eds Douglas, R. H. and Djamgoz, M.B.A. pp. 159–210. London: Chapman & Hall,.

Dogiel, A. (1895). Ein besonderer Typus von Nervenzellen in der mittleren gangliosen Schicht der Vogel-Retina. *Anat. Anzeiger*, 10, 750–760.

Dowling, J. E. (1987). *The Retina: an Approachable Part of the Brain*. Harvard: Belknap.

Dowling, J. E. and Ripps, H. (1970). Visual adaptation in the retina of the skate. *J. Gen. Physiol.*, 56, 491–520.

Dowling, J. E. and Ripps, (1972). Adaptation in skate photoreceptors. *J. Gen. Physiol.*, 60, 698–719.

Drager, U. C. and Hofbauer, A. (1984). Antibodies to heavy neurofilament subunit detect a supopulation of damaged ganglion cells in retina. *Nature Lond.*, 309, 624–626.

Drager, U. C. and Olsen, J. F. (1980). Origins of crossed and uncrossed retinal projections in pigmented and albino mice. *J. Comp. Neurol.*, 191, 383–412.

Dreher, B. and Robinson, S. R. (1988). Development of the retinofugal pathway in birds and mammals: evidence for a common 'timetable'. *Brain, Behav. Evol.*, 31, 369–390.

Duke-Elder, S. (1958). The eye in evolution. In *System of Ophthalmology, Vol. I*. St. Louis: Mosby.

Ebbesson, S. O. E. (1984). Evolution and the ontogeny of neural circuits. *Behav. Brain Sci.* 7, 321–366.

Eldredge, N. and Cracraft, J. (1980). *Phylogenetic Patterns and the Evolutionary Process*. New York: Columbia University Press.

Engbretson, G. A. and Anderson, K. H. (1990). Neuronal structure of the lacertilian parietal eye, I: A retrograde label and electron-microscopic study of ganglion cells in the photoreceptor layer. *Vis. Neurosci.*, 5, 395–404.

Engstrom, K. (1963). Cone types and cone arrangements in teleost retinae. *Acta Zool.*, 44, 179–243.

Famiglietti, E. V. and Kolb, H. (1976). Structural basis for on- and off-centre responses in retinal ganglion cells. *Science*, 194, 193–195.

Famiglietti, E. V., Kaneko, A. and Tachibana, M. (1977). Neuronal architecture of on and off pathways to ganglion cells in carp retinae. *Science*, 198, 1267–1269.

Fite, K. V. and Rosenfield-Wessels, S. (1975). A comparative study of deep avian foveas. *Brain, Behav. Evol.*, 12, 97–115.

Fritzsch, B. and Collin, S. P. (1990). Dendritic distribution of two populations of ganglion cells and the retinopetal fibres in the retina of the silver lamprey (*Ichthyomyzon unicuspis*). *Vis. Neurosci.*, 4, 535–545.

Frost, B. J., Wise, L. Z., Morgan, B. and Bird, D. (1990). Retinotopic representation of the bifoveate eye of the kestrel (*Falco sparverius*) on the optic tectum. *Vis. Neurosci.*, 5, 231–239.

Fuortes, M. G. F. and Simon, E. J. (1974). Interactions leading to horizontal cell responses in the turtle retina. *J. Physiol.*, 240, 177–198.

Gallego, A. (1985). Advances in horizontal cell terminology since Cajal. In *Neurocircuitry of the Retina, a Cajal Memorial*. Eds Gallego, A. and Gouras, P. pp. 123–140. New York: Elsevier.

Gallego, A. (1986) Comparative studies on horizontal cells and a note on microglial cells. *Progr. Retinal Res.*, 5, 165–206.

Gould, S. J. (1977). *Ontogeny and Phylogeny*. Harvard: Belknap.

Gould, S. J. and Lewontin, R. C. (1979). The spandrels of San Marco and the Panglossian paradigm: a critique of the adaptationist programme. *Proc. R. Soc. Lond., B.*, 205, 581–598.

Greferath, U, Grunert, U. and Wassle, H. (1990). Rod bipolar cells in the mammalian retina show protein kinase C-like immunoreactivity. *J. Comp. Neurol.*, 301, 433–442.

Guthrie, S. D. M. (1990). The physiology of the teleostean optic tectum. In *The Visual System of Fish*. Eds Douglas, R. H. and Djamgoz, M. B. A. pp. 279–343. London: Chapman & Hall.

Halpern, M. (1973). Retinal projections in blind snakes. *Science Wash. DC*, 182, 390–391.

Hamasaki, D. I. and Elder, D. J. (1977). Adaptive radiation of the pineal system. In *The Visual System in Vertebrates: Handbook of Sensory Physiology, Vol. VII/5*. Ed. Crescitelli, F. pp. 497–548. Berlin, Heidelberg: Springer.

Hartline, H. K. (1938). The response of single optic nerve fibres of the vertebrate eye to illumination of the retina. *Am. J. Physiol.*, 121, 400–415.

Hayes, B. P. and Brooke, M. L. (1990). Retinal ganglion cell distribution and behaviour in procellariiform seabirds. *Vision Res.*, 30, 1277–1289.

Hayes, B. P. and Hodos, W., Holden A. L. and Low, J. C. (1987). The projection of the visual field upon the retina of the pigeon. *Vision Res.*, 27, 31–40.

Hennig, W. (1979). *Phylogenetic Systematics*. Urbana: University of Illinois Press.

Hodos, W., Bessette, B. B., Macko, K. A. and Weiss, S. R. B. (1985). Normative data for pigeon vision. *Vision Res.*, 25, 1525–1528.

Holden, A. L. (1977a). Responses of directional ganglion cells in the pigeon retina. *J. Physiol.*, 270, 253–269.

Holden, A. L. (1977b). Concentric receptive fields of pigeon ganglion cells. *Vision Res.*, 17, 545–554.

Holmberg, K. (1977). The cyclostome retina. In *The Visual System in Vertebrates: Handbook of Sensory Physiology, Vol. VII/5*. Ed. Crescitelli, F. pp. 47–66. Berlin, Heidelberg: Springer.

Hoskins, S. G. (1989). Routing of axons at the optic chiasm: ipsilateral projections and their development. In *Development of the Vertebrate Retina*. Eds Finlay, B. L. and Sengelaub, D. R. pp. 113–148. New York: Plenum.

Hughes, A. (1977). The topography of vision in mammals of contrasting life styles: comparative optics and retinal organisation. In *The Visual System in Vertebrates: Handbook of Sensory Physiology, Vol. VII/5*. Ed. Crescitelli, F. pp. 613–756. Berlin, Heidelberg: Springer.

Ikushima, M., Watanabe, M. and Ito, H. (1986). Distribution and morphology of retinal ganglion cells in the Japanese quail. *Brain Res.*, 376, 320–334.

Illing, R.-B. and Wassle, H. (1981). The retinal projection to the thalamus in the cat: a quantitative investigation and a comparison with retinotectal pathway. *J. Comp. Neurol.*, 202, 265–285.

Ito, H. and Murakami, T. (1984). Retinal ganglion cells in two teleost species, *Sebastiscus marmoratus* and *Navodon modestus*. *J. Comp. Neurol.*, 229, 80–96.

Jacobs, G. H. (1990). Duplicity theory and ground squirrels: linkages between photoreceptors and visual function. *Vis. Neurosci.*, 5, 311–318.

Johns, P. R. (1982). The formation of photoreceptors in the growing retinas of larval and adult goldfish. *J. Neurosci.*, 2, 178–198.

Johns, P. R. and Easter, S. S. (1977). Growth of the adult goldfish eye. II. Increase in retinal cell number. *J. Comp. Neurol.*, 176, 331–342.

Johnson, J. I. (1986). Mammalian evolution as seen in visual and other neural systems. In *Visual Neuroscience*. Eds Pettigrew, J. D., Sanderson, K. J. and Levick, W. R. pp. 196–207. Cambridge University Press: Cambridge.

Kaas, J. H., Allman J. and Guillery, R. W. (1972) Some principles of organization in the dorsal lateral geniculate nucleus. *Brain, Behav. Evol.*, 6, 253–299.

Kaplan, E., Lee, B. B. and Shapley, R. M. (1989). New views of primate retinal function. *Progr. Retinal Res.*, 8, 273–334.

Karten, H. J., Fite, K. V. and Brecha, N. (1977). Specific projections of displaced retinal ganglion cells upon the accessory optic system in the pigeon (*Columba livis*). *Proc. Natl. Acad. Sci. USA*, 74, 1753–1756.

Kock, J.-H. and Reuter, T. (1978). Retinal ganglion cells in the crucian carp (*Carassius carassius*). I. Size and number of somata in eyes of different size. *J. Comp. Neurol.*, 179, 535–548.

Kolb, H. Nelson, R. and Mariani, A. (1981). Amacrine cells, bipolar cells and ganglion cells of the cat: a Golgi study. *Vision Res.*, 21, 1081–1114.

Koontz, M. A. and Hendrickson, A. E. (1987). Stratified distribution of synapses in the inner plexiform layer of primate retina. *J. Comp. Neurol.*, 263, 581–592.

Kuffler, S. W. (1953). Discharge patterns and functional organization of mammalian retina. *J. Neurophysiol.*, 16, 37–68.

Lasater (1990). Neurotransmitters and neuromodulators of the fish retina. In *The Visual System of Fish*. Eds Douglas, R. H. and Djamgoz, M. B. A. pp. 211–238. Chapman & Hall: London.

Lettvin, J. Y., Maturana, H. R., Pitts, W. H. and McCulloch, W. S. (1961). Two remarks on the visual system of the frog. In *Sensory Communication*. Ed. Rosenblith, W. A. pp. 757–776. Cambridge, MA: MIT Press.

Leventhal, A. G., Rodieck, R. W. and Dreher, B. (1985). Central projections of cat retinal ganglion cells. *J. Comp. Neurol.*, 237, 216–226.

Levick, W. R. (1986). Sampling of information space by retinal ganglion cells In *Visual Neuroscience*. Eds Pettigrew, J. D., Sanderson, K. J. and Levick, W. R. pp. 33–43. Cambridge: Cambridge University Press.

Lia, B., Williams, R. W. and Chalupa, L. M. (1987). Formation of retinal ganglion cell topography during prenatal development. *Science Wash. DC*, 26, 848–851.

Linden, R. and Perry, V. H. (1983). Massive retinotectal projection in rats. *Brain Res.*, 272, 145–149.

Livingstone, M. S. and Hubel, D. H. (1984). Anatomy and physiology of a color system in the primate visual cortex. *J. Neurosci.*, 4, 309–356.

Loew, E. R. and Lythgoe, J.N. (1978). The ecology of cone pigments in teleost fish. *Vision Res.*, 18, 712–722.

Long, K. O. and Fisher, S. K. (1983). The distributions of photoreceptors and ganglion cells in the California ground squirrel, *Spermophilus beecheyi*. *J. Comp. Neurol.*, 221, 329–340.

Luck, C. P. (1965). The comparative morphology of the eyes of certain African suiformes. *Vision Res.*, 5, 283–297.

Lund, R. D. and Lund, J. S. (1965). The visual system of the mole *Talpa europea*. *Exp. Neurol.*, 13, 302–316.

Lythgoe, J. N. (1979). *The Ecology of Vision*. Oxford: Oxford University Press.

Mariani, A. P. (1990). Amacrine cells of the rhesus monkey retina. *J. Comp. Neurol.*, 301, 382–400.

Martin, R. D. (1989). *Primate Origins and Evolution*. London: Chapman & Hall.

Masland, R. H. (1988). Amacrine cells. *Trends Neurosci.*, 11, 405–410.

Maturana, H. R. and Frenk, S. (1963). Directional movement and horizontal edge detectors in the pigeon retina. *Science Wash. DC* 142, 977–979.

Maturana, H. R., Lettvin, J. Y., Pitts, W. H. and McCulloch, W. S. (1960). Physiology and anatomy of vision in the frog. *J. Gen. Physiol.*, 43 Suppl. 129–175.

Maynard Smith, J., Burian, R., Kauffman, S., Alberch, P., Campbell, J., Goodwin, B. Lande, R., Raup, D. and Wolpert, L. (1985). Developmental constraints and evolution. *Q. Rev. Biol.*, 50, 265–287.

Meek, H. (1990). Tectal morphology: connections, neurones and synapses. In *The Visual System of Fish*. Eds Douglas, R. H. and Djamgoz, M. B. A. pp. 239–277. London: Chapman & Hall.

Meyer, D. B. (1977). The avian eye and its adaptations. In *The Visual System in Vertebrates: Handbook of Sensory Physiology*, Vol VII/5, Ed. Crescitelli, F. pp. 549–612. Berlin, Heidelberg: Springer.

Miles, F. A. (1972). Centrifugal control of the avian retina. I. Receptive field properties of retinal ganglion cells. *Brain Res.*, 48, 65–92.

Miller, W. H. and Snyder, A. W. (1977). The tiered vertebrate retina. *Vision Res.*, 17, 239–255.

Mollon, J. D. (1990). The club-sandwich mystery. *Nature Lond.*, 343, 16–17.

Munz, F. W. and McFarland, W. N. (1977). Evolutionary adaptations of fishes to the photic environment. In *The Visual System in Vertebrates: Handbook of Sensory Physiology, Vol. VII/5*. Ed Crescitelli, F. pp. 193–274. Berlin, Heidelberg: Springer.

Naka, K.-I. (1980). A class of catfish amacrine cells responds preferentially to objects which move vertically. *Vision Res.*, 20, 961–965.

Nathans, J., Thomas, D. and Hogness, D. S. (1986). Molecular genetics of human color vision: the genes encoding blue, green and red pigments. *Science Wash. DC*, 232, 193–202.

Negishi, K., Teranishi, T., Kuo, C.-H. and Miki, N. (1987). Two types of lamprey retina photoreceptors immunoreactive to rod- or cone-specific antibodies. *Vision Res.*, 27, 1237–1241.

Negishi, K., Kato, S. and Teranishi, T. (1988). Dopamine cells and rod bipolar cells contain protein kinase C-like immunoreactivity in some vertebrate retinas. *Neurosci. Lett.*, **94**, 247–252.

Nelson, R., Famiglietti, E. V. and Kolb, H. (1978). Intracellular staining reveals different levels of stratification for on-centre and off-centre ganglion cells in cat retina. *J. Neurophysiol.*, **41**, 472–483.

Ohman, P. (1976). Fine structure of photoreceptors and associated neurones in the retina of *Lampetra fluviatilis* (Cyclostomi). *Vision Res.*, **16**, 659–662.

O'Leary, D. D. M., Gerfen, C. R. and Cowan, W. M. (1983) The development and restriction of the ipsilateral retinofugal projection in the chick. *Dev. Brain Res.*, **10**, 93–109.

Oster, G. F., Shubin, N., Murray, J. D. and Alberch, P. (1988). Evolution and morphogenetic rules: the shape of the vertebrate limb in ontogeny and phylogeny. *Evolution*, **42**, 862–884.

Pedler, C. (1965). Rods and cones—a fresh approach In *Physiology and Experimental Psychology of Colour Vision. Ciba Foundation Symposium.* Eds Wolstenholme, G. E. W. and Knight, J. pp. 52–58. London: Churchill.

Peichl, L. and Wassle, H. (1981). Morphological identification of on- and off-centre brisk transient (Y) cells in the cat retina. With an appendix, 'Neurofibrillar staining of cat retinae' by B. B. Boycott and L. Peichl. *Proc. R. Soc. Lond. B*. **212**, 139–156.

Peichl, L., Ott, H. and Boycott, B. B. (1987). Alpha ganglion cells in mammalian retina. *Proc. R. Soc. Lond. B.*, **231**, 169–197.

Perry, V. H. (1979). The ganglion cell layer of the retina in the rat. *Proc. R. Soc. Lond. B.*, **204**, 363–375.

Perry, V. H. and Cowey, A. (1984). Retinal ganglion cells that project to the superior colliculus and pretectum in the macaque monkey. *Neurosci.*, **12**, 1125–1137.

Perry, V. H. and Cowey, A. (1985). The ganglion cell and cone distributions in the monkey's retina: implications for central magnification factors. *Vision Res.*, **25**, 1795–1810.

Perry, V. H., Henderson, Z. and Linden, R. (1983). Postnatal changes in retinal ganglion cell and optic axon populations in the pigmented rat. *J. Comp. Neurol.*, **219**, 356–368,.

Perry, V. H., Oehler, R. and Cowey, A. (1984). Retinal ganglion cells that project to the dorsal lateral geniculate nucleus in the macaque monkey. *Neurosci.* **12**, 1101–1123.

Pettigrew, J. D. (1979). Binocular visual processing in the owl's telencephalon. *Proc. R. Soc. Lond. B.*, **204**, 435–454.

Pettigrew, J. D. (1986). Flying primates? Megabats have the advanced pathway from eye to midbrain. *Science Wash. DC*, **231**, 1304–1306.

Pettigrew, J. D., Dreher, B., Hopkins, C. S., McCall, M. J. and Brown, M. (1988). Peak density and distribution of ganglion cells in the retinae of microchiropteran bats: implications for visual acuity. *Brain Behav. Evol.*, **32**, 39–56.

Pettigrew, J. D., Jamieson, B. G. M., Robson, S. K., Hall, L. S., McKnally, K. I. and Cooper, H. M. (1989). Phylogenetic relations between microbats, megabats and primates (Mammalia: Chiroptera and Primates). *Phil. Trans. R. Soc. Lond. B.*, **325**, 489–559.

Polyak, S. L. (1941). *The Retina*. Chicago: University of Chicago Press.

Pomeranz, B. and Chung, S. H. (1970). Dendritic-tree anatomy codes form-vision in tadpole retina. *Science Wash. DC*, **170**, 983–984.

Prasada Rao, P. D. and Sharma, S. C. (1982). Retinofugal projections in juvenile and adult channel catfish, *Ictalurus (Ameirus) punctatus*: an HRP and autoradiographic study. *J. Comp. Neurol.*, **210**, 37–48.

Provis, J. M. and Watson, C. R. R. (1981). The distribution of ipsilaterally and contralaterally projecting ganglion cells in the retina of the pigmented rabbit. *Exp. Brain Res.*, **44**, 82–92.

Provis, J. M., Driel, D. van, Billson, F. A. and Russell, P. (1985). Development of human retina: patterns of cell distribution in the ganglion cell layer. *J. Comp. Neurol.*, **233**, 429–451.

Rakic, P. and Riley, K. P. (1938). Regulation of axon number in primate optic nerve by prenatal binocular competition. *Nature Lond.*, **305**, 135–137.

Rager, G. and Rager, U. (1978). Systems-matching by degeneration. I. A quantitative electron microscopic study of the generation and degeneration of retinal ganglion cells in the chicken. *Exp. Brain Res.*, **33**, 65–78.

Reperant, J., Miceli, D., Rio, J. P. and Weidner, C. (1987). The primary optic system in a microphthalmic snake (*Calabria reinhardti*). *Brain Res.*, **408**, 233–238.

Rodieck, R. W. (1973). *The Vertebrate Retina*. San Francisco: Freeman.

Rodieck, R. W. (1988). The primate retina. *Comp. Primate Biol.*, **4**, 203–278.

Rodieck, R. W. and Brening, R. K. (1938). Retinal ganglion cells: properties, types, genera, pathways and trans-species comparisons. *Brain, Behav. Evol.*, **23**, 121–164.

Rowe, M. H. and Stone, J. (1977). Naming of neurones: classification and naming of cat retinal ganglion cells. *Brain, Behav. Evol.*, **14**, 185–216.

Ryder, J. A. (1895). An arrangement of retinal cells in the eyes of fishes partially simulating compound eyes. *Proc. Natl. Acad. Sci. USA*, 161–166.

Saito, H. A. (1983). Morphology of physiologically identified X-, Y- and W-type retinal ganglion cells of the cat. *J. Comp. Neurol.*, **221**, 279–288.

Sakai, H. M., Naka, K. I., Chappell, R.l. and Ripps, H. (1986). Synaptic contacts in the outer plexiform layer of the skate retina. *Biol. Bull. Mar. Biol. Lab., Woods Hole*, **170**, 497–498.

Samorajski, T., Ordy, J. M. and Keefe, J. R. (1966). Structural organisation of the retina in the shrew (*Tupaia glis*). *J. Cell Biol.*, **28**, 489–504.

Sarnat, H. B. and Netsky, M. G. (1981). *Evolution of the Nervous System*, 2nd edn. Oxford: Oxford University Press.

Scholes, J. H. (1975). Colour receptors, and their synaptic connections, in the retina of a cyprinid fish. *Phil. Trans. R. Soc. Lond. B*, **270**, 61–118.

Schultze, M. (1866). Zur Anatomie und Physiologie der Retina. *Arch. Mikrosk. Anat. Entw. Mech.*, **2**, 175–286.

Schultze, M. (1873). The retina. In *Manual of Human and Comparative Histology*. Ed. Stricker, S. Transl. Power, H. pp. 218–298. London: New Sydenham Society.

Shapley, R. M. and Perry, V. H. (1986). Cat and monkey retinal ganglion cells and their visual functional roles. *Trends Neurosci.*, **9**, 229–235.

Silveira, L. C. L., Picanco-Diniz, C. W. and Oswaldo-Cruz, E. (1989a) Distribution and size of ganglion cells in the retinae of large Amazon rodents. *Vis. Neurosci.*, **2**, 221–235.

Silveira, L. C. L., Picanco-Diniz, C. W., Sampaio, L. F. S. and Oswaldo-Cruz, E. (189b). Retinal ganglion cell distribution in the Cebus monkey: a comparison with cortical magnification factors. *Vision Res.*, **29**, 1471–1483.

Springer, A. D. and Landreth, G. E. (1977). Direct ipsilateral projections in goldfish (*Carassius auratus*). *Brain Res.*, **124**, 533–537.

Stell, W. K., Kretz, R. and Lightfoot, D. O. (1982). Horizontal cell connectivity in goldfish. In *The S-Potential*. pp. 51–75. New York: Liss.

Sterling, P. (1990). Retina. In *The Synaptic Organization of the Brain*. Ed. Shepherd, G. M. pp. 170–213. Oxford: Oxford University Press.

Stone, J. (1981). *The Wholemount Handbook*. Sydney: Maitland.

Stone, J. (1983). *Parallel Processing in the Visual System*. London: Plenum.

Stone, J. and Halasz, P. (1989). Topography of the retina in the elephant *Loxodonta africana*. *Brain, Behav. Evol*, **34**, 84–95.

Stone, J., Rapaport, D. H., Williams, R. W. and Chalupa, L. (1982). Uniformity of cell distribution in the ganglion cell layer of prenatal cat retina: implications for mechanisms of retinal development. *Dev.*

Svaetichin, G. and MacNichol, E. F. (1958). Retinal mechanisms for chromatic and achromatic vision. *Ann. NY Acad. Sci.*, **74**, 388–404.

Szel, A. and Rohlich, P. (1989). Colour vision and immunologically identifiable photoreceptor subtypes. In *Neurobiology of Sensory Systems*. Eds Singh, R. N. and Strausfeld, N. J. pp. 275–293. New York: Plenum.

Tiao, Y.-C. and Blakemore, C. (1975). Regional specialisation in the golden hamster's retina. *J. Comp. Neurol.*, **168**, 439–458.

Tomita, T. (1965). Electrophysiological study of the mechanisms subserving color coding in the fish retina. *Cold Spring Harbor Symp. Quant. Biol.*, **30**, 559–566.

Tomlinson, A. (1988). Cellular interactions in the developing *Drosophila* eye. *Development*, **104**, 183–193.

Toyoda, J.-I., Kujiraoka, T. and Fujimoto, M. (1982). The opponent color process and interactions of horizontal cells. In *The S-Potential*. Eds Drujan, B. D. and Laufer, M. pp. 151–160. New York: Liss.

Vesselkin, N. P., Reperant, J., Kenigfest, N. B., Rio, J. P., Miceli, D. and Shupaliakov, O. V. (1989). Centrifugal innervation of the lamprey retina. Light- and electron-microscopic and electrophysiological investigations. *Brain Res.*, **493**, 51–65.

Voneida, T. J. and Slingar, C. M. (1976). A comparative neuroanatomic study of retinal projections in two fishes: *Astyanax hubbsi* (the blind cave fish) and *Astyanax mexicanus*. *J. Comp. Neurol.*, **165**, 89–106.

Wagner, H.-J. (1978). Cell types and connectivity patterns in mosaic retinas. *Adv. Anat. Embryol. Cell. Biol.*, **55**, 1–81.

Wagner, H.-J. (1990). Retinal structure of fishes. In *The Visual System of Fish*. Eds Douglas, R. H. and Djamgoaz, M. B. A. pp. 109–157. London: Chapman & Hall.

Wagner, H.-J. and Wagner, E. (1988). Amacrine cells in the retina of a teleost fish, the roach (*Rutilus rutilus*). A Golgi study on differentiation and layering. *Phil. Trans R. Soc. Lond. B.*, **321**, 263–324.

Wakakuwa, K., Watanabe, M., Sugimoto, T., Washida, A. and Fukada, Y. (1987). An electron microscopic analysis of the optic nerve of the eastern chipmunk (*Tamais sibiricus asiaticus*): total fibre count and retinotopic organisation. *Vision Res.*, **27**, 1891–1901.

Walls, G. L. (1934). The reptilian retina. I. A new concept of visual-cell evolution. *Am. J. Ophthalmol.*, **17**, 892–915.

Walls, G. L. (1942). *The Vertebrate Eye and its Adaptive Radiation*. Bloomfield Hills, Missouri: The Cranbrook Institute of Science.

Walsh, C., Polley, E. H., Hickey, T. L. and Guillery, R. W. (1983). Generation of cat retinal ganglion cells in relation to central pathways. *Nature Lond.*, **302**, 611–614.

Ward, R., Reperant, J., Rio, J. P. and Peyrichoux, J. (1987). Étude quantitative du nerf optique chez la vipère aspic (*Vipera aspis*). *CR Acad. Sci. Paris*, **304**, 331–336.

Wassle, H. and Illing, R.-B. (1980). The retinal projection to the superior colliculus in the cat: a quantitative study with HRP. *J. Comp. Neurol.*, **159**, 419–438.

Wassle, H., Peichl, L. and Boycott, B. B. (1983). A spatial analysis of ON- and OFF-ganglion cells in the cat retina. *Vision Res.*, **23**, 1151–1160.

Wathey, J. C. and Pettigrew, J. D. (1989). Quantitative analysis of the retinal ganglion cell layer and optic nerve of the barn owl *Tyto alba*. *Brain, Behav. Evol.*, **33**, 279–292.

Webb, S. V. and Kaas, J. H. (1976). The sizes and distribution of ganglion cells in the retina of the owl monkey *Aotus trivirgatus*. *Vision Res.*, **16**, 1247–1254.

West, R. W. and Dowling, J. E. (1975). Anatomical evidence for cone and rod-like receptors in the gray squirrel, ground squirrel, and prairie dog retinas. *J. Comp. Neurol.*, **159**, 439–460.

Wikler, K. C. and Finlay, B. L. (1989). Developmental heterochrony and the evolution of species differences in retinal specialisations. In *Development of the Vertebrate Retina*. Eds Finlay, B. L. and Sengelaub, D. R. pp. 227–246. New York: Plenum.

Wikler, K. C. and Rakic, P. (1990). Distribution of photoreceptor subtypes in the retina of diurnal and nocturnal primates. *J. Neurosci.*, **10**, 3390–3401.

Williams, D. R. (1986). Seeing through the photoreceptor mosaic. *Trends Neuroscience*, **9**, 193–198.

Witpaard, J. and ter Keurs, H. E. D. J. (1975). A reclassification of retinal ganglion cells in the frog, based on tectal endings and response properties. *Vision Res.*, **15**, 1333–1338.

Wong, R. (1989). Morphology and distribution of neurons in the retina of the American garter snake (*Thamnophis sirtalis*). *J. Comp. Neurol.*, **283**, 587–601.

Young, H. M. and Vaney, D. I. (1990). The retinae of Prototherian mammals possess neuronal types that are characteristic of non-mammalian retinae. *Vision Res.*, **5**, 61–66.

Young, R. W. (1971). An hypothesis to account for basic distinction between rods and cones. *Vision Res.*, **11**, 1–5.

8 Functions of Subcortical Visual Systems in Vertebrates and the Evolution of Higher Visual Mechanisms

David J. Ingle

Introduction

Strictly speaking, we cannot describe states in the 'evolution' of vertebrate visual systems because our knowledge of the relevant ancestral species is limited to superficial views of brain topography. Despite our growing knowledge of visual processes in certain species at different phylogenetic levels (such as goldfish, frogs, rats, cats and monkeys), none of these is ancestral to any of the others. Nevertheless, we may set a useful goal in trying to draw generalized pictures of the main features of the visual system (and of behaviour) shared by each vertebrate class (fish, amphibians, reptiles, birds and mammals) in order to answer two broad questions. First, we should attempt to identify homologous visual cell groups and their connections which appear at various levels throughout the vertebrate phylogeny, and ask whether these homologous subsystems retain similar behavioural functions among diverse groups. Second, we can try to identify those advanced behavioural capacities of birds and mammals, which are not shared with the fishes or the amphibians, and which may reflect specialized or novel functions of an enlarged telencephalic visual system. In this chapter I will review some common functions among homologous subcortical components, but the difficult problem of comparing functions of cortical subregions among mammals and birds will be bypassed here. Rather, I will present data and hypotheses on the more general issue of which visual abilities depend upon the evolution of visual cortex in mammals. In reviewing these issues, I will draw most heavily on my own experimental work. Much of the detailed work of others is reviewed and better referenced in other reviews which I have listed below.

All inframammalian groups with good vision have large laminated optic tecta which dominate other visual nuclei in size, while all mammals have evolved a visual cortex whose size dominates their own optic tectum (or superior colliculus). Among all vertebrate classes projections from the eye terminate within five apparently homologous zones: (a) dorsal anterior thalamus, (b) ventral anterior thalamus, (c) two or three regions of pretectum, (d) the optic tectum, and (e) one or more accessory optic systems. I will review data from lower vertebrates (teleost fishes and anuran amphibians) and from selected mammals (mostly rodents) to support the idea that the visuomotor functions of each of the five retinofugal projection areas remain very similar throughout vertebrate phylogeny (see Fig. 8.1). This review sketches a basic vertebrate 'bauplann' upon which the many variations and evolutionary advances may be understood. I will then contrast the known limitations of this subcortical visual system (as inferred from studies of fishes and amphibians) with the new visual capacities of mammals which appear to depend upon visual cortex. In taking this broad overview I will omit discussion of the possible functional significance of the reptilian telencephalic visual regions, and will bypass the intriguing question as to how far convergent evolution has provided the birds with functional equivalents to mammalian neocortex.

Studies of inframammalian vision ought to be of interest to students of mammalian vision for at least two reasons. First, our knowledge of subcortical mechanisms is discouragingly incomplete for mammals, while the functions of homologous structures in fishes or in amphibians are often easier to sort out from behavioural and physiological studies, where they are uncomplicated by interactions with a visual cortex. The present assignment of specific visuomotor functions of tectum, pretectum and accessory optic pathways in lower vertebrates has already influenced research on mammalian vision, and some of the findings reviewed below provide a set of testable hypotheses for new mammalian studies. Second, our documentation of the limitations of fish and amphibian visual abilities helps to generate hypotheses as to the particular advantages in having a visual cortex. We give some examples where ablation of part of the mammalian visual cortex eliminates just the advanced abilities, leaving the animal able to perform simpler versions of the same task in the manner of fish or amphibian. On the other hand, we shall briefly review the case of 'colour constancy' where a comparison of fish and monkey suggests a radical reorganization of the neural basis of colour perception during evolution. While the understanding of many issues in visual perception requires a direct attack on the complex interleaved processes within the multiple visual cortical areas of primates, studies of lower vertebrates allow us to pose and sometimes solve fundamental questions as to how 'seeing' is coupled to brain mechanisms of motivation and movement.

Visuomotor Systems in Amphibia

An Overview of Visual Behavioural Taxonomy

In general, comparative psychologists have focused on the abilities of lower vertebrates (especially the fishes) to perform recognition or classification of visual patterns and colours, in order to compare these with mammalian discrimination and generalization abilities. By contrast, zoologists (and ethologists in particular) have stressed the natural uses of vision in capturing prey, avoiding predators and recognizing conspecifics. A key step in linking visual structures to their behavioural functions is to classify natural behaviours in terms of the kinds of responses elicited or guided by vision. I began this task in a broad review of vertebrate visuomotor behaviours (Ingle, 1982). A key first step was the demonstration (Ingle, 1973a) that the frog's optic tectum mediates at least two distinct classes of orienting behaviour (catching prey and avoidance of predators), while the pretectum mediates the negotiation of stationary barriers and optokinetic responses to global motion of the visual surround (Ingle, 1980; Fig. 8.1). It seems that tectum plays its chief role in responding to moving animate objects, while pretectum mediates adjustments of body posture or locomotor routes to rigid structures of the inanimate world. It is probable (but not yet demonstrated in frogs) that tectum also mediates approach to conspecifics during mating behaviour, and that pretectum is needed to detect chasms (to be

jumped over) as well as horizontal surfaces (to be stepped upon) and vertical surfaces (to be circumnavigated). Furthermore, recording studies in frogs (Muntz, 1962) have revealed a blue-light-sensitive group of cells within the

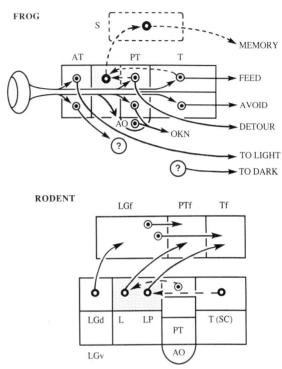

Fig. 8.1 *A comparison of the retinal projection targets in frogs and in rodents. In both groups the eye projects mainly contralaterally to the AT: two regions of the anterior thalamus; to PT: two or three regions of the pretectum, to T: the optic tectum with two major output systems; and to AO: the accessory optic system. For the frog, outputs from these various subsystems are shown as arrows, which activate specific classes of orienting behaviour. Still questionable are the behavioural function of the ventral thalamic target, and the retinal target region needed for elicitation of dark-hole seeking. Dashed arrows show the flow of visual information from tectum and pretectum to the striatum, relayed by cells in the middle thalamus (stippled). In mammals, the outputs to the brain stem are similar, so no redundant arrows are here depicted. Rather, we show the evolved ascending projections to cortex via thalamic cell groups. From the lateral geniculate dorsalis (LGd) comes a geniculofugal pathway to striate cortex (LGf). From the stippled zone (the apparent homologue of the stippled thalamic zone in frogs) cells of the lateral group (L) receive axons from pretectum (PT) and themselves project to visual cortex medial to the striate region, here labelled as pretectofugal (PTf). A parallel projection from tectum to the lateral posterior nucleus (LP) feeds a projection to regions just lateral to the striate cortex, labelled here as tectofugal (Tf). In this review, we argue that 'pretectal functions' seen in frogs are evolutionary precursors of advanced visual abilities of rodents mediated by the PTf zone of cortex.*

anterior thalamus, linked to the finding (Kicliter, 1973) that thalamic ablations abolish the frog's preference for approaching blue light over white or other coloured lights. More recently a tendency of frogs to approach dark areas within a lighted arena as 'hiding places' was described by Ingle (1983a) and it was suggested that this chronotaxic behaviour might be mediated by one thalamic visual region, while phototaxis is mediated by the second thalamic region. We shall now consider tectal and pretectal mechanisms in more detail and examine similarities of their function between frogs and mammals.

Orienting Functions of the Optic Tectum

In frogs and toads, ablation of the tectum unilaterally leaves the animal 'blind' to both prey and visual threat moving within the opposite monocular visual field. Yet as soon as prey or predator enters the intact field of vision, normal behaviours are triggered. Yet the motor sequences underlying these behaviours do not depend upon integrity of the tectum: both feeding and evasive movements can readily be triggered in the atectal frog by appropriate tactile stimulation of head or legs. The dependence of certain visually elicited behaviours on optic tectum was most dramatically illustrated by a method I have called 'rewiring' of the retinotectal projection (Ingle, 1973a). After ablation of one tectum results in a novel reconnection of the cut optic tract to the remaining ipsilateral tectum, the frog will turn and snap in a direction mirror-symmetrical to the actual location of prey (Ingle, 1973a) when prey moves within the field of the regenerated tract. We also found that threat approaching within the same field will elicit wrong-way turning and jumping into the ipsilateral field, often colliding with the looming black disc (Ingle, 1976). On the other hand, detours around barriers and optokinetic turns within a rotating drum are normally oriented. Unpublished studies in my laboratory also indicated that retinopretectal projections could be rewired to the ipsilateral side after a unilateral ablation of pretectum, and each of three animals with such a wrong-way projection to pretectum showed persistent wrong-way turns in tests of barrier avoidance, while preserving normal prey-catching.

Further exploitation of this unique rewiring paradigm to dissect visuomotor functions seems possible both in fish and amphibians (where optic nerve regeneration is well established) and among mammals when disruption of a pathway can be carried out before such regenerative capacity is lost. Using the golden hamster (a species born relatively prematurely), Schneider (1973) found that ablation of tectum prior to innervation from the opposite eye would result in the cut optic tract recrossing the dorsal midbrain and making functional synapses with cells of the medial ipsilateral tectum. As adults, these rewired hamsters would typically respond to moving food (sunflower seeds) within the upper field by turning initially in the wrong direction. Unlike the rewired frogs, these hamsters would not complete the full orientation to the mirror-symmetrical direction: perhaps because the tectum was quickly inhibited via the wrong-direction motion of the visual surround. Also unlike frogs, the rewired hamsters learn to inhibit all signs of wrong-way turning when such novel behaviour is not reinforced with food rewards. One may speculate that the dominant role of neocortex suppresses such maladaptive behaviour, and wonder whether the hamster would remain frog-like after removal of visual cortex.

Subdivision of Tectal Visuomotor Systems

The frog's tectum is a duplex system, containing output cells for activation of turns *towards* prey and other cells which activate turns *away* from threat (Fig. 8.2). When the crossed tectobulbar pathway is severed by splitting the ventral tegmentum (Ingle, 1982, 1983a) the frog loses its ability to orient the body towards a lateral prey object, but preserves the ability to turn sharply away from approaching threat. These split-tegmentum frogs will respond to nearby prey by leaping straight forwards and snapping, and to more distant prey by making short forward hops. These behaviours show that the disconnection between tectum and the motor-organizing circuitry of the opposite brain stem does not abolish feeding motivation or sensitivity to prey size and prey distance. Ingle (1983a) also described somewhat complementary effects of

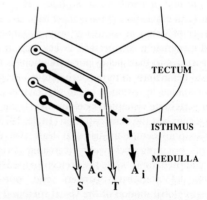

Fig. 8.2 *A schematic diagram of some descending tectal projections in frogs. The white neurones activate brain stem systems for snapping (S) and turning (T) towards prey via ipsilateral and contralateral pathways, respectively. The dark neurones elicit contraversive avoidance turns (A_C) via direct projections to the medulla, and ipsilateral avoidance turns (A_i) via an indirect projection, crossin the isthmus midline (dashed arrow.)*

severing the ipsilateral tectobulbar tract with cuts just behind one tectum through the lateral isthmus (and confirmed by lack of HRP transport from tectum to ipsilateral medulla). These frogs made no turn-and-snap responses to nearby crickets or dummy objects moved within the contralateral monocular field, but on most trials when the prey was moved just beyond the snapping zone the frogs promptly oriented to the same prey. Thus we could clearly dissociate the fixation-of-distant-prey response from snapping at prey, but could not dissociate with this lesion the orienting turn normally associated with snapping from the snapping mechanism. It seems that while the crossed tectobulbar path is necessary and sufficient for prey fixation, activation of the turn-and-snap movement requires integrity of both tectobulbar routes.

A similar picture contrasting ipsilateral and contralateral tectal efferents for the Mongolian gerbil is reported by Ellard and Goodale (1988). Their animals also failed to turn sharply towards stimuli associated with food following transection of the crossed tectobulbar trace at the predorsal bundle, but they showed normal patterns of avoidance elicited by overhead dark objects. Other gerbils with lesions of the lateral tegmentum which interrupted the ipsilateral tectobulbar projection turned normally to stimuli for food reward, but were impaired in fleeing from overhead threat. Deficits in turning to pursue prey were also reported by Dean et al. (1989) for rats with lesions of the predorsal bundle. Dean et al. add more information regarding parcellation of the ipsilateral tectal efferent system, noting that electrical stimulation of one midbrain target (the cuneiform nucleus) produces freezing or aversive running, a subset of behaviours elicited by direct tectum stimulation.

Northmore et al. (1988) have added more detail in hamsters to the specificity of the tectal system for avoidance behaviour. They find that stimulation of the superficial grey (SG) lamina at low current levels sufficed to elicit ducking, freezing or explosive jumping behaviours, while equivalent stimulation of the adjacent intermediate lamina did not. Because the SG has no projections to the brain stem, it seems likely that SG outputs relay upon cells of the deep tectum. Both Northmore et al. (1988) and Dean et al. (1989) agree that stimulation effects (chemical or electrical) on avoidance behaviour are optimal within the dorsal optic tectum, which receives projections from the upper visual field. Stimulation of the lateral tectum typically results in turning of the head contralaterally, as in food approach. This fact is in line with the anatomical finding in rats (Dean et al., 1989) that tectal cells giving rise to the crossed projection are located chiefly within the lateral region of the intermediate lamina.

The distinction between upper and lower fields was noted for gerbils by Ingle (1982) and by Ellard and Goodale (1988) and was studied in rigorous detail for hamsters by Ayers and Schneider in a study which has unfortunately remained unpublished. They found that hamsters while foraging would seldom flee back to their entry holes when confronted with floor-level stimuli but would frequently freeze or flee promptly during the appearance of stimuli at elevations of more than 45° from the floor. In unpublished studies, I carefully examined the effect of stimulus elevation on the frog's avoidance behaviour and found no sign of such an effect. In fact, frogs sitting at the edge of a table readily avoided looming stimuli approaching from 30° below the horizontal meridian.

In unpublished studies with gerbils, we found that floor-level stimuli would elicit stopping or swerving as they suddenly moved toward a free-running animal. However, gerbils were not afraid of such objects and usually pursued them to investigate them after their trajectory was disrupted. It was instructive to find that gerbils with large ablations of occipital cortex never responded to these floor-level looming stimuli and allowed the invariable collision, while the same animals foraging within an open arena would freeze or dart away in response to a surprising overhead black disc much as would normal gerbils. Thus it appears that gerbils and hamsters resemble frogs in depending upon retinotectal projections but not visual cortex for elicitation of unlearned avoidance behaviour by novel stimuli.

Subdivisions of the Tectal Avoidance Mechanism in Frogs

We carefully examined the avoidance pattern of frogs with lateral isthmus cuts which severed the ipsilateral tectobulbar pathway. In addition to loss of snapping toward contralateral prey, such frogs always lost the ability to jump away from looming black discs that were seen opposite the side of the cut. While the frequency of avoidance jumps was reduced well below the usual 100% prelesion level, vigorous jumps were elicited on more than half of the test trials for all of these lesioned animals. These jumps were now directed either straight forwards or actually into the rostral field on the same side as the threat approached. What could be the teleological reason for the existence of circuitry for turning a frog toward looming threat? Examination of normal animals revealed that in one situation they also made predominantly ipsiversive jumps: tests in which the stimulus approaches obliquely toward the frontal midline, just off a collision course. In such trials frogs usually (85% of 200 filmed trials) evade the crossing stimulus by cutting back parallel to the trajectory of the threat, much as a skilled football carrier might do to evade a tackler trying to intercept him. Further film studies (Ingle and Hoff, 1990) showed that even stimuli angling rostrally but located in the lateral field at the time of triggering escape tend to elicit rear-field escape directions, while lateral stimuli on a collision course or lateral stimuli angling caudally both elicit similar distributions of rostralwards jumps. We concluded that tectal cells sensitive to rostral motion interact with tectal cells sensitive to looming motion (both types described by Grusser and Grusser-Cornhels, 1976, in a review of frog tectal recording studies) in order to determine the actual escape direction taken.

In the case of ipsiversive jumps, these are either abolished or much reduced in frequency for all frogs with midline splits in the isthmic region (Ingle and Hoff, 1990). As control cases, frogs with the tegmental decussations severed and unable to orient toward prey were able to turn normally in either direction to make ipsiversive escape jumps. Because there are no tectal efferent fibres seen to cross the isthmic midline, we assume that tectofugal fibres contact cells in the tegmental or isthmus region ipsilaterally and these cells send axons across the midline to activate avoidance circuitry in the opposite medulla. It is our hypothesis that cells within that output pathway are sensitive to rostrally moving edges, while cells within the output pathway leading to contraversive escape are activated mainly by looming stimuli. Here a detailed analysis of escape strategies provides a possible explanation as to why large-field cells in the frog's tectum have selectivities to either rostralwards or looming trajectories but not to caudalward trajectories. It seems likely that mammals adjust their escape strategies to the perceived trajectories of approaching predators, but such analyses have not yet been reported.

Subdivision of Tectal Outputs for Feeding

The frog's feeding behaviour is composed of two distinct response components: orienting and snapping. These are often elicited separately and sequentially, but they may be tightly coordinated when the frog makes a 90° body turn ending with a tongue extrusion at the prey. Ingle (1982) first noted that split-tegmentum frogs lost the ability to turn toward contralateral prey but would continue to emit the snap response while lunging forwards. These two efferent systems arise from distinctive cell populations, as seen by back-filling tectal cells from implants of horseradish peroxidase (HRP) into the ventral medulla (Ingle, 1983a; Lazar et al., 1983). Cells giving rise to the contralateral tectomedullary pathway have wide dendrites which arborize only in the uppermost layer of tectum, where the small-field class-1 and class-2 retinal fibres terminate (Lettvin et al., 1959). These cells are shown schematically in Fig. 8.3. Cells of origin of the ipsilateral tectobulbar pathway are of at least three types, but all have dendrites which mainly arborize in deeper tectal layers, where the class-3 and class-4 retinal axons terminate. This arrangement strongly suggests that the crossed tectobulbar

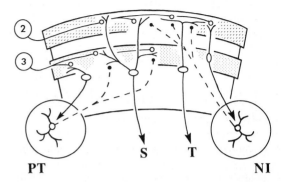

Fig. 8.3 *A schematic model of feedback modulation of tectal neurones by two satellite systems: lateral pretectum (PT) and nucleus isthmus (NI). Class-3 retinal fibres terminate (within the stipped zone) on dendrites of both tectopretectal neurones and T5-2 cells which are presumed to activate snapping (S). Some PT neurones feed back to inhibit the lower-level dendrites of T5-2 cells, making these tectal output cells temporarily responsive only to class-2 inputs accessing their upper-level dendrites. The T5-2 cells and also the cells giving rise to the crossed tectobulbospinal projection, which activates turning (T) toward prey, are each facilitated by feedback cells from the nucleus isthmus (NI). These NI cells are assumed to be activated from small-field tectal cells which themselves receive mainly (or only) class-2 retinal input. Thus while motion of large visual stimuli can inhibit T5-2 cells (via the PT loop) successive movements of small stimuli can escalate the T5-2 cell response via the NI loop. Indeed, Lettvin et al. (1961) have described escalation of response of one type of tectal cell (which we propose to be the crossed tectobulbar projection neurone) by just such repetitive motion of small spots.*

system, which determines the amplitude of orientation turns to prey, is triggered mainly by class-2 retinal fibres, themselves sensitive to small moving stimuli. By contrast, tectofugal cells needed for snapping at prey are activated also by fibres sensitive to large moving edges. This physiological dichotomy makes sense in terms of the finding that frogs snap preferentially at large-angle stimuli when nearby, but orient preferentially toward small-angle stimuli (as small as 2–3°) when set at a distance of 15 cm or more.

The tendency of frogs and toads to adjust their feeding preference to the real size of the stimulus, preferring different angular sizes at different distances, is called 'size-distance constancy' and was first reported by Ingle (1968) for frogs. Later Ewert and Gebauer (1973) and Ingle and Cook (1977) described the constancy phenomenon in more detail for toads and for frogs. Although functional differences between the two major tectal output pathways for feeding behaviour are established, nothing is yet known as to how either pathway can be selectively activated, or how the two can sometimes be precisely synchronized in time during the turn-and-snap sequence. Anatomical studies of the frog's tectum reveal many small interneurones which might function either to inhibit or to recruit one output system while the other is active.

Modulation of Tectal Functions by Satellite Visual Structures

Modulation of Feeding Selectivity by the Pretectum

Ewert (1970) made an important discovery for our understanding of feeding selectivity by amphibians when he noted that large lesions of the pretectal region (or cuts which disconnected pretectum from tectum) produced a remarkable kind of disinhibited feeding syndrome in toads. These animals would now orient to or snap at large objects that would have elicited avoidance or protective behaviour in intact toads. They would snap at their own feet, stationary objects as they walked by them, or at other toads moving nearby. They generally lost their usual tendency to habituate feeding attempts on repeatedly moved dummy objects, and would remain actively pursuing dummy stimuli for hours, instead of habituating within a few minutes.

Ingle (1973b) first recorded from such disinhibited frogs, finding that individual tectal neurones in the uppermost cell lamina had lost their usual habituation properties, and could sometimes be escalated into seizure-like activity by repetitive stimulation with prey-like stimuli. Furthermore, most cells in this layer had receptive fields about twice the typical size in intact frogs. Ewert (1984) carried out a series of experiments to show that the equivalent class of cells in toads (designated initially by the Grussers as 'T5-2' neurones) lost their normal size selectivity after either acute or chronic pretectal damage, and would (like the organism) respond abnormally to very large visual stimuli. Similar experiments (reviewed by Ewert, 1984) were carried out with salamanders following pretectum lesions, yielding similar evidence for pretectal modulation of tectum. Unfortunately, such studies have not yet been reported for teleost fish or reptiles, although many species in these groups have similar feeding preferences for small moving objects and show high rates of habituation towards dummy stimuli.

Ingle (1983b) proposed that tectal modulation by pretectal fibres act to preserve size-distance constancy effects during feeding. Our unpublished studies show that implantation of small pieces of HRP into the lateral pretectal nucleus (where damage produces feeding disinhibition) results in enzyme transport into that sublamina of tectum where the class-3 retinal fibres terminate (see Fig. 8.3). Pretectal activity could thus inhibit axodendritic depolarization of T5-2 cells in response to class-3 inputs, but not the cell's response to the more superficial class-2 inputs. Such inhibition would reduce activity of the T5-2 cells but make them more selective to small-size prey (or more distant prey) at that moment. Because size constancy operates within the lateral visual fields of frogs and toads (where binocular cues are not available), an animal's depth estimation is likely to be based upon monitoring of accommodative effort, as proposed by Collett (1977). Ingle (1983b) has proposed a model of size-constancy in which the brain's command to increase accommodative tension (in order to focus upon near prey) simultaneously reduces the activity of inhibitory pretectal axons within the tectum, and allows the T5-2 cells (or their equivalent in frogs) to respond optimally to large-angle moving stimuli. This mechanism for automatic modulation of tectal cell activity during changes in accommodation provides a mechanistic alternative to the supposition that a frog 'knows' that an object receding must subtend a smaller visual angle. Such a mechanism might still operate in mammals as a first stage of constancy which assists the learning of more flexible constancy rules.

Other Examples of Shifts in Prey Selection in Intact Animals

The idea that inhibition of prey-sensitive tectal cells can enhance size selectivity predicts a correlation between 'readiness to respond' to prey and the size preferred. Ingle (1973c) divided normal frogs into groups of fast-latency and slow-latency feeders, and found that the former group preferred 16°-high to 6°-high dummy prey (at the same distance) while the slow-latency animals showed the opposite preference. Ingle (1983b) made a second prediction from the pretectal-inhibition model described above: that when pretectal inhibition is high, frogs and toads will select prey more closely resembling optimal stimuli for activation of class-2 retinal fibres. In toads (but not in frogs), class-2 fibres are best activated by *white* spots or wormlike stimuli moving on a black background, while class-3 fibres are somewhat better activated by black objects than by white objects. Since recording studies (Ewert, 1971; Brown and Ingle, 1973) show that pretectal cells are best activated by large black objects, we tried to produce a 'hyperinhibition' of tectum by moving 15°-high black squares near a toad's midline on each side for 5 s before giving the animal a feeding choice between small black and small white prey dummies moved on a grey background. The results of this manipulation confirmed our prediction: the preference for white prey found on control trials was sharply reversed on trials where the 'frightening' black stimuli preceded each feeding preference trial (Ingle, 1983b).

Modulation of Tectum by Feedback from the Nucleus Isthmus

One potential problem with our model of pretectal modulation of size preference is that a strong inhibition of T5-2 cells might reduce excitability too far, such that small or brief stimuli would never excite the cell to discharge into the tectobulbar pathway. Yet Ingle (1975) has provided evidence of a facilitatory process which may serve to counteract effects of pretectal inhibition. When frogs were given dummy stimuli which moved for only 0.3 s (by a stepping motor), even the best feeding animals responded on only 10% of the test trials. Yet when the stimulus was programmed to make a second brief motion after a delay of 3 s the same frogs responded by snapping on 85% of 100 test trials. Further unpublished studies showed that the facilitation effect of double movements worked after delays of 4 s but not after delays of 6 or 8 s. This short time window for facilitation matched the time course of a class of non-retinal tectal unit recorded near the tectal surface which gave a few seconds after-discharge following a brief entry of a 2° spot into its small receptive field. These after-discharging units were triggered only by small spots (2–4° optimal size) and generally gave no response to 15°-high squares. Ingle (1975, 1983b) proposed that such after discharge serves to prime prey-sensitive cells within the tectum such that a second barrage of class-2 inputs to the same cell now elicits an efferent command to orient or snap (see Fig. 8.3).

Certain features of these after-discharging units (their precise retinotopy, their selectivity for small stimuli and their confinement to the uppermost tectal lamina) match characteristics of inputs to tectum from nucleus isthmus (Gruberg and Udin, 1978; Gruberg and Lettvin, 1980). Furthermore, in unpublished studies I found that this class of units could not be found within anterior tectum after a knife cut had divided the two tectal halves, but that vigorous afterdischarge units could still be recorded in the opposite tectum. Finally, a report by Caine and Gruberg (1985) describes 'scotomata' for feeding behaviour within the visual field of a frog that is opposite a focal lesion of nucleus isthmus. The relative weakness of tectal response within an area lacking facilitatory input from nucleus isthmus fits our model. In my own unpublished studies, total ablation of nucleus isthmus on one or both sides does not abolish good feeding behaviour, and similar observations are reported by Kostyk et al. (1987). What is needed for better confirmation of my model is evidence that after-discharge units are absent in just that region of this tectum corresponding to a 'behavioural scotoma' as described by Caine and Gruberg.

Modulation of Tectal Functions During Detour Behaviour

As we noted earlier, frogs with ablation of pretectum will crash into striped barriers set before them during escape from pinches of the dorsum from behind (Ingle, 1980). A variety of detour movements can also be elicited in frogs and toads viewing a moving prey behind a semitransparent, fence-like barrier (Ingle, 1970, 1971, 1983a). If the obstacle is too high to leap over but not too wide, the frog or toad initiates a three-stage sequence: (a) rotating the body to point just past the terminal edge of the barrier, (b) stepping or hopping forwards to clear the barrier, and (c) turning back toward the prey in preparation for a direct strike. The solution of the detour task thus requires at least three steps: inhibition of the tendency to strike directly at a nearby prey, addition of extra turning distance to clear the barrier, and a short-term memory for prey location to guide the final reorientation.

Ingle (1983a) has presented a two-stage neural model of detour behaviour, which bypasses an explanation for the final turn-back response. Studies on detour behaviour following vertical knife cuts (Ingle, unpublished) indicate that disconnection of the posterior nucleus of the pretectum from its laterally entering retinal input is sufficient to 'blind' the frog to barriers, but will not produce the disinhibited feeding described above which follows lesions to the lateral pretectal cell group. When HRP is placed locally within the posterior group anterograde transport is found into deeper tectal layers where it is presumed that pretectal axons modify efferent cells' responses to retinal input. Our unpublished studies (Ingle, 1983a) indicate that it is possible through lesion experiments to dissociate the role of these axons in inhibiting the tectal snapping mechanism, from the role of other descending pretectal efferents which modify the amplitude of orientation at a brainstem level. We filmed frogs with the isthmus region split at the midline, and found that they fail to turn beyond a barrier interposed between them and a moving prey (a cricket) but that they inhibit the temptation to strike at the prey and instead merely orient or take a half-step towards it. If the barrier is placed on one side, so that the adaptive detour response is to jump forwards past the rostral barrier edge (see Ingle, 1983a), the split-isthmus frog behaves normally!

A second experiment leading to the same conclusion regarding convergence of pretectal and tectal efferents in the medulla involves detour behaviour in frogs with caudal tectum ablated. Although such a frog can only 'see' prey up to a distance of about 60° from the midline, it can rotate by 90° in order to bypass a barrier edge at 70° after seeing a moving prey at 50° from the midline. We conclude from this demonstration that the tectum is not itself modified by pretectal afference to change its turn from 50° (at the prey) to the 90° detour turn, but that the tectal command to turn 50° is modified at the level of the medulla by descending pretectal output. Since the crossed tectobulbar system controls turn amplitude, the pretectal system must also cross the midline (at the isthmus level) to 'catch up' with the tectal-orienting command.

Short-term Memory for Barriers by Frogs

We have recently reported new experiments which reveal a surprising ability in frogs to remember the location of recently seen barriers (Ingle and Hoff, 1990). This is simply demonstrated by sudden removal of a vertically striped barrier from near the animal, and, after a delay, activating avoidance jumps with a looming black disc. As Fig. 8.4 shows, the directions of evasive jumps vary widely in order to avoid the region where the barrier was recently seen, even when the approach direction of threat is standardized. We have found good memory for a large barrier up to 60 s after its removal, providing that the frog remains stationary during that interval. However, if the frog spontaneously jumps away or is picked up for 5 s and replaced in its initial direction then the memory effect is erased. Barrier location memory is preserved after a passive rotation of 45° and avoidance tests show that these frogs remember the barrier in its real-world location, not in respect to its original retinotopic location. Given this level of complexity, it was not surprising to find that barrier memory depends upon integrity of the striatal region of the frog's telencephalon, an area homologized with the mammalian caudate–putamen complex. The frog's striatum is the main target of ascending thalamic projections from visual and auditory relay nuclei (Wilczynski and Northcutt, 1983) and it may function like the mammalian striatum to maintain a 'preparatory response set' to approach or avoid spatial locations, rather than to maintain a 'visual image' of the remembered object, which we presume to be a function of visual cortex. While it is rewarding to at last identify a non-olfactory function for the amphibian telencephalon, we have no hints from frog studies as to where the evolutionary precursor of visual cortex originates.

Mechanisms of Barrier Detour Behaviour in Mammals

I shall review some experiments with gerbils which fulfil our hypothesis that rodents share some visuomotor mechanisms with amphibians but that cortical mechanisms have provided new kinds of visuomotor competence beyond the abilities of lower vertebrates. Only one experiment has thus far been reported to test the hypothesis that mammalian pretectum plays a critical role in detection of barriers and programming of detour sequences. Goodale and Milner (1982) review a preliminary study of gerbils

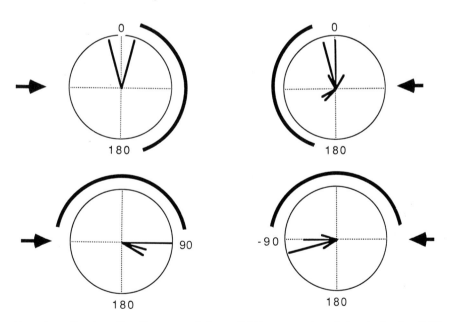

Fig. 8.4 *The avoidance jump directions of one frog are graphed as radial histograms where the length of each radial line is proportionate to the relative number of jumps within a 15°-wide bin. In each test trial the looming dark disc approaches from a direction 90° to left or right of the midline (arrow). Dark semicircles indicate the location of a barrier removed 10 s before evasive jumps are elicited. Note that these jumps usually clear the remembered barrier locations. For frontal barriers the frog can jump laterally in either direction, so that during the memory interval the animal must remember the location of both edges.*

with electrolytic lesions of pretectum, who showed clear deficits in their ability to swerve around a large striped barrier while running toward a dark aperture for food reward. The same animals oriented accurately toward foot-baited stimuli in the peripheral visual field, suggesting that their visuomotor deficit was a selective one. Gerbils with tectum lesioned did fail to orient promptly or accurately towards peripheral visual stimuli, but they evaded barriers as well as did intact gerbils. Unfortunately, the portion of rostral field where tectumless animals could still orient accurately (using visual cortex, according to Ingle, 1982) included that sector where the barrier was seen in detour tests. Tests of tectumless gerbils with wider barriers, with edges falling into the lateral field would be more decisive here.

Film studies by Ingle (1981) showed that gerbils readily turn and detour around barrier edges located 90° from their midline in order to obtain food that appears unpredictably at some location behind the semitransparent barrier. However, unlike frogs, gerbils do not make detours on initial trials: rather, they push at the barrier or run back and forth for a few trials until suddenly acquiring the adaptive detour trajectory. After this brief learning period, gerbils are accurate in choosing the shorter route to the food-baited disc, and will make their first head-orienting response (elicited by food appearance) toward the appropriate barrier edge rather than to the food itself. Moreover, gerbils can make this direction decision within 300 ms of the food's appearance, while frogs and toads take the same 2–3 second latency to begin detour responses, whether on the first trial or after days of experience. We suggest that while detour responses must be learnt in gerbils (but not frogs) they also learn to 'anticipate' after seeing a barrier that only two directions of turning past it are relevant, so that sight of food immediately selects one of the prepared turns. We suspect that frontal cortex plays a key role in this 'spatial planning', since ablations of frontal cortex in gerbils abolished the ability of these animals to plan the shortest route to the food; their best solution was to adopt long, stereotyped and inefficient circling trajectories to circumvent the barrier (Ingle, unpublished studies).

In a second unpublished study, we found that gerbils receiving massive occipital lesions at birth (removing all of visual cortex) could learn to detour around stationary barriers in order to reach a dark aperture on the other side, whose location remained fixed. This would be a good preparation to test the idea that a further lesion of pretectum would totally abolish barrier detection (as in frogs) but spare orienting to prey and avoidance of overhead threat.

That visual cortex plays an important role in a more demanding detour test is clear from experiments described by Ingle and Shook (1984). These animals received lesions either of frontal cortex (controls) or of posterior medial cortex, including areas 18b and 29c,d. These two medial cortical zones receive projections from the posterior nucleus of the thalamus, which in turn receives projections (as demonstrated in rats and hamsters) from the pretectum. This so-called 'pretectofugal' zone of rodent visual cortex contrasts with areas 18a and 19 (lateral to area 17) which receive projections from the lateral posterior nucleus, itself a recipient of tectal projections. We found that ablations of the 'prectal' zone of cortex abolished the ability of monocular gerbils to detect the nearer of two offset barriers as they tried to swerve and snake their way between them to reach a dark aperture on the other side. Frontal gerbils, by contrast, quickly recovered the ability to discriminate the nearer barrier, apparently using information about the asymmetry of visual flow gradients. The gerbils with medial visual cortex ablations were apparently normal in other tests of moving spot detection, and could avoid a single barrier set in front of the target aperture; they seemed selectively deficient in their use of 'expanding motion gradient' information to compute the distance to a surface while running briskly forwards. It would be important in future studies of this kind to evaluate the ability of such lesioned animals to utilize motion parallax for depth judgements.

Comparisons of Motion Sensitivity Between Fish and Mammals

Uses of Direction Selectivity in Visual Neurones

The finding of directional selectivity among retinal ganglion cells is ubiquitous among vertebrates, yet explanations of the observed variations among species in regard to the most commonly encountered directionality are lacking (see review by Ingle, 1973d). The fact that tectal units in goldfish are mostly selective for small objects moving in a nasalward direction (Cronly-Dillon, 1964) led Ingle (1967) to predict that at least one class of goldfish visual behaviour should be biased for the detection of nasalwards moving stimuli. Ingle reported that when confined goldfish were classically conditioned to reduce heart rate during the horizontal motion of dark spots in the lateral field, all fish proved more sensitive to nasalward motion than to caudalwards motion. Ingle proposed that such selectivity would be useful for a fish with side-to-side foraging movements, since small, food-like objects would then be best detected on the side toward which the fish is turning (Ingle, 1973d). That kind of explanation might hold for rabbits or pigeons as well, since their habit of side-to-side scanning while feeding coincides with a preponderance of tectal cells with nasalward sensitivity. In testing this hypothesis, it would be useful to compare the pigeon with the owl, in which a fixed head and eye direction is maintained until prey is sighted (or heard).

Perception of Relative Motion and Uses of Motion Parallax

The gerbils' sensitivity to relative motion shows the same lower velocity threshold as does sensitivity to real motion (unpublished Ph.D. thesis of B. Shook, reported by Ingle and Shook, 1984). In fact, gerbils show the classic 'induced-motion' effect when a slowly moving frame leads the animal to select an enclosed motionless seed over a moving seed coupled with a moving frame. However, ablation of striate cortex grossly elevated the threshold for simple motion detection, and in the test for induced motion the lesioned animals now preferred to approach the combined motion of seed and frame rather than that of the frame alone. We suggest that in this test the gerbil was dependent upon the detection of motion via the tectum and this mechanism (unaided by visual cortex) could not detect relative motion. It is interesting that striate ablation produced such a conspicuous reduction of motion sensitivity, since the same lesioned gerbils performed essentially normally in a set of pattern discrimination tests (Ingle and Shook, 1984). These experiments should make us guard against the untested assumption of many reviewers that tectum plays a more important role than does visual cortex in motion perception in lower mammals.

The discovery that the gerbil striate cortex (and no doubt other associated regions of visual cortex) is essential to detection of relative motion prompts the question as to whether fishes and amphibians (without any neocortex) are capable of detecting relative motion. While there are no relevant studies of that question with fishes, a careful study of toads (Honigman, 1944) found no induced motion effects in prey-catching behaviour. Although his toads snapped readily at worms on a slowly moving belt, they would not respond to a stationary worm viewed above a slowly moving patterned belt. On the other hand, studies of Ewert (1984) show that toads do see such movement of a background pattern. Toads will snap at dark, worm-like shapes while walking forwards on a homogeneous white floor, but are inhibited from feeding if the background consists of a black–white randomized texture. Indeed, there are recording studies which indicate that the toad's tectal 'T5-2' cells are inhibited in response to moving worms by such background texture motion. This study did not report situations where moving worm and background in opposite directions might enhance detection of the moving spots as was found for cells in the pigeon tectum (Frost *et al.*, 1981). The so-called 'double-opponency' properties of tectal cells in pigeon resemble those of many visual cortical cells in mammals, and may be derived from descending telencephalic visual projections in the pigeon rather than being created by intrinsic tectal circuitry.

Lower vertebrates might gain the most from detection of relative motion of stationary edges in the environment as they walk past them (rather than detecting relative motion of prey against background). Although no relevant tests have been yet reported with fishes or amphibians, there is good evidence that the up-and-down, head-bobbing behaviour of the gerbil is used to judge the absolute distances of surfaces preparatory to leaping a gap (Ellard and Goodale, 1990). When gerbils are deprived of binocular disparity cues (by sewing one eye shut) they compensate by making more of the larger-amplitude head bobs before jumping, and will make more such parallax-gaining movements for wider gaps. Further studies are needed to determine whether sensitivity to motion parallax is missing even in spatially adept amphibians (such as the tree-frogs) and whether among mammals this ability depends upon integrity of the visual cortex.

Visual Tracking Depends upon Visual Cortex

Although popular belief holds that a frog can strike a fast-moving fly on the wing, our unpublished films of frogs and toads reveal that these animals strike persistently at the location where the target was seen about 200 ms prior to initiation of the lunge. In other words, we have as yet no evidence that an amphibian can 'lead' its prey by anticipating its trajectory. A film study of a fast-moving predatory fish (Lancaster and Mark, 1975) reached the conclusion that fish would lunge each time at the instantaneous location of prey, but could not anticipate the trajectory of falling prey. On the other hand, gerbils are able to predict the trajectory of a food-baited disc moving at 30°/s after less than 300 ms of viewing time (Ingle *et al.*, 1979). Single-frame reconstructions of stimulus and gerbil head positions indicate that the animals turn the head about 10° ahead of the target, and in a second forward-running approach aim a further 10–15° ahead of the moving target, so as to intercept it. This ability is seen only for targets within the rostral field, up to about 40° from the rostral midline: an area where visual cortex seems to dominate orienting behaviour (Ingle, 1982). Of course, still more impressive skill is seen in dogs trained to catch Frisbees, which move in a decidedly curvilinear path. With the gerbil, the capacity for anticipatory orientation to moving discs is abolished by ablations which include cortical areas 17 and 18b (Ingle and Shook, 1984). Since the tectum seems programmed to turn head or eyes directly to a target's immediate location, the act of turning *ahead* of a target may reflect dominance of a cortical system over the optic tectum. Our candidate for such a forward-looking visuomotor system is the frontal cortex, which could inhibit the tectum via direct connections and bypass the tectum to activate motor control mechanisms in the brain stem. It is of great interest that gerbils, like rats, have direct projections from striate cortex to frontal area 8 (Ingle and Vogt, unpublished observations) and this direct connection is what may be needed for rapid activation of anticipatory orientation.

Shape Recognition in Vertebrates: Advantages of a Visual Cortex

Failure of Shape Constancies in the Fish

As humans, it is our common experience that distinctive shapes are easily recognized and labelled despite large variations in their size or in their spatial orientation. Thus we speak of shape recognition as invariant with transformations of size and of spatial orientation, although we recognize a few interesting exceptions such as the difficulty in recognizing upside-down faces. Studies with monkeys and with pigeons have provided clear evidence for shape-recognition invariance (or shape constancy) as stimuli are rotated in the frontal plane or during moderate changes in size. However, my review of other mammalian tests (Ingle, 1978), finds inconclusive evidence for good shape constancy, probably because of methodological weaknesses and lack of prolonged training procedures. Although one suspects that all highly visual birds and mammals have good shape constancy, the question needs to be more thoughtfully explored so that abilities of various species can be compared judiciously with those of primates.

A number of prewar European studies of teleost fish failed to demonstrate rotation invariance using relatively simple planar shapes, and a variety of generalization tests following discrimination training (reviewed by Ingle, 1978). A typical finding is that after fish learn to select a triangle (but not a square) for food reward, a generalization test with presentation of a diamond *vs* a rotated version of the same triangle will reveal the fish's preference for the diamond. The inference from this (and from similar tests) is that the fish has learned to identify the rewarded shape by noting the presence of a 'point' on its top, and not by attending to the entire configuration as such. Therefore, a diamond (with point at the top) is taken as similar to the initial triangle, while the rotated triangle (now with a flat top) looks very different.

Ingle (1971a, 1978) used a different training procedure to see whether similar results would be obtained if the fish discriminated shapes from a constant distance, and did not have to approach the shape. Fish were trained in a shuttlebox to avoid shock during presentation of one shape (by swimming forwards into a second box) and to inhibit avoidance in the presence of a second shape never associated with shock punishment. Using this successive discrimination, we found that after training goldfish to discriminate small squares from small circles (each 7° high) there was no transfer of the discrimination to presentations of large (14° high) versions of the same shapes. On the contrary, fish easily learned to avoid small circles (not small squares) and to avoid large squares (not large circles) without showing apparent conflict. It appeared that the size change altered the coding rules used to discriminate circles from squares. Ingle suggested that the mechanism for discriminating small shapes corresponds to an innate mechanism of food selection, and demonstrated that goldfish will spontaneously snap at small circles affixed to the side of the tank in preference to small squares or triangles of the same size.

Visuomotor Foundations of Shape Discrimination in Lower Vertebrates

On the other hand, Ingle (1978) suggested that the mechanism for discrimination among larger shapes might reflect an inborn mechanism biasing the negotiation of barriers and apertures found in the environment (Fig. 8.8).

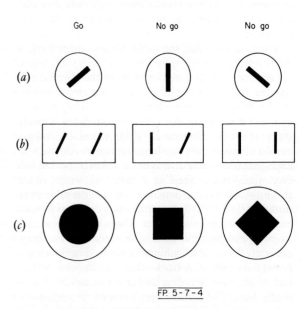

Fig. 8.5 *Discriminanda used by Ingle (1971a, 1976) to explore the 'relative difficulty' of shape discriminations in goldfish. For each of the three sets of shapes (a, b and c), fish were trained to escape from the leftmost stimulus, but to withhold the escape response during presentation of the other two stimuli. The middle shape was always easier to discriminate from the 'Go' shape than was the rightmost shape.*

A series of training studies will be described to illustrate this viewpoint. First, goldfish trained to avoid either 14° wide squares or circles were found to confuse diamonds more with circles than with squares. Ingle suggested that the fish noticed that the sides of a square were parallel while left vs right sides of a circle or a diamond were not (they were each curved). A second test showed that goldfish classified a pair of vertical line segments (set apart by 15°) as similar to a pair of 30°-tilted segments, but as different from a pair of non-parallel segments (0° + 30°). Had the fish been classifying each line segment individually then the non-parallel pair would have been intermediate in distinctiveness to the two parallel pairs. For the fish 'parallelness' is probably not an abstraction from prior labelling of orientation in each segment, but rather corresponds to noticing that distances between the respective top and bottom ends of the segments are equal. This symmetry has a distinctive visuomotor correlate: with apertures of parallel sides it does not matter if the fish swims higher or lower while passing through. But with apertures of non-parallel sides the fish trajectory through the aperture is pushed upwards or downwards.

Detailed studies of the frog's discrimination among apertures of varying widths (Ingle and Cook, 1977) reveals these animals to avoid leaping through holes whose width approaches their actual head width. Unpublished studies show (predictably) that the height of a frog's jump through an aperture (to avoid a pinch from the rear) depends upon whether the upper or lower region of the aperture is wider. Ingle and Cook (1977) further showed that in judgement of aperture width, frogs do not show size–distance constancy, but rather measure the visual angle between opposite edges. Thus the tendency to discriminate parallel and non-parallel pairs of edges may operate only within a limited size range in frogs (and in fish) and may not reflect an ability for abstract recognition of the actual configuration of the aperture.

A final example of goldfish discrimination among rectangles of different orientations will reinforce our hypothesis concerning the naturalistic basis of shape recognition in lower vertebrates. Mackintosh and Sutherland (1963) first demonstrated that goldfish confuse mirror image oblique rectangles (oriented at 45° and 135°) but easily discriminate verticals from horizontals. Ingle (1971a, 1978) reinforced this finding in showing that goldfish regard a 45° bar as more similar to a 135° bar than to either a horizontal or a vertical bar. As an obstacle, a vertical bar elicits a horizontal detour movement in fish, and a horizontal bar elicits either upwards or downwards detour directions. By contrast, oblique bars, are ambiguous and each elicits some horizontal and some vertical movements. Detour studies with frogs as well (Ingle, 1971b) show that vertically striped barriers elicit reliable detour attempts, while horizontal barriers do not 'repel' frogs effectively. Oblique stripes of either orientation are moderately effective. Thus we believe that line orientation is classified by fishes or frogs in a very concrete, 'environment-based' manner'.

References

Brown, W. and Ingle, D. (1973). Receptive field changes produced in frog thalamic units by lesions of the optic tectum. *Brain Res.*, 59, 405–409.
Caine, H. S. and Gruberg, E. R. (1985). Ablation of nucleus isthmi leads to loss of specific visually-elicited behaviors in the frog, *Rana pipiens. Neurosci. Lett.*, 54, 307–312.
Collett, T. (1977). Stereopsis in toads. *Nature Lond.*, 267, 349–351.
Cronly-Dillon, J. R. (1964). Units sensitive of direction of movement in goldfish optic tectum. *Nature Lond.*, 203, 214–215.
Dean, P., Redgrave, P. and Westby, G. W. (1989). Event or emergency? Two response systems in the mammalian superior colliculus. *Trends Neurosci.*, 12, 137–147.
Ellard, C. and Goodale, M. A. (1988). A functional analysis of the collicular output pathways: a dissociation of deficits following lesions of the dorsal tegmental decussation and the ipsilateral collicular efferent bundle in the Mongolian gerbil. *Exp. Brain Res.*, 71, 307–319.
Ellard, C. and Goodale, M. A. (1990). The role of image size and retinal motion in the computation of absolute distance by the Mongolian gerbil. *Vision Res.*, 30, 399–413.
Ewert, J.-P. (1970). Neural mechanisms of prey-catching and avoidance behavior in the toad, *Bufo bufo. Brain, Behav. Evol.*, 3, 36–56.
Ewert, J.-P. (1971). Single unit responses of the toad's (*Bufo americanus*) caudal thalamus to visual objects. *Z. Vergl. Physiol.*, 74, 81–102.
Ewert, J.-P. (1984). Tectal mechanisms that underlie prey-catching and avoidance behaviors in toads. In: *Comparative Neurology of the Optic Tectum.* ed. Vanegas, H. pp. 247–416. New York: Plenum.
Ewert, J.-P. and Gebauer, L. (1973). Grössenkonstanzphänomene im Beutefangverhalten der Erdkröte (*Bufo bufo*). *J. Comp. Physiol.*, 85, 303–315.
Frost, B. J., Scilley, P. L. and Wong, S. C. P. (1981). Moving background patterns reveal double opponency of directionally specific pigeon tectal neurons. *Exp. Brain Res.*, 43, 173–185.
Goodale, M. A. and Milner, A. D. (1982). Fractionating orienting behavior in rodents. In: *Analysis of Visual Behavior.* eds. Ingle, D., Goodale, M. and Mansfield, R. pp. 267–300. Cambridge, MA: MIT Press.
Gruberg, E. R. and Lettvin, J. Y. (1980). Anatomy and physiology of a binocular system in the frog, *Rana pipiens. Brain Res.*, 192, 313–325.
Gruberg, E. R. and Udin, S. B. (1978). Topographic projections between the nucleus isthmi and tectum of the frog, *Rana pipiens. J. Comp. Neurol.*, 179, 487–500.
Grüsser, O.-J. and Grüsser-Cornehls, U. (1976). Neurophysiology of the anuran visual system. In: *Frog Neurobiology.* eds. Llinas, R. and Precht, W. pp. 298–385. Berling: Springer.
Honigman, H. (1944). The visual perception of movement by toads. *Proc. R. Soc. Lond. B.*, 291–307.
Ingle, D. (1967). Two visual mechanisms underlying the behavior of fish. *Psychol. Forsch.*, 31, 44–51.
Ingle, D. (1968). Visual releasers of prey-catching in frogs and toads. *Brain, Behav. Evol.*, 1, 500–518.
Ingle, D. (1970). Visuomotor functions of the frog optic tectum. *Brain, Behav. Evol.*, 3, 57–71.

Ingle, D. (1971a). Discrimination of edge orientation by frogs. *Vision Res.*, 11, 1365–1367.

Ingle, D. (1971b). The experimental analysis of visual behaviour. In *Fish Physiology*, vol. 5. eds. Hoar, W. S. and Randall, D. J. 5, pp. 59–71. New York: Academic.

Ingle, D. (1973a). Two visual systems in the frog. *Science Wash. DC*, 181, 1053–1055.

Ingle, D. (1973b). Disinhibition of tectal neurons by pretectal lesions in the frog. *Science Wash. DC*, 180, 422–424.

Ingle, D. (1973c). Size preference for prey-catching in frogs: relationship to motivational state. *Behav. Biol.*, 9, 485–491.

Ingle, D. (1973d). Evolutionary perspectives on the function of the optic tectum. *Brain, Behav. Evol.*, 8, 211–237.

Ingle, D. (1975). Selective visual attention in frogs. *Science Wash. DC*, 188, 1033–1035.

Ingle, D. (1976). Spatial vision in Anurans. In: *The Amphibian Visual System*. ed. Fite, K. pp. 119–140. New York: Academic.

Ingle, D. (1978). Shape recognition by vertebrates. In: *Handbook of Sensory Physiology*, vol. 8. *Perception*. eds. Held, R., Leibowitz, H. W. and Teuber, H.-L., pp. 267–296. Berlin: Springer.

Ingle, D. (1980). Some effects of pretectum lesions on the frog's detection of stationary objects. *Behav. Brain Res.*, 1, 139–163.

Ingle, D. (1981). New methods for analysis of vision in the gerbil. *Behav. Brain Res.*, 3, 151–173.

Ingle, D. (1982). The organization of visuomotor behaviors in vertebrates. In: *The Analysis of Visual Behavior*. eds. Ingle, D., Goodale, M. and Mansfield, R. pp. 67–109. Cambridge, Mass: MIT Press.

Ingle, D. (1983a). Brain mechanisms of localization in frogs and toads. In: *Advances in Vertebrate Neuroethology*. eds. Ewert, J. P., Capranica, R. and Ingle, D. pp. 177–226. New York: Plenum.

Ingle, D. (1983b). Prey-selection in frogs and toads: a neuroethological model. In *Handbook of Behavioral Neurobiology*, vol. 6: *Motivation*. eds. Satinoff, E and Teitelbaum, P. pp. 235–264. New York: Plenum.

Ingle, D. and Cook, J. (1977). The effect of viewing distance upon size preference of frogs for prey. *Vision Res.*, 17, 1009–1014.

Ingle, D. and Hoff, K. (1990). Visually-elicited evasive behavior in frogs. *Bio-Science*, 40, 284–291.

Ingle, D. and Shook, B. (1984). Action oriented approaches to visuospatial functions. In: *Brain Mechanisms and Spatial Vision*. eds. Ingle, D., Jeannerod, M. and Lee, D. pp. 229–258. The Hague, Netherlands: Nijhof.

Ingle, D., Cheal, M. and Dizio, P. (1979). Cine analysis of visual orientation and pursuit by the Mongolian gerbil. *J. Comp. Physiol. Psychol.*, 93, 919–928.

Kicliter, E. (1973). Flux, wavelength and movement discrimination in frogs: forebrain and midbrain contributions. *Brain, Behav. Evol.*, 8, 340–365.

Kostyk, S. K. and Grobstein, P. (1987). Neuronal organization underlying visually elicited prey orienting in the frog. *Neurosci.*, 21, 57–82.

Lancaster, B. S. and Mark, R. F. (1975). Pursuit and prediction in the tracking of moving food by a teleost fish. *J. Exp. Biol.*, 63, 627–645.

Lazar, G., Toth, P., Csank, G. and Kicliter, E. (1983). Morphology and location of tectal projection neurons in the frog: a study with HRP and cobalt-filling. *J. Comp. Neurol.*, 215, 108–120.

Lettvin, J. Y., Maturana, H. R., McCulloch, W. S. and Pitts, W. H. (1959). What the frog's eye tells the frog's brain. *Proc. Inst. Radio Engrs NY*, 43, 1940–1951.

Lettvin, J. Y., Maturana, H. R., Pitts, W. H. and McCulloch, W. S. (1961). Two remarks on the visual system of the frog. In: *Sensory Communications*. ed. Rosenblith, W. A. pp. 757–776. Cambridge, MA: MIT Press.

Mackintosh, N. J. and Sutherland, N. S. (1963). Visual discrimination by goldfish: the orientation of rectangles. *Animal Behav.*, 11, 135–141.

Muntz, W. R. A. (1962). Microelectrode recordings from the diencephalon of the frog (*Rana pipiens*) and a blue-sensitive system. *J. Neurophysiol.*, 25, 699–711.

Northmore, D. P. M., Levine, E. S. and Schneider, G. E. (1988). Behavior evoked by electrical stimulation of the hamster superior colliculus. *Exp. Brain Res.*, 73, 595–605.

Schneider, G. E. (1973). Early lesions of the superior colliculus: factors affecting the formation of abnormal retinal projections. *Brain, Behav. Evol.*, 8, 73–109.

Wilczynski, W. and Northcutt, R. G. (1983). Connections of the bullfrog striatum: afferent organization. *J. Comp. Neurol.*, 214, 321–332.

9 Neural Control of Pursuit Eye Movements

Rolf Eckmiller

Introduction

During the evolution of vertebrates, various needs for controlled movements of an eye within the orbita arose, eventually leading to the development of several quite different oculomotor subsystems.[373] The latest version in old-world primates is the oculomotor system of humans (and several non-human primate species), which can be considered a 'federation' of oculomotor subsystems, sharing the eye as the object of movement and (to a large extent) the extraocular eye muscles, but using separate parts of the visual and vestibular systems as sensory input.

Various lines of evidence lead to the hypothesis that the phylogenetically oldest oculomotor subsystem (known as vestibulo-ocular reflex, VOR) was developed to reduce retinal slip of the projection of the visual world for moving animals (e.g. fish) down to tolerable values so as to improve vision. For this purpose, extraocular muscles were developed and then connected in an orderly fashion to the various vestibular organs. It has also been claimed that the phylogenetically most recent oculomotor subsystem is the pursuit system, which is only poorly developed in cats[125] but which evolved at some point in time in non-human primates.[373] In fact, smooth pursuit eye movements serve to continuously maintain sharp vision of a target as occurs during fixation, not only if it moves in front of the stationary head but also if the head alone or both target and head move. In other words, the pursuit system can also perform the task originally assigned to the VOR, and it can perform it even better since only the retina (and not the vestibular organs, which receive only head accelerations) receives the appropriate sensory information to assure maintenance of vision equivalent to fixation during various combinations of sensory stimulation.

The pursuit system as a separate oculomotor entity began to draw scientific attention toward the end of the nineteenth century, when the first observations by Ewald Hering in Prague[168] and studies with afterimages[202] indicated that, in humans, the line of gaze can exactly follow a moving target by means of smooth eye movements (*langsame Augenfolgebewegungen*) that cannot be voluntarily generated by most subjects without a target. Hering also clearly associated these pursuit movements with the process of visual fixation independent of whether the target or the head moved.

> *Eine langazsmere und doch stetige Bewegung der Augen lässt sich nur dadurch erzielen, dass man den eben fixirten Punkt langsam und stetig verschiebt, wobei die Augen getreu der Ortsveränderung des fixirten Objectes folgen, oder dadurch, dass man während man einen Punkt fixirt, den Kopf langsam wendet oder im Raume verschiebt.* [A smoother and yet steady eye movement can only be achieved by moving the just fixated point slowly and steadily, whereby the eyes follow precisely the dislocation of the fixated object; or by slowly turning or moving the head in space while fixating a point.]
> Hering [168, p 21]

Brown,[48] who also noticed the difference between saccades and fixation or pursuit ('If we keep our eyes fixed on the moving object, and this is possible if it does not move too fast or too irregularly, then we see it fixed and really fixed things moving.'), pointed out yet another oculomotor subsystem that generates 'compensatory movements' (VOR) or rotatory nystagmus of largely vestibular origin in response to head movements.

At about the same time, various clinical reports began to appear describing patients with blocked voluntary eye movements who were still able to perform involuntary eye movements. Specifically, these patients could not voluntarily look around by means of saccades but were able to move their eyes when a fixated visual object was being slowly moved.[17, 28, 319, 377] These clinical observations, which were already associated with dysfunctions of specific brain regions, clearly confirmed the view that the pursuit system is indeed a separate oculomotor subsystem.

It should be noted that pursuit movements are usually considered to be involuntary or reflective (e.g., 'optischer Bewegungsreflex'[17, 84, 202, 309], but sometimes they are also described as voluntary movements.[81, 177] Probably there is some truth to both views.[394]

Landolt[202] found that pursuit movements could not be generated voluntarily, even when subject were required to continuously move the line of gaze along a straight line. (Instead, using the afterimage method, he found alternating sequences of saccades and fixation episodes.) This finding was confirmed and elaborated in an influential study on eye movements during reading by Erdmann and Dodge.[123] Upon his return to the USA, the second author set up his own oculomotor laboratory and presented the first detailed description of five different oculomotor subsystems.[100] He suggested the term 'pursuit movements' for that type of eye movement in which the line of regard follows a target moving across the field of vision. These behavioural studies confirmed the fundamentally important hypothesis that humans 'see' (in the sense of recognition of visual objects) only during fixation, and they see those objects best that are projected onto the central retina.

This hypothesis, which was further elaborated by Öhrwall,[273] led to two further assumptions. (a) The visual recording process for a large structured visual scene (e.g., a book page or a painting) requires some sort of scanning process,[269] which means a sequential intake of visual information at a number of fixation points that are interconnected by saccadic eye movements,[394] and (b) pursuit movements try to 'lock' moving visual objects onto the fovea as occurs during fixation. Such a concept makes sense, in that smooth, pursuit-like eye movements cannot normally (without special training and predisposition) be generated (in contrast to voluntary smooth limb movements), since they would only blur vision in the absence of a visual target moving along with the eye.

Since pursuit movements were thought to serve fixation of a moving target, it was quite appropriate to compare the miniature eye movements[338, 346] during fixation of a stationary target with those during pursuit.[273] These eye movements (at that time known as *Elementarfixationen*), which were measured by means of afterimages and by observing bulbus movements under the microscope, appeared to be similar during pursuit and fixation, thereby supporting the established view of the function of pursuit movements. A very detailed and influential report on pursuit movements (they were called *Führungsbewegungen*) and their saccadic disturbances in various pathological cases was given by Cords[84] who also reviewed the older literature on pursuit movements and nystagmus. The author concluded that the pursuit system must partly use neural pathways and structures separate from those for the other types of eye movements. This conclusion which was partly based on clinical observations following Wernicke's[377] first report, will receive more confirmation in this review.

The oculomotor pattern of optokinetic nystagmus (OKN), which was also incorporated in the study on pursuit movements by Cords,[84] consists of a quick phase (saccade) in one direction and a slow, pursuit-like phase in the opposite direction. Optokinetic nystagmus was used as a powerful diagnostic tool by the use of a small rotatable drum with stripes.[17] Soon it became necessary to distinguish between at least two main types of OKN: (a) Schau (look) nystagmus, in which the subject deliberately fixates one of the moving stripes (or any large object that is part of the visual world and is moving relative to the unaccelerated head), and (b) Stier (stare) nystagmus, which leaves vision rather blurred, owing to considerable retinal slip movements.[355] The slow phase of Schau nystagmus, which is assumed to be equivalent to a pursuit movement, will not be considered further since the stimulus pattern typically covers large portions of the retina. Quantitative descriptions and measurements of the stimulus pattern with regard to different retinal locations are rather complex, as is the relationship between the size, pattern and location of the OKN stimulus and the OKN (Schau or Stier) pattern. The reader is referred to a number of detailed papers on this subject.[3, 58, 71, 79, 107, 359, 395].

After the Second World War, research on the control of pursuit movements was reinitiated by Westheimer,[379] who was the first person to apply linear systems theory to the pursuit system. He also postulated a mechanism of 'anticipation' or 'prediction' to explain how the pursuit system could overcome the considerable time delay between retinal stimulation and oculomotor pursuit response.

Two main approaches, which have been designed to study the dynamic behaviour of the human pursuit system, are the frequency analysis of sinusoidal pursuit[106, 126, 348] and pursuit movements in response to ramp and step-ramp target movements.[302, 306]

Several recent reviews are witnesses of the gradual development of our knowledge in this field.[118, 215, 307, 309, 398]

As there are considerable differences among the various types of smooth eye movements with regard to the physical stimulus, retinal events, neural control, and the movement time course,[118] the scope of this review will be confined to pursuit eye movements (PEM) in the original sense of Hering[168] and Dodge.[100] The hope is that once we understand how primates keep a small target as a moving fixation spot continuously on the fovea (or parafovea) against a homogeneous background during PEM, we will be better prepared to deal with more complex stimulus conditions. Such stimulus paradigms, can range from PEM against a structured background via pursuit of a perceptually completed contour that was extrapolated from peripheral retinal information[344] to Sigma-pursuit movements on the basis of a stroboscopically illuminated stationary stimulus pattern.[24]

Sensory Inputs for Smooth Pursuit System

Retinal Events Leading to Pursuit Eye Movements

Pursuit eye movements in primates are assumed to serve continuous maintenance of vision of a moving target, as occurs during fixation. It has proven useful to distinguish between *initiation* of PEM and *maintenance* of PEM.[112] Since the visual system typically recognizes and identifies a new target within the first few seconds of fixation, both maintenance of fixation and maintenance of PEM over a longer time span are employed mainly to provide information about where the target is rather than what it is. Obviously the current position of a moving target (prey or enemy) in space is crucial for primates in generating the appropriate motor response (prey catching or defence). Initiation of PEM can be studied by presenting a moving target within the field of gaze during fixation. Although very little is known about the details of the required signal processing, it is assumed that the neural network within the retina and the afferent visual system monitors the trajectory of the target movement on the retina. The various parameters of the target movement relative to the centre of the fovea (e.g., position-error vector $r(t)$ and slip velocity vector $v(t)$) are evaluated in a still unknown fashion, to generate the time course of pursuit initiation, which consists of a foveating saccade superimposed on the initial portion of smooth pursuit eye movements.[11] Numerous behavioural studies with step-ramp movements in humans[302, 306] as well as monkeys[20, 221] support the hypothesis that velocity and direction of the target movement on the initially motionless retina form the decisive stimulus for initiation of PEM. Simultaneously, the position error is reduced by a foveating saccade. One might speculate that velocity-sensitive and direction-selective afferent visual neurones with receptive fields throughout the retina (see 'Occipital and Temporal Cortex' below) provide the neural eye velocity signal for initiation of PEM, which actually begins with a delay of 100–150 ms after the beginning of the target movement on the retina (see reference 309). However, the fact that PEM can also be initiated in primates by means of a pure retinal position-error stimulus[197, 290, 306] argues against the notion that slip velocity is the sole stimulus for initiation of PEM. Although position errors larger than 10–20 min of arc typically lead to correctional saccades during fixation, this position error threshold for saccade elicitation can be increased during PEM such that position errors up to several degrees relative to the foveal centre can be tolerated.[81, 112, 298] This feature of a flexible position-error threshold is probably crucial for pursuit of highly unpredictable, manoeuvrable target trajectories. Initiation of PEM by means of pure position-error stimuli has been demonstrated using eccentric afterimages[161, 197] or various open-loop techniques.[290, 306]

Once PEM has been initiated within 400–500 ms after target movement onset, the target projection moves in an irregular fashion on the central retina in the vicinity of the foveal centre. This irregular movement is the result of miniature eye movements (with amplitudes of a few minutes of arc as during maintenance of fixation; see references 99, 346) and of differences between the target trajectory and the internally generated pursuit movement trajectory. In fact, the stabilization process during fixation[309] may share the same neural mechanisms that generate maintenance of pursuit. This hypothesis was already proposed by Gertz,[145] who suggested that maintenance of pursuit and the stabilization process during fixation share the same neural system, which he calles *Stellungsapparat* in contrast to a *Blickapparat* for rapid eye movements. For theoretical reasons it is extremely difficult to reconstruct the target trajectory in space from the position-error time course $r(t)$ during PEM.

Maintenance of PEM can best be studied with target trajectories, such as sinusoids (or constant-velocity triangle waves), which allow continuous PEM for at least several seconds back and forth within the normal oculomotor range. Ramp trajectories at higher velocities may not be optimal for this task, since the pursuing eye reaches the limits of the physiological oculomotor range very soon after the end of the initiation phase. Various behavioural studies on the dynamic properties of the pursuit system during maintenance of PEM in humans[81, 126, 348] and monkeys[112, 119, 218] have yielded a number of puzzling results with regard to retinal events during PEM, especially with regard to position error and slip velocity. These parameters are usually obscured in linear systems descriptions of gain and phase as a function of frequency. Position errors were found to increase with decreasing predictability of the target trajectory and reached values of up to several degrees. Detailed and quantitative measurements of the target path within the central retina during maintenance of PEM have not yet been reported, although several highly accurate eye movement recording systems (see, e.g., references 82, 87, 305) are available. An analysis of the target path relative to the foveal centre and its correlation with the pursuit eye movement time course could provide important information regarding the spatio-temporal stimulation pattern of afferent visual neurones with receptive fields within the central retina during PEM. Such data could specifically address the question of to what extent position error in contrast to slip velocity or direction of the slip movement can account for the resulting pursuit movement. For theoretical reasons, slip velocity has to be minimal and of varying direction during maintenance of PEM becauses the target and eye must move with approximately the same velocity or else the target projection would slip outside the fovea. Furthermore, the position error has to be continuously minimized, since a pure velocity servo system would often result in stabilization of the target somewhere on the peripheral retina rather than on the fovea. Probably the first person to become aware of this problem was the physiologist Johannes von Kries.[368] He suggested that maintenance of PEM might be achieved by feeding the continuous position-error signal into the control system of PEM. However, at the present time, the controversy concerning the importance of position error versus slip velocity signals for maintenance of PEM is still alive.[118, 221, 240, 291, 393] This may be due partly to the fact that data from quite different experimental paradigms (which will be further discussed below) are often compared. Furthermore, exact behavioural, as well as neurophysiological, measurements of the retinal events during PEM are still missing.

A number of structural and functional features of the eye and the afferent visual system can be listed as preconditions for the generation of pursuit eye movements. (a) The afferent visual system is arranged in retinotopic coordinates with its coordinate centre within the central retina. (b) A certain retinal area centred around the centre of the retinotopic coordinate system is responsible for maintaining vision during PEM. In other words, as long as the target projection stays within this retinal area (the pursuit area, which may be several degrees in diameter[119]), the various elements of the visual process such as pattern recognition and perception of depth of colour are assured. The concept of a pursuit area is related to similar descriptions of retinal events during fixation[273] and a positionally insensitive zone in the central fovea (see reference 126). (c) The afferent visual system can select visual objects as pursuit targets; it can distinguish between retinal events of the selected pursuit target and those of the other non-target objects.

If indeed the pursuit system is phylogenetically as young as primates, could further evolutionary steps lead to an even better pursuit system? What comes to mind is the functional limitation of the pursuit system owing to luminance adaptation in macaques and humans to the upper mesopic and photopic range of luminance adaptation. This limitation is based on the fact that, in this species, the fovea centralis is almost exclusively restricted to cones, rendering the foveal visual system functionally blind during dark adaptation. Therefore the luminance range of operation of the pursuit system could be expanded into the scopic range by means of a retina with a functional centre but with an even distribution of rods and cones. Interestingly, such an arrangement exists in the retina of certain nocturnal monkeys. However, the oculomotor performance of these monkeys has not been extensively tested.[270]

The following paragraphs are largely restricted to studies in the species *Macaca mulatta* and *M. fasicularis*, of the same macaque monkey genus[131] as well as *M. nemestrina* and *M. speciosa*. The monkeys have a clear fovea and oculomotor abilities quite similar to those of humans. The afferent visual system of macaques (as well as humans) seems to be functionally divided into a foveal (including parafovea) and an extrafoveal system.[96] Accordingly, initiation of PEM typically uses the extrafoveal system, whereas maintenance of PEM is more or less restricted to the foveal system.

The discussion of retinal events during PEM in macaques and humans is divided into psychophysical, morphological and neurophysiological considerations. It seems necesary to pay special attention to the retina, which serves as a transducer and filter of the visual stimulus pattern, since many of its features have often been ignored in studies of the pursuit system.

Psychophysics

Visual acuity and adaptation luminance

Both retinal location with respect to the fovea and adaptation luminance influence the static visual acuity.[12, 88, 194] The total luminance range of the primate visual system is usually divided into a scotopic range (dark adaptation) for adaptation luminances (L_a) below $L_a = 10^{-3}$ cd m^{-2}, a mesopic range between 10^{-3} and ~ 5 cd m^{-2}, and a photopic range (light adaptation) above $L_a = 5$ cd m^{-2}.[208] Whereas visual acuity at the foveal centre is maximal in the photopic range, it becomes sharply reduced with decreasing adaptation luminance in the mesopic range and drops to zero in the lower mesopic and the scotopic range.[12, 194] In the scotopic range, visual acuity reaches its maximum (although at a much lower level than in the photopic range) at retinal locations corresponding to the parafovea and perifovea at retinal eccentricities of ~ 5–$15°$.[12, 88, 172, 378] In other words, the central retina including the fovea and parts of the parafovea[293, 347, 367] is functionally blind for small targets in the scotopic and the lower part of the mesopic range. This fact has long been known as *Zentrales Dämmerskotom*, or central scotoma.[194, 367] The fovea and parafovea are, respectively, $\sim 3°$ and $6°$ in diameter in macaques[347] with slightly larger diameters of $5°$ and $8°$, respectively, in humans.[293]

Contrast

The photometric contrast $\Delta L/L = (L_t - L_a)/L_a$ refers to the luminance (L_t) of a pursuit target and the background

L_a. The range of operation for brightness perception is ~2–3 log units at any given L_a.[162, 191]

The relative luminance–difference threshold ($L/\Delta L$), which is the inverse of the just perceivable photometric contrast at threshold in humans, is highest in the fovea, with a gradual decrease toward the peripheral retina in the photopic range.[12] As would be expected from the visual acuity data, $L/\Delta L$ decreases with decreasing L_a and drops to zero in the fovea in the lower mesopic and scotopic range, whereas maximum values at these low adaptation levels are reached in the parafoveal and perifoveal region at retinal eccentricities between 5 and 15°. For example, at $L_a = 1$ cd m^{-2}, the contrast threshold ($\Delta L/L$) for small targets (diameter <0.5°) ranges from ~10^{-2} to 1.

As long as L_t remains within the given range of operation (2–3 log units around L_a), the target is visible without leading to dazzle (L_t too large relative to L_a). However, if L_t is too small relative to L_a then the target surface is only perceived as black. In those cases where L_t is far above the range of operation, the retinal events become rather undefined, since such a bright target leads not only to the perception of dazzle but also to local adaptation and afterimages wherever it moves.[12, 194] In fact, such a stimulus was probably very common in many behavioural pursuit studies in which a small, bright target was presented in a totally dark room (see, e.g., 302).

These psychophysical findings consequently require a distinction between at least three types of eye movements to pursue a single target: (a) photopic foveal pursuit of small targets (1–10 min of arc in diameter) at a photopic background L_a (or at least in the upper mesopic range) and L less than 10 times L_a to avoid dazzle, (b) scotopic foveal pursuit of small targets at scotopic L_a, but with L_t in the mesopic or photopic range to stimulate foveal cones, and (c) scotopic parafoveal pursuit of larger targets (0.5–2° diameter) at scotopic L_a and with L_t less than 10 times L_a (L_t within the range of operation of the retina at L_a).

It is evident that only photopic foveal pursuit can use the foveal centre for minimizing position errors, whereas scotopic foveal pursuit drives the central retina into a functionally undefined transient state, since L_t is far above the range of operation of the retina as given by L_a. This will be elaborated on in the following subsections on the morphology and neurophysiology of retinal events. Only fragmentary data exist on the psychophysics of dynamic local adaptation. This phenomenon is caused by a bright target that induces afterimage sensations and temporarily increases L_a levels of the stimulated parts of the central retina to a much greater extent than those of the peripheral retina.[12, 73, 74, 144, 194]. On the other hand, scotopic parafoveal pursuit certainly yields quite different eye movements as compared with photopic foveal pursuit, simply because the target can best be 'seen' on the parafovea (during dark adaptation), which forms a wide ring around the central scotoma of the fovea. Depending on the direction from which the target projection approaches the parafovea, maintenance of scotopic parafoveal pursuit can be attempted on various portions of the parafoveal ring and will therefore lead to considerable variations in the position error.

The significant functional difference among these three types of PEM have received little attention so far.[118] Several recent reports indicate the crucial dependence of PEM on the adaptation level.[382, 386] However, systematic measurements of the influence of L_a, L_t, target size and target movement on various parameters of PEM, including the retinal position error, are not yet available.

Morphology

Numerous anatomic studies show that the inner portion of the fovea, at least in humans (rod-free area 1.7–2° in diameter) and macaques (rod-free area at least 1° in diameter), is exclusively occupied by cones with the highest packing density compared with the retinal periphery.[281] Rods, in contrast, are most densely packed in the parafoveal region and have a higher density in the periphery than do cones.[2, 88, 172, 276, 293, 317] An interesting exception to this division into a foveal retina for the photopic range and a para- and extrafoveal retina for both day and night vision is found in a nocturnal primate, the owl monkey. The ratio of rods and cones is relatively unchanged across the retina of the owl monkey. Both rod and cone density are slightly increased toward the retinal centre of this afoveate primate.[270]

During the evolution of the oculomotor system, the ability to pursue slowly moving targets presumably evolved only in those non-human primates[373] having a fovea centralis or a functional equivalent in the form of a favoured location as the coordinate centre on the central retina (see reference 387). This phylogenetically youngest oculomotor subsystem is not developed in afoveate animals such as goldfish,[109] rabbit,[80] or cat,[125] although an optokinetic system for OKN (responding to large stimuli) is generally available in all vertebrates with motile eye balls. Many non-human primate species with or without a fovea[316, 387] have not been systematically examined with respect to their ability to generate PEM.

An afferent visual system with a functional organization centred around a specialized region in the central retina is necessary for the development of a foveal pursuit system with minimal position error, but the morphological existence of a fovea does not guarantee foveal PEM. In fact, many vertebrate species with no evidence of a foveal pursuit system have a well-developed fovea within the central retina; some of them (e.g., many predatory birds or anoline lizards) even develop a temporal fovea in addition to the central fovea.[299, 373, 387]

It is tempting to speculate that the oculomotor pursuit system evolved in primates because their binocular visual field was large enough and the functional and morphological specialization of the central retina was sufficiently advanced.

Several morphologically distinct types of retinal ganglion cells (P_α, P_β, P_γ and P_ε) have recently been described in the macaque retina.[280, 282] The diameter of the dendritic field of $P\alpha$ cells, which are likely to represent Y cells (see next subsection), ranges from $\sim 25\,\mu m$ (0.1°) at the fovea to $> 250\,\mu m$ (1°) in the periphery. P_β cells incorporate $\sim 80\%$ of all ganglion cells and are likely to represent X cells. Their dendritic field diameter is only $10\,\mu m$ (0.04°) throughout the central retina up to eccentricities of 10°.

Neurophysiology

Systematic neurophysiological studies of the primate retina during various types of PEM are not available. However, some of the retinal events during PEM can be pieced together from reports concerning the electrophysiology of the retina in anaesthetized animals (cats and monkeys).

The level of L_a significantly changes the functional antagonistic centre–surround organization of the receptive fields of retinal ganglion cells. These receptive field changes may lead to a complete disappearance of the antagonistic receptive field surround and to a considerable increase in the size of the receptive field centre during dark adaptation,[19] as well as to changes in the neural background activity and other dynamic properties of afferent visual neurones with receptive fields in the central and peripheral retina[154, 324, 397] As expected, these striking effects of the luminance level on the receptive organization also occur at more central levels, such as in the lateral geniculate nucleus of striate cortex.[300]

It is generally assumed that these receptive field changes are caused mainly by the photochemical and electrophysiological adaptation of the two main types of photoreceptors that are responsive to luminances either in the scotopic and mesopic range (rods) or in the mesopic and photopic range (cones). All vertebrate photoreceptors seem to exhibit similar features in this regard.[49, 268, 345, 356] The local electroretinogram (ERG) was also measured in various parts of the monkey retina to demonstrate the dramatic functional changes that accompany a shift from light to dark adaptation.[21, 49] For example, studies utilizing ERG recordings[49] or single-unit recordings from retinal ganglion cells with receptive fields in the parafovea of the monkey[155] have shown that cone responses to photopic stimuli have a considerably shorter latency than rod responses to scotopic stimuli. A number of recent single-unit studies of retinal ganglion cells in macaques established the existence of several functionally different cell types.[90, 91, 327]

X Cells or colour-opponent cells

X cells or colour-opponent cells are cells with sustained responses to stationary targets and with very small receptive field centres of only 0.06° in diameter (in the photopic range). These cells are the most common cell type found throughout the fovea and parafovea and are found only rarely in the peripheral retina.

Y cells or broad-band cells

Y cells or broad-band cells are cells with only transient response patterns and are found throughout the central and peripheral retina. The diameter of their receptive field centres ranges from at least 0.12° within the fovea, to 0.25° at $\sim 5°$ eccentricity, to more than 1° at eccentricities $> 20°$.

Non-concentric or rarely encountered cells

Non-concentric or rarely encountered cells are cells that form $< 10\%$ of all retinal ganglion cells. They typically do not show a well-defined centre–surround receptive field organization and exhibit only transient responses.

Let us now consider the task of retinal ganglion cells during PEM. According to the current view of the pursuit system, the following three parameters must be monitored by the visual input: (a) retinal target position, (b) retinal target slip velocity and (c) direction of the slip movement. Because of their concentration within the central retina, their small receptive field centres and their sustained response properties, X cells seem to be optimally suited for the first stage of the visual input for PEM. Theoretically at least, all three parameters can be deduced from X-cell activity. Target position is directly available from the retinotopic location of the activated X cell (or a few neighbouring X cells being activated by a larger target); slip velocity can be deduced from analysing the transient impulse rate increase of a newly activated X cell (cell B) in correlation with the transient impulse rate decrease of the previously activated neighbouring X cell (cell A); direction of slip movement follows directly from the retinotopic locations of X cells A and B.

In contrast, Y cells seem to be considerably less useful for monitoring these three target parameters on the retina because their receptive fields are much larger and only transient responses are possible. Further studies will determine whether Y cells, which could play an essential role in the phase of pursuit initiation on the peripheral retina, also play a role in the maintenance of pursuit on the fovea.

The spectral sensitivity of these retinal cells, which led to a distinction between colour-opponent and broad-band cells, will be disregarded in the following discussion, although there may be a phylogenetic correlation between the ability to perceive colour and to pursue small targets.

In summary, these psychophysical, morphological and neurophysiological data demonstrate that the neural control of PEM depends on the functional state of the primate retina, which can be shifted over a wide range by means of luminance adaptation and which is defined by a number of stimulus parameters. Consequently, the contribution of other brain regions to the control of PEM depends on the neural input that they receive from the various parts of the retina. The retina functions as a powerful filter of the spatial, temporal and intensity features of any given pursuit target and its corresponding background.

Neural Activity in the Visual System

Lateral Geniculate Body

The most prominent target for retinal ganglion cell axons is the lateral geniculate body, a part of the thalamus.[55, 282, 314] It consists of two subdivisions, the dorsal lateral geniculate nucleus (LGN) and the ventral lateral geniculate nucleus, which is also known as the perigeniculate nucleus (PGN) in primates (see references 37 and 294). Thus far attention has been focused on the LGN with its clearly laminated structure, whereas only little information is at present available on the PGN. These nuclei are discussed separately.

Dorsal Lateral Geniculate Nucleus

That portion of the LGN in macaques and humans that represents the central visual field consists of six layers with a retinotopic organization in register. The ventral magnocellular layers 1 and 2 can be morphologically distinguished from the adjacent dorsal parvocellular layers 3, 4, 5 and 6. The following picture concerning the representation of various cell types and regions of the ipsilateral or contralateral retina in these layers begins to emerge.

1. The layers are as follows: layer 1 (magnocellular) with contralateral input, layer 2 (magnocellular) with ipsilateral input, layer 3 (parvocellular) with ipsilateral input, layer 4 (parvocellular) with contralateral input, layer 5 (parvocellular) with ipsilateral input and layer 6 (parvocellular) with contralateral input.[83, 201]

2. The over-representation of the central retina is more pronounced in the parvocellular layers than in the magnocellular layers,[83] with the result that the ratio of parvocellular to magnocellular neurones may be 10 times larger in the fovea than in the retinal periphery.

3. All X cells of the morphological P_β type (80% of all ganglion cells) and most Y cells of the morphological P_α type (10% of all ganglion cells) in the macaque retina project to one of the six LGN layers.[282] With respect to the earlier classification of LGN neurones into type I–IV cells,[383] type I and type II cells can be classified as X cells, whereas type III and type IV cells have the characteristic features of Y cells.

4. Most authors agree that X cells project predominantly to one of the four parvocellular layers and that Y cells project predominantly to one of the two magnocellular layers.[90, 93, 105, 239, 328] One study[179] found considerably more X cells than Y cells in the magnocellular layers.

5. In contrast to the cat LGN, only very few cells with binocular receptive fields have been described in the macaque LGN.[238, 315]

6. Magnocellular neurones (magno-X cells and magno-Y cells) have a high contrast sensitivity (reciprocal of the just detectable contrast: pattern luminance–background luminance at a given spatial frequency) comparable to behavioural sensitivity values in humans and macaques,[96] whereas the spatial resolution is poor (up to 10 cpd), owing probably to the larger receptive field size. In contrast, parvocellular neurones (mostly parvo-X cells) with small receptive fields close to the fovea reach spatial resolution values of >40 cpd at the expense of contrast sensitivities 5 to 10 times lower than those for magno-Y cells.[93, 179]

These data on the LGN are by no means complete. For example, the large synaptic inputs from the reticular formation[103] and superior colliculus, as well as those reciprocal connections with areas V1 and V2, may play an important role in the alert state[22] when the visual system is actually being used to provide continuous information concerning the following three main questions with regard to PEM:

1. What is target and what is background?
2. Where is the target relative to the foveal centre?
3. What does the target mean?

Nevertheless, the concept of a separation of the afferent visual system into morphologically and functionally separate pathways[216, 314] is clearly preserved and even elaborated on at the LGN level. In fact, one single point in the visual space that has its first neural representation in the two retinas is represented in 12 separate LGN layers. There must be a good reason for this multiplication of neural representation of the visual world throughout the central visual system. Each LGN has three pairs of layers, representing similar stimulus properties as they appear on either eye. The X-cell pathway with its high spatial resolution and strong emphasis on foveal representation in the two pairs of parvocellular layers, is clearly separated from the Y-cell pathway with its high contrast sensitivity in the pair of magnocellular layers. In addition, a further subdivision of the X-cell pathway has been described[328]: into an on-centre branch projecting mainly to the most dorsal pair (parvocellular layers 5 and 6) and an off-centre branch projecting to the other pair (layers 3 and 4).

This detailed knowledge of the various parallel pathways not only is important for our understanding of the visual input into the pursuit system but may also influence the selection of the various stimulus parameters in further studies of the pursuit system.

Perigeniculate Nucleus

The ventral lateral geniculate or perigeniculate nucleus (PGN) is located adjacent to the LGN[37, 294] and is surprisingly large in monkeys and humans. Morphological studies indicate a direct retinal input from the ipsilateral and contralateral fovea by way of axon collaterals from those optic tract axons whose principal termination is the LGN.[165, 294] The main efferent PGN pathway is the pregeniculomesencephalic tract, which terminates in the vicinity of the oculomotor nuclear complex in the brain stem[294]. Neurophysiological studies[57, 231] have shown that some PGN neurones generate bursts of activity during saccades in the dark. However, these neurones were apparently not modulated by vestibular stimulation or during PEM. Further details concerning the suggested role of the PGN in the pupillary reflex[294] and in eye movements, especially in PEM, are not available. Such a short latency connection between the fovea centralis and the vicinity of the oculomotor nuclear complex, however, suggests various interesting possibilities for a contribution to the control of PEM.

Superior Colliculus

Because the superior colliculus (SC) has long been known to be involved not only in vision but also in eye movements, a substantial body of data exists with regard to morphology and single-unit activity in anaesthetized as well as in alert trained monkeys (see reference 390). Typically, the SC is associated with saccadic eye movements, although its partial involvement in PEM cannot be ruled out.

About 10% of all retinal ganglion cells project to the SC, predominantly (~70%) to the contralateral side, and have receptive fields spread over the entire retina including the fovea.[85, 280, 292] These direct retinal afferents are composed of very few Y cells and of P_γ and P_ε cells [280] that probably correspond to the functional type of 'atypical'[90] or 'rarely encountered'[327] cells having large receptive fields and lacking a clear centre–surround organization. There is general agreement that no X cells project to the SC.

Both the small percentage of direct retinal afferents and their large receptive fields seem to indicate that the SC would not be optimally suited for maintaining vision of a moving small target while keeping it continuously on the fovea. However, various indirect retinal inputs, for example, via the visual cortex,[32, 223, 224] may be available to provide very precise information about the visual target during PEM. Some neurones in the intermediate layers of the SC were found to discharge during PEM, but they appeared to be related to the visual target guiding the pursuit movement, rather than to PEM itself.[391] At present, more detailed single-unit studies during PEM with or without structured background in the trained monkey are not available for evaluating the possible role of the SC in one of the three tasks of the visual system during PEM: target background discrimination, target localization and target recognition.

Detailed studies on the influence of the SC lesions on pursuit performance are not available. However, generally the eye position error for saccades to visual targets was significantly increased. This increase corresponds to a reduction in fixation accuracy.[4, 329] Such a deficit should also impair initiation of PEM, which is dependent on the precision of foveating saccades. Furthermore, the range over which the eyes could be deviated during OKN was found to be drastically restricted following SC lesions.[329] Both electrophysiological[303] and neuroanatomical studies have clearly demonstrated projections from SC to both mesencephalic and pontine reticular formation,[75] as well as the dorsolateral pons and the nucleus reticularis tegmenti pontis.[164] A well-defined projection to the inferior olivary complex (especially the medial accessory olive) as part of a tecto-olivocerebellar pathway has also been found.[134, 164]

In summary, although the SC is probably not an essential structure for the neural control of maintenance of PEM, it may play a role in its initiation.

Accessory Optic System and Other Structures with Direct Retinal Projections

In macaques, several fasciculi of thin fibres (probably axon collaterals from optic tract axons having their principal termination in the LGN), mainly from the contralateral eye, branch off from the optic tract and form the accessory optic tract (AOT). The superior (posterior) accessory optic fasciculus projects to the lateral terminal nucleus of the accessory optic tract (LTN or NAOT) as described by various groups[165, 217] and even to the medial terminal nucleus of the accessory optic tract (MTN).[176] Another fasciculus[165] terminates in a region that possibly coincides with the dorsal terminal nucleus of the accessory optic tract (DTN), in which parts of the superior accessory optic fasciculus also terminate.[217]

The non-human primate accessory optic system appears to be capable of performing basic visual functions for the control of OKN (and possibly also PEM), at least after removal of the striate cortex.[279] A study in the alert macaque showed pursuit-related single-unit activity in the nucleus of the transpeduncular tract,[380] which is assumed to receive direct input from LTN (see reference 236). These neurones, which could be visually activated from either eye, had very large receptive fields, and their impulse rate increased with movement of a target (subtending at least several degrees) in a preferred direction (contralateral to recording site or vertical). These neurones could also be activated during pursuit of a large, structured background moving in the same direction, which led to an excitation by target movement. The functional properties of this neural population were interpreted by the authors with regard to a retinal slip signal for OKN or PEM. More detailed neurophysiological or lesion studies concerning the primate accessory optic system are not yet available.[37, 314]

The nucleus suprachiasmaticus hypothalami[314] and the nucleus supraopticus hypothalami[37] both receive direct retinal projections (probably as axon collaterals branching from the optic tract) but are not considered to play a role in the pursuit system.

Pretectum

The pretectal complex (see reference 314) with direct retinal input may be a visual oculomotor link of underestimated importance. According to the more recent terminology, it consists of the following nuclei:[173] nucleus of the optic tract (NOT), pretectal olivary nucleus (ON), anterior pretectal nucleus (NPA), medial pretectal nucleus (NPM) and posterior pretectal nucleus (NPP, homologous to sublentiform nucleus). Despite a considerable confusion with regard to terminology, there is now general agreement that retinal ganglion cells in macaques and other primates project bilaterally to the ON and NPP.[26, 98, 173]

Bilateral retinal projections to NOT in macaques and squirrel monkeys have also been described.[148, 173] Given its projections to area 19 probably including the middle temporal visual area (MT),[98] area 8,[211] the pontine nuclei including the dorsolateral pontine nucleus (DLPN), and the SC,[375] as well as to MRF,[62, 375] and the immediate vicinity of the IIIrd and IVth nerve nuclei,[343, 375] the pretectal complex is in a strategic position to participate in the control of PEM and to provide short latency links between the retina and the oculomotor nuclei. Area 8 has projections to NOT.[211] It has been recently shown that macaques with NOT lesions fail to generate the slow component of OKN in response to OKN stimulation toward the lesioned side. In contrast, PEM as well as saccades appear to remain undisturbed.[181] This interesting finding could be taken as another piece of evidence for two separate mechanisms for the control of PEM versus the slow component of OKN. Single-unit recordings in the NOT of anaesthetized squirrel monkeys and macaques[170] demonstrated a class of visually sensitive, direction-selective neurones with very large receptive fields. In alert macaques, about half of the single units recorded in the pretectum were found to be activated in relation with PEM (but not during saccades). The remaining neurones were either visually activated or both visually activated and pursuit-related.[258] More neurophysiological studies in the alert monkey are needed to evaluate the possible influence of the pretectal complex, an area that has been associated with pupillary construction and OKN,[314] on PEM.

Pulvinar

The pulvinar thalami optici (see reference 178), the largest (cushion-shaped) thalamic nuclear complex in primates (its relative size in humans is even bigger than in macaques), is usually subdivided into the inferior pulvinar (PI), lateral pulvinar (PL), and medial pulvinar (PM) nuclei.[37] Direct bilateral retinal projections to the PI were clearly demonstrated in macaques,[64, 176, 260] whereas one group[176] also reported a predominantly contralateral retinal projection to PM.

The fine structure of PI and PM[271] in macaques, as well as their extensive reciprocal connections to visual cortical areas 17, 18 and 19, was recently confirmed and elaborated.[254, 272, 364] Furthermore, connections from area 17 were found to establish two independent, topographically organized projections. One projection primarily within the PI represents the entire contralateral visual field, whereas the other projection within PL is limited to the contralateral central visual field.[364]

In the light of the projections from the retina, visual cortex, SC (see reference 364), NOT, MT, 7a and precentral cortex, as well as their efferent projections to V1, V2, MT[341] and area 7a, these pulvinar subdivisions have been suggested to function as integrators of visual cortical information.[272] However, lesion studies suggest that visual information from the visual cortex probably reaches the pursuit system in the brainstem via more than one pathway through the pulvinar. Monkeys with extensive and bilateral pulvinar lesions did not exhibit any deficits in pursuit eye movements.[362]

In the alert squirrel monkey and rhesus monkey, single-unit recordings in the pulvinar revealed the existence of both visually activated and eye-movement-related types of neurones.[283, 284] The visual receptive field properties of single units in the PI and PL were also studied in anaesthetized or at least sedated cebus monkeys.[143] Single-unit recordings in the pulvinar of macaques during PEM are not yet available.

Central Thalamus

The internal medullary lamina (IML) in the central thalamus is a region containing fibres of passage interspersed with several nuclei[37, 178] and receives substantial input from area 7 in the posterior parietal cortex.[396] The central thalamus projects to the visual cortex, area V1.[254] Recent pioneering studies in alert macaques[330, 332, 333] clearly demonstrated that several types of visually activated neurones (e.g., eye-position units) with large receptive fields as well as oculomotor-related neurones were present, especially in the rostral sectors of IML. None of these neurones was selectively active for PEM. In contrast to the previously mentioned parts of the thalamus (lateral geniculate body, LGN and PGN, and pulvinar), IML does not receive direct retinal input. Taking into consideration the dynamic properties of these IML neurones, the anatomically established inputs from cerebellum, reticular formation, SC, pretectum, parietal cortex and frontal eye fields to this part of IML, and IML projections to the striate cortex, parietal cortex, precentral cortex and the striatum, these authors[330, 332, 333] put forward the interesting hypothesis that the IML serves as part of a central controller rather than simply processing data signals. In fact, central thalamic lesions[275, 374] caused a neglect syndrome, as would be expected from the impairment of a putative central controller. Specific details concerning a possible role of the central thalamus in the control of PEM may come from future studies.

Neural Activity in the Vestibular System

Vestibular Nuclear Complex

The vestibular nuclei in primates form a nuclear complex that consists of four main nuclei, the medial (MVN), descending or inferior (IVN), superior (SVN) and lateral (LVN, or nucleus of Deiters) vestibular nuclei, as well as several minor cell groups (groups y, z, x, f and l) and nucleus Sv (supravestibularis) close to MVN.[39, 66, 177, 385] The vestibular nuclear complex has received considerable attention in oculomotor research for several reasons. (a) Many vestibular nuclei serve as relay nuclei in the direct vestibulo-ocular reflex (VOR) between primary vestibular neurones and the oculomotor nuclei. These nuclei would therefore be a possible location for superposition of any phylogenetically more recent modulatory visual inputs (e.g. for PEM) into the oldest oculomtor subsystem. (b) Some vestibular nuclei can be considered as misplaced cerebellar nuclei of the vestibulocerebellum owing to their extensive connections with flocculus, paraflocculus, nodulus, uvula and parts of the vermis.

A possible modulation of vestibular nerve activity by visual stimuli or eye movements (e.g., via the efferent vestibular system) has been ruled out by single-unit recordings in alert monkeys.[59, 185] Several neurophysiological studies in the vestibular nuclear complex of alert monkeys clearly demonstrated a long-latency, quasi-visual input during OKN stimulation.[34, 52, 166, 371] However, it was concluded from these studies that the visual input refers only to the gradual response to long-term, large OKN stimulation and that the visual signals controlling PEM are not being processed by vestibular nuclear neurones.[370] On the other hand it was also shown that many neurones (particularly those located in the SVN and MVN) had certain features of eye-movement-related neurones[136, 187, 249] and thus may be partially motor. It seems appropriate to summarize these findings with the statement that the visual input to the pursuit system (in contrast to OKN) is probably not being added to the VOR pathway at the level of the vestibular nuclear complex but reaches the oculomotor system along different pathways. However, a possible role for certain areas of the vestibular nuclear complex in the control of PEM will be discussed later (pp. 179–184).

Possible Role of Cerebral Cortex in Smooth Pursuit Control

Occipital and Temporal Cortex

Striate Cortex, Area V1 (or Area 17)

Most morphological and neurophysiological studies on area V1 in macaques have not specifically addressed its role in the control of PEM. Therefore one has to select those findings that might be helpful in tracing the specific afferent visual limb of the pursuit system and in evaluating the kind of information processing that might take place specifically for the purpose of PEM.

Bilateral ablation of area V1 in macaques has been shown to abolish PEM almost completely,[92, 252] whereas unilateral V1 lesions impaired only PEM of targets in the affected hemifield and left PEM of targets in the intact hemifield relatively unimpaired.[151] Similar pursuit deficits were induced by large, unilateral cortical lesions that spared only the frontal pole and the orbital and ventral surface of the temporal lobe.[361] In contrast, extensive lesions in areas 18 and 19 that completely spared the striate cortex left PEM intact.[92] In addition, morphological studies indicate that the vast majority of afferents from the LGN to the visual cortex terminate in the ipsilateral area V1 (reference 384; see also 365). These findings are usually taken as strong evidence that area V1 is as essential for PEM as is the LGN.

Morphology
The microcircuitry of area V1 in macaques as well as its afferent and efferent connections has been studied extensively, and most of the recent findings do not fit into the scope of this review. Several detailed summary articles on this subject are available.[32, 146, 171, 353, 365] Area V1 can be subdivided into at least nine cortical layers: 1, 2, 3, 4A, 4B, 4Cα, 4Cβ, 5 and 6. Area V1 contains ocular

dominance columns and orientation columns perpendicular to the stack of cortical layers in V1[171] It is also worth mentioning that the horizontal spread of intracortical neurones extends over > 5 mm corresponding with several degrees of visual field angle, for example in layer 4B.[129, 146] Furthermore, there seems to be a regular pattern of interlaminar connections from layer 4 to the superficial layers 1–3, from layers 2 and 3 to layer 5, from layer 5 to layer 6, and from layer 6 back to layer 4. The knowledge of this pattern of interlaminar connections within V1 has recently been found to be more prevalent than previously thought.[32, 130] These connections may constitute interlaminar feedback loops or provide only optional connections. Only a few main points are emphasized.

1. X cells from parvocellular LGN layers project to layers 4A and 4Cβ. The stellate cells in 4A and 4Cβ project to layer 3 and also to layer 6. Y cells from magnocellular LGN layers terminate exclusively in layer 4Cα of V1. The stellate cells in 4Cα project to layer 4B and also to layers 5A and 6. Therefore, these functionally different parallel visual pathways for X and Y cells are kept separate within V1, a feature important for further signal processing.[130, 223]

2. One important feature of the connectivity pattern among the various areas of the visual system is the morphological evidence for a large number of reciprocal connections. In addition to intracortical 'loops', one finds LGN to V1 to LGN loops, pulvinar to V1 to pulvinar loops, V1 to MT to V1 loops, or V1 to V2 to V1 loops, and so on. The functional importance of these possible loops is presently not well understood. It is quite possible, for example, that the contribution of a single neurone in V1 to the control of PEM cannot be deduced from the measurement of its receptive field properties under anaesthesia[286] but will reveal itself only while pursuit eye movements are actually occurring. In other words, one can hypothesize that these reciprocal connections define (select) different functions of a given neurone for different tasks. A possible role of the central thalamus in such a controlling task has already been mentioned.[330]

Projections from V1 to the pontine nuclei in macaques are quite small and arise primarily from the striate representation of the far peripheral field.[42, 149] Projections from the dorsomedial and midline tegmental regions of the pons to V1 have also been reported.[254]

Neurophysiology
Given the multilayered structure of area V1, it is clear why most single-unit recordings cannot be identified as cortical output activity in contrast to input or intracortical activity. Therefore the specific signal-processing features of V1, the best-studied part of the primate neocortex, can be deduced only on the basis of secondary evidence. Various recent single-unit studies in V1 of alert macaques confirmed the existence of binocular neurones[286, 288] and of orientation-sensitive as well as direction-selective neurones.[286, 287, 389] However, the foveal part of V1 seems to contain a majority of non-oriented neurones, neurones that respond equally well to a stationary or moving stimulus at any orientation and in any direction, whereas oriented neurones dominate in the parafoveal part.[286] Neurones with a non-adapting response to a stationary stimulus could be distinguished from rapidly adapting neurones.[389] In addition, most neurones in V1 were found to be sensitive to the location of the stimulus in depth. [288] A recent study regarding velocity sensitivity of neurones in the foveal representation of V1 in anaesthetized macaques[274] revealed the following pattern: 10% were velocity-tuned cells (preferred velocities ranging from 1.5 to $39°\,s^{-1}$; 70% were velocity low-pass cells (upper cutoff velocities ranging from ~ 1 to $80°\,s^{-1}$); 30% were velocity broad-band cells. In general, most cells were very sensitive to low stimulus velocities. The direction selectivity, when present, diminished with decreasing velocity.

In one of the very few neurophysiological studies on the function properties of V1 in alert macaques during PEM,[139] the response characteristics of oriented versus non-oriented neurones were compared using two paradigms. In *paradigm A*, the monkey fixated a small target while another optimally oriented stimulus was slowly moved over the parafoveal receptive field. In *paradigm B*, the monkey pursued the small target in the opposite direction so as to move the receptive field over the stationary visual stimulus. This study was designed to separate standard visual neurones, which cannot distinguish between stimulus movement during fixation (real motion) and movement of the eye in front of a stationary stimulus, from real-motion neurones and yielded the following interesting results. About 10% of all V1 neurones, particularly the oriented neurones with antagonistic receptive fields, could be classified as real-motion neurones, since they clearly responded during *paradigm A* but only weakly if at all during *paradigm B*. None of the non-oriented neurones exhibited the behaviour of real-motion neurones. These findings may be related to an earlier notion[35] that some neurones in V1 of alert macaques were excited only during PEM against a stationary background. These neurones did not respond when their receptive fields were swept across a stationary target. This important distinction between target movement and eye movement, which seemingly begins to evoke at the level of V1, may also play an important role in the necessary task of suppressing neural signals from background motion on the retina during PEM. Another recent study in alert macaques demonstrated directionally selective neurones (monocular or binocular), as well as neurones with special sensitivity to slow retinal image velocity in V1.[289] From these preliminary reports it appears that V1 makes various important contributions to the control of PEM (e.g., target-background discrimination; monitoring of target position, target velocity, and target movement direction relative to the foveal centre), but its role in PEM still needs to be systematically studied.

Middle Temporal Visual Area

Area V1 has particularly extensive reciprocal connections with area MT (within area 19), at least in macaques. Area MT had previously been termed[402] the motion area in the posterior bank of the superior temporal sulcus (STS) and is now commonly known as the middle temporal visual area (MT).[141, 366] The effects of chemical lesions in the MT on oculomotor performance have recently been examined in trained macaques.[261, 263] Saccades to stationary targets were found to be essentially normal, whereas initiation of PEM using a step-ramp paradigm was clearly impaired. Specifically, smooth eye velocity following the first foveating saccade was significantly lower than target velocity 24 h after the lesion. However, maintenance of PEM appeared normal as soon as the target projection had reached the fovea. Furthermore, the deficits in the initiation of PEM disappeared completely within a couple of days. Considering the fact that lesions in V1 induce a more or less permanent deficit in both initiation and maintenance of PEM (whereas lesions in MT do not), it seems reasonable to suggest that other parallel pathways, between V1 (possibly via V2 and precentral cortex) and the oculomotor brain stem regions, exist for the control of PEM.

Morphology

Area MT in macaques contains a complete, topographically organized representation of the contralateral visual hemifield even though its average surface area is $<3\%$ of the size of V1. The topography is cruder than in V1 and contains remarkable discontinuities that may serve as a structural basis for as yet unknown functional properties. These discontinuities can be envisaged as a mosaic of the (non-linearly distorted) retinal map, which got 'shuffled around' and reassembled in the 'wrong' order, whereby topography was maintained within each portion (mosaic stone) of the map.[366] The most prominent projection from V1 to the MT arises in layer 4B of V1 and reaches the MT at layer 4. This suggests that the 'Y-cell pathway' (which reaches area V1 at layer $4C\alpha$ and projects within V1 mainly to the adjacent layer 4B) connects the retina with MT.[130] The reciprocal projection from MT to V1 is considerably weaker (for review and original data, see reference 241). A projection from NOT to MT is likely.[98]

Area MT has strong connections with various subdivisions of the pulvinar. It is particularly noteworthy that these connections, whose function is not well understood, are reciprocal and topographically organized.[156, 241, 341, 364] Another output from area MT leads to some of the ipsilateral pontine nuclei (including the dorsolateral pontine nucleus (DLPN)), as confirmed by several recent studies.[149, 156, 241, 363] The frontal eye field (area 8) receives a projection from MT.[18, 234] In addition, MT has reciprocal connections with a newly identified cortical area, the medial superior temporal area,[241, 320, 354] as well as projections to the SC and the PGN.[156, 241, 353, 363]

Neurophysiology

The following neurophysiological data were obtained from paralysed and anaesthetized macaques. The receptive field size of visually sensitive MT neurones increases more rapidly with increasing retinal eccentricity and is significantly larger at a given eccentricity than in both V1 and V2. Receptive fields in the foveal centre can be as small as $1-1.5°$, but at eccentricities between 1 and $5°$ the average diameter is $\sim 10°$.[142, 354, 366, 402] Most MT neurones were found to receive approximately equal inputs from the two eyes.[5, 243] More than 80% of all MT neurones are direction selective in that the response to stimulus motion in the preferred direction is at least twice as large (and ~ 11 times larger on the average) as that in the null direction.[5, 6, 242, 354] By comparison, only 48% of neurones in V1 of anaesthetized macaques met this criterion of direction selectivity.[326] However, layer 4B in V1, from which the projection to the MT arises, was found to contain the highest proportion of direction-selective neurones in V1.[104] Therefore it has been suggested that the particularly high concentration of direction-selective neurones in MT reflects a highly selective projection from the large area V1 to the much smaller area MT.

By means of stationary, flashed stimulus bars it was shown that most MT neurones are also orientation-selective. The orientation preference is typically either perpendicular (type I MT neurones) or parallel (type II MT neurones) to the preferred direction of motion.[5, 242] Interestingly, the bandwidths of direction tuning and orientation tuning are significantly narrower for neurones in V1 in comparison with those in the MT.[5] Most MT neurones are clearly tuned to a certain preferred stimulus velocity,[242] a feature that has also been found (though to a lesser extent) in velocity-sensitive neurones in layers 4B and 6 of area V1.[274] In the MT, a unimodal distribution of preferred stimulus velocity was found with a peak at $32° s^{-1}$. The total range of preferred velocities extends from 2 to $250° s^{-1}$.

In summary, most MT neurones in anaesthetized

macaques are velocity-tuned, direction-selective, and orientation-selective, with wider bandwidths and larger receptive fields than neurones in V1. V1 neurones with features similar to type II MT neurones have not been found.

The few neurophysiological data on the dynamic properties of MT neurones in the alert monkey[262, 392] do not indicate functional differences between the anaesthetized and the alert state. These authors found no influence of eye movements on the activity of MT neurones.

Area MT may serve the function (as indicated by the lesion studies mentioned above) of monitoring the velocity and direction of target motion on the retina. However, information concerning the target position relative to the foveal centre is considerably degraded in area MT in comparison with V1 because of the significant increase in receptive field size, and therefore a minimization of the position error during PEM can hardly be achieved by means of MT neurones alone.

Medial Superior Temporal Area

A small cortical area lying in the superior temporal sulcus medial to area MT with a massive direct input from MT was recently defined as the medial superior temporal area (MST).[241] So far, reciprocal connections of MST have been described for area 7a in the posterior parietal cortex, area MT, and area 8 in the frontal cortex (see reference 241). Single-unit recordings in the MST of anaesthetized macaques[320, 354] revealed the following interesting functional properties: (a) receptive fields are very large, independent of eccentricity; (b) some cells with large excitatory fields respond more strongly to small moving bars than to wide field movements; and (c) $\sim 55\%$ of the cells are direction-selective (as compared with more than 75% in the MT).

In a preliminary study on neural activity in MST of alert macaques,[262, 392] the finding of very large receptive fields for cells in the MST was confirmed. Receptive fields typically cover a full quadrant and in some cases the contralateral hemisphere. These authors also confirmed that MST cells are predominantly direction-selective. Neural activity was also found to be pursuit-related during PEM of a small target in an otherwise dark environment. However, the neural activity level increased when a homogeneous background illumination was turned on, suggesting that these cells in the MST receive a retinal and an extraretinal input during PEM. Neurones with similar functional properties[323] have also been described in the posterolateral part of area 7a (see pp 178–179).

Other Areas of Occipital and Temporal Cortex

Since it is generally accepted that area V1 processes all essential afferent visual signals for PEM control, and that area MT may be important but not essential for PEM (as indicated by the finding of complete recovery of PEM following chemical lesions in MT), other visual areas with direct input from the LGN or from V1 have to be considered.

Area V2 (Within Area 18) in Macaques

Morphological studies showed that V2 surrounds V1 and is almost as large as V1.[142] Area V2 receives a topographically organized projection from V1[313, 403] containing a representation of the entire contralateral visual field. Areas V1 and V2 are reciprocally interconnected. Projections from V1 arise in layers 2 and 3, in layer 4B, and in the border region between layers 5 and 6, reaching layer 4 in V2.[190, 241] Area V2 has received even more attention after the recent finding that it receives direct projections from the interlaminar zones and the S layers of the LGN, as demonstrated by retrograde transport of fluorescent labels.[54, 190] The corresponding reciprocal pathway also exists.[see 224] The microcircuitry within area V2 is quite distinct from that in V1 and is currently under investigation in several laboratories.[224, 312, 358] In general, V2 appears to have considerably more connections with other visual cortical areas than does V1.[241] Area V2 projects to area 8 and possibly to other relevant areas in the precentral cortex.[18] Among the various subcortical connections of V2 are reciprocal connections with pulvinar and efferents to SC.[224, 390] In addition, V2 projects to the pontine nuclei, although this seems to represent only a very small portion of the corticopontine pathway.[149]

Neurophysiological data on area V2 are presently still quite limited and have been obtained from anaesthetized rather than alert macaques. The size of receptive fields in V2 at a given eccentricity is considerably larger than in V1.[142] Preferred spatial frequency for cells in V2 ranges from 0.2 to 2 cpd and is considerably lower than in V1.[132] These authors[132] also reported that 70% of V2 cells exhibit band-pass characteristics in response to temporal stimulation frequencies. Another recent study[274] found that V2 cells with receptive fields close to the fovea have a preference for slow stimulus movements (as is also true of most cells in V1) and have velocity–response curves with velocity low-pass characteristics. The portion of direction-selective cells is low (as in V1). These authors emphasized that most V2 cells (like those in V1 are sensitive to very low stimulus velocities below $1° s^{-1}$. Cells in V2 with receptive fields within $2°$ of the fovea were subdivided into 10% velocity-tuned cells (optimal velocity $0.9 = 23° s^{-1}$), 70% velocity low-pass cells and 20% velocity broad-band cells.

Single-unit recordings from V2 during PEM in trained macaques are not yet available.

Other regions of the occipital and temporal cortex (see reference 241) including V4 within area 18[127, 128, 232,

325, 404] and area TE in the inferior temporal cortex[94, 392] may be influential in the control of PEM, although this is not evident from the available data. Interestingly, both V4 and TE have recently been implicated in the gating of visual processing by selective attention.[256] Here is another indication (see also pp 173–174) that the disturbing number of morphologically possible neural connections among cortical regions may in fact be sharply reduced by selecting only a few connections for each specific visuomotor function.

Parietal Cortex

Inferior Parietal Cortex, Area 7

Area 7 in the inferior parietal cortex of primates has been suggested to be an important centre for neural control of directed visual attention, of spatial perception, and of spatially oriented movements including PEM. Several recent reviews on the still controversial functions of area 7 are available.[89, 174, 226] As originally proposed by the Vogts, area 7 is subdivided into area 7a (mediocaudal part of 7) and area 7b (rostrolateral part of 7; see reference 174).

Dysfunction of the parietal cortex caused by various syndromes in human patients[16] or induced by surgical lesions in trained macaques[229] has been associated with pursuit deficits and with OKN deficits toward the side of the lesion. However, such deficits could not be consistently linked to a specific cortical subregion within the inferior parietal cortex.

Morphology

Afferent connections to area 7 from the occipital cortex, area MST, frontal eye field, pulvinar and central thalamus, as well as efferent projections from area 7 to area 5, the occipital cortex, MST, frontal eye field, central thalamus, pulvinar, SC and lateral pontine nuclei have been reviewed.[174, 241] Several more recent neuroanatomic studies are mentioned below. Projections from various parts of area 7 to specific regions of the thalamus in macaques were described in detail by Yeterian and Pandya[396] (see also reference 376). These authors concluded that area 7a forms a clearly separate cortical region, with exclusive projections to the nucleus paracentralis and strong projections to the nucleus centralis lateralis of the IML within the central thalamus. In addition, area 7a projects to the PM and PL.[376] Projections from area 7a to three regions within the frontal cortex, area 8a and areas 45 and 46, have been described in detail.[8, 285]

Interestingly, it was found that the newly defined lateral intraparietal area (LIP adjacent to area 7a in the lateral wall of the intraparietal sulcus[8] projects much more strongly to the frontal cortex than does area 7a. In contrast to area 7a, area LIP as part of area POa[174] has strong and reciprocal connections with the SC in contrast to area 7a.[8, 228] Additional importance is given to area LIP (or area POa) by the earlier finding that it receives a strong input projection from area V4;[336] this could be one of the main inputs from visual to inferior parietal cortex. Area 7 also has a substantial projection to the pontine nuclei,[42, 149, 243a, 376] including the lateral pontine nuclei (see above).

Neurophysiology

A considerable number of single-unit studies in the posterior parietal cortex (areas 5 and 7) of alert, trained macaques have accumulated since about 1970 (for reviews, see references 89, 174, 226). In summary, different classes of neurones in area 7 could be distinguished that were activated by visual stimuli and/or during eye movements.[175, 230, 257, 311, 322] These classes were named visual fixation (VF) neurones, visual tracking (VT) neurones, saccade neurones and light-sensitive neurones.[see 174] The functional interpretation of these neural classes as visual afferents, attention and motor command neurones, or oculomotor efferents led to a very interesting controversy, which is still alive.

The functional properties of VT neurones in the anterior part of area 7a include the following features:

1. Impulse rate increases from about zero during fixation to 20–50 impulses $(imp)s^{-1}$ during pursuit of an interesting target in one direction.
2. The majority of VT neurones are direction-selective (ipsilateral, contralateral, up, or down), whereas some can be activated by PEM in either of two directions.[257, 322]
3. VT neurones also respond to movement of a visual target over the large stationary receptive field during fixation.[311] In other words, VT neurones are not only vision-related but also pursuit-related. Another interesting feature that was recently described[182, 183] is that many VT neurones are also modulated by purely vestibular rotatory stimulation in the dark. These authors found that most VT neurones (> 80%) are direction-selective for PEM and head rotation in the same direction (e.g., PEM to right or head rotation to right).

Two related studies focusing on the dynamic features of VT neurones[183, 323] revealed a number of further details.

1. About 30% of VT neurones are in fact located in the adjacent area MST (see pp 167–171), whereas the majority were located in the posterolateral part of area 7a. By comparison, only a few VT neurones were found in the anterior part of 7a.
2. The response of the majority of VT neurones to PEM in the dark was considerably weaker than to PEM against a light background.
3. The large majority respond not only to PEM but also to moving stimuli during fixation, thus confirming

earlier findings.[311] About 50% of these VT neurones prefer PEM in one direction, but they preferred stimulus movement during fixation in the opposite direction. The other half prefers the same movement direction during PEM and fixation.

4. The average impulse rate during PEM increases with increasing target velocity but reaches a saturation at velocities of $10-20°\,s^{-1}$. Furthermore, certain similarities and differences between VT neurones and gaze velocity neurones in the cerebellum were pointed out (see pp 180–182).

Visual tracking neurones typically continue to be active after sudden target disappearance during PEM as long as smooth eye movements are being generated.[183, 257] This feature, which is also typical for pursuit neurones[118, 120] in the brain stem (see pp 179–186), clearly indicates an extraretinal input to VT neurones.

Precentral Cortex

Frontal Eye Field, Area 8

Both its proximity to several motor cortical areas, i.e. 4 and 6 in the precentral cortex, and the old notion of elicitation of eye movements by electrical stimuli provide support for the assumption that the frontal eye field (area 8 in the prearcuate sulcus in macaques) may constitute a part of an oculomotor cortex.

Circumscribed unilateral and bilateral lesions of area 8 in macaques yield temporary oculomotor deficits regarding the accuracy and frequency of saccades to targets in the peripheral visual field. Recovery from these lesions was found to be complete within 1–4 weeks.[278, 329] A recent report described persisting pursuit deficits following bilateral lesions of area 8 in macaques.[227] Effects of frontal eye field lesions on the pursuit system in humans have not been described.[253] Interestingly, however, bilateral ablation of both frontal eye fields and the SC in macaques leads to permanent oculomotor deficits.[329] Although PEM was not tested specifically, it was noted that the oculomotor range of OKN was dramatically restricted. These authors interpreted their findings as evidence for two parallel channels (one via the frontal eye field, the other via the SC) for visually induced eye movements. Such a concept of several parallel channels may in fact explain various other examples of functional recovery after lesions of the pursuit system (see p 176).

Stimulation within the frontal eye field of alert macaques often yields saccades.[237, 310] Interestingly, the number of saccades induced by stimulation was found to be considerably reduced during PEM, suggesting an interaction between the saccadic and pursuit system at the level of area 8.[237] Conjugate smooth eye movements have been electrically induced in lightly anaesthetized monkeys[33, 310] and recently also in the alert state at certain locations deep within the arcuate sulcus.[51]

Morphology

Afferent visual signals can reach area 8 in macaques via several pathways from the central thalamus, including the IML,[18, 335] the pretectum including the NOT,[211] PM, area V2 and inferior parietal cortex,[8, 18, 285] as well as from areas MT[234] and MST. Efferent pathways from area 8 have been traced to numerous relevant structures including the inferior parietal cortex (area 7),[277] pretectum with NOT,[211] PM,[210] SC,[195, 200, 212] central thalamus with IML,[10, 210] area MST (see reference 241), pontine nuclei probably including DLPN, [10, 149, 199a] and nucleus prepositus hypoglossi,[334] as well as the paramedian pontine reticular formation (PPRF)[210, 213, 334] and even the oculomotor nuclear complex.[209, 214] A subdivision of area 8 on morphological and functional grounds is in progress (see reference 18)

Neurophysiology

Single-unit studies in alert macaques showed that many neurones in area 8 have large visual receptive fields and respond to moving as well as stationary stimuli independent of orientation or movement direction.[51, 255] An interesting feature of these vision-related neurones is their long visual response latency (>60 ms) in comparison with either the striate cortex or SC (typical latencies of 25–50 ms[50]).

Saccade-related neurones (type I neurones), which exhibit an activity increase following the beginning of a saccade[29, 30] or even exhibit a presaccadic activity increase[50, 51] for visually guided saccades, have been localized especially in a subregion comprising area 8a and area 45 of Walker.

Pursuit-related neurones (type II neurones) form a third major class of neurones within the frontal eye field.[29, 30] These neurones have recently been localized deep within the arcuate sulcus[51] and are known to respond to PEM or the slow phase of OKN in a preferred direction and to exhibit steady activation during fixation of gaze in preferred eccentric position. Quantitative data on the functional properties of pursuit-related neurones are not yet available.

Other Areas of Precentral Cortex

Area 46 of Walker in macaques surrounds the principal sulcus adjacent to area 8 and area 45 (see references 8, 138). Lesioning of area 46 may lead to impairment of complex tasks such as spatial delayed alternation;[56] however, specific oculomotor deficits were not mentioned by these authors. Area 46 has afferent and efferent connections with relevant brain regions and therefore deserves some attention. Specifically, it has been shown that area

78a in the inferior parietal cortex[8, 285] and parts of area 8[18] have substantial projections to area 46. Efferent projections from area 46 to the SC are also established.[152, 212] Single-unit studies in alert macaques (see 138) found activity enhancement during fixation of a visual target[349] or with a visual stimulus initiating a lever-pressing trial.[199, 321] A possible participation of such neurones in area 46 in the tasks of initiation or maintenance of PEM in the presence of other stimuli that are competing for attention has not yet been explored.

Another interesting oculomotor zone in the prefrontal cortex is located in Vogt's area 6aβ and is referred to as frontal dorsomedial cortex or as the supplementary eye field.[331, 388] The supplementary eye field appears to project to the region of the oculomotor complex[214] and the pontine tegmentum, including the PPRF.[213] Neurophysiological evidence for oculomotor-related activity in the supplementary eye field is gradually being accumulated.[36, 331] In the first single-unit study using trained macaques,[331] saccade-related neurones were found to be activated before the onset of spontaneous saccades in a preferred direction. Microstimulation at the recording site elicited saccades in this preferred direction. Some pursuit-related neurones have also been found in this region (J. Schlag, personal communication).

These preliminary findings are quite encouraging and suggest that future studies may lead to the establishment of several pursuit-related oculomotor zones in the precentral cortex.

Neural Activity in Cerebellum and Brain Stem

Several parts of the cerebellum are known to participate in the control of eye movements[297] as evidenced by oculomotor deficits associated with cerebellar dysfunction (see references 97, 147 and 401). In recent years it has become more and more clear that cerebellar cortical inputs typically project in parallel to the corresponding target neurones of Purkinje cells in the vestibular or cerebellar nuclei. Several possible wiring diagrams for these corticonuclear microcomplexes (CNMC) have been discussed (see reference 177). From this CNMC concept it follows that any lesion of the cerebellar cortex may modify (often only temporarily) the signal flow and processing in the cerebellum, whereas lesions in the vestibular or cerebellar nuclei result in a total, permanent interruption of this signal flow. Also, an evaluation of cerebellar function requires a comparison of the cerebellar input signals (from climbing and mossy fibres) with the output signals of the corresponding portion of the vestibular or cerebellar nuclei. In contrast, Purkinje cell activity, which represents the cerebellar cortical output, can be regarded as an intrinsic cerebellar signal rather than its output. To remain within the scope of this review, cerebellar and brain stem studies related to the control of OKN or VOR plasticity are not included (see reference 27).

The cerebellar hemispheres (especially crus I and II) are likely to be involved in PEM as indicated by stimulation studies[318] and by their massive cerebral cortical input (including area 8) via the pontine nuclei.[43] However, owing to a lack of specific data, the further discussion will be restricted to the flocculus and vermis.

Flocculus Cerebelli

Cerebellar cortical lesions involving the flocculus and paraflocculus in adult macaques disturb the ability of these monkeys to hold the gaze in eccentric positions as well as to generate PEM.[381, 400] These deficits, however, are only transient in young monkeys, and recovery is complete in adult macaques that received cerebellar lesions as neonates as long as the vestibular and cerebellar nuclei were spared.[121, 122] The characteristic pursuit deficit following flocculectomy is a reduction in the ratio of eye velocity to target velocity (from ~ 1 in normal primates to ~ 0.65 for PEM.[400] This implies that smooth eye movements that are related to the time course of the visual target movement can still be generated. However, the position error relative to the fovea and retinal slip velocity cannot be properly minimized. The finding of complete recovery from pursuit deficits following neonatal or juvenile cerebellar lesions indicates that the flocculus is not essential (in the sense of irreplaceable) for PEM. However, it seems premature to conclude from these lesion results that the flocculus normally generates the neural signal for minimization of position error and slip velocity.

Stimulation of the flocculus in alert macaques was shown to evoke smooth eye movements towards the stimulated side.[318] Smooth eye velocity increased with increasing stimulus intensity until a complete nystagmus pattern developed. A more detail microstimulation study[15] revealed several floccular zones distinguishable on the basis of the direction of the elicited smooth disconjugate eye movement (lateral, downward or rotatory movement of the ipsilateral eye).

Morphology
Several detailed histological studies on afferents to the macaque flocculus described a large number of different input sources, which probably project to selected microzones within the flocculus (see reference 177). The inferior olive projects to the contralateral flocculus.[47, 206] Massive bilateral and topographically organized projections arise in the medial, inferior, and vestibular nuclei, and the abducens nucleus, as well as the (mostly) ipsilateral y

group.[41, 206] Other relevant input structures with bilateral projections include the nucleus reticularis tegmenti pontis (NRTP), parts of the pontine nuclei including DLPN,[45, 206] and the nucleus prepositus hypoglossi (PH), including the supragenu nucleus (sg).[40, 206] In addition, bilateral projections could be identified arising from a broadly distributed set of neurones underlying the cerebellar nuclei, which has been named the basal interstitial nucleus of the cerebellum.[204, 206]

Efferent projections from the macaque flocculus were demonstrated to be exclusively ipsilateral and to terminate in the following nuclei: y group; medial, inferior and superior vestibular nuclei; and the basal interstitial nucleus of the cerebellum.[69, 205] All these nuclei, which have reciprocal connections with the various microzones of the flocculus, can thus be regarded as components in a set of floccular CNMCs (see reference 177), some of which participate in the control of horizontal or vertical PEM. A recent report indicated the existence of projections from the paraflocculus (but not the flocculus) to the PH in squirrel monkeys.[25]

Neurophysiology

To evaluate the function of the flocculus and its corresponding target nuclei, it seems appropriate to first review the input to both flocculus and target nuclei, subsequently the floccular Purkinje cell signals, and finally the output from the target nuclei. As would be expected from the many input projections, various types of neural activity that represent visual, vestibular and oculomotor signals have been identified in the granular layer of the flocculus in alert macaques. One type of presumed mossy fibre activity was classified as burst-tonic units[220, 251, 266, 267, 372] because of their characteristic impulse rate bursts (or transients) related to saccades and their tonic impulse rate values proportional to eye position in a preferred direction. During sinusoidal PEM, these burst-tonic units exhibited phase lead values relative to eye position ranging from ~ 25–$80°$. Phase lead increased with increasing pursuit frequency (in the range of 0.1–1 Hz) and was quite similar for the paradigms: smooth pursuit (target moves, head is stationary) and VOR (light) (head moves, target is stationary), whereas no modulation occurred during VOR suppression (target moves with head). These findings were taken as evidence that burst-tonic units in the granular layer of the flocculus represent oculomotor signals arising somewhere in the brainstem. Other presumed mossy fibres were classified as vestibular-only units, resembling vestibular primary afferents, or as vestibular-plus-saccade units, which exhibited additional impulse rate pauses for saccades in one or more directions.[220, 251] These vestibular units, which were not modulated by PEM, exhibited a phase lead relative to chair position of $>90°$ during sinusoidal oscillations.

These authors also described vestibular-plus-position units, which showed tonic impulse rates that were linearly related to horizontal eye position in the absence of vestibular stimulation. Phase lead during smooth pursuit ranged from ~ 25 to $75°$ and was similar to that for burst-tonic units. However, during VOR (light), phase lead was comparable to that of the other vestibular units, namely leading head velocity. These findings indicate the superposition of primary vestibular signals with pursuit eye velocity and eye position signals at some prefloccular locations.

Visual information seems to reach the flocculus mainly via mossy fibres but also via climbing fibres (from the inferior olive). Climbing-fibre activity at average impulse rate values of only 1 imp s^{-1} was found to be poorly if at all modulated by large moving visual stimuli[251] and unrelated to vestibular or oculomotor activity. However, a small number of presumed mossy fibre inputs responded, after considerable delay (mean 87 ms), to rapid shifts of the visual background in one or more directions.[251] These visual units were also modulated during PEM against a structured background and in total darkness. Visual units were also demonstrated to respond to sinusoidal movement of the visual background during fixation.[264] Some visual units showed only directional selectivity, whereas the impulse rate of others was well correlated with the sinusoidal stimulus movement in one direction with a phase lag of $\sim 45°$ at 0.5 Hz. These visual units of unknown origin monitor retinal image motion with a slip velocity and slip acceleration component.

In summary, the available single-unit data suggest that the macaque flocculus is supplied with a head velocity signal (via vestibular units), pursuit eye velocity and eye position signals (via burst-tonic units) and retinal slip velocity signals (via visual units). These signals have not yet been traced back to their origin or to the corresponding target nuclei of the flocculus (perhaps with the exception of vestibular units). One of the key questions for understanding the pursuit system is how the motor program of pursuit eye velocity could be generated on the basis of visual events. In this respect the neurophysiological evidence suggests that the pursuit eye velocity signal is not generated by the flocculus (unless one can prove that its target nuclei provide and eye velocity feedback signal), but rather it is fed into the flocculus via mossy fibres of non-cerebellar origin.

Floccular Purkinje cell activity in trained macaques during PEM has been analysed by several groups, yielding several types of distinctive dynamic responses. Purkinje cells provide the only output from this part of the cerebellar cortex and are assumed to inhibit neurones in the corresponding target nuclei, representing the relevant cerebellar output. Tonic activity in a number of Purkinje cells was related to eye position in a preferred direction,[219, 251, 265] but this position sensitivity was considerably weaker than for burst-tonic units at the floccular input. The tonic impulse rate in the primary eye position for all Purkinje cells (with or without position sensitivity) was larger than 10 and lower than 160 imp s^{-1} (mean ~80 imp s^{-1}). During saccades, the activity of most Purkinje cells exhibited either a pause or burst in all directions or a pause and burst in opposite directions (similar to burst-tonic units).

During smooth pursuit, most Purkinje cells exhibited an additional eye velocity component for movements in either the ipsilateral or the contralateral direction.[219, 251, 265, 267] These authors also found that the phase lead values of maximum impulse rate relative to pursuit eye position ranged from 30 to 150°. A quantitative analysis of the phase spectra of position-related Purkinje cells revealed that their individual phase lead values at a given frequency formed a continuum with a range of ~45°.[267] Values ranged from ~20 to 64° phase lead at 0.3 Hz and shifted upward with increasing frequency, thus indicating an increase of the velocity component (see also reference 219).

The superposition of an eye velocity and eye position component was further demonstrated by the finding that the phase lead value at a given pursuit frequency shifted (e.g., from 78 to 40°) when the position range of pursuit was shifted into the on direction (e.g., from right to left field of gaze) of a given position-related Purkinje cell.[267]

During VOR (light)[250, 251] or during VOR in the dark,[219] Purkinje cells were not (or only weakly) modulated. In contrast, a strong modulation could be elicited during VOR suppression. The phase lead (relative to head position) was on the average slightly larger than phase lead relative to eye position during smooth pursuit and ranged from ~40 to 140°.[219, 251] For a given Purkinje cell, neural activity increased for target movement (smooth pursuit) or head movement (VOR suppression) in the same direction. The characteristic modulation pattern (disregarding a possible eye position component) during smooth pursuit, VOR (light) and VOR suppression led to the hypothesis that floccular Purkinje cells might be gaze velocity cells.[251] However, this hypothesis has recently been challenged by Büttner and Waespe[60] on the basis of their finding that Purkinje cells were modulated during smooth pursuit (e.g., eye velocity of 30° s^{-1}), but not during constant OKN slow phase at the same eye velocity.

Gaze velocity-type activity has not been found in the main target nuclei of the flocculus, the medial, inferior, or superior vestibular nuclei of alert macaques[188, 372] (see also p 174). These vestibular nuclei and their corresponding floccular microzones may not play a major role in PEM control. However, pursuit-related neurones whose activity was modulated approximately in phase with upward eye velocity[72, 357] have been described in the y group of the vestibular nuclear complex. The dynamic properties of these neurones resembled those of vertical gaze velocity neurones in the flocculus, and some of them exhibited an additional vertical eye position component. Neurophysiological data on the cell types in the newly defined basal interstitial nucleus of the cerebellum[204] are at present not available.

Vermis Cerebelli

Several lesion studies established the finding that ablation of the cerebellar vermis involving lobule V and VII in macaques disturbs the control of saccades,[9, 31, 304] but not of PEM (see, however, reference 304). Microstimulation within this part of the vermis yielded only saccades[248, 318] unless the stimulating current reached the paravermis, where smooth eye movements could be elicited. These findings suggest that the vermis is specifically involved in the control of saccades. However, single-unit data (see below) indicate that the vermis also plays a role in PEM control.

Morphology

Afferent projections to lobule V and VII (and to parts of the lobulus VI) in macaque vermis have thus far been traced from various sources including the accessory olives,[46, 133] the vestibular nuclear complex,[41] the PH including the sg,[40, 41] the pontine nuclei,[43] and specifically to lobulus VII from the NRTP,[44, 133] the DLPN[133] and the abducens nucleus.[41]

Efferent projections are known to reach the ipsilateral (except for a narrow midline region of the vermis) fastigial nucleus, interpositus nucleus, nucleus prepositus hypoglossi and vestibular nuclear complex.[37, 163, 177] However, detailed tracing studies of the vermal output connections in macaques are presently not available.

Neurophysiology

Several recent single-unit studies in the posterior vermis (lobule VI and VII) of trained macaques revealed that the activity of both mossy fibre afferents and Purkinje cells was correlated not only with saccades but also with pursuit eye velocity, target velocity, retinal slip velocity and even head velocity. The impulse rate of mossy fibres was activated approximately in phase with eye velocity in one direction during smooth pursuit.[352] Most mossy fibres were also vision-related in that impulse rate increased in phase with velocity of a moving light spot (2° in diameter) in the same direction (as during smooth pursuit), while the monkey fixated a stationary target. Some of them also responded to movement of a random-dot background moving in the opposite direction during fixation. These findings indicate that this mossy-fibre input to the posterior vermis carries two signals important for the control of PEM; these are the pursuit eye velocity signal and a retinal slip velocity signal. The origin of these signals remains a topic for speculation (however, see eye velocity neurones in the next section). The behaviour of these mossy fibres was not analysed during VOR (light) and VOR suppression.

Some Purkinje cells were modulated only in phase with pursuit eye velocity, whereas others were also activated by purely visual stimuli during fixation.[180, 352] The vision-related Purkinje cells responded to retinal slip of a light spot moving in the same direction (as during smooth pursuit) while fixating a stationary target. Modulation of the impulse rate by movement of a random-dot background during fixation (either in the pursuit direction or the opposite direction) also occurred. More recently it was found that some Purkinje cells in this part of the vermis (probably including Purkinje cells of the previously described types) were also modulated approximately in phase with head velocity during VOR suppression,[350] which is similar to what occurs in vestibular-related neurones in the fastigial nucleus.[140]

Premotor Neural Activity in the Brainstem

Several recent reviews have focused on the role of the brainstem in PEM.[301, 309] These reviews have emphasized the possible involvement of the flocculus and the need for a neural integrator for converting the premotor pursuit eye velocity signal into the required eye position signal (with additional eye velocity component) for oculomotor motorneurones.

In the course of the present review, various thalamic, tectal, cerebral, and cerebellar brain regions, which might contribute to the control of PEM, have been mentioned. Control of PEM requires first of all the generation of a motor control signal for pursuit eye velocity. There is evidence for pursuit-related motor activity (in addition to vision-related sensory activity) in various parts of the cerebral cortex (MST, 7a, 8, 45, and supplementary eye field) and the pretectum (NOT). These findings suggest several possibilities for the generation of the pursuit velocity signal. (a) The pursuit signals at the level of the cerebral cortex, which represents a first step of sensory-motor transformation, reach the brain stem via several parallel pathways. A neural network involving brain stem and cerebellum (and possibly other structures such as the basal ganglia) generates the desired motor control signal for the pursuit eye velocity time course by 'refining' the incoming pursuit-related cerebral output and comparing the motor control signal with sensory (visual and vestibular) signals. (b) Another possibility is that the cerebral cortex provides no pursuit motor signal but rather only visual signals that serve as a basis for the generation of the pursuit velocity signal at the level of brain stem and cerebellum. The brain stem in turn sends pursuit-related activity to the cerebral cortex.

The relevant structures (as judged from the present literature) within the brain stem are subdivided here into four groups:

1. 'Precerebellar' nuclei, including the pontine nuclei (especially the DLPN, NRTP, PH including the sg, and the inferior olivary nuclei.
2. Pontine and mesencephalic reticular formation.
3. Cerebellar nuclei and vestibular nuclear complex.
4. Oculomotor nuclei (IIIrd, IVth, VIth).

The precerebellar nuclei project not only to the cerebellar cortex (in addition to non-cerebellar regions) but also in parallel to the corresponding cerebellar or vestibular nuclei (see references 37, 177). Nearly all neurones of the pontine nuclei (for exceptions, see reference 254) and of the inferior olivary nuclei appear to project to the cerebellum (this does not rule out the presence of parallel projections to non-cerebellar regions via axonal branching). This means that the functional impact on PEM control of all signals travelling along the corticopontine pathways has to be judged by the activity at the cerebellar output, namely the cerebellar and vestibular nuclei. Pursuit-related components of these cerebellar output signals, however, seem to be very rare (see pp 174, 180–183). This suggests that non-cerebellar pathways may form more important links between neocortex and brain stem. In this regard it is noteworthy that a substantial number of neurones in both the PH and NRTP (see reference 37) do not reach the cerebellum. In this way a certain potential for cerebellar-independent signal processing is retained.

Visual responses during fixation or PEM were found by a single-unit study in the DLPN of macaques.[351] Some DLPN neurones were modulated by the movement of a random-dot background pattern in a specific direction while the monkey fixated a stationary target. A second class of DLPN neurones responded during fixation to movement of small stimuli. The majority of these neurones exhibited increased activity when the stimulus crossed the fovea in a specific direction, and they remained active at a reduced level as long as the stimulus continued its movement away from the fovea in the same direction. These findings suggest a superposition of two separate visual inputs, one from the fovea and the other from visual neurones with large receptive fields. During episodes of imperfect PEM, these neurones seemed to monitor stimulus movement relative to the fovea in the same fashion as during fixation. Thus these DLPN neurones appear to be correlated only with visual rather than oculomotor parameters. Given that the DLPN receives inputs from the pretectum,[375] SC,[164] area MT,[149, 241, 363] and area 7a[241] and that it projects to the cerebellar flocculus and posterior vermis,[45, 133, 206] Suzuki and Keller[351] suggested that the vision-related DLPN neurones convey both direction and retinal slip velocity signals of targets in the foveal range during PEM from the afferent visual system to the cerebellum. At present it is unclear whether these neurones also project to non-cerebellar targets, which cerebellar or vestibular nuclei they reach, and what their possible contribution is to the required generation of the pursuit velocity signal. Another study of single-unit activity in the DLPN reported the presence not only of vision-related neurones but also of a substantial number of pursuit-related neurones and mixed vision- and pursuit-related neurones.[259]

Low-current (25–70 μA) microstimulation within the DLPN of alert macaques was recently reported to elicit pursuit eye velocity changes, but only during ongoing PEM.[244]

Another interesting class of 'multimodal' neurones was found in the dorsomedial paramedian basilar pontine grey, also known as the rostromedial dorsal pontine nucleus (MDPN) adjacent to NRTP.[186] These neurones responded to visual and vestibular stimuli and were also modulated during PEM. The MDPN appears to receive inputs from area 8[10, 213] and possibly also from subcortical structures such as the pretectum.[375] A substantial percentage of MDPN neurones, all of which were activated by movement of a whole-field stimulus in a specific direction during fixation of a stationary target, exhibited a pursuit-related activity modulation during sinusoidal PEM of a small target against a dark background. The on direction of these neurones was the same for whole-field stimulation and PEM. These neurones with the additional pursuit-related component could not, however, be activated during fixation by movements of the small target anywhere in the visual field. During VOR suppression or VOR in the dark, many of these multimodal neurones were modulated in phase with head velocity in the direction opposite to the on direction during PEM (in contrast to gaze velocity neurones in the cerebellum or pursuit neurones in PH and PPRF). Their maximum impulse rate reached saturation at a velocity of only $10° s^{-1}$ for target, whole field or head movement.

Single-unit activity in the NRTP of alert macaques[86] was found to be modulated by visual stimuli and/or during saccades. However, no correlation between neural activity and PEM or OKN could be detected in the NRTP, and this area will not be considered any further.

Nucleus Prepositus Hypoglossi

Several lines of evidence indicate that the PH,[38] including the sg,[206] is probably one of the essential structures for PEM control in the brain stem. Bilateral chemical microlesions involving the PH and the adjacent MVN in macaques caused severe pursuit deficits (in addition to gaze-holding deficits), which was interpreted as partial destruction of the neural integrator.[65] However, these monkeys exhibited a substantial recovery from their pursuit deficits within several days.

Morphology

Detailed morphological studies of the PH in the cat (see references 246 and 247) have demonstrated afferent projections from SC, NOT, vestibular nuclear complex, NRTP, pontine reticular formation, cerebellar cortex, fastigial nucleus and extraocular motor nuclei, as well as efferent projections to SC, pretectum, PGN, central thalamus, vestibular nuclear complex, pontine reticular formation, cerebellar cortex (flocculus and vermis), inferior olive and extraocular motor nuclei. This morphological picture is presently less complete in macaques.[38] Afferent projections have been reported to reach the PH from the NOT,[230a] pontine reticular formation,[63] and area 8.[334] The PH was found to project to the abducens nucleus,[207] oculomotor nucleus,[67, 343] flocculus,[40, 206] the posterior vermis[40, 41] and the dentate nucleus.[70]

Neurophysiology

Single-unit recordings in the PH of alert cats (see reference 222) have revealed the existence of many neurones with eye velocity and eye position components similar to burst-tonic neurones in monkeys. The activity of these neurones was consistently related to all types of eye movements in a preferred direction. Pursuit was not specifically studied, but during the slow phase of nystagmus the impulse rate gradually changed and was highly regular, as is also seen in ocular motoneurones. These authors suggested that the PH participates in neural integration (transformation of eye velocity into eye position signals) and provides a 'corollary discharge' for eye movement signals throughout the brain stem.

Similar activity patterns probably exist in the PH of alert macaques.[184, 225] The recording sites in these two studies involved a large area of the brain stem, including the PH and the pontine reticular formation.

More recently, several studies on single-unit activity in the vicinity of the abducens nucleus during PEM in trained macaques revealed the existence of three classes of premotor neurones with features relevant for pursuit control. Unit activity was recorded in stereotaxically and histologically verified sites in the PH and the sg.[115, 118, 120] Monkeys were trained to pursue a small visual target (8-min arc in diameter) in the horizontal plane against a homogeneous background (1 cd/m²) under three standard conditions: smooth pursuit during sinusoidal target movement (10° amplitude at frequencies between 0.2 and

1.4 Hz), VOR (light) with the same amplitude and frequency range and VOR suppression (fixation of the target, which moved with the sinusoidally rotating chair).

Pursuit Neurones

Pursuit neurones (PU), the first class of neurones,[118, 120] were located in the area of the sg dorsal to the abducens nucleus and ventral to the abducens nucleus in the caudal PPRF. These neurones had the following dynamic properties:

1. During spontaneous eye movements (saccades and fixation), the activity of pursuit neurones was found to be only very loosely, if at all, correlated with eye movement parameters. Therefore one could easily fail to detect these neurones unless the monkey performed PEM.

2. Activity modulation was approximately in phase with eye velocity ipsilateral to the recording site during smooth pursuit.

3. No or only small modulation (approximately in phase with contralateral eye velocity eqivalent to ipsilateral head velocity) occurred during VOR (light).

4. Pronounced modulation (usually larger than during smooth pursuit) occurred approximately in phase with head velocity ipsilateral to the recording site during VOR suppression.

5. Bode plots indicated a wide range of phase values independent of frequency for individual pursuit neurones between $\sim 30°$ lead and $30°$ lag relative to velocity.

In summary, pursuit neurones appeared to have dynamic properties similar to gaze velocity neurones in the cerebellar cortex except that no eye position sensitivity was found for pursuit neurones.

Eye-Velocity Neurones

The second class of premotor neurones is made up of eye-velocity neurones (EV).[118] These neurones were located just dorsal to the abducens nucleus in part of the sg and had the following dynamic features:

1. The impulse rate was often loosely correlated with saccades in a preferred direction but was low and exhibited little or no eye position sensitivity during fixation.

2. Modulation was in phase with horizontal eye velocity ipsilateral or contralateral to the recording site during smooth pursuit.

3. During VOR (light), modulation in phase with eye velocity was very similar to that observed during smooth pursuit.

4. Little or no modulation occurred during VOR suppression.

5. Bode plots indicated a considerably narrower range of phase values (as compared with the phase range of pursuit neurones) of individual neurones between $\sim 15°$ lead and $15°$ lag relative to eye velocity.

In summary, EV neurones with negligible eye position sensitivity exhibit a strong correlation with smooth eye velocity independent of the stimulus condition. Thus far they are the most likely candidate for a neural control signal specifically for pursuit eye velocity (equivalent to the control signal of medium lead burst neurones specifically for saccadic eye velocity) that would provide an input into the required neural integration stage.

Intermediate-Phase Neurones

The third class consists of intermediate-phase neurones (IN).[115, 117] These neurones were located dorsal to the abducens nucleus in the sg nucleus, but also at the ventral rim of the abducens nucleus, within the medial longitudinal fasciculus (MLF) rostrodorsal to the abducens nucleus, and in the central mesencephalic reticular formation (cMRF) close to the oculomotor nuclear complex. Intermediate phase neurones had the following dynamic properties:

1. During spontaneous eye movements, the impulse rate exhibited brief positive transients for saccades in the on direction followed by slow negative drifts. This pattern was reminiscent of a leaky integrator and quite different from the corresponding activity time course of ocular motor neurones.

2. Eye position sensitivity during fixation was small for those IN neurones with phase lead values close to $90°$ (similar to EV neurones) but larger for IN neurones with smaller intermediate phase values close to $30°$ at 0.8 Hz (similar to ocular motor neurones).

3. The impulse rate of individual IN neurones was modulated during smooth pursuit at a given frequency with intermediate phase values ranging from $90°$ lead to $30°$ lead relative to eye position.

4. Impulse rate modulation during VOR (light) was similar to that during smooth pursuit except for slightly larger phase lead values.

5. During episodes of perfect VOR suppression, no impulse rate modulation occurred.

6. Bode plots indicated a gradual increase of phase lead with increasing frequency (for smooth pursuit or VOR (light)). Superimposed phase-vs-frequency plots of a number of IN neurones yielded a continuum between the phase plots of eye-velocity neurones and those of ocular motor neurones.

In summary, IN neurones are modulated during PEM and other types of eye movements, as would be expected of premotor neurones representing intermediate results in the process of neural integration. These findings support the hypothesis that the neural integration of eye-velocity signals into eye position signals involves these IN neurones within the region of PH and sg and that the integrator network consists of a cascade of IN neurones.[115]

Further detailed neurophysiological studies in alert primates are needed to clarify the connectivity pattern of PH with its several subdivisions and their functional role in PEM.

Pontine and Mesencephalic Reticular Formation
The PPRF is a brain stem region (for localization in the monkey and cat, see references 150 and 158) with premotor functions essential for the saccadic system. Surgical PPRF lesions in macaques[150] generated permanent paralysis of conjugate horizontal gaze toward the side of the lesion. However, a more recent study on the effects of unilateral and bilateral chemical PPRF lesions[203] found that compensatory slow eye movements toward the hemifield ipsilateral to the lesion could still be elicited during head rotation in the contralateral direction. Furthermore, some ability to pursue moving objects was preserved, even following bilateral PPRF lesion. These findings emphasize the role of PPRF in generating horizontal quick eye movements.

Interestingly, electrical stimulation of the PPRF was found to elicit smooth, ipsilateral, horizontal, conjugate eye movements, independent of eye position and with a constant eye velocity that was linearly related to the frequency of stimulation.[77]

Afferent projections from the precentral cortex, including area 8,[210, 213, 334] the SC,[75, 303] and the cMRF,[76] were identified in macaques; additional afferents from the NOT and PH were also demonstrated in cats (see reference 246). The PPRF in macaques was found to project to the rostral interstitial nucleus of the MLF (iMLF),[62] the ipsilateral abducens nucleus,[63, 157] IML, the PH, and possibly also to the ipsilateral oculomotor nuclear complex.[63] Single-unit recordings in the PPRF of alert macaques revealed several classes of premotor neurones including burst neurones, pause neurones, burst-tonic neurones and tonic neurones during spontaneous or visually induced saccadic eye movements.[167, 184, 225, 303] Whereas burst neurones and pause neurones are specifically related to rapid eye movements in a preferred direction, burst-tonic and tonic neurones exhibit a smooth impulse rate modulation during PEM.[184, 339] It was found that some of the tonic neurones with significant eye position sensitivity in a preferred on direction exhibited additional eye velocity sensitivity; however, this occurred only during PEM in the on direction.[184] These tonic neurones seem to encode eye position plus eye velocity in the on direction. More recently, a new class of pursuit neurones, which had already been found to occur in the sg nucleus of the PH,[118, 120] was also found in the caudal PPRF ventral to the abducens nucleus.[118, 120] Pursuit neurones had no eye position sensitivity and seemed to be specifically correlated with pursuit similar to what is found in gaze-velocity neurones.

Two areas within the region of the MRF laterally adjacent to the oculomotor nuclear complex and extending further rostrally along the MLF have recently been distinguished on histological and functional grounds. These areas are referred to as the rostral interstitial nucleus of the MLF (rostral iMLF)[62] and the central MRF (cMRF).[75]

Unilateral lesions in the MRF induced deficits in the slow phase of contralateral OKN,[196] without producing an impairment in rapid eye movments or VOR. The slow-phase velocity of OKN toward the side of lesion was slowed, indicating a possible interruption of the visual input (presumably from the NOT) for OKN. These lesions left the pursuit and gaze-holding systems normal. A more recent study found that bilateral MRF lesions did not impair horizontal eye movements but caused substantial deficits in vertical rapid eye movements.[198] Vertical pursuit movements were initially also slightly impaired but soon recovered.

Unilateral electrical stimulation of the cMRF generated conjugate saccadic eye movements toward the contralateral side.[78, 196] In contrast, bilateral stimulation in other parts of the MRF was demonstrated to induce downward binocular movements.[198]

The cMRF receives afferent projections from the SC,[75] whereas the rostral iMLF receives inputs from the NOT,[62, 375] the PPRF[62] and the SVN.[69] The cMRF projects to the SC and PPRF,[76] whereas the rostral iMLF projects to the oculomotor nuclear complex.[62]

A single-unit study of the MRF (presumably including the rostral iMLF)[193] of trained macaques described two classes of neurones exclusively related to vertical (upward or downward) eye movements. Burst-tonic neurones exhibited linear relationships between impulse rate and vertical eye position during fixation and exhibited an additional eye-velocity component during smooth pursuit or VOR in the dark. Phase lead values relative to eye position for individual burst-tonic neurones during smooth pursuit ($\pm 10°$ at 1 Hz) ranged from 24 to 62° and during VOR in the dark from 18 to 50°. During VOR suppression, the impulse rate was more or less unmodulated, depending on the monkey's performance. The second class consisted of irregular tonic neurones that exhibited irregular activity that occasionally appeared to be related to spontaneous eye movements but often changed to a reliable static characteristic for impulse rate vs eye position during fixation of a stationary target of interest. These neurones were consistently modulated during vertical PEM. Phase lead values for individual irregular tonic neurones during smooth pursuit ranged from 17 to 75° and, during VOR in the dark, from 18 to 76° with average phase values slightly higher than those of burst-tonic neurones. Bode plots and dynamic data during VOR (light) were not presented. However, the wide range of phase lead values in both neural populations indicates interesting similarities between these rostral MRF neurones with vertical on directions and IN neurones with horizontal on directions,[115] which were located not only in the PH (for details, see review of the PH) but also in the cMRF close to the oculomotor nuclear complex. Accordingly, the suggested role for IN neurones in neural integration might also be applicable to these rostral MRF neurones with regard to vertical eye movements.

'Final Common' Pathway

It is generally assumed that all ocular motor neurones located in the oculomotor nuclear complex (IIIrd nerve nucleus), the trochlear nucleus (IVth nerve nucleus) and the abducens nucleus (VIth nerve nucleus), structures that innervate the six extraocular muscles of each eyeball, participate in the control of all types of eye movements. The structural and functional properties of this final common pathway incorporating the motor neurones and the corresponding end organ have been studied in great detail (for a review, see reference 308). Two recent neurophysiological studies in alert macaques suggest that the neural control signal of individual ocular motor neurones for a given eye movement time course can vary slightly when this eye movement is elicited by superimposed visual and rotatory vestibular[337] or superimposed visual and linear vestibular stimuli,[113] rather than purely visual pursuit stimulation. Thus in a strict sense the final common pathway consists only of orbit plant, where the individual contributions of motor neurones and their corresponding fast-twitch or slow-twitch muscle fibres[245] are combined.

The two horizontal eye muscles (medial rectus and lateral rectus) in humans and macaques move the line of gaze rather precisely along a horizontal line (with the head in the upright position). Also, the two downward-pulling pairs of eye muscles, with upward and downward components acting synergistically, produce purely vertical eye movements.[360] These mechanical features of the orbit plant seem to support the available neurophysiological evidence for separate horizontal and vertical oculomotor systems. However, some neurophysiological[167] and behavioural data[95] on the saccadic oculomotor system in macaques have emphasized the possibility that oblique eye movements into tertiary positions may be encoded in polar coordinates rather than cartesian coordinates. Corresponding data on oblique pursuit eye movements are not available.

Morphology of Oculomotor Nuclei

The oculomotor nuclei comprise ocular motor neurones and intranuclear interneurones.[137, 169] It should be noted that axon collaterals have recently been found to also occur in ocular motor neurones,[124, 340] at least in cats. The following data on afferent and efferent projections refer to macaques unless otherwise specified.

The IIIrd nerve nucleus receives afferent input from the NOT,[375] area 8 and the supplementary eye field (in area 6aβ),[209, 214] the contralateral VIth nerve nucleus,[61, 68, 343] the vestibular nuclear complex,[68, 69, 343] the y group,[342, 343] the PH[67, 343] and possibly also from the ipsilateral PPRF.[63] Efferents from this nucleus innervate four extraocular muscles (ipsilateral inferior rectus, medial rectus, inferior oblique and contralateral superior rectus) and project to the VIth nerve nucleus. In cats (see reference 246) additional efferents to the PH have been described.

The IVth nerve nucleus was found to receive afferents from the NOT,[375] the vestibular nuclear complex, especially the SVN,[66, 69] and from the y group.[342] Efferent projections innervate the contralateral superior oblique muscle. In cats, reciprocal connections with the PH have also been described (see reference 246).

The VIth nerve nucleus receives afferent projections from the SC, the NRTP, the PH,[207] the y group[342] and the PPRF,[63, 157, 207] the vestibular nuclear complex[66, 207] and the IIIrd nerve nucleus. Efferents from this nucleus innervate the ipsilateral lateral rectus muscle and project to the flocculus[41, 206] and the contralateral IIIrd nerve nucleus. In cats, efferent projections to the PH have also been found (see reference 246).

Neurophysiology of Oculomotor Nuclei

The dynamic properties of ocular motor neurones in alert macaques have been approximated by a first-order differential equation that gives the impulse rate for all types of eye movements as the sum of an eye position component (for positions above an individual threshold position) and an eye-velocity component (see reference 309). The number of active motor neurones increase with movement of the eyeball further and further into the on direction (recruitment). Further quantitative studies have revealed a number of differences in the static and dynamic properties for eye movements in the on direction (agonist phase) vs off direction (antagonist phase). The finding that the impulse rate for encoding a given fixational eye position is always slightly larger if this position is reached by a movement in the on direction (horizontal or vertical) rather than the off direction[110, 153] was described as static hysteresis. For many ocular motor neurones, the change from spontaneous eye movements to the mode of PEM was typically found to be accompanied by a shift in the neural activity level.[111, 119] The impulse rate of individual motor neurones either increased or decreased during PEM relative to the static characteristic during fixation. This finding was taken as evidence for the hypothesis that individual motor neurones make different contributions to the neural control of eye movements in the modes of fixation vs PEM. A quantitative comparison of the positive eye velocity component during PEM in the on direction with the negative eye-velocity component in the off direction at a given eye position revealed significant differences in oculomotor neural activity in the IIIrd and VIth nerve nuclei.[119] Some neurones had a larger (positive) eye-velocity component in the agonist phase than in the antagonist phase, whereas this relationship was reversed for other motor neurones (or intranuclear interneurones). This finding of different-velocity components was interpreted as evidence that ocular motor neurones receive their eye-velocity components for PEM in opposite directions from two separate premotor sources.

These various functional differences between ocular motor neurones led to a more specific quantitative description by means of two separate first-order differential equations, one for the agonist phase and one for the antagonist phase during PEM.[119]

The frequency response of the maximum impulse rate and of the phase lead of ocular motor neurones during horizontal PEM[119] indicated a continuous increase in both parameters with increasing frequency. However, the maximum impulse rate dropped for frequencies above ~1 Hz, thereby encoding the gain drop (ratio of eye movement amplitude and target movement amplitude) of the overall pursuit system. Evidently the neural control system of PEM uses both amplitude and phase lead increase to compensate the frequency-dependent mechanical load of the orbit plant. Typical phase lead values relative to eye position in the on direction ranged from 10° at 0.2 Hz to 30° at 1 Hz for these motor neurones. During sinusoidal PEM (at 10° amplitude) at various frequencies, the target was turned off for 800 ms at random phases within the movement cycle. While the monkeys performed slow postpursuit eye movements (PPEM) for more than 1 s after target disappearance, the corresponding impulse rate of ocular motor neurones was smoothly modulated.[119] Thus PPEM, which typically continued at the most recent frequency before target disappearance (but with gradually reduced amplitude), was clearly controlled by motor neurones. This is an important finding, supporting the hypothesis that the neural control of PEM is based on an internally generated motor programme with optional updating while the target is visible, rather than on a reflective movement that depends entirely on retinal events. The impulse rate of ocular motor neurones with vertical on directions was also slightly modulated during pure horizontal PEM. This might correspond with the old finding that the four vertical eye muscles have small additional pull components in the horizontal plane.

The dynamic properties of ocular motor neurones and intranuclear interneurones in the IIIrd and IVth nerve nuclei with vertical on directions[193] were found to be comparable to those with horizontal on directions, but the range for the phase lead values (13–63°) was larger for the former.

In summary, ocular motor neurones carry the superposition of an eye position signal (in the on direction of the corresponding extraocular muscle) and eye-velocity signals, which are positive in the on direction and negative in the off direction during PEM. During VOR (light), the phase lead values of individual ocular motor neurones at a given frequency may be slightly different from those during smooth pursuit, thus indicating individual rather than pooled premotor inputs. Impulse rate modulation during PPEM, VOR suppression, and VOR in the dark reflects the corresponding eye movement.

Proprioceptive Afferents from Extraocular Muscles

Theoretically, stretch receptors in the eye muscles are in a unique position to monitor otherwise unobtainable signals involving the length–tension status within the orbit plant. However, no positive evidence presently exists for an influence of such proprioceptive afferents on eye movements, especially on PEM. A monosynaptic stretch reflex does not exist in the oculomotor system of macaques.[189] Since stretch receptors of various kinds have been identified in the extraocular muscles and their activity has been traced to several relevant brain regions in primates and other vertebrates, some of this literature will be briefly reviewed.

The morphology and dynamic properties of proprioceptors in the eye muscles have been studied particularly in humans, macaques, cats, and ungulates.[13, 23, 160, 233, 235, 296, 369] In macaques, the cell bodies of primary afferents of individual extraocular muscles were localized in the ipsilateral semilunar ganglion.[296] These eye muscle afferents were recently demonstrated to terminate in the caudal brain stem in the ipsilateral spinal trigeminal nucleus and the ipsilateral cuneate nucleus[295] These two nuclei project to the flocculus and vermis of the cerebellum and to the SC (for a review of this literature, see reference 295). In the cat, responses to passive stretch or electrical stimulation of extraocular muscles were traced to the LGN,[101] SC,[1, 102], visual cortex,[53] frontal eye fields,[108] and vermis.[135] A detailed study of the modulation of Purkinje cells in the flocculus of anaesthetized rabbits during stepwise or sinusoidal passive eye movements[192] revealed various types of neural activity with direction selectivity and eye position as well as eye velocity sensitivity. Comparable data on macaques are not at present available.

In summary, the role of stretch receptors as potential monitors of the execution of motor programmes in the final common pathway, a pathway that might be important during early stages in ontogeny and following surgically or chemically induced changes in eye muscle function, is still unclear and requires future research.

Missing Links and Model Considerations

Schematic Summary of Physiological and Anatomical Data

The circuit plan in Fig. 9.1 summarizes the described connections between the various neural processing stages and indicates the corresponding types of neural activity in the alert macaque. The layout of this circuit plant is deliberately open-ended to allow the addition of new connections and neural processing stages without changing the existing circuit. The chosen sequence of stages from the retina at the top via thalamus, tectum, cerebral cortex, pons and cerebellum to the oculomotor nuclei is artibrary. Connections are always arranged in clockwise direction to distinguish 'downstream' connections on the right (e.g., from the LGN to V1) from 'upstream' connections on the left (e.g., from the MRF to the SC). The different symbols at the various stages indicate their characteristic types of neural activity. By means of these symbols, stages with vision-related neurones (cross), oculomotor-related (unrelated with pursuit) neurones (o), pursuit-related (but not encoding pursuit velocity) neurones (p) and with neurones correlated with gaze velocity ($\dot{g}$), or even with pursuit eye velocity ($\dot{e}$) or eye position (e) can easily be recognized. This circuit plan necessarily ignores the fact that each neural processing stage consists of several neural subdivisions that may be the exclusive source or the exclusive target of individual pathways. However, such a simplification can be particularly helpful in a discussion of missing links and models of the pursuit system. Three points will be addressed.

Firstly in a consideration of the fundamental difference between vision-related (on the sensory side) and pursuit-related (on the oculomotor side) neurones, the question arises at what stage pursuit-related signals are being generated as the basis for the required pursuit eye velocity signal. Figure 9.1 shows pursuit-related activity (p) in the NOT, the MST, 7a, the precentral cortex (area 6, 8) and the pontine nuclei including the DLPN. It is tempting to speculate that the 'first draft' of the required motor programme, which may be represented by pursuit-related activity, is being generated within the precentral cortical areas 6, 8 (also including areas 45 and 46). This stage (6 and 8) has a unique location within this circuit plan (as judged by the presently known connections) in that it contains pursuit-related neurones and has direct connections to all other stages in which pursuit-related activity has been described. Considerable work is still required to find supporting evidence for this hypothesis. However, it is clear that the precentral cortex is involved in the contol of PEM to a much greater degree than had previously been assumed (see pp 179–180).

Secondly, with the assumption that the generation of the required eye velocity signal in the brain stem is based on pursuit-related signals arising in the precentral cortex (possibly also involving other stages of the cerebral cortex), the brain stem and cerebellum have to perform the necessary signal transformation steps. The available data do not allow definite statements to be made about the specific contributions of the cerebrocerebellar pathways via the pontine nuclei *vs* the non-cerebellar pathways from the cerebrum to MRF, PPRF and PH. However, the neurophysiological findings suggest a certain signal processing sequence: (a) generation of pursuit-related signals in the desired movement direction, (b) generation of a gaze velocity signal, (c) superposition of the gaze velocity signal and a head velocity signal to generate the required pursuit eye velocity signal and (d) neural integration of the eye velocity signal to generate the required eye position signal for the ocular motor neurones.

Thirdly, our present knowledge of the primate pursuit system is still rather preliminary and incomplete. Several major questions (including those listed below) still need to be answered, although the oculomotor system, including the pursuit system, is probably much better understood than most skeletomotor systems: (a) What is the neural mechanism of motor programme generation? (b) Do stretch receptor afferents from the extraocular muscles participate in the neural control of PEM in ontogeny or following mechanical changes within the orbit plant?

Models of Neural Network for Pursuit Eye Movements

Most recent models of the pursuit system include the hypothesis of internally generated velocity signals to explain the various dynamic properties of PEM. This internal signal generator is referred to as the target-velocity regenerator (for the perceived target velocity). It uses an efference copy signal of eye velocity[399] as an adaptive controller[14, 159] or as a motor programme generator for eye velocity.[112, 114] Several authors agree that the retinal error signals (position error or slip velocity) during PEM, which are available to modify (or update) the internally generated velocity signal, can be treated as command signals for eye acceleration.[112, 218] It is generally assumed that the goal of PEM is the continuous

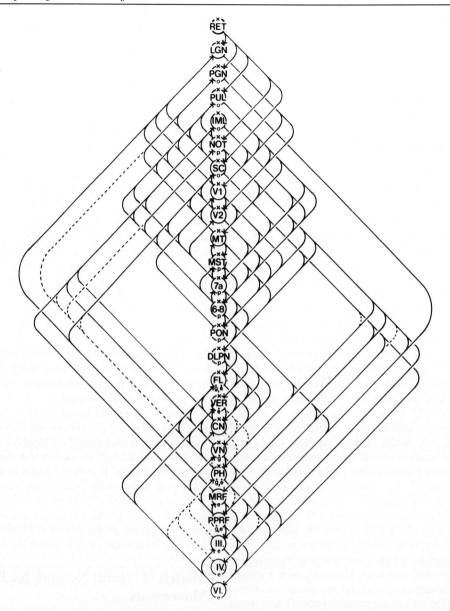

Fig. 9.1 *Schematic circuit plan of the pursuit system in macaques. Connections so far only confirmed in cats are indicated by dotted lines. Multiple projections from a given neural processing stage to several other stages below (connections on right) or above (connections on left) may consist of independent parallel pathways or of axon collaterals. Characteristic types of neural activity at various stages are indicated as follows: vision-related neurones ($\times$), oculomotor-related neurones unrelated to pursuit ($\bigcirc$), pursuit-related neurones that did not encode gaze or eye velocity (p), neurones correlated with gaze velocity ($\dot{g}$), neurones correlated with eye velocity ($\dot{e}$), neurones correlated with eye position in addition to eye velocity (e). Afferents from eye muscle proprioceptors were not included. Abbreviations of neural processing stages: RET, retina; LGN, dorsal lateral geniculate nucleus; PGN, perigeniculate nucleus; PUL, pulvinar; IML, internal medullary lamina; NOT, nucleus of optic tract; SC, superior colliculus; V1, area V1; V2, area V2 within area 18; MT, middle temporal visual area within area 19; MST, medial superior temporal area in superior temporal sulcus; 7a, area 7a including lateral intraparietal area (LIP); 6,8, frontal eye field in area 8 plus supplementary eye field in area 6aβ plus area 45 and area 46; PON, pontine nuclei except for DLPN (dorsolateral pontine nucleus); FL, flocculus cerebelli; VER, lobule V–VII of vermis cerebelli; CN, deep cerebellar nuclei; VN, vestibular nuclear complex; PH, nucleus prepositus hypoglossi including supragenu nucleus (sg); MRF, mesencephalic reticular formation including rostral interstitial nucleus of MLF and central MRF; PPRF, parmedian pontine reticular formation; III, oculomotor nuclear complex; IV, trochlear nucleus; VI, abducens nucleus.*

minimization of the retinal position error (see pp 165–167. In a model of the corresponding neural network for PEM,[112, 114] a class of afferent visual neurones was postulated that would monitor the current position error or eccentricity of the target projection relative to the fovea and serve as an acceleration signal (in the sense of a signal to change the ongoing eye velocity) for minimization of the position error. Such neurones with features of eccentricity detectors in a preferred direction are likely to exist but have not yet been found. It was pointed out[112] that this model of the fovea pursuit system, which basically transforms a displacement at the sensory level (target displacement on the fovea or cupula displacement) into a corresponding rotation of the eyeball, underscores striking similarities between the phylogenetically oldest (VOR) and youngest (PEM) oculomotor subsystem.

Models of the motor programme generator in the pursuit system are particularly difficult to deduce from the available neurophysiological data, if one requires not only the simulation of zero-latency pursuit of targets with certain predictable movement trajectories[14] but also postpursuit eye movements following sudden target disappearance. A highly theoretical model of a motor programme generator has recently been proposed,[116, 116a] which is neurobiologically plausible and accounts for zero-latency pursuit, neural prediction and postpursuit eye movements. This model consists of a neural net that is arranged in space as a triangular lattice on a circular surface. An activity peak (like a soliton) travels over this lattice with a constant velocity and defines the generated velocity time course at the output by its trajectory on the lattice. This model not only is capable of simulating the generation of any velocity trajectories with or without sensory updating (acceleration) signals, but also can learn, memorize and later retrieve various trajectories.

Conclusions

The earlier parts of this chapter devote special attention to a proper definition of the sensorimotor task of pursuit eye movements and to the corresponding neural events in the afferent visual system. Although it is safe to conclude that the retina, LGN and area V1 are essential stages for PEM (see Fig. 9.1), the specific contribution of X cells and Y cells is still unclear. The possible role of the IML, SC or area V2 is entirely open.

The involvement of the MST in the superior temporal sulcus, area 7a in the posterior parietal cortex, and various parts of the precentral cortex (including area 6 with the supplementary eye field, area 8 with the frontal eye field, and areas 45 and 46) in the neural control of PEM is apparent. A first draft of the motor control signal for PEM, as represented by pursuit-related neurones, may arise in one of these stages, but this hypothesis needs considerably more work for its substantiation.

The available data on the pursuit system in macaques can be tentatively summarized by a hypothesis with the following components. (a) The primary motor programme for PEM is represented by gaze velocity signals in the brain stem (PH, y group) and cerebellum (FL), which may be internally generated on the basis of pursuit-related activity and visual activity related to retinal events (position error and slip velocity). (b) Eye velocity signals for PEM in a preferred direction are generated by comparing gaze velocity with head velocity signals and are represented by eye velocity neurones in the PH. (c) Eye position signals are being generated by means of neural integration of eye velocity signals; this process partly involves the PH and cMRF.

From a more general perspective, the smooth pursuit system in primates may serve as a model for various oculomotor and skeletomotor systems in that it reflects the present state of the art in our knowledge of sensorimotor brain functions. In this regard, it is useful to take a critical look at the methods and concepts presently available and to consider their intrinsic limitations.

The available knowledge of the neural control of the pursuit system in primates originates from the correlation of lesions, stimulations and single-unit recordings in loosely circumscribed brain regions with the overall ability of the brain to 'fixate and see' small moving targets. Particular attention should be paid to the ever-increasing number of newly discovered neural connections between various brain regions and to newly described details of their microcircuitry and subdivisions, as well as to the standard finding that circumscribed lesions (e.g. in the MT or in the PH) cause only temporary pursuit deficits.

We are so deeply influenced by the concept of flow diagrams for the programs of our serially operating computers that we may be automatically looking for a serially operating pursuit system, such as retina to LGN to V1 to MT to MST to 7a to DLPN to FL to VN to PPRF to VI. This approach is reminiscent of the once very popular concept of a sequence of cells from simple cells to 'grandmother' cells for visual pattern recognition.

1. What if the pursuit system requires parallel (rather than serial) signal processing structures?

2. The network architecture consisting of a *few* neural connections that may have been only temporarily selected by 'command neurones' for the purpose of smooth pursuit, may be hidden in the jungle of the *many possible* histologically established neural connections.

3. Essential aspects of the neural signal processing events may occur by means of subthreshold activity that

could be detected only by intracellular DC recordings rather than the usual extracellular AC recordings.

4. Other processing components may use ensembles of neurones scattered over several stages (Fig. 9.1), which could only be monitored by multielectrode recording methods.

This list of possible limitations of the presently used methods is by no means complete.

Brain research has often advanced in steps. It is quite possible that the present and still rather patchy knowledge of various neural components of the pursuit system will lead to more integrated studies, both experimental and theoretical, which describe the function of the pursuit system as *one complex system* and lead to the development of appropriate new theoretical tools for its description.

Acknowledgements

This chapter was originally published in *Physiological Reviews*, **67**, 797–857 and has been reproduced with the kind permission of the American Physiological Society.

I thank those scientists who kindly provided helpful comments, criticism and advice on an early version of this review, particularly Joachim Grüsser, Edward Keller and John Schlag. I especially thank Elke Jaworski and Anneliese Thelen for their help with literature searches and their secretarial expertise.

This work was supported by the Deutsche Forschungsgemeinschaft, SFB 200/A-1.

Abbreviations for Brain Regions

AOT	accessory optic tract
cMRF	central mesencephalic reticular formation
DLPN	dorsolateral pontine nucleus
DTN	dorsal terminal nucleus of accessory optic tract
FL	flocculus cerebelli
IML	internal medullary lamina in the central thalamus
IVN	inferior or descending vestibular nucleus
LGN	dorsal lateral geniculate nucleus
LTN	lateral terminal nucleus of accessory optic tract equivalent to NAOT
LVN	lateral vestibular nucleus or nucleus of Deiters
MDPN	rostromedial dorsal pontine nucleus
MLF	medial longitudinal fasciculus
MRF	mesencephalic reticular formation
MST	medial superior temporal area
MT	middle temporal visual area (within area 19)
MTN	medial terminal nucleus of the accessory optic tract
MVN	medial vestibular nucleus
NAOT	nucleus of accessory optic tract equivalent to LTN
NOT	nucleus of the optic tract
NPA	anterior pretectal nucleus
NPM	medial pretectal nucleus
NPP	posterior pretectal nucleus equivalent to sublentiform nucleus
NRTP	nucleus reticularis tegmenti pontis
ON	pretectal olivary nucleus
PGN	ventral lateral geniculate or perigeniculate nucleus
PH	nucleus prepositus hypoglossi
PI	inferior pulvinar nucleus
PL	lateral pulvinar nucleus
PM	medial pulvinar nucleus
PPRF	paramedian pontine reticular formation
rostral iMLF	rostral interstitial nucleus of the MLF
SC	superior colliculus
sg	supragenu nucleus of nucleus prepositus hypoglossi
SVN	superior vestibular nucleus
V1	area V1 or area 17, striate cortex
V2	area V2 (within area 18)
III	oculomotor nuclear complex
IV	trochlear nucleus
VI	abducens nucleus
7a, 7b	area 7, inferior parietal cortex

References

1. Abrahams, V. C. Proprioceptive influence from eye muscle receptors on cells of the superior colliculus. In *Reflex Control of Posture and Movement*. Eds. Granit, R. and Pompeiano, O. (1979) vol. 50, pp. 325–334. Amsterdam: Elsevier. (Prog. Brain Res. Ser.)
2. Adams, C. K., Perez, J. M. and Hawthorne, M. N. Rod and cone densities in the rhesus. *Invest. Ophthalmol.*, **13**, 885–888 (1974).
3. Adler, B. and Grüsser, O.-J. Sigma-movement and optokinetic nystagmus elicited by stroboscopically illuminated stereopatterns. *Exp. Brain Res.*, **47**, 353–364 (1982).
4. Albano, J. E. and Wurtz, R. H. Deficits in eye position following ablation of monkey superior colliculus, pretectum and posterior-medial thalamus. *J. Neurophysiol.*, **48**, 318–337 (1982).
5. Albright, T. D. Direction and orientation selectivity of neurones in visual area MT of the macaque. *J. Neurophysiol.*, **52**, 1106–1130 (1984).
6. Albright, T. D., Desimone, R. and Gross, C. G. Columnar organization of directionally selective cells in visual area MT of the macaque. *J. Neurophysiol.*, **51**, 16–31 (1984).
7. Allman, J., Miezin, F. and McGuinness, E. Stimulus specific

responses from beyond the classical receptive field: neurophysiological mechanisms for local-global comparisons in visual neurones. *Ann. Rev. Neurosci*, **8**, 407–430 (1985).
8. Andersen, R. A., Asanuma, C. and Cowan W. M. Callosal and prefrontal associational projecting cell populations in area 7a of the macaque monkey: a study using retrogradely transported fluorescent dyes. *J. Comp. Neurol.*, **232**, 443–455 (1985).
9. Aschoff, J. C. and Cohen, B. Cerebellar ablations and spontaneous eye movements in monkey. *Bibl. Ophthal.*, **82**, 169–177 (1972).
10. Astruc, J. Corticofugal connections of area 8 (frontal eye field) in *Macaca mulatta*. *Brain Res.*, **33**, 241–256 (1971).
11. Atkin, A. Shifting fixation to another pursuit target: selective and anticipatory control of ocular pursuit initiation. *Exp. Neurol.*, **23**, 157–173 (1969).
12. Aulhorn, E. and Harms, H. Visual perimetry. In: *Handbook of Sensory Physiology. Visual Psychophysics*. Eds. Jameson, D. and Hurvich, L. M. vol. VII/4. pp. 102–145. Berlin: Springer (1972).
13. Bach-Y-Rita, P. Structural-functional correlations in eye muscle fibers. Eye muscle proprioception. In: *Basic Mechanisms of Ocular Motility and Their Clinical Implications*. Eds. Lennerstrand, G. and Bach-Y-Rita, P. pp. 91–111. Oxford: Pergamon (1975).
14. Bahill, A. T. and McDonald, J. D. Model emulates human smooth pursuit system producing zero-latency target tracking. *Biol. Cybern.*, **48**, 213–222 (1983).
15. Balaban, C. D. and Watanabe, E. Functional representation of eye movements in the flocculus of monkeys (*Macaca fuscata*). *Neurosci. Lett.*, **49**, 199–205 (1984).
16. Baloh, R. W., Yee, R. D. and Honrubia, V. Optokinetic nystagmus and parietal lobe lesions. *Ann. Neurol.*, **7**, 269–276 (1980).
17. Barany, R. Die Untersuchung der reflektorischen vestibulären und optischen Augenbewegungen und ihre Bedeutung für die topisch Diagnostik der Augenmuskelähmungen. *Münch. Med. Wochenschr.*, **54**, 1072–1075 (1907).
18. Barbas, H. and Mesulam, M.-M. Organization of afferent input to subdivisions of area 8 in the rhesus monkey. *J. Comp. Neurol.*, **200**, 407–431 (1981).
19. Barlow, H. B., Fitzhugh, R. and Kuffler, S. W. Change of organization in the receptive fields of the cat's retina during dark adaptation. *J. Physiol. Lond.*, **137**, 338–354 (1957).
20. Barmack, N. H. Modification of eye movements by instantaneous changes in the velocity of visual targets. *Vision Res.*, **10**, 1431–1441 (1970).
21. Baron, W. S., Boynton, R. M. and Van Norren, D. Primate cone sensitivity to flicker during light and dark adaptation as indicated by the foveal local electroretinogram. *Vision Res.*, **19**, 109–116 (1979).
22. Bartlett, J. R., Doty, R. W., Pecci-Saavedra, J. and Wilson, P. D. Mesencephalic control of lateral geniculate nucleus in primates. III. Modifications with state of alertness. *Exp. Brain Res.*, **18**, 214–224 (1973).
23. Batini, C. Properties of the receptors of the extraocular muscles. In: *Progress in Brain Research. Reflex Control of Posture and Movement*. Eds. Granit, R. and Pompeiano, O., vol. 50. pp. 301–314. Amsterdam: Elsevier (1979).
24. Behrens, F. and Grüsser, O.-J. Smooth pursuit eye movements and optokinetic nystagmus elicited by intermittently illuminated stationary patterns. *Exp. Brain Res.*, **37**, 317–336 (1979).
25. Belknap, D. B. and McCrea, R. A. Anatomy and physiology of cerebellar efferents to the vestibular complex and nucleus prepositus hypoglossi of the squirrel monkey. *Soc. Neurosci. Abstr.*, **11**, 1034 (1985).
26. Benevento, L. A., Rezak, M. and Santos-Anderson R. An autoradiographic study of the projections of the pretectum in the rhesus monkey (*Macaca mulatta*): evidence for sensorimotor links to the thalamus and oculomtor nuclei. *Brain Res.*, **127**, 197–218 (1977).
27. Berthoz, A. and Melvill Jones, J. *Adaptive Mechanisms in Gaze Control. Facts and Theories*. Amsterdam: Elsevier (1985).
28. Bielschowsky, A. Das klinische Bild der assoziierten Blicklähmung und seine Bedeutung für die topische Diagnostik. *Münch. Med. Wochenschrift*, **50**, 1666–1670 (1903).
29. Bizzi, E. Discharge of frontal eye field neurones during saccadic and following eye movements in unanesthetized monkeys. *Exp. Brain Res.*, **6**, 69–80 (1968).
30. Bizzi, E. and Schiller, P. H. Single unit activity in the frontal eye fields of unanesthetized monkeys during eye and head movement. *Exp. Brain Res.*, **10**, 151–158 (1970).
31. Blair, S. and Gavin, M. Modification of the macaque's vestibulooocular reflex after ablation of the cerebellar vermis. *Acta Otolaryngol.*, **88**, 235–243 (1979).
32. Blasdel, G. G., Lund, J. S. and Fitzpatrick, D. Intrinsic connections of macaque striate cortex: axonal projections of cells outside lamina 4C. *J. Neurosci.*, **5**, 3350–3369 (1985).
33. Blum, B., Kulikowski, J. J., Carden, D. and Harwood D. Eye movements induced by electrical stimulation of the frontal eye fields of marmosets and squirrel monkeys. *Brain Behav., Evol.*, **21**, 34–41 (1982).
34. Boyle, R., Büttner, U. and Markert, G. Vestibular nuclei activity and eye movements in the alert monkey during sinusoidal optokinetic stimulation. *Exp. Brain Res.*, **57**, 362–369 (1985).
35. Bridgeman, B. Visual receptive fields sensitive to absolute and relative motion during tracking. *Science Wash. DC*, **178**, 1106–1108 (1972).
36. Brinkman, C. and Porter, R. Supplementary motor area in the monkey: activity of neurones during performance of a learned motor task. *J. Neurophysiol.*, **42**, 681–709 (1979).
37. Brodal, A. *Neurological Anatomy*. New York: Oxford University Press (1981).
38. Brodal, A. The perihypoglossal nuclei in the macaque monkey and the chimpanzee. *J. Comp. Neurol.*, **218**, 257–269 (1983).
39. Brodal, A. The vestibular nuclei in the macaque monkey. *J. Comp. Neurol.*, **227**, 252–226 (1984).
40. Brodal, A. and Brodal, P. Observations on the projection from the perihypoglossal nuclei onto the cerebellum in the macaque monkey. *Arch. Ital. Biol.*, **121**, 151–166 (1983).
41. Brodal, A. and Brodal, P. Observations on the secondary vestibulocerebellar projections in the macaque monkey. *Exp. Brain Res.*, **58**, 62–74 (1985).
42. Brodal, P. The corticopontine projection in the rhesus monkey. Origin and principles of organization. *Brain*, **101**, 251–283 (1978).
43. Brodal, P. The pontocerebellar projection in the rhesus monkey: an experimental study with retrograde axonal transport of horseradish peroxidase. *Neuroscience*, **4**, 193–208 (1979).
44. Brodal, P. The projection from the nucleus reticularis tegmenti pontis to the cerebellum in the rhesus monkey. *Exp. Brain Res.*, **38**, 29–36 (1980).
45. Brodal, P. Further observations on the cerebellar projections from the pontine nuclei and the nucleus reticularis tegmenti pontis in the rhesus monkey. *J. Comp. Neurol.*, **204**, 44–55 (1982).
46. Brodal P. and Brodal, A. The olivocerebellar projection in the monkey. Experimental studies with the method of retrograde tracing of horseradish peroxidase. *J. Comp. Neurol.*, **201**, 375–393 (1981).
47. Brodal, P. and Brodal, A. Further observations on the olivocerebellar projection in the monkey. *Exp. Brain Res.*, **45**, 71–83 (1982).

48. Brown, A. C. The relation between the movements of the eyes and the movements of the head. *Nature Lond.*, **52**, 184–189 (1895).
49. Brown, K. T., Watanabe, K. and Murakami, M. The early and late receptor potentials of monkey cones and rods. *Cold Spring Harbor Symp. Quant. Biol.*, **30**, 457–482 (1965).
50. Bruce, C. J. and Goldberg, M. E. Primate frontal eye fields. I. Single neurones discharging before saccades. *J. Neurophysiol.*, **53**, 603–635 (1985).
51. Bruce, C. J., Goldberg, M. E., Bushnell, M. C. and Stanton, G. B. Primate frontal eye fields. II. Physiological and anatomical correlates of electrically evoked eye movments. *J. Neurophysiol.*, **54**, 714–734 (1985).
52. Buettner, U. W. and Büttner, U. Vestibular nuclei activity in the alert monkey during suppression of vestibular and optokinetic nystagmus. *Exp. Brain Res.*, **37**, 581–593 (1979).
53. Buisseret, P. and Singer, W. Proprioceptive signals from extraocular muscles gate experience-dependent modifications of receptive fields in the kitten visual cortex. *Exp. Brain Res.*, **51**, 443–450 (1983).
54. Bullier, J. and Kennedy, H. Projection of the lateral geniculate nucleus onto cortical area V2 in the macaque monkey. *Exp. Brain Res.*, **53**, 168–172 (1983).
55. Bunt, A. H., Hendrickson, A. E., Lund, J. S., Lund, R. D. and Fuchs, A. F. Monkey retinal ganglion cells: morphometric analysis and tracing of axonal projections, with a consideration of the peroxidase technique. *J. Comp. Neurol.*, **164**, 265–286 (1975).
56. Butters, N., Pandya, D., Sanders, K. and Dye, P. Behavioral deficits in monkeys after selective lesions within the middle third of sulcus principalis. *J. Comp. Physiol. Psychol.*, **76**, 8–14 (1971).
57. Büttner, U. and Fuchs, A. F. Influence of saccadic eye movements on unit activity in simian lateral geniculate and pregeniculate nuclei. *J. Neurophysiol.*, **36**, 127–141 (1973).
58. Büttner, U., Meienberg, O. and Schimmel-Pfennig, B. The effect of central retinal lesions on optokinetic nystagmus in the monkey. *Exp. Brain Res.*, **52**, 248–256 (1983).
59. Büttner, U. and Waespe, W. Vestibular nerve activity in the alert monkey during vestibular and optokinetic nystagmus. *Exp. Brain Res.*, **41**, 310–315 (1981).
60. Büttner, U. and Waespe, W. Purkinje cell activity in the primate flocculus during optokinetic stimulation, smooth pursuit eye movements and VOR-suppression. *Exp. Brain Res.*, **55**, 97–104 (1984).
61. Büttner-Ennever, J. A. and Akert, K. Medial rectus subgroups of the oculomotor nucleus and their abducens internuclear input in the monkey. *J. Comp. Neurol.*, **197**, 17–27 (1981).
62. Büttner-Ennever, J. A. and Büttner, U. A cell group associated with vertical eye movements in the rostral mesencephalic reticular formation of the monkey. *Brain Res.*, **151**, 31–47 (1978).
63. Büttner-Ennever, J. A. and Henn, V. An autoradiographic study of the pathways from the pontine reticular formation involved in horizontal eye movements. *Brain Res.*, **108**, 155–164 (1976).
64. Campos-Ortega, J. A., Haghow, W. R. and Clüver, R. F. A note on the problem of retinal projections to the inferior pulvinar nucleus of primates. *Brain Res.*, **22**, 126–130 (1970).
65. Cannon, S. C. and Robinson, D. A. The final common integrator is in the prepositus and vestibular nuclei. In: *Adaptive Processes in Visual and Oculomotor Systems*. Eds. Keller, E. L. and Zee D. S. pp. 307–311. New York: Pergamon (1986).
66. Carleton, S. C. and Carpenter, M. B. Afferent and efferent connections of the medial, inferior and lateral vestibular nuclei in the cat and monkey. *Brain Res.*, **278**, 29–51 (1983).
67. Carpenter, M. B. and Batton, R. R. Abducens internuclear neurones and their role in conjugate horizontal gaze. *J. Comp. Neurol.*, **189**, 191–209 (1980).
68. Carpenter, M. B. and Carleton, S. C. Comparison of vestibular and abducens internuclear projections to the medial rectus subdivision of the oculomotor nucleus in the monkey. *Brain Res.*, **274**, 144–149 (1983).
69. Carpenter, M. B. and Cowie, R. J. Connections and oculomotor projections of the superior vestibular nucleus and cell group 'y'. *Brain Res.*, **336**, 265–287 (1985).
70. Chan-Palay, V. *Cerebellar Dentate Nucleus. Organization, Cytology and Transmitters.* Berlin: Springer (1977).
71. Cheng, M. and Outerbridge, J. S. Optokinetic nystagmus during selective retinal stimulation. *Exp. Brain Res.*, **23**, 129–139 (1975).
72. Chubb, M. C. and Fuchs, A. F. Contribution of y group of vestibular nuclei and dentate nucleus of cerebellum to generation of vertical smooth eye movements. *J. Neurophysiol.*, **48**, 75–99 (1982).
73. Cibis, P. Zur Pathologie der Lokaladaptation. I. Mitteilung: Physiologische und klinische Untersuchungen zur quantitativen Analyse der örtlichen Umstimmungserscheinungen des Licht; und Farbensinnes unter besonderer Berücksichtigung hirnpathologischer Fälle. *Graefes Arch. Clin. Exp. Ophthalmol.*, **148**, 1–92 (1948).
74. Cibis, P. Zur Pathologie der Lokaladaptation. II. Mitteilung: Konstruktive Darstellung des Ablaufs der Erregungsvorgänge im normalen und erkrankten Sehorgan bei konstanter und phasischer Reizung umschriebener Sehfeldstellen. *Graefes Arch. Clin. Exp. Ophthalmol.*, **148**, 216–257 (1948).
75. Cohen, B. and Büttner-Ennever, J. A. Projections from the superior colliculus to a region of the central mesencephalic reticular formation (cMRF) associated with central saccadic eye movements. *Exp. Brain Res.*, **57**, 167–176 (1984).
76. Cohen, B., Büttner-Ennever, J. A., Waitzman, D. and Bender, M. B. Anatomical connections of a portion of the dorsolateral mesencephalic reticular formation of the monkey associated with horizontal saccadic eye movements. *Soc. Neurosci. Abstr.*, **7**, 776 (1981).
77. Cohen, B. and Komatsuzaki, A. Eye movements induced by stimulation of the pontine reticular formation: evidence for integration in oculomotor pathways. *Exp. Neurol.*, **36**, 101–117 (1972).
78. Cohen, B., Matsuo, V., Fradin, J. and Raphan, T. Horizontal saccades induced by stimulation of the central mesencephalic reticular formation. *Exp. Brain Res.*, **57**, 605–616 (1985).
79. Cohen, B., Matsuo, V. and Raphan, T. Quantitative analysis of the velocity characteristics of optokinetic nystagmus and optokinetic after-nystagmus. *J. Physiol. Lond.*, **270**, 321–344 (1977).
80. Collewijn, H. An analog model of the rabbit's optokinetic system. *Brain Res.*, **36**, 71–88 (1972).
81. Collewijn, H. and Tamminga, E. P. Human smooth and saccadic eye movements during voluntary pursuit of different target motions on different backgrounds. *J. Physiol. Lond.*, **351**, 217–250 (1984).
82. Collewijn, H., Van Der Mark, F. and Jansen, T. C. Precise recording of human eye movements. *Vision Res.*, **15**, 447–450 (1975).
83. Connolly, M. and Van Essen, D. The representation of the visual field in parvicellular and magnocellular layeres of the lateral geniculate nucleus in the macaque monkey. *J. Comp. Neurol.*, **226**, 544–464 (1984).
84. Cords, R. Zur Pathologie der Führungsbewegungen. *Graefes Arch. Clin. Exp. Ophthalmol.*, **123**, 173–218 (1930).
85. Cowey, A. and Perry, V. H. The projection of the fovea to the superior colliculus in rhesus monkeys. *Neuroscience*, **5**, 53–61 (1980).

86. Crandall, W. F. and Keller, E. L. Visual and oculomotor signals in nucleus reticularis tegmenti pontis in alert monkey. *J. Neurophysiol.*, **54**, 1326–1345 (1985).
87. Crane, H. D. and Steele, C. M. Accurate three-dimensional eye-tracker. *Appl. Optics*, **17**, 691–714 (1978).
88. Crawford, M. L. J. Central vision of man and macaque: cone and rod sensitivity. *Brain Res.*, **119**, 345–356 (1977).
89. Darian-Smith, I., Johnson, K. O. and Goodwin, A. W. Posterior parietal cortex: relations of unit activity to sensorimotor function. *Annu. Rev. Physiol.*, **41**, 141–157 (1979).
90. De Monasterio, F. M. Properties of concentrically organized X and Y ganglion cells of macaque retina. *J. Neurophysiol.*, **41**, 1394–1417 (1978).
91. De Monasterio, F. M. and Gouras, P. Functional properties of ganglion cells of the rhesus monkey retina. *J. Physiol. Lond.*, **251**, 167–195 (1975).
92. Denny-Brown, D. and Chambers, R. A. Physiological aspects of visual perception. I. Functional aspects of visual cortex. *Arch. Neurol.*, **33**, 219–227 (1976).
93. Derrington, A. M. and Lennie, P. Spatial and temporal contrast sensitivities of neurones in lateral geniculate nucleus of macaque. *J. Physiol. Lond.*, **357**, 219–240 (1984).
94. Desimone, R. and Gross, C. G. Visual areas in the temporal cortex of the macaque. *Brain Res.*, **178**, 363–380 (1979).
95. Deubel, H. Adaptivity of gain and direction in oblique saccades. In *Eye Movements: From Physiology to Cognition*. Eds. O'Regan, J. K. and Levy-Schoen, A. pp. 181–190. Amsterdam: North-Holland (1986).
96. De Valois, R. L., Morgan, H. C., Polson, M. C., Mead, W. R. and Hull, E. M. Psychophysical studies of monkey vision. I. Macaque luminosity and color vision tests. *Vision Res.*, **14**, 53–67 (1974).
97. Dichgans, J. and Jung, R. Oculomotor abnormalities due to cerebellar lesions. In *Basic Mechanisms of Ocular Motility and Their Clinical Implications*. Eds. Lennerstrand G. and Bach-Y-Rita, P. pp. 281–302. Oxford: Pergamon (1975).
98. Dineen, J. T. and Hendrickson, A. Overlap of retinal and prestriate cortical pathways in the primate pretectum. *Brain Res.*, **278**, 250–254 (1983).
99. Ditchburn, R. W. and Ginsborg, B. L. Involuntary eye movements during fixation. *J. Physiol. Lond.*, **119**, 1–17 (1953).
100. Dodge, R. Five types of eye movement in the horizontal meridian plane of the field of regard. *Am. J. Physiol.*, **8**, 307–329 (1903).
101. Donaldson, I. M. L. and Dixon, R. A. Excitation units in the lateral geniculate and contiguous nuclei of the cat by stretch of extrinsic ocular muscles. *Exp. Brain Res.*, **38**, 245–255 (1980).
102. Donaldson, I. M. L. and Long, A. C. Interactions between extraocular proprioceptive and visual signals in the superior colliculus of the cat. *J. Physiol. Lond.*, **298**, 85–110 (1980).
103. Doty, R. W., Wilson, P. D., Bartlett, J. R. and Pecci-Saavedra, J. Mesencephalic control of lateral geniculate nucleus in primates. I. Electrophysiology. *Exp. Brain Res.*, **18**, 189–203 (1973).
104. Dow, B. M. Functional classes of cells and their laminar distribution in monkey visual cortex. *J. Neurophysiol.*, **37**, 927–946 (1974).
105. Dreher, B., Fukada, Y. and Rodieck, R. W. Identification, classification and anatomical segregation of cells with x-like and y-like properties in the lateral geniculate nucleus of old-world primates. *J. Physiol. Lond.*, **258**, 433–452 (1976).
106. Drischel, H. Über den Frequenzgang der horizontalen Folgebewegungen des menschlichen Auges. *Pflügers Arch.*, **268**, 34 (1958).
107. Dubois, M. F. W. and Collewijn, H. Optokinetic reactions in man elicited by localized retinal motion stimuli. *Vision Res.*, **19**, 1105–1115 (1979).
108. Dubrovsky, B. O. and Barbas, H. Frontal projections of dorsal neck and extraocular muscles. *Exp. Neurol.*, **55**, 680–693 (1977).
109. Easter, S. S. Pursuit eye movements in goldfish. *Vision Res.*, **12**, 673–688 (1972).
110. Eckmiller, R. Hysteresis in the static characteristics of eye position coded neurones in the alert monkey. *Pflügers Arch.*, **350**, 249–258 (1974).
111. Eckmiller, R. Differences in the activity of eye-position coded neurones in the alert monkey during fixation and tracking movements. In *Basic Mechanisms of Ocular Motility and Their Clinical Implications*. eds. Lennerstrand, G. and Bach-Y-Rita, P. pp. 447–451. Oxford: Pergamon (1975).
112. Eckmiller, R. A model of the neural network controlling foveal pursuit eye movements. In *Progress in Oculomotor Research*. Eds. Fuchs, A. F. and Becker, W. pp. 541–550. Amsterdam: Elsevier (1981).
113. Eckmiller, R. Concerning the linear acceleration input to the neural oculomotor control system in primates. In: *Physiological and Pathological Aspects of Eye Movements*. Eds. Roucoux, A. and Crommelinck, M. pp. 131–137. Boston: Junk (1982).
114. Eckmiller, R. Neural control of foveal pursuit versus saccadic eye movements in primates – single unit data and models. *IEEE Trans. Syst. Man Cybern.*, **13**, 980–989 (1983).
115. Eckmiller, R. The transition between pre-motor eye velocity signals and oculomotor eye position signals in primate brain stem neurones during pursuit. In: *Adaptive Processes in Visual and Oculomotor Systems*. Eds. Keller, E. L. and Zee, D. S. pp. 301–305. New York: Pergamon (1986).
116. Eckmiller, R. Computational properties of a neural net with a triangular lattice structure and a travelling activity peak. In: *Proceedings International Conference of Systems, Man and Cybernetics Atlanta 1986*, pp. 633–637. Washington: McGregor & Werner.
116a. Eckmiller, R. Topological and dynamical aspects of a neural network model for generation of pursuit motor programs. In: *Proceedings Annual Conference of the Cognitive Science Society 8th 1986*. pp. 645–651. Hillsdale: Erlbaum.
117. Eckmiller, R. and Bauswein, E. Pre-motor neurones in monkey oculomotor system with intermediate phase values between those of eye velocity neurones and oculomotor motoneurones during pursuit. *Neurosci. Lett.* Suppl. 22, S256 (1985).
118. Eckmiller, R. and Bauswein, E. Smooth pursuit eye movements. In: *Oculomotor and Skeletalmotor System*. Eds. Freund, H.-J., Büttner, U., Cohen, B. and Noth, J. vol. 64, pp. 313–323. Amsterdam: Elsevier (1986). (Prog. Brain Res. Ser.)
119. Eckmiller, R. and Mackeben, M. Pursuit eye movements and their neural control in the monkey. *Pflügers Arch.*, **377**, 15–23 (1978).
120. Eckmiller, R. and Mackeben, M. Pre-motor unit activity in the monkey brain stem correlated with eye velocity during pursuit. *Brain Res.*, **184**, 210–214 (1980).
121. Eckmiller, R., Meisami, E. and Westheimer, G. Neuroanatomical status of monkeys showing functional compensation following neonatal cerebellar lesions. *Exp. Brain Res.*, **56**, 59–71 (1984).
122. Eckmiller, R. and Westheimer, G. Compensation of oculomotor deficits in monkeys with neonatal cerebellar ablations. *Exp. Brain Res.*, **49**, 315–326 (1983).
123. Erdmann, B. and Dodge, R. *Psychologische Untersuchungen über das Lesen*. Halle: Niemeyer (1898).
124. Evinger, C., Baker, R. and McCrea, R. A. Axon collaterals of cat medial rectus motoneurones, *Brain Res.*, **174**, 153–160 (1979).
125. Evinger, C. and Fuchs, A. F. Saccadic, smooth pursuit and optokinetic eye movements of the trained cat. *J. Physiol. Lond.* **285**, 209–229 (1978).

126. Fender, D. H. and Nye, P. W. An investigation of the mechanisms of eye movement control. *Kybernetik*, **1**, 81–88 (1961).
127. Fischer, B. and Boch, R. Peripheral attention versus central fixation: modulation of the visual activity of prelunate cortical cells of the rhesus monkey. *Brain Res.*, **345**, 111–123 (1985).
128. Fischer, B., Boch, R. and Bach, M. Stimulus versus eye movements: comparison of neural activity in the striate and prelunate visual cortex (A17 and A19) of trained rhesus monkey. *Exp. Brain Res.*, **43**, 69–77 (1981).
129. Fisken, R. A., Gary, L. J. and Powell, T. P. S. The intrinsic, association and commissural connections of area 17 of the visual cortex. *Phil. Trans. R. Soc. Lond. B. Biol. Sci.*, **272**, 487–536 (1975).
130. Fitzpatrick, D., Lund, J. S. and Blasdel, G. G. Intrinsic connections of macaque striate cortex: afferent and efferent connections of lamina 4C. *J. Neurosci.*, **5**, 3329–3349 (1985).
131. Fooden, J. Rhesus and crab-eating macaques: intergradation in Thailand. *Science Wash. DC*, **143**. 363–365 (1964).
132. Foster, K. H., Gaska, J. P., Nagler, M. and Pollen, D. A. Spatial and temporal frequency selectivity of neurones in visual cortical areas V1 and V2 of the macaque monkey. *J. Physiol. Lond.*, **365**, 331–363 (1985).
133. Frankfurter, A., Weber, J. T. and Harting, J. K. Brain stem projections to lobule VII of the posterior vermis in the squirrel monkey: as demonstrated by the retrograde axonal transport of triated horseradish peroxidase. *Brain Res.*, **124**, 135–139 (1977).
134. Frankfurter, A., Weber, J. T., Royce, G. J. Strominger, N. L. and Harting, J. K. An autoradiographic analysis of the tecto-olivary projection in primates. *Brain Res.*, **118**, 245–257 (1976).
135. Fuchs, A. F. and Kornhuber, H. H. Extraocular muscle afferents to the cerebellum of the cat. *J. Physiol. Lond.*, **200**, 713–722 (1969).
136. Fuchs, A. F. and Kimm, J. Unit activity in vestibular nucleus of the alert monkey during horizontal angular acceleration and eye movement. *J. Neurophysiol.*, **38**, 1140–1161 (1975).
137. Furuya, N. and Markham, C. H. Arborization of axons in oculomotor nucleus identified by vestibular stimulation and intra-axonal injection of horseradish peroxidase. *Exp. Brain Res.*, **43**, 289–303 (1981).
138. Fuster, J. M. Prefrontal cortex in motor control. In: *Handbook of Physiology. The Nervous System*. sect. 1, vol. 2, pt. 2, chap. 25, pp. 1149–1178. Bethesda, Md: Am. Physiol. Soc. (1981).
139. Galletti, C., Squatrito, S., Battaglini, P. P. and Maioli, M. G. 'Real motion' cells in the primary visual cortex of macaque monkeys. *Brain Res.*, **301**, 95–110 (1984).
140. Gardner, E. P. and Fuchs, A. F. Single-unit responses to natural vestibular stimuli and eye movements in deep cerebellar nuclei of the alert rhesus monkey. *J. Neurophysiol.*, **38**, 627–649 (1975).
141. Gattass, R. and Gross, C. G. Visual topography of striate projection zone (MT) in posterior superior temporal sulcus of the macaque. *J. Neurophysiol.*, **46**, 621–638 (1981).
142. Gattass, R., Gross, C. G. and Sandell, H. Visual topography of V2 in the macaque. *J. Comp. Neurol.*, **201**, pp. 519–539 (1981).
143. Gattass, R., Oswaldo-Cruz, E. and Sousa, A. P. B. Visual receptive fields of units in the pulvinar of cebus monkey. *Brain Res.*, **160**, 413–430 (1979).
144. Geisler, W. S. Adaptation, afterimages and cone saturation. *Vision Res.*, **18**, 279–289 (1978).
145. Gertz, H. Über die gleitende (langsame) Augenbewegung. *Z. Sinnesphysiol.*, **49**, 29–58 (1916).
146. Gilbert, C. D. Microcircuitry of the visual cortex. *Annu. Rev. Neurosci.*, **6**, 217–247 (1983).
147. Gilman, S., Bloedel, J. R. and Lechtenberg. R. *Disorders of the Cerebellum*. Philadelphia, PA: Davis (1981).
148. Giolli, R. A. and Tigges, J. The primary optic pathways and nuclei of primates. In: *The Primate Brain. Advances in Primatology*. Eds. Noback, C. and Montagna, W. vol. 1, pp. 29–54. New York: Appleton-Century-Crofts (1970).
149. Glickstein, M., May, J. G. and Mercier, B. E. Corticopontine projection in the macaque: the distribution of labeled cortical cells after large injections of horseradish peroxidase in the pontine nuclei. *J. Comp. Neurol.*, **235**, 343–359 (1985).
150. Goebel, H. H., Komatsuzaki, A., Bender, M. B. and Cohen, B. Lesions of the pontine tegmentum and conjugate gaze paralysis. *Arch. Neurol.*, **24**, 431–440 (1971).
151. Goldberg, M. E., Bruce, C. J., Ungerleider, L. and Mishkin, M. Role of the striate cortex in generation of smooth pursuit eye movements. *Ann. Neurol.*, **12**, 113 (1982).
152. Goldman, P. S. and Nauta, W. J. H. Autoradiographic demonstration of a projection from prefrontal association cortex to the superior colliculus in the rhesus monkey. *Brain Res.*, **116**, 145–149 (1976).
153. Goldstein, H. P. and Robinson, D. A. Hysteresis and slow drift in abducens unit activity. *J. Neurophysiol.*, **55**, 1044–1056 (1986).
154. Gouras, P. The effects of light-adaptation on rod and cone receptive field organization of monkey ganglion cells. *J. Physiol. Lond.*, **192**, 747–760 (1967).
155. Gouras, P. and Link, K. Rod and cone interaction in dark-adapted monkey ganglion cells. *J. Physiol. Lond.*, **184**, 499–510 (1966).
156. Graham, J., Lin, C.-S. and Kaas, J. H. Subcortical projections of six visual cortical areas in the owl monkey. *Aotus trivirgatus*. *J. Comp. Neurol.*, **187**, 557–580 (1979).
157. Graybiel, A. M. Anatomical pathways in the brain stem oculomotor system. In: *Eye Movements and Movement Perception Symposium Center for Visual Science 9th Rochester 1975*, pp. 37–38.
158. Graybiel, A. M. Direct and indirect preoculomotor pathways of the brainstem: an autoradiographic study of the pontine reticular formation in the cat. *J. Comp. Neurol.*, **175**, 37–78 (1977).
159. Greene, D. E. and Ward, F. E. Human eye tracking as a sequential input adaptive process. *Biol. Cybern*, **33**, 1–7 (1979).
160. Greene, T. and Jampel, R. Muscle spindles in the extraocular muscles of the macaque. *J. Comp. Neurol.*, **126**, 547–549 (1966).
161. Grüsser, O.-J. The effect of gaze motor signals and spatially directed attention on eye movements and visual perception. In: *Progress in Brain Research. Oculomotor and Skeletalmotor System*. Eds. Freund, H.-J., Büttner, U., Cohen, D. and Noth, J. vol. 64, pp. 391–404. Amsterdam: Elsevier (1986).
162. Haberich, F. J. and Lingelbach, B. Psychophysical measurements concerning the range of visual perception. *Pfluegers Arch.*, **386**, 141–146 (1980).
163. Haines, D. E. Cerebellar corticovestibular fibers of the posterior lobe in a prosimian primate, the lesser bushbaby (*Galago senegalensis*). *J. Comp. Neurol.*, **160**, 363–398 (1975).
164. Harting, J. K. Descending pathways from the superior colliculus: an autoradiographic analysis in the rhesus monkey (*Macaca mulatta*). *J. Comp. Neurol.*, **173**, 583–612 (1977).
165. Hendrickson, A., Wilson, M. E. and Toyne, M. J. The distribution of optic nerve fibers in *Macaca mulatta*. *Brain Res.*, **23**, 425–427 (1970).
166. Henn, V., Young, L. R. and Finley, C. Vestibular nucleus units in alert monkeys are also influenced by moving visual fields. *Brain Res.*, **71**, 144–149 (1974).
167. Hepp, K. and Henn, V. Spatio-temporal recoding of rapid eye movement signals in the monkey paramedian pontine reticular formation (PPRF). *Exp. Brain Res.*, **52**, 105–120 (1983).
168. Hering, E. Über Ermüdung und Erholung des Sehorgans. *Graefes Arch. Clin. Exp. Ophthalmol.*, **37**, 1–36 (1891).
169. Highstein, S. M., Karabelas, A., Baker, R. and McCrea, R. A.

Comparison of the morphology of physiologically identified abducens motor and internuclear neurones in the cat: a light microscopic study employing the intracellular injection of horseradish peroxidase. *J. Comp. Neurol.*, **208**, 369–381 (1982).
170. Hoffmann, K.-P. and Distler, C. The role of direction-selective cells in the nucleus of the optic tract of cat and monkey during optokinetic nystagmus. In: *Adaptive Processes in Visual and Oculomotor Systems*. Eds. Keller, E. L. and Zee, D. S. pp. 261–266. New York: Pergamon (1986).
171. Hubel, D. H. and Wiesel, T. N. Ferrier lecture. Functional architecture of macaque monkey visual cortex. *Proc. R. Soc. Lond. Ser. B Biol. Sci.*, **198**, 1–59 (1977).
172. Hughes, A. The topography of vision in mammals of contrasting life style: comparative optics and retinal organisation. In: *Handbook of Sensory Physiology. The Visual System in Vertebrates*. Ed. Crescitelli, F. vol. VII/5, pp. 613–756. Berlin: Springer-Verlag (1977).
173. Hutchins, B. and Weber, J. T. The pretectal complex of the monkey: a reinvestigation of the morphology and retinal terminations. *J. Comp. Neurol.*, **232**, 425–442 (1985).
174. Hyvärinen, J. *The Parietal Cortex of Monkey and Man*. Berlin: Springer (1982).
175. Hyvärinen, J. and Poranen, A. Function of the parietal associative area 7 as revealed from cellular discharges in alert monkeys. *Brain*, **97**, 673–692 (1974).
176. Itaya, S. K. and Van Hoesen, G. W. Retinal axons to the medial terminal nucleus of the accessory optic system in old-world monkeys. *Brain Res.*, **269**, 361–364 (1983).
177. Ito, M. *The Cerebellum and Neural Control*. New York: Raven (1984).
178. Jones, E. G. Functional subdivision and synaptic organization of the mammalian thalamus. *Int. Rev. Physiol.*, **25**, 173–245 (1981).
179. Kaplan, E. and Shapley, R. M. X and y cells in the lateral geniculate nucleus of macaque monkeys. *J. Physiol. Lond.*, **330**, 125–143 (1982).
180. Kase, M., Noda, H., Suzuki, D. A. and Miller, D. C., Target velocity signals of visual tracking in vermal purkinje cells of the monkey. *Science Wash. DC*, **205**, 717–720 (1979).
181. Kato, I., Harada, K., Hasegawa, T., Igarashi, T., Koike, Y. and Kawasaki, T. Role of the nucleus of the optic tract in monkeys in relation to optokinetic nystagmus. *Brain Res.*, **364**, 12–22 (1986).
182. Kawano, K., Sasaki, M. and Yamashita, M. Vestibular input to visual tracking neurones in the posterior parietal association cortex of the monkey. *Neurosci. Lett.*, **17**, 55–60 (1980).
183. Kawano, K., Sasaki, M. and Yamashita, M. Response properties of neurones in posterior parietal cortex of monkey during visual-vestibular stimulation. I. Visual tracking neurones. *J. Neurophysiol.*, **51**, 340–351 (1984).
184. Keller, E. L. Participation of medial pontine reticular formation in eye movement generation in monkey. *J. Neurophysiol.*, **37**, 316–332 (1974).
185. Keller, E. L. Behavior of horizontal semicircular canal afferents in alert monkey during vestibular and optokinetic stimulation. *Exp. Brain Res.*, **24**, 459–471 (1976).
186. Keller, E. L. and Crandall, W. F. Neuronal responses to optokinetic stimuli in pontine nuclei of behaving monkey. *J. Neurophysiol.*, **49**, 169–187 (1983).
187. Keller, E. L. and Daniels P. D. Oculomotor related interaction of vestibular and visual stimulation in vestibular nucleus cells in alert monkey. *Exp. Neurol.*, **46**, 187–198 (1975).
188. Keller, E. L. and Kamath, B. Y. Characteristics of head rotation and eye movement-related neurones in alert monkey vestibular nucleus. *Brain Res.*, **100**, 182–187 (1975).
189. Keller, E. L. and Robinson, D. A. Absence of a stretch reflex in extraocular muscles of the monkey. *J. Neurophysiol.*, **34**, 908–919 (1971).
190. Kennedy, H. and Bullier, J. A double-labeling investigation of the afferent connectivity to cortical areas V1 and V2 of the macaque monkey. *J. Neurosci.*, **5**, 2815–2830 (1985).
191. Kern, E. Der Bereich der Unterschiedsempfindlichkeit des Auges bei festgehaltenem Adaptationszustand. *Z. Biol.*, **105**, 237–245 (1953).
192. Kimura, M. and Maekawa, K. Activity of flocculus Purkinje cells during passive eye movements. *J. Neurophysiol.*, **46**, 1004–1017 (1981).
193. King, W. M., Fuchs, A. F. and Magnin, M. Vertical eye movement-related responses of neurones in midbrain near interstitial nucleus of Cajal. *J. Neurophysiol.*, **46**, 549–562 (1981).
194. Kohlrausch, A. Tagessehen, Dämmersehen, Adaptation. In: *Handbuch der normalen und pathologischen Physiologie*. Eds. Bethe, A., Bergmann, G. V., Embden, G. and Ellinger, A. vol. 12/2, pp. 1499–1594. Berlin: Springer (1931).
195. Komatsu, H. and Suzuki, H. Projections from the functional subdivisions of the frontal eye field to the superior colliculus in the monkey. *Brain Res.*, **327**, 324–327 (1985).
196. Komatsuzaki, A., Alpert, J., Harris, H. E. and Cohen, B. Effects of mesencephalic reticular formation lesions on optokinetic nystagmus. *Exp. Neurol.*, **34**, 522–534 (1972)
197. Kommerell, G. and Klein, U. Über die visuelle Regelung der Okulomotorik: die optomotorische Wirkung exzentrischer Nachbilder, *Vision Res.*, **11**, 905–920 (1971).
198. Kömpf, D., Pasik, T., Pasik, P. and Bender, M. B. Downward gaze in monkeys: stimulation and lesion studies. *Brain*, **102**, 527–558 (1979).
199. Kubota, K., Iwamoto, T. and Suzuki, H. Visuokinetic activities of primate prefrontal neurones during delayed-response performance. *J. Neurophysiol.*, **37**, 1197–1212 (1974).
199a. Künzle, H. and Akert, K. Efferent connections of cortical area 8 (frontal eye field) in *Macaca fascicularis*. A reinvestigation using the autoradiographic technique. *J. Comp. Neurol.*, **173**, 147–164 (1977).
200. Künzle, H., Akert, K. and Wurtz, R. H. Projection of area 8 (frontal eye field) to superior colliculus in the monkey. An autoradiographic study. *Brain Res.*, **117**, 487–492 (1976).
201. Kupfer, C., Chumbley, L. and Downer, J. de C. Quantitative histology of optic nerve, optic tract and lateral geniculate nucleus of man. *J. Anat.*, **101**, 393–401 (1967).
202. Landolt, E. *Festschrift zur Feier des 70. Geburtstages von Hermann von Helmholtz*. Stuttgart (1891).
203. Lang, W., Henn, V. and Hepp, K. Gaze palsies after selective pontine lesions in monkeys. In: *Physiological and Pathological Aspects of Eye Movements*. Eds. Roucoux, A. and Crommelinck, M. pp. 209–218. Boston, Mass: Junk (1982).
204. Langer, T. P. Basal interstitial nucleus of the cerebellum: cerebellar nucleus related to the flocculus. *J. Comp. Neurol.*, **235**, 38–47 (1985).
205. Langer, T., Fuchs, A. F., Chubb, M. C., Scudder, C. A. and Lisberger, S. G. Floccular efferents in the rhesus macaque as revealed by autoradiography and horseradish peroxidase. *J. Comp. Neurol.*, **235**, 26–37 (1985).
206. Langer, T., Fuchs, A. F., Scudder, C. A. and Chubb, M. C. Afferents to the flocculus of the cerebellum in the rhesus macaque as revealed by retrograde transport of horseradish peroxidase. *J. Comp. Neurol.*, **235**, 1–25 (1985).
207. Langer, T., Kaneko, C. R. S., Scudder, C. A. and Fuchs, A. F. Afferents to the abducens nucleus in the monkey and cat. *J. Comp. Neurol.*, **245**, 375–400 (1986).
208. Le Grand, Y. Spectral luminosity. In: *Handbook of Sensory Physiology. Visual Psychophysics*. Eds. Jameson, D. and Hur-

vich, L. M. vol. VII/4. pp. 413–433. Berlin: Springer (1972).
209. Leichnetz, G. R. An anterogradely-labeled prefrontal cortico-oculomotor pathway in the monkey demonstrated with HRP gel and TMB neurohistochemistry. *Brain Res.*, **198**, 440–445 (1980).
210. Leichnetz, G. R. The prefrontal cortico-oculomotor trajectories in the monkey. *J. Neurol. Sci.*, **49**, 387–396 (1981).
211. Leichnetz, G. R. Connections between the frontal eye field and pretectum in the monkey: an anterograde/retrograde study using HRP gel and TMB neurohistochemistry. *J. Comp. Neurol.*, **207**, 394–402 (1982).
212. Leichnetz, G. R., Spencer, R. F., Hardy, S. G. P. and Astruc, J. The prefrontal corticotectal projection in the monkey: an anterograde and retrograde horseradish peroxidase study. *Neuroscience*, **6**, 1023–1041 (1981).
213. Leichnetz, G. R., Smith, D. J. and Spencer, R. F. Cortical projections to the paramedian tegmental and basilar pons in the monkey. *J. Comp. Neurol.*, **228**, 388–408 (1984).
214. Leichnetz, G. R., Spencer R. F. and Smith, D. J. Cortical projections to nuclei adjacent to the oculomotor complex in the medial dienmesencephalic tegmentum in the monkey. *J. Comp. Neurol.*, **288**, 359–387 (1984).
215. Leight, R. J. and Zee, D. S. Smooth pursuit and ocular stabilization. In: *The Neurology of Eye Movements*. Eds. Leigh, R. J. and Zee D. S. pp. 69–88. Philadelphia, Pa: Davis (1983).
216. Lennie P. Parallel visual pathways: a review. *Vision Res.*, **20**, 561–594 (1980).
217. Lin, H. and Giolli, R. A. Accessory optic system of rhesus monkey. *Exp. Neurol.*, **63**, 163–176 (1979).
218. Lisberger, S. G., Evinger, C., Johanson, G. W., and Fuchs, A. F. Relationship between eye acceleration and retinal image velocity during foveal smooth pursuit in man and monkey. *J. Neurophysiol.*, **46**, 229–249 (1981).
219. Lisberger, S. G. and Fuchs, A. F. Role of primate flocculus during rapid behavioral modification of vestibuloocular reflex. I. Purkinje cell activity during visually guided horizontal smooth-pursuit eye movements and passive head rotation. *J. Neurophysiol.*, **41**, 733–763 (1978).
220. Lisberger, S. G. and Fuchs, A. F. Role of primate flocculus during rapid behavioral modification of vestibuloccular reflex. II. Mossy fiber firing patterns during horizontal head rotation and eye movement. *J. Neurophysiol.*, **41**, 764–777 (1978).
221. Lisberger, S. G. and Westbrook, L. E. Properties of visual inputs that initiate horizontal smooth pursuit eye movements in monkeys. *J. Neurosci.*, **5**, 1662–1673 (1985).
222. Lopez-Barneo, J., Darlot, C., Berthoz, A. and Baker, R. Neuronal activity in prepositus nucleus correlated with eye movement in the alert cat. *J. Neurophysiol.*, **47**, 329–352 (1982).
223. Lund, J. S. and Boothe, R. G. Interlaminar connections and pyramidal neurone organization in the visual cortex, area 17 of the macaque monkey. *J. Comp. Neurol.*, **159**, 305–334 (1975).
224. Lund, J. S., Hendrickson, A. E., Ogren, M. P. and Tobin, E. A. Anatomical organization of primate visual cortex area V2. *J. Comp. Neurol.*, **202**, 19–45 (1981).
225. Luschei, E. S. and Fuchs, A. F. Activity of brain stem neurones during eye movements of alert monkeys. *J. Neurophysiol.*, **35**, 445–461 (1972).
226. Lynch, J. C. The functional organization of posterior parietal association cortex. *Behav. Brain Sci.*, **3**, 485–534 (1980).
227. Lynch, J. C. The interaction of prefrontal and parieto-occipital cortex in the control of purposive eye movements in rhesus monkeys (abstr.). In: *IUPS Satellite Meeting. Developments in Oculomotor Research, Gleneden Beach, Oregon, 1986*, p. 37.
228. Lynch, J. C., Graybiel, A. M. and Lobeck, L. J. The differential projection of two cytoarchitectonic subregions of the inferior parietal lobule of macaque upon the deep layers of the superior colliculus. *J. Comp. Neurol.*, **235**, 241–254 (1985).
229. Lynch, J. C. and McLaren, J. W. The contribution of parieto-occipital association cortex to the control of slow eye movements. In: *Functional Basis of Ocular Motility Disorders*. Eds. Lennerstrand, G., Zee, D. S. and Keller, E. L. pp. 501–510. New York: Pergamon (1982).
230. Lynch, J. C., Mountcastle, V. B., Talbot, W. H. and Yin, T. C. T. Parietal lobe mechanisms for directed visual attention. *J. Neurophysiol.*, **40**, 362–389 (1977).
230a. Magnin, M. and Baleydier, C. Possible non cerebellar visuo-vestibular pathways in primates (abstr.). In: *IUPS Satellite Meeting. Developments in Oculomotor Research, Gleneden Beach, Oregon*, p. 44 (1986).
231. Magnin, M. and Fuchs, A. F. Discharge properties of neurones in the monkey thalamus tested with angular acceleration, eye movement and visual stimuli. *Exp. Brain Res.*, **28**, 293–299 (1977).
232. Maguire, W. M. and Baizer, J. S. Visuotopic organization of the prelunate gyrus in rhesus monkey. *J. Neurosci.*, **4**, 1690–1704 (1984).
233. Maier, A., Deasntis, M. and Eldred, E. The occurrence of muscle spindles in extraocular muscles of various vertebrates. *J. Morphol.*, **143**, 397–408 (1974).
234. Maioli, M. G., Squatrito, S., Galletti, C., Battaglini, P. P. and Sanseverino, E. R. Cortico-cortical connections from the visual region of the superior temporal sulcus to frontal eye field in the macaque. *Brain Res.*, **265**, 294–299 (1983).
235. Manni, E. and Bortolami, R. Proprioception in eye muscles. In: *Functional Basis of Ocular Motility Disorders*. Eds. Lennerstrand, G., Zee, D. S. and Keller, E. L. pp. 55–64. Oxford: Pergamon (1982).
236. Marg, E. Neurophysiology of the accessory optic system. In: *Handbook of Sensory Physiology. Visual Centers in the Brain*. Ed. Jung, R. vol. VII/3, part B, chapt. 15, pp. 103–111. Berlin: Springer (1973).
237. Marrocco, R. T. Saccades induced by stimulation of the frontal eye fields: interaction with voluntary and reflexive eye movements. *Brain Res.*, **146**, 23–34 (1978).
238. Marrocco, R. T. and McClurkin, J. W. Binocular interaction in the lateral geniculate nucleus of the monkey. *Brain Res.*, **168**, 633–637 (1979).
239. Marrocco, R. T., McClurkin, J. W. and Young, R. A. Spatial summation and conduction latency classification of cells of the lateral geniculate nucleus of macaques. *J. Neurosci.*, **2**, 1275–1291 (1982).
240. Martins, A. J., Kowler, E. and Palmer, C. Smooth pursuit of small-amplitude sinusoidal motion. *J. Opt. Soc. Am. Ser. A*, **2**, 234–242 (1985).
241. Maunsell, J. H. R. and Van Essen, D. C. The connections of the middle temporal visual area (MT) and their relationship to a cortical hierarchy in the macaque monkey. *J. Neurosci.*, **3**, 2563–2586 (1983).
242. Maunsell, J. H. R. and Van Essen, D. C. Functional properties of neurones in middle temporal visual area of the macaque monkey. I. Selectivity for stimulus direction, speed, and orientation. *J. Neurophysiol.*, **49**, 1127–1147 (1983).
243. Maunsell, J. H. R. and Van Essen, D. C. Functional properties of neurones in middle temporal visual area of the macaque monkey. II. Binocular interactions and sensitivity to binocular disparity. *J. Neurophysiol.*, **49**, 1148–1167 (1983).
243a. May, J. G. and Andersen, R. A. Different patterns of cortico-pontine projections from separate cortical fields within the inferior parietal lobe and dorsal prelunate gyrus of the macaque. *Exp. Brain Res.*, **63**, 265–278 (1986).
244. May, J. G., Keller, E. L. and Crandall, W. F. Changes in eye velocity during smooth pursuit tracking induced by microsti-

mulation in the dorsolateral pontine nucleus of the macaque. *Soc. Neurosci. Abstr.*, **11**, 79 (1985).
245. Mayr, R. Functional morphology of the eye muscles. In: *Disorders of Ocular Motility. Neurophysiological and Clinical Aspects.* ed. Kommerell, G. pp. 1–16. Munich: Bergmann (1978).
246. McCrea, R. A. and Baker, R. Anatomical connections of the nucleus prepositus of the cat. *J. Comp. Neurol.*, **237**, 377–407 (1985).
247. McCrea, R. A., Baker, R. and Delgado-Garcia, J. Afferent and efferent organization of the prepositus hypoglossi nucleus. In: *Progress in Brain Research. Reflex Control of Posture and Movement.* Eds. Granit, R. and Pompeiano, O. pp. 653–665. Amsterdam: Elsevier (1979). vol. 50.
248. McElligott, J. G. and Keller, E. L. Cerebellar vermis involvement in monkey saccadic eye movements: microstimulation. *Exp. Neurol.*, **86**, 543–558 (1984).
249. Miles, F. A. Single unit firing patterns in the vestibular nuclei related to voluntary eye movements and passive body rotation in conscious monkeys. *Brain Res.*, **71**, 215–224 (1974).
250. Miles, F. A. and Fuller, J. H. Visual tracking and the primate flocculus. *Science Wash. DC*, **189**, 1000–1002 (1975).
251. Miles, F. A., Fuller, J. H., Braitman, D. J. and Dow, B. M. Long-term adaptive changes in primate vestibuloocular reflex. III. Electrophysiological observations in flocculus of normal monkeys. *J. Neurophysiol.*, **43**, 1437–1476 (1980).
252. Miller, M., Pasik, P. and Pasik, T. Extrageniculostriate vision in the monkey. VII. Contrast sensitivity functions. *J. Neurophysiol.*, **43**, 1510–1526 (1980).
253. Milner, B. Some cognitive effects of frontal-lobe lesions in man. *Phil. Trans. R. Soc. Lond. Ser. B Biol. Sci.*, **298**, 211–226 (1982).
254. Mizuno, N., Uchida, K., Nomura, S., Nakamura, Y., Sugimoto, T. and Uemura-Sumi, M. Extrageniculate projections to the visual cortex in the macaque monkey: an HRP study. *Brain Res.*, **212**, 454–459 (1981).
255. Mohler, C. W., Goldberg, M. E. and Wurtz, R. H. Visual receptive fields of frontal eye field neurones. *Brain Res.*, **61**, 385–389 (1973).
256. Moran, J. and Desimone, R. Selective attention gates visual processing in the extrastriate cortex. *Science Wash. DC*, **229**, 782–784 (1985).
257. Mountcastle, V. B., Lynch, J. C. Georgopoulos, A., Sakata, H. and Acuna, C. Posterior parietal association cortex of the monkey: command functions for operations within extrapersonal space. *J. Neurophysiol.*, **38**, 871–908 (1975).
258. Mustari, M. J., Fuchs, A. F. and Wallman, J. Visual and oculomotor response properties of single units in the pretectum of the behaving rhesus macaque. *Soc. Neurosci. Abstr.*, **11**, part 1, 78, (1985).
259. Mustari, M. J., Fuchs, A. F. and Wallman, J. The physiological response properties of single pontine units related to smooth pursuit in the trained monkey. In: *Adaptive Processes in Visual and Oculomotor Systems.* Eds. Keller, E. L. and Zee, D. S. pp. 253–260. New York: Pergamon (1986).
260. Nakagawa, S. and Tanaka, S. Retinal projections to the pulvinar nucleus of the macaque monkey: a re-investigation using autoradiography. *Exp. Brain Res.*, **57**, 151–157 (1984).
261. Newsome, W. T., Dürsteler, M. R. and Wurtz, R. H. The middle temporal visual area and the control of smooth pursuit eye movements. In: *Adaptive Processes in the Visual and Oculomotor Systems.* Eds. Keller, E. and Zee, E. pp. 223–230. Oxford: Pergamon (1986).
262. Newsome, W. T. and Wurtz, R. H. Identification of architectonic zones containing visual tracking cells in the superior temporal sulcus (STS) of macaque monkeys (abstr.). *Invest. Ophthal. Vis. Sci.*, **22**, Suppl.: 238 (1982).
263. Newsome, W. T., Wurtz, R. H., Dursteler, M. R. and Mikami, A. Deficits in visual motion processing following ibotenic acid lesions in the middle temporal visual area of the macaque monkey. *J. Neurosci.*, **5**, 825–840 (1985).
264. Noda, H. Visual mossy fiber inputs to the flocculus of the monkey. *Ann. NY Acad. Sci.* **374**, 465–475 (1981).
265. Noda, H. and Suzuki, D. A. The role of the flocculus of the monkey in fixation and smooth-pursuit eye movements. *J. Physiol. Lond.*, **294**, 335–348 (1979).
266. Noda, H., and Suzuki, D. A. Processing of eye movement signals in the flocculus of the monkey. *J. Physiol. Lond.*, **294**, 349–364 (1979).
267. Noda, H. and Warabi T. Eye position signals in the flocculus of the monkey during smooth-pursuit eye movements. *J. Physiol. Lond.*, **324**, 187–202 (1982).
268. Normann, R. A. and Werblin, F. S. Control of retinal sensitivity. I. Light and dark adaptation of vertebrate rods and cones. *J. Gen. Physiol.*, **63**, 37–61 (1974).
269. Noton, D. and Stark, L. Scanpaths in saccadic eye movements while viewing and recognizing patterns. *Vision Res.*, **11**, 929–942 (1971).
270. Ogden, T. E. and receptor mosaic of *Aotes trivirgatus*: distribution of rods and cones. *J. Comp. Neurol.*, **163**, 193–202 (1975).
271. Ogren, M. P. and Hendrickson, A. E. The structural organization of the inferior and lateral subdivisions of the macaca monkey pulvinar. *J. Comp. Neurol.*, **188**, 147–178 (1979).
272. Ogren, M. P. and Hendrickson, A. E. The morphology and distribution of striate cortex terminals in the inferior and lateral subdivisions of the macada monkey pulvinar. *J. Comp. Neurol.*, **188**, 179–200 (1979).
273. Öhrwall, H. Die Bewegungen des Auges während des Fixierens. *Scand. Arch. Physiol.*, **27**, 304–314 (1912).
274. Orban, G. A., Kennedy, H. and Bullier J. Velocity sensitivity and direction selectivity of neurones in areas V1 and V2 of the monkey: influence of eccentricity. *J. Neurophysiol.*, **56**, 462–480 (1986).
275. Orem, J., Schlag-Rey, M. and Schlag, J. Unilateral visual neglect and thalamic intralaminar lesions in the cat. *Exp. Neurol.*, **40**, 784–797 (1973).
276. Osterberg, G. Topography of the layer of rods and cones in the human retina. *Acta Ophthalmol.* suppl. 6, 3–103 (1935).
277. Pandya, D. N. and Vignolo, L. A. Intra- and interhemispheric projections of the precentral, premotor and arcuate areas in the rhesus monkey. *Brain Res.*, **26**, 217–233 (1971).
278. Pasik, P. and Pasik, T. Oculomotor functions in monkeys with lesions of the cerebellum and the superior coliculus. In: *The Oculomotor System.* Ed. Bender, M. B. chapt. 3. pp. 40–80. New York: Harper & Row (1971).
279. Pasik, T. and Pasik, P. The visual world of monkeys deprived of striate cortex: effective stimulus parameters and their importance of the accessory optic system. *Vision Res.*, **11**, suppl. 3, 419–435 (1971).
280. Perry, V. H. and Cowey, A. Retinal ganglion cells that project to the superior colliculus and pretectum in the macaque monkey. *Neuroscience*, **12**, 1125–1137 (1984).
281. Perry, V. H. and Cowey, A. The ganglion cell and cone distributions in the monkey's retina: implications for central magnification factors. *Vision Res.*, **25**, 1795–1810 (1985).
282. Perry, V. H., Oehler, R. and Cowey, A. Retinal ganglion cells that project to the dorsal lateral geniculate nucleus in the macaque monkey. *Neuroscience*, **12**, 1101–1123 (1984).
283. Perryman, K. M., Lindsley, D. F. and Lindsley, D. B. Pulvinar neurone responses to spontaneous and trained eye movements and to light flashes in squirrel monkeys. *Electroencephalogr. Clin. Neurophysiol.*, **49**, 152–161 (1980).
284. Petersen, S. E., Robinson, D. L. and Keys, W. Pulvinar nuclei of the behaving rhesus monkey: visual responses and their

modulation. *J. Neurophysiol.*, **54**, 867–886 (1985).
285. Petrides, M. and Pandya, D. N. Projections to the frontal cortex from the posterior parietal region in the rhesus monkey. *J. Comp. Neurol.*, **228**, 105–116 (1984).
286. Poggio, G. F., Baker, F. H., Mansfield, R. J. W. Sillito, A. and Grigg, P. Spatial and chromatic properties of neurones subserving foveal and parafoveal vision in rhesus monkey. *Brain Res.*, **100**, 25–59 (1975).
287. Poggio, G. F., Doty, R. W. Jr and Talbot, W. H. Foveal striate cortex of behaving monkey: single-neurone responses to square-wave gratings during fixation of gaze. *J. Neurophysiol.*, **40**, 1369–1391 (1977).
288. Poggio, G. F. and Fischer, B. Binocular interaction and depth sensitivity in striate and prestriate cortex of behaving rhesus monkey. *J. Neurophysiol.*, **40**, 1392–1405 (1977).
289. Poggio, G. F. and Talbot, W. H. Mechanisms of static and dynamic stereopsis in foveal cortex of the rhesus monkey. *J. Physiol. Lond.*, **315**, 469–492 (1981).
290. Pola, J. and Wyatt, H. J. Target position and velocity: the stimuli for smooth pursuit eye movements. *Vision Res.*, **20**, 523–534 (1980).
291. Pola, J. and Wyatt, H. J. Active and passive smooth eye movements: effects of stimulus size and location. *Vision Res.*, **25**, 1063–1076 (1985).
292. Pollack, J. G. and Hickey, T. L. The distribution of retinocollicular axon terminals in rhesus monkey. *J. Comp. Neurol.*, **185**, 587–602 (1979).
293. Polyak, S. *The Retina*. Chicago: University of Chicago Press (1941).
294. Polyak, S. *The Vertebrate Visual System*. Chicago: University of Chicago Press (1957).
295. Porter, J. D. Brainstem terminations of extraocular muscle primary afferent neurones in the monkey. *J. Comp. Neurol.*, **247**, 133–143 (1986).
296. Porter, J. D., Guthrie, B. L. and Sparks, D. L. Innervation of monkey extraocular muscles: localization of sensory and motor neurones by retrograde transport of horseradish peroxidase. *J. Comp. Neurol.*, **218**, 208–219 (1983).
297. Precht, W. Cerebellar influences on eye movements. In: *Basic Mechanisms of Ocular Motility and their Clinical Implications*. Eds. Lennerstrand, G. and Bach-Y-Rita, P. pp. 261–280. Oxford: Pergamon (1975).
298. Puckett, J. and Steinman, R. M. Tracking eye movements with and without saccadic correction. *Vision Res.*, **9**, 695–703 (1969).
299. Pumphrey, R. J. The theory of the fovea. *J. Exp. Biol.*, **25**, 299–312 (1948).,
300. Ramoa, A. S., Freeman, R. D. and Macy, A. Comparison of response properties of cells in the cat's visual cortex at high and low luminance levels. *J. Neurophysiol.*, **54**, 61–72 (1985).
301. Raphan, T. and Cohen, B. Brainstem mechanisms for rapid and slow eye movements. *Ann. Rev. Physiol.*, **40**, 527–552 (1978).
302. Rashbass, C. The relationship between saccadic and smooth tracking eye movements. *J. Physiol. Lond.* **159**, 326–338 (1961).
303. Raybourn, M. S. and Keller, E. L. Colliculoreticular organization in primate oculomotor system. *J. Neurophysiol.*, **40**, 861–878 (1977).
304. Ritchie, L. Effects of cerebellar lesions on saccadic eye movements. *J. Neurophysiol.*, **39**, 1246–1256 (1976).
305. Robinson, D. A. A method of measuring eye movement using a scleral search coil in a magnetic field. *IEEE Trans Biomed. Eng.*, **10**, 137–145 (1963).
306. Robinson, D. A. The mechanics of human smooth pursuit eye movement. *J. Physiol. Lond.*, **180**, 569–591 (1965).
307. Robinson, D. A. The physiology of pursuit eye movements. In: *Eye Movements and Psychological Processes*. Eds. Monty, R. A., Senders, J. W. part I.2. pp. 19–31. New York: Wiley (1976).
308. Robinson, D. A. The functional behaviour of the peripheral oculomotor apparatus: a review. In: *Disorders of Ocular Motility. Neurophysiological and Clinical Aspects*. Ed. Kommerell, pp. 43–61. Munich: Bergmann (1978).
309. Robinson, D. A. Control of eye movements. In: *Handbook of Physiology. The Nervous System*. vol. II, sect. 1, part 2, chapt, 28, pp. 1275–1320. Bethesda: Am. Physiol. Soc. (1981).
310. Robinson, D. A. and Fuchs, A. F. Eye movements evoked by stimulation of frontal eye fields. *J. Neurophysiol.*, **32**, 637–648 (1969).
311. Robinson, D. L., Goldberg, M. E. and Stanton, G. B. Parietal association cortex in the primate: sensory mechanisms and behavioral modulations. *J. Neurophysiol.*, **41**, 910–932 (1978).
312. Rockland, K. S. A reticular pattern of intrinsic connections in priamte area V2 (area 18). *J. Comp. Neurol.*, **235**, 467–478 (1985).
313. Rockland, K. S. and Pandya, D. N. Laminar origins and terminations of cortical connections of the occipital lobe in the rhesus monkey. *Brain Res.*, **179**, 3–20 (1979).
314. Rodieck, R. W. Visual pathways. *Ann Rev. Neurosci.*, **2**, 193–225 (1979).
315. Rodieck, R. W. and Dreher, B. Visual suppression from non-dominant eye in the lateral geniculate nucleus: a comparison of cat and monkey. *Exp. Brain Res.*, **35**, 465–477 (1979).
316. Rohen, J. W. and Castenholz, A. Über die Zentralisation der Retina bei Primaten. *Folia Primatol.*, **5**, 92–147 (1967).
317. Rolls, E. T. and Cowey, A. Topography of the retina and striate cortex and its relationship to visual acuity in rhesus monkeys and squirrel monkeys. *Exp. Brain Res.*, **10**, 298–310 (1970).
318. Ron, S. and Robinson, D. A. Eye movements evoked by cerebellar stimulation in the alert monkey. *J. Neurophysiol.*, **36**, 1004–1022 (1973).
319. Roth, W. C. Demonstration von Kranken mit Ophthalmoplegie. *Neurol. Zentralbl.*, **20**, 921–923 (1901).
320. Saito, H., Yukie, M., Tanaka, K., Hikosaka, K., Fukada, Y. and Iwai, E. Integration of direction signals of image motion in the superior temporal sulcus of the macaque monkey. *J. Neurosci.*, **6**, 145–157 (1986).
321. Sakai, M. Prefrontal unit activity during visually guided level pressing reaction in the monkey. *Brain Res.*, **81**, 297–309 (1974).
322. Sakata, H., Shibutani, H. and Kawano, K. Spatial properties of visual fixation neurones in posterior parietal association cortex of the monkey. *J. Neurophysiol.*, **43**, 1654–1672 (1980).
323. Sakata, H., Shibutani, H. and Kawano, K. Functional properties of visual tracking neurones in posterior parietal association cortex of the monkey. *J. Neurophysiol.*, **49**, 1364–1380 (1983).
324. Sakmann, B. and Creutzfeldt, O. D. Scotopic and mesopic light adaptation in the cat's retina. *Pfluegers Arch.*, **313**, 168–185 (1969).
325. Schein, S. J., Marrocco, R. T. and De Monasterio, F. M. Is there a high concentration of color-selective cells in areas V4 of monkey visual cortex? *J. Neurophysiol.*, **47**, 193–213 (1982).
326. Schiller, P. H., Finlay, B. L. and Volman, S. F. Quantitative studies of single-cell properties in monkey striate cortex. I. Spatiotemporal organization of receptive fields. *J. Neurophysiol.*, **39**, 1288–1319 (1976).
327. Schiller, P. H. and Malpeli, J. G. Properties and tectal projections of monkey retinal ganglion cells. *J. Neurophysiol.*, **40**, 428–445 (1977).
328. Schiller, P. H. and Malpeli, J. G. Functional specificity of lateral geniculate nucleus laminae of the rhesus monkey. *J. Neurophysiol.*, **41**, 788–797 (1978).
329. Schiller, P. H., True, S. D. and Conway, J. L. Deficits in eye movements following frontal eye-field and superior colliculus ablations. *J. Neurophysiol.*, **44**, 1175–1189 (1980).

330. Schlag, J. and Schlag-Rey, M. Visuomotor functions of central thalamus in monkey. II. Unit activity related to visual events, targeting, and fixation. *J. Neurophysiol.*, 51, 1175–1195 (1984).
331. Schlag, J. and Schlag-Rey, M. Unit activity related to spontaneous saccades in frontal dorsomedial cortex of monkey. *Exp. Brain Res.*, 58, 208–211 (1985).
332. Schlag, J. and Schlag-Rey, M. Role of the central thalamus in gaze control. In: *Progress in Brain Research. Oculomotor and Skeletalmotor System.* Eds. Freund, H.-J., Büttner, U., Cohen, B., and Noth, J. vol. 64. pp. 191–202. Amsterdam: Elsevier (1986).
333. Schlag-Rey, M. and Schlag, J. Visuomotor functions of central thalamus in monkey. I. Unit activity related to spontaneous eye movements. *J. Neurophysiol.*, 51, 1149–1174 (1984).
334. Schnyder, H., Reisine, H., Hepp, K. and Henn, V. Frontal eye field projection to the paramedian pontine reticular formation traced with wheat germ agglutinin in the monkey. *Brain Res.*, 329, 151–160 (1985).
335. Scollo-Lavizzari, G. and Akert, K. Cortical area 8 and its thalamic projection in *Macaca mulatta*. *J. Comp. Neurol.*, 121, 259–269 (1963).
336. Seltzer, B. and Pandya, D. N. Converging visual and somatic sensory cortical input to the intraparietal sulcus of the rhesus monkey. *Brain Res.*, 192, 339–351 (1980).
337. Skavenski, A. A. and Robinson, D. A. Role of abducens neurones in vestibuloocular reflex. *J. Neurophysiol.*, 36, 724–738 (1973).
338. Skavenski, A. A., Robinson, D. A. Steinman, R. M. and Timberlake, G. T. Miniature eye movements of fixation in rhesus monkey. *Vision Res.*, 15, 1269–1273 (1975).
339. Sparks, D. L. and Sides, J. P. Brain stem unit activity related to horizontal eye movements occurring during visual tracking. *Brain Res.*, 77, 320–325 (1974).
340. Spencer, R. F., Evinger, C. and Baker, R. Electron microscopic observations of axon collateral synaptic endings of cat oculomotor motoneurones stained by intracellular injection of horseradish peroxidase. *Brain Res.*, 234, 423–429 (1982).
341. Standage, G. P. and Benevento, L. A. The organization of connections between the pulvinar and visual area MT in the macaque monkey. *Brain Res.*, 262, 288–294 (1983).
342. Stanton, G. B. Afferents to oculomotor nuclei from area 'y' in *Macaca mulatta*: an anterograde degeneration study. *J. Comp. Neurol.* 192, 377–385 (1980).
343. Steiger, H.-J. and Büttner-Ennever, J. A. Oculomotor nucleus afferents in the monkey demonstrated with horseradish peroxidase. *Brain Res.*, 160, 1–15 (1979).
344. Steinbach, M. J. Pursuing the perceptual rather than the retinal stimulus. *Vision Res.*, 16, 1371–1376 (1976).
345. Steinberg, R. H. Rod–cone interaction in S-potentials from the cat retina. *Vision Res.*, 9, 1331–1344 (1969).
346. Steinman, R. M., Haddad, G. M., Skavenski, A. A. and Wyman, D. Miniature eye movement. The pattern of saccades made by man during maintained fixation may be refined but useless motor habit. *Science Wash. DC*, 181, 810–819 (1973).
347. Stone, J., Leicester, J. and Sherman, S. M. The naso-temporal division of the monkey's retina. *J. Comp. Neurol.*, 150, 333–348 (1973).
348. Sünderhauf, A. Untersuchungen über die Regelung der Augenbewegungen. *Klin. Monatsbl. Augenheilkd.*, 136, 837–852 (1960).
349. Suzuki, H. and Azuma, M. Prefrontal neuronal activity during gazing at a light spot in the monkey. *Brain Res.*, 126, 497–508
350. Suzuki, D. A. and Keller, E. L. Vestibular signals in the posterior vermis of alert monkey cerebellum. *Exp. Brain Res.*, 47, 145–147 (1982).
351. Suzuki, D. A. and Keller, E. L. Visual signals in the dorsolateral pontine nucleus of the alert monkey: their relationship to smooth-pursuit eye movements. *Exp. Brain Res.*, 53, 473–478 (1984).
352. Suzuki, D. A., Noda, H. and Kase, M. Visual and pursuit eye movement-related activity in posterior vermis of monkey cerebellum. *J. Neurophysiol.*, 46, 1120–1139 (1981).
353. Swadlow, H. A. Efferent systems of primary visual cortex: a review of structure and function. *Brain Res. Rev.*, 6, 1–24 (1983).
354. Tanaka, K., Hikosaka, K., Saito, H., Yukie, M., Fukada, Y. and Iwai, E. Analysis of local and wide-field movements in the superior temporal visual areas of the macaque monkey. *J. Neurosci.*, 6, 134–144 (1986).
355. Ter Braak, J. W. G. Untersuchungen über optokinetischen Nystagmus. *Arch. Neerl. Physiol.*, 21, 309–376 (1936).
356. Tomita, T. Electrical response of single photoreceptors. *Proc. IEEE*, 56, 1015–1023 (1968).
357. Tomlinson, R. D. and Robinson, D. A. Signals in vestibular nucleus mediating vertical eye movements in the monkey. *J. Neurophysiol.*, 51, 1121–1136 (1984).
358. Tootell, R. B. H., Silverman, M. S., De Valois, R. L. and Jacobs, G. H. Functional organization of the second cortical visual area in primates. *Science Wash. DC*, 220, 737–739 (1983).
359. Trincker, D., Sieber, J. and Bartual, J. Schwingungsanalyse der vestibulär, optokinetisch und durch elektrische Reizung ausgelösten Augenbewegungen beim Menschen. 1. Mitteilung: Stetige Augenbewegungen: Frequenzgänge und Ortskurven. *Kybernetik*, 1, 21–28 (1961).
360. Tschermark, A. *Physiologische Optik.* Berlin: Springer (1929).
361. Tusa, R. J., Zee, D. S. and Herdman S. J. Recovery of oculomotor function in monkeys with large unilateral cortical lesions. In: *Adaptive Processes in Visual and Oculomotor Systems.* Eds. Keller, E. and Zee, D. pp. 209–216. Oxford: Pergamon.
362. Ungerleider, L. G. and Christensen, C. A. Pulvinar lesions in monkeys produce abnormal eye movements during visual discrimination training. *Brain Res.*, 136, 189–196 (1977).
363. Ungerleider, L. G., Desimone, R., Galkin, T. W. and Mishkin, M. Subcortical projections of area MT in the macaque. *J. Comp. Neurol.*, 223, 368–386 (1984).
364. Ungerleider, L. G., Galkin, T. W. and Mishkin, M. Visuotopic organization of projections from striate cortex to inferior and lateral pulvinar in rhesus monkey. *J. Comp. Neurol.*, 217, 137–157 (1983).
365. Van Essen, D. C. Visual areas of the mammalian cerebral cortex. *Annu. Rev. Neurosci.*, 2, 227–263 (1979).
366. Van Essen, D. C., Maunsell, J. H. R. and Bixby, J. L. The middle temporal visual area in the macaque: myeloarchitecture, connections, functional properties and topographic organization. *J. Comp. Neurol.*, 199, 293–326 (1981).
367. von Kries, J. Die Gesichtempfindungen. In: *Handbuch der Physiologie des Menschen.* Ed. Nagel, W. vol. 3, chapt. 3, pp. 109–282. Braunschweig: Vieweg (1904).
368. von Kries, J. Zusätze von v. Kries. In: *Handbuch der Physiologischen Optik. 3. Band: Die Lehre von den Gesichtswahrnehmungen.* Ed. von Helmholtz, H. pp. 226–233. Hamburg: Voss (1910).
369. Voss, H. Beiträge zur mikroskopischen Anatomie der Augenmuskeln des Menschen. (Faserdicke, Muskelspindeln, Ringbinden.) *Anat. Anz.*, 104, 345–355 (1957).
370. Waespe, W. and Cohen, B. Flocculectomy and unit activity in the vestibular nuclei during visual-vestibular interactions. *Exp. Brain Res.*, 51, 23–35 (1983).
371. Waespe, W. and Henn, V. Vestibular nuclei activity during optokinetic after-nystagmus (OKAN) in the alert monkey. *Exp. Brain Res.*, 30, 323–330 (1977).
372. Waespe, W., Rudinger, D. and Wolfensberger, M. Purkinje cell

activity in the flocculus of vestibular neurectomized and normal monkeys during optokinetic nystagmus (OKN) and smooth pursuit eye movements. *Exp. Brain. Res.,* **60**, 243–262 (1985).
373. Walls, G. L. The evolutionary history of eye movements. *Vision Res.,* **2**, 69–80 (1962).
374. Watson, R. T. and Heilman, K. M. Thalamic neglect. *Neurology,* **29**, 690–694 (1974).
375. Weber, J. T. and Harting, J. K. The efferent projections of the pretectal complex: an autoradiographic and horseradish peroxidase analysis. *Brain Res.,* **194**, 1–28 (1980).
376. Weber, J. T. and Yin, T. C. T. Subcortical projections of the inferior parietal cortex (area 7) in the stump-tailed monkey. *J. Comp. Neurol.,* **224**, 206–230 (1984).
377. Wernicke, C. Herderkrankung des unteren Scheitelläppchens. *Arch. Psychiatr.,* **20**, 243–275 (1888).
378. Wertheim, T. Über die indirekte Sehschärfe. *Z. Psychol.,* **7**, 172–187 (1894).
379. Westheimer, G. Eye movement responses to a horizontally moving visual stimulus. *Am. Med. Assoc. Arch. Ophthalmol.,* **52**, 932–941 (1954).
380. Westheimer, G. and Blair, S. M. Unit activity in accessory optic system in alert monkeys. *Invest. Ophthalmol.,* **13**, 533–534 (1974).
381. Westheimer, G. and Blair, S. M. Functional organization of primate oculomotor system revealed by cerebellectomy. *Exp. Brain. Res.,* **21**, 463–472 (1974).
382. Wheeless, L. L., Cohen, G. H. and Boynton, R. M. Luminance as a parameter of the eye-movement control system. *J. Opt. Soc. Am.,* **57**, 394–400 (1967).
383. Wiesel, T. N. and Hubel, D. H. Spatial and chromatic interactions in the lateral geniculate body of the rhesus monkey. *J. Neurophysiol.,* **29**, 1115–1156 (1966).
384. Wilson, M. E. and Cragg, B. G. Projections from the lateral geniculate nucleus in the cat and monkey. *J. Anat.,* **101**, 677–692 (1967).
385. Wilson, V. J. and Melvill Jones, G. *Mammalian Vestibular Physiology.* New York: Plenum (1979).
386. Winterson, B. J. and Steinman, R. M. The effect of luminance on human smooth pursuit of perifoveal and foveal targets. *Vision Res.,* **18**, 1165–1172 (1978).
387. Wolin, L. R. and Massopust, L. C. Jr. Characteristics of the ocular fundus in primates. *J. Anat.,* **101**, 693–699 (1967).
388. Woolsey, C. N., Settlage, P. H., Meyer, D. R., Spencer, W., Hamy, T, P. and Travis, A. M. Patterns of localization in the precentral and 'supplementary' motor area and their relation to the concept of a premotor area. *Res. Publ. Assoc. Res. Nerv. Ment. Dis,* **30**, 238–364 (1952).
389. Wurtz, R. H. Visual receptive fields of striate cortex neurones in awake monkeys. *J. Neurophysiol.,* **32**, 727–742 (1969).
390. Wurtz, R. H. and Albano, J. E. Visual-motor function of the primate superior colliculus. *Annu. Rev. Neurosci.* **3**, 189–226. (1980).
391. Wurtz, R. and Goldberg, M. E. Activity of superior colliculus in behaving monkey. III. Cells discharging before eye movements. *J. Neurophysiol.,* **35**, 575–586 (1972).
392. Wurtz, R. H., Richmond, B. J. and Newsome, W. T. Modulation of cortical visual processing by attention, perception, and movement. In: *Dynamic Aspects of Neocortical Function.* Ed. Edelman, G. M., Gall, W. E. and Maxwell, C. W. pp. 195–217. New York: Wiley (1984).
393. Wyatt, H. J. and Pola, J. Smooth pursuit eye movements under open-loop and closed-loop conditions. *Vision Res.,* **23**, 1121–1131 (1983).
394. Yarbus, A. L. Eye movements during perception of moving objects. In: *Eye Movements and Vision.* Ed. Yarbus, A. L. pp. 159–170. New York: Plenum (1967).
395. Yasui, S. and Young L. R. On the predictive control of foveal eye tracking and slow phases of optokinetic and vestibular nystagmus. *J. Physiol. Lond.,* **347**, 17–33 (1984).
396. Yeterian, E. H. and Pandya, D. N. Corticothalamic connections of the posterior parietal cortex in the rhesus monkey. *J. Comp. Neurol.,* **237**, 408–426 (1985).
397. Yoon, M. Influence of adaptation level on response pattern and sensitivity of ganglion cells in the cat's retina. *J. Physiol. Lond.,* **221**, 93–104 (1972).
398. Young, L. R. Pursuit eye tracking movements. In: *The Control of Eye Movements.* Ed. Bach-Y-Rita, P. and Collins, C. C. pp. 429–443. London: Academic (1971).
399. Young L. R. Pursuit eye movement – what is being pursued? In: *Control of Gaze by Brain Stem Neurones.* Ed. Baker, R. and Berthoz, A. pp. 29–36. Amsterdam: Elsevier (1977).
400. Zee, D. S., Yamazaki, A., Butler, P. H. and Gücer, G. Effects of ablation of flocculus and paraflocculus on eye movements in primate. *J. Neurophysiol.,* **46**, 878–899 (1981).
401. Zee, D. S., Yee, R. D., Cogan, D. G., Robinson, D. A. and Engel, W. K. Ocular motor abnormalities in hereditary cerebellar ataxia. *Brain,* **99**, 207–234 (1976).
402. Zeki, S. M. Functional organization of a visual area in the posterior bank of the superior temporal sulcus of the rhesus monkey. *J. Physiol. Lond.,* **236**, 549–573 (1974).
403. Zeki, S. M. The cortical projections of foveal striate cortex in the rhesus monkey. *J. Physiol. Lond.,* **277**, 227–244 (1978).
404. Zeki, S. Colour coding in the cerebral cortex: the reaction of cells in monkey visual cortex to wavelengths and colours. *Neuroscience,* **9**, 741–765 (1983).

10 Patterns of Function and Evolution in the Arthropod Optic Lobe

D. Osorio

> *These animals [the insects] have an extraordinarily complex and differentiated nervous system, with an intricacy of construction which extends to the realms of the ultra-microscopic. In comparing the visual or brain-like ganglia from a bee or a dragonfly with those from a fish or an amphibian one is extraordinarily surprised. The quality of the psychic machine does not increase with the zoological hierarchy; on the contrary it has to be acknowledged that in fishes and amphibians the nervous centres have undergone an unexpected simplification. Certainly the grey matter has grown considerably in bulk; but when its structure is compared with that from the Apidae or Libellulidae [bees and dragonflies], it appears to us exceedingly coarse, rough and rudimentary. It is as if we are attempting to equate the qualities of a great wall clock with those of a miniature watch, a marvel of detail, delicacy and precision. As always, the genius of life, building its astonishing artifices, shines in the small far more than in the large.*   Cajal and Sanchez, 1915, translation.

Introduction

The three ganglia of the optic lobe are responsible for retinotopic visual processing beneath the compound eyes of insects and crustaceans. The anatomy of the optic lobe is well known; its orderly structure of repeated columns divided into many tangential layers has been admired for nearly a century (Cajal and Sanchez, 1915; Hanström, 1928; Strausfeld and Nässel, 1981). These anatomical studies give a picture of visual processing characterized by a divergence of receptor output to many different classes of small-field cells followed by a convergence upon wide-field cells. The optic lobe contains over half of the cells in the insect brain, indicating the cost of low-level visual processing (Strausfeld, 1976). Yet the general anatomy of the visual system is shared by arthropods with widely differing lifestyles and phylogenies, such as flies and crabs (Figs 10.1 and 10.7), hinting that a common plan of neural processing may underlie their retinotopic vision.

This chapter gives an overview of the function and evolution of insect and crustacean visual processing. The first part, 'General Anatomy', outlines the principal stages of retinotopic visual processing, while the second, 'Evolution', looks at the phylogeny of these structures. Retinotopic ganglia occupy much of the brain in arthropods and in vertebrates, but we have no clear idea what has been the driving force behind the investment in this neurone-intensive stage of visual processing when the small-field cells in these ganglia do not themselves project directly to muscles, but their outputs are integrated by wide-field cells that arise at several points in the visual pathway. It is interesting to ask which particular responses of wide-field cells require prior processing by small-field cells and so have been the mainspring for the evolution of retinotopic processing, and which responses could be obtained directly at a single synapse from, say, a photoreceptor. Can a class of synaptic processes be identified that has to be performed locally before integration of the visual signal by wide-field cells? Many studies over the past 40 years have focused on mechanisms of directional selectivity, and we look at the sites of wide-field directionally selective cells in insects to see what insights the description of these mechanisms may give into the evolutionary origins of the optic lobe.

The second part tries to combine our knowledge of anatomy and physiology with an appreciation of phylogeny, and looks at how visual processing has evolved over the past 600 million years. Arthropods are a highly diverse group, and there has been considerable argument as to whether they are monophyletic. Some authors have seen differences between insects, crustaceans and arachnids which appear to preclude the possibility that a jointed exoskeleton has been acquired only once in evolution (Manton, 1977). Others see similarities which could not have arisen by convergence (Boudreaux, 1979). In the light of these varied opinions, what can we say about the evolution of arthropod (or more specifically insect and crustacean) compound eyes? Do the common features of insect and crustacean visual systems arise because there is only one way to implement visual processing beneath a compound eye, or are they a reflection of phylogenetic conservatism in a nervous system that has been able to accommodate a wide variety of lifestyles without major change? Can we trace the evolution of particular neural faculties by comparing modern species with simple visual systems to those with more complex vision, or have examples of ancestral arthropod visual systems vanished?

Organization of Visual Processing

We now describe the retinotopic visual pathway and visual processing, outlining first the overall anatomy and then the physiology at various levels in this anatomical framework. Most of this section refers to the fly (Diptera) and, later, to the locust. However, many common features are shared by insects and malacostracan crustaceans, the two groups whose evolutionary relationship is discussed on pp 214–219. As well as generalizing from the fly's eye, the description here tends to imply that the retina and visual ganglia are uniform across the visual field, whereas in fact regional specializations are well known. For example, the frontal visual field of the male *Musca* eye has several features which appear to be adaptations for pursuit of small targets (Hardie, 1985, 1986; Strausfeld, 1989), and in many insects the dorsal rim of the eye is specialized for the analysis of sky polarization patterns (Rossel and Wehner, 1986; Rossel, 1989).

General Anatomy

The retina of the compound eye is divided into ommatidia, one for each of the several thousand facet lenses. The ommatidium contains eight coaxial photoreceptors and various accessory cells (the fly's 'neural superposition' eye differs slightly from this common type of organization, but this is not important here (see Meinertzhagen, this volume, chapter 17; Kirschfeld, 1967; Hardie, 1986). There are two anatomical receptor types, short visual fibres (SVFs) and long visual fibres (LVFs) (Fig. 10.1). The fly's six SVFs subserve a variety of behaviours, including motion detection and the landing response, but less is known about the roles of the two LVFs, although they must contribute to colour vision since the SVFs share a common spectral sensitivity (Heisenberg and Buchner, 1977; Tinbergen and Abeln, 1983: Hardie, 1985, 1986).

Receptors directed to each point in space project to a specific column (or cartridge) of cells in the first two visual ganglia, imposing a strictly retinotopic organization on the optic lobe (Fig. 10.1; Horridge and Meinertzhagen, 1970). Just as the ommatidium is the basic unit of the compound eye so the column of cells is the basic unit in the optic lobe. The SVFs end in the first optic ganglion, the lamina, whereas the LVFs pass through the lamina to end in the second optic ganglion, the medulla (Fig. 10.1a). The third ganglion, the lobula, is also retinotopic and columnar but has a coarser structure. Beyond the lobula, columnar organization and retinotopic projection appear to be absent, and higher visual centres are designated here simply as 'the brain' (Braitenberg, 1972; Strausfeld and Nässel, 1981; Strausfeld, 1989).

Retinotopic Ganglia and the Divergence of Visual Pathways

The three ganglia in the optic lobe, the lamina, the medulla and the lobula, are easily recognizable because they are separated by two antero-posterior chiasms (Figs 10.1 and 10.7). Each neural column in these ganglia is divided into tangential layers as different sets of neurites arborize at different depths. The medulla and the lobula are divided into distinct proximal and distal halves. In some holometabolous insects (Coleoptera, Diptera, Lepidoptera) the main projection from the medulla enters between the two parts of the lobula. The posterior half is then called the lobula plate (Fig. 10.7(b,d)). However, cell classes and connectivity patterns seem to be unaffected by this rerouting of the fibres (Strausfeld, 1976; 1989).

In vertebrates and in arthropods retinotopic visual processing proceeds first by dividing information among many different cell types with small receptive fields, and subsequently by integration of the outputs of these small-field cells (DeYoe and Van Essen, 1988; Strausfeld, 1989; Horridge, this volume, chapter 12). This strategy is reflected in the anatomical organization of the fly's optic lobe. Each lamina cartridge contains six afferent cells of which three, called the large monopolar cells (LMCs), receive direct inputs from the receptors, while the others are driven either by the LMCs or by amacrine cells (Shaw, 1984; Fig. 10.8; see also 'Lamina' below). These six lamina cells, together with the coaxial LVFs (and hence the signals from all of the receptors directed to a given point in space), reach a single column in the medulla where each cell type ends at a specific depth. The medulla is much more complex than the lamina and in the fly *Drosophila* a set of over 40 distinct cell types forms each medulla column (Strausfeld, 1976; Fischbach and Dittrich, 1989). From the medulla some cells project directly to the brain (proto- and deuterocerebrums) or to the contralateral optic lobe, but most cross the second optic chiasm to end in the lobula. Here the columnar structure is coarser than that imposed on the medulla by the retinal array, and in the fly there are a quarter as many lobula columns as there are points in visual space (Braitenberg, 1972). Both halves of the lobula receive direct inputs from the medulla, and they are also connected by columnar cells (Strausfeld, 1976).

In addition to their columnar elements, all three ganglia contain wide-field (tangential) cells and amacrine cells. There are only one or two tangential cells in the lamina but many more in the medulla and lobula. These cells arise at various levels in the optic lobe between the distal medulla and the lobular and form distinct 'serpentine layers', the largest of which divides the two halves of the medulla (Figs 10.1, 10.7 and 10.9a; Strausfeld and Nässel, 1981; Strausfeld, 1989). Some of the wider-field cells are probably efferent, providing feedback to the early stages of visual processing, while others are afferent.

Anatomy shows how distinct pathways may be formed as information diverges from the receptors and LMCs among many small-field cells in the medulla, and in medium or wide-field cells in the lobula and beyond (Strausfeld, 1976, 1989). The best candidate 'pathway' running through the optic lobe may subserve the optomotor response (Buchner, 1984; Strausfeld, 1989). One class of lamina cell (L2), which is driven by the SVFs, ends in a layer of the distal medulla near to the inputs to two classes of columnar cell, Tm1 and Tm-sub. These medulla columnar cells end in a layer of the proximal medulla where two further types of medulla cell, T4 and T5, arborize. T4 and T5 project in turn to the lobula plate, which contains wide-field directionally selective cells. The implication that this is a 'motion-detection pathway' is supported by the finding, made by 2-deoxyglucose labelling, that the tangential layers of the medulla containing the T4 and T5 inputs are active during stimulation with motion, but are quiescent under static flicker (Buchner *et al.*, 1984). At present the responses of the LMCs and the lobula plate cells are known, but none of the medulla columnar cells have been recorded. According to Strausfeld (1976, p 148), the set of interneurones L2, Tm1, T4 and T5 are found in all insect species studied (fly, bee, dragonfly, etc.) and they end in layers of the lobular near to cells analogous to the giant cell of the fly lobula plate (see pp 210–211). The reason why a multi-step pathway, incorporating three sets of interneurones, is used here might be a key to understanding the origins and function of the medulla, a question to which we return on pp 210–211.

Physiology

Recordings have been made from few of the cells in the optic lobe. Most papers describe the responses either of the three LMCs, or of various wide-field cells in the lobula. Best known in the latter group are the directionally selective cells of the fly lobula plate (Hausen and Egelhaaf, 1989), and the locust's lobula giant movement detector (LGMD) and its postsynaptic element, the descending contralateral movement detector (DCMD) (Rowell *et al.*, 1977; Rind, 1987). The function of the lamina cells and the way in which their outputs are used, together with speculation about local processing on the input pathways to the wide-field cells, inspires much current research on the optic lobe, but supporting data are sparse owing to the shortage of recordings from small-field cells in the medulla (see Rowell *et al.*, 1977; Coombe *et al.*, 1989; Franceschini *et al.*, 1989; Horridge, this volume, chapter 11).

LMC Function: Optimization of Information Capacity, an Essential First Step in Visual Processing?

The LMCs are non-spiking cells which give a phasic hyperpolarization to a step intensity increment and are subject to lateral inhibition (Fig. 10.2; Laughlin and Osorio, 1989). Identifiable on the basis of their morphology, and especially by the positions of their outputs in the medulla (Fig. 10.8), the fly's three LMCs are physiologically similar (but see Laughlin and Osorio, 1989). The LMCs are the sole afferent cell class to receive direct inputs from the SVFs (Shaw, 1984), and because this single physiological class contributes to a wide variety of behaviours its responses are probably not tailored to any particular behaviour (Heisenberg and Buchner, 1977; Tinbergen and Abeln, 1983; but cf. Coome *et al.*, 1989). Rather, it has been suggested that the LMCs encode all of the information in the receptors with equal reliability, stripping redundancy from the retinal signal and thereby maximizing the information capacity of the channel from the receptors to the medulla (Srinivasan *et al.*, 1982;

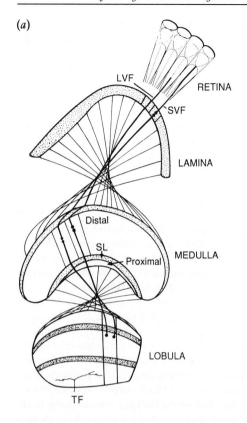

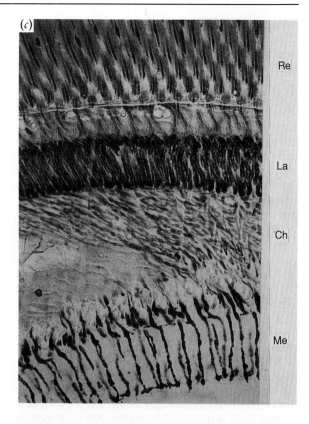

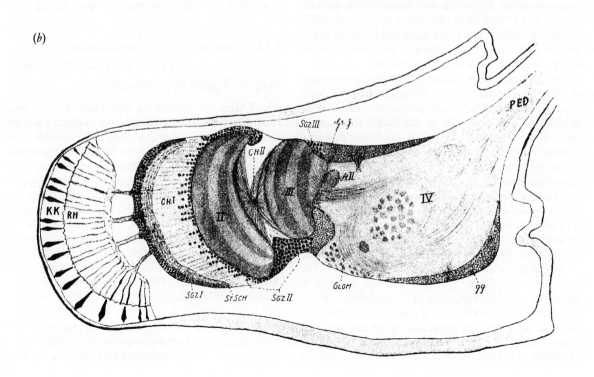

Laughlin, 1988, 1989).

The responses of LMCs differ from those of receptors in two main respects. Firstly, they are more responsive to contrast fluctuations; the receptor to LMC synaptic gain is about six (Laughlin *et al.*, 1987). Secondly, LMCs are more phasic than photoreceptors and are subject to lateral inhibition, which is absent in the retina (Fig. 10.2(c); Dubs, 1982; Laughlin and Osorio, 1989). Lateral and self inhibitions remove DC components in the signal (in space and in time) and, more generally, decorrelate the response at any one instant or location from its neighbours. This decorrelation in principle maximizes the information transmitted by the LMC array (save for the triplication of cells in each cartridge) by producing a random signal. That is, in natural conditions, the response voltage at any one instant in any one cell cannot (in principle) be predicted by looking at the response history either of that cell or of its neighbours, and so no redundant (repeated) information is transmitted to the medulla (Barlow, 1961; Srinivasan *et al.*, 1982; Laughlin, 1988, 1989).

If they are adapted to strip redundancy from the signal, the LMCs can have two possible functions: firstly, to amplify this signal to 'protect' it from noise introduced during later processing, and secondly to distribute the retinal signal to several sites in the medulla (Laughlin, 1988, 1989). The former function is consistent with the observation that the synapse from receptors to LMCs is particularly large (Nicol and Meinertzhagen, 1982), so that many synaptic vesicles can be released by the receptors, and the contribution of synaptic (quantization) noise to the LMC response will be small (Laughlin *et al.*, 1987).

Despite its appeal as a quantitative approach to neural function, maximization of information capacity should perhaps be treated circumspectly. The visual system did not evolve to transmit signals with minimal corruption, but to abstract and analyse specific features in the image. Selection might minimize reaction time or enhance object recognition, but it will not act to provide a 'perfect picture' of the world to a homunculus in the brain. None the less, the maximization of information capacity is a 'null hypothesis' against which any more specific hypothesis of function can be tested. For example, it has recently been suggested that (when light-adapted) LMCs are optimized to detect moving edges (Srinivasan *et al.*, 1990). This implies that features other than moving edges could be shown to be relatively poorly encoded (i.e. with a lower signal-to-noise ratio).

Given that the LMCs have a very general role in vision, it is not surprising that cells with similar properties have evolved elsewhere, and, equally, they can be expected to retain their properties even as visual behaviour changes during evolution. For example, the LMCs were compared to the vertebrate bipolar cells by the anatomists Cajal and Sanchez (1915), an analogy which has since been found to extend to physiological responses (Attwell, 1986; Laughlin, 1988, 1989). The laminas of different arthropods such as the fly and the crayfish are also closely comparable; whether this similarity is due to convergent evolution or conserved function is discussed on pp 215–219.

The Medulla: Multiple Representations of the Retinal Image

LMCs (and the LVFs) drive the many separate cell types in the medulla directly and others indirectly via other medulla or lamina cells (see Fig. 11.8). Records from small-field columnar cells in the locust's medulla reveal a range of properties, but most fall into two main classes which, following Honegger (1978, 1980), and hence the classification of cat retinal ganglion cells (Levick and Thibos, 1983), are called either 'sustaining' or 'transient' (Fig. 10.3; Osorio, 1987a,b). (It is impossible to say what proportion of the 40 or so columnar cells have been recorded, and it is quite likely that a few cell types with large neurites make up the bulk of records.) 'Sustaining' does not mean tonic and sustaining cells may be either phasic or tonic, depending on the background light intensity. (It is difficult to compare the insect 'sustaining' cells with those of the crustaceans which are said to be tonic (Wiersma *et al.*, 1982; Kirk *et al.*, 1982).) Sustaining cells

Fig. 10.1 *(a) Diagram showing the main features of the arthropod optic lobe, based on the crab optic lobe illustrated in Fig. 10.7(c). Note how receptors from adjacent facets project to separate columns in the lamina and medulla, but converge to a single lobular column. LVF: long visual fibre; SVF: short visual fibre; SL: serpentine layer (indicated by stippled regions in the medulla and the lobula but not the lamina); TF: tangential fibre (diagrammatic illustration of a fibre which can be seen in Fig. 10.7. (b) Diagram of the crab (Pachygrapsus) eyestalk (From Hanström, 1924). KK: crystalline cone (optical apparatus); RH: receptors. I: Lamina. II: Medulla: III: Lobula; IV: protocerebrum/medulla interna. CH: chiasm. In II and III paler areas correspond to serpentine layers, where tangential cells arborize. Note that in many malacostracans the eyestalk is distant from the main part of the brain and so the tract between the eye and the brain forms a long optic nerve; in insects this tract is very short, concealing an analogy between the most proximal ganglion in the crustacean eyestalk, which does not have a columnar structure, and was formerly called the medulla interna, and the lateral protocerebrum of insects (see Strausfeld and Nässel, 1981). (c) The fly (Calliphora) visual system stained with an antibody to histamine, which labels seven of the eight photoreceptors (Courtesy of M. Holmqvist). The six short visual fibres (SVFs) terminate in the lamina and a single class of long visual fibres (LVF), R8, can clearly be seen projecting to the medulla. The other LVF, R7, appears to be GABAergic (Datum et al., 1986; Nässel et al., 1988). Re: retina; La: lamina, Ch: first chiasm; Me: medulla.*

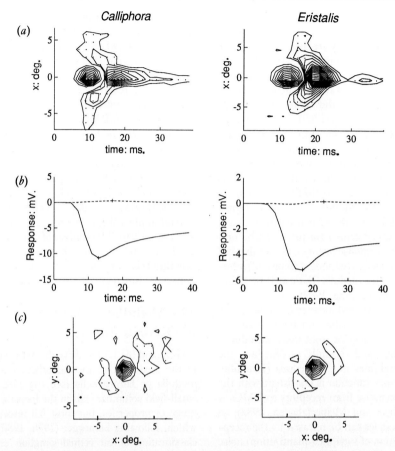

Fig. 10.2 *Linear components of response to spatiotemporal stimulation for light adapted ($500 \, cd \, m^{-2}$) large monopolar cells of two families of fly,* Calliphora *(left) and* Eristalis *(Syrphidae) (right) (Courtesy of A. C. James). The least-squares estimates of the linear spatiotemporal kernels were estimated from intracellular responses to stimulation with a two-dimensional chequerboard pattern of white noise. (a) The horizontal marginal kernels, obtained by integrating over the vertical dimension. These describe the sensitivity over horizontal position and time to a stimulus extending uniformly in the vertical dimension. Cross-sections along time are similar to impulse responses; cross-sections across space are similar to receptive field profiles. Contour levels are relative to the extreme value, and are at 10% intervals for the region of primary, negative values (unstippled), and at 2% intervals for the regions of antagonistic, positive values (stippled). Note the presence of strong temporal antagonism, the weaker spatial antagonism. Note also that the kernels are not space–time separable, with the time course differing on and off axis. This contrasts with a difference-of-Gaussian spatial receptive field model of spatial organization (Marr and Hildreth, 1980). (b) The step responses corresponding to the spatiotemporal kernels, obtained by taking the cumulative integral along time. This stimulates the response in millivolts to a step of contrast one over an area of one square degree. Shown are the curves for a stimulus on-axis, and at the position of maximum spatial antagonism. The crosses indicate the extreme values of response. Note again the greater size of temporal antagonism, and the later extreme value for the off-axis position. (c) The extreme value of the simulated step-responses, plotted as a function of two-dimensional stimulus position. Contour levels are relative to overall extreme value, as in (a). The presence of two distinct antagonistic regions lying diagonally relative to receptive field centre was seen in many cells. This antagonism is reminiscent of the much stronger lateral antagonism reported in spiking lamina cells (Arnett, 1972). Results for the two families of fly are similar, but with* Calliphora *having a consistently wider field of lateral antagonism.*

are essentially linear; they give opposite responses to brief intensity increments and decrements and show linear spatial summation within their receptive fields (Fig. 10.3a). Some sustaining cells are also spectrally and/or spatially opponent: i.e. the response at one wavelength or at one position is opposite to that in another region of the spectrum or of space (Osorio, 1968a, 1987a, unpublished observations). The transient cells, by comparison, are highly non-linear; they are responsive to motion in any direction within their receptive visual fields but less so to static flicker, and give very similar responses to increments and decrements, and also to pulses and steps (Fig. 10.3b). These responses typically comprise one or two spikes which occur at a fixed latency after any suprathreshold

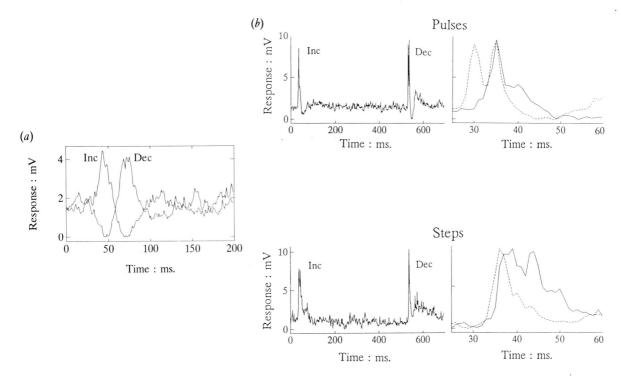

Fig. 10.3 *Responses of small-field cells from the locust medulla. Showing some of the characteristics of the sustaining and the transient cell types. Data shown are obtained by intracellular recording and signal averaging the responses to 30 stimulus presentations. (a) Sustaining cell response to brief (10-ms) increments and decrements of contrast 0.5 about a mean background intensity. Note that the responses to increments and decrements are antisymmetric and triphasic. The cell is quite sharply tuned to temporal frequencies of 15 Hz with a bandwidth of below one octave. (The locust's receptor flicker fusion frequency is about 60 Hz, which is similar to man's.) When dark-adapted the impulse response of this cell is purely monophasic, indicating that the cell is tonic. (b) Transient cell responses to brief (10-ms) and step increments and decrements of equal contrast about a mean background. The responses to all four stimuli are similar, indicating that this unit would give the same responses to a dark border and to a pale line moving past the eye. Note that the spiking response contains one or two spikes which occur at a precise time after stimulation. The left- and right-hand figures show the same data at different time scales. On the right the responses to increments are indicated by solid and those to decrements by dashed lines. The abscissa indicates the time after stimulus onset.*

stimulus.

Medulla Function

It is not surprising that medulla cells have widely divergent properties when low-level image features are extracted or information is gathered relevant to specific behavioural outputs, but at present we have no clear idea of how visual information is divided among the many medulla columnar cells (see Strausfeld, 1989; Horridge, this volume, chapter 12). On the basis of the responses obtained in the locust one can suggest only that sustaining cells encode general information about intensity distributions. The transient cells appear to be more 'specialized' and could abstract information about specific features in the image at an early stage in visual processing. In addition to being insensitive to stimulus polarity or to the difference between edges and lines, the transient cells can adapt to a mean background 'contrast', such as a uniform texture moving past the eye. We have suggested that the transient cells will fire in response to the relatively high contrasts which mark object borders, and so provide the basis for image segmentation where it is not necessary to encode the structure or the polarity of the boundaries (Osorio *et al.*, 1987; Osorio, 1990).

A useful alternative to asking what is 'represented' by the neural image in cells at any one stage of vision may be to look at the response specificities of those at the next stage. One can then ask: how is the lower-order representation suited to building the higher order responses? In the optic lobe this approach is attractive because data from the medulla are so scarce, and several suggestions have been made about medulla cell properties on the basis of recordings from the lobula-complex (Rowell et al., 1977; Egelhaaf et al., 1989, 1990; Franceschini et al., 1989). On the basis of available medulla recordings from the locust it is difficult to suggest how the sustaining cells might be used, since small-field linear cells can subserve almost any behaviour (cf. LMC properties, above). In contrast, the transient cells 'discard' so much information that it is unlikely that they are used (at least alone) for processes like directional motion detection in which signal polarity is preserved. The transient cells could drive the LGMD, which is a wide-field, non-directionally selective 'movement-detector', and superficially their responses resemble small-field versions of the LGMD (Osorio, 1987b). Both types of cell give similar responses to ON and OFF stimuli and both show local adaptation. However, the responses of the LGMD cannot be obtained simply by integrating the outputs of transient cells (Osorio, 1990), and wide-field inhibition, long-term adaptation or habituation and efferent inputs seem to be crucial in shaping the lobula cells' responses but are not apparent in the medulla.

Computational Mechanisms and the Integration of Small-field Cells' Responses

Small-field medulla cells do not project directly to muscles, and so the key to understanding local processing in the medulla will be to show how retinal information is converted into a form which can be used to build the responses of higher-order, wide-field cells. Computations must be performed among small-field cells which could not be achieved directly at the synapses from receptors or LMCs upon wide-field units. It is not clear what these computations might be, and so the *raison d'être* of the medulla and hence half of the insect brain is a mystery. This mystery cannot easily be solved by modelling neuronal interactions because we do not understand fully the range of local integrative capabilities of the nervous system, and hence the operations which could be implemented at, say, the synapse of an LMC upon a wide-field neurone (Bullock, 1976). In practice, large-field afferents receive their inputs at every level of the optic lobe: at the distal edge of the insect medulla lie directionally selective cells (Osorio, 1986a; Ibbotson et al., 1990a), while in the crayfish medulla there are 14 sustaining cells whose overlapping receptive fields tonically encode light intensity over the entire eye, and outputs from the medulla provide the visual input to the lobula and brain (Rowell et al., 1977; Kirk et al., 1982; Rind, 1987; Pfeiffer and Glantz, 1989; Strausfeld, 1989). It will be of interest to identify differences between the responses of tangential cells which receive their inputs distally from those which arise in the proximal lobula, since such a comparison may indicate how columnar cells in the medulla can transform the retinal image.

What is the Site and the Mechanism of the Computation of Directional Motion?

The difficulties in establishing which kinds of 'neural computation' require small-field cells are well illustrated by the example of directional motion detection in the fly. Arthropods measure motion flowfields to stabilize locomotion and the algorithms describing stabilization behaviour (the optomotor response) are well known. The optomotor response is probably mediated by the wide-field directionally selective cells in the lobula plate, and the responses of these cells to motion can be described by algorithms which account for the behaviour. However, it has been difficult to establish how these algorithms are implemented by the neural machinery of the optic lobe (Reichardt, 1969; Heisenberg and Buchner, 1977; Buchner, 1984; Egelhaaf et al., 1988; Hausen and Egelhaaf, 1989; but cf. Heisenberg and Wolf, 1988).

Direction is measured locally, and so requires interactions in the medulla or between the outputs of medulla cells. Essentially, the outputs from adjacent receptors are correlated after one has been delayed for a short time (Buchner, 1984; Egelhaaf et al., 1989). However, despite the use of various techniques – electrophysiological (DeVoe, 1980), pharmacological (Egelhaaf et al., 1990), anatomical (Buchner et al., 1984; Strausfeld, 1989) and optical (Franceschini, 1989) – it is not known whether the measurement of direction requires small-field medulla cells to be directionally selective, or if the directional computation is performed directly at the synapse from small-field elements onto the wide-field cells (Torre and Poggio, 1978). Going beyond the question of where the first directionally selective cell lies, one can ask how the small-field medulla cells should filter their input to permit the computation of direction. Some argue that separation of signals into ON and OFF channels is a necessary step before the outputs from adjacent receptors are correlated (Franceschini et al., 1989), whereas others imply that the responses of the wide-field cells can be obtained at a single synapse by correlation of the outputs of two linear cells with different temporal tuning (Borst and Egelhaaf, 1989; Egelhaaf et al., 1989).

Butterfly and Locust: Directional Selectivity Outside the Lobula Plate

The character of the medulla inputs to the lobula plate may soon be known, either from direct recording or by inference, and it would be reasonable to assume that neural processing which takes place on the 'directional-motion pathway' *en route* to the lobula plate exemplifies the interactions for which the medulla evolved. But it now appears that the medulla may be unnecessary for the detection of visual flowfields, because wide-field direction-sensitive cells emerge in the distal part of the medulla (Fig. 10.4).

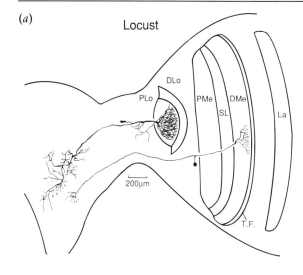

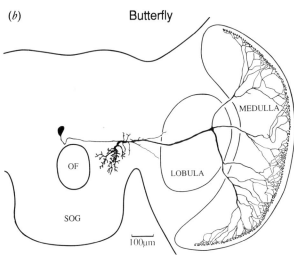

Fig. 10.4 *Three anatomically distinct classes of directionally selective cells found in the optic lobe; all cells were filled with dye after intracellular recording. (a) Two cells in the locust, one in the distal medulla (below) and the other in the proximal lobula LDSMD(F) neurone (above; adapted from Rind, 1990). The more distal cell is drawn from sections, while the lobula cell was originally drawn in whole mount. La: lamina; T.F.: tangential fibre layer covering the distal edge of the medulla; DMe: distal medulla; SL: serpentine layer (tangential fibres); PMe: proximal medulla; DLo: distal lobula; PLo: proximal lobula. (b) Cell with a preference for downward motion arborizing in the distal part of the butterfly's medulla. SOG: sub-oesphageal ganglion. OF: oesophageal foramen (Courtesy of M. R. Ibbotson).*

Like flies and bees, locusts and butterflies have wide-field direction-sensitive cells in the proximal lobula–lobula plate (Fig. 10.4a, Hausen and Egelhaaf, 1989; Ibbotson, 1990; Rind, 1990; Maddess *et al.*, 1990). Rind (1990) notes several anatomical and physiological differences between the locust lobula cells and those in the fly, but none the less it seems likely these cells are homologues (see pp 204–205). Apart from these lobula directionally selective cells, both the latter species have medium- and wide-field directional cells in the distal part of the medulla.

In the butterfly the direction-sensitive cells in the medulla (Fig. 10.4(b)), at least some of which are afferent (R. A. Dubois and M.R. Ibbotson, personal communication), receive their inputs in the most distal layer of the medulla close to the terminals in the butterfly lamina monopolar cells L1 and L4 (Ribi, 1987). At least two cells are sensitive to vertical and three to horizontal motion. These distal directional cells terminate in the same part of the brain (deuterocerebrum) as the lobula plate cells of the fly (and the butterfly) and have similar physiological properties. For example, they measure contrast frequency (not velocity) and they adapt rapidly in response to a continuous movement (Maddess *et al.*, 1990; Ibbotson *et al.*, 1990a,b).

Another kind of directionally selective cell which arborizes in the distal part of the medulla is found in the locust (Fig. 10.4(a); Osorio, 1986b). These cells are excited by upward motion and have relatively small receptive fields (about 10° across), but like the wider field units seem to measure motion locally (between the outputs of adjacent receptors). A comparison of the responses of these locust cells with those of the fly lobula plate indicates that the underlying mechanisms of computation of directional selectivity may differ (Fig. 10.5). An apparent motion (phi) stimulus, comprising two successive brief (10-ms) pulses at 50-ms intervals and separated by one inter-receptor angle, elicits a directional response both in the fly lobula plate cell H1 and in the locust medulla cells (Osorio, 1986b, 1990; Franceschini *et al.*, 1989). In H1 the basis of this directional response seems to be a *facilitatory* interaction so that the response induced by the second flash is much larger than that to the first. By comparison, in the locust medulla directional interaction seems to be based on an *inhibition*, so that when the apparent motion is in the preferred direction both the first and the second stimuli elicit equal 'flicker' responses, but if the stimulus stimulates motion in the anti-preferred direction the response to the second flash is suppressed. This veto, or AND-NOT, mechanism of directional selectivity was first proposed for retinal ganglion cells in the rabbit (Barlow and Levick, 1965).

A different type of directionally selective cell has been found in the medulla of the fly *Eristalis* (Collett and King, 1975). These cells are characterized by strong spatial antagonism, so that they respond best to movement of targets about 2.5° across (i.e. one receptor width), and may be used for tracking and pursuit.

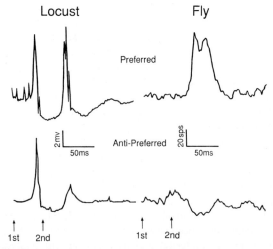

Fig. 10.5 *Responses of two classes of directionally selective cells, indicating that the mechanisms underlying their responses may differ. In each case the stimulus is a pair of brief (10-ms) pulses (arrowed), separated by one interreceptor angle and 50 ms. In the locust medulla cell (right) these were given in the far field (normal vision), whereas in the fly the stimulus was applied ophthalmoscopically to two individual receptors by Franceschini et al. (1989). The directional interaction in the fly's cell H1 is a facilitation of the flicker response to motion in the preferred direction, whereas in the locust there is a suppression of the response to antipreferred motion (see text). Experiments at other intervals indicate that the time course of the directional interaction is similar in both cell types, with a maximum directional response when the interval was about 30 ms (locust) or 50 ms (fly), declining to zero after 100 ms (locust) or 200 ms (fly) (Franceschini et al., 1989; Osorio, 1990) (Fly adapted from Franceschini et al., 1989).*

Evolutionary considerations

Directional selectivity is a clearly recognizable, well characterized and widespread feature of low-level visual processing that may be a useful basis for studies of the phylogeny of visual processing mechanisms (see below). At this stage three points emerge from finding that there are several classes of directionally selective cells in the medulla, perhaps using different neural mechanisms to achieve their responses:

1. The characterization of several distinct sites of directional selectivity in the optic lobe seems to vindicate Strausfeld's (1989) suggestion that: 'A reasonable supposition is that because so many computational tasks are based upon motion sensitivity, circuits for motion detection may be legion'. Directional cells may emerge wherever they are needed, with synaptic inputs from nearby small-field lamina or medulla cells evolving accordingly. As in invertebrates (Grüsser and Grüsser-Cornehls, 1973), we need not expect to find just one location or mechanism of directional selectivity in arthropods.

2. The corollary of the first point is that even if the properties of the columnar cells whose outputs are integrated by any one class of wide-field cells are known, this does not necessarily establish the fundamental function of the medulla ganglion. The inference from the butterfly and the locust is that directional tuning can be achieved at the distal edge of the medulla, perhaps with direct inputs from lamina monopolar cells. This implies that the two sets of medulla columnar cells (Tm1 and S-ub, and T4 and T5) on the pathway to the lobula plate (see pp 204–205; Strausfeld, 1989) need be no more than relays of lamina cell inputs. It is most unlikely that the trans-medullary cells are simple relays, but we are left to wonder why the medulla evolved.

3. The final conclusion to be drawn from the finding that directional selectivity can appear independently in several parts of one optic lobe has a bearing on questions which are addressed in the next part, where we try to identify evolutionary homologies and convergences in arthropod vision. If analogous cells found in two different animals are directionally selective and also use the same synaptic mechanisms, these similarities are not due inevitably to constraints on neural signal analysis which oblige the nervous system to implement a fixed set of local interactions to measure direction. Instead, one might argue that the two animals have a common ancestor which had homologous directionally selective cells (see p 219).

Evolution

Behaviour and Neural Processing

Invertebrate behaviours can be hierarchically classified according to their complexity. For example kineses, where a stimulus elicits random motion, are seen as more primitive than taxes, where the movement is oriented by the stimulus. More advanced still are visual behaviours where the direction of motion and then the velocity of motion are taken into account, or where targets are fixated (see Fraenkel and Gunn, 1961; Poggio and Reichardt, 1976; Horridge, 1987; Horridge, this volume, chapter 12; Srinivasan, 1991). Several algorithms which describe the behaviours (and the responses of single neurones) have been discovered; notably the autocorrelation model of directional motion selectivity in flies (Reichardt, 1969; Poggio and Reichardt, 1976; Borst and Egelhaaf, 1989). The impression that there is a hierarchy of complexity is reinforced by these mathematical descriptions of behaviour, because the complexity of the algorithms tend to be higher in behaviours which appear to be more complex. For example, phototaxis can be mediated by linear mechanisms, whereas directional-motion detection requires second-order, and fixation (based on relative motion cues) fourth-order, non-linearities (Poggio and Reichardt, 1976). It seems plausible that this behavioural hierarchy is due to evolutionary progression, and that this progression has been underpinned by a concomitant evolution of neural processing from rudimentary beginnings (Bullock and Horridge, 1965; Horridge, this volume, chapter 12).

This part tries to discern the evolutionary origins and radiation of crustacean and insect visual systems associated with the diversification of lifestyles that has occurred since arthropods first appeared 600 million years ago. What was ancestral arthropod vision like and how have particular physiological mechanisms of visual processing evolved? For example, the suggested title for this chapter was: 'Evolutionary response to visual shadows: lateral inhibition and contrast sensitivity and the subsequent development of movement and directional sensitivity.' Accepting that the evolutionary path implied by this title is plausible, where should we look for the primitive animals which perform lateral inhibition using structures homologous to the directional motion detectors of more advanced species? (Reasons why the LMCs, which are the primary site of lateral inhibition in the optic lobe, would be unlikely to evolve directional selectivity have already been mentioned.)

Convergence and Homology

Palaeontological data pertinent to neural processing will never be abundant, so how can inferences about evolutionary history be drawn from contemporary forms? For our present purposes, several different points of comparison may be drawn between organs in any two groups. Firstly, the organs may be homologous and similar, owing to conservation of a character. Secondly, they may be homologous and similar due to parallel evolution from a common ancestor. Thirdly, two characters may be homologous but differ because they represent different 'stages' in the evolution of a character (horses' feet are a familiar example). If reliably identified homologous characters at different stages of evolutionary development are very valuable for reconstructing phylogeny. In the optic lobe the evolution of fly lamina synapses from 'primitive' nematoceran to 'advanced' cycloraphid families can be reconstructed using this reasoning in conjunction with careful comparative study of many contemporary groups (Shaw, 1989, 1990; Shaw and Moore, 1989). Homologous characters can also diverge to suit different functions. Finally, convergent evolution of non-homologous structures can occur. How can we disentangle the threads of arthropod evolution to distinguish convergences, which are the raw material of comparative study, from homologies which are the keys to phylogeny and the reconstruction of ancestral types?

The Origins of Arthropods

Arthropoda is a large and diverse phylum of animals characterized by having jointed legs (exoskeleton) and includes the following classes/subphyla: Trilobita, Merostomata (e.g. *Limulus*), Arachnida (e.g. spiders, scorpions and mites), Crustacea, Insecta and Chilopoda (centipedes) as well as various minor classes. This chapter is concerned with a group sometimes known as Mandibulata (Snodgrass, 1938), which includes, among others, the last three named classes. The second main subdivision of the (living) Arthropoda is Chelicerata, which includes spiders, scorpions and *Limulus*. Taxonomic nomenclature of Crustacea follows Schram (1982), who assigns the rank of class to the main divisions within the subphylum Crustacea, e.g. Branchiopoda, and Malacostraca. Insect classification follows Hennig (1981).

A reconstruction of the history of visual processing should be based on a known phylogeny, but the phylogenetic relationships between the arthropods is controversial, and many issues remain unresolved. For example, it is not known whether 'arthropodization' occurred independently on several occasions or if it was a unique event (i.e. whether 'Arthropoda is polyphyletic or monophyletic; Manton, 1977; Gupta, 1979; Briggs and Fortey, 1989; Willmer, 1990).

Compound eyes do not automatically imply that the body has an articulated exoskeleton, but these eyes have long been recognized as providing good evidence for monophyly, especially in insects and crustaceans. Even those authors who – on the strength of other evidence – believe that these two taxons have different soft-bodied ancestors 'hesitate in admitting convergence in production of the compound eye' (Tiegs and Manton, 1958), and despite their rejection of this possibility, subsequent authors have used the compound eye as evidence for monophyly of insects and crustaceans (Paulus, 1979; Meinertzhagen, this volume, chapter 17). Hanström (1926) is perhaps the latest author to have discussed the evolutionary relationships between the arthropods basing arguments on the anatomy of the visual centres, and his work (1926, 1928) remains a most comprehensive comparative anatomy of the invertebrates. Figure 10.6 summarizes Hanström's views, and it is salutary to compare these to those expressed here on the barest shreds of additional evidence.

As long as phylogenetic relationships are unresolved it will be uncertain whether characters like the compound eyes of the crab and the fly are homologous, or whether they are the products of convergent evolution. Thus arguments of the kind addressed here could be stood on their heads: a convergence full of comparative potential, according to one phylogenetic history, may be homology shedding little light on function according to another. The

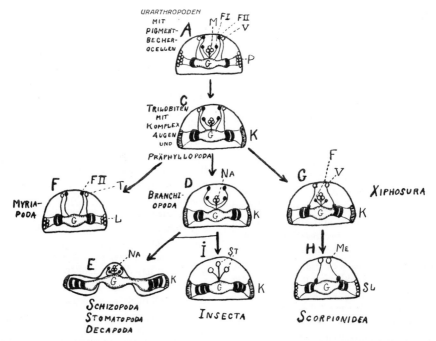

Fig. 10.6 *Figure from Hanström (1926), illustrating the phylogenetic relationships between arthropods based on his study of their visual systems. Note the implication that evolution is marked by a consistent increase in complexity, and that arthropod compound eyes are monophyletic.*

next section compares the optic lobes of insects and malacostracan crustaceans presenting evidence for and (some) against the idea that they are monophyletic. (The Malacostraca includes large crustaceans such as crabs, crayfish, krill, mantis shrimps, and woodlice. Other crustacea are mentioned later. The argument here is similar to that made elsewhere (Paulus, 1979; Meinertzhagen, this volume, chapter 16), and these references should be consulted for more detailed information, especially on the anatomy of the optics and retina.

Similarities Between Insect and Malacostracan Visual Conservatism or Convergence?

Compound eye optics, where many-facet lenses produce an erect image on the retina, have appeared independently on about half-a-dozen occasions, and subsequently evolved along similar lines (Land, 1981; Nilsson, 1989, 1990; Meinertzhagen, this volume, chapter 17). Convergences occur because the properties of biological materials and the physics of light mean that a limited number of optical 'designs' can evolve. Moreover, the performance of optics can be quantified by taking into account their size, light-gathering capacity, and the image quality, and so it is possible to compare different eyes (Land, 1981;

Snyder, 1979). The same judgments cannot be made for neural processing, where we have little idea of the function of most neurones or of the constraints under which they have evolved, and even less of how to quantify their performance (optimal coding is an interesting exception; see pp 205–207). Consequently, it is difficult to say whether analogous neural structures or processes in two different species are likely to have arisen as a result of convergent or of parallel evolution, or are simply a reflection of evolutionary conservatism and have remained unchanged since their two lines of descent diverged.

Optics and Retina

The compound eyes of crustaceans and insects exhibit some obvious evolutionary convergences or parallelism, like the acquisition of superposition optics (Nilsson, 1989). However, the similarities between insect and crustacean visual systems extend beyond the optical mechanisms. These start with several common features in the cellular architectures of the optics: the corneal lens is composed of two cells, and the crystalline cone of four cells (Meinertzhagen, this volume, chapter 17; Paulus, 1979), and in addition a specific glial antigen occurs in the crystalline cone of insects and crustaceans (Edwards and Meyer, 1990). Beneath each facet lens the ommatidium contains (usually) 8 photoreceptors, of which 6 or 7 short visual fibres (SVFs) end in the lamina while 1 or 2 long visual fibres (LVFs) project to the medulla (see pp 204–205; Strausfeld and Nässel, 1981; Trujillo Cenoz, 1985; Meinertzhagen, this volume, chapter 16). The precise numbers of these cells varies somewhat; often crustaceans have 7 SVFs and 1 LVF and insects 6 and 2 respectively, but in cockroach and in 1 region of the male fly eye there are 7 SVFs, and *Daphnia* (Branchiopoda) has the 'typical' insect arrangement (Macagno and Levinthal, 1975; Ribi, 1977; Hardie, 1985). The LVFs are often UV-sensitive, whereas the SVFs' photopigments are normally most sensitive to longer wavelengths (Menzel and Blakers, 1976; Cummins and Goldsmith, 1981; Hardie, 1985; Osorio, 1986a; Smith and Macagno, 1990). A notable point about the variation seen in the retina, as elsewhere (such as the numbers of lamina cells – see 'Lamina' below) is that the degree of variation in the numbers of receptor cell types between comparatively closely related species is comparable to that between insects and crustaceans (Meinertzhagen in this volume, chapter 16, describes variants on the generalization made here).

By contrast with the common pattern shared by insects and crustaceans, *Limulus* (a chelicerate) compound (lateral) eye has a very different organization: there are variable numbers of receptors in each ommatidium, an eccentric cell which has no obvious homologue in the Mandibulata, and no equivalent of the LVS–SVF dichotomy (see Meinertzhagen, this volume, chapter 16). Another distinguishing feature of *Limulus* is that efferent neurones extend to the retina (Barlow *et al.*, 1989), whereas in insects and crustaceans they do not extend beyond the lamina. However, screening pigments in the crustacean retina do respond to neural or neurohormonal modulators like serotonin, while in insects this does not seem to be the case. None the less, the movement of screening pigment in beetles and in moths is subject to a circadian rhythm independent of illumination, implying that hormonal influences play some part, and insects have wide-field aminergic fibres in their laminas analogous to those which may mediate pigment migration in the crayfish (see below and Eloffson and Klemnn, 1972; Fleissner, 1982; Weyrauther,1989; Aréchiga *et al.*, 1990; E. J. Warrant, personal communication). The extent of this apparent difference between the two groups needs verification.

Lamina

As Strausfeld and Nässel (1981, p 49) say in their comprehensive description of insect and crustacean optic lobes: 'Retinotopically organized neuropils are arranged beneath the insect compound eye much in the same way as ... the crayfish.' These authors use the same terminology for describing the two visual systems (Fig. 10.7), both for overall organization and for naming individual cells, a practice which we follow here but should not preempt resolution of the question as to whether the two systems are homologous (Rowell, 1989). The similarities between insects and crustaceans are most obvious and best documented in the lamina, where, in addition to anatomical data, there are now known to be analogies between the physiologies of crayfish and fly lamina ganglion cells.

The complement of lamina cells has been catalogued in several species of insects (Strausfeld, 1976; Kral, 1987) and of crustaceans (Strausfeld and Nässel, 1981; Elofsson and Hagberg, 1986). The synaptic connections in the fly lamina have been described in exquisite detail, as have those of various other species of insect and also the crayfish (Meinertzhagen, this volume, chapter 17; Nässel and Waterman, 1977; Shaw, 1984, 1989; Kral, 1987). Figure 10.8 shows the columnar neurones from *Drosophila* and (partially diagrammatically) from the crayfish. The monopolar (M) cells and a T(= T-shaped) cell, which has its soma in the distal medulla, are afferent. One or two C (= centrifugal) cells have their somas in the proximal medulla and are probably efferent, and together with one or two lamina amacrines these complete the complement of cells found in each lamina cartridge. In addition there is a medium-field tangential cell (La wf1 in fly, Tan-1 in crayfish) which has its cell body in the distal medulla and may have a lower periodicity than the retinal array. Finally, there are one or two giant wide-field tangential cells, which are efferents that arborize throughout the optic lobe (not illustrated) and in fly a lamina-intrinsic wide-field cell. At least one of the tangential cells is probably efferent in both species (Pfeiffer and Glantz, 1989) and are aminergic (possibly serotonergic) in fly and in crayfish (Nässel and Klemm, 1983; Aréchiga, *et al.*, 1990; but cf. Elofsson, 1983). It should be noted that the somas of arthropod neurones are usually electrically isolated from the active neurites (as was recognized by Cajal and Sanchez (1915)!) and so there is no obvious reason why convergent evolution should give sets of cells with cell bodies in similar positions, as in the laminas of fly and crayfish.

Of the lamina cells the physiology of those monopolar cells which receive their main inputs directly from the photoreceptors (L1–L3 in fly, M2–M4 in crayfish) is best known because they have large-diameter axons for transmitting graded potential signals to the medulla, and so are easily recorded (Fig. 10.2; Wang-Bennett and Glantz, 1987a). In the crayfish and in the fly these cells give nonspiking hyperpolarizing responses to intensity increments, they are non-directional, they show centre–surround spatial antagonism, and they give phasic responses to step intensity changes. Unlike the fly, but like various other insects, including the bee and the dragonfly (Ribi, 1981; Meinertzhagen *et al.*, 1983), all six receptors do not converge onto each one of the M cells in crayfish, but instead

216 *Evolution of the Eye and Visual System*

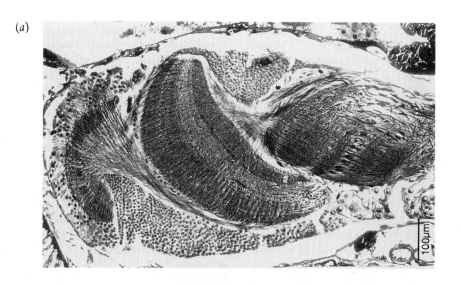

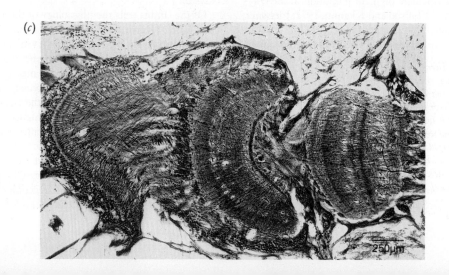

(d)

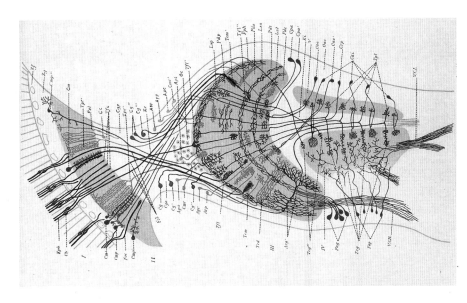

(e)

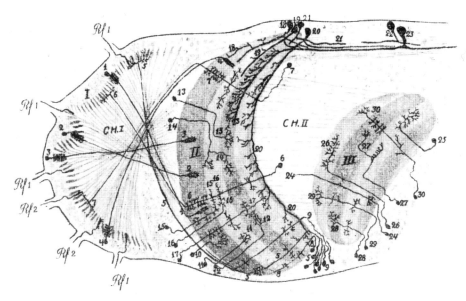

Fig. 10.7 *Micrographs of the optic lobes of two insects, (a)* Notonecta *and (b)* Calliphora, *and the malacostracan crustacean* Carcinus *(c), silver stained horizontal sections showing the chiasms. The columnar structure of the medulla is especially obvious in* Notonecta *and* Calliphora. *A tangential cell is visible in the proximal part of the lobula of* Carcinus. *The division of the medulla by a serpentine layer is generally more clear and consistent in insects; see also Fig. 11.9(a). Re: retina; La: lamina; Me: medulla; Lo: lobula; LoPl: lobula plate (labels on blowfly only). (a) and (c) Reduced-silver stains using the Fraser–Rowell technique (Courtesy of G. A. Horridge). (b) Silver stained using a modified Holmes–Blest technique with lutidine (Courtesy of A. D. Blest. From Blest and Davie, 1980). (d) Optic lobe of* Calliphora vomitoria *from Cajal and Sanchez (1915), based on Golgi stained material. Compare the lamina cells with those from* Drosophila *shown in Fig. 10.8(a). (e) Optic lobe of a crab,* Pachygrapsus *(from Hanström, 1924). Hanström studied a wide range of invertebrates and his studies (1926, 1928) remain unmatched as a source of comparative data. Few Golgi studies of the medulla and lobula of Crustacea have been published since, whereas studies of insects have thrived during the past 20 years. It should be noted that figures of the kind shown here are partly diagrammatic and may not have been intended to show the precise morphologies of cells.*

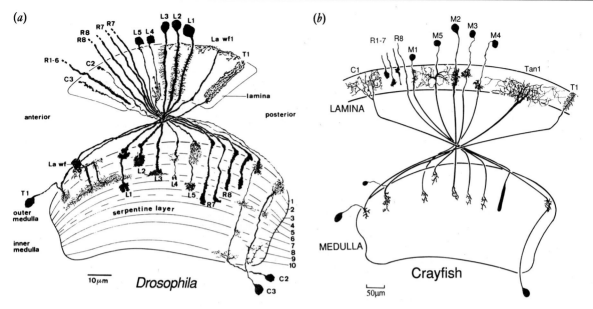

Fig. 10.8 *Cell types present in the laminas of* Drosophila melanogaster *(from Fischbach and Dittrich, 1989) and crayfish, mainly* Pacifastacus lenisculus *(also* Procambarus clarkii*) (Adapted from Hanström, 1924; Nässel, 1977 and Wang-Bennett and Glantz, 1978a,b). Those parts of the cells in the crayfish lamina are accurately drawn, whereas apart from M3, M4 and Tan1 the parts of the medulla are diagrammatic, illustrating the positions of the cell bodies and primary neurites, but not the fine arborizations. The cells postsynaptic to receptors (SVFs) are M2–M4 in the crayfish and in the fly L1–L3 (LMCs) and also one or two amacrine cells (not illustrated) (Strausfeld, 1976; Nässel and Waterman, 1977; Shaw, 1984). The cells La wf1 and Tan1 are analogous 'medium-field' cells of unknown periodicity. The small crayfish monopolar cell M1 is duplicated in each cartridge. Wang-Bennett and Glantz (1987a,b) filled cells with dye (Lucifer Yellow), while the other studies used Golgi staining.*

the receptor outputs are kept separate in the lamina (Nässel and Waterman, 1977). This isolation of receptor outputs may be an adaptation to polarization or spectral vision (Sabra and Glantz, 1985).

The other lamina cells, which do not receive direct receptor inputs, are less well known in insects, but in crayfish one monopolar (M5) arborizes in several columns. It is a spiking cell which gives a fairly tonic *depolarization* to axial illumination, and is subject to surround antagonism (Wang-Bennett and Glantz, 1987b). These physiological properties are reminiscent of one of the spiking units in the fly lamina (Arnett, 1972), and the anatomy of M5 is analogous to that of the fly monopolar cell L4, in that both cells arborize in more than one column, and have a thin axon (compared to the LMCs or M2–M4). Other cell types in the crayfish lamina have been recorded and marked (Wang-Bennett and Glantz, 1987) and have anatomical but no known physiological analogues in the fly. These include the T-cell, two wide-field tangential cells and an amacrine cell. The similar anatomies of the T-cells in a decapod (*Hippolyte?*) and *Drosophila* have been interpreted as indicating that these neurones 'perform a basic and conserved function' (Fischbach and Dittrich, 1989; see also Meinertzhagen, this volume, chapter 17). (Moreover, Hanström, 1928, from where the (partly diagrammatic) illustration of *Hippolyte* is taken, illustrates a very similar T-cell, and general anatomy, in the first optic ganglion of *Limulus*; so perhaps there is conserved function here also!)

Pharmacology may be useful for establishing evolutionary relationships between different taxa as well as for investigating signal processing in single species (Pfeiffer-Linn and Glantz, 1989), but so far has had little to say on the question at hand. For example, histamine is the receptor transmitter in the compound and in the simple eyes of various members of Arthropoda, including insects and *Limulus*; this is evidence that all (compound and simple) arthropod eyes have a common origin, but says little about relationships within this group (Hardie, 1989a). However, the LVF, R7, of flies is not histaminergic but may be GABAergic; this could be a derived character useful for establishing relationships within the Arthropoda (Fig. 10.1(c); Datum *et al.*, 1986; Nässel *et al.*, 1988). Unfortunately, the LMC transmitter, which is probably glutamate in the bee (Schaefer, 1987, cited in Hardie, 1989b), is not known in Crustacea. Similar distributions of cholinergic activity are seen in the crayfish and in the fly laminas, but have been attributed to different cells. In the crayfish a cell class other than the LMCs is cholinergic, perhaps the 'terminal fibres' of the C cells or the tangential cells (one of which appears to be serotonergic in insects and in crayfish (Nässel and Klemm, 1983; Aréchiga et al., 1990)). Formerly, cholinergic activity in the fly was also associated with the C cells, but it is now attributed to amacrines (Datum et al., 1989; Wang-Bennett et al., 1989). This apparent discrepancy should be confirmed, and further comparisons drawn between the pharmacologies of insect and malacostracan laminas.

The fly and the crayfish laminas have closely analogous anatomies and, apparently, physiologies. Given the difference in habits of the two animals (one is aerobatic and diurnal, the other aquatic and nocturnal), these similarities are surprising. It is likely that some convergence has occurred, for more closely related species differ in their synaptology and numbers of cells (see Meinertzhagen, this volume, chapter 17). Nevertheless, this is striking evidence both for homology and for conservatism in the lamina ganglion of the crustacean and the insect.

Higher Ganglia

The basic anatomy of medulla and lobula of malacostracan crustaceans and of insects is similar (Fig. 10.7). In both groups (as described above), the entire afferent projection from the lamina (also called the lamina ganglionaris in Crustacea) crosses the first optic chiasm to the medulla (also called the medulla externa in Crustacea). In insects, the medulla is conspicuously divided into proximal and distal parts by a 'serpentine layer' which contains the arborizations of large-field tangential cells; an analogous division is said to occur in malacostracans, but is less obvious (Strausfeld and Nässel, 1981). The same general columnar cell types can be recognized in crustaceans and in insects (Strausfeld and Nässel, 1981) but the anatomy of individual cells in the crustacean medulla is insufficiently known (from Golgi studies) to justify drawing comparisons. Projections from the medulla are again similar in insects and crustaceans; most outputs cross the second chiasm to the lobula (also called the medulla interna in Crustacea), which is the final retinotopic ganglion, but has a lower periodicity, with one column of cells for every four facets of the compound eye in insects (Braitenberg, 1972), and an unknown periodicity in Malacostraca. Some medulla axons bypass the lobula and project directly to the brain or cross to the opposite optic lobe (Strausfeld and Nässel, 1981). A fourth ganglion which is situated at the base of the eyestalks in Crustacea called the medulla interna is not columnar and is perhaps analogous or homologous to the lateral protocerebrum of insects (Strausfeld and Nässel, 1981).

There are many similarities between visual responses of the wide-field afferents from the optic lobes of insects and of crustaceans (Wiersma et al., 1982; DeVoe, 1985). For example, both have directional cells, sustaining cells and non-directional, motion-sensitive cells or jittery-movement detectors. There are at present no grounds for homologizing cells on the basis of such broadly defined response properties, given the conclusions reached above (see pp 210–211), which argue that particular response properties such as directional selectivity can evolve independently at many different loci. However, it would be of interest to know whether the proximal lobula of crustaceans is dominated by wide-field directionally selective cells analogous to those in insects, and what mechanism of directional selectivity is used. Similarities here would show that the most proximal parts of the optic lobes of insects and crustaceans resemble the laminas in sharing common neural properties (see p 212).

Homology or Convergence?

It is difficult to make any generalization about the neurophysiology or neuroanatomy of the insect visual system which does not also apply to malacostracan crustaceans. Constraints imposed by being an arthropod with compound eyes may partly dictate the structure of the visual system, but there are many similarities where no obvious reason for convergence is at hand and which might give valuable insights into the function of many structures which are so far unexplained. Why do (typically) the long-wavelength photoreceptor cells end in the lamina, whereas the short-wavelength photoreceptors project directly to the medulla? Why are there three columnar ganglia separated by two anteroposterior chiasms, with the first two arranged with a column of cells repeated for each facet in the compound eye, whereas the third has a coarser structure? And finally, why are the laminas so similar? If, on the other hand, the crustacean and the insect visual systems are largely homologous it is probable that they diverged long ago, and we are unlikely to be able to deduce much about the evolution of contemporary arthropod visual systems by searching for surviving ancestral forms.

The similarities between the eyes and the visual systems of insects and malacostracan crustaceans suggests to others, especially Paulus (1979), that they are homologous (see also Meinertzhagen, this volume, chapter 16), and much of the succeeding argument follows from the acceptance of this conclusion. Homologies are also suspected elsewhere in the nervous system: for example a similar 'identifiable' neurone occurs in the thoracic ganglion of *Drosophila* and crayfish (Thomas et al., 1984), and in the sensory periphery the chordotonal organs (mechanoreceptors) of crustaceans and insects are similar down to their anatomical details, and thought unlikely to be the products of convergent evolution (Hoffmann, 1964). (Chordotonal organs are absent from Chelicerates.)

None the less, one cannot hope to provide a decisive

answer to the question of whether the common ancestor of crustaceans and insects was an arthropod (with compound eyes), for as Manton (1977) sternly advises: 'Reliable results are not gained quickly or by guessing or by considering too few examples.' And much contrary evidence can be adduced; for example from Manton's studies of locomotion and from embryology (Anderson, 1973).

The use of monoclonal antibodies and tissue *in situ* hybridization will provide new data, which, when combined with data from older approaches such as comparative anatomy and physiology, may serve to resolve ambiguities and provide greater confidence in drawing conclusions. Specific molecular markers may help to establish whether the optic lobes of different arthropods are truly homologous. For example, the *Drosophila* mutant 'small optic lobes' shows specific defects in the medulla (Fischbach et al., 1989). The gene for small optic lobes has now been cloned, and it will be of interest to establish the pattern of expression of its product in various arthropod taxa. In addition, the choreography of the developing *Drosophila* retina has now been analysed, showing how the eight retinula cell types differentiate, a demonstration of similar processes in crustacean retinas would be good evidence for homology (Meinertzhagen, this volume, chapter 16; Ready, 1989).

The History of Arthropod Vision

A reasonable case has been made that compound eyes and their visual ganglia are homologous so now we turn to explore the consequences of this conclusion. What can be said about the vision of ancestral arthropods, and how might visual processing have evolved in the 600 million years or so since the insect and crustacean lines diverged? The simplest explanation for homology (and similarity) of an organ in two distinct taxons is that it evolved in their common ancestor and has since retained its form and function. However, a possible alternative is that genetic and embryological constraints may be operating that have no direct bearing on the function in question. For example, the mechanisms of body segmentation (Akam, 1987) might dictate the number of ganglia in the optic lobe, and likewise general morphogenetic processes might be responsible for the presence of chiasms between these ganglia. The conservatism of morphogenesis is exemplified by the suggestion that homologous genes control development in vertebrates and insects (Wilkinson, 1989). Due to their pleiotropic effects (i.e. the phenotypic effects felt outside the visual system), genetic constraints could cause similarities to arise even if the common ancestor of insects and crustaceans did not have compound eyes.

Myriapod Vision

One means of establishing the significance of 'non-visual' constraints as a cause of the similarities between crustaceans and insects is to compare their visual systems with those of a third related group. Myriapods (including centipedes and millipedes, classes Chilopoda and Diplopoda respectively) are thought to be allied to insects and both are placed in the major taxon Tracheata alongside the Crustacea.

Most myriapods have poor vision and some possess only ocelli (simple eyes). This might seem to be a fatal flaw in the hypothesis that the compound eyes of insects and crustaceans are homologous, since the typical myriapod state could be argued to represent the primitive state of the Tracheata from which compound eyes have evolved. But eyes can easily be lost when they are not needed, and perhaps because myriapods normally live in moist, dark places their vision has degenerated (Paulus, 1979). However, centipedes in the order Scutigera, which are notable for their cursorial habits, have (re-evolved?) good compound eyes. (Scutigerids have several primitive myriapod characters, and so their compound eyes might also be primitive (Hennig, 1981).) Paulus (1979) argues that the anatomy of the optics and retina of scutigerids differs from the common anatomy of insects and crustaceans (see also Meinertzhagen, this volume, chapter 16). A similar pattern can be seen in the anatomy of the optic lobe, so far as it is known, with only superficial similarities between the visual systems of centipedes and insects. The scutigerid centipede *Thereuopoda* has two well-defined optic ganglia (named the lamina and medulla), however there is no optic chiasm between the first and second ganglia, and (unlike in insects) a direct projection from the lamina bypasses the medulla (Hanström, 1928; Fahlander, 1938; cited in Joly and Deschamps, 1987). The millipede *Julus* has a small compound eye and, like *Thereuopoda*, also has two optic ganglia and no chiasm, but no lamina outputs bypass the medulla (Hanström, 1928). In fact, the visual system of *Julus* is rather like that of *Daphnia* (see pp 222–223).

More studies are needed, but insect and scutigerid visual systems appear to have less in common than those of crustaceans and insects, notwithstanding the generally agreed affinity of myriapods and insects. This argues that similarities between the visual systems of insects and crustaceans are not due mainly to embryological or genetic constraints that are general to arthropods and are independent of the necessity for good vision, but instead, unlike scutigerids, neither insects nor malacostracans have passed through a 'partially blind' phase during their evolution (Paulus, 1979).

Primitive Arthropod Vision and the Cambrian Explosion

If homology and stasis rather than convergence do explain

most of the similarities between the two best-known compound-eye visual systems their ancestors will probably not be found, since no known arthropod is a plausible common ancestor for the Tracheata (insects, myriapods) and the Crustacea. This ancestor must have lived before the mid-Cambrian, when the first malacostracan crustaceans appear in the fossil record (see below), and to justify a complex visual system one can imagine that it was a large and mobile animal (the vision of small and sessile animals is generally inferior; see pp 222–223). The widespread occurrence of UV-sensitive LVFs argues that this animal lived in shallow water, where UV is present! Other likely characteristics of the common ancestor of insects and crustaceans or Arthropoda in general are outlined elsewhere (Hennig, 1981; Briggs and Fortey, 1989).

The first part of the Cambrian was a period of great diversification in animal morphologies, known as the 'Cambrian explosion', and several major groups including the Arthropoda emerge in the fossil record. Large arthropods (with compound eyes) seem to have evolved comparatively rapidly from soft-bodied ancestors at least to the level of complexity of some modern malacostracans (see below; Cisne, 1982; Briggs and Fortey, 1989; Gould, 1989). Maybe the visual system appeared early during the period of 'phyletic experimentation' and was inherited by several classes which went on to diversify radically in other ways (limb structure, for instance) for a longer period (Gould, 1989).

Ancestral Forms

The suggestion that the visual system has remained unchanged in the major arthropod lineages is supported by modern forms which are believed (from general comparative anatomy) to have changed least from the ancestor. The Crustacea have primitive arthropod characters, when compared with other Cambrian taxa, like trilobites (Briggs and Fortey, 1989), and the malacostracan subclass Phyllocarida is not only primitive but is also represented by one

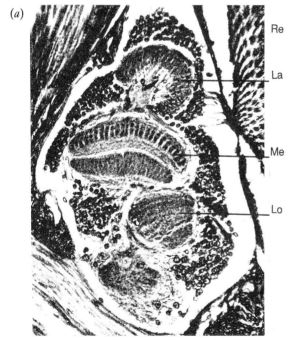

Machiloides : (Archaeognatha)

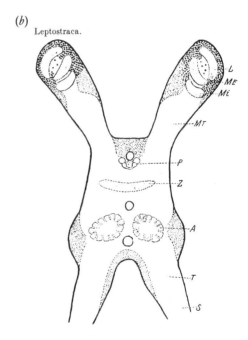

Nebalia : (Phyllocarida: Leptostraca)

Fig. 10.9 *Illustrations (from Hanström, 1928, 1940) showing the optic lobes of two groups which are believed to possess ancestral characters. Neither show any marked differences from other modern insects and malacostracan crustaceans. (a) Optic lobe of* Machiloides *(Archaeognatha), an apterygote insect. Note the presence of two chiasms, and the columnar distal medulla separated from the proximal medulla by a broad serpentine layer. The lobula is also divided into proximal and distal parts as in the locust (Fig. 10.4). Labelling as in Fig. 10.7(b). (b) Diagram of the visual system of a modern representative of the Phyllocarida,* Nebalia *(order Leptostraca). Note the presence of stalked eyes, which appear to be a primitive character in the malacostraca but are generally absent from insects. L: lamina; ME: medulla externa (= medulla); Mi: medulla interna (= lobula); MT: medulla terminalis (= protocerebrum?); P: protocerebral commissure; Z: central body; A: antennal glomeruli; T: tritocerebrum; S: circumoesophageal connective.*

of the earliest known fossil crustaceans, *Canadaspis* (Briggs, 1978). *Canadaspis* was about 60 mm long, it probably lived on the floor of a shallow sea and, like most malacostracans, had stalked eyes (Briggs, 1978). A modern representative of Phyllocardia (order Leptostraca), *Nebalia bipes*, also has stalked eyes and the three optic ganglia typical of Malacostraca (Fig. 10.9(b); Hanström, 1928).

Similarly, in the insects the Archaeognatha (bristletails) are wingless insects believed to be a relict stock that have diverged little from the original insects that lived prior to the early Devonian period soon after the initial colonization of land (Hennig, 1981). Modern Archaeognatha, like *Machiloides* and *Petrobius*, have good compound eyes with a clearly recognizable lamina, medulla and lobula separated by two chiasms (Fig. 10.9(a); Hanström, 1940). The medulla is divided into two parts by a serpentine layer just as in flying insects and the lobula also has distinct proximal and distal parts.

Clearly it will not be possible to discern the evolution of the neural structures and physiology underlying visual processing by looking at primitive forms. The neural processes and overall 'computational organization' underlying arthropod vision, as reflected in the general anatomy of the optic lobe, has changed little since the Cambrian.

Patterns of Evolution

Genetic Constraints on the Evolution of Compound Eyes

Arthropod evolutionary history seems to differ from that of vertebrates where there has been a tendency for overall brain size and the relative sizes of different parts of the brain to vary considerably according to life style, especially among mammals and birds. Indeed the absence of any gross changes in the anatomy of the optic lobe, and hence the amount of brain devoted to retinoptic visual processing, between crustaceans or apterygote insects (order Archaeognatha) and flying insects is surprising; apparently the basic plan of the optic lobe did not need elaboration when insects took wing (compare Fig. 10.7(a,b,d) with Figs 10.7(c,e) and 10.9(a). Flying insects have large eyes, but there is no disproportionate elaboration of any optic lobe ganglion. By comparison, birds have a much larger cerebellum than reptiles or mammals, which is often attributed to the need for manoeuvrability in flight, and other regions of the vertebrate brain also vary in size according to the animal's habits (Szarski, 1980; Harvey and Krebs, 1990). However, the apparent conservatism in insect and malacostracan visual system anatomy and (maybe) physiology does not reflect a general evolutionary stasis in arthropods. For example, the 'limbs' and their associated thoracic musculature vary greatly between locusts and flies, not to mention crayfish (Chapman, 1971; Manton, 1977; see also below).

It is an interesting possibility that the requirement for precise wiring, necessary to give reliable visual behaviour in individuals, has been responsible for phylogenetic conservatism in the compound eye. The fly's eye in particular has a very regular or 'neurocrystalline' organization (Ready et al., 1976). This regularity is not merely superficial but is exemplified by the accuracy of the projection of receptors to the lamina and in the precise numbers of cells in each lamina cartridge (Horridge and Meinertzhagen, 1970; Meinertzhagen, 1974). By comparison *Daphnia*, which has inferior vision to, and hence perhaps less need for reliability than a fly, shows high variability in the number of cells in a lamina cartridge, even between members of a single clone (Macagno et al., 1973); similarly, the number of L-cells (second order interneurones) in the grasshopper ocellus (simple eye) is quite variable (Goodman, 1977).

While the basic plan of large compound eyes may have been conserved for almost the entire history of the Arthropoda, two main kinds of evolutionary change can be seen. Firstly, there is a tendency to specialize or simplify, typified by the small crustaceans, and secondly there is flexibility within the constant ground-plan of the visual system. These two types of evolution are introduced briefly in the final parts of this chapter.

Simple Eyes and Neoteny in Crustacean Evolution

The preceding discussion has stressed the similarity of the compound eye and its visual ganglia in insects and crustaceans. The close resemblance of the two visual systems, and evidence from the myriapods, argue that their common ancestor also had good vision, and that this was not lost in either lineage. To pursue these arguments Malacostraca (specifically decapods) are taken as a representative, implying that theirs is the primitive crustacean visual system. But many crustacean classes, generally smaller and less familiar animals such as *Daphnia*, copepods, ostracods and barnacles (classes Branchiopoda, Copepoda, Ostracoda and Cirripeda respectively), do not have this common type of visual system. *Daphnia* has compound eyes and a familiar retinal and lamina anatomy (with six SVFs projecting to five lamina monopolar cells and two LVFs which end in the (much-reduced) medulla), but lacks a lobula and a first optic chiasm. The other subclasses lack compound eyes entirely and they often have reduced visual systems and visual behaviour (Bullock and Horridge, 1965; Macagno et al., 1973; Macagno and Levinthal 1975; Strausfeld and Nässel 1981; Sandeman, 1982; Land, 1984).

Instead of compound eyes many of the smaller crustaceans retain the nauplius eye, the (optically) simple eye used by the larvae. Crustaceans possessing nauplius eyes as adults are often small or live deep in the ocean (Land,

1984). Under these conditions, simple visual systems arise because the eye cannot make a useful image. Moreover, many small crustaceans are transparent, an excellent form of camouflage that is inevitably compromised by visual pigments. The vision of a pelagic copepod is described by Gregory (this volume, chapter 3). Barnacles, on the other hand, are sessile, their visual behaviour seems to be restricted to a 'shadow-reflex' and they have a rudimentary visual nervous system (Gwilliam and Millecia, 1975). (An intermediate, mobile, phase in the barnacle's life cycle does have compound eyes (Land, 1984).)

If crustacean compound eyes are primitive the persistence of nauplius eyes in adults is a neotenous character. 'Neotenous' is used here when strictly the term should be '*paedomorphic*'. Paedomorphy can take two forms: neoteny when juvenile features are retained in the adult, and progenesis when juveniles acquire gonads (Schram, 1982).) For reasons that have nothing to do with vision *per se*, perhaps because crustaceans have evolved in unstable environments where rapid sexual maturation and the production of many small offspring are beneficial, it has been suggested that paedomorphy has been a major feature of crustacean evolution (Schram, 1982). The absence of compound eyes in several orders (but not barnacles), especially where the adults are small, may be a corollary of this type of evolution as well as being due to the direct selective pressures.

Neoteny is not an important feature of insect evolution (Schram, 1982), but there is clear evidence for convergence with the Crustacea in small, flightless (usually parasitic) orders where compound eyes have been lost secondarily. These include fleas (Siphonaptera), scale insects (Hemiptera), various 'lice' (Psocoptera, Mallophaga, and Anoplura), and also the springtails (Collembola), which although belonging to an ancient flightless lineage are a 'very derived dwarf form' (Hennig, 1981).

Specializations and simplifications engendered by sedentary life styles, small size, dim habitats or a need for accelerated maturation, and not increases in complexity, may have been the mainspring of major change in arthropod visual anatomy. If the retinal image is dim or blurred there will be no justification for costly neural processing machinery and so simplicity should not be equated with the ancestral state in attempts to reconstruct the evolutionary processes leading to the development of particular 'advanced' faculties. The view that compound eyes with three optic ganglia are a primitive feature is consistent with the finding that representatives of ancient orders have such visual systems (see pp 221–222), and can be contrasted with that of the anatomist Hanström (1926), who based a phylogeny of the Arthropoda on the basis (among others) that evolution was marked by a consistent increase in the number of visual ganglia (Fig. 10.6).

The observation that compound eyes are absent only from adult crustaceans and insects which are small or sessile raises a well-known 'unsolved problem' in physiolgical optics: Why do compound eyes exist when simple eyes seem to be superior, giving better resolution for a given size of eye (Kirschfeld, 1974; Land, 1981)? Why have copepods with advanced simple eyes not displaced crabs as the predominant large crustaceans? Similarly, in insects the larvae in some (holometabolous) orders have simple eyes (stemmata) which are as 'good' as the adults' compound eyes (Land, 1981; Waterman, 1981): why are stemmata not retained after pupation? (For a discussion of the evolution of stemmata from compound eyes, see Paulus, 1979.) In contrast, many spiders have simple eyes with far higher acuity than any known compound eye (Land, 1985).

Substrates for Evolution: Flexibility in a Conservative Framework

Conservatism does not pervade every aspect of the 'typical' compound eye. If the general neural architecture has indeed been conserved then this has been a framework for small-scale flexibility. The visual system must have changed as insects acquired first flight and then flower recognition, but what are the substrates for this evolution?

One area of flexibility is in the pharmacology of the optic lobes. Related species often have very different distributions of neurotransmitters, especially of aminergic transmitters, which are used by (efferent?) tangential cells (Nässel and Klemm, 1983; Nässel and Elofsson, 1987; Hardie, 1989b; Nässel, 1989). Another area where variability has been found is in the synaptology of the lamina. The complement of cells in each lamina cartridge is fairly constant and some basic rules of connectivity apply across the insects and malacostracan crustaceans (e.g. monopolar cells are postsynaptic to receptors), but there is flexibility in the precise wiring of the lamina between and within orders, as shown by the studies of flies and other insects (Meinertzhagen, this volume, chapter 17; Shaw and Meinertzhagen, 1986; Kral, 1987; Shaw and Moore, 1989; Shaw, 1989, 1990).

A future challenge is to relate the variability seen in wiring and pharmacology to the evolution of particular faculties. A good candidate for this approach is spectral information processing and colour vision, where there is great diversity within the Arthropoda (Menzel, 1979; Menzel and Backhaus, 1989; Goldsmith, and Menzel and Backhaus, volume 6, chapters 5 and 14 of this series). The primitive compound eye may have had 6 or 7 longwavelength ($c.$ 500 nm) SVFs, and 1 or 2 short-wavelength LVFs (see pp 204–205, 219–220 and 222–223). Species like the locust and the crayfish (Cummins and Goldsmith, 1981; Osorio, 1986a and unpublished observations) have (retain?) this organization. Others have departed from this simple plan: the mantis shrimp (Stomatopoda) has 10 or more spectral receptor types, although it is not known what type of 'colour vision' is supported (Cronin and Marshall, 1989). The honeybee, on the other hand, appears to have trichromatic colour vision, comparable to that of man (Menzel and Backhaus, 1989 and volume 6, chapter 14 of this series). The presence of 'colour vision' is not necessarily a corollary of a large or complex visual system, for example *Daphnia* has four spectral receptor types in each ommatidium (Smith and Macagno, 1990; Goldsmith, volume 6, chapter 5 of this series).

Unfortunately, to close on a pessimistic note, there is a considerable gulf between studies of neurophysiology, anatomy and behaviour, which will have to be bridged if we are to understand the neural basis of evolutionary change in arthropod vision. Recordings from the honeybee optic lobe have provided the basis for modelling spectral information processing (Kien and Menzel, 1977; Hertel and Maronde, 1987; Menzel and Backhaus, 1989 and volume 6, chapter 14 of this series), but comparable studies of animals with simpler 'colour vision' such as the locust (Osorio, 1986a) are needed for us to begin to take an evolutionary approach. In the fly lamina, our knowledge of physiology and comparative anatomy is much greater than elsewhere in the optic lobe, but even here we cannot discern any significant physiological corollary of the observed changes in synaptic connectivity, either between cell classes in the same eye or between species (Laughlin and Osorio, 1989). The most detailed description of LMCs available (a two-dimensional spatio-temporal white-noise analysis (James, 1990; James and Osorio, in preparation)) reveals no marked differences between the lamina cell (LMC) responses of a Syrphid (*Eristalis*) and *Calliphora* (Fig. 10.2), despite differences in synaptic connectivity (Shaw, 1989), although lateral inhibition is somewhat stronger in the latter.

Further Reading

The references given here are at best selective, but to assist access to the primary literature recent works and more specialized reviews have been cited where possible. Stavenga and Hardie (1989) give an authoritative introduction to many current areas of compound eye research. Other useful reviews include Ali (1984), Singh and Strausfeld (1989) and Kerkut and Gilbert (1985). DeVoe's (1985) review in the latter volume covers much of the literature on the electrophysiology of insect compound eyes and Wiersma *et al.* (1982) provide a comparable overview of the crustacean optic lobe. Autrum (1980–1981) is unsurpassed in its depth of coverage and contains several key reviews. Finally, the comparative anatomy from early this century should not be forgotten, especially the work of Cajal and Sanchez (1915) on insects, and Hanström (1926, 1928) on general comparative anatomy. Much of this early work was summarized by Bullock and Horridge (1965). The present chapter should be read in conjunction with Horridge, Chapter 11 and Meinertzhagen, Chapter 16. Neumeyer, Chapter 13 and Goldsmith, Volume 6 Chapter 5 discuss the evolution of colour vision.

Acknowledgements

I thank A. D. Blest, G. A. Horridge, M. R. Ibbotson, A. C. James, D. R. Nässel and E. J. Warrant, who provided unpublished data and specimens, and J. Vaughan, who did the illustrations. In addition, many of my colleagues were helpful, especially W. Backhaus, G. L. G. Miklos, D. C. O'Carroll, P. Gullan and S. J. Stowe. I also thank I. A. Meinertzhagen for discussion of his work. The literary resources and knowledge of G. A. Horridge were invaluable.

References

Akam, M. (1987). The molecular basis for metameric pattern in the *Drosophila* embryo. *Development*, 101, 1–22.

Ali, M. A. ed. (1984). *Photoreceptors and Vision in Invertebrates*. New York: Plenum.

Anderson, D. T. (1973). *Embryology and Phylogeny in Annelids and Arthropods*. Oxford: Pergamon.

Autrum, H.-J. ed. (1980–1). *The Handbook of Sensory Physiology*, vol. VII. 6: *Invertebrate Visual Centres and Behaviour*. I. Berlin: Springer.

Aréchiga, H., Bañuelos, E., Frixione, E., Picones, A. and Rodriguez-Sosa, L. (1990). Modulation of crayfish retinal sensitivity by 5-hydroxytryptamine. *J. Exp. Biol.*, 150, 123–143.

Arnett, D. W. (1972). Spatial and temporal integration properties of units in the first optic ganglion of dipterans. *J. Neurophysiol.*, 35, 429–444.

Attwell, D. (1986). Ion channels and signal processing in the outer retina. *Q. J. Exp. Physiol.*, 71, 497–536.

Barlow, H. B. (1961). Possible principles underlying the transformations of sensory messages. In: *Sensory Communication*. ed. Rosenblith, W. A. pp. 217–234. Cambridge, MA: MIT.

Barlow, H. B. and Levick, W. R. (1965). The mechanism of directionally selective units in the rabbit's retina. *J. Physiol.*, 178, 477–504.

Barlow, R. B., Chamberlain, S. C. and Lehman, H. K. (1989). Circadian rhythms in the invertebrate retina. In: *Facets of Vision*. eds. Stavenga, D. G. and Hardie, R. C. pp. 257–280. Berlin: Springer.

Blest, A. D. and Davie, P. S. (1980). Reduced silver impregnations derived from the Holmes technique. In: *Neuroanatomical Techniques: Insect Nervous System*. eds. Strausfeld, N. J. and Miller, T. A. pp. 98–119. Berlin: Springer.

Borst, A. and Egelhaaf, M. (1989). Principles of visual motion detection. *Trends Neurosci.*, 12, 297–306.

Boudreaux, H. B. (1979). *Arthropod Phylogeny with Special Reference to Insects*. New York: Wiley.

Braitenberg, V. (1972). Periodic structures in the visual system of the fly. In: *Information Processing in the Visual Systems of Arthropods*. ed. Wehner, R. pp. 3–15. Berlin: Springer.

Briggs, D. E. G. (1978). The morphology, mode of life and affinities of *Canadaspsis superba* (Crustacea: Phyllocarida), Middle Cambrian, Burgess Shale, British Columbia. *Phil. Trans. R. Soc. Lond. B*, 281, 439–487.

Briggs, D. E. G. and Fortey, R. A. (1989). The early radiation and relationships of the major arthropod groups. *Science Wash. DC.*, **246**, 241–243.

Buchner, E. (1984). Behavioural analysis of spatial vision in insects. In: *Photoreception and Vision in Invertebrates*. ed. Ali, M. A. pp. 561–622. New York: Plenum.

Buchner, E., Bader, R., Buchner, S., Cox, J., Emson, P. C., Flory, E., Heizmann, C. W., Hemm, S., Hofbauer, A. and Oertel, W. H. (1988). Cell-specific immunoprobes for the brain of normal and mutant *Drosophila melanogaster*. I. Wildtype visual system. *Cell Tissue Res.*, **253**, 357–370.

Buchner, E., Buchner, S. and Bülthoff, I. (1984). Deoxyglucose mapping of nervous activity induced in *Drosophila* brain by visual movement. *J. Comp. Physiol. A*, **155**, 471–483.

Bullock, T. H. (1976). In search of principles in neural integration. In: *Simpler Networks and Behavior*. ed. Fentress, J. C. pp. 52–60. Sunderland: Sinauer.

Bullock, T. H. and Horridge, G. A. (1965). *Structure and Function of the Nervous Systems of Invertebrates. Pt II*. San Francisco: Freeman.

Cajal, S. R. and Sanchez y Sanchez, D. (1915). Contribucion al conocimiento de los centros nerviosos de los insectos. Parte I. Retina y los centros opticos. *Trab. Lab. Invest. Biol Univ. Madr.*, **13**, 1–168.

Chapman, R. F. (1971). Movement and control of wings. In: *The Insects*. pp. 184–206. London: Hodder and Stoughton.

Cisne, J. L. (1982). Origin of the Crustacea. In: *The Biology of Crustacea*. ed. Abele, L. G. vol. 2 *Systematics the Fossil Record, and Biogeography*, pp. 65–92. New York: Academic.

Collett, T. and King, A. J. (1975). Vision during flight. In: *The Compound Eye and Vision in Insects*. ed. Horridge, G. A. pp. 437–466. Oxford: Oxford University Press.

Coombe, P. E. and Heisenberg, M. (1986). The structural brain mutant *Vacuolar medulla* of *Drosohila melanogaster* with specific behavioral defects and cell degeneration in the adult. *J. Neurogen.*, **3**, 135–158.

Coombe, P. E., Srinivasan, M. V. and Guy, R. G. (1989). Are the large monopolar cells of the insect lamina on the optomotor pathway? *J. Comp. Physiol. A*, **166**, 23–35.

Cronin, T. W. and Marshall, N. J. (1989). A retina with at least ten spectral types of photoreceptors in a mantis shrimp. *Nature Lond.*, **339**, 137–140.

Cummins, D. and Goldsmith, T. H. (1981). Cellular identification of the violet receptor in the crayfish eye. *J. Comp. Physiol.*, **142**, 199–202.

Datum, K.-H., Weiler, R. and Zettler, F. (1986). Immunocytochemical demonstration of γ-amino butyric acid and glutamic acid decarboxylase in R7 photoreceptors and C2 centrifugal fibres in the blowfly visual system. *J. Comp. Physiol. A*, **159**, 241–249.

Datum, K.-H., Rambold, I. and Zettler, F. (1989). Cholinergic neurons in the lamina ganglionaris of the blowfly *Calliphora erythrocephala*. *Cell Tissue Res.*, **256**, 153–158.

DeVoe, R. D. (1980). Movement sensitivities in the fly's medulla. *J. Comp. Physiol. A*, **138**, 93–119.

DeVoe, R. D. (1985). The eye: electrical acitivity. In: *Comprehensive Insect Physiology, Biochemistry and Pharmacology, Vol. 6: Nervous System: Sensory*. eds. Kerkut, G. A. and Gilbert, L. I. pp. 277–354. Oxford: Pergamon.

DeYoe, E. A. and Van Essen, D. C. (1988). Concurrent processing streams in monkey visual cortex. *Trends Neurosci.*, **11**, 219–226.

Dubs, A. (1982). The spatial integration of signals in the retina and lamina of the fly compound eye under different conditions of luminance. *J. Comp. Physiol. A*, **146**, 321–334.

Edwards, J. S. and Meyer, M. R. (1990). Conservation of Antigen 3G6: a crystalline cone constituent in the compound eye of arthropods. *J. Neurobiol.*, **21**, 441–452.

Egelhaaf, M., Borst, A. and Pilz, G. (1990). The role of GABA in detecting visual motion. *Brain Res.*, **509**, 156–160.

Egelhaaf, M., Borst, A. and Reichardt, W. (1989). Computational structure of a biological motion detection system as revealed by local detector analysis in the fly's nervous system. *J. Opt. Soc. Am. A*, **6**, 1070–1087.

Egelhaaf, M., Hausen, K., Reichardt, W. and Wehrhahn, C. (1988). Visual course control in flies relies on neuronal computation of object and background motion. *Trends Neurosci.*, **11**, 351–358.

Elofsson, R. (1983). 5-HT like immunoreactivity in the central nervous system of the crayfish *Pacifastacus leniusculus*. *Cell Tissue Res.*, **232**, 221–236.

Elofsson, R. and Hagberg, M. (1986). Evolutionary aspects on the construction of the first optic neuropile (lamina) in Crustacea. *Zoomorphology*, **106**, 174–178.

Elofsson, R. and Klemm, N. (1972). Monoamine-containing neurons in the optic ganglia of crustaceans and insects. *Z. Zellforsch.*, **133**, 475–499.

Fahlander, K. (1938). Beiträge zur Anatomie und systematischen Einteilung der Chilopoda. *Zool. Bidr. Upps.*, **17**, 1–148.

Fischbach, K.-F. and Dittrich, A. P. M. (1989). The optic lobe of *Drosophila melanogaster*. I. A Golgi analysis of wild type structure. *Cell Tiss. Res.*, **258**, 441–475.

Fischbach, K.-F. and Heisenberg, M. (1984). Neurogenetics and behaviour in insects. *J. Exp. Biol.*, **112**, 65–93.

Fischbach, K.-F., Barleben, F., Boschert, U., Dittrich, A. P. M., Gschwander, G., Hoube, B., Jaeger, R., Kaltenbach, E., Ramos, R. G. P. and Schlosser, G. (1989). Developmental studies on the optic lobe of *Drosophila melanogaster* using structural brain mutants. In: *Neurobiology of Sensory Systems*. eds. Singh, R. N. and Strausfeld, N. J. pp. 171–194. New York: Plenum.

Fleissner, G. (1982). Isolation of an insect circadian clock. *J. Comp. Physiol. A*, **149**, 311–316.

Fraenkel, G. S. and Gunn, D. L. (1961). *The Orientation of Animals*. New York: Cover.

Franceschini, N., Riehle, A. and LeNestour, A. (1989). Directionally selective motion detection by insect neurons. In: *Facets of Vision*. eds. Stavenga, D. G. and Hardie, R. C. pp. 360–390. Berlin: Springer.

Goodman, C. S. (1977). Neuron duplications and deletions in locust clones and clutches. *Science Wash. DC*, **197**, 1384–1386.

Gould, S. J. (1989). *Wonderful Life: The Burgess Shale and the Nature of History*. New York: Norton.

Grüsser, O. J. and Grüsser-Cornehls, U. (1973). Neuronal mechanisms of visual movement perception and some psychophysical and behavioural correlates. In: *Handbook of Sensory Physiology*, vol. VII/3: *Central Information Processing*. ed. Jung, R. pp. 333–429. Berlin: Springer.

Gupta, A. P. (1979). *Arthropod Phylogeny*. New York: Van Nostrand: Reinhold.

Gwilliam, G. F. and Millecia, R. J. (1975). Barnacle photoreceptors: their physiology and role in the control of behaviour. *Progr. Neurobiol.*, **4**, 71–102.

Hanström, B. (1924). Untersuchengen ueber das Gehirn, insbesondere die Sehganglien der Crustaceen. *Arch. Zool.*, **16**, 1–119.

Hanström, B. (1926). Eine genetische studie über die Augen und Sehzentren von Turbellarien, Anneliden und Arthropoden (Trilobiten, Xiphosuren, Eurypteriden, Arachnoiden, Myrispoden, Crustaceen und Insecten). *K. Svenska Vetensk Akad. Handl.*, **4** (1), 1–176.

Hanström, B. (1928). *Vergleichende Anatomie des Nervensystems der wirbellosen Tiere*. Berlin: Springer.

Hanström, B. (1940). Inkretorische organe, sinnesorgane und nervensystem des kopfes einiger niederer insektenordnungen. *K. Svenska Vetensk Akad. Handl.*, **18**(8), 1–265.

Hardie, R. C. (1985). Functional organisation of the fly retina. In: *Progress in Sensory Physiology, vol. 5*. ed. Ottson, D. pp. 1–79. Berlin: Springer.

Hardie, R. C. (1986). The photoreceptor array of the dipteran retina. *Trends Neurosci.*, **9**, 419–423.

Hardie, R. C. (1989a). A histamine-activated chloride channel involved in neurotransmission at a photoreceptor synapse. *Nature Lond.*, **339**, 704–706.

Hardie, R. C. (1989b). Neurotransmitters in compound eyes. In: *Facets of Vision*. eds. Stavenga, D. G. and Hardie, R. C. pp. 235–356. Berlin: Springer.

Hardie, R. C. and Stavenga, D. G. (eds) (1989). *Facets of Vision*. Berlin: Springer.

Harvey P. H. and Krebs, J. R. (1990). Comparing brains. *Science Wash DC*, **249**, 140–146.

Hausen, K. and Egelhaaf, M. (1989). Neural mechanisms of visual course control in insects. In: *Facets of Vision*. eds. Hardie, R. C. and Stavenga, D. G. pp. 391–424. Berlin: Springer.

Heisenberg, M. and Buchner, E. (1977). The role of retinula cell types in visual behaviour of *Drosophila melanogaster*. *J. Comp. Physiol.*, **117**, 127–162.

Heisenberg, M. and Wolf, R. (1988). Reafferent control of optomotor yaw torque in *Drosophila melanogaster*. *J. Comp. Physiol. A*, **163**, 373–388.

Hennig, W. (1981). *Insect Phylogeny*. Chichester: Wiley.

Hertel, H. and Maronde, U. (1987). Processing of visual information in the honeybee brain. In: *Neurbiology and Behaviour of Honeybees*, eds. Menzel, R. and Mercer, A. pp. 141–157. Berlin: Springer.

Hoffmann, C. (1964). Bau und Vorkommen von propriozeptiven Sinnesorganen bie den Arthropoden. *Ergebn. Biol.*, **27**, 1–38.

Honegger, H-W. (1978). Sustained and transient responding units in the medulla of the cricket *Gryullus campestris*. *J. Comp. Physiol.*, **125**, 259–266.

Honegger, H-W. (1980). Receptive fields of sustained medulla neurons in crickets. *J. Comp. Physiol., A*, **136**, 191–201.

Horridge, G. A. (1987). The evolution of visual processing and the construction of seeing systems. *Proc. R. Soc. Lond. B*, **230**, 279–292.

Horridge, G. A. and Meinertzhagen, I. A. (1970). The exact neural projection of the visual fields upon the first and second ganglia of the insect eye. *Z. Vergl. Physiol.*, **66**, 369–378.

Ibbotson, M. R. (1990). Wide-field motion-sensitive neurons tuned to horizontal movement in the honeybee, *Apis mellifera*. *J. Comp. Physiol. A*, (in press).

Ibbotson, M. R., Maddess, T. and DuBois, R. A. (1990a). A system of insect neurons sensitive to horizontal and vertical image motion connects the distal medulla and midbrain: I Anatomy. *J. Comp. Physiol. A*, (in preparation).

Ibbotson, M. R., Maddess, T. and DuBois, R. A. (1990b). A system of insect neurons sensitive to horizontal and vertical image motion connects the distal medulla and midbrain. II Directional tuning properties. *J. Comp. Physiol. A*, (in preparation).

James, A. C. (1990). White-noise Studies in the Fly Lamina. Ph.D. Thesis, Australian National University, Canberra.

Joly, R. and Deschamps, M. (1987). Histology and ultrastructure of the myriapod brain. In: *Arthropod Brain: Its Evolution, Development, Structure, and Functions*. ed. Gupta, A. P. pp. 135–158. New York: Wiley.

Kerkut. G. A. and Gilbert, L. I. eds. (1985). *Comprehensive Insect Physiology, Biochemistry and Pharmacology*, vol. 6: *Nervous System: Sensory*. Oxford: Pergamon.

Kien, J. and Menzel, R. (1977). Chromatic properties of interneurons in the optic lobes of the bee. II. Narrow band and colour opponent neurons. *J. Comp. Physiol.*, **113**, 17–34.

Kirk, M. D., Waldrop, B. and Glantz, R. M. (1982). The crayfish sustaining fibers. I. Morphological representation of visual receptive fields in the second optic neuropil. *J. Comp. Physiol. A*, **146**, 175–179.

Kirschfeld, K. (1967). Die projektion der optischen Umwelt auf das Roster der Rhabdomere im Komplexange von *Musca*. *Exp. Brain Res.*, **3**, 248–270.

Kirschfeld, K. (1974). The absolute sensitivity of lens and compound eyes. *Z. Naturforsch.*, **29c**, 592–596.

Kral, K. (1987). Organisation of the first optic neuropil (or lamina) in different insect species. In: *Arthropod Brain: Its Evolution, Development, Structure and Functions*. ed. Gupta, A. P. pp. 181–202. New York: Wiley.

Land, M. F. (1981). Optics and vision in invertebrates. In: *Handbook of Sensory Physiology*. ed. Autrum, H. vol. *VII/6B: Invertebrate Visual Centres and Behaviour. I*. pp. 471–592. Berlin: Springer.

Land, M. F. (1984). Crustacea. In: *Photoreception and Vision in Invertebrates*. ed. Ali, M. A. pp. 401–438. New York: Plenum.

Land, M. F. (1985). The morphology and optics of spider eyes. In: *Neurobiology of Arachnids*. ed. Barth, F. G. pp. 53–78. Berlin: Springer.

Laughlin, S. B. (1988). Form and function in visual processing. *Trends Neurosci.*, **10**, 478–483.

Laughlin, S. B. (1989). Coding efficiency and design in visual processing. In: *Facets of Vision*. eds. Hardie, R. C. and Stavenga, D. G. pp. 213–234. Berlin: Springer.

Laughlin, S. B., Howard, J. and Blakeslee, B. (1987). Synaptic limitations to contrast coding in the retina of the blowfly *Calliphora*. *Proc. R. Soc. Lond. B*, **231**, 437–467.

Laughlin, S. B. and Osorio, D. (1989). Mechanisms for neural signal enchancement in the blowfly compound eye. *J. Exp. Biol.*, **144**, 113–146.

Levick, W. R. and Thibos, L. N. (1983). Receptive fields of cat ganglion cells: classification and construction. In: *Progress in Retinal Research*, vol. 2 eds. Osborne, N. and Chader, G. pp. 267–319: Oxford: Pergamon.

Macagno, E. R. and Levinthal, C. (1975). Computer reconstruction of the cellular architecture of the *Daphnia magna* optic ganglion. In: *Proceedings E.M. Soc. Am. 33rd Annual Meeting*, ed. Bailey, G. W. pp. 284–285. Baton Rouge: Claitor's.

Macagno, E. R., Lopresti, V. and Levinthal, C. (1873). Structure and development of neuronal connections in isogenic organisms: variations and similarities in the optic system of *Daphnia magna*. *Proc. Natl Acad. Sci. USA*, **70**, 57–61.

Maddess, T., DuBois, R. A. and Ibbotson, M. R. (1990). Response properties and adaptation of neurons sensitive to image motion in the butterfly *Papilio aegeus*. *J. Exp. Biol.*, in press.

Manton, S. M. (1977). *The Arthropoda: Habits, Functional Morphology and Evolution*. Oxford: Clarendon.

Marr, D. and Hildreth, E. (1980). A theory of edge detection. *Proc. R. Soc. Lond B.*, **207**, 187–217.

Menzel, R. (1979). Spectral sensitivity and colour vision in invertebrates. In: *Handbook of Sensory Physiology*, vol. VII/6A: Invertebrate Visual Centres and Behaviour. I. ed. Autrum, H, pp. 503–580. Berlin:Springer.

Menzel, R. and Backhaus, W. (1989). Colour vision in honey bees: phenomena and physiological mechanism. In: *Facets of Vision*. eds. Stavenga, D. G. and Hardie, R. C., pp. 281–297.

Menzel, R. and Blakers, M. (1976). Colour receptors in the bee eye: morphology and spectral sensitivity. *J. Comp. Physiol.*, **108**, 11–33.

Nässel, D. R. (1977). Types and arrangements of neurons in the crayfish optic lobe. *Cell Tiss. Res.*, **179**, pp. 45–75.

Nässel, D. R. (1989). Chemical neuroanatomy of the insect visual system. In: *Neurobiology of Sensory Systems*. eds. Singh, R. N. and Strausfeld, N. J. pp. 295–318. New York: Plenum.

Nässel, D. R. and Elofsson, R. (1987). Comparative anatomy of the crustacean brain. In: *Arthropod Brain: Its Evolution Development Structure and Functions*. ed. Gupta, A. P. pp. 111–133. New York: Wiley.

Nässel, D. R., Holmqvist, M., Hardie, R. C., Hakanson, R. and Sundler, F. (1988). Histamine-like immunoreactivity in photoreceptors of the compound eyes and ocelli of the flies *Calliphora erythrocephala* and *Musca domestica*. *Cell Tiss. Res.*, 253, 639–646.

Nässel, D. R. and Klemm, N. (1983). Serotonin like immunoreactivity in the optic lobes of three insect species. *Cell Tissue Res.*, 232, 129–140.

Nässel, D. R. and Waterman, T. H. (1977). Golgi EM evidence for visual information channeling in the crayfish lamina ganglionaris. *Brain Res.*, 130, 556–563.

Nicol, D. and Meinertzhagen, I. A. (1982). An analysis of the number and composition of the synaptic populations formed by photoreceptors of the fly. *J. Comp. Neurol.*, 207, 29–44.

Nilsson, D.-E. (1989). Optics and evolution of the compound eye. In: *Facets of Vision*. eds. Hardie, R. C. and Stavenga, D. G. pp. 30–73. Berlin: Springer.

Nilsson, D.-E. (1990). From cornea to retinal image in invertebrate eyes. *Trends Neurosci.*, 13, 55–64.

Osorio, D. (1986a). Ultraviolet sensitivity and spectral opponency in the locust. *J. Exp. Biol.*, 122, 193–208.

Osorio, D. (198b). Directionally selective cells in the locust medulla. *J. Comp. Physiol. A*, 159, 841–847.

Osorio, D. (1987a). Temporal and spectral properties of sustaining cells in the medulla of the locust. *J. Comp. Physiol. A*, 161, 441–448.

Osorio, D. (1987b). The temporal properties of non-linear transient cells in the locust medulla. *J. Comp. Physiol. A.*, 161, 431–440.

Osorio, D. (1990). Inhibition, spatial pooling and signal rectification in the subunits of non-directional motion sensitive cells in the locust optic lobe. *Vis. Neurosci*, submitted.

Osorio, D., Snyder, A. W. and Srinivasan, M. W. (1987). Bi-partitioning and boundary detection in natural scenes. *Spatial Vision*, 2, 191–198.

Paulus, H. F. (1979). Eye structure and the monophyly of the arthropoda. In: *Arthropod Phylogeny*. ed. Gupta, A. P. pp.299–383. New York: Van Nostrand Reinhold.

Pfeiffer, C. and Glantz, R. M. (1989). Cholinergic synapses and the organization of contrast detection in the crayfish optic lobe. *J. Neurosci.*, 9, 1873–1882.

Pfeiffer-Linn, C. and Glantz, R. M. (1989). Acetylcholine and GABA mediate opposing actions on neuronal chloride channels in crayfish. *Science Wash. DC*, 245, 1249–1251.

Poggio, T. and Reichardt, W. (1976). Visual control of orientation behaviour in the fly. II. Towards the underlying neural mechanisms. *Q. Rev. Biophys.*, 9, 377–483.

Ready, D. F. (1989). A multifacetted approach to neural development. *Trends Neurosci.*, 12, 102–110.

Ready, D. F., Hanson, T. E. and Benzer, S. (1976). Development of the *Drosophila* retina, a neurocrystalline lattice. *Dev. Biol.*, 53, 217–240.

Reichardt, W. (1969). Movement perception in insects. In: *Processing of Optical Data by Organisms and By Machines*. ed. Reichardt, W. pp. 465–493. New York: Academic.

Ribi, W. A. (1977). Fine structure of the first optic ganglion of the cockroach *Periplaneta americana*. *Tissue Cell*, 9, 57–72.

Ribi, W. A. (1981). The first optic ganglion of the bee. IV. Synaptic fine structure and connectivity patterns of receptor axons and first order interneurones. *Cell Tissue Res.*, 236, 443–464.

Ribi, W. A. (1987). Anatomical identification of spectral receptor types in the retina and lamina of the Australian orchard butterfly, *Papilio aegeus aegeus* D. *Cell Tissue Res.*, 247, 393–407.

Rind, F. C. (1987). Non-directional, movement sensitive neurons of the locust optic lobe. *J. Comp. Physiol. A*, 161, 477–494.

Rind, F. C. (1990). Identification of directionally selective motion-detecting neurones in the locust lobula and their synaptic connections with an identified descending neurone. *J. Exp. Biol.*, 149, 21–43.

Rossel, S. (1989). Polarization sensitivity in compound eyes. In: *Facets of Vision*. eds. Hardie, R. C. and Stavenga, D. G. pp. 298–316. Berlin: Springer.

Rossel, S. and Wehner, R. (1986). Polarization vision in bees. *Nature Lond.*, 323, 128–131.

Rowell, C. H. F., O'Shea, M. and Williams, J. L. D. (1977). The neuronal basis of a sensory analyser, the acridid movement detector system. III. Control of response amplitude by tonic lateral inhibition. *J. Exp. Biol.*, 65, 617–625.

Rowell, C. H. F. (1989). The taxonomy of invertebrate neurons: a plea for a new field. *Trends Neurosci.*, 12, 169–174.

Sabra, R. and Glantz, R. M. (1985). Polarisation sensitivity of crayfish photoreceptors is correlated with their termination sites in the lamina ganglionaris. *J. Comp. Physiol. A*, 156, 315–318.

Sandeman, D. C. (1982). Organization of the central nervous system. In: *The Biology of Crustacea, vol. III: Neurobiology: Structure and Function*. eds. Atwood, H. L. and Sandeman, D. C. pp. 1–61. New York: Academic.

Schaefer, S. (1987). Immunocytologische Untersuchungen am Bienengehirn. Ph.D. Thesis, Freie Universität Berlin.

Schram, R. F. (1982). The fossil record and evolution of Crustacea. In: *The Biology of Crustacea, vol. 2: Systematics, the Fossil Record, and Biogeography*. ed. Abele, L. G. pp. 93–147. New York: Academic.

Shaw, S. R. (1984). Early visual processing in insects. *J. Exp. Biol.*, 112, 225–251.

Shaw, S. R. (1989). The retina–lamina pathway in insects, particularly Diptera, viewed from an evolutionary perspective. In: *Facets of Vision*. eds. Hardie, R. C. and Stavenga, D. G. pp. 186–212. Berlin: Springer.

Shaw, S. R. (1990). The photoreceptor axon projection and its evolution in the neural superposition eyes of some primitive brachyceran Diptera. *Brain Behav. Evol.*, 35, 107–125.

Shaw, S. R. and Meinertzhagen, I. A. (1986). Evolutionary progression at synaptic connections made by identified homologous neurons. *Proc. Natl Acad. Sci., USA*, 83, 7961–7965.

Shaw, S. R. and Moore, D. (1989). Evolutionary remodelling in a visual system through extensive changes in the synaptic connectivity of homologous neurons. *Vis. Neurosci.*, 3, 405–410.

Singh, R. N. and Strausfeld, N. J., eds. (1989). *The Neurobiology of Sensory Systems*. New York: Plenum.

Smith, K. C. and Macagno, E. R. (1990). UV photoreceptors in the compound eye of *Daphnia magna* (Crustacea, Branchiopoda). A fourth spectral class in single ommatidia. *J. Comp. Physiol. A*, 166, 597–606.

Snodgrass, R. E. (1938). Evolution of the Annelida, Onychophora, and Arthropoda. *Smithson Misc. Collect.*, 97, 1–59.

Snyder, A. W. (1979). The physics of vision in compound eyes. In: *Handbood of Sensory Physiology*, vol. VII/6A: *Invertebrate Photoreceptors*. ed. Autrum, H. pp. 225–314. Berlin: Springer.

Srinivasan, M. V. (1991). How insects infer range from visual motion. In: *Visual Motion and Its Role in the Stabilization of Gaze*. eds. Miles, F. A. and Wallman, J. Amsterdam: Elsevier. in press.

Srinivasan, M. V., Laughlin, S. B. and Dubs, A. (1982). Predictive coding: a fresh view of inhibition in the retina. *Proc. R. Soc. Lond. B.*, 216, 427–459.

Srinivasan, M. V., Pinter, R. B. and Osorio, D. (1990). Matched filtering in the visual system of the fly: large monopolar cells of the lamina are optimized to detect moving edges and blobs. *Proc. R. Soc. Lond. B*, (in press).

Strausfeld, N. J. (1976). *An Atlas of an Insect Brain*. Berlin: Springer.

Strausfeld, N. J. (1989). Beneath the compound eye: neuroanatomical analysis and physiological correlates in the study of insect vision. In: *Facets of Vision*. eds. Hardie, R. C. and Stavenga, D. G. pp. 317–359. Berlin: Springer.

Strausfeld, N. J. and Nässel, D. R. (1981). Neuroarchitecture of brain regions that subserve the compound eyes of Crustacea and insects. In: *Handbook of Sensory Physiology*, vol. VII/6B: *Invertebrate Visual Centres and Behaviour. I.* ed. Autrum, H. pp. 1–132. Berlin:

Springer.

Szarski, H. (1980). A functional and evolutionary interpretation of brain size in vertebrates. *Evol. Biol.*, **13**, 149–174.

Thomas, J. B., Bastiani, M. H., Bate, M. and Goodman, C. S. (1984). From grasshopper to *Drosophila*: a common plan for neuronal development. *Nature Lond.*, **310**, 203–207.

Tiegs, O. W. and Manton, S. M. (1958). The evolution of the Arthropoda. *Biological Reviews*, **33**, 255–337.

Tinbergen, J. and Abeln, R. G. (1983). Spectral sensitivity of the landing blowfly. *J. Comp. Physiol. A*, **150**, 319–328.

Torre, V. and Poggio, T. (1978). A synaptic mechanism possibly underlying directional selectivity to motion. *Proc. R. Soc. Lond. B*, **202**, 409–416.

Truillo Cenoz, O. (1985). The eye: development structure and neural connections. In: *Comprehensive Insect Physiology Biochemistry and Pharmacology*, vol. 6: *Nervous System: Sensory*. eds. Kerkut, G. A. and Gilbert, L. I. pp. 171–223. Oxford: Pergamon.

Wang-Bennett, L. T. and Glantz, R. M. (1987a). The functional organization of the crayfish lamina ganglionaris. I. Nonspiking monopolar cells. *J. Comp. Physiol. A*, **161**, 131–145.

Wang-Bennett, L. T. and Glantz, R. M. (1987b). The functional organization of the crayfish lamina ganglionaris. II. Large-field spiking and nonspiking cells. *J. Comp. Physiol. A*, **161**, 147–160.

Wang-Bennet, L. T. Pfeiffer, C., Arnold, J. and Glantz, R. M. (1989). Acetylcholine in the crayfish optic lobe: concentration profile and cellular localisation. *J. Neurosci.*, **9**, 1864–1871.

Waterman, T. H. (1981). Polarisation sensitivity. In: *Handbook of Sensory Physiology*, vol. VII/6B: *Invertebrate Visual Centres and Behaviour. I*. ed. Autrum, H. pp. 281–470. Berlin: Springer.

Weyrauther, E. (1989). Requirements for screening pigment migration in the eye of *Ephestia kuehniella* Z. *J. Insect Physiol.*, **35**, 925–934.

Wiersma, C. A. G., Roach, J. L. M. and Glantz, R. M. (1982). Neural integration in the optic system. In: *The Biology of Crustacea*, vol. 4. eds. Atwood, H. and Sandeman, D. C. pp. 1–31. New York: Academic.

Wilkinson, D. G. (1989). Homeobox genes and development of the vertebrate CNS. *BioEssays*, **10**, 82–85.

Willmer, P. (1990). *Invertebrate Relationships: Patterns in Animal Evolution*. Cambridge: Cambridge University Press.

11 Evolution of Visual Processing

G. Adrian Horridge

Introduction

The distribution of eyes, their anatomy and the visual behaviour that they make possible, are well documented in comprehensive texts, of which several are referenced with an asterisk (*) at the end of this article. Here I will endeavour to draw attention to a few general principles that apply to the simpler examples of natural visual processing so far as we know it in invertebrates, and to our primitive efforts to copy low-level natural vision into artificial systems. These principles apply to a very large number of scattered examples of eye types and to a wide variety of visual behaviour, but most of the data actually refer to insect vision, which is reasonably well known.

The Anthropocentric View of Vision

The first hurdle to be overcome, and the most difficult, is the acceptance of the view of primitive vision as necessarily resembling human vision. Humans create a marvellous visual world which arrives in consciousness already endowed with meaning. We see trees, dogs and faces already given names and nouns by deep unconscious interaction with language and other memories. We overcome these preconceptions with the aid of comparative anatomy and comparative visual behaviour, which immediately reveals the limitation of the view that (say) insect vision is like all VISION which is like human VISION. I prefer to call the visual performance of animals like insects *semivision*, because we have no evidence that they have our kind of visual world, although they obviously see very well.

No Comprehensive Theory

The evolution of eyes and visual processing encompasses so many topics that it is not easy to see the significant aspects, especially because numerous disciplines are involved. Our relatively isolated example of technology which mimics natural vision, the camera, gives a false impression that 'vision' sees 'pictures', but the camera only *transmits* the picture. In Canberra in the 1970s we elaborated a general theory to explain how the dimensions of eyes and the angles between visual axes are related to the limits of lens resolution, as fixed by the diffraction of light, and the noise limits arising from the random arrival of photons. The theory referred to the sensitivity which depends on receptor size and lens diameter and to the spatial resolution which also depends on receptor size and spacing. Basically, it was a sampling theory (Kirschfeld, 1976; Snyder, 1975, 1979). However, the advent of the new architectures of computers with parallel distributed processing brings another framework, which can now bind together a new synthesis to span the gap downstream between the array of photoreceptors, the functions of single neurones and the coordinated processing of visual behaviour. Indeed, for almost a century there have been discussions about the emergent properties of nervous systems; in the case of the visual system we can now see clearly that emergent properties reveal themselves because in cooperative parallel processing the neurones are individually ambiguous and inadequate exactly because they have evolved to operate in groups. However, I anticipate my conclusion and must return to the evolution of vision.

Importance of Broad Absorption Spectra

Pigment Spots and Visual Pigments

The commonest visual pigments (called rhodopsins) are combinations of a protein (often called an *opsin*) with a carotenoid, which is lipophilic and has a broad absorption spectrum. Carotenoid molecules have a long hydrocarbon chain of alternating double and single carbon–carbon bonds which resonate in many ways and cause a broad absorption spectrum. The position of the peak, that is, the colour of the visual pigment, depends on the constitution of the protein, so that pigments can readily evolve to suit the visual world and the visual tasks in that world. Pigments are known with a peak in the range from 350 mm in the ultraviolet to about 620 mm in the mid-red. Most of the broad-spectral sensitivity curves have a shape that agrees with the Dartnall (1953) nomogram. Visual pigments apparently evolved independently many times from related molecules already in the cell. Eyes with 3 photoreceptor colour types are common in several groups of animals; eyes with a single pigment are also widespread, and a few are known with 2 (cockroach), 4 (butterflies and some birds) or 5 (dragonflies).

The carotenoid part of the molecule is held by the cell membrane and the absorption of a photon opens channels in the membrane that cause a polarization or depolarization of the cell via an amplifying cascade. In many photoreceptors the single photon captures cause individual miniature potentials (called 'bumps'), best known in arthropods. An important consequence of this mechanism is that the magnitude of the response is independent of the wavelength of the photon. Photoreceptors therefore act as photon counters irrespective of the colour of the captured photon, and the discrimination of colour must depend on an accurate measurement of the ratios of photon numbers captured by photoreceptors with different spectral peaks. The spectral sensitivity of a photoreceptor is its field of view in the colour dimension: these fields are broad and they overlap a great deal where there are several spectral types. Colour discrimination is excellent because it depends on overlapping spectral sensitivity of different receptors which are excited in different ratios. This principle is important and we will return to colour discrimination as a model for processing in other dimensions.

The Laws of Optics Govern Eye Evolution

Absorbing Light Guides and Fused Rhabdomeres

The rhabdomere is the organelle composed of tightly packed microvilli in the photoreceptor cell. The microvilli contain the visual pigment molecules and, being composed of a high proportion of lipid membrane, they have a relatively high refractive index (of about 1.39 compared with 1.34 in the adjacent cytoplasm). The visual pigment molecules in a layer 1 µm thick absorb only about 1% of the light, and therefore there has been strong evolutionary pressure to increase the depth of the absorbing layer. At the same time the cross-sectional area of the rhabdomere must remain small in order to optimize the resolution and match the cross-section of the absorbing organelle to the resolution of the lens (see Fig. 11.3). As a result, we find that many excellent eyes have the rhabdomere in the form of a long, thin rod which points towards the nodal point of the optical system. Examples occur in coelenterates, annelids, arthropods and molluscs and in the rods and cones of vertebrate eyes. The rod acts as a light guide which conducts the light falling on its end, absorbing the light as it passes along the rod (Figs 11.1 and 11.2). Where these receptor rods occur we can infer that resolution and sensitivity are being optimized so that they do not act against each other, as they would if the rhabdomeres were merely increased in size, like the silver grains in a sensitive film.

Increasing sensitivity by increase in the length of the rod soon leads to the self-absorption effect. Absorption in the rod is about 1% per µm at the absorption peak, so that most of this peak light has been absorbed at 100 to 200 µm along the rod. The off-peak photons are absorbed also, although less effectively, with the nett effect that the longer the receptor rod the broader the absorption curve. The ultimate result is that when all the light is absorbed the receptor is *black*. This achieves a maximum sensitivity but rules out the possibility of colour vision by receptors of different colours. This is presumably why the cones of vertebrates are shorter than the rods. The above considerations mean that the absorption must be below about 50% and the receptor rod must not exceed 30 to 50 µm in length, depending on the exact numerical values of the coefficients (Snyder, 1975, p 197).

The fused rhabdom is an anatomical solution to the problem of increasing sensitivity and retaining colour vision. The rhabdomere is the contribution of a receptor cell to the whole rhabdom, which is now the light guide. The light is therefore absorbed by each rhabdomere as it is carried down the rhabdom as a unit (Fig. 11.2). If the rhabdomeres differ in their absorption peaks, having for example peaks in the ultraviolet, blue and green, each absorbs a different fraction of the total light and it is now possible to absorb all of the light and retain colour vision. Most of the larger insects have 2, 3 or 4 spectral types of rhabdomeres fused into rhabdoms more than 100 µm long. Elongated *fused* rhabdoms are therefore a sign of colour vision.

The same principle, combining 100% absorption with complete discrimination, applies to the plane of polarization of polarized light. Crustacea commonly have interleaved layers of microvilli arranged so that microvilli of one set of receptor cells are at right angles to those of the other cells. Insects, spiders, molluscs and others commonly have rhabdomeres of adjacent cells pressed together side by side with microvilli of different orientations. This arrangement is not necessarily proof that discrimination of the polarization plane is important for animals that have it, because absorbing light in all polarization planes is also a way of increasing sensitivity in general.

Matching Receptor Resolution and Lens Resolution

Following a seminal paper by Kirschfield (1976), and inspired by Allan Snyder, a long series of papers have

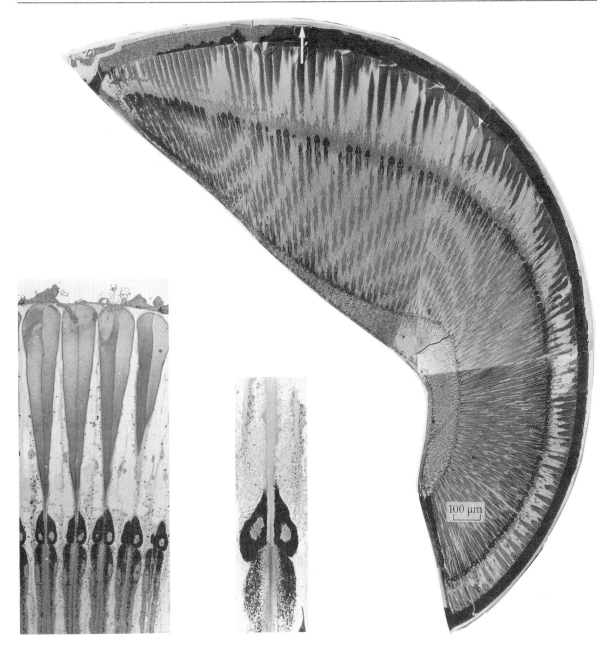

Fig. 11.1 *The anatomical basis of the regional differences in spatial resolution in the eye of the praying mantis. This is a horizontal section through the compound eye of* Tenodera. *The forward-looking part of the eye has smaller angles between visual axes (smaller $\Delta\phi$), longer cones and therefore longer focal length (smaller $\Delta\rho$) and larger facets than the side of the eye. The insets show the critical region where the cone meets the rhabdom, where the ray optics is matched to the light-guide optics.*

appeared from Canberra on the geometry of receptors, the dimensions of eyes and the diffraction limitation of light. Essentially, the optics and receptors must be matched to make the best use of the lens resolution. All eyes are sampling devices in angular coordinates. To resolve contrast with least blurring, the focus must be perfect, the optics of the lens must be as good as diffraction allows, and then to detect as small a contrast as possible, the size of the receptors must match the optical resolution of the finest detail by the lens. These conditions of matching spatial resolution, least-motion resolution and lens resolution, are met by working at the limit set by the wavelength of light, and

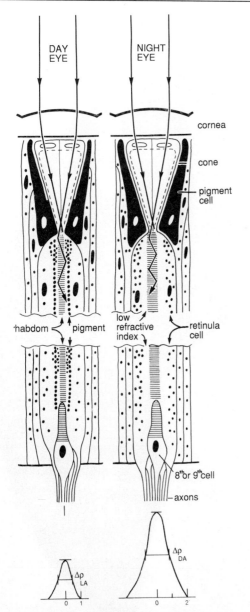

Fig. 11.2 *One ommatidium in vertical section: (left) the day state, with pigment around the cone tip and around the narrow rhabdom; (right) the night state, with a clear space around the track of the light and a broader rhabdom. As shown by the angular sensitivity fields plotted at the foot of each figure, the day eye is less sensitive and has a narrower field at the 50% sensitivity contour. This diagram includes features of several of the groups of large diurnal insects.*

the best compromise is found when the photoreceptors are only 1 to 2×10^{-6} μm in diameter, which is 2 to 4 times the wavelength of green light (see below). A similar grain size is found in photographic film for similar reasons,

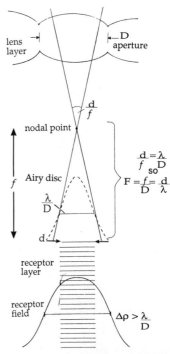

Fig. 11.3 *A lens aperture D generates a blur circle of width λ/D radians at 50% intensity on a receptor of width d in the focal plane. For the blue circle width to match the subtense of the receptor, $\lambda/D = d/f$ radians. The F-number of the lens is f/D. Therefore $d = F/2$ μm, because $\lambda = 0.5$ μm. The field width $\Delta\rho$ of the receptor cannot be narrower than the blur circle on account of the finite width of the receptor.*

because an F-number of about 2 to 4 is similar for eyes and cameras. This relation follows because the width of the circle of confusion (Airy disc) is λ/D at 50% intensity (Fig. 11.3), the F-number is f/D and the subtense of the photoreceptor of diameter d is d/f (all measured in radians), where f is the focal length and D is the lens aperture (Horridge, 1978a). Therefore, to match the diffraction properties of light, receptors must be only a few μm wide, and their fields must touch or overlap. Larger facets are accompanied by narrower rhabdoms (Fig. 11.4) and nocturnal eyes have huge rhabdoms that throw away resolution in exchange for sensitivity (Fig. 11.5).

Related Light-Guide Dimensions

From the relation $d/f = \lambda/D$ (above) we can write $\lambda/d = D/f = 1/F$ so that for $\lambda = 0.5$ μm and $F = 2$ it follows that the receptor has a diameter of 1 μm. Very conveniently, given the refractive index of the rhabdom material, raised above that of the surrounding cytoplasm by a high percentage of lipid, it can be shown that 1 μm is also approximately the minimum diameter for a rhabdom to function as a light guide. Therefore we could say that the absorbing

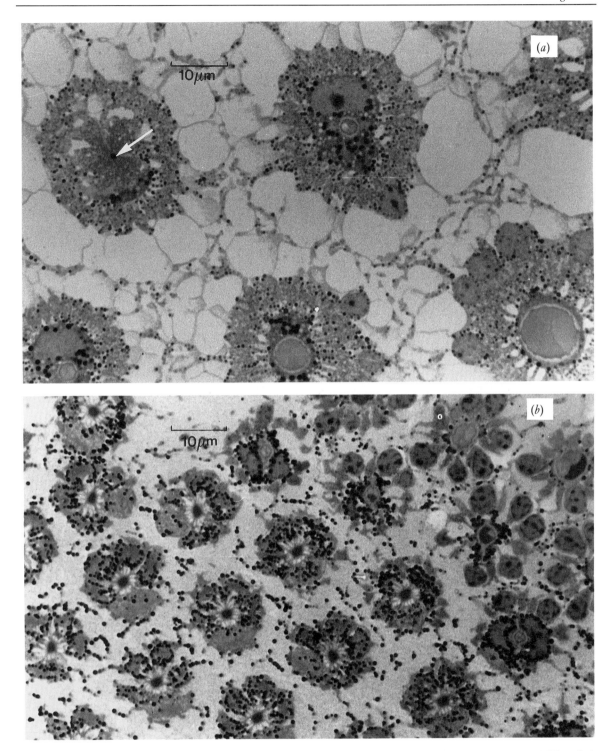

Fig. 11.4 *Transverse sections through the region of the cone tip (see Fig. 11.1) of a dragonfly eye (*Austrogomphus*). (a) Through the foveal region looking forwards and upwards. The facets are large (from centre to centre) but the rhabdom (arrow) is only about 1 μm in diameter. (b) In the ventral part of the eye the facets have an area a quarter of that in the fovea and the rhabdom area is correspondingly four times as large as in (a), giving a nett uniform sensitivity to a diffuse source.*

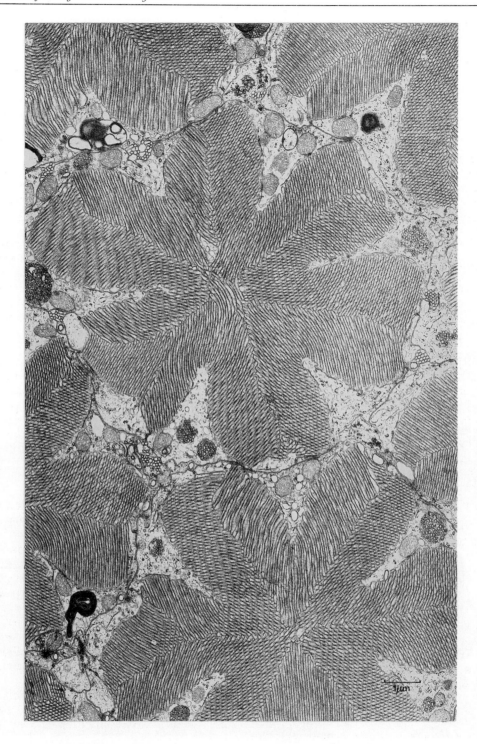

Fig. 11.5 *The rhabdoms of a nocturnal eye fill the cross-section within the eye to maximize the sensitivity to a diffuse source. This is a horizontal section through the eye of* Chrysopa *(Neuroptera). There are six large rhabdomeres to each rhabdom, but two other rhabdomeres at another level. The visual pigment molecules lie within the rhabdomere microvilli.*

light guides in eyes are admirably adapted to the *F*-number of the lenses that commonly occur. The geometry of the eye is governed by the laws of physics.

Modulation

The main feature of a photoreceptor that governs the ability of the eye to see is the *field* of each individual receptor, which looks through this window at the outside world. The actual field is measured experimentally either by a moving, flashing point source (Fig. 11.6) or from the modulation of the potential caused by a range of regular striped patterns of different periods (Fig. 11.7). The stimulus to the receptor is the modulation of the light falling upon it as integrated in its field: the response is the modulated receptor potential, which is propagated to the second-order neurones (Fig. 11.10).

Matching the Spacing of Visual Axes to the Lens Resolution

An array of receptors can resolve *in space* a spatial frequency up to a limit that is determined by the angle between the visual axes. Regular stripes, of period $\Delta\theta$ subtended at the eye, can be distinguished when the angle between receptor axes $\Delta\phi$ is smaller than the limit set by $\Delta\theta = 2\Delta\phi$, as illustrated in Fig. 11.8. In the same eye the stripes of period $\Delta\theta$ can be resolved by the lens when $\Delta\theta = \lambda/D$ from diffraction theory. Therefore $\lambda/D = 2\Delta\phi$ when $\lambda = 0.5\,\mu\text{m}$ (Fig. 11.9).

We now have a relation between the spatial sampling resolution and the lens resolution, which in turn is linked to the receptor diameter, as applied to eyes with reasonable *F*-number and which can afford to operate near the diffraction limit because there is plenty of light. Of course, the processing mechanism behind the retina must be able to deal with this spatial resolution, which is the reason for postulating (later in this review) motion-detection templates *at this resolution*.

The above relation can be used to calculate the diameter of a compound eye for a given resolution. To obtain a spatial resolution of 1° we have $\lambda/D = 1°$, so that D is approximately 30 μm minimum (as in eyes of large dragonflies). The eye radius (R) is therefore given by $D/R = 1°$ if the facets are adjacent, so that R is approximately 1.8 mm.

Sacrifice of Resolution for Sensitivity

Many animals live in situations, or emerge at times of day, when vision is limited by lack of light, not resolution. Photons arrive at random, and so cause shot noise in the receptors at low light levels. Sunlight gives about 10^{14} useful photons cm^{-2} s^{-1}, moonlight about 10^7 of them. We can calculate that if 1% of the photons are reflected from objects in full moonlight, a facet 30 μm in diameter on an insect eye catches about 1 photon s^{-1}, which is hardly sufficient for vision even at 50% capture efficiency. Under the same illumination a camera-type eye with a lens 3 mm in diameter and low *F*-number has sufficient light in moonlight, even with receptors 1 μm in diameter, e.g. rods in human eye.

The sensitivity of a receptor behind a lens is proportional to the area of the lens and the solid angle of the receptor area as subtended at the nodal point of the lens:

$$\text{sensitivity} = \frac{kd^2 D^2}{f^2} = \frac{kd^2}{F^2}$$

where k depends on the efficiency of photon capture in the rhabdomere, and is approximately 1% per μm of depth.

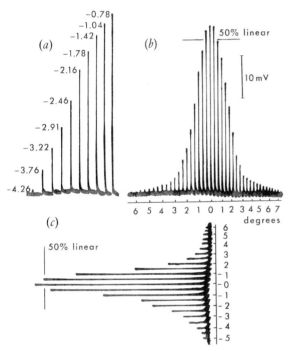

Fig. 11.6 *The way that the visual field of a receptor is measured with a microelectrode in the receptor as a graded potential at each flash of light. (a) Calibration by means of a neutral density series. A light was flashed on axis at the log I values shown. (b) A constant small point source subtending less than half a degree at the eye was flashed at each half-degree step in the horizontal plane across the axis of the receptor. The angle subtended by this source at the eye is important. The brightness was selected to be always below saturation on axis. (c) A similar run through the axis in the vertical plane. From (b) and (c), angular sensitivity curves can be calculated from the calibrations in (a). From the angular sensitivities one measures $\Delta\rho$ at the 50% sensitivity level for the light adaptation employed (Fig. 11.2), but note how close to the peak of the responses the 50% linear reduction in effective intensity is found.*

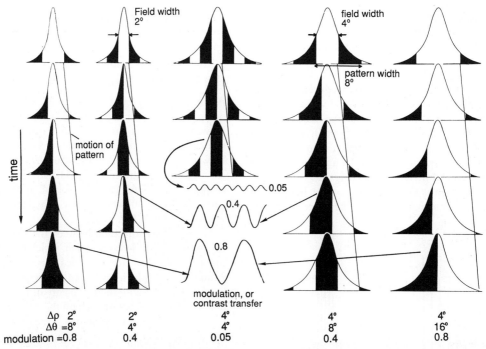

Fig. 11.7 *The way that changes in intensity of the light falling on the photoreceptor are converted into modulation of the potential of the receptor, in relation to the field width. In this diagram the fields $\Delta\rho$ are 2° or 4° width at 50% sensitivity. The stripe pattern of period $\Delta\theta$ subtends 4°, 8°, or 16° at the eye and moves to the right. As the individual bright and dark areas pass the field, which acts like a shaped window, they cause modulation of 0.05 (5%), 0.4 or 0.8 in the receptor. This in turn causes an oscillation in the potential of the cell at the frequency at which the stripes go past. A modulation of 0.05 is near the cut-off for fine patterns. This principle is the basis of the loss of resolution of fine detail as a result of increased field widths.*

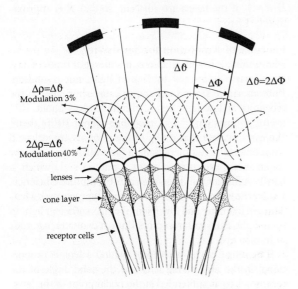

Fig. 11.8 *The relation between the width and overlap of the fields of receptor cells in typical eyes of large diurnal flying insects such as dragonflies, flies, butterflies and bees, and the minimum period of a regular pattern which such an eye can resolve. Let us first examine the relation between the visual axes and the repeat period of the pattern. To resolve a striped pattern of period $\Delta\theta$, the angle between the visual axes must be less than $\Delta\theta = 2\Delta\phi$ (at top right). To generate sufficient modulation for vision (3%) the field width of the photoreceptors must be less than $\Delta\rho = \Delta\theta$, as explained in Fig. 11.4. To generate 40% modulation, we have $\Delta\rho = \Delta\theta/2 = \Delta\phi$, which is a commonly occurring compromise that requires better lens resolution than that set by the diffraction limit. This principle is the basis of the compromise of spatial resolution as limited by the density of the visual axes and the overlap and width of the receptor fields.*

Therefore the only way to increase sensitivity at constant *F*-number and capture efficiency is to increase receptor area (Fig. 11.5). This is why, in a camera for a given film type, more sensitive film has larger grains and therefore poorer spatial resolution. For the same reason, the evolution of eye performance is almost entirely related to the *number* of receptors of a given size that can be packed in.

The lower the light level is, the poorer is the spatial resolution that can be achieved, and the larger the lens aperture needed to catch the incident light. Therefore the

The Evolution of Vision

The Old Problem of Eyes and Brains

For 130 years there has been discussion as to how the eye (of man) could have evolved together with the visual centres of the brain when every small improvement in one would be useless without an equivalent improvement in the other. The problem was well known to Darwin and the later proponents of evolution, and also to those who attacked the theory of evolution using arguments of this kind. A related problem, how the eye of a squid or octopus comes to resemble that of a man or elephant, was regarded as evidence that evolution limited to small steps selected from random mutations could never fully account for the facts. It was also clear that the theory of selection could *not at that time* be shown to be the only mechanism of evolution, even though the best scientists of Europe were convinced, and therefore the difficult example of the evolution of eye and brain was repeatedly quoted as support for alternative explanations of animal structure.

In hindsight we see that the problem was caused partially by the assumption that an eye with the structure of a camera must produce a two-dimensional picture, and because it was not realized that the evolution of an excellent eye is governed by the laws of optics, while the progressive evolution of the visual centres can follow along independently later, as the visual tasks become more complicated.

Eyes Without Brains

Extensive texts describe the structure and histology of numerous eyes in primitive animals. Descriptions up to 1963 will be found in Bullock and Horridge (1965); the best special textbook is still Plate (1924), while modern work is reviewed in Ali (1984) and by Meinertzhagen and by Osorio (this volume).

Comparative anatomy tells us that excellent eyes, with numerous small photoreceptors and an apparently good optical system, evolved independently many times in lower invertebrates such as medusae and worms (Fig. 11.11) in the absence of a nervous system which could achieve much visual processing. A great many of these eyes generate no complex visual behaviour. The eye was clearly an organ of spatial resolution long before the nervous processing structures evolved to make more use of its potential. Many of these primitive eyes detect motion while they are stationary, with no suggestion that a scanning eye like that of *Copilia* (Gregory, 1967) or the fly lava (Fraenkel and Gunn, 1940) preceded them in evolution. The selective advantage of the lens resolution and spatial sampling resolution of these lower-invertebrate eyes is that an approaching predator or the direction of a mate is perceived as the smallest possible motion of the least pos-

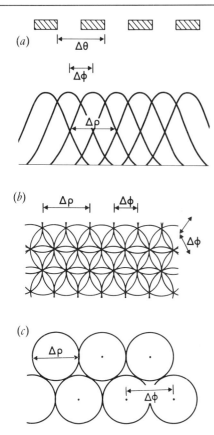

Fig. 11.9 *(a) Optimum sampling in one dimension in bright light, unrestricted by photon noise. There must be at least one visual axis looking at each peak and trough in the image, so that $\Delta\phi = \Delta\theta/2$. To provide sufficient modulation in the receptor, the field must have a maximum width of $\Delta\rho$ at 50% sensitivity, where $\Delta\rho = 2\Delta\phi = \Delta\theta$. (b) Putting together these relations, we obtain a hexagonal pattern of overlapping fields. Most large day-flying insects do not achieve this density of sampling: instead they have their receptor fields touching, as shown in (c), and in Fig. 11.8, where the same spatial resolution is achieved by collaboration between adjacent rows of receptors.*

ratio of lens resolution to spatial resolution must increase as the eye evolves for vision in dim light. At one extreme, matching the diffraction limit to the spatial resolution in bright light (Fig. 11.9(b)), we had $1/D = 2\Delta\phi$ so that $D\Delta\phi = 0.25$ μm. $D\Delta\phi$ is called the *eye parameter* and it is half of the ratio of the spatial resolution to the lens resolution. A larger eye parameter, sacrificing some resolution for sensitivity, is shown in Fig. 11.9(c), where $D\Delta\phi = 0.5$ μm. To maintain the optimum resolution and sensitivity, the eye parameter must increase for low light levels to values between 3 and 5 μm (Snyder, 1979, p 249). We see that much of the geometry of the eye is governed by the laws of physics.

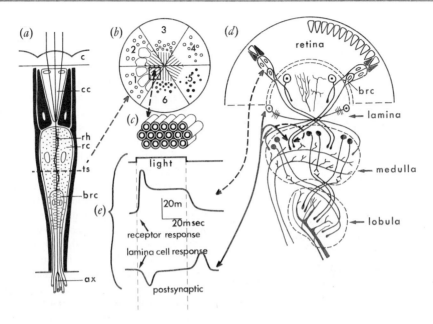

Fig. 11.10 *Synopsis of retina–lamina relations. (a) Light is absorbed in the rhabdom. (b) Ommatidium in transverse section, with six receptor cells. (c) Rhabdomere microvilli. (d) Optic lobe in horizontal section showing the relation between the retina and the layers below. (e) Lamina cell responses are approximately inverted temporal derivatives of retinula cell responses.*

sible contrast at the appropriate average light intensity.

The common visual behaviour of the medusae and worms which have excellent eyes is simply withdrawal from a moving shadow. Little can be learnt from them about visual processing without data on neurone functions. There are some examples where the behaviour depends on the angle at which the moving contrast lies, and many where motion of the whole visual field has little effect compared with local motion, but few where the directionality of motion is important until we come to the insects and crustaceans. For the practical reason that we know most about them, the insects illustrate very well a level of complexity where there are several different visual responses in one animal, and we also find a variety of visual behaviours when we look at different insects. It has become obvious over the past few decades that insects have several neuronal mechanisms in parallel behind every visual axis. Recording from insect optic lobe neurones shows that these mechanisms for visual detection of colour, direction of motion, polarity of an edge, stimulus orientation, flicker frequency, range and angle at the eye are subsequently channelled into partially separate but overlapping neuronal pathways. This is clearly a result of the history of their staged evolution. The model to be developed depends mainly on our understanding of insect visual mechanisms because this group is the best understood, but many of the concepts apply to all vision.

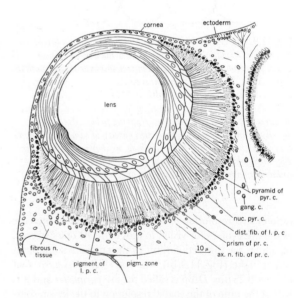

Fig. 11.11 *The eye of the Cubomedusan, Charybdea (Coelenterata), in section, showing numerous photoreceptors and the typical arrangement of a camera eye, which is ideally a set of sampling stations of optimum spatial resolution and lens resolution. The only visual response of this animal is a reaction to moving shadows by inverting and swimming downwards away from the surface (Berger, 1900).*

The Basis of Vision

Whether they are stationary or move, eyes primarily see transients and motions, not steady states or intensities. A few sessile animals have stationary eyes which detect a small movement of a distant predator and trigger a retreat into a shell or tube. An excellent example is the giant clam, *Tridacna*, with dozens of eyes embedded in the edge of its mantle: tubiculous polychaete worms sometimes have eyes on their head or tentacles with a similar function, associated with a rapid contraction controlled by giant neurones of the nerve cord. The barnacle eye is another well-studied detector of a passing shadow (reviewed by Laughlin, 1981). A general principle is that these eyes rapidly habituate to the average level of background movement and they progressively become more sensitive when no motion is visible. This is the general rule for sensory systems everywhere.

Most animals, however, move forward in a stabilized posture so that their visual systems operate in the context of a predictable flow field. In a three-dimensional world this has several important consequences.

1. At any one moment there is a predictable one-dimensional motion at each point on the retina so that processing need not be two-dimensional.

2. The angular velocity at which contrasts move across the retina (excluding eye rotation) is inversely proportional to their range for a known eye motion, giving an idea of the three-dimensional structure of the surroundings if this angular velocity can be measured.

3. As the eye moves, nearer objects move across the background, causing sharp discontinuities of the flow field at their edges, here called parallax. It is tempting to think that visual processing mechanisms of freely moving animals have evolved to see parallax as a primary feature which provides information about the separation of objects in three dimensions. If so, parallax detection takes over from motion detection as the driving force for evolution of better resolution towards hyperacuity.

4. Vision is based on the motion of the image across the retina, which is projected like a map upon the fixed arrays of processing neurones inside. Therefore the moving features that the neurones at early stages detect at any one place (Fig. 11.15) are transients and there is a major problem as to how any part of the image is 'captured' as it moves across the neural array within the optic lobe.

Field Sizes

The size of a neurone's field has evolved to match what the neurone is trying to see, especially at threshold. With reference to Fig. 11.12, if we search for an object that differs slightly from background, the signal-to-noise ratio is optimum when the field size matches the object size. In natural situations there has to be a compromise because

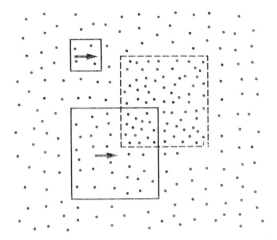

Fig. 11.12 *If a neurone responds to an object that differs slightly from background, the signal-to-noise ratio is optimized if the neurone's field is the same size as the expected object. This principle applies to all features of interest in the image in space and time, so that all visual neurones fields must be evolved towards a spatio-temporal optimum that is related to the visual task.*

the geometry of the image is never so predictable, and to detect the object location there must be many separately line-labelled fields. Therefore, fields cannot be specific. This is a basic principle for the distributed parallel processing within optic lobes.

Neurone field sizes in general in insects are as diverse as they can be, ranging from a single sensory cell input on the one hand to the whole of both eyes in the case of vision. Both extremes raise questions, in that fields corresponding to single receptors somehow have to be integrated spatially in order to abstract the significance of their combinations. Small fields miss the stimulus: large fields fail to locate it: moderate overlap is the usual compromise, One way to explain large fields is to suppose that they are alerting or inhibitory at their outputs, signalling a freeze. Possibly they act only in combination with small-field neurones which require a stimulus that is not specific with reference to direction, so that high resolution at every point is combined with a wide field. Large-field units sometimes relate to behavioural sensitivity towards the whole background in the visual world, as in optomotor responses, but they do not have to do so. Large fields may be a crude way to keep a moving contrast within the field of one neurone. On the other hand, small-field units, even though clearly related to high-resolution tasks, must always be tested on a variety of backgrounds because the detection of a moving contrast of any kind is essentially a detection against background.

Insect Vision

The analysis of visual processing by electrophysiological identification of neurone functions, fields and anatomy (surveyed by De Voe, 1985, and by Osorio in this volume) has suffered from the lack of a theoretical framework for 40 years. The Reichardt model of motion perception is concerned with the computation of a single output by a mathematical operation on the stimulus pattern, and therefore fails to relate to the variety of neurone fields and their interactions and does not approach the subtlety of vision in freely flying insects. Since the early days of Burtt and Catton (1960) empirical data on optic lobe neurones have been collected in terms of ON, OFF, ON–OFF and directional motion-sensitive units, small-object detectors, etc., but the chief feature of sensory processing, the adaptation, has been neglected. It has never been possible to fit the electrophysiological data into a logical scheme and the known fields of neurones only increase the difficulty in understanding how the moving image is captured (if indeed it is) as it moves fleetingly across the eye (Fig. 11.15).

The High-Resolution Tracks Through the Optic Lobe

Referring to Figs. 11.8 and 11.9, we observe that the visual fields of width $\Delta\rho$ of adjacent receptors are narrow and overlap; and the angle $\Delta\phi$ between visual axes sets the limit of spatial resolution of the eye as an angle-labelled sampling array. The eye cannot distinguish the separate bars of a striped pattern of period less than $2\Delta\phi$ subtended at the eye, and single receptors receive no modulation from (are not stimulated by) a striped pattern of period narrower than about $\Delta\rho$ subtended at the eye. We find that many eyes have evolved compromise optics and anatomy such that the field width $\Delta\rho$ approximately equals the receptor spacing $\Delta\phi$. As more complex subretinal mechanisms evolve, the receptors are able to map the visual scene into them with the high spatial resolution that was evolved for directional and non-directional motion detection.

In insects three further stages of neurones as far as the lobula retain this topographical map at the highest spatial resolution (Fig. 11.13). First, the function of the lamina monopolar ganglion cells, so far as we know it, can be summarized as transforming intensity into the temporal derivative of intensity (Fig. 11.10) with minimal latency while retaining the maximum spatial resolution (Laughlin, 1981, 1987). The next level in insects, the medulla, is where the excitation spreads into a large number of small neurones arranged in columns on each visual axis. These neurones presumably respond in different combinations to different local details of the stimulus pattern. In the small fly Drosophila there are about 30 intrinsic neurones in each of the vertical columns which correspond to a visual axis in the eye and a greater number of column neurones with axons to the lobula (Fischbach and Dittrich, 1989). This anatomical diversification into many small neurones recently led me to formulate the template model in which there are on each visual axis a number of

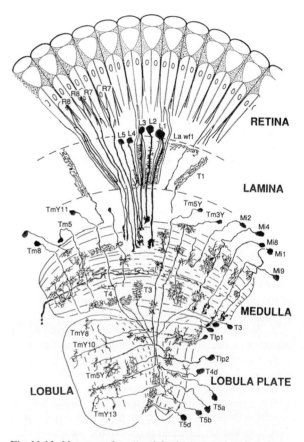

Fig. 11.13 *Neurones of restricted dendritic spread in the insect optic lobe, mainly based on the fly* Drosophila. *Narrow field receptors of the retina feed into columns of the lamina. From here the axons of the lamina ganglion cell types (L1–L5) transmit high-resolution information to columns of the medulla. The long retinal fibres of receptors R 7 and R8 run direct to the medulla. The chiasma between lamina and medulla has been omitted for clarity. So far all these neurones have no spikes, only electrotonic spread. In the medulla are 30–50 local neurone types in each column, which corresponds to a visual axis. Some of these project to the lobula, preserving much of the spatial resolution but the mapping of the retinal projection continues no further. Every level in this columnar projection has tangential fibres with long arborizations in specific layers and also medium-field neurones of several types. Large-field and small-field directional and non-directional motion detectors occur in the medulla and lobula but object detection appears to begin in the lobula (Redrawn from Fischbach and Dittrich, 1989).*

bers of the more peripheral units. They are directionally motion-sensitive, some vertically, some horizontally, have large fields, and appear to detect the direction and changes in visual field motion for the optomotor responses for control of stability in flight. Neurones for other unknown behaviour at this level (called object detectors) are non-directionally sensitive to motion of a small object anywhere in a large field. Some of the latter, acting in groups, may be for collision avoidance or for chasing the specific patterns of mates or prey. The number of combinations of inputs that make possible the variety of outputs are severely limited at this level by restricting the output task. The search for subtle properties of high-level neurones and the mechanisms by which they are achieved has advanced only slowly since 1950.

The Tangential or Horizontal Neurones

Insects and crustaceans are typical in not having efferent fibres to the retina ending within the retina on the receptors. Insects have tangential fibres of unknown function spreading through the lamina, sometimes in more than one stratum. There have been suggestions that efferent fibres to the lamina control the relative gain of the inputs from different colour receptors, or local sensitivity, but no effects of this kind have yet been recorded from the lamina monopolar cells.

Numerous different neurones run in horizontal strata of the medulla, lobula and lobula plate (Fig. 11.14), some of them efferent. So far, all those recorded in the medulla have been wide-field collector neurones from large numbers of local motion-detectors, either directional or non-directional, with high spatial and temporal resolution. The directional neurones have a maintained response to continued motion, and are related to optomotor responses rather than to feature vision. Possibly they are gating neurones which allow other circuits to function only when the general motion is in the specified direction. These large-field optomotor neurones are especially well known from the lobula plate of the fly (Hausen and Egelhaaf, 1989; Strausfeld, 1989) where they respond as if they collect from a large number of high-resolution directional detectors separately for horizontal and vertical motions.

So far as we know, object-detector neurones tend to be complexly ramifying, not limited to one stratum, and are located in the lobula or brain. Almost nothing is known about identified neurones with small or medium-sized arborizations in the deep optic lobes (Osorio, Chapter 10).

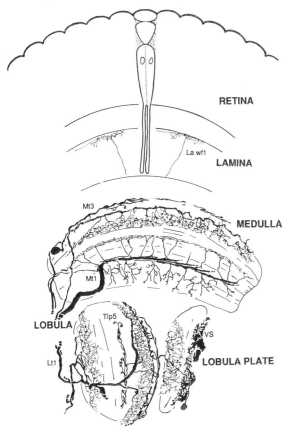

Fig. 11.14 *Horizontal fibres with widely spread arborizations in tangential layers. There is a striking similarity to the collector neurones in the ensemble processing system of Fig. 11.28. Labels on neurones here and in Fig. 11.13 after Fischbach and Dittrich (1989).*

templates that respond differently to the different local spatio-temporal patterns in the visual scene (see below). At various horizontal strata in the medulla there are also wide-field neurones (Fig. 11.14) which are sensitive to motion, usually directionally. It is important to stress that the full retinal resolution appears to be used only for motion detection and not for colour vision, object vision, polarized light, or any other visual function that we know of. Only two types of simple behaviour, both obviously not restricted to insects, retain the full spatio-temporal resolution of the eye. They are the directional and non-directional responses to motion. All other vision has poorer resolution, as if many templates are involved. There is no evidence that the *positions* of edges are located by anything resembling a 'zero-crossing' mechanism.

Corresponding to the visual axes, anatomically small-field neurones arranged in columns continue into the third neuropile of the optic lobe, the lobula (Fig. 11.13). At this level the most obvious neurones collect from large num-

The Evolutionary Approach

Evolution of Visual Processing

The progressive evolution of visual processing is the progressive addition of new neurones into a mechanism that already operates at high spatial resolution with many receptors in parallel. Possibly, in compound eyes, there was also a reduplication of visual axes after some mechanisms of processing had appeared. The first stage, found in many lower invertebrates, has many photoreceptors backed by many local detectors of moving shadows, all feeding into one circuit that withdraws the animal into deeper water, or into a hole. Next, we have small-field motion-detectors that are a little more subtle than shadow detectors; then we find *directional* motion detectors which are asymmetrical. At about this level of evolution, we find an indifference to total background motion and different colour sensitivities appear, although directional motion-detection is apparently colour blind in insects.

Never can we contemplate a primitive starting point in which the visual processing mechanism was able to analyse or 'see' the whole picture or discriminate large numbers of patterns. Vision obviously evolved by progressive addition of task-directed ways of analysing the visual world, and the addition of functions implies the addition of new neurones. The first requirement was to detect a small number of relevant features, a task that can be done by a small number of neuronal templates that cover these features. Nor was it ever necessary for a primitive visual system to analyse the flow field caused by its own motion. Only a few aspects of the flow field are useful. It is tempting to imagine that the directional neurones first acted as non-adapting gates so that other channels function only during motion in the appropriate direction. This would fit in with their large fields and wide arborizations. A useful ability for a moving eye is the measurement of local angular velocity along one expected line of motion, to estimate range, to interpret the three-dimensional world and distinguish solid objects from patches of light and shade by use of parallax. Looming is also a convenient signal for impending collision, but, again, the analysis can be one-dimensional in each region of the eye.

The main problem with introducing templates matching the biologically significant features in the visual scene is that all template responses are 'event-driven', by which I mean that the template responds every time its combination of contrasts in the visual field passes its visual axis. If template responses are summed in any way in the animal or in an experiment (as in Fig. 11.17), the total response is dependent on the temporal frequency of the appropriate trigger features. At a higher level, a butterfly with templates that respond to flowers would need a mechanism to cope with a whole bed of flowers. This event-driven property has been known for many years in directional motion-detector neurones of the fly (Fig. 11.17), which respond to the passing of each edge: they therefore respond more to the movement of groups of edges than to single edges, so that the summed response detects direction of motion but cannot measure velocity independently of pattern. They

respond to contrast frequency even though they may be tuned to a particular range of angular velocities. The simplest way to avoid being event-driven in this way is to adapt rapidly and respond only to the first presentation, as commonly occurs with object detector neurones. Another way is to follow the example of colour vision, and take the output as the *ratio* of the numbers of responses of different templates, and so eliminate the cumulative effects of event-driven responses. Taking ratios makes it possible to detect qualities of features independently of the number of times they move across the visual axes, i.e. irrespective of pattern, and this implies 'fast' and 'slow' motion-detectors to measure velocity.

Neural Adaptation at Every Stage

Photoreceptors adapt by many mechanisms, but they can still be considered as photon counters with a calibration

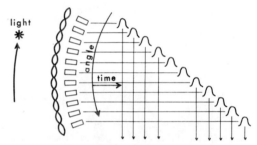

Fig. 11.15 *Any stimulus that moves relative to the eye generates a series of responses spread out in time and space in the sampling array. Therefore visual processing has to be considered in spatio-temporal coordinates, and one of the main questions in vision studies is how the image is tracked across the spatio-temporal array within the nervous system.*

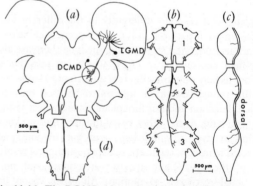

Fig. 11.16 *The DCMD (descending contralateral motion detector) neurone of the locust and its feeder neurone the LGMD (lateral giant). These neurones are detectors of non-directional movement of any small contrasting object at the full spatial resolution of the retina but are little influenced by a large background motion. The peripheral connections run to flight motor neurones, and the probable function of this neurone is the rapid avoidance of obstacles when in flight.*

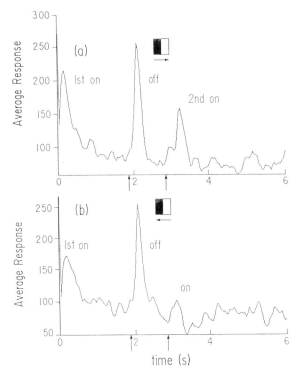

Fig. 11.17 *A sample of the data behind the template model. This is the directional high-resolution response of a wide-field motion-detector neurone (the H1 neurone of the fly lobula plate). These are post-stimulus accumulations of nerve impulses (average responses). On the left at* t = 0 *the pattern with a single edge appears on the screen. At* t = 2 s *the pattern disappears and is replaced by plain grey, then 1 s later the pattern reappears but displaced by 1.5°. The response is shown (a) in the preferred direction, (b) in the antipreferred direction (note the inhibition of background impulses). This effect shows that the first position of the pattern is 'remembered' somehow for at least 1 s although there is a large OFF response at* t = 2 s. *Therefore opposite effects are propagated in opposite directions from the edge, causing the directionality, and in the absence of the stimulus these effects persist for a short time at the spatial resolution of the visual axes. Compare the templates in Fig. 11.18.*

that can change. Some of the adaptation, caused by membranes and ions, is rapid; other adaptation, caused by movement of the receptor cells or of screening pigment, is slower, but all adaptation helps to keep the photoreceptor near the middle of the intensity range of the ambient light at the appropriate time of day. Photoreceptors can be modelled as photon counters with adaptation.

Lamina ganglion cells adapt so rapidly that they act like differentiating circuits with an output proportional to the temporal derivative of the intensity modulation in the photoreceptors on the same visual axis. Some photoreceptor axons bypass this stage and run direct to the next neuropile, the medulla. So far as they are known (Mimura, 1972; Honegger, 1978; Osorio, 1987, 1990;), many of the small-field cells of the medulla also adapt rapidly (Osorio, this volume, chapter 10).

The higher-order neurones adapt even more obviously, especially those sensitive to motion of a small object anywhere in a larger field. Some respond only to a novel stimulus and, at least in restrained preparations, require a long recovery period before they respond again. The DCMD and LGMD neurones of the locust (Fig. 11.16) are examples of this type (Rowell, 1971; Rind, 1987). The directionally sensitive optomotor neurones of large insects (e.g. Hausen, in Ali, 1984; Rind, 1990) continue to respond for long periods to a steadily maintained motion, but they often respond much better to a change in velocity.

The Template Model

The Template Model of the Optic Medulla

Visual processing necessarily involves temporal and spatial correlations (Fig. 11.15) between photoreceptors that are adjacent on the retina, otherwise the temporal and spatial resolution is wasted. The second-order neurones of the insect lamina effectively take the temporal derivative of the photoreceptor output on each axis, at high gain and minimum latency. We also know that in motion detection there is a rapid saturation of contrast (Horridge and Marcelja, 1990b), although there may also be other separate mechanisms which actually measure contrast. Therefore we take a threshold temporal contrast of 0 ± 0.008 on each visual axis, and base the model upon these threshold changes at adjacent visual axes at successive instants of time (Sobey and Horridge, 1990). The model is one-dimensional in space because the insect's own stereotyped motion generates one-dimensional motion at each point on the retina. An increase of intensity over the threshold at a single photoreceptor is ($\uparrow$), a decrease is ($\downarrow$) and indeterminate or 'no change' is ($-$). There are nine possible pairs at adjacent axes in one dimension, namely ($--$), ($\downarrow -$), ($\downarrow\downarrow$), ($\downarrow\uparrow$), ($-\uparrow$), ($\uparrow\uparrow$), ($\uparrow -$), ($\uparrow\downarrow$), ($-\downarrow$). When we take nine pairs at two successive times (Fig. 11.20) we obtain 81 spatiotemporal 2×2 templates which are the smallest possible primitives for spatio-temporal analysis of the visual scene when contrast is thresholded. This is the simplest possible way that all quantized contrast changes can be included in time and space with full resolution.

The templates differ from neurones in that every group of two adjacent pixels at adjacent times causes the response of only one of the templates (Fig. 11.18), but real neurones in the columns behind one visual axis can respond in parallel in different numbers in various combinations simultaneously, and can yield graded responses as well. In

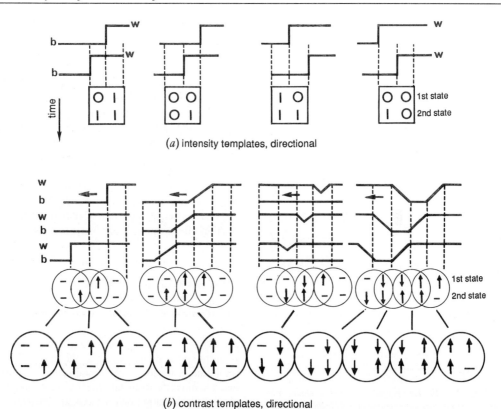

Fig. 11.18 *The way that the template notation relates to the motion of a contrast in the outside world. b, black or dark level. w, white or light level. (a) Intensity templates; 0 represents dark and 1 represents light. The template consists of the* state *(either 0 or 1) at two adjacent visual axes at two successive instants. The diagonal 3 : 1 symmetry of the spatiotemporal (a) template is characteristic of directional sensitivity. (b) Contrast templates are preferred because they cope with the wide range of background intensity. Any increase in intensity is (↑), any decrease (↓) and 'no change' is (–). Sharp edges, as on the left, are uncommon in real eyes and natural images consist mainly of gradient edges. All of the templates shown here are directional for motion to the left, as shown by the symmetry about a diagonal (see Fig. 11.23, 11.24). A group of template responses are necessary to convey the nature of the moving contrast. The template idea shows how spatial image structure in motion is fed into an array of processing units.*

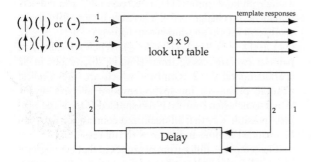

Fig. 11.19 *One way to correlate adjacent pairs at successive instants is to arrange a state machine with a delay in the feedback loop. A neuronal circuit would presumably have a persistent transmitter with the delayed effect.*

fact, I expect that many of the small neurones of the optic neuropiles are non-spiking neurones with mutual inhibitory interactions similar to those described by Burrows (1980) and his associates in the control of locust walking. The templates in the present model are therefore very simplified representations of parts of neurones or groups of neurones, but they illustrate the principles of parallel processing in discrete time steps by an artificial or natural system of this type. One way to visualize the state machine on each visual axis is shown in Fig. 11.19.

The Significance of 'No Change' (–) and 'No Response'

In the development of the theory, the idea of 'no change' (–) arose naturally from the indeterminate region below

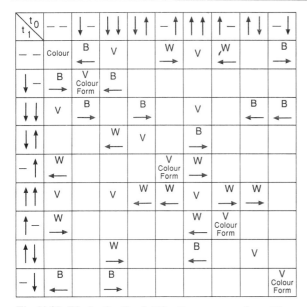

Fig. 11.20 *The 9 × 9 table of all 2 × 2 contrast templates. The symbols are: (−), no change; (↓), decreasing light level; (↑), increasing light level. Directionally sensitive templates are always those with 3 : 1 symmetry about a diagonal. Many of the templates rarely responded in our conditions. Meaning of other symbols: (→), motion to the right; (←), motion to the left; (B), dark follows light in direction of motion; (V), templates that indicate velocity by their ratios.*

Fig. 11.21 *The primary data of the one-dimensional image at successive times. This spatio-temporal printout is 200 pixels wide, covering a field of view of 10.5°, and 180 successive instants from top to bottom, over a period of 7.2 s at a sampling rate of 25 frames per second. In the picture are five targets plus a background, located at 180, 380, 740, 1210, 1640 and 2660 mm from the camera. This picture is a portion of the whole image 512 pixels wide.*

threshold. Subsequently, because not all templates are employed, we may have many points in the spatio-temporal map with no template response. In general, absence of activity is significant in behaviour in three ways: firstly, so that a behaviour pattern can proceed in the absence of a veto or inhibitory gating; secondly, in the progressive increase in sensitivity with time that occurs in the absence of stimulation; thirdly, in leaving processing channels undisturbed for other types of discriminations to occur. These generalities apply to behaviour patterns and to single neurones. Conversely, repeated stimulation raises thresholds, sometimes precipitately, and can cause habituation for long periods, especially in higher-order neurones.

Also, 'no change' is significant because colour vision depends on the ratios of graded responses derived from different colour types of photoreceptors, and the ratios are more easily measured in undisturbed channels separate from those conveying temporal contrast.

Implementation of the Model

To obtain data we take a real scene and scan it at 25 Hz with a single horizontal line of pixels in a moving CCD camera. The scene changes as a result of the horizontal motion of the camera, so that successive scans by the same line of pixels generate a spatio-temporal picture of graded intensity in which motions of contrasts are seen as diagonal lines (Fig. 11.21). At this stage we can, if we wish, introduce some lateral inhibition which influences later processing (see below). To obtain temporal contrast we subtract the intensity at each pixel from the intensity in the same pixel in the previous scan and take a threshold contrast level so that

(↑) is greater than $+0.008$, (↓) is less than -0.008 and (−) is 0 ± 0.008.

This threshold is arbitrarily found by adjustment above the noise level of the camera, so that small contrasts are detected but spurious responses are minimized.

The resulting digitized spatio-temporal map of contrast (Fig. 11.22) is then scanned with selected templates for particular primitives (Fig. 11.23). Let us examine the mechanism in detail. The motion of a graded contrasting edge to the right is represented pixel by pixel in x, t coordinates as a series of steps (Fig. 11.24(a)). The template (−−)/(↓−) responds to exposed corners on this profile and the template (↓−)/(↓↓) to inside corners: both of these templates have light–dark polarity and are directional because these particular corners occur normally with the motion of an edge with decreasing intensity towards the right. Directional templates are event-driven and therefore their total number of responses measures the number

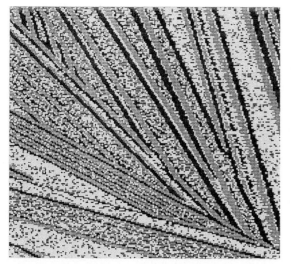

Fig. 11.22 *The same spatio-temporal data as in Fig. 11.21, with the 8-bit intensity at each pixel subtracted from that at the previous instant on the same pixel. The threshold at a contrast of 0 ± 0.0008 is then applied and the three resulting states are plotted; (·) 'no change' or indeterminate; (■) decreasing and (▶) increasing intensity. This spatio-temporal map is then compared, pixel by pixel, with the templates in Fig. 11.20. Note that gradients at edges are often 3 to 5 pixels wide as a result of the preprocessing.*

of edges that pass irrespective of pattern. They indicate the direction of motion reliably and respond instantly to transients. Directional templates such as $(-)/(\downarrow -)$ and $(\downarrow -)/(\downarrow \downarrow)$ respond in equal numbers (Fig. 11.24(a)), and their ratios do not measure velocity.

The templates $(--)/(\downarrow \downarrow)$ and $(\downarrow -)/(\downarrow -)$ respond differently, and are not directional. We see from Fig. 11.24 that the local ratio of the numbers of responses of these, and related non-directional templates, is a measure of the angular velocity. Setting the threshold low reduces the count of responses of templates such as $(--)/(\downarrow -)$ and $(-\downarrow)/(\downarrow \downarrow)$ with a 'no change' symbol. When this symbol is not used, vision is still possible, although only 16 possible template types remain. These 16 include directional ones such as $(\downarrow \uparrow)/(\downarrow \downarrow)$ and non-directional ones such as $(\uparrow \uparrow)/\downarrow \downarrow)$, so that measurement of direction and velocity in separate later channels is still possible. Also, templates for moving dark–light edges are necessarily the mirror images of those for light–dark edges, so that edges of opposite polarity are separately processed.

This model has many points of similarity to natural visual systems as seen in insects. Only a few types of templates need be utilized for limited vision. Evolution of such a system is easy because additional templates can be brought into use while others drop out. The observed greater number of neurones as evolution progresses is interpreted as the evolution of more templates. Template responses can be gathered up in subsequent combinations to suit more complex visual tasks. The way to keep down the numbers of combinations of template responses in processing of this type is to make use of only the combinations needed for the behavioural output, the properties of the visual world, and the features to be abstracted.

Properties of Templates

Template responses are highly non-linear, and the range of their necessary variety is compatible with our knowledge of optic lobe neurones. The effect of converting sharp edges to gradients by convolution with the visual fields of the receptors, and of emphasizing contrasts by adding some lateral inhibition, is to change the selection of templates that subsequently respond. These mathematical operations upon the image should be regarded as biologically significant, not because they reject high or low frequency in the spatial information of the visual scene, as stated for the past century or so, but because they allow better scope for separating template responses deeper in the processing mechanism.

The responses of individual templates are insufficient to convey much information and are ambiguous with reference to their stimulus, but this is an essential feature of a distributed processing mechanism. Groups of templates evolve together and templates are therefore incomplete in isolation. A system with a few essential templates on each visual axis could generate the behaviour that we observe in insects but, of course, would not analyse flow fields or stationary patterns in two dimensions. The response of a template is a unit of energy which is fed into the next stage of processing: a directional template generates a unit vector impulse at that point in the visual field of the whole eye. Preformed templates are an ideal mechanism for saving time in responses to transient presentations because all pathways act in parallel and any number can add their vector input at the same time. Templates avoid computation, do not take averages, and improve signal and reject noise by use of a greater number of templates, not by deeper computation.

To illustrate templates in action, we recorded an actual scene with a moving camera and one line of pixels so that the spatio-temporal data can be represented on a page with responses of certain templates superimposed on the primary data (Fig. 11.23). Templates of the form $(-\downarrow)/(\downarrow \downarrow)$ and $(\uparrow \downarrow)/(\downarrow \downarrow)$ respond directionally to moving edges of one or other polarity. Templates of the form $(--)/(\downarrow \downarrow)$ and $(-\downarrow)/(-\downarrow)$ give a measure of velocity by the ratio of their responses. Templates of the form $(-\downarrow)/(\downarrow -)$ rarely respond because sharp edges have become gradients (Fig. 11.18).

Counting Line-Labelled Template Responses

The template operation separates significant primitives in the visual scene into different lines – a process called *line*

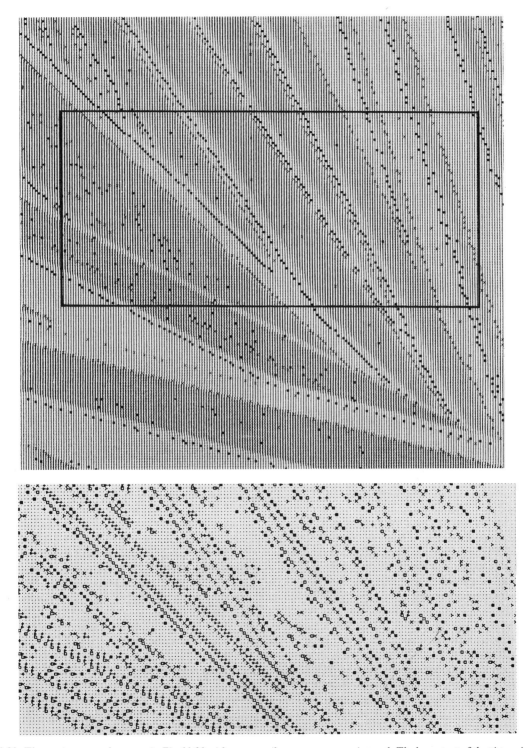

Fig. 11.23 *The spatio-temporal scene as in Fig. 11.21 with some template responses superimposed. The lower part of the picture is the outlined area at higher magnification and with lateral inhibition added before the threshold stage. We observe that different edges have different ratios of template responses. Templates are as follows:* ■ = (↑↑/–↑), X = (↓–/↓↓), ➤ = (↓↓/–↓); □ = (↓↓/↑↓), ▶ = (↓↑/↓↓) *and* + = (↓↓/↓↑).

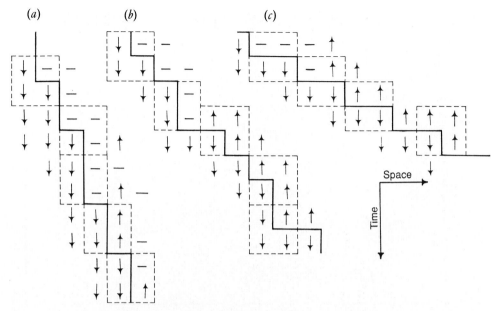

Fig. 11.24 *The motion to the right of a contrast (an increased darkening) as seen at high magnification by a one-dimensional eye in spatio-temporal coordinates. (↑) increasing intensity, (↓) decreasing intensity, (−) no change. Three different velocities are shown (a) slow, (b) medium and (c) fast. The directional templates with 3 : 1 diagonal symmetry detect the corners (which are an arbitrary consequence of the pixel–interval pattern). The ratios of the number of responses of the non-directional templates (↑↑/↓↓) and (−−/↓↓) on the horizontals, to the (↓−/↓−) and (↓↑/↓↑) on the verticals, gives a measure of the velocity but at a lower resolution in space and time. The whole spatio-temporal map can be broken down to template responses in this way.*

labelling in sensory systems – and each line carries digitized responses which can be counted. To reduce all these responses in numerous lines to a simple set of decisions or actions, we can summarize them in various ways. We can count all responses of the same kind at the same time (along horizontal lines in Figs. 11.25 and 11.26), or at the same place (along vertical lines in Figs. 11.25, 11.26), or we can count them along the diagonal lines in spatio-temporal coordinates by use of the Hough transform (Fig. 11.33) (Sobey and Horridge, 1990). We can also take running ratios or differences of the local numbers of responses of particular templates, or count significant groupings in medium-sized spatio-temporal fields. Taking ratios gets away from the event-driven property of all feature detectors, and can then be followed by logical AND to accommodate two-dimensional features. Loss of spatio-temporal resolution inevitably accompanies this process. Essentially the template responses are countable, which makes them easy to process further. Noise in the stimulus generates complementary template responses that tend to neutralize each other when ratios are taken. Even the thresholded contrasts, (↑) (↓) and (−), are countable in the same way. The higher-level neurones are now postulated to be leaky counters of template responses, just as the photoreceptors are leaky counters of photons. The number of different templates on each visual axis is a measure of the ability to discriminate structural diversity by the visual processing. The variety of the templates counted and number of alternative outputs is determined by the required sophistication of the behaviour, which in turn governs what is worth processing, and so semivision mechanisms are designed for a particular set of input and output tasks, and are never universal. Of course, bringing together the template responses into higher-level fields means that their identities and order of occurrence are lost, so that the spatio-temporal relations within the higher field are exchanged in favour of a decision relating to a particular visual task. Likely locations for these convergences are the dorsal and ventral optic glomeruli (Strausfeld, 1989).

That photoreceptors are counters of photons has long been a respectable idea, in particular when colours are represented as a colour triangle with different proportions of input from three receptor types. Photons are line-labelled and counted according to which (line-labelled by colour) receptor responds, and the colour of the stimulus is identified by a ratio of these responses in different lines. The theory for colour vision preceded the identification of the spectral types of receptors and their quantitative description. If we follow the same method to analyse vision of motion and form, by template response ratios, the next step is to identify which templates are used. We

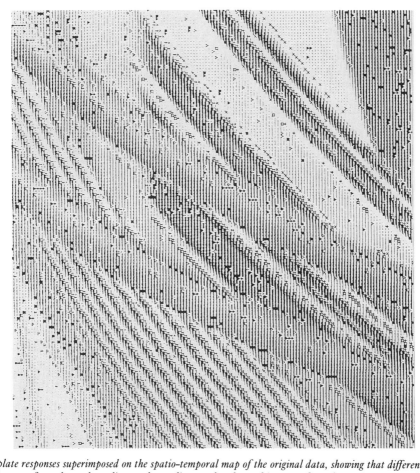

Fig. 11.25 *Template responses superimposed on the spatio-temporal map of the original data, showing that different objects yield characteristic mixtures of templates, depending on the gradient at the edge and its range from the moving camera. In this example the camera is moving towards the objects and passing by one of them; therefore the diagonal lines of motion of contrasts through space and time are curved. At bottom left the eye is passing an object with a regular texture. The meanings of the symbols are as follows:*
■ $= (\downarrow\downarrow/\uparrow\uparrow)$, ➤ $= (-\downarrow/-\downarrow)$, ▶ $= (\uparrow\uparrow/--)$; $+ = (\uparrow\uparrow/\downarrow\uparrow)$, □ $= (\downarrow\uparrow/\downarrow\downarrow)$, × $= (\uparrow-/\uparrow\uparrow)$.

can represent four colour types or equally well four related template types at the corners of a tetrahedron (Fig. 11.27). Points at different distance from the corners then represent different ratios of template responses, which is a pictorial way of saying that the numbers of template responses can be processed in similar ways to numbers of photons in receptors, as well as by adaptation, antagonistic interaction, and temporal summation at subsequent stages. In the colour system we find receptors and neurones with responses that depend on light intensity, although final colour discriminations are independent of intensity. Similarly, the early directional neurones that respond to moving edges are strongly dependent on contrast frequency, being driven by each edge event that passes. Nevertheless the freely moving insect is able to discriminate features of edges irrespective of contrast frequency. In both cases the final visual behaviour depends on the taking of ratios at a location *after* these neurones. Down this track we will find a lot of room for exploration of mechanisms of semivision in both natural and artificial systems.

Across-fibre Information Processing

Theories of chemoreception long ago introduced the idea that smells and tastes depend on simultaneous stimulation of clusters of specific combinations of different neurones (Erickson, 1963, 1982). Theories of the function of the mammalian hippocampus or cortex have also been presented in terms of ensembles of neurones in distributed processing mechanisms which carry a complex pattern in chemosense, touch or vision, by having information distributed in parallel in many neurones. The models usually envisage physically orthogonal arrays of two sets of neurones, usually considered to be in columns and in layers

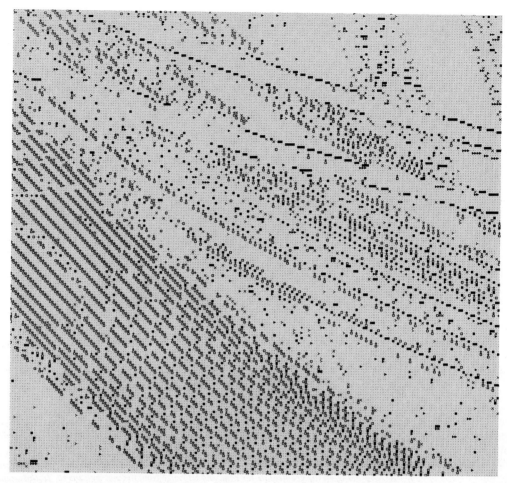

Fig. 11.26 *Template responses in a spatiotemporal plot as in Fig. 11.25, but for clarity the background of the original data (Fig. 11.21) is omitted. The meanings of the symbols are as follows:* □ = (↓↓/↑↓), ⊳— = (↓↓/–↓), ▶ = (↓↑/↓↓); × = (↓–/↓↓), + = (↑↑/↓↑), ■ = (↑↑/–↑).

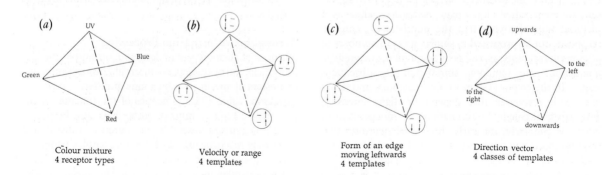

Fig. 11.27 *Because template responses are line-labelled and can be counted, groups of them can be represented in the same way as photon counts in receptors of different colour sensitivity. Here a few possibilities are illustrated ((a)–(d)).*

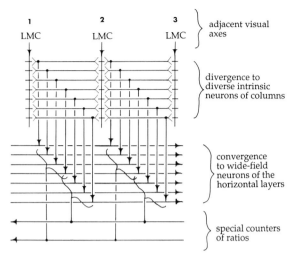

Fig. 11.28 *Ensemble processing by an array in the insect visual system. The inputs are the lamina monopolar cells which carry information about temporal contrast on each visual axis (L1 to L5 in Fig. 11.13). In each column these diverge to a number of intrinsic neurones which presumably abstract different aspects of the temporal contrast in various spatio-temporal combinations (templates). Templates of a given type then converge upon collector neurones, and other neurones are connected to selected groups as special counters of particular ratios.*

running cross the columns (Fig. 11.28). Where the two sets cross there are synapses which could be modified by a conditioning stimulus or by usage. Such a matrix responds with a specific output pattern of line-labelled neurons for each input pattern (reviewed in Rolls, 1987) and has the appropriate properties of tolerance to change in the input pattern, but still has hard-edged outputs. Such systems can complete an incomplete input pattern and can generalize stimulus patterns within limits. The insect optic neuropiles appear to be arrays of this type (Figs. 11.13, 11.14 and 11.28). Similar ideas are proposed by Strausfeld (1989).

The template model has been elaborated at length because it illustrates how a natural visual system could abstract spatial and temporal correlations from the moving image by a distributed mechanism. This is the new framework which allows us to think concretely about vision, both artificial and natural, and its evolution, in new ways.

Other Models

The Autocorrelation Model
For many years, data on motion perception in insects have been tested against the autocorrelation theory of Reichardt (1961), partly because the mathematics of systems analysis and filter theory have been conveniently available, partly because no other likely system has been investigated, and partly because in the early work there was some positive evidence pointing towards multiplication of inputs. The basis of the original model was a series of observations on the optomotor response of the beetle *Chlorophanus* making choices at Y junctions in its path (Hassenstein, 1959).

1. The response was proportional to the square of the contrast in the stimulus.
2. During motion of a regular striped pattern, reversal of contrast caused the perception of direction to be reversed. In the original experiments, this was perhaps no more than a natural effect of the phase change.
3. Pairs of adjacent facets were sufficient for directional motion perception. (Later it was found in the fly that additional facets participate in the processing mechanism.)
4. The response to velocity was a bell-shaped curve, showing that time constants control the upper and lower limits of velocity.
5. Later it was shown that the relative phase of the first and third Fourier components of a regular pattern can be shifted without influence on the steady-state response.
6. It was also shown later that the response depends on the temporal frequency of the passing of stripes, independent of spatial frequency, i.e. independent of pattern. This, however, is a feature of any 'event-driven' mechanism that detects edges if the responses are summed, and is not a test of a particular processing mechanism.

On the above experimental basis, Reichardt (1961) proposed a mathematical model in which the overall response and its direction could be calculated by making an autocorrelation between the modulation in a receptor and the same modulation shifted in time to correspond with a receptor on the adjacent visual axis (Fig. 11.29).

Deficiencies of the Correlation Model
The general difficulties are:

1. It is a mathematical operation in which the filtered image is multiplied with itself, whereas visual processing is an operation in which many neurone fields respond to a spatio-temporal pattern of input.
2. Motion perception involves much more than optomotor or fixation behaviour, and the correlation model has not proved useful for object vision or artificial systems.
3. This model does not suggest how the visual processing actually operates, or how to interpret the numerous neurones of the optic medulla with ON, OFF and ON-OFF properties.
4. It gives a single quantitative output where we know that even for optomotor behaviour there is a great deal of parallel processing by neurones, and we are interested in the mechanism rather than a calculation of the final summed effect.

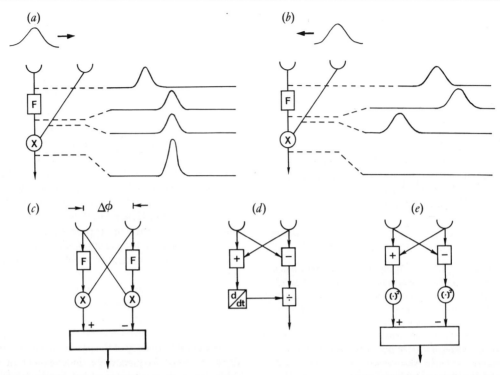

Fig. 11.29 *Classical algorithms for motion detection. (a) Motion in the preferred direction over one-half of the Reichardt multiplication model (c); the filter F causes a phase delay which causes a coincidence with the spatially delayed signal. (b) In the antipreferred (null) direction there is no coincidence, so that multiplication yields zero. (c) The full multiplicative system which eliminates spurious responses to intensity change. (d) The gradient model. Velocity and its direction are obtained by dividing temporal contrast by spatial contrast, irrespective of pattern. (e) The energy model. The difference between the squares of the sum and the difference for two adjacent intensities. The final result is similar to that in (c) for some stimulus patterns.*

5. Actual calculations from whole-animal responses yield arbitrary time and filter parameters that could have arisen from properties of the receptors, muscles or mechanical inertia, and are not necessarily related to the processing of motion.

6. In systems analysis, an equal number of gradients of opposite polarity are presented at the same time, generating the second harmonic in the responses.

In particular, there are other difficulties:

1. The finding that summed responses of large-field optomotor neurones or optomotor behaviour depend on the contrast (drift) frequency of the stimulus (e.g. the frequency of passing of edges) is not helpful in distinguishing processing mechanisms, and it makes us wonder how flying insects are independent of spatial frequency in the visual world.

2. The correlation theory predicts that white–black and black–white edges are treated together, but in fact we find that edges of opposite polarity are processed separately and differently (Kien, 1975; Franceschini *et al.*, 1989; Horridge and Marcelja, 1990a).

3. It predicts that (response) is proportional to (contrast)^2, which is difficult to demonstrate in neurone responses. In fact, we find that the response is more nearly proportional to (contrast)^1 in flies, and that the response saturates at low contrast levels (Horridge and Marcelja, 1990b).

4. Responses to visual stimuli in early processing neurones are essentially phasic and immediate: they are not averaged responses.

5. Responses of visual processing neurones have overlapping response patterns in time and space which are not specific for individual features of the stimulus, suggesting that groups of neurones respond together in spatiotemporal combinations, as in other sensory and motor pathways in all nervous systems.

6. The behavioural responses and the individual-motion-sensitive neurones rapidly adapt: the high-resolution systems in insects appears to be phasic and designed to see direction and non-directional motion, change in velocity, temporal frequency and especially change in contrast. Similarly, in the colour domain, the opponent cells of the optic lobes of locus (Osorio, 1986)

and bee (Hertel, 1980) apppear to be detectors of colour contrast over a narrow spectral range.

7. A direct test for multiplication of inputs, made by reversal of the contrast of a single bar as it jumps by one inter-receptor angle, fails to evoke a directional response in either direction (Horridge and Marcelja, 1990a). Unless a moving edge preserves its polarity when moving, only OFF or ON responses are given (Franceschini *et al.*, 1989).

The autocorrelation model, with a low-pass filter in the arm from one receptor and a high-pass filter in the other, depends on the phase delay between these two filters to generate a diagonal spatio-temporal sensitivity. The low-pass filter is able to pass the 'no change' signal which is necessary for a response to a single black–white edge which jumps by one receptor spacing, as in Fig. 11.18a. The spatio-temporal models illustrate two points with reference to directionality: (a) the diagonal spatio-temporal directional templates can be generated in many ways and (b) all that is needed for directionality similar to that given by the autocorrrelation model is a 3:1 structure with diagonal spatio-temporal symmetry in any neurone field which can be simplified to a 2×2 template.

Two-Dimensional Vision

Let us try to work out what is meant by two-dimensional vision with distributed parallel processing as outlined by the model with templates as crude mimics of local neurones.

Making Two-dimensional Templates

The basis of the one-dimensional model is that the flow-field is a predictable consequence of locomotion or scanning while the animal moves on an even keel, so that semi-vision in insects is concerned with the detection and discrimination of contrasts moving in predetermined directions. We can extend this to the detection of corners and edges by the simplest two-dimensional templates, which have a $2 \times 2 \times 2$ structure (Fig. 11.30), but we can immediately anticipate several problems.

1. There are 6561 of these $2 \times 2 \times 2$ templates, of which only a few may be useful, as determined by running tests with natural scenes.

2. The $2 \times 2 \times 2$ templates appear to be insufficiently specific to detect two-dimensional features, and at the same time too numerous to collaborate together for discriminations because there would be far too many possible groupings of them.

3. The $2 \times 2 \times 2$ templates are readily fooled by selec-

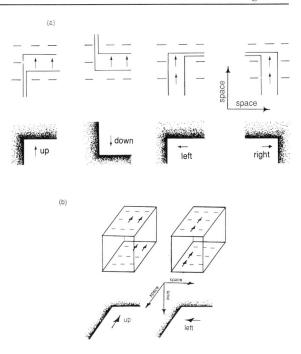

Fig. 11.30 *How to construct templates for two dimensions in space: (a) corners that move in predictable ways generate patterns of intensity change; (b) adding the time dimension to (a) gives us three-dimensional templates.*

ted one-dimensional stimuli, as may be checked by testing with a variety of edge inclinations.

When we implemented this model, as before, but with successive two-dimensional images in spatio-temporal coordinates, we were met by complete failure with natural scenes. There are so many possible templates with combinations of three states at four adjacent axes at two successive times that the responses of almost all of the templates are scattered like noise across the scene. A few $2 \times 2 \times 2$ templates show up moving edges that happen to be sharp and in line, but most edges turn out to be stepped or notched at the detailed level of individual visual axes, so that there is a chance for a variety of templates to respond here and there without revealing significant template groupings.

The situation is far worse when we try templates of the $3 \times 3 \times 2$ type, of which there would be 387 420 489 on each visual axis. Many of these templates are reasonably specific in detecting moving corners, but they are so numerous that we cannot map any feature of a moving visual scene with the responses of any of them.

The situation can be simplified by combining templates together so that any one of several gives a response. This process is equivalent to making simpler templates such as $3 \times 2 \times 2$ type, or reducing the number of states from 3 to

2. In fact, it is also equivalent to combining together a few of the simplest one-dimensional 2 × 2 templates by a logical AND. This line of thought shows that although correlation of changes at adjacent visual axes is essential, one soon reaches a practical limit in the variety of simultaneous correlations. The conclusion is that it is impossible to operate even reasonably specific templates in two spatial dimensions simultaneously, because there are so many of them. At the same time it is clear that the spatial resolution has evolved for a good reason, namely that the high-resolution unit detectors of edge motion are based on adjacent axes at successive times.

Abandon Two-Dimensional Purity

A solution to the impasse caused by the combinatorial explosion outlined in the two paragraphs above is provided by the insect visual system, if we are prepared to accept that insect semivision is *one-dimensional in two dimensions*. By this I mean that the processing mechanisms (neurones) that are equivalent to templates have one spatial and one time dimension, and that they are oriented along the lines of the predictable flow field so that motion and processing are one-dimensional in each channel at each point on the retina. The line can be vertical up the front and along the top of the eye. There are additional templates aligned vertically at the side of the eye, for optomotor correction of roll in flight; in fact, there can be a superimposed orthogonal or a three-axis array for one-dimensional vision along two or three axes in any part of the eye, but no early templates to detect features in two spatial dimensions. Templates in the vertical plane may abstract vertically moving contrasts as postulated for others in the horizontal plane, but there is no reason to take all combinations of the 81 possible horizontal and 81 possible vertical templates and get back to the 6561 combinations in two dimensions, which individually respond as frequently to noise as they do to real scenes. There is no reason why the 50 or 60 small local neurones on each visual axis in the medulla column should not be sufficient for some vertical one-dimensional vision during head nodding (as in butterfly flight) and also for some horizontal vision in head scanning.

Presumably we can devise tests presenting trained bees with moving targets to see whether all their visual discriminations are accounted for by scanning along predetermined lines by one-dimensional vision dependent on the bee's motions or whether simultaneous correlations in two spatial dimensions have to be accepted. If vision truly in two spatial dimensions simultaneously (whatever that means) occurs in insects, it sets a big problem for evolution and for implementation with a restricted visual system, but so far the difficulty seems to have been in deciding what we mean by tests for vision in two spatial dimensions.

Semivision Performance

Semivision is the kind of vision that is inferred from the functional analysis of visual neurones and from visual behaviour of lower animals, with no suggestion of a conscious visual world or categorization of objects, but none the less providing excellent vision for mobility and recognition of features of mates, prey or obstacles in flight.

What Insects See

Freely flying insects appear to see the angle on the eye of moving contrasting edges and manoeuvre relative to them. Somehow, as a result of their own eye motion, they get a measure of the range of contrasting edges at each angle on the eye (Lehrer *et al.*, 1988; Kirchner and Srinivasan, 1989). This gives them a crude representation of the surrounding three-dimensional world just at the time that they most need that information. Their vision of moving intensity gradients, shades of grey and smooth shading has hardly been studied, either at neuronal or at behavioural level. Insect colour vision is related to object detection and fixation rather than to motion. Object and colour discrimination seems to be associated with poorer spatial resolution than motion perception and (so far as we know) is usually associated with groups of visual axes feeding into specialized processing neurones in the part of the eye looking forward, together with visual fixation behaviour, sometimes fixating while scanning.

Parallax

The word is from a Greek intransitive verb ($\pi\alpha\rho\alpha\lambda\lambda\alpha\chi\theta\eta\nu$) meaning 'to pass by one another' but for centuries has been used in astronomy to mean the angular displacement of an object relative to other stars as a result of the annual movement of the earth. Here, parallax means object motion against background as a result of eye motion.

Besides being able to measure the range of contrasts in different directions as a result of their own motion (Fig. 11.31), and to recognize the differences between contrasting edges, some insects can detect the three-dimensional structure of their surroundings (Fig. 11.32) by seeing the motion of edges against a structured background (Srinivasan *et al.*, 1990). In our spatio-temporal maps (Figs 11.23 and 11.25), parallax is detected as the place where a diagonal line of templates suddenly starts or stops. A general-purpose template to detect parallax is not easy to construct because the foreground and background may be moving in either direction across the eye, and the point of parallax may itself be moving in a different direction, but the sudden termination of a familiar mixture of template responses reveals the closing parallax where a distant contrast goes behind a nearer one, and the sudden novelty of a new mix of templates is a good sign of opening

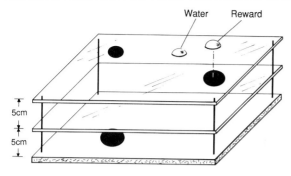

Fig. 11.31 *The apparatus used to demonstrate the ability of a bee to measure the range of objects irrespective of absolute size or position on the eye as they fly along. The bee flies over the three (or more) tiers of clear perspex sheet upon which discs of various sizes can be placed. The bees can readily discriminate* and *land over the discs of shortest, intermediate, or furthest range irrespective of their randomized sizes and locations. The target for the bee is the drop of sugar solution over the desired disc and other drops are only water to teach the bee not to look for the drop of sugar solution alone. The final discrimination tests are done with no reward drop present. Bees act as if colour blind in the measurement of range.*

Fig. 11.32 *The apparatus used to demonstrate the ability of a bee to detect the raised patterned platform by the parallax generated against the patterned background as the bee moves. By use of a variety of patterns it can be shown which combinations of foregound and background are most effective for the bees. In these tests the bees come in to land at right angles to the near edge, which therefore generates closing parallax and would cause responses in a specific ensemble of templates (Photo kindness of Dr M. Srinivasan).*

parallax. If the insect assumes that distant large objects are stationary, parallax provides a powerful measure of range during a predictable movement of the eye. Parallax is a reliable signal that distinguishes a solid or separate object from a shadow or flat patch of colour, and it looks very much as if parallax is the driving force that has led to the evolution of vernier resolution or hyperacuity in man by integration along the whole edge.

Object-motion Detection

We might well ask, after so much theory of mechanisms and illustrative but hypothetical templates, what actually responds when small, contrasting moving objects on a background are presented to the visual processing neurones. A class of neurones in dragonflies that detect motion of a small contrast but do not respond to motion of a large background of moving contrasts has been defined (Oldberg, 1986) as object-motion detectors. The DCMD neurone of the locust (Fig. 11.16), the earliest such object-motion detector to be described (Rowell, 1971; Pinter, 1979), probably functions in the avoidance of crashing into obstacles while in flight. The input of these object detectors appears to be numerous non-directional small-field neurones that pick up a moving contrast and feed into a pooling neurone at synapses which rapidly adapt at each location. As a result, the stimulus soon fails to excite at that point and must again become novel in some way to renew the response. This looks like an adaptation to one particular mix of templates, but a renewed excitation as soon as the stimulus is shifted or a new mixture appears. Opening parallax continually generates new mixtures. At present we have no better way to interpret the electrophysiology of the detection of moving objects against a textured background.

A great deal of work remains to be done in uncovering the properties of anatomically identified neurones before we can make sense of insect visual processing at the lobula level where object vision may be based if it exists. One of the main difficulties is to identify the real visual fields and significant backgrounds because systems analysis, with stimuli containing black–white and white–black edges *always together* over a range of temporal and spatial frequencies, is not yielding the data we need for inferring neurone functions. If there are medium-level templates, how do we delineate them? So far, the known object-detector neurones do not reveal by their individual response profiles what specific targets they select, possibly because they have to work in groups. However, they suggest that insects do not have a relatively few complicated templates in visual processing, but have instead numbers of (ratios of) less specific template responses in parallel at every level. That makes the task of analysis even harder.

Fixation, Foveas and Object Recognition

A major problem in visual processing is that the relative motion of the eye and the visual world causes the image to move across the receptor array, and therefore the representation of any particular contrast moves across the central map (Fig. 11.15). To detect this motion of contrasts across the eye it is necessary to look along diagonal

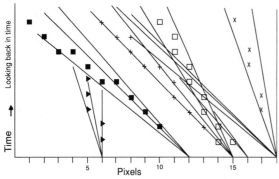

Fig. 11.33 *The Hough transform. Counting the different template responses along the diagonal lines in spatio-temporal coordinates for a moving eye. The slope of the diagonal line of template responses is a measure of the range if the eye scans laterally and does not rotate. So, from any point in time, we look back over the template responses and find the angles at which large counts of responses occur. These angles give the ranges of contrasting features in front of the eye. A vertical line of response represents a feature at infinity or at the fixation point.*

Fig. 11.34 *Fixation. The primary one-dimensional image (plotted along) at successive times (plotted downwards) in spatio-temporal coordinates when the eye scans and also fixates upon one feature in the visual scene. The fixation point is the apparently stationary vertical band down the middle. Objects in front of the fixation point move to the right and down, while those behind the fixation point move to the left and down; the steeper the slope the nearer the objects are to the fixation point.*

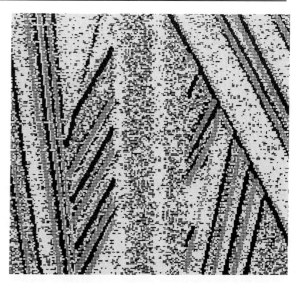

Fig. 11.35 *The spatio-temporal map of thresholded contrasts, pixel by pixel, from Fig. 11.34. (■) decrease in intensity, (▶) increase, (−) no change. We now see that the object which is fixated upon generates a lot of (−) 'no change', although it contains contrasts.*

lines in spatio-temporal coordinates and to count the template responses to any particular visual feature as they move across the central projection (Fig. 11.33). A serious difficulty arises when these diagonal lines are not straight, i.e. when motions are not constant (Fig. 11.25). The difficulty is increased by changes in the appearance of solid objects, and their relations to each other, as the eye moves.

Insects have evolved a strategy to reduce these difficulties. We find many examples of visual fixation by insects upon objects, especially of prey, food sources, nest sites, mates or any novelty which moves, and especially when landing from flight. More and more examples have been described over the past few decades, from all groups of active insects (Van Praagh et al., 1980; Wehner, 1981; Section D; Rossel, 1986; Lehrer et al., 1990; Zhang et al., 1990; Zeil et al., 1989).

Inseparable from this visual fixation behaviour we find the repeated independent evolution of acute zones or foveas which have selective advantage only if there is visual fixation. When held at a predetermined position on the eye with the help of a fixation mechanism, a single contrasting feature can be processed over a reasonable time within a single neuronal field of reasonable size instead of giving fleeting phasic responses as it moves across the internal projection of the retina. However, the fixation behaviour requires some kind of attention-directing circuit so that the fovea is 'locked on' to one particular object and other objects are rejected. Foveas and fixation behaviour certainly evolved independently many times over, so that the advantage must be worth the investment in the control circuits. Acute zones are found in many groups of arthropods (Horridge, 1978a) and always associated with fixation. Predetermined neurones behind the fovea have been described in the male housefly

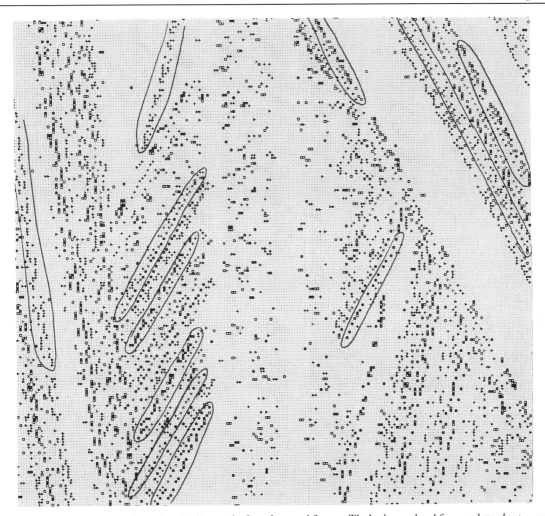

Fig. 11.36 *Responses of directional templates looking at the fixated scanned feature. The background and foregound stand out, moving in opposite directions. Fixation knocks out the directional templates from the fixated object. Each moving edge in the scene has its own characteristic mixture of template responses, some of which are outlined. Templates are as follows:* ■ = (−↑/↑↑) *and* (↑↑/−↑); □ = (↑−/↑↑) *and* (↑↑/−↑); × = (↓↑/↑↑); + = (↓↓/↓−) *and* (−↓/↓↓); ▶ = (↓↓/−↓) *and* (↓−/↓↓); ≻ = (↑↓/↑↑).

and are related to the chasing behaviour by the males (Strausfeld and Nässel, 1981, p 102).

Fixation upon one particular contrast does not necessarily imply that the insect is stationary relative to the contrast. It may be moving forwards or scanning sideways at that time, as if coming in to land, pursuing a target or inspecting a contrasting feature. Fixation usually includes a relative movement of the eye as it wanders about the target. Therefore parallax becomes more important when combined with fixation. On a spatio-temporal mapping, the contrasting feature which is fixated upon generates a vertical band (Figs 11.34–11.37). Considering higher-level neurones as leaky counters of template responses, with the advantage of fixation they can count over longer times and with greater spatial resolution, i.e by neurones with smaller fields than would be possible if the image were moving across the retina.

The effect of fixation upon the counts of template responses opens up further interpretation of the function of foveas in the visual process. Moving the eye sideways while fixating upon a contrast in the middle distance causes contrasts in the foreground and background to move in opposite directions (Figs 11.34–11.36) and they move faster the further they are from the point of fixation (see Sandini and Tistarelli, 1990, for recent references).

Fixation also assists colour vision. For obvious reasons few templates for motion respond around the fixation point. Instead, templates carrying two or more adjacent

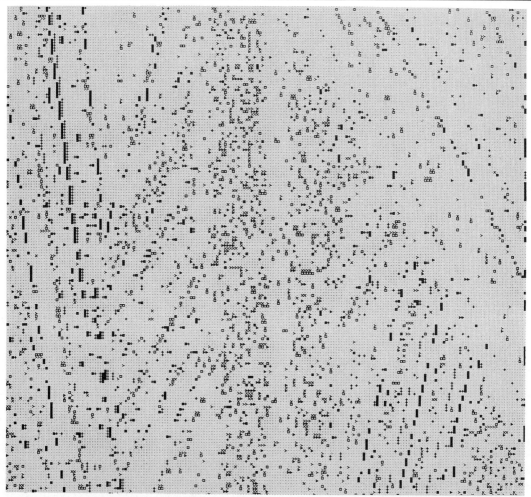

Fig. 11.37 *Responses of non-directional templates to a scene which includes the fixated, scanned feature in Fig. 11.34. The jitter in the fixation causes noise which appears as non-directional motion within the fixated object. Combinations of these templates detect certain velocities but are not particularly useful for identifying other features. Symbols:* ■ = (−↓/−↓) *and* (↓−/↓−); ▶ = (↓↓/↑↑); × = (↑↑/↓↓); + = (−↑/−↑) *and* (↑−/↑−); ≻ = (−−/↓↓); □ = (↓↓/−−).

'no change' pixels respond frequently in the blank areas of the spatio-temporal map (Fig. 11.35). Regions of the visual field where adjacent pixels are not changing are exactly those where opponent colour mechanisms and spatial interactions for colour discrimination can function undisturbed by motion or flicker. We already know that the high-resolution motion pathway is colour blind (Lehrer et al., 1990); we will probably find that pattern discrimination and colour vision are 'motion blind' and that colour vision has its own channels with templates containing adjacent 'no change' pixels.

Why Templates Are Proposed

This theory, based on small, high-resolution motion-detection templates in early vision, extends beyond previous theories. The optic lobes of insects which process the motion information do not function by multiplication of the stimulus intensity with itself: they function by convolution of the pattern as it moves over the eye, with the fields of many neurones located behind every visual axis. There are 50 or more of them in each column of the *Drosophila* medulla (Fischbach and Dittrich, 1989); see Fig. 11.13). There is essentially an instantaneous phasic response to the local spatiotemporal features from the neurones within, as the image of the visual world flows across the central projection.

Insects execute active manoeuvres, especially when flying, which give them information on range, parallax and, in some cases, some kind of rudimentary form of vision and memory of individually distinct landmarks in different directions. The neurophysiologist in search of the optimum fields and functions of small optic lobe neurones must have a framework of ideas about what the neurones might be doing, so that electrophysiological experiments to determine neurone patterns can be put into action as soon as each neurone is found. Systems analysis, steady-state theories or mathematical operations on the stimulus pattern are not useful in this experimental context. They are not much use, either, when making inferences about neurone fields.

The effort to proceed beyond the anecdotal stage in studies of visual behaviour must be based on a theory which is useful in so far as it suggests experiments and appropriate controls to make the experimental results definitive. We can readily see, by study of old descriptions of visual discrimination tests with bees, for example, a lot of experimental effort wasted for lack of a useful theory. There is also a lot of speculation in vision for want of useful data.

These broad generalizations led me to search at the interface between electrophysiology and visual behaviour for a theory which is based upon large numbers of small-field neurones in parallel. We expect to find numbers of small-field neurones with properties that reveal which templates have in fact been selected from the complete range of possible spatio-temporal combinations. If vision depends on contrast changes at adjacent pairs of receptors at successive instants, and we have a thresholding mechanism, then 2×2 spatio-temporal templates are a simple way of generating responses that can be counted like photons. Actual neurones are, of course, more flexible and overlapping than templates, but we now have a scheme which exemplifies the parallel distributed processing in operation with real scenes. The template theory provides us with a notation (Fig 11.23–11.26) to assist further thinking about visual processing. The template theory also turns out to be a useful way to think about the evolution of visual processing, and it generates new ideas about colour vision, discrimination of features independently of their number, the effect of fixation and the significance of one-dimensional vision with predictable motion of the eye.

The old lessons from biological processing of information are that neural mechanisms are sloppy and inaccurate, that they respond rapidly to transients by preformed expectant sensors and few successively higher level neurones, and that they avoid extensive sequential computation. The new lessons are that several very simple, specialized pathways can work in unison to achieve specificity and discrimination, that parallel distributed systems cannot be analysed as if they are single channels and that prior knowledge of the task is built into every aspect of structure and function by progressive evolution of earlier processing mechanisms.

The idea of templates helps us to visualize the mechanisms and is a useful theory for many approaches to the problem of how vision works. The templates show how the spatial organization of contrast can be fed into an array of neurones.

How Semivision Avoids Problems in Artificial Vision

A number of problems have been identified by those who simulate computer vision or try to build artificial seeing systems. With the insights gained from the template theory, we can now see how insect semivision has overcome or avoided these difficulties.

1. Motion of the eye is assumed to be horizontal and forward. This yields a one-dimensional analysis of motion at each point on the eye. To control roll and pitching in flight there are additional systems in the vertical planes, but no evidence that these are used for two-dimensional picture analysis of the visual world.

2. Contrast is saturated by means of a high gain in the channels for motion perception. This reduces the information load with little loss of the image structure in return for seeing any moving contrast at high spatial and temporal resolution.

3. The visual world is composed of edges so that 2×2 spatio-temporal templates are useful in groups. This generates a useful compromise between ambiguity of responses and variety of templates. Templates that are more specific are too numerous.

4. The template system for motion is colour blind: colour vision is in separate channels. Maybe colour vision is directionally motion blind.

5. The high spatial resolution of the retina and lamina continues into directional and non-directional motion detectors of the medulla. Each of these motion pathways is then separately summed into large-field neurones which control vital locomotory reflexes. There is no evidence of other kinds of vision, e.g., of objects, at the highest resolution.

6. In particular, there is no evidence that the position of edges are *located* with any kind of mechanism downline that can give the coordinates of a 'zero crossing', so such operations are irrelevant.

7. To obtain a simple three-dimensional map of the world, the range in each direction is measured from the relative motion on the eye caused by the predictable flow field at each point. There is no need to analyse the whole flow field.

8. The three-dimensional map of the world is augmented by the parallax caused by motion of nearby edges over a patterned background. Parallax is detected where different motions of edges meet. Parallax detection is incompatible with the 'smoothness constraint' that is sometimes assumed in the analysis of flow fields in artificial visual systems.

9. The variety of outputs is limited and visual behaviour relies on a few specific visual cues for each task. This helps reduce the enormous number of template combinations that would be essential for vision in two dimensions.

10. Picture analysis and categorization of objects are replaced by a reliance on trigger features or predictable cues.

11. If possible, during a difficult discrimination task, the image is held stationary in the central projection by fixation while scanning (Figs 11.34–11.37).

12. The use of one-dimensional processing along predictable flow lines means that the classical problems of two-dimensional vision do not arise.

13. The 'aperture problem' arises because the true motion of a moving straight edge cannot be determined when it is seen through a single-bounded window such as a neurone field. The 'smoothness constraints' assume that edges of objects are continuous, that velocity is constant over small areas of the image, and that objects are not elastic. These assumptions are not relevant to one-dimensional vision.

14. Contrary to the smoothness constraint, discontinuities in the flowfield are useful indicators of parallax, and freely flying insects interacting in flight cannot assume that the surrounding world is rigid.

15. The enormous number of combinations of inputs that makes picture analysis computationally heavy and slow is reduced by one-dimensional vision, built-in templates, limitation of input and output tasks, ensemble processing and the use of large fields for high-level channels.

16. Learning is a separate process which changes the weighting in ensemble processing, i.e. changes the preferred *ratios* of template responses, which applies equally to vision of features and colours.

Larger Templates

Recognition of Significant Images by Semivision

An obvious feature of insect vision is the ability to recognize and chase a mate or prey, fly away and return to the same twig or leaf, repeatedly visit flowers of one kind, or use landmarks to return to a nest hole. How are these features recognized?

So far I have outlined a theory of elementary colour-blind templates which abstract direction of motion, polarity of contrasts and non-directional motion at the maximum level of spatial resolution that the design of the visual axes allows (Fig. 11.18). Those templates drive the retina towards the evolution of better resolution. The same templates can provide ratios to give measures of contrast qualities, just as is done by receptor inputs in colour vision (Fig. 11.27). After that, the ratios, now independent of pattern, can be put together by logical AND to trigger deeper functions for specific tasks to avoid the combinatorial explosion. Is there more?

Certainly, at the whole-eye level there is more, as outlined in examples below; but we should carefully consider a group of mechanisms of intermediate scale that detect biologically significant images which are commonly called 'trigger features' or 'sign stimuli' following early popularization of these terms with work on the visual behaviour of fishes, spiders and birds (Tinbergen, 1951). To be as critical as possible, I would like to see discriminations that cannot be accounted for by ratios or logical AND of responses of templates such as those in Fig. 11.18 and 11.20. Let us examine a few examples before trying to analyse the possible processing mechanisms.

Recognition of Mate, Prey or Predator

Examples abound; one is the preference of stick insects which, when they fall on the ground, walk towards any object that looks like a bush (Fig. 11.38). Another is the threatening posture which is displayed by praying mantis of the genus *Stagmatoptera* when encountering an insectivorous bird. The bird is recognized visually. The behaviour is restricted to this genus of mantis which have large imitation eyes on the prothoracic femur, and the response is fairly specific to certain species of birds (Crane, 1952), but is apparently not learnt. The mantis rears up, exposes

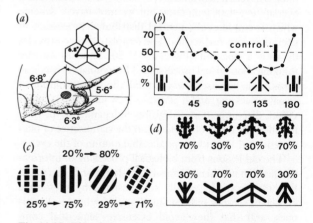

Fig. 11.38 *The relative attractiveness of different images to a hungry stick insect when given a choice. (a) The large interommatidial angles of the eye. (b) The angle at which the side branches project has a large effect on the visual preference. (c) Preference is always to the right-hand member of the pair. (d) Roughening the branches makes them more attractive when they are pointing upwards. Inbuilt visual behaviour of this kind suggests that predetermined groups of appropriate templates occur in the optic lobes (Modified from Horridge, 1978, after Jander and Volk-Heinrichs, 1970).*

the mock eyes, elevates its wings, stridulates violently and sways from side to side, often frightening away the bird. With one compound eye covered over, the mantis can still recognize a bird, or a film of one, but cannot estimate its range (Maldonado, 1970).

The significance of visual shape in arthropod behaviour patterns is reviewed at length, with numerous examples, by Wehner (1981). The main categories of images are the shapes of flowers, the recognition of mates and prey, courtship behaviour (especially in hunting spiders and fiddler crabs) and the pattern on the wings of butterflies, dragonflies and others. The basic problem is that we are unable to obtain much information about mechanisms of visual processing in these cases; therefore we turn either to the analysis of object-detector neurones (if they can be found) or to the visual behaviour of the trained bee.

Eidetic Images

The word is from the Greek root ΕΙΔΩ (latin, *video*) or Σϊδωλον, meaning an image in the mind, which has come into English as 'idol'. The word 'eidetic' means an imprint of the image in the visual processing mechanism, but authors are usually not clear whether the eidetic image can float about in the spatial array of the visual projection or whether it is burnt in at one location, requiring congruence at the same location before recognition can occur.

As mentioned above, experimental analysis is almost restricted to bees. The honeybee can be trained to come to a food source that is recognized visually, and then, being trained for one target pattern, can be tested with the same or other patterns to reveal something about what has been learned. Three warnings are essential when considering this topic. Firstly, the bee must be given a frame of reference in the form of her own motion, including horizontal scanning, and the direction of gravity; she is then able to discriminate many shapes, angles of inclination and locations relative to patterns or landmarks. Most experiments before Wehner's work, starting about 1970, were done with patterns laid flat, probably copying Von Frisch's early experiments with colours. The bees, having no reference axis while flying over the flat patterns, showed that they could discriminate *degrees of disruption* in the patterns, but little more. Only experiments on vertical surfaces are useful. Secondly, the bee must be forced to make her decision at some distance from the target, and target positions must be randomized regularly during training. Bees that are allowed to examine targets closely before making a choice can concentrate on regions of the target, e.g. the top corner, so that it is then impossible to know what part of the target the bee is looking at. Thirdly, all aspects of the pattern except the detail that is the topic of the experiment must be randomized repeatedly during the experiment to teach the bee what features *not* to look at and learn. This experimental design of controls is so frequently ignored that it is impossible to be sure whether the bees were looking out for some totally irrelevant detail. Many other controls are sometimes essential, such as restricting the approach path of the bee, or photographing the bees in flight, because the cue that the bee uses to locate the reward may be one that the experimenter does not anticipate. By randomizing all other features, Lehrer *et al.* (1988) showed that bees can measure range to objects independently of size, and can discriminate a selected range (Fig. 11.31) to get a reward; Van Hateren *et al.* (1990) and Srinivasan (unpublished data) showed that bees can discriminate a difference of 45° in the slope of parallel lines in a pattern of random parallel stripes (Fig. 11.39), but they cannot discriminate one random pattern from a similar one at the same slope; Srinivasan *et al.* (1990) showed that bees can use parallax (Fig.11.32) to get an idea of the separateness of an object from background. These experiments can all be interpreted readily by templates responding to horizontal and vertical motion with scanning (Fig. 11.20). Apparently, pattern disruption interferes with pattern discrimination by bees, as would be expected from a theory based on ratios of template responses. To make inferences about more complex tem-

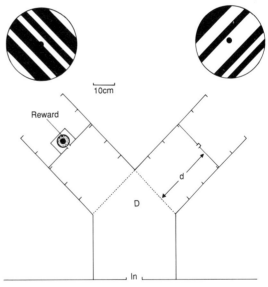

Fig. 11.39 *The apparatus used to test visual discrimination of forms and angles. The bee enters through the hole, and in the middle of the Y must make a choice between the two targets which can be placed at various distances (d) in the arms of the Y. The reward is found in the hole in the centre of the pattern. The two examples shown were from a selection of randomly striped patterns. It was found that from a distance the bee can discriminate a difference of 45° angle irrespective of pattern but bees cannot discriminate these two patterns if they are inclined at the same angle (Redrawn from Van Hateren et al., 1990; Srinivasan, unpublished data).*

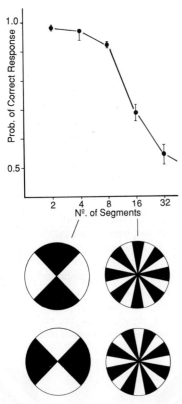

Fig. 11.40 *One of the original experiments on a vertical surface suggesting that bees can learn a two-dimensional pattern by something like an eidetic image. The bee is rewarded inside a hole at the centre of the disc, and is tested on a discrimination between the upper and the lower pattern of each pair, one rotated by a segment relative to the other. Bees can do this with discs up to about 32 segments (After Wehner, 1981). The question still remains whether it is necessary to postulate an eidetic image, and is so, whether that can be done by one-dimensional visual processing.*

plates for shapes requires very careful design of control with patterns on vertical surfaces.

Early experiments (Fig. 11.40) were done 20 years ago by Rudiger Wehner (review, 1981) who showed by cinephotography that the bees faced the target as they hovered or scanned in front of it. He later found (with Flatt) that they could not recognize the target if the eye region that learnt it was painted over, and suggested that something is fixed in the map behind the eye. For this to be so it is evident that the image cannot be at the full resolution of the eye – it must be fuzzified by neurones with large fields so that the image can be recognized *within a region*, albeit less precisely. Taking ratios of template responses would be one way to enlarge those spatio-temporal fields.

Over the past decade the standard way to think of pattern recognition in insects, by a combination of old and new results, was to imagine accurate measurement of flicker sequences as images swept across the retina and also as a less precise recognition of a pattern that is briefly lined up with an internal representation like an eidetic image (e.g. Collett and Cartright, 1983; Gould, 1985). The fovea of the insect eye was seen as the most effective region for picking up the eidetic images and responding to their reccurrence, so that fixation behaviour was tied into object discrimination.

Template Theory Applied to Object Vision

First, setting aside colours, let us refer to the colour-blind high-resolution templates in Fig. 11.20, noting that most respond to *polarity* and directional or non-directional *motion* at adjacent visual axes. As discussed already, making larger templates *at this resolution* is futile on account of the combinatorial explosion, but taking ratios of template counts, followed by logical AND in local spatio-temporal regions, looks promising, although two-dimensional shape is thrown away in the process. We add the possibility that there is also a crude, one-dimensional semivision of this type in the vertical plane, even if it is less effective.

Within a brief period in any local region of the eye, as the bee scans horizontally with one-dimensional horizontal and vertical directional detectors, the ratio of template responses for upwards, downwards, left-and-right motion (Fig. 11.27(d)) could be a measure of the angle of inclination of the edge, independent of form, and it could also distinguish edge polarity. If the direction of the bee's own scanning motion is monitored by large-field directional neurones, then a stationary angle could be discriminated by a freely flying bee.

Bees can in fact discriminate the angle of inclination of randomly arranged stripes (Fig. 11.39) when the form of the pattern presented is randomized in the tests so that the bee is taught to look only at the angle. In these tests, however, bees cannot discriminate different examples of random stripes except by trivial properties such as average brightness or contour density. When the tests are controlled in this way, the bee cannot make an eidetic image *from a distance* (Fig. 11.39).

The experiments that most strongly support the idea of an eidetic image are those of Wehner (1981, p 477) done in 1972. Bees can detect the change from white to black at the top of a disc with 16 alternating black and white segments (Fig. 11.40). At first sight it is hard to explain the bee's ability to discriminate a small angular shift of the radial pattern, because the sum of all the angles is independent of the angular displacement. In these experiments, however, the bee comes close to the hole in the centre of the pattern, which therefore occupies a large part of the eye. The experiment seems designed to tell the bee that she should look to see whether the horizontal midline of the two eyes

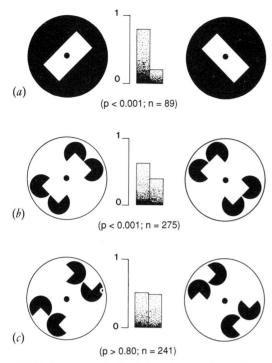

Fig. 11.41 *Bees can easily discriminate between the two figures in (a) and fairly easily between those in (b), but cannot distinguish the two patterns in (c). It is an interesting question whether their failure to discriminate in (c) is caused by the disruption in the patterns or by the loss of the visual illusion (Redrawn from Van Hateren et al., 1990).*

is in line with black or white. Another puzzling discrimination test, done recently in Canberra, is illustrated in Fig. 11.41. Bees easily discriminate between the patterns in (a). The patterns in (b) are more difficult, but are still distinguished correctly more often than not. The same patterns as (b), but rearranged as in (c), so as to eliminate the visual illusion, cannot be discriminated by bees. Whether the failure to distinguish in (c) is simply because the pattern is more disrupted than (b), as seen in horizontal scans, or whether the bees see the illusion as we do, cannot be decided. Certainly it is hard to design crucial tests for eidetic images which have only one interpretation.

All of the patterns recently employed for tests for pattern vision on bees, except the randomized use of selections of random patterns (Van Hateren *et al.*) can be criticized in the light of the discussion of proper controls. It is certainly possible to devise template ratios that discriminate the patterns and provide alternatives to eidetic images in the supposedly definitive studies (e.g. Gould, 1985, 1986). However, there are other significant natural situations, akin to patterns, that may depend solely on spatial (possibly one-dimensional) correlations over large

visual fields: for example landmarks, and whole-eye or two-eye templates.

Landmarks

Insects apparently use visual landmarks so that without retracing their steps they can return to a known site that they have previously learned, often having made exploratory flights. The extensive literature on navigation by landmarks is reviewed by Wehner (1981). Experiments with movable landmarks show that bees learn what the distribution of landmarks should look like, as seen from the point where they desire to be. They search around until the bearings of the landmarks each lie on the retina where they have been accustomed to seeing them and at their normal apparent size subtended at the eye. However, because bees look forward but fly in any direction we are faced with the question of how the eidetic images of the landmarks can rotate within the bee. The best solution offered is that the bee has several internal snapshots which can be used when looking in the appropriate directions (Collett and Cartright, 1983). If this is so, we have no idea where they are located (Strausfeld, 1989).

Whole-Eye Templates

Viewing vision through human eyes leads to the error that lower animals see a picture of the outside world as we do. Consideration of the evolution of visual processing, bottom up, from receptors to simple templates involving pairs of receptors and temporal sequences, leads to the further error that an evolution of increasing complexity has produced hierarchical structures in vision – generating templates looking at templates and so on, up to categories like 'dog' and 'chair'. It doesn't work out so simply. Indeed, there probably was a progressive addition of templates in parallel at each level, and a progressive increase in the number of levels, during the evolution of vertebrate visual centres, or within the Crustacea from primitive ones with few optic neuropiles up to crabs with about five of them (This volume, Fig. 9.6; Bullock and Horridge, 1965, Fig. 19.2). However, when we actually examine a variety of examples of complex visual discriminations done by lower animals, it turns out that the sampling array itself is structured around a restricted set of tasks, or dedicated to a single task, so that eye structure and at least one channel of processing act together like a single template (Wehner, 1987). The following are three examples.

Size Constancy in an Object-Motion Neurone
The pond backswimmer *Notonecta* hangs in a predetermined position below the water surface, waiting for prey such as damaged insects that struggle as they float within

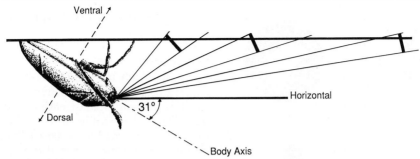

Fig. 11.42 *The backswimmer* Notonecta *hangs in a predictable posture below the water surface. The angles between visual axes looking horizontally to the distant water surface are less than those looking at objects lying closer on the water surface. As a result the angular subtense, measured in number of visual axes, is similar for an object of one size at any range within limits. The responses of an object-detector neurone of the optic lobe show that this gradient in the spatial magnification factor tends to make the neurone sensitive to objects of a given absolute size irrespective of range (Redrawn after Schwind, 1978).*

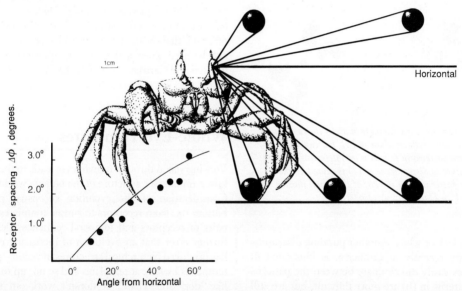

Fig. 11.43 *The ghost crab* Ocypode *has a vertically elongated eye with small angles between the visual axes looking at the horizon, grading into larger angles between axes looking up or down (inset). This built-in gradient of magnification in sampling could have the effect that an object approaching along the ground cuts the same number of visual axes irrespective of range. Similarly, an object above the horizontal is sampled by a constant number of visual axes irrespective of its distance, over a limited range, but the absolute size cannot be measured because the altitude is not known (Redrawn from Wehner, 1987).*

range. A large neurone in the optic lobe responds to motion of any small, black object within a field which includes the forward-looking part of both eyes. The neurone's response is reduced if only one eye sees the object. Motion of a large patterned background elicits no response and reduces that to a superimposed small object. The interesting feature is that objects of the size that produces the maximum response do so irrespective of their distance along the water surface. The preferred object size of the eye region looking at the horizon is smaller than that looking at a point near to the animal (Fig. 11.42). In the figure, the angular subtense of an object decreases as the object recedes, but the neurone (and perhaps the retina) compensates in such a way that the preferred absolute dimension of an object (at the water surface) is constant. In fact, the most preferred object is one-fifth the size of the *Notonecta*, a convenient size for a prey (Schwind, 1978). The same principle has been extended to ghost crabs (*Ocypode*) and other flat-world crabs, and related in a specific way to the eye anatomy (Fig. 11.43). The visual axes pointing to the horizon are closer together than those looking down to the ground, with intervals between axes graded in such a way that an object subtends a similar number of visual axes whatever its range, over limited distances (Zeil *et al.*,

1989). Clearly, this simple mechanism for size constancy functions only when the posture of the animal is predictable and the surface is level. The angle of the eyestalks is stabilized visually with reference to the horizon, which is therefore always relatively magnified vertically.

The Celestial Compass of the Honeybee

The clear blue of the sky everywhere carries a sun compass in the form of the plane of polarization of the ultraviolet rays. Elongated dust particles tend to float horizontally in the atmosphere and the light scattered by them is partially polarized, an effect which is strongest for the ultraviolet. This light originates at the sun and strikes the dust particles, where it is scattered before it arrives at the eye. As a result, imaginary lines drawn on the sky at right angles to the plane of polarization all point to the position of the sun. Even if the sun is behind a cloud, its position can be found if there is a patch of blue sky. In conjunction with an internal clock this is sufficient to act as a compass.

The template of the eye has a corresponding pattern of sensitivity (Wehner, 1987). The dorsal margin of the compound eye of the bee (and of many other insects) has ommatidia unlike those in the rest of the eye. They have poor optics but the photoreceptor cells are sensitive to the polarization plane in the ultraviolet. The axis of maximum sensitivity points towards the dorsal pole of the eye, in the same way (radially) as the lines at right angles to the planes of polarization in the blue of the sky point to the sun. Therefore, as the bee turns round, each blue part of the sky will appear maximally bright at two positions 180° apart. The pole of the eye then gives the direction of the sun (or the direction away from the sun). The ambiguity of the sun's position is overcome by use of the general brightness of the sky and by the integrative action of the bee's eye as a whole, because the sun is not expected to be below the horizon.

Fly-Grabbing by a Hungry Mantis

The praying mantis has a foreleg modified to flick out and catch a fly in its tarsal–tibial joint. To be effective the mantis must have the fly at the range that suits the length of its leg. Normally the mantis lies in wait and looks

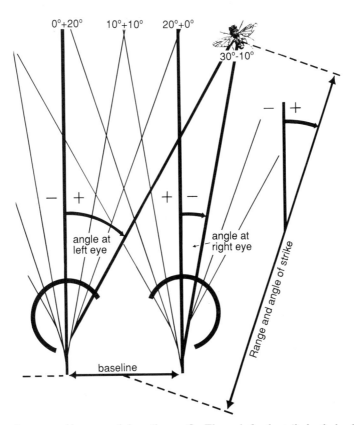

Fig. 11.44 *The praying mantis eyes are able to control the strike at a fly. The angle for the strike by the leg depends on the mean of the angles subtended from the midlines of the two eyes (with signs as shown). The difference between these angles gives the range and is approximately constant for a given range, as shown. An interesting corollary is that as the mantis grows and its leg lengthens, the visual angles and processing must also change to maintain the accuracy. (Redrawn after Rossel, 1986).*

towards an approaching fly. The visual fields of mantid eyes have a binocular overlap that is abnormally large for an insect – about 45°, with a wide, shallow, forward-looking fovea (see Fig. 11.1). In a series of ingenious experiments with prisms placed in front of the mantis's eyes, Rossel (1986) has found that the mantis strikes at the apparent range of the fly as seen through the prisms by both eyes. The visual axes radiating out from the two fixed eyes therefore form a fixed lattice (Fig. 11.44) which is able to locate the fly and indicate its range. The mechanism still functions quite well when the fly is up to about 20° from the mid-line. An approximate range can be calculated by adding together the angles subtended by the fly at the two eyes and taking the reciprocal, but knowing the way insect nervous systems function, it seems likely that there are internal templates which are set for just those combinations of angles that give the correct range, as suggested in Fig. 11.44.

Conclusions from Whole-Eye Templates

Although directional motion detection has the full spatial resolution of the retina, there is no evidence of invertebrate semivision mechanisms that locate the position of a stationary edge with the same accuracy. Although the minimum displacement of a moving edge is less than the interommatidial angle in crabs and insects, the only indications that the location is measured relative to the eye are the very ones where a fixed eye geometry measures the direction for tasks such as grabbing prey (Fig. 12.45g,h).

A new principle shows up clearly in these whole-eye

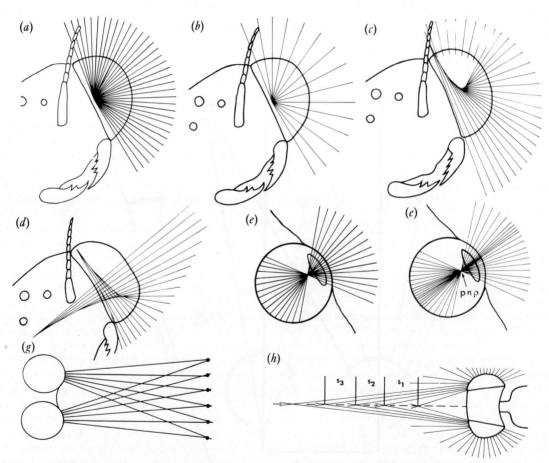

Fig. 11.45 *The template model is stimulated with eyes of equal-sized pixels and equal angles between visual axes. However, as illustrated in this diagram, natural vision evolves in the context of acute zones which generate unequal spatial magnification in the central projections, clearly improving the visual processing of the fixated object. (a) A diurnal eye with small facets and dense sampling array. (b) A nocturnal eye with larger facets and fewer visual axes. (c) A forward-looking fovea with increased eye radius. (d) Two foveas, as in some dragonflies. (e) A camera-type eye without an acute zone. (f) A camera-type eye with a fovea of increased sampling density and therefore narrower receptors. (g) Mantis type of binocular overlap for control of the strike of the foreleg off the midline. (h) Binocular overlap along the midline, as in dragonfly larvae, for control of the strike by the mouthparts.*

templates. There is a contribution from each part of the eye acting as a sampling array. When the whole of the bee's polarization mechanism sees the sky, it all functions together, but if only a small patch is visible then that is sufficient to give the sun's direction. There are many examples in insect visual behaviour, especially in chasing or turning to look at an object, or when fixating upon an object, when the strength of the response depends on the sum of a number of inputs subtended at different points on the eye.

These whole-eye templates, *which allow stationary animals to measure range*, clearly depend on hard-wired circuitry that is already fully functional when the imago insect emerges from its pupa or larva, and yet we find this exact correspondence to the structure of the outside world and the visual task. A job for the future is to combine the template model with the acute zone, and implement artificial visual systems with gradients of angle between the visual axes as illustrated in Fig. 11.45. In terms of evolutionary adaptations, the insect or crab displays its visual behaviour in its sampling array. Such adaptations are more characteristic of arthropods than (say) mammals, where the eye tends to be general-purpose and the behaviour to be perfected by learning. We might draw an analogy with modern technology: if special-purpose computers were as easy to evolve as arthropod eyes, we would make a special one for every task: in fact, we use rapidly adaptable software in general-purpose computers which have flexible behaviour, like mammals.

Beyond Semivision

Adding to the Information Processed

The template model illustrates clearly how the adaptations of the neurones to the visual task of trying to see, and to the action of the ensuing visual behaviour, actually add their contribution to the information that is processed. Darwinian selection promotes templates that represent the prior knowledge of the visual tasks. They lie waiting for their spatio-temporal fields to be excited, with a threshold. The templates do not 'pick up information' from the visual world. In their clusters *they create a new internal visual world* in their own domain by interaction of only certain aspects of the input with their own preformed fields of sensitivity. The preformed templates are the 'prior knowledge' put into vision. Each template response is definite and it contributes its own line-label, although its stimulus feature must have been fuzzy. The same process of specifying what combinations of low-level templates can exceed threshold takes place at the next higher level of templates, even though the mixture of low-level templates was only approximately satisfactory. This process is repeated from one layered neuropile to the next and repeated in parallel, with a threshold at each level down to the motor neurones. The output actions are clear-cut and sharp, although the message is a wobbly transient at every level, and noise is prevented from evoking error by having many circuits in parallel. There is no reason why these principles should not apply to both man and bee.

Responses of the whole organism can depend on mechanisms which count large numbers of template responses in parallel to obtain reliable detailed information, but which can also respond to a single template response that threatens survival, or any intermediate between these extremes. Each different insect species has its own selection of simple templates which are counted by higher-level neurones and diversity of visual behaviour is presented to us by a diversity of insects. The mammalian cortex appears to have an additional mechanism which makes its own templates by flexible circuitry.

The Wider Context of Adaptive Low-Level Vision

Studies of visual behaviour and eye structure in a variety of invertebrates show that to a large degree visual systems are special-purpose designs; but clearly in any one group of animals they have common features upon which a diversity of sophisticated and rapidly evolved examples are based. My guess is that in insects this basis is the set of high-resolution templates that go with the visual control of locomotion (Fig. 11.20). Given these, the rest can be done with low-resolution ratios followed by logical coincidences of particular ratios. Then, built upon this foundation is another layer of fewer, more specialized templates for biologically significant features, either genetically built in or assembled by learning for specific tasks. That was the basis of semivision.

At a low level of complexity, invertebrate visual systems illustrate very clearly that they process only the information necessary for their actions. If we extend this line of thought, we see that all visual systems evolve by natural selection in the context of the animal's activity and there will be no extra structure or process which is not needed for its specialized normal behaviour. Each of the templates in the visual processing mechanism represents a prior knowledge of just those visual tasks of selective importance. We are not used to thinking in this way because we have our own marvel of apparently universal vision 'before our open eyes'. When we see things they are already endowed with meaning based on memory; we have size constancy, colour constancy, visual illusions and hallucinations of visual images. Here we leave the realm of semivision.

Addition of a Huge Memory with Immediate Access

Meaning in human vision or hearing depends on categories, which in turn depend on associations and a long history of learning. The process of mixing, filtering and abstracting pattern from sense data starts at the eye or ear. The recognition of categories is done with the aid of an enormous recall memory before the level of consciousness. This is the only way to explain how thoughts and sensory perceptions arrive in consciousness already coloured by everything which memory recalls for the occasion. The visual or auditory categorization is an unconscious action of the brain, learned in childhood, clearly cultural but also related to the empirical world against which it is continually rechecked. Although it is operationally successful and based on data and reality, much of what we humans think we see is a series of hallucinations generated by memory, in which the detail is regularly updated by eye movements, and we learn to see objects which are then given names and significance before they arrive in our conscious visual world. An excellent example is reading, as you are now doing. Clearly the idea of successive layers of templates which create a new visual world in their own domain is still useful for explanations of vision up to the highest level, but we are no longer concerned with on-line semivision.

In primitive eyes where the relation between complexity of processing and complexity of visual behaviour is more direct, we see that they cannot be expected to 'see' the visual world as we do, or 'see' the whole visual flow field as they move. Although an insect may have 360° vision, perhaps only the nearest relative motion, or one set of contrasts resembling a mate or prey, is of interest to it at any one time, and whatever is of interest has already been installed as the prior knowledge embodied in mechanisms of processing via Darwinian evolution of something like templates. In human vision the prior knowledge is installed by far more complex processes of maturation and learning.

Conclusion

In the study of natural visual processing at a higher level, we find that prior knowledge has already been incorporated in a way that is impossible with mathematical operations on the stimulus alone. In the vernacular we say that you have to know the relevant features of the visual world before you can see what you are looking at. In man, these 'top-down' effects include long training in infancy, memory and rational inference, but in primitive vision the equally essential requirement for prior knowledge is met by the choice of processing structures that are the result of Darwinian selection – what I have called the choice of templates. Therefore, to design a simple artificial visual system it is essential to start with what we want to see and the directions in which we expect contrasts to move at each point on the retina. Then we insert in parallel in repeated columns only as many simple templates as are required. The choice of templates represents the prior knowledge. In deciding what to do with the template responses, we count only sufficient combinations of them to generate the variety of outputs needed for the ensuing action and we count in fields that are as large as possible. Visual fixation helps by limiting the processing to one place on the map behind the retina. By copying natural vision in these ways it may be possible to overcome some of the problems of low-level artificial vision, but we will only begin to understand vision by drawing upon many different approaches involving many scientific disciplines.

Acknowledgements

I would like to thank many who work or have worked on these and related questions in Canberra over the past few years while these ideas have been fermenting. The comprehensive theory has depended on research on the visual behaviour of the bee, much of it done by Miriam Lehrer, on the electrophysiology of insect optic lobes studied by Ljerka Marcelja over many years, and on computer modelling done recently by Peter Sobey. There are many others besides, working on insect visual processing. Especially, however, I would like to thank my colleague Mandyam Srinivasan for his interest and unfailing support, and for the ideas and critical comments that emerge from our discussions. Numerous drafts of the MS have been typed by Elizabeth Watson. The work on object vision in insects is supported by a grant from the Fujitsu Company of Japan, and by funds from the Centre for Visual Sciences and the Centre for Information Science in ANU.

References

(Works marked with an asterisk (*) are general accounts or parts thereof.)

*Ali, M. A. (1984). *Photoreception and Vision in Invertebrates*. London and New York: Plenum.

Berger, E. W. (1900). Physiology and histology of the Cubomedusae, including Dr F. S. Conant's notes on the physiology. *Mem. Biol. Lab. Johns Hopkins Univ.*, 4 (4), 1–84.

*Bullock, T. H. and Horridge, G. A. (1965). *Structure and Function in the Nervous Systems of Invertebrates*. San Francisco and London: Freeman.

Burrows, M. (1980). The control of sets of motoneurones by local interneurones in the locus. *J. Physiol. Lond.*, 298, 213–233.

Burtt, E. T. and Catton, W. T. (1960). The properties of unit discharges in the optic lobe of the locust. *J. Physiol. Lond.*, **154**, 479–490.

Collett, T. S. and Cartright, B. A. (1983). Eidetic images in insects: their role in navigation. *Trends Neurosci.*, **6**, 101–105.

Crane, J. (1952). A comparative study of innate defensive behaviour in Trinidad mantids (*Orthoptera: Mantoidea*). *Zoologica*, **37**, 259–293.

Dartnall, H. J. A. (1953). The interpretation of spectral sensitivity curves. *Br. Med. Bull.*, **9**, 24–30.

*De Voe, R. (1985). The eye: electrical activity. In: *Comprehensive Insect Physiology, Biochemistry and Pharmacology*, vol. 4: *Nervous System: Sensory*. eds Kerkut, G. A. and Gilbert, L. I. pp. 277–354. Oxford: Pergamon.

Erickson, R. P. (1963). Stimulus encoding in topographic and non-topographic modalities: on the significance of the activity of individual sensory neurones. *Psychol. Rev.*, **75**, 447–465.

Erickson, R. P. (1982). The across-fiber pattern theory: an organizing principle for molar neural function. *Cont. Sens. Physiol.*, **6**, 79–110.

Fischbach, K. F. and Dittrich, A. P. M. (1989). The optic lobe of *Drosophila melanogaster*. I. A Golgi analysis of wild-type structure. *Cell Tiss. Res.*, **258**, 441–475.

*Fraenkel, G. S. and Gunn, D. O. (1940). *The Orientation of Animals*. Oxford: Clarendon.

*Franceschini, N., Riehle, A. and Le Nestour, A. (1989). Directionally-selective motion detection by insect neurones. In *Facets of Vision*. eds. Stavenga, D. and Hardie, R. pp. 360–390. Berlin: Springer.

Gould, J. L. (1985). How bees remember flower shapes. *Science Wash. DC*, **227**, 1492–1494.

Gould, J. L. (1986). Pattern learning by honeybees. *Animal Behav.*, **34**, 991–997.

Gregory, R. L. (1967). Origin of eyes and brains. *Nature Lond.*, **213**, 369–372.

Hassenstein, B. (1959). A cross-correlation process in the nervous center of an insect eye. *Nuovo Cimento*, X**13**, 617–619.

Hausen, K. and Egelhaaf, M. (1989). Neural mechanisms of visual course control in insects. *Facets of Vision*. eds. Stavenga, D. and Hardie, R. pp. 391–424. Berlin: Springer.

Hertel, H. (1980). Chromatic properties of identified interneurones in the optic lobes of the bee. *J. Comp. Physiol.*, **137**, 215–231.

Honegger, H. W. (1978). Sustained and transient responding units in the medulla of the cricket *Gryllus campestris*. *J. Comp. Physiol.*, **125**, 259–266.

*Horridge, G. A. (1978a). The separation of visual axes in apposition compound eyes. *Phil. Trans. R. Soc. Lond.*, **285**, 1–59.

*Horridge, G. A. (1978b). A different kind of vision: the compound eye. In: *Handbook of Perception*, vol. 8. eds. Carterette, E. C. and Friedman, M. P. pp. 3–82. London: Academic.

Horridge, G. A. and Marcelja, L. (1990a). Responses of the H1 neurone to jumped edges. *Phil. Trans. R. Soc. Lond B*, **329**, 65–73.

Horridge, G. A. and Marcelja, L. (1990b). Responses of the H1 neurone to contrast and moving bars. *Phil. Trans. R. Soc. Lond.*, **329**, 75–80.

Jander, R. and Volk-Heinrichs, I. (1970). Das strauschspezifische visuel Perceptorsystem der Stabheuschrecke (*Carausius morosus*). *Z. vergl. Physiol.*, **70**, 425–477.

*Kien, J. (1975). Motion detectors in locusts and grasshoppers. In: *The Compound Eye and Vision of Insects*. eds. Horridge, G. A. pp. 410–422. Oxford: Clarendon.

Kirchner, W. H. and Srinivasan, M. V. (1989). Freely flying honeybees use image motion to estimate object distance. *Naturwissen.*, **76**, 281–282.

*Kirschfeld, K. (1976). The resolution of lens and compound eyes. In: *Neural Principles in Vision*. eds. Zettler, F. and Weiler, R. pp. 354–370. Heidelberg: Springer.

*Laughlin, S. B. (1981). Neural principles in the visual system. In: *Handbook of Sensory Physiology*, vol. VII/6B: *Invertebrate Visual Centers and Behavior*. pp. 133–280. Heidelberg: Springer.

Laughlin, S. B. (1987). Form and function in retinal processing. *Trends Neurosci.*, **10**, 478–483.

Lehrer, M., Srinivasan, M. V., Zhang, S. W. and Horridge, G. A. (1988). Motion cues provide the bee's visual world with a third dimension. *Nature Lond.*, **332**, 356–357.

Lehrer, M., Srinivasan, M. V. and Zhang, S. W. (1990). Visual edge detection in the honeybee and its chromatic properties. *Proc. R. Soc. Lond. B*, **238**, 321–330.

Maldonado, H. (1970). The deimatic reaction in the praying mantis *Stagmatoptera biocellata*. *Z. Vergl. Physiol.*, **68**, 60–71.

Mimura, K. (1972). Neural mechanisms subserving directional selectivity of movement in the optic lobe of the fly. *J. Comp. Physiol.*, **80**, 409–437.

Oldberg, R. M. (1986). Identified target-selective visual interneurones descending from the dragonfly brain. *J. Comp. Physiol.*, **159**, 827–840.

Osorio, D. (1986). Ultraviolet sensitivity and spectral opponency in the locust. *J. Exp. Biol.*, **122**, 193–208.

Osorio, D. (1987). The temporal properties of non-linear transient cells of the locust medulla. *J. Comp. Physiol.*, **161**, 431–440.

Osorio, D. (1990). Inhibition, spatial pooling and signal rectification in the sub-units of non-directional motion-sensitive cells in the locust optic lobe. *J. Neuroscience*, submitted.

Pinter, R. B. (1979). Inhibition and excitation in the locust DCMD receptive field; spatial frequency, spatial and temporal characteristics. *J. Exp. Biol.*, **80**, 191–216.

*Plate, L. J. (1924). *Allgemeine Zoologie und Abstammungslehre*. Part 2: *Die Sinnesorgane der Tiere*. Jena: Gustav Fischer.

*Reichardt, W. (1961). Autocorrelation: a principle for evaluation of sensory information by the central nervous system. In: *Principles of Sensory Communication*. Rosenblith, W. A. pp. 303–317. New York: Wiley.

Rind, F. C. (1987). Non-directional, movement-sensitive neurones of the locust optic lobe. *J. Comp. Physiol. A*, **161**, 477–494.

Rind, F. C. (1990). A directionally selective motion-detecting neurone in the brain of the locust: physiological and morphological characterization. *J. Exp. Biol.*, **149**, 1–19.

*Rolls, E. T. (1987). Information representation, processing and storage in the brain: analysis at the single neurone level. In: *Neural and Molecular Bases of Learning*. eds. Changeux, J. P. and Konishi, M. Chichester and New York: Wiley.

Rossel, S. (1986). Binocular spatial localization in the praying mantis. *J. Exp. Biol.*, **120**, 265–281.

Rowell, C. F. H. (1971). The orthopteran descending movement detector (DMD) neurones: a characterization and review. *Zeit. vergl. Physiol.*, **73**, 167–194.

Sandini, G. and Tistarelli, M. (1990). Active tracking strategy for monocular depth inference over multiple frames. *IEEE Trans Patt. Anal. and Mach. Intel.*, **12**, 13–27.

Schwind, R. (1978). Visual system of *Notonecta glauca*. A neurone sensitive to movement in the binocular visual field. *J. Comp. Physiol.*, **123**, 315–328.

*Snyder, A. W. (1975). Optical properties of invertebrate photoreceptors. In: *The Compound Eye and Vision of Insects*. ed. Horridge, G. A. pp. 179–235. Oxford: Oxford University Press.

Snyder, A. W. (1979). The physics of vision in compound eyes. In: *Handbook of Sensory Physiology*, Vol VII/6A: *Vision in Invertebrates*, ed. Autrum, H. pp. 255–314. Heidelberg: Springer.

Sobey, P. J. and Horridge, G. A. (1990). Implementation of the template model of vision. *Proc. R. Soc. Lond. B*, **240**, 211–229.

Srinivasan, M. V., Lehrer, M. and Horridge, G. A. (1990). Visual figure-ground discrimination in the honeybee: the role of motion parallax at boundaries. *Proc. R. Soc. Lond. B*, **238**, 331–350.

*Strausfeld, N. J. (1989). Beneath the compound eye: neuroanatomical analysis and physiological correlates in the study of insect vision. In: *Facets of Vision*. eds. Stavenga, D. G. and Hardie, R. C. pp. 317–359.

*Strausfeld, N. J. and Nassel, D. R. (1981). Neuroarchitectures serving compound eyes in crustacea and insects. In: *Handbook of Sensory Physiology*, vol. VII/6B: *Vision in Invertebrates*. ed. Autrum, H. pp. 1–132. Berlin, Heidelberg: Springer.

*Tinbergen, N. (1951). *The Study of Instinct*. Oxford: Oxford University Press.

Van Hateren, J. H., Srinivasan, M. V. and Wait, P. B. (1990). Pattern recognition in bee: orientation discrimination. *J. Comp. Physiol.*, in press.

Van Praagh, J. P., Ribi, W., Wehrhahn, C. and Wittmann, D. (1980). Drone bees fixate the queen with the dorsal front part of their compound eyes. *J. Comp. Physiol. A*, 162, 159–172.

*Wehner, R. (1981). Spatial vision in arthropods. In: *Handbook of Sensory Physiology*, vol. VII/6C: *Vision in Invertebrates*. ed. Autrum, H. pp. 287–616. Berlin, Heidelberg: Springer.

Wehner, R. (1987). 'Matched filters' – neural models of the external world. *J. Comp. Physiol. A*, 161, 511–531.

*Zeil, J., Nalback, G. and Nalbach, H. O. (1989). Spatial vision in a flat world: optical and neural adaptations in arthropods. In: *Neurobiology of Sensory Systems*. eds. Singh, R. N. and Strausfeld, N. pp. 123–136. New York: Plenum.

Zhang, S. W., Wang, X, Liu, Z. and Srinivasan, M. W. (1990). Visual tracking of moving targets by freely flying honeybees. *Vis. Neurosci.* 4, 379–386.

12 Evolution of Binocular Vision

John D. Pettigrew

Introduction

We are all familiar with the coordinated use of both eyes to examine the same visual scene. After all, this is the habitual mode of operation of our own visual systems and, except for 5% of the population who are stereoblind (Richards, 1970), most of us are familiar with the fact that this mode of operation, binocular vision, enables us to extract subtle details which are not available to one eye alone. This is illustrated by random-dot stereograms, where an embedded figure is invisible to monocular inspection but stands out vividly in stereoscopic depth on binocular inspection (Julesz, 1971).

Binocular vision has evolved independently at least twice among the vertebrates, in both mammals and birds, and in the present account I shall attempt to reconstruct these two independent evolutionary events, largely from examination of the living forms and the way in which they achieve binocular vision. Since all mammals, with the exception of the Cetaceae, appear to have at least the rudiments of binocular vision, particular emphasis will be placed upon avian binocular vision. Comparison of the bird groups which appear to have achieved binocular vision with those where binocular vision does not seem to have been achieved, can give some insight into the evolutionary pressures which have given rise to binocular vision.

Hallmarks of Binocular Vision

Binocular Overlap

A first prerequisite in the achievement of binocular vision is clearly a satisfaction of the optical and oculomotor restraints which would otherwise prevent the eyes from cooperating. In other words, an animal with functional binocular vision would be expected to have at least some degree of binocular overlap in the visual fields of each eye. It must be emphasized, however, that a large degree of binocular overlap is neither necessary nor sufficient for the achievement of functional binocular vision. For example, the owl achieves a high degree of specialization in its binocular vision and yet the region of binocular overlap represents only half the total visual field of one eye and only a third of the total visual field of both eyes together (see Fig. 12.1).

Nor does the degree of binocular overlap need to be constant, since a number of species have oculomotor strategies which enable them to change from binocular visual processing to panoramic, monocular vision at will. For example, the tawny frogmouth (*Podargus strigoides*), in the camouflage posture which it adopts when threatened, completely eliminates binocular overlap between the visual fields of each eye to maximize its panoramic field of view. At other times, such as when hunting, the frogmouth can align its eyes frontally and achieve functional binocular vision of a grade equal to the owl (Wallman and Pettigrew, 1985). This kind of behaviour seems at first sight very strange to animals like ourselves in which a marked divergence of the visual axis is regarded as pathological and given the name *strabismus*. 'Voluntary strabismus' may not be confined to birds, however, since wide divergence of the visual axes occurring under some circumstances has been described for ungulates (cf. the whites-of-the-eyes posture sometimes adopted by frightened horses and the large divergent saccades shown by sheep).

That binocular overlap is not sufficient for functional binocular vision is shown by a number of bird species which have significant overlap of the visual fields of each eye but yet show no evidence of functional binocular vision. Such species include the swift and the oil bird (see below).

Area Centralis

Perhaps more important for binocular vision than the optical presence of binocular overlap *per se* is the development of retinal specializations in the temporal retina sub-

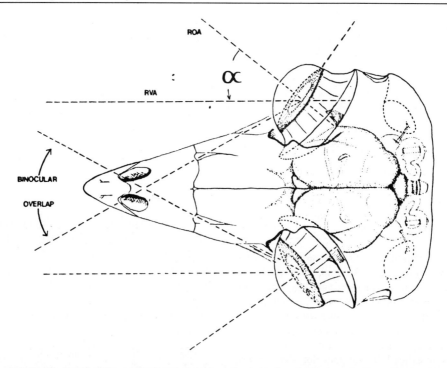

Fig. 12.1 *Schematic dorsal view of a bifoveate bird to show* angle alpha *(α) subtended between the binocular visual axis of the right eye (RVA), passing through the fovea located on the temporal retina and the optical axis (ROA), passing through the fovea located on the central retina (monocular visual axis). Anatomical and physiological studies confirm that functional binocular vision exists along the binocular visual axis, despite the small proportion of the total visual field which is involved in binocular overlap and despite the high visual acuity possible along the monocular visual axis.*

serving the binocular field. Since binocular vision demands a degree of visual acuity sufficient for the small differences between the two retinal images to be measured, a universal feature of animals with functional binocular vision is the presence of a retinal specialization with increased cell densities in the area of the retina which projects forwards into the binocular visual field. In other words, the presence of a temporal retinal specialization in each eye is a strong indication of the presence of functional binocular vision. Defining the presence of such an area of retinal specialization may require extra information about the oculomotor strategies of the species concerned (as already pointed out for those species which sometimes adopt extremely divergent eye position) and is not a surefire guarantee since there are some birds which appear not to make stereoscopic use of foveas which project forward into binocular space (e.g. the swift). The degree of specialization required of the temporal *area centralis* or fovea will obviously depend upon such species-specific factors as the distance at which binocular vision is normally used and the size of the targets to be detected. For these reasons, the increase in density there may not be dramatic and escape attention without special techniques, as in the case of the rabbit. The low degree of speciali-

zation in its temporal area is perhaps related to the low demands placed upon the rabbit's need for binocular vision as it feeds on targets at very close range with a consequent larger angular target size and retinal disparity. In rabbits, the temporal area specialization is limited to one cell class which, in ordinary preparations without special labelling, tends to be overshadowed by the high density of neurones in the monocular, central retina (Provis, 1979).

Bifoveate Retina: *Angle Alpha*

In contrast to the monofoveate condition of mammals, we may cite some of the bifoveate birds which have two foveal specializations with high cell densities, one looking into the monocular field and one into the binocular field. This dual foveal specialization in the same eye raises an important point about what can be called *angle alpha*, viz. that there need not be an alignment of the optical axis of the eye and the visual axis of the eye. Since the optical axis and the visual axis are separated only by a very small amount in man (*angle alpha* = 6°) there is a common tendency to treat these two axes as the same. For this reason it is commonly assumed that animals with a very marked degree of divergence of their *optical* axes will have a correspondingly high degree of divergence of their *visual* axes. That this is not the case is well illustrated by the kingfishers, which have the largest *angle alpha* so far measured in the vertebrate kingdom (around 40–50°). As shown in Fig. 12.1, the binocular visual axis which passes through the nodal point of the eye and the binocular fovea is some 40 degrees away from the optical axis of the eye which is also aligned with the monocular fovea. As described below, kingfishers achieve full functional binocular vision despite this very wide separation of the binocular visual axis from the optical axis of the eye.

The question might reasonably be asked, since it appears to be possible to have simultaneous panoramic vision and binocular vision as shown by so many birds, what are the selection pressures which have operated to bring such a close alignment of the optical axis and visual axis in primates? This question has been addressed elsewhere and it appears to be a complicated one involving perhaps the optical constraints imposed by both the demands of binocular vision and the demands of a nocturnal niche, with its accompanying requirements for a large-aperture optical system (Allman, 1977; Pettigrew, 1978).

In summary, we may say therefore that if we see a considerable degree of specialization in the temporal retina subserving binocular vision, then there is a good chance this animal achieves functional binocular vision. Perhaps the best example of this is the owl, where there is a high degree of specialization with increased ganglion cell densities in the temporal retina. On the other hand, if there is a strong emphasis on the acquisition of high visual acuities in the monocular field, the specializations in temporal retina may be swamped and it may take special techniques to demonstrate the temporal specialization. This is perhaps well illustrated in the case of the burrowing owl, which has a high-density visual streak of retinotectal ganglion cells which tend to overshadow the peak retinothalamic-ganglion cell density in temporal retina. Separate labelling of these two populations of ganglion cells can make the binocular specialization more evident (Bravo and Pettigrew, 1981), as in the case of the rabbit (Provis, 1979).

Convergence of the Visual Pathways from Each Eye

Sir Isaac Newton in his treatise on optics (1730) was one of the first scientists to point out the necessity, in functional binocular vision, for information from both eyes to converge on the same site in the brain. Newton reasoned that a partial decussation of the fibres in the optic nerves would be a necessary consequence in a system with functional binocular vision, and his anatomical prediction was verified not long afterwards in gross dissections of the mammalian optic chiasm (see Polyak, 1957). That a partial decussation of the optic chiasm is not the only way to achieve binocular integration is shown by the birds, among which all so far described have a total decussation. This total decussation has led many authors (e.g. Walls, 1942) to conclude that functional binocular vision is impossible in birds, but recent work has shown that avian binocular vision is accomplished by different anatomical means. In the owls, some nightjars, kingfishers and diurnal birds of prey, the visual thalamus projects bilaterally to the visual wulst where binocular integration occurs (Pettigrew and Konishi, 1976; Pettigrew, 1979; Bravo and Pettigrew, 1981).

It is rather difficult to conceive of some intermediate stage of evolution where there is both a partial decussation of the optic chiasm and a bilateral projection from the thalamus, if topographic relations are to be maintained. There seems, therefore, to be little chance that birds and mammals share a common predecessor with functional binocular vision. This may be regarded as an example of convergent evolution, both birds and mammals having independently achieved functional binocular integration by different mechanisms. Since the mammalian partial decussation and the avian double decussation by no means exhaust the possibilities for binocular interaction, it may be expected that further search may reveal other mechanisms for convergence of binocular information. For example, although there is little in the way of other evidence to support functional binocular vision in the eel and the shark, they both have rather unusual decussations from one side of the visual pathway to the other (Ebbesson and Schroeder, 1971; Ekström, 1982).

Binocular Nerve Cells

One of the last steps in the demonstration of functional binocular vision is a description of the properties of nerve cells which receive input from both eyes. Since the requirements of a functional binocular system capable of extracting the disparity information necessary for stereopsis have been well described, there appear to be a limited number of solutions to this problem. Evidence to support this view comes from the gradual evolution of algorithms for machine stereopsis which appear to be gradually approaching a solution rather like that achieved in the visual cortex of birds and mammals (see below). The essential two steps in the achievement of stereopsis from binocular inputs are (a) the solution of the 'matching' or 'correspondence' problem, whereby the parts of the two images belonging to the same outside object are identified, and (b) the measurement of the slight differences in these paired parts of the image (disparity detection).

The solutions to these two problems are, in a certain sense, incompatible, since the recognition of the matching elements involves at some level the detection of similarity whereas the disparity detection task involves the measurement and recognition of small differences. It therefore appears necessary for these two tasks to be carried out gradually, in parallel with one another, to avoid the difficulties which may arise, for example, when the matching process is taken too far before the disparity measurement. In physiological terms, what this means is that, despite the presence in the retina of highly specialized retinal ganglion cells which are capable of carrying out some of the first stages of the matching task by identifying common elements in the two images, these retinal ganglion cells are *not* used to construct binocular receptive fields. In preference, the concentrically organized retinal ganglion cells are used, which do not 'jump to conclusions' about the nature of the retinal stimulus except in so far as to identify the exact position of the local contrast change. Specialization for the exact form of the local part of the retinal image occurs only where there is convergence of information from the two eyes and appears to proceed gradually, hand in hand with the increasing degree of binocularity seen at higher levels in the visual pathway.

For example, at the lateral geniculate nucleus, where the dominant influence on the properties of the cells is from the concentrically organized retinal input, there are the beginnings of binocular interaction, with an inhibitory field in the non-dominant eye which has the beginnings of some selectivity for the form of the stimulus (Sanderson *et al.*, 1969). At the level of the visual cortex where fully fledged binocular neurones are seen, these binocular neurones are also highly selective for the nature of the visual stimulus, in contrast to the immediately preceding stage where they are selective only for location. These binocular neurones are also orientation-selective. This orientation selectivity is tightly matched for both receptive fields (one from each eye) within tightly controlled statistical limits (Nelson *et al.*, 1977). It is this close matching between the selectivity in the receptive fields in each eye which may go some way towards solving the matching problem. Since there are a variety of cells with the same matched receptive field properties for a variety of features, such as orientation, size and direction of motion, but with differing disparity selectivity, then we can see that at this stage in the visual pathway there are individual binocular neurones which are capable of providing the important elements for stereoscopic processing.

These highly selective binocular neurones, with tightly matched receptive field properties in each eye, are to be contrasted with the large, diffusely organized binocular receptive fields which have been described in the optic tectum of a variety of species (e.g. Gordon, 1973). There are a number of functions which these large binocular fields might subserve, such as the detection of motion-in-depth, but they seem quite unsuited to the demanding and subtle task of achieving stereopsis and therefore, in my view, do not constitute any compelling basis for the existence of binocular visual processing such as is involved in stereopsis. Of course the final proof should come with the behavioural demonstration of stereoscopic abilities. One must, however, be aware that the latency between the demonstration of neurones which are capable of stereoscopic depth discrimination and the demonstration of the same abilities in the whole, cantankerous animal may be quite long. In the case of the cat, over 10 years elapsed between the first demonstrations of disparity-selective binocular neurones (Barlow *et al.*, 1967; Pettigrew *et al.*, 1968) and the convincing behavioural demonstrations of the same abilities using the whole cat (e.g. Mitchell *et al.*, 1979).

Binocular Vision in Mammals

It has been possible to demonstrate all the hallmarks of functional binocular vision in each of the mammals which have been studied with the aim of investigating this aspect (e.g. cat: Bishop, 1973; sheep: Clarke *et al.*, 1976; monkey: see review in Poggio and Poggio, 1984; and rabbit: Hughes and Vaney, 1982). Since there is also anatomical evidence for binocular vision in all of the mammals which have been so far studied, with the possible exception of the Cetaceae, it seems unlikely that it will be possible to reconstruct an evolutionary scenario for binocular vision in mammals. It seems more likely that binocular vision was an attribute of the earliest mammals and may in fact have played a large part in the successful radiation of this group. The earliest fossil primates such as *Purgatorius* and *Tetonius* (e.g. Allman, 1977; Archer and Clayton, 1984) show evidence of large, frontally placed eyes and there is every reason to think that they had functional binocular visual systems as adequate as those possessed by prosimian primates living today. There are a number of special advantages of binocular vision which accrue to animals which occupy the nocturnal niche. These include increased signal-to-noise ratio, camouflage breaking, and the absence of the need to move in order to generate motion parallax. In view of the prevailing evidence that the early mammals were nocturnal, it may further be speculated that binocular vision was one of the features which contributed to their success (as pointed out by Polyak, 1957). The functional anatomy and physiology of binocular vision in mammals is remarkably similar in all the species which have been studied, such as the cat, various primates, the sheep and the rabbit. This is perhaps not so surprising in view of the strong likelihood that they all arose from a common mammalian ancestor with binocular vision, but it may be valuable to summarize quickly what has been found, in terms of the organization of binocular vision in these various mammals to help emphasize some of the general features of binocular visual systems.

Binocular Vision in the Rabbit

The rabbit forms a convenient counterpoint to other mammals, since it was long considered that functional binocular vision was not a feature of its visual system. Indeed the rabbit, like many birds, can eliminate its frontal region of functional binocular overlap by the appropriate oculomotor posture, such as the one it adopts when in the freeze position (Hughes and Vaney, 1982). Recent work has shown, however that rabbits do converge their eyes when feeding (Zuidam and Collewijn, 1979) and that there are binocular neurones in the visual cortex which have most of the properties enumerated above to indicate that they are involved in functional binocular vision. For example, these binocular neurones have orientation selectivity which is tightly matched for both receptive fields (Hughes and Vaney, 1982). Previous work had failed to demonstrate this nice selectivity, which is matched on both retinas, and had even led to the suggestion that the binocular connections of the rabbit might be maladaptive (Van Sluyters and Stewart, 1974). The conclusion that the binocular fields were not matched may well have been based upon inappropriate eye position, or failure to correct the eccentric optics of the rabbit, or both, and this serves as a good example of the need to establish optical and eye position parameters in any species for which binocular vision is being investigated (cf. the long line of baseline investigations of the cat's optics and binocular rest position of the eyes carried out by Peter Bishop during his career).

Retinal Trigger Features Not Used for Binocular Vision

There is another important message which can be gleaned from the work on binocular vision in the rabbit, and it concerns the way in which the visual system is organized to provide binocular neurones which have receptive fields on both retinae. It was in the rabbit retina that the highly complicated receptive field properties such as orientation selectivity, direction selectivity and local edge detection were first described for mammalian retinal ganglion cells (Barlow et al., 1964; Levick, 1967). Although the specialized receptive field properties were thought to be characteristic of the 'lower' vertebrates such as birds and anuran amphibians, their demonstration in the rabbit opened the way for the later experiments which demonstrated that such highly specialized receptive field properties are probably found in retinae right across the vertebrates. Differences in the proportion of specialized retinal ganglion cells turn out, then, to be quantitative rather than qualitative, and provide the following useful generalization: the proportion of concentrically organized, non-specialized retinal ganglion cells, as a function of the highly specialized retinal ganglion cells, tends to increase with the importance of binocular vision for the animal. This was first noticed and pointed out by Levick with respect to the rabbit, in which the number of concentrically organized retinal ganglion cells is rather small, as is the proportion of the visual cortex devoted to binocular vision. More recent work has extended this generalization further, since it can be shown that the ipsilateral input to the lateral geniculate nucleus which provides the basis for binocular interaction is contributed almost exclusively by the concentrically organized retinal ganglion cells (Y. Takahashi and T. Ogawa, personal communication). In other words despite the large numbers and variety of specialized retinal ganglion cells in the rabbit retina, including the orientation-selectivity variety which might be expected to be used for solving the matching problem with binocular vision, none of these can be shown to contribute to the binocular neurones' receptive field properties in the visual cortex. Instead, the minority population of concentrically organized retinal ganglion cells and their lateral geniculate relay cells are used to bring about binocular receptive fields with matched orientation selectivity in the two eyes.

That this is the case in primates and carnivores was perhaps not so surprising, in view of the fact that it was long felt that the only retinal ganglion cells available for the task were of the concentric variety. The recent evidence that primate and carnivore retinae may also contain significant proportions of the specialized retinal ganglion cells (see Stone, 1983, for a review), coupled with the fact that rabbits achieve a degree of functional binocular vision, serves to underline the generalization that, even when they are available, highly specialized retinal ganglion cells do not appear to make an important contribution to the genesis of binocular receptive fields with highly specialized properties on both retinas.

A moment's consideration will show why this might be the case. Firstly, if this were otherwise there would have to be an exceedingly large number of different retinal ganglion cells to cover all the permutations and combinations of orientation, direction, size, etc., at all of the different relative retinal locations to provide disparity selectivity. In addition, as pointed out below with reference to machine stereopsis, there may be underlying theoretical constraints upon the optimal solution to the matching problem which require that it avoid 'jumping to conclusions' about the important features for stereopsis which are present in the monocular stimulus. Whatever the underlying reasons, we can say with some generality, based on work on the binocular visual pathways of cat, various primates, sheep and goats, rabbits and various birds, that specialized receptive field construction is delayed in the visual pathways subserving binocular vision until information from both eyes converges.

Cats, Owls, Monkeys and Machines

In a facetious reference to disparity-selective binocular neurones in aardvarks, Mayhew and Frisby (1979) question the significance of physiological studies of disparity-selective binocular neurones carried out so far, because of their apparent failure to address important issues such as the particular computational strategy used for stereopsis. These demeaning remarks fall short on two counts, since (a) they fail, characteristically, to acknowledge the important stimulus to machine stereopsis which the preceding neurophysiological studies provided, and (b) they miss one of the major points of the comparative work on binocular vision, i.e. *that the computational strategies adopted for stereopsis by unrelated species are identical in the sense that all avoid feature extraction until information from both eyes has converged.* The hierarchical step from monocular, concentric organization to binocular, matched, feature-specific organization is the same in both mammals and birds, despite the considerable differences in their respective neural apparatus which might have led to other solutions, such as the synthesis of binocular receptive fields from monocular elements which were already feature-specific; an abundance of the latter exist in lower levels of the visual pathway, particularly in birds, so it is a fair question to ask why monocular concentric rather than feature-selective elements are used as inputs to the binocular neurones. The answer to such a question, while not illuminating the whole computational strategy, certainly eliminates some possibilities. The avoidance of a high degree of monocular preprocessing has also 'evolved' in the later machine algorithms for stereopsis, so it may be valuable to give some specific consideration to the machine–animal comparison.

Machine vs. Animal Stereopsis

In recent years there has been a variety of machine algorithms which successfully compute depth from a pair of Julesz random-dot stereograms (reviewed in Poggio and Poggio, 1984). These models have shown a process of evolution, since they have been subject to successive refinements which have improved their versatility and efficiency. For example, one of the early models was very sensitive to small-scale differences between the two images, and could not achieve stereopsis if one image had reversed contrast (Marr and Poggio, 1976; Grimson and Marr, 1979), neither of which problems present any difficulty for the human stereoscopic system (see Julesz, 1971). Later models have overcome the reversed-contrast problem (Marr and Poggio, 1979) but still hang up if there is size anisotropy or large vertical disparity.

It may be instructive to examine the continuing evolution of machine algorithms for stereopsis to gain further insights into the constraints operating upon the evolution of animal stereopsis. We may ask, for example, whether the striking similarities between the avian and mammalian stereoscopic systems which exist in spite of their different origins and structural components reflect a fundamental constraint upon a successful algorithm. Is there, perhaps, a single, most-efficient solution to the problem of stereopsis?

The designers of machine stereopsis algorithms are usually reluctant to attempt any measure of the efficiency of the successful ones, although this could be a desirable objective if they are to be measured against each other and against the performance of human subjects on a range of stereoscopic tasks. We may, nevertheless, infer that the algorithms so far devised fall short of some ideal because of the ferment in this area and the fairly steady stream of new versions (e.g. Frisby and Mayhew, 1980; Mayhew and Frisby, 1980).

One of the key areas of ferment concerns the choice of strategy for the solution of the *correspondence* or *matching* problem, viz. the identification in each image of the paired parts which correspond to the same outside objects. The human visual system shows extraordinary versatility in solving this problem, to a degree that makes it unlikely that a search for a particular primitive is always used to match up the two images. For example, Frisby and Julesz (1975) have shown that, under some circumstances, edge orientation can be ruled out as the local primitive for matching, since successful fusion and stereopsis can follow if the stereo pair is constructed of randomly oriented line segments. That this particular problem can be solved (this writer can do it only with the greatest difficulty) does not imply that the special resources brought to bear on this occasion are used as a matter of course in the operation of stereopsis under more normal circumstances. After all, a dancing poodle, however impressive its performance, does not necessarily lead to the conclusion that normal canine locomotion is bipedal. The experiment does imply, however, that if orientation is commonly used as a primitive for matching (as Barlow *et al.* (1967) suggested for the cat visual cortex), it must be done in a facultative, rather than an obligatory, fashion so that other strategies for matching can also operate in parallel or as alternatives. It is therefore of some interest that one of the more recent machine algorithms for stereopsis has some features which avoid the rigidity in the matching strategy which hampers some earlier models and at the same time comes closer in conceptual design to the avian and mammalian blueprints (Frisby and Mayhew, 1980).

The key strategy, then, may be the avoidance of commitment to a particular feature for matching before the disparity information is extracted. In this way we can understand the avoidance, in both avian and mammalian binocular pathways, of feature extraction before information from both eyes comes together. One reason for this has already been suggested in terms of the impossibly large number of monocular feature detector neurones which would have to be located in the retina, where size and space limitations would mitigate against it. A second reason may be that too early a commitment to a particular feature may preclude later solution of the disparity problem. After all, retinal disparity involves any measurable difference between the images related to depth and can involve slight mismatches in size, position, velocity of movement, orientation, or even (when one considers the rivalrous inputs from lustrous surfaces) brightness and colour. Matching the images, on the other hand, inevitably requires some estimation of similarity between parts of the images which could run the risk of losing information valuable to subsequent disparity detection task. It may be for this reason that avian, mammalian and the recent machine algorithms for stereopsis all avoid feature analysis at early stages before information from both eyes is compared. Once the inputs from both have converged on binocular neurones, a variety of features can then be used for matching, such as size (the X and Y streams), edge orientation, direction of movement, velocity, and end-stopping – a few properties which are known to be matched in the receptive field pairs of binocular disparity-sensitive neurones, whether these are found in cats (Barlow et al., 1967, Pettigrew et al., 1968), monkey (Poggio and Talbot, 1981), sheep (Clarke et al., 1976) or owls (Pettigrew and Konishi, 1976; Pettigrew, 1979).

Vertical Disparity

Vertical retinal disparities are as accurately detected by binocular cortical neurones as are horizontal disparities, despite the commonly accepted view that only the latter can contribute to stereopsis. This finding appears to be a general one, since all studies of disparity-selective neurones, from the pioneering work on the cat (Pettigrew et al., 1968) to the more recent work on the monkey (Poggio and Talbot, 1981), have failed to demonstrate any prominent difference between the respective codings for disparities in the two dimensions.

The absence of the expected anisotropy in favour of horizontal disparity was felt by some to be fatal to the proposition that the disparity-selective binocular neurones could form a basis for stereopsis. This was in spite of the arguments raised to the effect that vertical disparities were an inevitable consequence of both eye movements and the optics of the situation for which provision would have to be made by the system even if there were no obvious use for them. Recent work has revived interest in the question of vertical disparities by drawing attention to old psychophysical data which clearly demonstrated a role for them, at the same time as showing that they can make a contribution to depth perception which is independent of that made by horizontal disparities.

Mayhew and Longuet-Higgins (1982) proposed a scheme whereby a gradient of vertical disparities could be used to judge distances independent of the vergence position of the eyes, if these were surfaces within reasonable viewing distance and if vertical disparity detection were as accurate as horizontal. This scheme accurately predicts a number of psychophysical phenomena in binocular vision such as the 'induced effect' first described by Ogle (1950).

Although it still remains to be shown by the neurophysiologists how the horizontal and vertical disparity information is treated separately, the new insights help to account for the accurate preservation of *both* by binocular disparity-selective neurones.

Binocular Inhibition in the Lateral Geniculate Nucleus

Another important feature of all the mammalian binocular systems studied so far is the phenomenon of binocular inhibition in the lateral geniculate nucleus (LGN). Although true convergence of excitatory inputs from both eyes is first seen at the level of the visual cortex, limited binocular interaction also occurs at the LGN before the generation of binocular receptive fields with matched properties. This takes the form of an inhibitory influence from the non-dominant eye and is a subtle feature which has required special attention for its demonstration. Documentation is best in the cat where all laminae appear to be subject to binocular inhibition (Sanderson et al., 1969, 1971), but it has also been demonstrated in the rhesus monkey where the magnocellular laminae, but not the parvocellular laminae, show this influence (Dreher et al., 1976). This species difference is very likely due to the fact that there is a much higher proportion of concentrically organized, non-oriented receptive field properties in lamina IV of the monkey visual cortex and it seems very likely that further study of lamina IVc beta cells receiving parvocellular input will reveal significant binocular interaction there. One could regard the monkey's IVc beta with its enhanced segregation of inputs from both eyes, as representing comparable processing to that which takes place in the cat's LGN, except that it is 'encephalized' one synapse. The significance of this binocular inhibition lies in the way in which the orientation-selective properties of the cortex can subtly influence binocular processing at the very earliest stage before highly specialized binocular receptive fields are constructed. In theoretical terms this subtle influence may have its counterpart in recent machine algorithms for stereopsis, where it has been found that efficiency is increased by allowing a small degree of binocular interaction to occur at the very earliest stages, just before the matching problem is solved (Mayhew and Frisby, 1979).

Lateral Geniculate Lamination

A universal feature of mammalian LGNs is the segregation of inputs from the two eyes into separate laminae. The apparent paradox by which the inputs from both eyes appear to be segregated in such a precisely aligned fashion after so much trouble has been expanded to bring both eyes together, has been discussed extensively by Kaas *et al.* (1972). This paradox may be resolved partially in terms of the new findings, which indicate a significant degree of binocular interaction, of an inhibitory kind, within the LGN. These have already been discussed in the previous section. In addition, there is recent evidence which suggests that geniculate lamination patterns may be an important mechanism by which disparity selectivity is generated.

The new observations derive from the work showing that different retinal ganglion cell classes have different patterns of decussation within the optic chiasm (see Levick, 1977). For example, the alpha cells which have been identified with physiological class Y-cells in the cat have a decussation pattern which is significantly to the temporal side of the zero meridian through the centre of the area centralis. In contrast, the beta cells have a decussation line which lies nasal to that of the alpha cells. Recent evidence suggests that one of the W-cell subclasses has a decussation pattern which is even more nasal than that of the beta cells. The significance of these different decussation patterns when translated into the LGN is shown in Fig. 12.2, where it can be seen that if the medial edge of the LGN represents the most medial retinal ganglion cells within the decussation, then projection lines drawn through the different layers of the LGN will result in binocular pairings which have different retinal disparities. For example, because of the more temporal position of the decussation for Y-cells, the projection line passing through adjacent laminae representing the two eyes would intersect at a retinal correspondence which would give rise to a very convergent retinal disparity. In contrast, the epsilon class of W-cells would tend to code for divergent disparities and the beta cells for disparities close to the fixation plane. The finding that different retinal ganglion cell classes connect with specific targets within the LGN and that these specific targets, in turn, relay on to specific sublaminae and even specific visual cortical areas, suggests that one important aspect of parallel processing concerns retinal disparity. The nature of the decussation patterns the generation of binocular neurones connected to different and the pattern of retinogeniculate lamination will lead to classes of retinal ganglion cells which also have different retinal disparities.

The segregation of disparity processing to different laminae could also be relevant to the question of cortico-

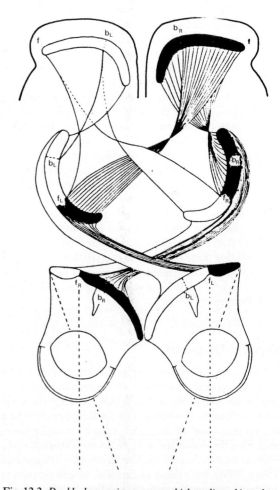

Fig. 12.2 *Double decussation pattern which mediates binocular interaction in owls and some other predatory birds. Despite the total decussation of the optic nerve fibres, information from corresponding parts of each retina converge in the brain by means of a second decussation. Note that the fibres projecting back to the opposite side of the brain arise from the representation in the thalamic relay nucleus of retina temporal to the fovea (subserving the region of binocular overlap). This pattern achieves a similar end result, at the final cortical destination, to the more familiar partial decussation of optic nerve fibres seen in mammals. f, the binocular fovea on the temporal retina and its representation in the central visual pathway; b, the limit of binocular overlap, which corresponds roughly to the position of the blind spot at the pecten/optic nerve head.*

fugal feedback, which could thereby be used to enhance a particular disparity plane at the expense of others.

Ipsilateral Visual Field Representation

A feature of both the ungulate and the feline binocular visual pathway not evident in the primate is a significant representation of the ipsilateral visual field. This representation is much more prominent in the sheep and goat than it is in the cat (Pettigrew et al., 1984) and can be seen at the level of the retina, where the decussation pattern for the alpha cells involves spread into the temporal retina of the contralateral eye, in the medial interlaminar nucleus, which is a prominent feature of both the cat and sheep visual thalamus and which has a representation of the ipsilateral visual field, and, finally, in the visual cortex in the boundary zone where areas 17 and 18 adjoin. The functional significance of this representation of the ipsilateral visual field can be suggested from the fact that the binocular neurones with receptive fields that have a large ipsilateral component tend to have atypically large and convergent disparities (Pettigrew et al., 1984). This fact, in combination with the large disjunctive eye movements which are known to occur, particularly for ungulates, suggests a role for this lack of differentiation in the medial edge of the descussation in allowing for a relative lack of precision of binocular alignment. In contrast, the very sharp decussation of the primate would, without a large callosal mechanism for mediating midline integration, result in a complete failure of binocular fusion when the eyes were misaligned following a disjunctive eye movement. This interpretation is supported by the finding of both very large convergent disparities and a large ipsilateral representation of the visual field within the optic tectum of the oppossum (Ramôa et al., 1983). Like ungulates, marsupials have a relatively poor degree of binocular coordination of eye movements, and large disparities across the vertical midline will occur as a result of the peculiar disjunctive eye position adopted by marsupials under some circumstances. The disparities created by large disjunctive eye movements contrast markedly with the tiny retinal disparities involved in fine binocular depth discriminations and one may speculate that the generation of binocular neurones coding for such large disparities would have a different mechanism. Such a mechanism, involving a large ipsilateral field representation and interhemispheric connections, will lead to the generation of very large field disparities. It is possible that a similar mechanism, operating from ocular dominance column to ocular dominance column instead of from hemisphere to hemisphere, may explain the generation of the small receptive field disparities which characterize the majority of binocular neurones so far described.

The Naso-temporal Decussation

One unsuspected feature of the naso-temporal descussation in both cats and ungulates was that it did not form a right-angle at its intersection with the horizontal meridian. In these two species as well as in the opossum there is an outward tilt of naso-temporal decussation such that the superior arm lies in more temporal retina. The significance of this tilt was first realized from work on the comparison of binocular visual systems in a terrestrial owl with that of the cat and can be explained in the context of Helmholtz's vertical horopter (Cooper and Pettigrew, 1979). In the vertical plane, as predicted by Helmholtz, the horopter is tilted to pass through the fixation point and the ground below the subject's feet. The tilt can be verified by a number of independent methods, both anatomical and physiological. Psychophysical studies in man also support the nature of the tilt. The magnitude of the tilt will be greatest for terrestrial animals which have a large pupillary separation in relation to the distance from the eyes to the ground and may help explain the fact that decussations in some of the small terrestrial mammals, such as the mouse, tend to have such an oblique inclination with respect to the horizon (e.g. Drager and Olsen, 1980).

The presence of the tilt has ramifications in a number of different areas, such as the area of binocular visual development. For example, in altricial animals born with a large pupillary separation in relation to height there will be a constant change with growth of angle θ. This constant change in θ will require some remarkable readjustments within the binocular visual system if binocular vision is to remain functional throughout the period of growth, and it is of great interest, therefore, to find that the dynamics of changes in θ correspond exactly to those for the critical period in those binocular species where data on both of these changes are available (Pettigrew et al., 1984). In addition to changing during development, θ may also change markedly during the daily life of an animal. During hunting, cats may adopt a posture with the head very close to the ground in which θ would have to be markedly increased. Is it possible that during this time, when a greater degree of excyclotorsion of the eyes would be necessary, the resulting change in the binocular correspondence would necessitate a switch to a form of visual processing which is different from that which occurs when the cat is in the normal posture? I suggest that one of the roles for the many different visual cortical areas may be to accommodate such a change and, further, that satellite nuclei, such as the visual claustrum, may play an important role in enabling the switch from one mode of visual processing to another to occur concomitant with changes in the total motor patterning adopted by the animal.

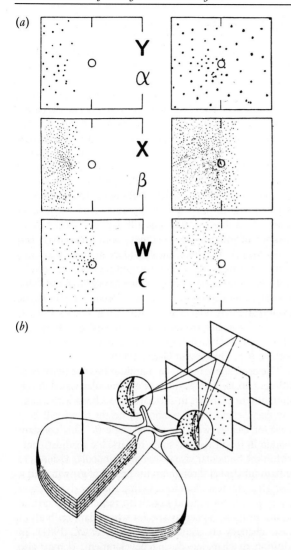

Fig. 12.3 (a) Schematic illustration of the postulated nasotemporal decussation for different retinal ganglion cell classes. Note that temporal extension is greatest for the Y-cell class. (b) Consequences of different ganglion cell decussation patterns for formation of receptive fields of binocular neurones in the geniculostriate pathway, if the edge of each decussation is aligned with the medial edge of the LGN. The three different ganglion cell classes are shown projecting to six different laminae with the LGN. The arrow indicates a 'projection line' along which lateral geniculate neurones representing each eye will converge onto the same binocular neurone. Binocular neurones with Y input will tend to have more convergent disparities and respond best to nearer depth planes than binocular cortical neurones receiving X input because of the different decussation patterns. The limited data on the epsilon W class suggest that they may have a decussation pattern and central connectivity (in area 19 of the cat visual cortex) appropriate to a depth plane at infinity, more distant than the fixation plane represented by the X-driven system.

Evolution of Avian Stereopsis

Owls have evolved an elaborate neural substrate for binocular vision which is comparable to that already described for cats, primates and ungulates. The parallels in physiological organization are quite striking when one considers that the anatomical organization is very much different, owls having a double decussation instead of the partial decussation of mammals (Fig. 12.3). Thus, the avian thalamic neurones which contribute to the construction of binocular receptive fields at the level of the visual cortex are all of the concentrically organized receptive field type, just as one finds in mammals, despite the fact that the avian retina has a very large number of the highly specialized receptive field types such as direction-selective, orientation-selective and edge-selective varieties. In other words, in the owl visual cortex, just as in the striate cortex of mammals, the process of binocular interaction proceeds hand in hand with the elaboration of highly specialized receptive field properties, and binocular neurons have tightly matched receptive field properties in each eye. Indeed, most of the features so far described for the mammalian binocular visual pathway have their counterparts in the binocular visual pathway of the owl. So far no functional equivalent of the visual claustrum has been described in the owl, nor does the owl have ocular dominance stripes in the normal situation. With regard to the latter, however, it must be remembered that ungulates likewise have no ocular dominance stripes, yet these appear following different forms of visual deprivation during early development. The latter is true also for the New World primates and also appears to be the case for owls, which do show the phenomenon of ocular dominance banding patterns if visual input is restricted during early development.

Binocular Visual Organization in Relatives of the Owl

The finding of a neural substrate for binocular vision in the owl prompted the investigation of close relatives of the owl in an effort to sketch some kind of evolutionary history of functional binocular vision in birds. All members so far studied in the families Strigidae and Tytonidae have comparable patterns of organization subserving binocular vision. In the diverse order Caprimulgiformes, there is some heterogeneity with respect to binocular vision however. For example, the oilbird *Steatornis caripensis* is a large, neotropical frugivorous bird which lives in totally dark caves to which it returns after nocturnal forays to feed on palm fruit. It has a highly developed visual system in addition to an echolocation system which enables it to avoid obstacles in the dark, so long as they are more than

20 cm or so in diameter. Studies of its visual system have failed to reveal any evidence for a neural substrate for binocular vision. Instead, the visual cortex has large numbers of orientation-selective cells which cluster around the horizontal and vertical axis and which can be driven only by the contralateral eye (Pettigrew and Konishi, 1984). Likewise, a number of members of different genera in the true nightjar families, Caprimulginae and Chordeilinae, have failed to provide any evidence for binocular vision after detailed electrophysiological and anatomical investigations like those carried out on the owl.

Some of the Caprimulgiformes have provided evidence that they can achieve binocular vision like the owls. Two such families are the Podargidae and Aegothelidae, both Australasian groups, which have highly developed visual systems which are used for predation, mostly on invertebrates, at night. Like the owls, these two families of nightjar-like birds have the ability to take prey from the substrate. This ability sets them apart from the other nightjar-like families which have neither the ability to take prey from the substrate nor any apparent mechanism for stereopsis.

In this large assemblage of nocturnal birds at least, it appears that the selection pressures operating for the emergence of stereopsis appear to be greatest in the niche involving predation from a substrate such as the ground. These pressures do not appear to be particularly strong in predators which are exclusively aerial, such as the true nightjars. One can imagine a number of reasons for this. First, it is easy to see that the consequences of mislocalizing prey are much greater for an avian predator hurtling towards the ground at high speed than they would be for a nightjar taking a moth in the air. Since stereopsis is such an accurate and direct means of judging depth, it is easy to see why it might be more important in the former case. Secondly, one can see that the powerful monocular depth cue of movement parallax is much more readily available to an aerial predator as it moves in relation to its prey than it is for a 'perch-and-pounce' kind of predator which may be at a disadvantage if it moves around too much and thereby reveals itself to the prey. Finally, as pointed out by Julesz (1971), it is possible for a predator with stereoscopic ability to 'break' camouflage since a target which is invisible to monocular inspection may be visible stereoscopically if it has a slightly different depth plane from the background. Breaking camouflage is not a pervasive problem for aerial predators, for which prey will be either silhouetted or highlighted against the background. Conspicuity of prey is much more of a problem for predators taking prey on a substrate and it seems likely that the advantage conferred upon the predator which could break camouflage would be great enough to increase greatly its survival value.

The acquisition of stereopsis can be gauged roughly in birds by looking at the size of the visual wulst which forms a prominent bulge on the dorsal surface of the brain. In the nightjars and swifts, which appear to use monocular cues exclusively for prey capture, the visual wulst is relatively small and the vallecular groove, which marks its lateral boundary, is quite close to the midline. In contrast, in the owls, podargids and aegothelids, which have developed stereopsis, the visual wulst is enormous and the vallecular margin extends almost out to the lateral margin of the brain.

The presence of the large wulst, in combination with the rather small tectum, allows one to infer the presence of specialization for binocular vision in an indirect way from fossil material. The relative paucity of fossil avian material makes it difficult to go far with this approach, but the existing evidence is consistent with the idea that the primitive condition amongst avian predators is stereoblindness and that the advanced condition involves the evolution of stereoscopic processing. For example, the fossil *Quipollornis* is an aegothelid nightjar which is intermediate between the existing aegothelids and the aerial insectivorous nightjars in that it has relatively longer wings (Rich and McEvey, 1977).

Stereopsis in Other Vertebrate Groups?

There is some evidence for the existence of binocular visual processing capable of subserving stereopsis in vertebrate groups apart from the birds and mammals considered above. A number of teleost fish and anuran amphibians have anatomical evidence for binocular integration within the tectum and this evidence, coupled with behavioural evidence for binocular depth discrimination in the toad (Collet, 1977), suggests that further work may reveal fully fledged binocular visual processing in these groups (Finch and Collett, 1983; Gaillard, 1985). Certainly the diverse visual adaptations of the teleosts make it likely that at least some of them have evolved the capability for binocular vision. Whether this has an anatomical substrate similar to that of mammals, or to that of birds, or to a third system remains to be discovered.

Acknowledgement

This chapter was originally published in *Visual Neuroscience*, eds. J. D. Pettigrew, K. J. Sanderson and W. R. Levick, 1986, pp. 208–222. Cambridge, UK: Cambridge University Press.

References

Archer, M. and Clayton, G. (1984). *Vertebrate Zoogeography and Evolution in Australasia*. Carlisle, Western Australia: Hesperian Press.

Allman, J. M. (1977). Evolution of the visual system in the early primates. In *Progress in Psychobiology and Physiological Psychology*, vol. 7. ed. Sprague, J. M. pp. 1–53. New York and San Francisco: Academic Press.

Barlow, H. B., Blakemore, C. and Pettigrew, J. D. (1967). The neural mechanism of binocular depth discrimination. *J. Physiol.*, 193, 327–342.

Barlow, H. B., Hill, R. M. and Levick, W. R. (1964). Retinal ganglion cells responding selectively to direction and speed of image motion in the rabbit. *J. Physiol.*, 173, 377–407.

Bishop, P. O. (1973). Neurophysiology of single vision and stereopsis. In: *Handbook of Sensory Physiology*, vol. VII/2. *Central Processing of Visual Information. A: Integrative Functions and Comparative Data*. ed. Jung, R. pp. 255–305. Berlin, Heidelberg and New York: Springer-Verlag.

Bravo, H. and Pettigrew, J. D. (1981). The distribution of neurons projecting from the retina and visual cortex to the thalamus and tectum opticum of the barn owl, *Tyto alba* and the burrowing owl, *Speotyto cunicularia. J. Comp. Neurol.*, 199, 419–441.

Clarke, P. G. H., Donaldson, I. M. L. and Whitteridge, D. (1976). Binocular visual mechanisms in cortical areas I & II of the sheep. *J. Physiol.*, 256, 509–526.

Collet, T. (1977). Stereopsis in toads. *Nature (Lond.)*, 267, 349–351.

Cooper, M. L. and Pettigrew, J. D. (1979). A neurophysiological determination of the vertical horopter in the cat and owl. *J. Comp. Neurol.*, 184, 1–26.

Drager, V. C. and Olsen, J. F. (1980). Origins of crossed and uncrossed retinal projections in pigmented and albino mice. *J. Comp. Neurol.*, 191, 383–412.

Dreher, B., Fukada, Y. and Rodieck, R. W. (1976). Identification, classification and anatomical segregation of cells with X-like and Y-like properties in the lateral geniculate nucleus of old world monkeys. *J. Physiol.*, 258, 433–452.

Ebbesson, S. O. E. and Schroeder, D. M. (1971). Connections of the nurse shark's telencephalon. *Science*, 173, 254–256.

Ekström, P. (1982). Retinofugal projections in the eel, *Anguilla anguilla* L. (Teleostei), visualised by the cobalt-filling technique. *Cell Tiss. Res.*, 225, 507–524.

Finch, D. J. and Collett, T. S. (1983). Small-field, binocular neurons in the superficial layers of the frog optic tectum. *Proc. R. Soc. Lond. B*, 217, 491–497.

Frisby, J. P. and Julesz, B. (1975). The effect of orientation difference on stereopsis as a function of line length. *Perception*, 4, 179–186.

Frisby, J. P. and Mayhew, J. E. W. (1980). Spatial frequency tuned channels: implications for structure and function from psychophysical and computational studies of stereopsis. *Phil. Trans. R. Soc. Lond. B*, 290, 95–116.

Gaillard, F. (1985). Binocularly-driven neurons in the rostral part of the frog optic tectum. *J. Comp. Physiol. A*, 157, 47–56.

Gordon, B. (1973). Receptive fields in deep layers of cat superior colliculus. *J. Neurophysiol.*, 36, 157–158.

Grimson, W. E. L. and Marr, D. (1979). A computer implementation of a theory of human stereo vision. In *Image Understanding Workshop. (April 1979, A. I. Lab.)*. Cambridge, Mass: MIT.

Hughes, A. and Vaney, D. I. (1981). Contact lenses change the projection of visual field onto rabbit peripheral retina. *Vision Res.*, 21, 955–956.

Hughes, A. and Vaney, D. I. (1982). The organisation of binocular cortex in the primary visual area of the rabbit, *J. Comp. Neurol.*, 204, 151–164.

Julesz, B. (1971). *Foundations of Cyclopean Perception*. Chicago: University of Chicago Press.

Kaas, J. H., Guillery, R. W. and Allman, J. M. (1972). Some principles of organisation in the lateral geniculate nucleus. *Brain. Behav. Evol.*, 6, 253–299.

Kato, H., Bishop, P. O. and Orban, G. A. (1978). Hypercomplex and simple/complex cell classifications in the cat striate cortex. *J. Neurophysiol.*, 41, 1071–1095.

Kaye, M., Mitchell, D. E. and Cynader, M. (1981). Selective loss of binocular depth perception after ablation of cat visual cortex. *Nature (Lond.)*, 293, 60–62

Levick, W. R. (1967). Receptive fields and trigger features of ganglion cells in the visual streak of the rabbit's retina. *J. Physiol.*, 188, 285–307.

Levick, W. R. (1977). Participation of brisk-transient retinal ganglion cells in binocular vision – an hypothesis. *Proc. Aust. Physiol. Pharmacol. Soc.*, 8, 9–16.

Luiten, P. G. M. (1981). Two visual pathways in the telencephalon of the nurse shark, *Ginglymostoma cirratum*, II. Ascending thalamotelencephalic connections. *J. Comp. Neurol.*, 196, 539–548.

Marr, D. (1982). *Vision*. San Francisco: Freeman.

Marr, D. and Poggio, T. (1976). Cooperative computation of stereo disparity. *Science*, 194, 283–287.

Marr, D. and Poggio, T. (1979). A computational theory of human stereo vision. *Proc. R. Soc. Lond. B*, 204, 301–328.

Mayhew, J. E. W. and Frisby, J. P. (1979). Surfaces with steep variations in depth pose difficulties for orientationally tuned disparity filters. *Perception*, 8, 691–698.

Mayhew, J. E. W. and Frisby, J. P. (1980). The computation of binocular edges. *Perception*, 9, 69–86.

Mayhew, J. E. W. and Longuet-Higgins, H. C. (1982). A computational model of binocular depth perception. *Nature (Lond.)*, 297, 376–378.

Mitchell, D. E., Kaye, M. and Timney, B. (1979). A behavioural technique for measuring depth discrimination in the cat. *Perception*, 8, 389–396.

Nelson, J. I., Kato, H. and Bishop, P. O. (1977). The discrimination of orientation and position disparities by binocularly-activated neurons in cat striate cortex. *J. Neurophysiol.*, 40, 260–264.

Newton, I. (1730). *Opticks*. 1952 issue based on 4th edn. New York: Dover Publications.

Ogle, K. N. (1950). *Researches in Binocular Vision*. New York: Saunders.

Pettigrew, J. D. (1979a). Comparison of the retinotopic organisation of the visual Wulst in nocturnal and diurnal raptors, with a note on the evolution of frontal vision. In *Frontiers of Visual Science*, ed. Cool, S. J. and Smith E. L. pp. 328–335. New York: Springer-Verlag.

Pettigrew, J. D. (1979b). Binocular visual processing in the owl's telencephalon. *Proc. R. Soc. Lond. B*, 204, 435–454.

Pettigrew, J. D. and Konishi, M. (1976). Neurons selective for orientation and binocular disparity in visual Wulst of the barn owl (*Tyto alba*). *Science*, 193, 675–678.

Pettigrew, J. D., Nikara, T. and Bishop, P. O. (1968). Binocular interaction on single units in cat striate cortex: simultaneous stimulation by single moving slit with receptive fields in correspondence. *Exp. Brain Res.*, 6, 391–410.

Pettigrew, J. D. and Konishi, M. (1984). Some observations on the visual system of the oilbird (*Steatornis caripensis*). *Natl. Geogr. Res. Rep.*, 16, 439–450.

Pettigrew, J. D., Ramachandran, V. S. and Bravo, H. (1984). Some neural connections subserving binocular vision in ungulates. *Brain, Behav. Evol.*, 24, 65–93.

Poggio, G. F. and Poggio, T. (1984). The analysis of stereopsis. *Ann. Rev. Neurosci.*, 7, 379–412.

Poggio, G. F. and Talbot, W. H. (1981). Mechanisms of static and dynamic stereopsis in foveal striate cortex of the rhesus monkey. *J. Physiol.*, **315**, 469–492.

Polyak, S. (1957). *The Vertebrate Visual System*. Chicago: University of Chicago Press.

Provis, J. M. (1979). The distribution and size of ganglion cells in the retina of the pigmented rabbit: a quantitative study. *J. Comp. Neurol.*, **185**, 121–139.

Romôa, A. S., Rocha-Miranda, C. E., Méndez-Otero, R. and Jousá, K. M. (1983). Visual receptive fields in the superficial layers of the opossums's superior colliculus. *Exp. Brain Res.*, **49**, 373–381.

Rich, P. V. and McEvey, A. (1977). A new owlet-nightjar from the early to mid-Miocene of eastern New South Wales. *Mem. Nat. His. Mus. Victoria*, **38**, 247–253.

Richards, W. (1970). Stereopsis and stereoblindness. *Exp. Brain Res.*, **10**, 380–388.

Sanderson, K. J., Darian-Smith, I. and Bishop, P. O. (1969). Binocular corresponding receptive fields of single units in the cat dorsal lateral geniculate nucleus. *Vision Res.*, **9**, 1297–1303.

Sanderson, K. J., Darian-Smith, I. and Bishop, P. O. (1971). The properties of binocular receptive fields of lateral geniculate neurons. *Exp. Brain Res.*, **13**, 178–207.

Stone, J. (1983). *Parallel Processing in the Visual System*. New York: Plenum Press.

Van Sluyters, R. C. and Stewart, D. L. (1974). Binocular neurons of the rabbit's visual cortex. Receptive field characteristics. *Exp. Brain Res.*, **19**, 166–195.

Wallman, J. and Pettigrew, J. D. (1985). Conjugate and disjunctive saccades in two avian species with contrasting oculomotor strategies. *J. Neurosci.*, **5**, 1418–1428.

Walls, G. L. (1942). *The Vertebrate Eye and its Adaptive Radiation*, 1967 fascimile of 1942 edn. New York and London: Hafner Publishing Co.

Zuidam, I. and Collewijn, H. (1979). Vergence eye movements of the rabbit in visuomotor behaviour. *Vision Res.*, **19**, 185–194.

13 Evolution of Colour Vision

Christa Neumeyer

Introduction

'Colour vision' means the capability of a visual system to respond differently to light differing in wavelength only. It is based on the existence of two or more photoreceptor types containing photopigments maximally absorbing in different spectral ranges. The absorption spectrum of a photopigment can be read as the probability with which a photon of a certain energy (inversely related to the wavelength of light) is going to be absorbed. If an absorption event takes place, the photoreceptor responds with the same change of excitation independent of the energy of the photon. This is in essence the 'principle of univariance', which can also be formulated as: 'The output depends upon quantum catch, but not upon what quanta are caught' (Rushton, 1972). It has the consequence that each photoreceptor is actually 'colour blind'. Therefore, the central nervous system can gain information about 'colour' only by comparing the outputs of the different photoreceptor types. The processing of colour-specific information in eye and brain gives rise to the perception of colour as a subjective phenomenon. As human beings we classify colours in terms of 'hue', 'brightness' and 'saturation', and name different hues with the basic colour terms 'blue', 'green', 'yellow', and 'red', as well as 'white' and 'black'. The number of colours we are able to discriminate is enormous, so that it is hard to imagine that any other colour vision system may be superior. But human colour vision and that of higher primates is only one realization of colour vision in the animal kingdom. It is generally assumed that colour vision was invented several times independently during evolution (Walls, 1942). If we consider only highly effective colour vision systems which use three or more photoreceptor types, there are at least three separate lines: one leading to primate colour vision, one to colour vision of fishes, amphibia, reptiles and birds, and one to colour vision of hymenopteran insects.

The question how these different colour vision systems may have been developed during evolution will never be answered with certainty, but, as demonstrated in primates during the last decade, there is hope of getting rather close by approaching the problem with different methods:

1. A refined behavioural analysis which clearly characterizes colour vision in several closely related species.
2. Microspectrophotometrical measurements of the photopigments involved.
3. Deciphering the genes decoding the amino-acid sequence of the opsins.

This approach, which is connected with the names of G. Jacobs, J. Bowmaker, and J. Nathans and their co-workers (see below), gives some insight into the steps of evolution from dichromatic forms of colour vision to its trichromatic realization.

Until there are more data, especially about the genetic basis of colour vision in other vertebrate and invertebrate animals, it is perhaps not unwise to follow Gordon Walls's approach to obtain some ideas about the phylogeny and evolution of colour vision by asking the three questions: 'what?', 'how?' and 'why?' (Walls, 1942, pp 462ff.). Thus, in the first part of this chapter, answering the 'what'-questions will be attempted by collecting the knowledge we have about colour vision found in different animals. Here one can move on the solid grounds of comparative colour vision. In this context, the following problems are of interest: what are the characteristics of colour vision realized in different animals? Are there basic or only gradual differences? Are all colour vision systems designed according to the same functional principle, even those that most probably evolved independently? Then, Gordon Walls's 'how' questions may be asked dealing with the neuronal realization of colour vison: which photoreceptor types are involved, and what are the steps in neuronal processing of colour-specific information? Can we discriminate between 'simple' and more 'complex' patterns of processing? Finally, the 'why' questions deal with the selective pressures that gave rise to the forms of colour vision that now exist. Here, one has to consider the physical limits of colour vision as well as the properties of the natural habitat, such as the spectral characteristics of natural daylight and its changes, the light conditions in air and under water, and the spectral remittance of natural objects.

It is not intended to give a complete review of comparative colour vision and related problems. The reader will find excellent summaries of the older literature about vertebrates in Walls (1942) and in Jacobs (1981); about invertebrates reference may be made to Menzel (1979). In addition, the other chapters in this volume may be consulted, as well as the relevant chapters in volume 6 of this series, especially that by Goldsmith.

Ability to Discriminate Wavelengths

The only way to show that an animal has colour vision and to characterize the system is to perform behavioural experiments. The most appropriate method consists in a training technique in which the animals are rewarded with food. This technique was introduced into comparative sensory physiology by Karl von Frisch (1913, 1915). In these early experiments, von Frisch was the first who was able to show convincingly that a cyprinid fish, the minnow *Phoxinus laevis*, and the honeybee possess colour vision. The evidence was given by demonstrating that the animals used only the 'colour' cue to discriminate between the different stimuli, and not 'brightness'. While in all experiments in which pieces of coloured paper are used as stimuli together with different shades of grey, colour vision can be shown only quantitatively (Walls, 1942; Jacobs, 1981), a qualitative method which allows a comparison between different colour vision systems requires monochromatic colour stimuli adjustable in wavelength and intensity. Most informative for the characterization of a colour vision system is the knowledge of its ability to discriminate light of different wavelengths which is adjusted in intensity to equal 'animal-subjective' brightness. In the following, only such colour vision systems are taken into consideration in which the wavelength discrimination function (the $\Delta \lambda$ function) is known.

To obtain a $\Delta \lambda$ function, the animal is trained to approach a testfield illuminated with light of a given wavelength, while a comparison testfield is illuminated by light of another, adjacent wavelength. Correct choices are rewarded with food. Discrimination ability is measured as the relative frequency with which the training wavelength is chosen. That specific wavelength is determined at which a threshold criterion of 70% choice frequency is reached. $\Delta \lambda$ is then the difference between the training wavelength and the wavelength interpolated at 70% choice frequency. The values of $\Delta \lambda$ have to be found for a series of training wavelengths throughout the spectrum in distances of 5–20 nm. Since, for each training wavelength, λ, two values of $\Delta \lambda$ are obtained, which can be very different, it is not appropriate to plot the mean of the two $\Delta \lambda$ values, but each value has to be plotted separately over $\lambda +/- \Delta \lambda/2$. It is essential for these measurements that the animal discriminates the monochromatic lights only because of hue, and not because of brightness. For this reason, the stimuli have to be adjusted to equal (animal subjective) brightness. This requires that the spectral sensitivity function be known for the same experimental conditions. Complete $\Delta \lambda$ functions which fulfil these criteria have so far been measured for only a few animals. The measurements are rather difficult and extremely time-consuming. Figs 13.1–13.7 show almost all examples of $\Delta \lambda$ functions found in the literature, plotted on the same scale to allow a more convenient comparison.

Fish

Fig. 13.1 shows the $\Delta \lambda$ function of the goldfish measured by Neumeyer (1986) with the method described above. The function implies that discrimination ability is highest in three spectral ranges: around 400 nm, at 500 nm and around 610 nm. In the ranges between these minima, around 450 nm and at 550 nm, discrimination ability is much worse. The smallest values of $\Delta \lambda$ were found at 500 nm and reached 4 nm. The goldfish seems to be the only fish in which a $\Delta \lambda$ function was determined. An earlier measurement by Yarczower and Bitterman (1965) in the goldfish did not include the short-wavelength spectral range, and the result differs to some extent from that in Fig. 13.1, probably because the monochromatic lights were not adjusted to equal brightness. This was also not the case in the early measurement by Wolff (1925) with the minnow *Phoxinus laevis*, which, however, yielded a high discrimination ability in the same three spectral ranges as in goldfish.

Amphibia

Complete $\Delta \lambda$ functions have not been measured. Only recently, the discrimination ability between colours produced by a colour monitor was determined in the tiger salamander *Salamandra salamandra* (Przyrembel and Neumeyer, 1990). Using a technique in which the salamanders had to detect a coloured 'worm' dummy against a differently coloured but equally bright background, we found that they were able to discriminate colours equivalent to 460 nm from 530 nm, as well as 530 nm from 590 nm. The discrimination between colours that appear to a human observer as 'green' and 'red' was very good and

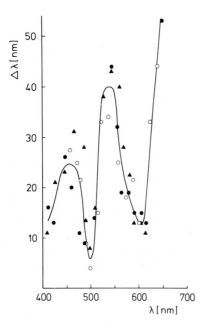

Fig. 13.1 *Wavelength discrimination* ($\Delta\lambda$) *function of the goldfish. The symbols stand for three individual fish.* (*After Neumeyer, 1986*).

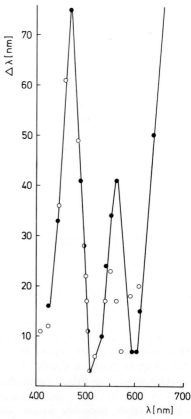

Fig. 13.2 *Wavelength discrimination* ($\Delta\lambda$) *function of the turtle* Pseudemys scripta elegans. *Two turtles tested.* (*After Arnold and Neumeyer, 1987.*)

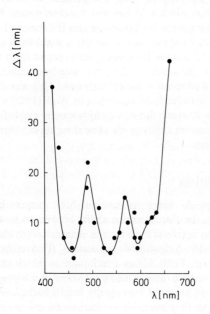

Fig. 13.3 *Wavelength discrimination* ($\Delta\lambda$) *function of the pigeon* (Columba livia). (*After Emmerton and Delius, 1980, from the mean values of* $\Delta\lambda$ *in their Fig. 2.*)

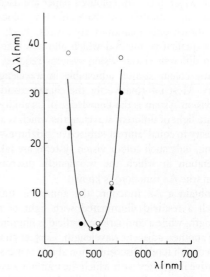

Fig. 13.4 *Wavelength discrimination* ($\Delta\lambda$) *function of the Western grey squirrel* Sciurus grieseus. (*After Jacobs, 1981, Fig. 5.2.*)

Colour Vision 287

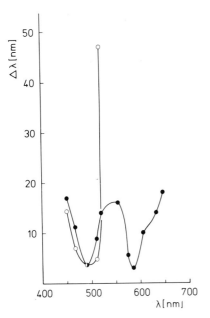

Fig. 13.5 *Wavelength discrimination (Δλ) functions of the squirrel monkey* Saimiri sciureus. *(After Jacobs and Neitz, 1985.) Open circles: function found in males and some females; filled circles: function found exclusively in females.*

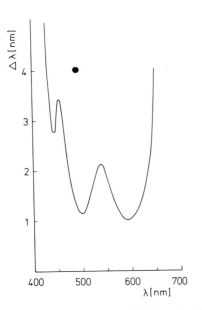

Fig. 13.6 *Wavelength discrimination (Δλ) function of humans (After Wyszecki and Stiles, 1982, from figure 1 (7.10.2) based on data by Wright and Pitt, 1934).*

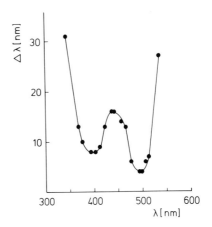

Fig. 13.7 *Wavelength discrimination (Δλ) function of the honeybee* Apis mellifica. *(After von Helversen, 1972.)*

reached Δλ values of about 15 nm, testing 530 nm and 590 nm. An even higher discrimination ability is to be expected in the spectral range between these two wavelengths.

Reptiles and Birds

The colour vision of diurnal birds and of reptiles is especially interesting because their cones possess brightly coloured oil droplets. They are located just below the outer segments which contain the photopigments, so that the incident light has to traverse them. The oil droplets act as cutoff filters which absorb ultraviolet and short-wavelength light. Therefore, they alter the spectral sensitivity functions of the cone types by steepening their short-wavelength limbs and by shifting maximal sensitivity towards longer wavelengths.

That turtles have an excellent wavelength discrimination ability was already shown by Wojtusiak (1933) in *Clemmys caspica*. In another freshwater turtle, *Pseudemys scripta elegans*, a Δλ function was measured (Arnold and Neumeyer, 1987). This function, shown in Fig. 13.2, is the only one known for reptiles. As in the goldfish (Fig. 13.1), there are three narrow spectral ranges in which wavelength discrimination is best; at 400 nm, at 510 nm and at 570 or 600 nm (in the long-wavelength range, the two subjects gave differing results). Deviating from the goldfish, the turtles showed no wavelength discrimination at all in the range between 450 nm and 510 nm which is indicated by the very high Δλ values at 460 nm in Fig. 13.2. The lack of wavelength discrimination can be explained by the filter effect of the oil droplets present in the mid-wavelength cones. They are so effective in absorbing short-wavelength light that this photoreceptor type does not respond at all when stimulated with wavelengths shorter than 510 nm.

In birds, the most thoroughly investigated colour vision system is that of the pigeon *Columba livia*. Its $\Delta\lambda$ function, measured by Emmerton and Delius (1980), is shown in Fig. 13.3. Best discrimination ability was found at 460 nm, at 530 nm and at 595 nm, with smallest values of $\Delta\lambda$ equal to 3 nm. Furthermore, Emmerton and Delius have shown in a separate experiment that the pigeon is even able to discriminate wavelengths in the ultraviolet range at 360–380 nm. Other investigators reported a good discrimination around 500 nm in addition (a summary of all wavelength discrimination data is given in Emmerton and Delius, 1980). A similar wavelength discrimination function was found in the hummingbird *Archilochus alexandri* (Goldsmith et al., 1981).

Mammals

Owing to the careful studies of mammalian colour vision by Jacobs and his co-workers, several wavelength discrimination functions are known for non-primate and primate mammals. The method by which the $\Delta\lambda$ functions in mammals were obtained differed slightly from the examples above, as here the so-called 'oddity' procedure was applied. Here, the animals have to make choices between three testfields, two of which are equally illuminated while the third is different. They have to select the different one. Another difference consists in the fact that, in the $\Delta\lambda$ function, the mean of the two $\Delta\lambda$ values is plotted over the training wavelength, which blurs the curves to some extent. Fig. 13.4 shows the $\Delta\lambda$ function of the western grey squirrel *Sciurus griseus* (after Jacobs, 1976). Here, best discrimination was found in one wavelength range only, around 500 nm. Towards longer and shorter wavelengths, discrimination ability fell off sharply, so that the animals were entirely unable to discriminate 550 nm from 600 nm. A similar U-shaped $\Delta\lambda$ function was found in the squirrel monkey *Saimiri sciureus*, a New World primate. Here, interestingly, this kind of $\Delta\lambda$ function was found only in a part of the population; the other part had a function with two minima, around 490 nm and around 590 nm (Jacobs, 1984). Both functions are shown in Fig. 13.5 (after Jacobs and Neitz, 1985). Testing 36 squirrel monkeys, it turned out that all males produced U-shaped $\Delta\lambda$ functions, while in most females (76%) the functions revealed two minima. Such a polymorphism of colour vision was also found in other neotropical monkeys such as *Saguinus fuscicollis* (Jacobs, Neitz and Crognale, 1987). The $\Delta\lambda$ functions for Old World monkeys are very similar to the two-minima type shown in Fig. 13.5 (for review see Jacobs, 1981, Table 5.1 and Fig. 5.4). The best performance here was in the range of 2–3 nm. The shape of the function in Fig. 13.5 is also similar to that of humans, which is shown in Fig. 13.6

(from Wyszecki and Stiles, 1982, p 570, after data from Wright and Pitt). The smallest value of $\Delta\lambda$ an experienced observer is able to discriminate is in the range of 1 nm and sometimes even smaller. Therefore, in Fig. 13.6 the scale of the ordinate differs from those in the other figures by a factor of 10. The two minima of the function are again in the range of 500 nm and 590 nm.

Invertebrates

Of all vertebrate and invertebrate species, the colour vision of the honeybee *Apis mellifica* is most extensively investigated. The reason is that they are ideal for training experiments in which a food reward is given. The bee behaves in the experiment in the same manner as in the natural situation when it collects nectar from flowers. It gathers sugar water on artificial colour stimuli and delivers it in the comb when its honey stomach is filled. Therefore, it never becomes 'saturated' and is willing to come to the lab for many hours per day. While Karl von Frisch, who developed the method, worked in the garden of the Zoological Institute with a crowd of bees collecting sugar water on a watch glass, the technique has been refined considerably during the last 30 years. Nowadays experiments use only single, individually trained bees which are rewarded several times during one visit with tiny drops of sugar water. Using a spectral apparatus with which ultraviolet light can be presented, the $\Delta\lambda$ function was measured by von Helversen (1972). The function is shown in Fig. 13.7. Note that here the abscissa is shifted for 100 nm into the short-wavelength range. Best wavelength discrimination was found in two spectral ranges around 400 nm and around 500 nm. The smallest $\Delta\lambda$ value was 4 nm.

The comparison of the different $\Delta\lambda$ functions in Figs. 13.1–13.7 may indicate some common properties which could be of selective significance during evolution:

1. The minima in these functions are pronounced, i.e. the ranges of best discrimination are rather narrow.

2. The ranges of best discrimination are restricted to certain spectral ranges: around 400 nm, around 500 nm, and at 600 nm. Only the pigeon seems to be an exception. Here, the minima are less clear, and best discrimination occurred at 460 nm, at 530 nm and at 600 nm. The pigeon seems to be able to discriminate very well in the spectral range which appears to us as 'green', a range in which other animals are rather impaired.

3. The best performance was always in the range of a $\Delta\lambda$ value of 3–4 nm. Only human beings do better. It is unclear whether this is due to a better discrimination ability, or to a more strict 'internal' criterion.

Dimensionality of Colour Vision

General Considerations and an Example

The wavelength discrimination function can be understood – but only in a first approximation – on the basis of the spectrally different photoreceptors that provide input into the colour vision system. As mentioned in the Introduction, a visual system can obtain information about 'hue' only from a comparison of the excitation values of the receptor types, by determining their ratios. If, in a given spectral range, different wavelengths excite exclusively one photoreceptor type, they always elicit the same excitation ratio. In this case, the wavelengths cannot be discriminated. The best discrimination ability has to be expected in those spectral ranges in which the limbs of the spectral sensitivity functions of the receptors are steep and cross each other. Here, the ratio of the excitation values changes rapidly with wavelength. Thus, if we consider the case of dichromatic colour vision which is based on two photoreceptor types, it can be expected that the wavelength discrimination function has only one minimum. In trichromatic colour vision with three photoreceptor types there should be two minima, and in a tetrachromatic one based on four receptor types, three minima will be found, and so forth. However, it is important to realize that the reversed inference need not necessarily hold: three minima in a $\Delta\lambda$ function can be due to the existence of four photoreceptor types, but they can also be due to three photoreceptors when one of them has a side maximum of spectral sensitivity in the short-wavelength range based on the β-band of the photopigment. Separate behavioural experiments have to be performed to decide between these two possibilities. The following example may illustrate this point. In the goldfish, it was a great surprise to find three minima in the $\Delta\lambda$ function (Fig. 13.1) because it was very well documented that goldfish and carp possess three cone types with maximal sensitivity at 450 nm, at 535 nm and at 620 nm (from microspectrophotometrical data by Marks, 1965, and Harosi, 1976; from intracellular recordings from single cones by Tomita, 1965; and from the determination of the cone fundamentals by analysing ganglion cell responses by 'silent substitution' by van Dijk and Spekreijse, 1984). Therefore, it seemed reasonable to assume that the high discrimination ability around 400 nm is due to a β-band of the long-wavelength cone photopigment. Model computations have shown that this would indeed explain the experimental findings (Neumeyer, 1986). The alternative possibility, which explains them equally well, is that there is a fourth photoreceptor type with a maximum in the near-ultraviolet range (at 350 nm–370 nm), and a sensitivity reaching up to 450 nm.

To decide between these two possibilities, colour mixture experiments have been performed (Neumeyer, 1985). The rationale was as follows. If, in the range of 400 nm the β-band of the long-wavelength cone type is involved, then light of 400 nm will stimulate the long-wavelength cone type and the short-wavelength one. Thus, this wavelength stimulates the photoreceptors at the same ratio as an additive mixture of 450 nm and 683 nm when the relative amount of quanta is appropriately chosen. Then, at a certain mixture relation, the fish should be unable to discriminate the monochromatic light from the mixture. This, however, was never the case. Instead, the fish completely confused 400 nm with an additive mixture of short-wavelength light (434 nm) and ultraviolet light (367 nm). This experiment provided the evidence that the goldfish has an ultraviolet cone as a fourth photoreceptor type involved in colour vision, and that its colour vision system is tetrachromatic.

The goldfish example taught us the lesson that the characterization of the photoreceptors by 'direct' methods analysing the cellular level can always be incomplete. In cyprinid fishes it became evident only from 1983 on that there are ultraviolet sensitive cones (Avery et al., 1982; Hárosi and Hashimoto, 1983; Hárosi, 1986). They are morphologically described as single miniature cones, the smallest of all cone types and distributed only sparsely. Until recently, the evidence for ultraviolet cones in goldfish came only from behavioural experiments (Neumeyer, 1985; Hawryshyn and Beauchamp, 1985). On the other hand, even if one could be certain of having found all photoreceptor types, it would not be possible to name the dimensionality of the colour vision system. This always requires a behavioural test which shows that all photoreceptor types are involved in colour vision, and that they are processed independently from each other. Theoretically, it is possible that the visual system sums up two receptor inputs or that one photoreceptor type is not used for colour vision but for other visual functions. So far, there is no example where colour vision does not use all existent receptor types (the rods are here not taken into consideration). Possible exceptions to this rule may be found in reptiles and birds where the oil droplets increase the number of cone types (see below). Therefore, to name a colour vision system with the terms 'dichromatic', 'trichromatic', 'tetrachromatic' or 'pentachromatic', or even higher chromatic, it is essential to perform colour mixture experiments in which the question is asked: how many spectral colours are necessary and sufficient in an additive mixture to be indiscriminable from the electromagnetic radiation similiar to that of the sun which appears to a human observer as 'white'. (Please note that such a 'white' light has to include the near-ultraviolet range, and should cover the range between 300 and 700 nm.)

Special Cases

Fish

Teleost fishes are the most numerous in species (about 22 000) of all vertebrates, and they are found in the most diverse habitats. They live in the deep sea which is never reached by the radiation of the sun as well as in the shallow waters of clear rivers, they inhabit the blue ocean as well as greenish-yellow ponds. The light environment is never stable but changes according to time of day, cloudiness, the state of the surface layers, and the movements of the fish into deeper or upper layers (Loew and McFarland, 1990). Therefore, it would not be surprising to find very different colour vision systems. Fishes that possess only rods but no cones are probably restricted to the deep-sea habitat. Here, in several species, only one photopigment type has been reported, while in others two types were found (for review see Bowmaker, 1990). Other possible *monochromats* which possess rods and only one cone type also seem to exist. Bowmaker (1990, Fig. 4.6) shows the absorption spectrum of the cone photopigment of a catfish, *Kryptopterus bicirrhis*, which has its maximum at 607 nm. Colour vision requires at least two types of photopigments. Such possible *dichromats* are obviously frequent in the coastal waters of the Atlantic Ocean (Lythgoe, 1984). Another potential dichromat is the perch *Perca fluviatilis*, which lives in fresh water. Here, a measurement of spectral sensitivity in a behavioural experiment revealed a function with only two maxima in the mid- and long-wavelength range (550 and 670 nm) (Cameron, 1982), and two photopigments were found by microspectrophotometry (see Bowmaker, 1990, Fig. 4.6(c)). A short-wavelength visual pigment seems to be missing. This seems to be also the case in other freshwater teleosts as reported by Loew and Lythgoe (1978). Provided that there is no ultraviolet-sensitive cone type in addition, the cichlid fish *Haplochromis burtoni* would be a good candidate for a *trichromat*. Here, three cone visual pigments have been identified with maximal absorbance at 454 nm, 523 nm and 562 nm (Fernald and Liebman, 1980). In the 'four-eyed' fish *Anableps anableps*, three types of photopigments were found, the absorption maxima are here at 409, 463 and 576 nm (Avery and Bowmaker, 1982). In cyprinids, four cone types were found in several species (for review see Bowmaker, 1990). The brown trout *Salmo trutta* is especially interesting. Here, the juveniles up to 1–2 years old possess ultraviolet-sensitive cones which disappear entirely when the fish become older (Bowmaker and Kunz, 1987).

The only direct evidence for a *tetrachromatic* colour vision was obtained in the goldfish by performing colour mixture experiments (Neumeyer, 1985 and 1988). The spectral sensitivity functions of the four cone types most probably underlying are shown in Fig. 13.8(a). The effect of a given spectral distribution of light on the four cone types can be represented as a single point in a four-dimensional space, with the excitation values of the four cone types as coordinates. If one is interested only in the *relative* excitation values which determine 'hue', abandoning information about 'brightness', the effect of all possible colour stimuli can be shown in a tetrahedron, a three-dimensional plot. In this tetrahedron (Fig. 13.8(b)), the corners represent an exclusive stimulation by one cone type (UV, S = short-, M = mid-, and L = long-wavelength cone). The symbols indicate the colour loci of the spectral colours. Between 680 nm and 450 nm all loci lie in the ground-plane (L, M, S) of the tetrahedron. Towards shorter wavelengths the loci approach the UV corner, and between 350 nm and 300 nm they turn back. This latter part of the spectral colour locus is highly uncertain. In the stereographic plot of Fig. 13.8(c) the spatial relationships can be seen more clearly. 'XW' indicates the locus of the white light emitted by a xenon-arc lamp. Provided the tetrahedron in Fig. 13.8 represents the actual situation of goldfish colour vision, then an additive mixture of 404 nm and 599 nm (marked by open circles) should be found which is equal or very similar to this white light, as the line connecting the two loci runs inside the tetrahedron and comes very close to the locus of XW. Indeed, a mixture proportion was found which was for the goldfish indistinguishable from xenon-white, which contains ultraviolet light, but not from the white light of a tungsten lamp. In this case, ultraviolet light (367 nm) had to be added to obtain a match. Furthermore, xenon-white was indiscriminable from a certain additive mixture of four monochromatic lights of the wavelengths 367, 434, 523 and 641 nm. Only three were not sufficient. Thus, the colour vision system of the goldfish is tetrachromatic, fulfilling the requirement that four spectral lights are necessary and sufficient to be equal to white light.

Amphibia

In amphibia, most of the studied species have rods and cones. Deviating from other vertebrates there are two types of rods, 'red' rods containing a 'normal' rhodopsin with maximal absorption at 500 nm, and 'green' rods which have a photopigment with maximal absorption around 430–450 nm. The latter could be involved in colour vision. The cone types (there are double and single cones) seem to contain the mid- and long-wavelength photopigments with maximal absorption at 500 nm and at 560–580 nm (for review see Donner and Reuter, 1976; Roth, 1987). In the European tiger salamander *Salamandra salamandra*, the absorption spectra of the rod and cone photopigments have not yet been measured. However, the results of our colour discrimination experiments and our measurements of spectral sensitivity could be explained by assuming a trichromatic colour vision based

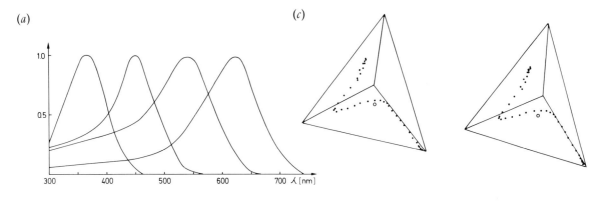

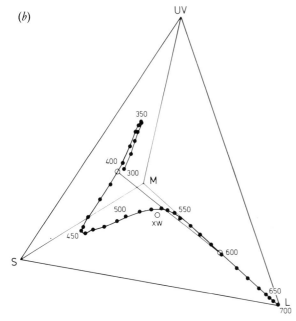

Fig. 13.8 *Tetrachromatic colour vision in the goldfish. (a) Spectral sensitivity functions of the four cone types underlying colour vision (UV: after Hárosi and Hashimoto (1983); S, M, L: 'cone fundamentals' from van Dijk and Spekreijse, 1984). (b) Colour tetrahedron of the goldfish calculated on the basis of the cone sensitivity functions in (a) (As described in Neumeyer, 1980). Each point in this trichromatic space represents the ratio in which the four cone types are excited by a given colour stimulus (the sum of the coordinate values of each point is equal to 1). UV, S, M, L: corners of the tetrahedron, representing colours which excite this respective cone type exclusively. Filled circles: colour loci of the spectral colours between 300 and 700 nm. Open circles: loci of the monochromatic colours (404 and 599 nm) which were confused with Xenon-'white' light (XW) when they were presented in an additive mixture. (c) Three-dimensional plot to be viewed in a stereoscope. The 'white' point XW should be seen suspended above the ground plane SML of the tetrahedron (Calculations and graphic presentation in B and C by Dr J. Schramme).*

on green rods with maximal sensitivity at 450 nm, and on two cone types maximally sensitive at 500 and 580 nm (Neumeyer *et al.*, 1988; Przyrembel and Neumeyer, 1990).

Reptiles and Birds

Colour vision in reptiles and birds is complicated by the fact that the cones contain different brightly coloured oil droplets which act as colour filters. The oil droplets were already discovered in the last century (Hannover, 1840; cited after Peiponen, 1964). It was assumed that the spectrally different cone types underlying colour vision were created by combining the same photopigment with different oil droplets. That this cannot be the case was clearly shown by Donner (1953) in his early recordings from retinal ganglion cells in the pigeon eye. In the turtle *Pseudemys scripta elegans*, microspectrophotometric measurements revealed three different visual pigments of the porphyropsin type with maximal absorption at 450, 518 and 620 nm (Liebman and Granda, 1971). The oil droplets alter the shape of the cone spectral sensitivity functions as shown in electrophysiological measurements (Baylor and Hodgkin, 1973). According to Lipetz (1985), clear oil droplets are combined with the short-wavelength photopigment, yellow oil droplets with the mid-wavelength one, and oil droplets of as many as three different colours – orange, red and pale green – are found in cones with the long-wavelength photopigment. Thus, between 400 and 700 nm there are 5 different cone types. Does this mean that colour vision is pentachromatic in this range? The model computations analysing the $\Delta\lambda$ function of *Pseudemys* led us to the conclusion that the turtle uses only *one* long-wavelength cone type: the result from one turtle could be explained by assuming a combination with red oil droplets (dark symbols in Fig. 13.2), that from the other one by assuming orange oil droplets (open circles). Thus, the long-wavelength cones with different oil droplets seem not to be processed separately, so that between 400 nm and 700 nm there are only three cone types which play a role in colour vision. However, to explain the high discrimination ability in the range of 400 nm, we had to assume that there is an additional photoreceptor maximally sensitive in the ultraviolet range (Arnold and Neumeyer, 1987). The colour mixture experiments performed were the same as described above for the goldfish, and they indicated that the turtle must have an ultraviolet-sensitive cone type in addition. Till now, such a UV receptor has not been identified in turtles. Despite the fact that we did not perform extensive colour mixture experiments in other spectral ranges, we are fairly sure that the turtle *Pseudemys* has a tetrachromatic colour vision system similar to that found in goldfish.

In the pigeon, the situation is similar. Here, three visual pigments of the rhodopsin type were identified with maximal absorption at 460, 514 and 567 nm which are combined with different oil droplets in such a manner that there are three cone types containing the 567 nm pigment which have different maximal sensitivities between 567 and 620 nm (Bowmaker and Knowles, 1977). To find out whether all three long-wavelength cone types are relevant for colour vision, Palacios et al. (1990) performed colour mixture experiments in the range between 580 and 640 nm. However, the result was not clear, as it could be explained by assuming two as well as three cone types. It seems possible to explain the $\Delta\lambda$ function of the pigeon shown in Fig. 13.3 by three cone types with the maximal effective sensitivity at 485, 575 and 619 nm as proposed by Bowmaker and Knowles (1977), and an ultraviolet or violet receptor type in addition, as shown by Govardovskii and Zueva (1977).

In other birds, the situation seems to be less complicated. Behavioural investigations in the jackdaw, *Corvus monedula* L., revealed four maxima of spectral sensitivity at about 420, 480, 550 and 620 nm, which may reflect four cone types as a basis of a tetrachromatic colour vision (Wessels, 1974). Four maxima were also found in the spectral sensitivity function of a passerine bird, the Pekin robin, *Leiothrix lutea* (Burkhardt and Maier, 1989). They were located at 380, 470, 530 and around 600 nm. Absolute maximal sensitivity was highest in the ultraviolet, and decreased with longer wavelengths in the ratio 10 : 2.5 : 2 : 1. The presence of four types of cones was also shown by recording transretinal voltages in eye-cup preparations of a variety of passerine birds (Chen and Goldsmith, 1986). Sensitivity maxima were found at 370, 450, 480 and 570 nm. In the duck (*Anas platyrhynchos* L.), four different photopigments were shown in microspectrophotometric examination which were absorbing maximally at about 420, 450, 502 and 570 nm (Jane and Bowmaker, 1988). Summing up all these findings, it seems that tetrachromatic colour vision is rather widespread among diurnal birds. So far, there is no evidence that colour vision in the species studied is higher-dimensional. Tri- or dichromatic colour vision systems have not yet been found.

Mammals

Earlier it was doubted that colour vision is a widespread visual capacity in non-primate mammals (Walls, 1942). Thanks to careful studies, mainly by Gerald Jacobs and his coworkers, our knowledge about colour vision in quite a number of species is now well founded (for review of the older literature see Jacobs, 1981). In all non-primate species studied (with one exception, the rat; see below) colour vision was found, and was characterized as dichromatic, based on two types of cones. Dichromatic colour vision systems are characterized by U-shaped $\Delta\lambda$ functions as shown in Fig. 13.4, and by the existence of a 'neutral' point in the spectral range. Both properties can be understood on the basis of the underlying two cone types, as indicated in Fig. 13.9. The absorption spectra shown in Fig. 13.9(a) have maxima at 433 nm and at 550 nm. The colour space determined by the excitation values of the two cone types is two-dimensional, while 'hues' can be represented on a line as shown in Fig. 13.9(b). Here, the spectral loci for loci between 400 and 460 nm nearly coincide in a very narrow range, as well as the loci for wavelengths between 540 and 700 nm. Thus, wavelength discrimination is possible only in the range in between. If one assumes that the 'white' light of a xenon-arc lamp stimulates the two cone types in equal ratios, this would also be the case for one single wavelength around 480 nm. This wavelength should not be discriminated from 'white' light. The spectral locus of this wavelength marks the 'neutral' point of the spectrum.

Colour vision in American squirrels has been especially well studied (Jacobs, 1981). This group is interesting from the evolutionary point of view as it includes diurnal as well as nocturnal species. Ground-dwelling squirrels have their 'neutral' point in the range around 505 nm (Jacobs, 1978; see also Fig. 5.1 in Jacobs, 1981). In the same range, wavelength discrimination was best as shown in Fig. 13.4 for the grey squirrel (Jacobs, 1976). The two cone photopigments seem to have maximal absorption at 530–540 and at about 450 nm. In another rodent, the rat, the situation was rather unclear (see Jacobs, 1981), until a re-examination showed that this animal obviously does not have any colour vision (Neitz and Jacobs, 1986a). Until now, the rat has been the only animal for which the non-existence of colour vision could be demonstrated. Other examples on the list in Jacobs (1981, Table 5.3) await re-examination. For the pig, this was done recently, and a

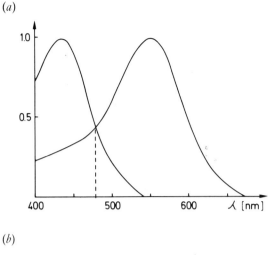

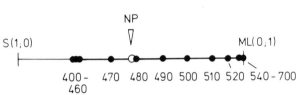

Fig. 13.9 *Dichromatic colour vision as found in mammals. (a) Spectral sensitivity functions of the two cone types (similar to those in dichromatic squirrel monkeys). The broken line represents an equally strong excitation of the two cone types. (b) One-dimensional line on which all 'hues' can be represented. S: point for all colour stimuli which excite the short-wavelength cone exclusively (excitation ratio: 1 : 0). (No real colours are found that are represented by this point because of the overlap of the two cone sensitivities in the short-wavelength range.) ML: corresponding point for the second cone type (excitation ratio: 0 : 1). Filled circles: loci of the spectral colours. Open circle and arrow: 'neutral' point (NL) of the spectrum that appears in the same hue as 'achromatic' light.*

dichromatic colour vision was found (Neitz and Jacobs, 1989). The domestic cat probably possesses colour vision, as it can be inferred from electrophysiological data that there must be at least two cone types. But until now it has not been possible to characterize it. The reason is that it is surprisingly difficult to train a cat on a colour stimulus, while there is no problem with other stimuli like shape. It seems that, as stated by Jacobs (1981), this animal is more 'predisposed to operate on cues other than colour', or that 'its colour vision is not highly developed'. It is odd that no-one took a closer look at the dog for many years. Only recently has it been shown convincingly that dogs have a dichromatic colour vision with a maximal wavelength discrimination ability (smallest $\Delta\lambda = 3$ nm), and a 'neutral' point at 480 nm (Neitz, Geist and Jacobs, 1989). It is probably based on two visual pigments with maximal absorption at 429 and 555 nm. Similar cone types probably underlie the colour vision of the tree shrew *Tupaia belangeri*. This animal has attracted much interest in visual research, as it is generally regarded as a very primitive primate. However, its actual taxonomic status is still disputed; there are many morphological and behavioural characters pointing to a position closer to primitive insectivores (for the present state of discussion, see Martin, 1990). The colour vision of the tree shrew is clearly dichromatic with a neutral point at 505 nm (Jacobs and Neitz, 1986).

In prosimian primates, the colour vision of the ringed-tailed lemur *Lemur catta* was investigated (Blakeslee and Jacobs, 1985). Here the situation was not entirely clear, and only one cone type could be identified with maximal sensitivity at 543 nm. The New World monkeys are especially interesting because within the individuals of the same species (shown in the squirrel monkey *Saimiri sciureus*, and in the tamarin *Saguinus fuscicollis*) there are significant variations in colour vision (Jacobs, 1983, 1984, 1990; Jacobs and Neitz, 1985; Jacobs et al., 1987). Within each species, dichromatic as well as trichromatic colour vision was identified. As mentioned above, this difference is sex-related, with the males exclusively being dichromats and the females being either di- or trichromats (the ratio for the squirrel monkey is about 1 : 3). This was tested by measuring wavelength discrimination (the functions show one or two minima; Fig. 13.5), by determining the neutral points and by performing colour mixture experiments. Furthermore, there is variation within the groups of dichromatic and trichromatic animals, respectively, which is based on a polymorphism of the cone photopigments. Whereas the short-wavelength visual pigment seems to be the same in all animals, there are three photopigment types in the mid- and long-wavelength range with maximal absorption at 545 nm, 557 nm and 562 nm. From these each individual has either one (when it is a dichromat) or two (in the case of a trichromat), as revealed in behavioural experiments and microspectrophotometry (Mollon et al., 1984; Jacobs et al., 1987). It is assumed that the genes for these different photopigments are alleles of the same gene located at the X-chromosome.

Colour vision in Old World monkeys seems to be exclusively trichromatic. As summarized in Jacobs (1981), all wavelength discrimination functions known so far have two minima of the type shown in Fig. 13.5 and 13.6. Cone photopigments were investigated in macaques and some other species and have maximal absorption at 430, 535 and 565 nm (Bowmaker et al., 1978; Hárosi, 1987; Baylor et al., 1987).

That human colour vision is trichromatic was suggested already by Thomas Young (1802), and was shown in colour mixture experiments performed by Maxwell and

Helmholtz in the middle of the last century. Compared with this knowledge, based on psychophysical experiments, the three cone types were identified only recently by analysing the absorption spectra of the photopigments (Marks *et al.*, 1964; Rushton, 1965; Bowmaker and Dartnall, 1980), which revealed maximal absorption at 420, 534 and 563 nm. In highly refined electrophysiological recordings from single mid- and long-wavelength cones, maximal sensitivities were found at 530 and 560 nm (Schnapf *et al.*, 1987). However, there are considerable variations, which suggests a polymorphism of photopigments in humans also (Dartnall *et al.*, 1983; Neitz and Jacobs, 1986b).

Invertebrates

The colour vision system of the honeybee is clearly trichromatic. This can be inferred from the wavelength discrimination function shown in Fig. 13.7, together with the results of colour mixture experiments by Daumer (1956). Three spectral types of retinula cells were found in electrophysiological recordings (Autrum and von Zwehl, 1964; Menzel *et al.*, 1986) with maximal sensitivity at 335, 435 and 540 nm. In a trichromatic colour vision, each colour stimulus can be represented by a point in a two-dimensional plot of the colour triangle shown in Fig. 13.10. Here, each point stands for the *ratio* in which the three receptor types are excited, and represents 'hue'. The colour triangle was calculated on the basis of the three spectral sensitivity functions shown in Fig. 13.10(a). The results of the colour mixture experiments by Daumer (1956), which provide evidence for the honeybee's trichromacy, can be inferred from Fig. 13.10(b). Here, the colour loci of the wavelengths on which the bees were trained (triangles) are shown, together with the loci of the mixtures (marked by stars) which the bees confused. The comparison reveals that the corresponding loci come very close, which indicates that the assumption of three receptor types underlying colour vision is valid. To match 'white' light, so-called 'bee-white', containing ultraviolet, it was necessary to mix three monochromatic lights in a certain proportion ('white' = 15% 360 nm + 30% 440 nm + 55% 590 nm). For a similar representation of Daumer's data see Menzel and Backhaus (1989, Fig. 13.2). Other hymenopteran insects also seem to have a trichromatic colour vision. The ant *Cataglyphis bicolor* was suspected of even four types of retinula cells on the basis of behavioural training experiments (Kretz, 1979). However, colour mixture experiments were never performed, and in electrophysiological recordings only two types of photoreceptors could be identified (Labhart, 1986). In butterflies, however, at least four retinula cell types seem to exist. Long-wavelength photopigments with maximal absorption at about 610 nm were shown in different species (Bernard, 1979). This was very surprising, as rho-

(a)

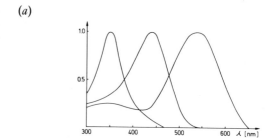

(b)

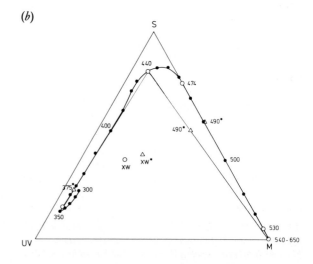

Fig. 13.10 *Trichromatic colour vision in the honeybee. (a) Spectral sensitivity functions of the retinula cells of the honeybee (From Neumeyer, 1980). (b) Colour triangle calculated on the basis of the functions in (a), which were weighted by the ratio 5.6 : 2.6 : 1 (UV) : S : M). Filled circles: loci of the spectral colours. Open circles: spectral colours used by Daumer (1956) in his colour mixture experiments. Triangles: loci of the colour mixtures which gave a match with the spectral colours marked by the star. For example: wavelength 490 nm was matched by the bees with an additive mixture of 440 nm (40%) and 540 nm (60%). 490 nm was also confused with a mixture of 474 nm (55%) and 530 nm (45%). (Note that point 490* is not located in the middle between 474 and 530 because of the weighting of the functions.) Wavelength 375 nm was for the bee equal to the mixture from 360 nm (80%) and 440 nm (20%). Xenon-'white' (XW) could be matched with a mixture from 360 nm (15%), 440 nm (30%) and 588 nm (55%).*

dopsin photopigments with maximal absorption at longer wavelengths than 560–580 nm were thought not to exist. In recordings from retinula cells in a Japanese swallowtail butterfly, *Papilio xuthus*, even five types of photoreceptors were identified with maximal spectral sensitivities at 360, 400, 460, 520 and 600 nm (Arikawa *et al.*, 1987). It seems to be unclear whether they contain five different photopigments or whether two of them are due to filter effects of substances within the retinula cells or due to a screening by other receptor types. Behavioural evidence for tetra- or pentachromacy is still lacking. The most elaborate recent behavioural study was performed on the butterfly *Pieris brassicae* L. by Scherer and Kolb (1987). Here, action spectra for spontaneous behavioural reactions were measured, as all attempts to train the butterflies on colour stimuli were entirely unsuccessful. For the 'feeding reaction' (landing and probing the stimulus with unrolled proboscis), a maximum of the action spectrum was found at 447 nm with a side maximum at 600 nm. 'Drumming' (rapid movements with the first leg pair shown before egg-laying), and 'egg-laying' (induced by a certain chemical) had a maximum of their action spectra at 558 and 542 nm, respectively. The 'open-space' reaction was induced by ultraviolet light (maximal reaction at 370 nm) and by white light (including UV). All four action spectra were rather narrow, and resemble the sensitivity functions measured by Paul *et al.*, (1986) and by Arikawa *et al.* (1987). Each of the four behavioural reactions of the butterfly seem to get input from only one or two of the different photoreceptor types. Whether the information of the four photoreceptor types is processed in such a manner that a true tetrachromatic colour vision results is still an open question. That there must be a comparison between the output of the different retinula cells was indicated by the fact that a stimulation of the short- and long-wavelength retinula cells as such is not sufficient to elicit feeding behaviour: white light is not effective. The same holds for the mid-wavelength retinula cells and the egg-laying response.

In the retinae of jumping spiders, the photoreceptors are arranged in four layers (Land, 1969). It is not known whether they are of different spectral types, or whether the spiders possess colour vision. Highly exciting is a more recent finding in crustaceans. Here, in two species of gonodactyloid stomatopods (mantis shrimps) at least 10! different photoreceptor types were found in the spectral range between 400 and 700 nm which contain different photopigments (Cronin and Marshall, 1989). Does this mean that these shrimps have a decachromatic colour vision?

Phylogeny of Colour Vision

Gordon Walls (1942; Fig. 156) presented a hypothetical phylogenetic tree of colour vision in vertebrates, which is shown in an updated version in Fig. 13.11. Here, only those colour vision systems are indicated that have been clearly characterized, whereas in Walls' scheme they were marked as 'being present', or as 'possibly' or 'perhaps' being present. As shown in Fig. 13.11, there are tetrachromats (and probably also tri- and dichromats) in teleost fishes, in turtles and in birds. Trichromats were found in urodeles and in primates, and dichromats in some primates, and in non-primate mammals. Differing from Walls's presentation, colour vision in lizards is not marked, as it was certainly shown to exist by Wagner (1933) but was not characterized in more detail. It seems to be at least trichromatic. Two new branches can be added to Walls's colour vision tree: that of Urodeles where a trichromatic colour vision was shown recently in *Salamandra salamandra*, and that of higher placentals where colour vision was characterized as dichromatic in various species. For all other branches 'down' to primitive chondrosteans or further, it may be assumed that colour vision exists at least in its dichromatic expression. Exceptions may be found in species which are strictly nocturnal (as the rat *Rattus norvegicus*) or living in dark environments, and which are probably 'younger' from the evolutionary point of view. Dichromatic forms of colour vision have perhaps been invented several times during the evolution of vertebrates. In their oldest expressions, they were probably based on two-cone photopigments with maximal absorptions in the short-wavelength range, around 430–450 nm, and in the mid-wavelength range, between 520 and 580 nm. All colour vision systems which could be clearly characterized (see above) used photopigments with absorption maxima in these ranges (solely or in addition to others). But there may be also exceptions, such as the possible dichromats in teleost fishes living in fresh water, which have two photopigments maximally absorbing in the middle- and long-wavelength range (Loew and Lythgoe, 1978).

The course of evolution from dichromatic and trichromatic realizations of colour vision can be traced to some extent in primates, by comparing the different colour vision systems of Old and New World monkeys, and the genes coding for the different opsin molecules of human cone photopigments. While the gene coding for the human rod photopigment is very similar to the gene for the bovine rod pigment analysed earlier, all three human cone opsins show a correspondence of only about 40% with the rod photopigment (Nathans *et al.*, 1986; Nathans, 1987). The correspondence of the genes for the human short-wavelength cone opsin with that for the mid- and the long-wavelength opsins was also about 40%. However, there was a very high similarity of 96% when the mid- and long-wavelength opsins were compared. As they are located on the same branch of the X-chromosome, it was concluded that they originate from the same gene by duplication. However, in the New World monkeys the behavioural analysis of colour vision indicated that the genes coding for the mid- and long-wavelength opsins are alleles of the same gene (Jacobs and Neitz, 1985). Three alleles are known so far; two of them can be expressed in females (located on each of the two X-chromosomes), but

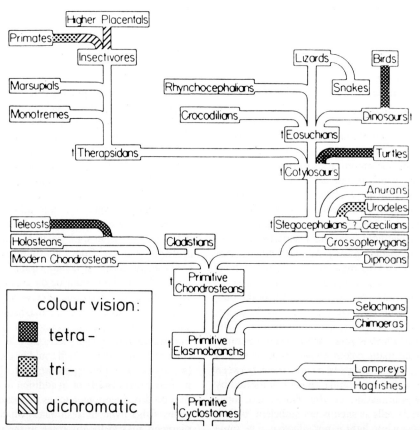

Fig. 13.11 *Gordon Walls' phylogenetic tree of colour vision in vertebrates, updated (After Walls, 1942; Fig. 156). Here, only the clearly characterized colour vision systems with known dimensionality are marked. Each branch is marked according to the highest dimensionality found, which does not imply that there are not also colour vision systems with lower dimensionality.*

only one in males, which explains the phenotypes of di- and trichromatic colour vision found in the two sexes (Jacobs and Neitz, 1985; 1987). As the geographic separation of New and Old World monkeys occurred about 30–40 million years ago, it may be assumed that the duplication event and the generation of two separate genes for the mid- and long-wavelength photopigments occurred after that time (Nathans et al., 1986).

To get more general insight into the evolution of colour vision in vertebrates, it would be highly interesting to compare the human cone opsins at least with those found in cyprinid fishes. Are the genes coding for mid- and long-wavelength opsins similarly recent aquisitions, or are they rather different in fishes, indicating an old age? What about the ultraviolet cone photopigment? Is it closely related to the short-wavelength one? What is the relationship between short-wavelength cone photopigments in different vertebrate classes, and between mid-wavelength ones? Are they derived from a common ancestor?

The hypothetical scheme for the evolution of the photoreceptor cells given by Walls (1942, Plate I) assumes one common ancestor with one type of cone from which a diversification into rods and cones, as well as into different cone types, could have proceeded. Apart from the question of whether a monophyletic origin for all vertebrates is reasonable to assume, this scheme is interesting as it emphasizes that double cones, and also oil droplets, are present already in the 'oldest' vertebrates. Double cones – their functional significance in colour vision seems still not to be clear – are found in all classes of vertebrates and are missing only in the higher placentals. Oil droplets, which play an important role in the colour vision of birds and reptiles, also seem to be a very old acquisition. In 'primitive' fishes such as sturgeons in which only cone-like photoreceptors are found, the oil droplets are clear (L. Zueva, personal communication). Their function is probably to increase quantum catch, as they bundle the light into the outer segment of the cones (Ives et al., 1983).

For some invertebrates, such as *Drosophila melanogaster*, the genes coding for photopigments were also analysed (see the appropriate chapters of this volume).

Comparative Colour Perception

The Peculiarities of Human Colour Vision

Colour perception is the result of neuronal processing in the peripheral and central nervous system. The human brain assigns sensations of colour to the objects in the 'outer world', so that 'colours' are experienced like physical properties. Accordingly, our thinking about 'colour' vision in animals is guided by our own introspection, and we assume that the world appears to them in the same way as to us. It was Jakob von Uexküll (1928) who clearly stated that each animal lives in its own subjective world (its 'Umwelt'), determined by its senses and its motoric actions. Keeping this in mind, one has to be very careful in extrapolating our sensations to those that animals may have. In this context, one should be aware of the specific characteristics of human colour perception. There are four colours that are regarded as 'pure' or 'unique' colours, and are named as 'blue', 'green', 'yellow' and 'red', as well as 'white' and 'black', which are perceived as 'achromatic' or 'neutral'. These six sensations seem to be universal for all human beings, which can be concluded from the fact that in different languages there are 'basic colour terms' used to name them (Berlin and Kay, 1969). They have been invented exclusively to designate colours, in contrast to other terms which are derived from names for objects, such as 'purple', 'orange', 'violet', 'turquoise' and so on. The 'unique' or 'pure' colours can be found in the spectrum by asking questions like: is this green a 'pure' green, or does it contain traces of yellow or blue? The ranges in which 'pure' or 'unique' colours are named are rather narrow, and their location differs slightly between different individuals. 'Derived' colour terms, on the other hand, are used for 'mixed' sensations; thus 'orange' designates a colour which is 'yellow' as well as 'red'. The unique colours are organized in pairs of 'opponent colours': red and green, blue and yellow, which exclude each other perceptually (as was stated explicitly for the first time by Ewald Hering, 1920). This means that it is impossible to imagine 'bluish-yellow' or 'greenish-red'. A transition from yellow to blue is perceptually possible only via green by making yellow successively greenish and green bluish, or via unsaturated shades of blue and yellow which become finally 'white'. For the pair black and white, 'opponency' is also assumed, despite the fact that there are transitions which we call 'grey'.

These peculiarities of human 'opponent' colour perception are the result of a specific neuronal information processing, and cannot be explained on the receptor level. This becomes very obvious if one reflects on a presentation of the colour triangle in which the colours are painted (for example in Cornsweet, 1970, Fig. 10.3). It is perhaps not so surprising that colour stimuli exciting one of the three cone types maximally (three corners of the colour triangle) elicit three different sensations: red, green and blue. But it is very strange that colour stimuli exciting the mid- and the long-wavelength cones equally strongly give rise to a sensation of 'yellow', and not to a sensation of 'red-green', as in the other cases where we find 'blue-green', and 'purple'. The second peculiarity is the fact that colour stimuli which excite all three cone types simultaneously about equally strongly, located in the middle of the colour triangle, do not have any colour character, but are perceived as 'white' or 'neutral', and not as 'blue-green-red'. However, the 'opponent' property of sensations that exclude each other, and the existence of 'complementary colours', are only consequences of the existence of 'yellow' and 'white', and can be understood in the colour triangle: here, a straight line between 'red' and 'green' traverses 'yellow', and a line connecting 'blue' and 'yellow' traverses 'white', whereas a line between 'green' and 'blue' indicates a 'direct' transition called 'blue-green'.

Colour Perception in Animals

It is obvious that statements such as 'this animal is able to discriminate between red and green' can have only the meaning that the animal is able to discriminate between colours which appear to *us* 'red' and 'green'. It is probably reasonable to assume that colour stimuli which excite maximally each of the different photoreceptor types give rise to different perceptions. However, it is not self-evident that an animal perceives 'white' as 'neutral' or 'achromatic'. The experimental findings of 'neutral' points in the spectrum as in dichromatic systems, or of 'complementary colours' (which are defined as colours which are in an additive mixture equal to 'white'), indicate only that these colour stimuli have the same effect as a broad spectral distribution such as sunlight, which appears to *us* as 'achromatic'. However, this can be understood already on the basis of the photoreceptor types underlying colour vision, as both stimuli give rise to the same excitation ratio. That there is 'white' in the sense of 'neutral' or 'achromatic' has to be shown experimentally, which is a very difficult task. The only experiment from which it can perhaps be inferred that 'white' is perceived as 'achromatic' was performed in the honeybee (Menzel, 1981). In this experiment, the bees showed the same response to monochromatic light as to 'white' light of a xenon-arc lamp, when the former was presented at intensities below a certain level. The two stimuli were equally answered despite the fact that they caused a different excitation ratio of the receptor types. This result can be understood on the assumption that the monochromatic light was perceived under these conditions as 'colourless' in the same way as the 'achromatic' xenon-'white' light. The other important question, still open in all (non-human) colour vision systems, is how many categories of hue the perceived colours can be divided into. Does the number of 'unique' colours exceed the number of receptor types involved? Is there, for example, an analogue for the (human) 'yellow'? Are there other 'unique' colours, perhaps located near 500 nm, or around 400 nm? To answer such questions, experiments are required which test the degree of generalization in a similar manner as in the ingenious training experiments performed by Wright and Cumming (1971) with pigeons (see also the detailed description in Jacobs, 1981, p 117). The results indicate that there are two points in the spectral range between 450 and 650 nm at which a transition from one type of perceived hue to another type takes place: at 540 and at 595 nm. The transition point at 540 nm does not have an analogue in human colour vision, where one point, the transition between 'blue' and 'green', is located at about 495 nm. In humans, there are two transition points in the long-wavelength range: one between 'green' and 'yellow' at 565–585 nm, and one between 'yellow' and 'red' at 610 nm. Whether the transition point in pigeons at 595 means that the pigeons' perceived hue changes into an analogue of 'yellow', or whether it indicates a direct transition from 'green' into 'red', cannot be decided on the basis of these experiments.

In all colour vision systems in which an ultraviolet photoreceptor is involved, like those of honeybees, goldfish, turtles and birds, we are entirely lost when we try to image how 'ultraviolet' may appear. But it seems reasonable to assume that there is a specific colour sensation elicited by ultraviolet light, qualitatively different from the others.

For all animals in which colour vision could be investigated with a training technique using food rewards, it may be assumed that there is an 'internal representation' of colour which is in principle similar to our own. Another possibility would be that the outputs of the different photoreceptor types are more 'directly' connected with behaviour. This could be the case in the above-described example of the butterfly *Pieris brassicae* L., where the action spectra of different behavioural responses indicated the influence of different single-receptor types. (This is similar to the situation in the honeybee where the optomotor response and the orientation to polarized light are determined by one receptor type only.) However, the findings in the butterfly by Scherer and Kolb (1987) do not exclude the possibility that there are 'real' colour sensations.

The Significance of Opponent Colour Coding

To explain the 'opponent' character of human colour vision, Ewald Hering (1920, see in Wasserman, 1978, p 90) formulated the hypothesis that each of three photoreceptors may assume two opposite metabolic states: dissimilation and assimilation, depending on the spectral composition of incident light. The result of early recordings from the retina of a cyprinid fish by Svaetichin (1953) fitted nicely into this idea, as it was found that the same cell responded with depolarization to light of one spectral range, but with hyperpolarization to light of another range. These S-potentials were later shown to arise in horizontal cells, whereas the cones respond only with hyperpolarization (Tomita, 1963; 1965). So-called colour-opponent cells are found in the retinae of all lower vertebrates at the level of the horizontal cells, and at the level of ganglion cells in primates. As there are three response types in primates, characterized as red/green, blue/yellow, and one spectrally non-opponent type (see for review Zrenner, 1983), it is very tempting to relate them directly to the perceived opponent colours. That colour-opponent neurones play an important role in the processing of colour-specific information may be inferred from the fact that they are found not only in vertebrates but also in the honeybee (for review see Menzel and Backhaus, 1989). However, it may be doubted that the existence of colour-opponent neurones implies that colour perception is realized as an opponent-colour system. Instead, it can be assumed that they have primarily a filter-function, narrowing cone spectral sensitivity (Gouras and Zrenner, 1981, p 164). This was also supposed in the goldfish to explain the specific shape of the behaviourally measured spectral sensitivity function (Neumeyer, 1984), as well as the wavelength discrimination function (Neumeyer, 1986).

The Question 'Why?'

Finally, we have to deal with Gordon Walls' most difficult but most interesting question about the reasons why colour vision was evolved, and why the specific realizations of colour vision were created during evolution. The question has many aspects, and probably we are aware only of some of them.

What Determines the Spectral Range Used for Colour Vision?

The range of the electromagnetic spectrum that can be used for vision is limited to the range between 300 and

800 nm. Radiation of shorter wavelengths damages proteins and nucleic acids severely in all living cells, owing to the process of photo-oxidation as a result of the interaction between oxygen and light. In photoreceptor cells, photo-oxidation is already a limiting factor in the near-ultraviolet range between 300 and 400 nm, as it may destroy the membranes containing the photopigments (Zhu and Kirschfeld, 1984). There are several strategies to prevent ultraviolet light from reaching these structures, the acquisition of coloured oil droplets being only one (Kirschfeld, 1982). In the long-wavelength spectral range the limiting factor seems to be mainly the thermal isomerization of the photopigments, which increases photoreceptor noise (Barlow, 1957; Aho et al., 1988; Barlow, 1988). So far, no photopigments (of the porphyropsin type with dehydroretinal as the prosthetic group) are known with maximal absorption at wavelengths longer than 620 nm, while rhodopsin pigments are maximally absorbing only up to 580 nm. Colour vision systems based on rhodopsin pigments were mainly evolved in land animals (insects, amphibia, reptiles, birds and mammals) and in marine fishes and turtles. While in land-living species temperature may be the decisive factor in making rhodopsin more advantageous than porphyropsin, the reduced transmittance of water in the long-wavelength range may be the reason in marine species.

The physical limits of photo-oxidation and thermal isomerization seem to determine the outer boundary of the visible range an animal can maximally cover. Whether an animal will have an ultraviolet receptor type depends probably on its ability to keep the costs for repairing damages due to photo-oxidation as small as possible. This can be achieved by the use of coloured oil droplets which absorb effectively ultraviolet radiation in all cone types maximally sensitive at longer wavelengths, so that repair is necessary only in UV receptors. In mammals there are no oil droplets, and the ultraviolet light is already filtered out by the optic media, mainly by the lens. So far, no ultraviolet receptors have been found. But also other colour vision systems do not use the entire, theoretically possible spectral range. The honeybee, for example, has a visible spectrum shortened at the long-wavelength side (at about 650 nm). In species living under water, the visible range might be determined mainly by the spectral absorption of water and the substances dissolved in it, and hence by the spectral 'window' in which light is transmitted.

What Determines the Dimensionality of Colour Vision?

Given a certain spectral range (between 400 nm and 700 nm), and a fixed half-bandwidth of the absorption spectra of visual pigments, an application of sample theory on human colour vision showed that *three* different photopigments are necessary and sufficient to transmit all spectral information (Barlow, 1982). For cases in which the half-bandwidths of the photoreceptors are narrowed by carotenoids that are either concentrated in oil droplets or diffusely distributed, such as in fly retinula cells (Hardie, 1985), Barlow predicted the invention of a fourth photoreceptor, and a tetrachromatic colour vision. Otherwise, the narrowing of cone spectral sensitivity functions would cause a loss of spectral information, which is indeed found in the turtle *Pseudemys*, as described above (Fig. 13.2). Here, wavelength discrimination was impossible in the range between 460 and 510 nm. However, no photoreceptor was added in this spectral range. (Whether the three different types of long-wavelength cones found in turtles and pigeons are significant for colour vision is not yet clear, as discussed above.) The impairment of wavelength discrimination due to oil droplets is probably selectively neutral, because colour vision has been evolved to discriminate object colours with a broad spectral distribution of reflected light and not spectral colours. For this task, colour vision systems with oil droplets may even be superior as they enlarge the photoreceptor space (Govardovskii, 1983; Govardovskii and Vorobyev, 1989). Tetrachromacy was also predicted from Barlow's theoretical analysis for the case that the visible range of the spectrum is extended into the ultraviolet (Barlow, 1986). This was found in cyprinid fishes, in turtles and in birds. However, the European tiger salamander does not have an ultraviolet receptor type, despite the fact that it is sensitive in this spectral range.

Regarding the above-mentioned cases of butterflies with 5 photoreceptor types in the range between 300 and 700 nm, and of mantis shrimps with at least 10 types between 400 nm and 700 nm (Arikawa et al., 1987; Cronin and Marshall, 1989), it is not known whether they have a 'real' colour vision. The sensitivity functions are rather narrow, which would be in line with Barlow's considerations. However, it is also possible that here an entirely different principle of colour vision is realized, in which spectral information is obtained not by comparing the output of the different photoreceptor types, but by a kind of frequency analysis like in hearing, with narrowly tuned receptors.

The two cone types found in many marine fishes (Loew and Lythgoe, 1978) are perhaps an adaptation to the specific light conditions with a more narrow spectral range; they are also in line with Barlow's analysis. However, why non-primate mammals have also only a dichromatic colour vision is not clear in this context. The same dichromatic, colour vision systems are found in species more active in twilight, which contains more energy in the short-wavelength range, and also in diurnal species.

What is the Selective Advantage of Having More Than One Photoreceptor Type?

The above discussion about why a certain number of photoreceptors may underly colour vision made the assumption that the visual system is optimized to extract spectral information from its environment. But is this really so important? Can't we do almost equally well when we see the world only in different shades of grey? Is it possible that the selective pressure to create a visual system with two or more simultaneously active photoreceptors has to be seen within another functional context? So did evolution of colour vision start perhaps from a 'pre-adaptive' level, in which two or more photoreceptor types had already been invented?

One reason to develop an additional photoreceptor type is simply to broaden the spectral sensitivity of the eye, and hence to increase quantum catch. Another idea, supported by detailed measurements of spectral distribution of light under water, and by microspectrophotometry of visual pigments of many species of teleost fishes, claims that two different photoreceptor types may have been evolved to optimize visual contrast (Lythgoe, 1968; Munz and McFarland, 1975; McFarland and Munz, 1975; Loew and Lythgoe, 1978; Lythgoe, 1979). For a fish swimming in bluish or greenish sea water, it is a question of survival to detect an object, whether prey or predator, from as large a distance as possible. It was pointed out that maximum visibility of objects under water requires two types of photoreceptors, one matched to and the other offset from the spectral distribution of background light against which the object has to be detected. Model computations performed recently have shown that the existing two photoreceptor types are optimal for the detection of contrast under the specific natural conditions of coastal water (Lythgoe and Partridge, 1991). Whether the fish living in this habitat are also able to discriminate wavelengths, and possess a dichromatic colour vision, has not yet been investigated.

For land-living animals, there is probably another severe disadvantage for having only one photoreceptor type, which was pointed out by von Campenhausen (1986). His argument takes into account the properties of object colours under natural conditions of illumination: firstly, biologically relevant objects have surfaces with a broad spectral reflectance; secondly, the incident natural light changes its spectral characteristics considerably during the day (Henderson, 1977; see also Wyszecki and Stiles, 1982, pp 145–146). This has the consequence that an animal with only one photoreceptor type cannot rely on contrast: an object with maximal reflectance in the mid- and short-wavelength range will reflect a high amount of quanta under a blue sky, but much less under a yellowish morning or evening illumination. The opposite will be the case when an object predominantly reflects in the long-wavelength range. Thus, the perceived lightness of an object is not constant, and the contrast between object and surround can be reversed (for example, a strawberry would appear darker than the leaves under one kind of natural daylight, but brighter in another). This was tested in an experiment in which human subjects had the task of classifying about 30 differently coloured papers when the papers were illuminated with yellowish or bluish light (von Campenhausen, 1986). As shown in Fig. 13.12, about the same sequence was obtained under 'normal' viewing conditions, but not when subjects were wearing dark goggles that transformed them from trichromats into (rod)-monochromats. Under this condition, classification was possible only by using the lightness cue.

The precondition for achieving lightness constancy, which must be of high selective advantage, is the existence of two photoreceptors differing in spectral sensitivity. The neuronal mechanisms which extract information about the spectral character in the visual field, and which allow a correction to a state in which the output of the two photoreceptors is about equally strong, is probably already sufficient to provide colour vision. But there is also the possibility predicted by von Campenhausen (1986) that vertebrates may exist with two cone types that show lightness constancy but no ability to discriminate colours. At the moment, the author's colleague Mark Tritsch is testing whether the cat is such an animal, as it is known (see above) that cats have more than one cone type, but nevertheless seem to be little interested in colour.

Colour Constancy

A compensation for the spectral changes of natural daylight is not only required to achieve lightness constancy; it is also necessary for perception of 'colour'. As explicitly stated by Hering (1920), an object would appear in different 'hues' under illuminations of different spectral distribution, and the selective advantage of using spectral information would disappear if there was no mechanism providing colour constancy. In human colour vision, colour constancy is a very powerful capacity, which has the effect of 'neutralizing' the colour of an illumination, so that the perceived hue of an object is always the same, despite the fact that the reflected light can be physically different. That perceived colour is not simply determined by the excitation ratio of the photoreceptor types, but has to be understood as the result of neural processing, was impressively demonstrated and provided with a theoretical framework by Edwin Land (e.g. Land and McCann, 1971; Land, 1983; 1986).

The necessity to compensate for the continuous

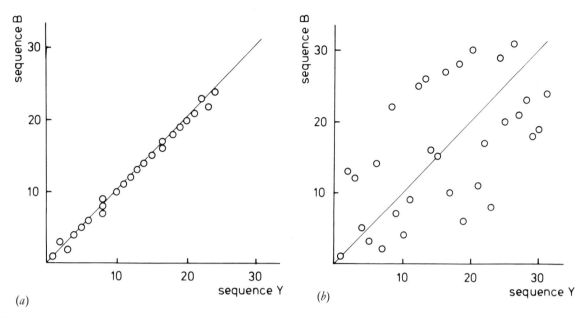

Fig. 13.12 *Perceptive sequence for a series of object colours under blue and yellow illumination. Abscissa: number in the sequence from the darkest (0) to the brightest colour (30) when illuminated with yellow light; ordinate: number in the sequence when illuminated with blue light. (a) Similarity sequences for a normal trichromat. (b) Lightness sequences for a rod monochromat. The points should fall on the line when the sequence is identical under the two illuminations. (After von Campenhausen, 1986.)*

changes in the spectral composition of natural daylight may be regarded not only as the driving force for the evolution of colour vision (as explained above); it is perhaps also the reason for the specific properties of (human) colour vision with 'white', 'complementary colours', 'colour circle' and 'colour contrast phenomena'. That these characteristics of human colour vision can be understood in a functional context with colour constancy was pointed out by Erich von Holst (1957). Von Holst approached the problem by reflecting on how an engineer would design an apparatus which corrects for eventual changes of illumination colour. For him, it is a precondition that sunlight does not have a colour property but is perceived as neutral. The colour of the illumination, which can only be derived from the light reflected from the objects, is assumed to correspond to that colour which is predominant in the visual field. This seems to be reasonable, as it is rather improbable that all objects in a natural surrounding will have the same colour. Then, a correction mechanism adds actively the complementary colour, to compensate the illumination colour to neutral. To ensure that this is possible for all colours, perceived colours have to be arranged in a colour circle.

Colour constancy must be of high selective value, and is a necessary condition for *each* colour vision system. In animals, it has to date been shown in the honeybee (Neumeyer, 1981; Werner *et al.*, 1988), and, only qualita- tively, in the goldfish (Ingle, 1985). The phenomena of simultaneous and successive colour contrasts which can be regarded as by-products of the mechanism providing colour constancy were also shown in honeybees (Neumeyer, 1980, 1981), and were even found in the fruit- fly *Drosophila melanogaster* by Fischbach (1979). 'Colour opponent' neurones of the type found in goldfish horizon- tal cells, or in the optic ganglia of the honeybee (see in Menzel and Backhaus, 1989), cannot account for these phenomena. Instead, neurones from the 'double- opponent' type are required, as was shown for the first time in the goldfish retina by Daw (1967).

Colour Vision and Coloration

Colour vision is adapted to the conditions of natural sky- light and the spectral reflectance of natural objects. It is surprising that the best discrimination abilities for wavelengths are found mainly in three narrow spectral ranges (Figs 13.1–13.7): around 400 nm, around 500 nm and between 570 and 600 nm. What might be the selective value of having the optima just there? Perhaps it has to do with the absorption spectrum of cholorophyll: it has two maxima, one around 450 nm and a second one at 630 nm (see for example in Kirk, 1983, Fig. 8.7) with flanks at 400, 500 and 600 nm. At least the ranges of best wavelength discrimination ability at 500 and 600 nm could have as their biological significance the optimal discrimination of colours just different from that of green leaves. (The UV-reflectance of chlorophyll in leaves seems to be reduced by the carotenoids.)

The colours of blossoms and fruit have to be regarded as the result of coevolution between plants and animal colour vision. Flowers, which have evolved to attract insects, very often have patterns which are invisible for us but can be seen by the honeybee (Daumer, 1958). For example, the petals of *Potentilla reptans* L., which look for us homogeneously yellow, are for the bee divided into an outer part reflecting yellow plus UV ('bee-purple'), and an inner zone near the stamen which reflects only yellow. UV patterns may also occur on flowers which are pollinated by birds. Furthermore, the colour 'character' may be entirely different: the red poppy *Papaver rhoeas*, for example, is for the bee ultraviolet, whereas the white narcissus *Narcissus poeticus* is 'bee-bluegreen' as it does not reflect in the ultra-violet range. As was shown by Burkhardt (1982), there are blue, black or green berries or fruits, which for us are sometimes hard to detect among the green leaves, but which may be highly visible for birds as they reflect in the ultraviolet range. White fruits, on the other hand, like those of the mistletoe *Viscum album*, do not reflect in the ultraviolet, and must appear to birds as coloured.

That the coloration of animals must be of high selective value was already discussed by Darwin, and many examples are given by Wallace (1889). Protection against predators can be provided either by a cryptic coloration, in combination with patterns, spots or stripes, which break up the outline of the body, or by very obvious colours, mostly red or yellow, which are used as warning signals. Now, as we know so much more about the colour vision systems of mammals, birds, reptiles, fishes and insects, it would be worthwhile to reinvestigate animal coloration under the aspect of coevolution between body colour and the colour vision systems of potential predators. Is it a specific predator for which an animal wants to be invisible, or is it of higher selective advantage to be cryptic for as many as possible?

Body coloration is also important in many species with good colour vision as a signal in communication. Furthermore, it may have the selective advantage of separating closely related species. Here, one has to take into consideration that animals with ultraviolet receptors may have patterns which are invisible to human eyes. This was shown to be the case in the plumage of birds (Burkhardt, 1989), and on the wings of butterflies (Meyer-Rochow and Eguchi, 1983).

Acknowledgements

For improving the readability of the English manuscript I thank Neil Beckhaus and Dr Mark Tritsch. I am also very grateful to Dr J. Schramme for calculating the tetrahedron.

References

Aho, A.-C., Donner, K., Hyden, C., Larsen, L. O. and Reuter, T. (1988). Low retinal noise in animals with low body temperature allows high visual sensitivity. *Nature Lond.*, 334, 348–350.

Arikawa, K., Inokuma, K. and Eguchi, E. (1987). Pentachromatic visual system in a butterfly. *Naturwissen.*, 74, 297–298.

Arnold, K. and Neumeyer, C. (1987). Wavelength discrimination in the turtle *Pseudemys scripta elegans*. *Vision Res.*, 27, 1501–1511.

Autrum, H. and Zwehl, V. von (1964). Spektrale Empfindlichkeit einzelner Sehzellen des Bienenauges. *Z. Vergl. Physiol.*, 48, 357–384.

Avery, J. A. and Bowmaker, J. K. (1982). Visual pigments in the four-eyed fish, *Anableps anableps*. *Nature Lond.*, 298, 62–63.

Avery, J. A., Bowmaker, J. K., Djamgoz, M. B. A. and Downing, J. E. D. (1982). Ultra-violet sensitive receptors in a freshwater fish. *J. Physiol. Lond.*, 334, 23P.

Barlow, H. B. (1957). Purkinje shift and retinal noise. *Nature Lond.*, 179, 255–256.

Barlow, H. B. (1982). What causes trichromacy? A theoretical analysis using comb-filtered spectra. *Vision Res.*, 22, 635–643.

Barlow, H. B. (1986). Why can't the eye see better? In *Visual Neuroscience*. eds. Pettigrew, J. D., Sanderson, K. J. and Levick, W. R. pp. 3–18. Cambridge: Cambridge University Press.

Barlow, H. B. (1988). The thermal limit to seeing. *Nature Lond.*, 334, 296.

Baylor, D. A. and Hodgkin, A. L. (1973). Detection and resolution of visual stimuli by turtle photoreceptors. *J. Physiol. Lond.*, 234, 163–198.

Baylor, D. A., Nunn, B. J. and Schnapf, J. L. (1987). Spectral sensitivity of cones in the monkey *Macaca fascicularis*. *J. Physiol. Lond.*, 390, 145–160.

Berlin, B. and Kay, P. (1969). *Basic Color Terms*. Berkeley: University of California Press.

Bernard, G. D. (1979). Red-absorbing visual pigment of butterflies. *Science Wash. DC*, 203, 1125–1127.

Blakeslee, B. and Jacobs, G. H. (1985). Color vision in the ringed-tailed lemur (*Lemur catta*). *Brain, Behav. Evol.*, 26, 154–166.

Bowmaker, J. K. (1990). Visual pigments of fishes. In *The Visual System of Fish*. eds. Douglas, R. H. and Djamgoz, M. B. A. pp. 81–107. London: Chapman & Hall.

Bowmaker, J. K. and Knowles, A. (1977). The visual pigments and oil droplets of the chicken retina. *Vision Res.*, 17, 755–764.

Bowmaker, J. K., Dartnall, H. J. A., Lythgoe, J. N. and Mollon, J. D. (1978). The visual pigments of rod and cones in the rhesus monkey, *Macaca mulatta*. *J. Physiol. Lond.*, 274, 329–348.

Bowmaker, J. K. and Dartnall, H. J. A. (1980). Visual pigments of rods and cones in a human retina. *J. Physiol. Lond.*, 298, 501–511.

Bowmaker, J. K. and Kunz, Y. W. (1987). Ultraviolet receptors, tetrachromatic colour vision and retinal mosaics in the brown trout (*Salmo trutta*): age-dependent changes. *Vision Res.*, 27, 2101–2108.

Burkhardt, D. (1982). Birds, Berries and UV. *Naturwissen.*, 69, 153–157.

Burkhardt, D. (1989). UV vision: a bird's eye view of feathers. *J. Comp. Physiol. A*, 164, 787–796.

Burkhardt, D. and Maier, E. (1989). The spectral sensitivity of a passerine bird is highest in the UV. *Naturwissen.*, 76, 82–83.

Cameron, N. E. (1982). The photopic spectral sensitivity of a dichromatic teleost fish (*Perca fluviatilis*). *Vision Res.*, 22, 1341–1348.

Campenhausen, C. von (1986). Photoreceptors, lightness constancy and color vision. *Naturwissen.*, **73**, 674–675.

Chen, D.-M. and Goldsmith, T. H. (1986). Four spectral classes of cone in the retinas of birds. *J. Comp. Physiol.*, **159**, 473–479.

Cornsweet, T. N. (1970). *Visual Perception*. New York: Academic.

Cronin, T. W. and Marshall, N. J. (1989). Multiple spectral classes of photoreceptors in the retinas of gonadactyloid stomatopod crustaceans. *J. Comp. Physiol. A*, **166**, 261–275.

Dartnall, H. J. A., Bowmaker, J. K. and Mollon, J. D. (1983). Human visual pigments: microspectrophotometric results from the eyes of seven persons. *Proc. R. Soc. Lond. B*, **220**, 115–130.

Daumer, K. (1956). Reizmetrische Untersuchungen des Farbensehens der Bienen. *Z. vergl. Physiol.*, **38**, 413–478.

Daumer, K. (1958). Blumenfarben, wie sie die Bienen sehen. *Z. vergl. Physiol.*, **41**, 49–110.

Daw, N. W. (1967). Goldfish retina: organization for simultaneous colour contrast. *Science Wash. DC*, **158**, 942–944.

Donner, K. O. (1953). The spectral sensitivity of the pigeon's retinal elements. *J. Physiol. Lond.*, **122**, 526–537.

Donner, K. O. and Reuter, T. (1976). Visual pigments and photoreceptor functions. In *Frog Neurobiology*. eds Llinas, R. and Precht, W. pp. 251–277. Berlin, Heidelberg, New York: Springer.

Emmerton, J. and Delius, J. D. (1980). Wavelength discrimination in the 'visible' and ultraviolet spectrum by pigeons. *J. Comp. Physiol.*, **141**, 47–52.

Fernald, R. D. and Liebman, P. (1980). Visual receptor pigments in the african cichlid fish, *Haplochromis burtoni*. *Vision Res.*, **20**, 857–864.

Fischbach, K. F. (1979). Simultaneous and successive colour contrast expressed in 'slow' phototactic behaviour of walking *Drosophila melanogaster*. *J. Comp. Physiol.*, **130**, 161–171.

Frisch, K. von (1913). Weitere Untersuchungen über den Farbensinn der Fische. *Zool. Jahrb. Abt. f. Zool. und Physiol.*, **34**, 43–68.

Frisch, K. von (1915). Der Farbensinn und Formensinn der Biene. *Zool. Jahrb., Abt. f. Zool. und Physiol.*, **35**, 1–188.

Goldsmith, T. H., Collins, J. S. and Perlman, D. L. (1981). A wavelength discrimination function for the hummingbird *Archilochus alexandri*. *J. Comp. Physiol.*, **143**, 103–110.

Gouras, P. and Zrenner, E. (1981). Color vision: a review from a neurophysiological perspective. In *Progress in Sensory Physiology*, vol. 1. ed. Ottoson, D. pp. 139–179. Berlin, Heidelberg, New York: Springer-Verlag.

Govardovskii, V. I. (1983). On the role of oil drops in colour vision. *Vision Res.*, **23**, 1739–1740.

Govardovskii, V. I. and Zueva, L. V. (1977). Visual pigments of chicken and pigeon. *Vision Res.*, **17**, 537–543.

Govardovskii, V. I. and Vorobyev, M. V. (1989). The role of colored oil drops of cones in the color vision. *Sensory Systems*, **3**, 150–159 (in Russian).

Hannover, A. (1840). Über die Netzhaut und ihre Gehirnsubstanz bei Wirbeltieren, mit Ausnahme des Menschen. *Müllers Arch. Anat. Physiol.*, **2**, 320–345.

Hardie, R. C. (1985). Functional organization of the fly retina. In *Progress in Sensory Physiology*, vol. 5. ed. Ottoson, D. pp. 1–79. Heidelberg: Springer.

Harosi, F. I. (1976). Spectral relations of cone pigments in goldfish. *J. Gen. Physiol.*, **68**, 65–80.

Harosi, F. I. (1986). Ultraviolet- and violet-absorbing vertebrate visual pigments: dichroic and bleaching properties. In *The Visual System*, eds. Fein, A. and Levine, J. S. pp. 41–55. New York: Alan R. Liss.

Harosi, F. I. (1987). Cynomolgus and rhesus monkey visual pigments. *J. Gen. Physiol.*, **89**, 717–743.

Harosi, F. I. and Hashimoto, Y. (1983). Ultraviolet visual pigment in a vertebrate: a tetrachromatic cone system in the dace. *Science Wash. DC*, **222**, 1021–1023.

Hawryshyn, C. W. and Beauchamp, R. (1985). Ultraviolet photosensitivity in goldfish: an independent u.v. retinal mechanism. *Vision Res.*, **25**, 11–20.

Helversen, O. von (1972). Zur spektralen Unterschiedsempfindlichkeit der Honigbiene. *J. Comp. Physiol.*, **80**, 439–472.

Henderson, S. T. (1977). *Daylight and Its Spectrum*. 2nd edn; Bristol: Hilger.

Hering, E. (1920). Grundzüge der Lehre vom Lichtsinn. In *Handbuch der gesamten Augenheilkunde*. eds. Graefe, A. von and Saemische, T. Berlin: Springer. Transl. Hurvich, L. M. and Jameson, D. (1964). Cambridge: Harvard University Press.

Holst, E. von (1957). Aktive Leistungen der menschlichen Gesichtswahrnehmung. *Studium Generale*, **10**, 231–243.

Ingle, D. J. (1985). The goldfish as a retinex animal. *Science Wash. DC*, **227**, 651–654.

Ives, T. J., Normann, R. A. and Barber, P. W. (1983). Light intensification by cone oil droplets: electromagnetic considerations. *J. Opt. Soc. Am.*, **73**, 1725–1731.

Jacobs, G. H. (1976). Wavelength discrimination in gray squirrels. *Vision Res.*, **16**, 325–327.

Jacobs, G. H. (1978). Spectral sensitivity and colour vision in the ground-dwelling sciurids: results from the gold-mantled ground squirrel and comparisons for five species. *Animal Behav.*, **26**, 409–421.

Jacobs, G. H. (1981). *Comparative Color Vision*. New York: Academic.

Jacobs, G. H. (1983). Within-species variations in visual capacity among squirrel monkeys (*Saimiri sciureus*): sensitivity differences. *Vision Res.*, **23**, 239–248.

Jacobs, G. H. (1984). Within-species variations in visual capacity among squirrel monkeys (*Saimiri sciureus*): color vision. *Vision Res.*, **24**, 1267–1277.

Jacobs, G. H. (1990). Discrimination of luminance and chromaticity differences by dichromatic and trichromatic monkeys. *Vision Res.*, **30**, 387–397.

Jacobs, G. H. and Neitz, J. (1985). Color vision in squirrel monkeys: sex-related differences suggest the mode of inheritance. *Vision Res.*, **25**, 141–143.

Jacobs, G. H. and Neitz, J. (1986). Spectral mechanisms and color vision in the tree shrew (*Tupaia belangeri*). *Vision Res.*, **26**, 291–298.

Jacobs, G. H. and Neitz, J. (1987). Polymorphism of the middle wavelength cone in two species of South American monkey: *Cebus apella* and *Callicebus molloch*. *Vision Res.*, **27**, 1263–1268.

Jacobs, G. H., Neitz, J. and Crognale, M. (1987). Color vision polymorphism and its photopigment basis in a callitrichid monkey (*Saguinus fuscicollis*). *Vision Res.*, **27**, 2089–2100.

Jane, S. D. and Bowmaker, J. K. (1988). Tetrachromatic colour vision in the duck (*Anas platyrhynchos* L.): microspectrophotometry of visual pigments and oil droplets. *J. Comp. Physiol. A*, **162**, 225–235.

Kirk, J. T. O. (1983). *Light and Photosynthesis in Aquatic Ecosystems*. Cambridge: Cambridge University Press.

Kirschfeld, K. (1982). Carotenoid pigments: their possible role in protecting against photooxidation in eyes and photoreceptor cells. *Proc. R. Soc. Lond. B.*, **216**, 71–85.

Kretz, R. (1979). A behavioural analysis of colour vision in the ant *Cataglyphis bicolor* (Formicidae, Hymenoptera). *J. Comp. Physiol..* **131**, 217–233.

Labhart, T. (1986). The electrophysiology of photoreceptors in different eye regions of the desert ant, *Cataglyphis bicolor*. *J. Comp. Physiol. A*, **158**, 1–7.

Land, E. H. (1983). Recent advances in retinex theory and some implications for cortical computations: color vision and the natural image. *Proc. Natl. Acad. Sci. USA*, **80**, 5163–5169.

Land, E. H. (1986). Recent advances in retinex theory. *Vision Res.*, **26**, 7–21.

Land, E. H. and McCann, J. J. (1971). Lightness and retinex theory.

J. Opt. Soc. Am., 61, 1–11.
Land, M. F. (1969). Structure of the retinae of the principal eyes of the jumping spiders (Salticidae: Dendryphantinae) in relation to visual optics. Exp. Biol., 51, 443–470.
Liebman, P. A. and Granda, A. M. (1971). Microspectrophotometric measurements of visual pigments in two species of turtle, *Pseudemys scripta* and *Chelonia mydas*. Vision Res., 11, 105–114.
Lipetz, L. E. (1985). Some neuronal circuits of the turtle retina. In *The Visual System*. eds Fein, A. and Levine, J. S. pp. 107–132. New York: Liss.
Loew, E. R. and Lythgoe, J. N. (1978). The ecology of cone pigments in teleost fishes. Vision Res., 18, 715–722.
Loew, E. R. and McFarland, W. N. (1990). The underwater visual environment. In *The Visual System of Fish*. eds Douglas, R. H. and Djamgoz, M. B. A. pp. 1–43. London; Chapman & Hall.
Lythgoe, J. N. (1968). Visual pigments and visual range underwater. Vision Res., 8, 997–1012.
Lythgoe, J. N. (1979). *The Ecology of Vision*. Oxford: Clarendon.
Lythgoe, J. N. (1984). Visual pigments and environmental light. Vision Res., 24, 1539–1550.
Lythgoe, J. N. and Partridge, J. C. Visual pigments of teleost dichromats in green coastal water. Vision Res., 31, 361–371.
Marks, W. B. (1965). Visual pigments of single goldfish cones. J. Physiol. Lond., 178, 14–32.
Marks, W. B., Dobelle, W. H. and MacNichol, E. F. Jr., (1964). Visual pigments of single primate cones. Science Wash. DC, 143, 1181–1183.
Martin, R. D. (1990). *Primate Origins and Evolution. A Phylogenetic Reconstruction*. London: Chapman & Hall.
McFarland, W. N. and Munz, F. W. (1975). Part III: the evolution of photopic visual pigments in fishes. Vision Res., 15, 1071–1080.
Menzel, R. (1979). Spectral sensitivity and color vision in invertebrates. In *Handbook of Sensory Physiology, vol. VII/A*. ed. Autrum, H. pp. 504–580. Berlin, Heidelberg, New York: Springer.
Menzel, R. (1981). Achromatic vision in the honeybee at low light intensities. J. Comp. Physiol., 141, 389–393.
Menzel, R., Ventura, D. F., Hertel, H., de Souza, J. and Greggers, U. (1986). Spectral sensitivity of photoreceptors in insect compound eyes: comparison of species and methods. J. Comp. Physiol. A, 158, 165–177.
Menzel, R. and Backhaus, W. (1989). Color vision in honey bees: phenomena and physiological mechanisms. In *Facets of Vision*. eds. Stavenga, D. G. and Hardie, E. C. pp. 281–297. Berlin: Springer.
Meyer-Rochow, V. B. and Eguchi, E. (1983). Flügelfarben, wie sie die Falter sehen — a study of UV and other colour patterns in Lepidoptera. Ann. Zool. Jpn, 56, 85–99.
Mollon, J. D., Bowmaker, J. K. and Jacobs, G. H. (1984). Variations in color vision in a New World primate can be explained by polymorphism of retinal photopigments. Proc. R. Soc. B., 222, 373–399.
Munz, F. W. and McFarland, W. N. (1975). Part I: presumptive cone pigments extracted from tropical marine fishes. Vision Res., 15, 1045–1062.
Nathans, J. (1987). Molecular biology of visual pigments. Ann. Rev. Neurosci., 10, 163–194.
Nathans, J., Thomas, D. and Hogness, D. S. (1986). Molecular genetics of color vision: the genes encoding blue, green and red pigments. Science Wash. DC, 232, 193–202.
Neitz, J. and Jacobs, G. H. (1986a). Reexamination of spectral mechanisms in the rat (*Rattus norvegicus*). J. Comp. Psychol., 100, 21–29.
Neitz, J. and Jacobs, G. H. (1986b). Polymorphism of the long-wavelength cone in normal human colour vision. Nature Lond., 323, 623–625.
Neitz, J. and Jacobs, G. H. (1989). Spectral sensitivity of cones in an ungulate. Vis. Neurosci., 2, 97–100.
Neitz, J., Geist, T. and Jacobs, G. H. (1989). Color vision in the dog. Vis. Neurosci., 3, 119–125.

Neumeyer, C. (1980). Simultaneous color contrast in the honeybee. J. Comp. Physiol., 139, 165–176.
Neumeyer, C. (1981). Chromatic adaptation in the honeybee: successive color contrast and color constancy. J. Comp. Physiol., 144, 543–553.
Neumeyer, C. (1984). On spectral sensitivity in the goldfish. Evidence of neural interactions between different 'cone mechanisms'. Vision Res., 24, 1223–1231.
Neumeyer, C. (1985). An ultraviolet receptor as a fourth receptor type in goldfish colour vision. Naturwissen., 72, 162–163.
Neumeyer, C. (1986). Wavelength discrimination in the goldfish. J. Comp. Physiol., 158, 203–213.
Neumeyer, C. (1988). *Das Farbensehen des Goldfisches*. Stuttgart: Thieme.
Neumeyer, C., Przyrembel, C. and Keller, B. (1988). The tiger salamander sees ultraviolet light without having an ultraviolet photoreceptor type. In *Sense Organs: Proceedings 16th Göttingen Neurobiology Conference*. eds Elsner, N. and Barth, F. G. p. 269. Stuttgart: Thieme.
Palacios, A., Martinoya, C., Bloch, S. and Varela, F. J. (1990). Color mixing in the pigeon. A psychophysical determination in the longwave spectral range. Vision Res., 30, 587–596.
Paul, R., Steiner, A. and Gemperlein, R. (1986). Spectral sensitivity of *Calliphora erythrocephala* and other insect species studied with Furier Interferometric Stimulation (FIS). J. Comp. Physiol. A, 158, 669–680.
Peiponen, V. A. (1964). Zur Bedeutung der Ölkugeln im Farbensehen der Sauropsiden. Ann. Zool. Fenn., 1, 1964, 281–302.
Przyrembel, C. and Neumeyer, C. (1990). Trichromatic colour vision in the tiger salamander. In *Brain — Perception and Cognition: Proceedings 18th Göttingen Neurobiology Conference*. eds Elsner, N. and Roth, G. p. 265. Stuttgart: Thieme.
Roth, G. (1987). *Visual Behaviour in Salamanders*. Berlin: Springer.
Rushton, W. A. H. (1965). The Newton Lecture: chemical basis of colour vision and colour blindness. Nature Lond., 206, 1087–1091.
Rushton, W. A. H. (1972). Pigments and signals in colour vision. J. Physiol., 120, 1–31P.
Scherer, C. and Kolb, G. (1987). Behavioral experiments on the visual processing of color stimuli in *Pieris brassicae* L. (Lepidoptera). J. Comp. Physiol. A, 160, 645–656.
Schnapf, J. L., Kraft, T. W. and Baylor, D. A. (1987). Spectral sensitivity of human cone photoreceptors. Nature Lond., 325, 439–441.
Svaetichin, G. (1953). The cone action potential. Acta Physiol. Scand., 29 (suppl. 106), 565–600.
Tomita, T. (1963). Electrical activity in the vertebrate retina. J. Opt. Soc. Am., 53, 49–57.
Tomita, T. (1965). Electrophysiological study of the mechanisms subserving color coding in the fish retina. Cold Spring Harbor Symp. Quant. Biol., 30, 559–566.
Uexküll, J. von (1928). *Theoretische Biologie*. Berlin: Springer. Frankfurt am Main: Suhrkamp. (1973).
van Dijk, B. W. and Spekreijse, H. (1984). Color fundamentals deduced from carp ganglion cell responses. Vision Res., 24, 211–220.
Wagner, H. (1933). Über den Farbensinn der Eidechsen. Z. vergl. Physiol., 18, 378–392.
Wallace, A. R. (1889). *Darwinism: an Exposition of the Theory of Natural Selection*. London. Reprint (1975) New York: AMS Press.
Walls, G. L. (1942). *The Vertebrate Eye and Its Adaptive Radiation*. Reprint (1967) New York: Hafner.
Wasserman, G. S. (1978). *Color Vision. An Historical Introduction*. New York: Wiley.
Werner, A., Menzel, R. and Wehrhahn, C. (1988). Color constancy in the honeybee. J. Neurosci., 8, 156–159.
Wessels, R. H. (1974). Tetrachromatic vision in the daw (*Corvus monedula* L.). Ph.D Thesis, University of Utrecht.
Wojtusiak, R. J. (1933). Über den Farbensinn von Schildkröten. Z.

vergl. Physiol., **18**, 393–436.

Wolff, H. (1925). Das Farbunterscheidungsvermögen der Ellritze. *Z. vergl. Physiol.*, **3**, 279–329.

Wright, A. A. and Cumming, W. W. (1971). Color-naming functions for the pigeon. *J. Exp. Anal. Behav.*, **15**, 7–17.

Wyszecki, G. and Stiles, W. S. (1982). *Color Science*. 2nd edn. New York: Wiley.

Yarczower, M. and Bitterman, M. (1965). Stimulus generalization in the goldfish. In *Stimulus Generalization*. ed. Mostofsky, D. J. pp. 179–192. Stanford: Stanford University Press.

Young, T. (1802). On the theory of light and colours. *Philos. Trans. R. Soc. Lond.*, 20–71.

Zhu, H. and Kirschfeld, K. (1984). Protection against photodestruction in fly photoreceptors by carotenoid pigments. *J. Comp. Physiol. A*, **154**, 153–156.

Zrenner, E. (1983). *Neurophysiological Aspects of Color Vision in Primates*. Berlin, Heidelberg, New York: Springer.

14 Uses and Evolutionary Origins of Primate Colour Vision

J. D. Mollon

Introduction

In his *Uncommon Observations About Vitiated Sight*, a collection of case reports published in 1688, Robert Boyle reproduces his notes on a gentlewoman whom he had examined some years earlier (Fig. 14.1). During an unidentified illness, apparently treated by fierce vesication, the young woman had transiently gone blind. Visual function had gradually returned, and Boyle satisfied himself that she could read from a book and had good acuity:

> ...having pointed with my Finger at a part of the Margent, near which there was the part of a very little Speck, that might almost be covered with the point of a Pin; she not only readily enough found it out, but shewed me at some distance off another Speck, that was yet more Minute....

But what was most singular and strange for Boyle was that she was left with a permanent loss of colour vision:

> ...she can distinguish some Colours, as Black and White, but it is not able to distinguish others, especially Red and Green....

Boyle's early case of acquired achromatopsia, and similar cases described in modern times (e.g. Meadows, 1974; Mollon *et al.*, 1980), bear on the celebrated question of whether there are submodalities within each of the major sensory modalities: if colour vision can be disproportionately impaired, while processing of spatial detail remains relatively intact, then we must suspect that there is some independence between the analysis of colour and the analyses of other attributes of the retinal image (Chisholm, 1869). But a second interest of Boyle's gentlewoman lies in the practical expression of her disability:

> ...that when she had a mind to gather Violets, tho' she kneeled in that Place where they grew, she was not able to distinguish them by the Colour from the neighbouring Grass, but only by the Shape, or by feeling them.

Her disability reminds us that we can best understand primate colour vision, and its evolution, by considering the price of not enjoying colour vision: what are the advantages that we should then forgo?

Advantages of Colour Vision

The Detection of Targets Against Dappled Backgrounds

In the current literature it is commonly suggested that colour vision serves particularly to detect edges between 'equiluminant' surfaces, surfaces of the same luminance but different chromaticity. Such a view was systematically expressed, for example, by Gouras and Eggers (1983). But, in fact, it is rare in the natural world for one surface to lie in front of another, in such a way that both have the same reflectance, both lie at the same angle to the incident illumination, and the nearer throws no shadow on the farther.

And if Nature finds it hard to generate equiluminant edges, so does Art: in the 1870s, when the first 'pseudoisochromatic' plates were constructed for identifying colour-deficient observers, attempts were made to print targets of one colour on fields of different colour, in such a way that dichromats could not detect them. The enterprise was not long pursued. For the printer could not so completely eliminate edge artifacts, and could not so exactly equate the lightnesses of figure and ground, as to deceive the average dichromat. But in the early pseudoisochromatic plates of Stilling (1877, 1883), like the later ones of Ishihara, the problem was neatly sidestepped by

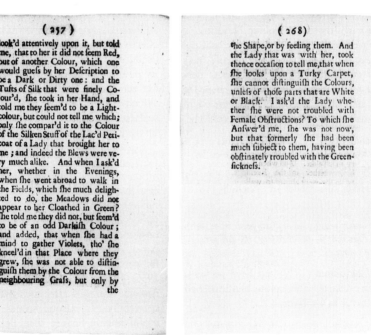

Fig. 14.1 *Title page and excerpts from Robert Boyle's* Vitiated Sight *of 1688. (Note: the original pagination is in error.)*

two ploys: first, the target and background were formed from discrete patches, each with its own limiting contour; and secondly, the lightnesses of the individual patches were varied rather than being equated (see Plate 1). Stilling himself was quite explicit in his insight into how to prepare successful pseudoisochromatic plates. He wrote:

Der störende Einfluss der Buchstabenconturen ist einfach dadurch beseitigt, dass die Conturen über das ganze Feld gleichmässig vertheilt und die Buchstaben aus discontinuirlich von Grund sich abhebenden Quadraten gebildet sind. Der zweite Fehler ist dadurch völlig eliminirt, dass die sämmtlichen Quadrate des Grundes abwechselnd dunklere und hellere Töne zeigen.

The pseudoisochromatic plates of Stilling and Ishihara, which have proved so successful in detecting colour-deficient observers, nicely show us the tests that defeat the colour-deficient in the natural world. The colour-deficient observer is not often confronted by a homogeneous target of one chromaticity that lies on an equiluminant surface of different chromaticity. Rather, his difficulty arises when the background is dappled and brindled, when, that is, luminosity is varying randomly (Mollon, 1987). Such a variation in luminosity can arise because the illuminant is interrupted by foliage; or it can arise because the background consists of component surfaces that lie at varying angles to the illuminant or (as in the case of new and old leaves) themselves vary in reflectance. Boyle's gentlewoman is poor at spotting violets among grass: she cannot use lightness to solve the task, and is forced to use shape, but under conditions such that the presence of similar, masking, contours slows down a search on the basis of shape. And in the early reports of colour blindness, a recurrent theme is difficulty in searching for natural targets against foliage, vegetation, or other visually busy backgrounds. The celebrated shoemaker Harris, described by Huddart in the *Philosophical Transactions* for 1777, had noticed as a child that other children could detect cherries on a tree at a greater distance than could he:

Large objects he could see as well as other persons; and even the smaller ones if they were not enveloped in other things, as in the case of cherries among the leaves.

Similarly, the observer described by Nicholl (1818) reports:

... the fruit on trees when red, I cannot distinguish from the leaves, unless when I am near it, and then from the difference of shape rather than colour.

The protanope of Colquhoun (1829) had been perverse enough to become an orchardist after colour blindness had stymied his earlier career as a weaver:

He cannot discern, even in a loaded bush, the existence of red gooseberries among the leaves, until he has almost approached so near as to be able to take hold of the branch. Rosy apples on a tree, which may be discovered by ordinary eyes, at a distance of from thirty to forty yards at least, are lost to his sense, until he has come within ten or twenty yards of the tree, when he can trace out the fruit by its form...

The colour-deficient brassfounder of Edinburgh, described by Combe (1824), was a keen shot, but could not 'discover game upon the ground, from the faintness of his perception of colours':

... last year, when a large covey of partridges rose within ten or twelve yards of him, the back ground being a field of Swedish turnips, he could not perceive a single bird.

Such complaints were familiar in the early literature on Daltonism. Today, in the developed countries, a smaller proportion of us live in agricultural societies, but a recent Australian study showed formally that many colour-blind observers still report difficulty in finding berries or fruit among foliage (Steward and Cole, 1989). However commonplace the complaint might seem, we should listen to it. For it tells us the disability of the Daltonic observer; and the complement of his disability must be the advantage of the normal trichromat. The essence of the Daltonian's disability is captured in Huddart's telling phrase '... if they were not enveloped in other things, as... cherries among the leaves.' *A primary advantage of colour vision is that it allows us to detect targets against dappled or variegated backgrounds, where lightness is varying randomly.*

Perceptual Organization by Colour: Two Senses of 'Segregation'

In the preceding section, we have considered the target as effectively punctate – a cherry amongst foliage. The pseudoisochromatic plates of Stilling or of Ishihara show us an additional and related way in which colour assists our perception: the shared hue of different elements in an extended visual array may serve as a basis for perceptual segregation and so allow us to identify a figure that is made up from discrete patches that vary in lightness.

Segregation of the visual field into elements that belong together is an important preliminary to the actual recognition of objects. This segregation corresponds to the process that the Gestalt psychologists called 'figure-ground differentiation' or 'perceptual organization'. Gestalt writers recognized that many features of the visual image – brightness, colour, local form, local texture, binocular disparity, direction of movement – can each serve as a basis for perceptual organization (Wertheimer, 1923; Koffka, 1935). Barlow (1981) used the term 'linking features' to describe stimulus attributes that can sustain perceptual segregation. In natural scenes, colour may be a more useful linking feature than lightness. For an extended object seldom consists of a single plane surface lying at a constant angle to the illuminant. Rather, its component surfaces may lie at different angles or may be curved or corrugated, and indeed the resulting variation in luminance allows the visual system to recover the object's shape – a process recently called 'shape from shading' but earlier known as the 'modelling effect of shadows' (Vallée, 1821; Norden, 1948). It is true that such variations in luminance are often accompanied by correlated changes in chromaticity, from which shape can in principle be recovered (Monge, 1791): for in so far as the reflection from an object has a specular component, the chromaticity of the light reflected from the object's surfaces will vary according to the orientation of those surfaces. Nevertheless, these variations are constrained to lie along a single line in chromaticity space (the line that connects the chromaticity of the illuminant with the non-specular component of the object's reflectance); and in general we might expect chromaticity to vary less freely than does luminance across the surface of the object, and thus chromaticity will be the more useful linking feature.

The argument of the preceding paragraph is in direct contradiction to the position of Livingstone and Hubel (1988), who make the remarkable claim that 'only luminance contrast, and not colour differences, is used to link parts together'. This claim should be understood exclusively in terms of the particular matter with which they are concerned, the incoherence of the visual field when edges are defined only by chromatic differences: under such conditions of 'equiluminance', it becomes difficult, for example, to relate different parts of colinear edges, and we lose the coordination that would otherwise impose a uniform interpretation on apparent movements in different parts of the field. But perceptual failures of this kind may have a special explanation: under the unusual conditions of equiluminance the long-wave receptors of the retina are signalling spatial and temporal transients of opposite, contradictory, sign from those signalled by the middle-wave receptors; and it is probably this *contradiction of normally-yoked signals* that impairs our perceptual performance (see Mollon, 1980, 1987, and below). The very existence of pseudoisochromatic plates shows that Livingstone and Hubel's statement cannot be generally true: normal observers can perfectly well use colour to link different elements in the visual field and extract a figure. Notice, in the case of the Stilling or Ishihara plates (Plate 1), that this is done under conditions where lightness is varying randomly – conditions that are very much more common in the natural world than are conditions where lightness remains constant across the field.

But Livingstone and Hubel's position has distinguished precedents in the Gestalt literature. Thus Koffka (1935) writes: '... a mere colour difference, has, to say the least, much less power to produce a segregation of ... two areas in the psychophysical field than a very small difference in luminosity'; and Koffka and Harrower (1932) write: '... mere colour difference without the aid of concomitant difference in brightness has surprisingly little organizing power.' Such remarks were explicitly inspired by the Liebmann effect, the perceptual melting of the edges of a coloured patch on a background of the same lightness (Liebmann, 1927). To understand the apparent contradiction between Koffka's position and the patent role of colour in the Stilling plates, we must distinguish between segregation in the sense of *articulation of contour* and segregation in the sense of *organization of the field*. The former is likely to be a largely retinal process, whereas the latter requires the linking of well-separated elements (as in Plate 1) and is likely to be a cortical process.

The Articulation of Contour

In the sense that edges are subjectively either weak or unstable, contours are indeed poorly articulated when equiluminant colours abut one another. Probably more than one factor is at work here. First we should note that the Liebmann effect is particularly strong when the two abutting colours are ones that would be confused by a tritanope, i.e. when the edge is indicated only by a change in the signal of the short-wave cones. This special case of 'soft' colours will be discussed below (pp 311–313) and is almost certainly due to the insensitivity of the short-wave cone system to high spatial frequencies.

When the edge is visible to the long- and middle-wave cones (i.e. when the two equiluminant colours do not lie on a tritan confusion line in colour space), a second factor impairs the representation of the contour: from the long-wave and from the middle-wave cones, the visual system receives contradictory accounts of the sign of the edge. The long-wave ON-centre midget ganglion cells indicate an edge opposite in sign to that signalled by the middle-wave ON-centre midget ganglion cells; and the long-wave OFF-centre cells similarly tell a different story from their middle-wave fellows, to whom they are usually yoked. Such contradictory signals may then impair secondary functions (e.g. movement perception, integration of elements from different parts of the field) that depend on the primary process of extracting edges from the visual array (Mollon, 1988).

The Organization of the Field

But these two effects of equiluminance on the local articulation of contour are quite a different matter from the central use of colour to link together low-frequency information in the spatial image. When we recognize a digit in the Ishihara plates, the contours of individual patches are detected by their luminance contrast, but the digit is discovered only by colour. The cortex must be able to collect together widely separated patches of similar chromaticity while yet preserving the local signs – the spatial addresses – that belong with these patches; and must be able to deliver this subset of spatial addresses as the input to a form-recognition system. We do not know in detail how this is done, although it is plausible to suppose (with Barlow, 1981) that the correlation of patches of similar chromaticity is achieved in the prestriate region in which colour is emphasized (Zeki, 1978) – V4 in the rhesus monkey and perhaps the region of the lingual and fusiform gyri in man (Meadows, 1974); and it is also plausible to suppose that the set of spatial addresses is ultimately delivered to a common form-recognition system, which is located in the inferotemporal lobe and is able to accept similar sets of spatial addresses defined by other linking features such as movement, texture, depth and orientation. But whatever cortical machinery collects common chromaticities while preserving spatial addresses, an important point for our present purpose is that such machinery must evolve before an animal can use colour for the particular purpose of perceptual segregation.

Is there any price to be paid for the advantage of being able to segregate the field on the basis of colour? Here again the Ishihara plates point to the answer. There exist plates (e.g. plates 18–21 of the 11th edition) where a digit is readily visible to the dichromat, on the basis of differences in lightness or in the short-wave component, but is masked for many normals, owing to a stronger, rival, organization suggested by red–green variation. Morgan *et al.* (1989) prepared a visual display in which one quadrant of the array differed from the rest in the size or orientation of texture elements; they found that the textural difference could be masked for normals if a random variation in colour was introduced, but that dichromats performed as well as they did in a monochrome condition. A fundamental limitation of the visual system is that it cannot concurrently entertain different perceptual organizations; and this in itself suggests that figural organizations based on colour, texture, orientation and other attributes are delivered to a common pattern-recognition system.

Identification by Colour

Finally, colour may serve to identify, that is, it may (a) help us assign an already segregated object to a given category or (b) indicate what lies beneath a given surface.

Colour may be one of the cues that monkeys use to identify the species of particular trees or particular plants. Certainly, it must be a major cue that catarrhine (Old World) monkeys use to identify conspecifics, the sex of conspecifics, and the sexual state of conspecifics. In many members of the genus *Cercopithecus*, the slate blue scrotum of the male contrasts with surrounding yellow fur; and often this contrast is echoed in facial markings, most notably in the moustached guenon (*Cercopithecus cephus*) whose vivid blue face is surrounded by a yellow ruff. Such colourings in *Cercopithecus* are often accompanied by species-specific 'flagging' movements and may serve to maintain species isolation in habitats where several species of guenon overlap (Kingdon, 1980, 1988). Other striking examples of the use of colour as sexual signals are offered by the bright-red ischial callosities of the female baboon and the blue, violet and scarlet patterns of the genitalia and muzzle of the male mandrill.

Colour offers a means to judge remotely the structure underlying a surface (Katz, 1935). In the case of fructivorous primates, one of the chief functions of trichromatic colour vision must be to discover the state of ripeness of fruit from the external appearance. Similarly, variations in the colour of ground-covering vegetation can reveal at a distance the presence of water; and in the semi-desert regions of the world, a splotch of green provides a remote indication of moisture. In this category of the use of colour, we can also include judgements of complexion: human observers have a finely developed (though little studied) capacity to use skin colour to estimate the health, or emotional state, of conspecifics, and this ability may have biological advantage in the selection of sexual partners, in the care of infants, and in social interactions. Small variations in complexion arise from dilatation of blood vessels, the density of erythrocytes, the ratio of oxyhaemoglobin to reduced haemoglobin, and the levels of bilirubin, melanin and carotenoids (Edwards and Duntley, 1939). In the survey of colour-deficient subjects by Steward and Cole (1989), deuteranopic respondents did report difficulty in detecting changes in skin coloration that accompanied illness.[1]

In so far as an animal uses colour to identify – in so far as it uses colour vision to recognize permanent attributes of an object or to make absolute judgments about the spectral reflectances of surfaces – we should expect it to exhibit colour constancy: it ought to be able to recognize the spectral reflectance of a surface independently of the spectral composition of the illuminant (see e.g. Katz, 1935; Land, 1983). Of the three uses of colour vision distinguished here, only the third necessarily requires colour constancy; but the distinction between the three functions is not absolute and there will be many cases, for example, where an animal is searching a complex array not merely for a salient stimulus but for a stimulus of a particular spectral reflectance.

For an alternative approach to the uses of colour vision, the discussion by Jacobs (1981) is recommended.

Trichromacy and Its Evolution

The Photoreceptors

In man, the three classes of retinal cone exhibit overlapping spectral sensitivities, with peak sensitivities lying close to 420, 530 and 560 nm (Plate 2). Estimates of these sensitivities have now been obtained by psychophysics (Smith and Pokorny, 1975; Boynton, 1979), by microspectrophotometry of the outer segments of individual cones (Dartnall *et al.*, 1983), and by electrophysiological measurements of isolated receptors that have been sucked into micropipettes (Schnapf *et al.*, 1987); and the different estimates agree to a first approximation. Notice the asym-

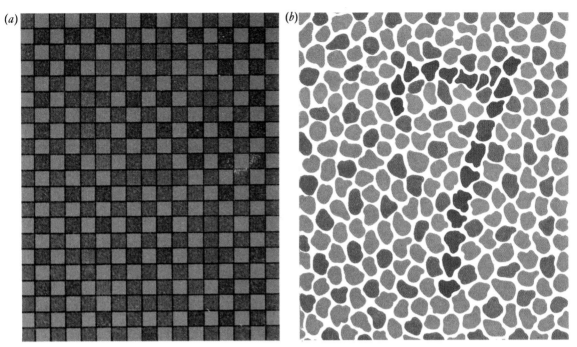

Plate 1 *Test patterns from the pseudoisochromatic plates of Stilling. Note that the figure and field are made up of discrete patches, varying in lightness and each having its own contour. Pattern (a) was the first design (Stilling, 1877) and the less regular pattern (b) was used in later editions (Stilling, 1883).*

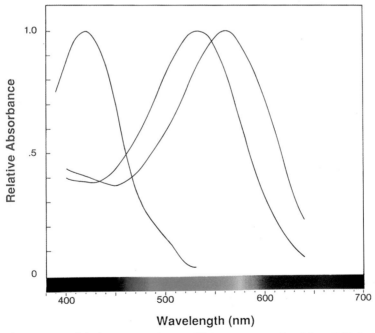

Plate 2 *The absorbance spectra of the human cone pigments. These curves are replotted from Table 2 of Dartnall* et al. *(1983) and were obtained by microspectrophotometric measurements of individual photoreceptors from human eyes removed in cases of ocular cancer. The ordinate represents the (normalized) logarithm of the factor by which monochromatic light is attenuated as it passes through the outer segment of an isolated receptor.*

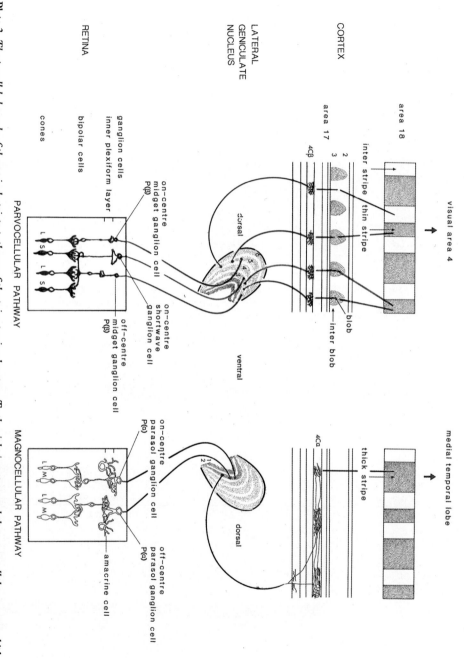

Plate 3 The parallel channels of the geniculostriate pathway of the primate visual system. To the right is represented the magnocellular system, which carries little chromatic information: the large Pα cells of the retina project to the magnocellular layers (1 and 2) of the lateral geniculate nucleus and thence to layer 4Cα of the visual cortex and the 'thick stripes' of area 18, which stain strongly for cytochrome oxidase. A second parallelism is represented within the parvocellular pathway on the left; and here the parallelism is between a channel that carries almost purely chromatic information (the primordial subsystem of colour vision) and a channel that carries information about both colour and spatial contrast (the second subsystem of colour vision). The two subsystems, represented by blue and red respectively, appear to remain independent at least as far as the cytochrome oxidase 'blobs' of Area 17 of the cortex.

As early as 1883, F. C. Donders proposed that our present trichromatic vision had evolved by natural (and possibly sexual) selection from an earlier dichromatic system; and he even anticipated recent discussions (Bowmaker *et al.*, 1987; Jacobs, 1990) by suggesting that there was an intermediate stage at which only females enjoyed trichromacy (Donders, 1883). After Donder's death, the belligerent Mrs Ladd-Franklin (Fig. 14.2) took for her own the idea that colour vision had successively evolved from monochromacy to a form of deuteranopia and thence to full trichromachy (Ladd-Franklin, 1892). But it is only fair to record that the prescience of Donders and Ladd-Franklin was incomplete: both held that the evolution of colour vision depended on elaborating a single photosensitive molecule (with different cleavage products) rather than on creating a family of slightly different molecules.

The Primordial Subsystem of Mammalian Colour Vision

For the main business of vision, for the detection of contours, movement and flicker, most mammals depend on a single class of cone, which has its peak sensitivity near the peak of the solar spectrum, in the range 510 to 570 nm (Jacobs, 1981; Bowmaker, this volume). A long time ago, a few short-wave (violet-sensitive) cones were sparsely added to the matrix of middle-wave cones. These short-wave cones, with their wavelength of peak sensitivity well removed from that of the middle-wave cones, provided mammals with a basic, dichromatic, system of colour vision by comparing the quantum catch of the short-wave cones with that of middle-wave cones, the visual system is able to estimate the sign and the slope of the change in spectral flux from one end of the spectrum to another. If the spectral distribution of the light reaching us from an object is itself considered as a waveform (Barlow, 1982), then this primordial colour-vision system seems designed to extract the lowest Fourier component of the waveform (Mollon *et al.*, 1990). Subjectively, it is the primordial subsystem that divides our colour sensations into warm, cool and neutral (McDougall, 1901). And it is this subsystem that survives in the common forms of human dichromacy.

The antiquity of the primordial subsystem is suggested by the molecular genetic results of Nathans *et al.* (1986a), who have sequenced the genes that code for the protein moieties of the four human photopigments. The amino-acid sequence for the short-wave cone pigment is as different from those for the long- and middle-wave pigments as it is from the sequence for the human rod pigment, rhodopsin: in a pairwise comparison the short-wave pigment shares 42% of its amino acids with rhodopsin, 43% with the long-wave cone pigment and 44% with the middle-wave cone pigment. Nathans and his colleagues

Fig. 14.2 *Christine Ladd-Franklin (courtesty of The Ferdinand Hamburger, Jr, Archives of the Johns Hopkins University).*

metry of the spectral positions of the receptors: there is an interval of 100 nm between the short-wave and middle-wave receptors, but an interval of only 30 nm between the middle-wave and long-wave receptors.

The response of any individual receptor obeys the Principle of Univariance: its electrical response varies only with the rate at which photons are captured by the outer segment, and all that changes with wavelength is the probability that any given photon will be absorbed. Thus, if the visual system is to learn about wavelength, independently of radiance, it must not only have receptors with different sensitivities but must also have the neural machinery to measure the relative rates of photon capture in different classes of receptor (Rushton, 1972).

Textbooks often represent the three types of human cone as equal elements in a trichromatic scheme, as if they evolved all at once and for a single purpose. However, all the evidence suggests that they evolved at different times and for different reasons; and our own colour vision seems to depend on two, relatively independent, subsystems – a phylogenetically recent subsystem overlaid on a much more ancient subsystem (Mollon and Jordan, 1988). This account, which is elaborated below, has long antecedents.

use the difference in sequence between bovine and human rhodopsin to estimate the rate at which photopigments diverge, and they conclude that the short-wave pigment diverged from the middle-wave cone pigment (or perhaps indeed from an ancestral rod pigment) more than 500 million years ago.

The Primordial Subsystem Considered as a Purely Chromatic Channel

It would be tempting to say that the short-wave cones have been given to us only for colour vision, and that the pathway which carries their signals is the nearest we come to a channel that conveys purely chromatic information. There is certainly good reason why the short-wave cones should not be used for the discrimination of spatial detail: the short-wave component of the retinal image is almost always degraded, since bluish, non-directional skylight dilutes the shadows of the natural world and since the eye itself is subject to chromatic aberration and usually chooses to focus optimally for yellow light. The short-wave cones were therefore added sparingly to the retinal matrix, and they remain rare: they constitute only a few per cent of all cones in every primate species where cones have been identified directly by microspectrophotometry (e.g. Hárosi, 1982; Dartnall, Bowmaker and Mollon, 1983; Mollon *et al.*, 1984; Bowmaker *et al.*, 1991). At a post-receptoral level, little sensitivity to spatial contrast is exhibited by those post-receptoral neurons that draw signals from the short-wave cones (Derrington *et al.*, 1984; Gouras, 1990): such cells are *chromatically* opponent but tend not to draw their opposed inputs from clearly distinct, spatially antagonistic, regions in the way that is usual for other cells in the retina and lateral geniculate nucleus.

It would, in fact, be careless to say that the information provided by the short-wave cones is *purely* chromatic. Their signals must carry with them enough of a local sign that the chromatic information can be associated with the corresponding object. And the local signs are adequate to support perceptual segregation (see pp 308–309 above), as is demonstrated by those specialized pseudoisochromatic plates that bear figures visible to the normal trichromat but invisible to tritanopes (the rare type of dichromats who lack the short-wave cones): examples of such plates are those of Willmer (see Stiles, 1952) and of Farnsworth (see Kalmus, 1955). Moreover, if the short-wave cone signals are isolated by presenting a violet target on an intense long-wave adapting field, then the contours formed directly by short-wave signals can support the perception of gratings of low spatial frequency (e.g. Brindley, 1954); such gratings can give an orientationally selective adaptation (Stromeyer *et al.*, 1980) and can provide an input to a motion-detecting mechanism that also draws inputs from the long-wave cones (Stromeyer and Lee, 1987). Webster *et al.* (1990) isolated the short-wave system by forming an equiluminant low-frequency grating of two colours that are confusable to a tritanope; they reported that the discrimination of orientation and spatial frequency was as good as it was for an equiluminant grating that was visible by virtue of variations in the ratio of the signals of the long-wave and middle-wave cones.

Nevertheless, in natural viewing, the short-wave cones must play little role in the extraction of high spatial frequencies from the image, that is, in the detection of sharp edges and the discrimination of spatial detail. Using a chequerboard pattern, Stiles (1949) first demonstrated the low values that are obtained for foveal acuity when vision depends only on the signals of the short-wave cones. And later measurements of spatial contrast sensitivity suggest that the short-wave receptors cannot sustain a resolution of much greater than $10\,c\,deg^{-1}$ (for references, see Mollon, 1982, Table 1); the maximal sensitivity for vision with these cones lies close to $1\,c\,deg^{-1}$. Tansley and Boynton (1976) required subjects to adjust the relative luminances of two hemifields of different chromaticity until the edge between them was minimally distinct; the subject then rated the salience of the residual edge on an eight-point scale. When the two half-fields differed only in the degree to which they excited the short-wave cones, when, that is, they lay on the same tritan confusion line in colour space, observers gave a rating of zero to the distinctness of the border between the two fields: the fields 'melted' into one another. And the rated distinctness of other (non-tritan) borders could be predicted simply by how much the two half-fields differed in the ratio with which they excited the long- and middle-wave cones (see also Thoma and Scheibner, 1980). In her classical description of 'hard' and 'soft' colours, Liebmann (1927) described how a blue figure perceptually merged ('*verschwimmt*') with an equiluminant green ground; and in retrospect we may suspect that these two colours were ones that came close to lying along the same tritan confusion line. In the recent literature on equiluminance the Liebmann effect has sometimes been cited, but often no distinction is made between the modest effects obtained when long- and middle-wave cones are modulated and the much more spectacular weakening of contours and of figural organization when the target and field differ only along a tritan confusion line, i.e. when they differ only in the degree to which they excite the primordial subsystem of colour vision.

The Morphological Basis of the Primordial Subsystem

There are growing hints that the primordial subsystem of colour vision has a morphologically distinct basis in the primate visual system (as indicated in blue in Plate 3).

Thus Mariani (1984) has described a special type of bipolar cell, which resembles the common 'invaginating midget' bipolar but which makes contact with two well-separated cone pedicles; he suggests that these bipolars are exclusively in contact with short-wave cones. Further evidence for the distinctness of this class of bipolar has come from an immunological study using antisera that recognize precursors of the neuropeptide cholecystokinin (Marshak *et al.*, 1990). Using micropipettes filled with the dye 'Procion-yellow', De Monasterio (1979) made intracellular recordings from, and stained, three ganglion cells that showed a maximal sensitivity for 440 nm increments in the presence of a long-wave field: the cell bodies of these short-wave ganglion cells were clearly larger than the standard 'midget' ganglion cell that draws opposed inputs from the long- and middle-wave receptors[2]. In the lateral geniculate nucleus, the cells that draw excitatory inputs from the short-wave cones appear to be found predominantly in the parvocellular layers 3 and 4 (Schiller and Malpeli, 1978; Michael, 1988).

The primordial subsystem may remain distinct as far as Area 17 of the primate cortex: Ts'o and Gilbert (1988) have examined the properties of individual cells in the so-called 'cytochrome oxidase blobs' of layers 2 and 3 (Livingstone and Hubel, 1984; see Plate 3) and report that cells within a given blob exhibited the same type of colour opponency, 'blue–yellow' or 'red–green', the latter type being three times as common as the former. It is likely that Ts'o and Gilbert's 'blue–yellow' cells correspond to what I am here calling the primordial subsystem; but it must be said that the cells were classified only by their responses to coloured lights, and the presence of short-wave inputs needs to be confirmed by the use of selection adaptation (e.g. Gouras, 1974) or of tritanopic substitution (Derrington *et al.*, 1984). Since the maximum ratio of middle- to long-wave cone sensitivity occurs near 460 nm (see e.g. Boynton and Kambe, 1980), it is possible for the spectral cross-over point of a cell (the wavelength where an excitatory response replaces inhibition) to lie at short wavelengths even though the cell draws its opposed inputs from the middle- and long-wave cones.

The Primordial Subsystem: Conclusions

Man and the Old World primates thus share with many mammals an ancient colour-vision channel, which depends on comparing the quantum catch of the short-wave cones with the quantum catch of a much more numerous class of cones, which have their peak sensitivity in the middle of the visible spectrum and which subserve the other functions of photopic vision. There may be a distinct morphological basis for this channel in the primate visual pathway, although the short-wave cones, and the cells that carry their signals, are always in the minority.

It would be foolish to say that this primordial subsystem is a purely chromatic channel, since the several uses of colour vision (see pp 306–310) themselves *require* that the chromatic signals carry with them a local sign; and the signals of this subsystem can sustain a number of spatial discriminations, provided only that the stimuli are of low spatial frequency. But the primordial colour channel has little role in the detection of edges and spatial detail.

The Second Subsystem of Colour Vision

As far as is known at present, the Old World primates are the only mammals that share our own form of colour vision (Jacobs, 1981; Jacobs and Neitz, 1986; Bowmaker, this volume), although the New World monkeys have found their own route to trichromacy (see below, pp 314–317). The recent ancestors of the Old World primates acquired a second subsystem of colour vision, overlaid on the first; and the two subsystems remain relatively independent, physiologically independent at early stages of the visual system and psychophysically independent in detection and discrimination (Mollon and Jordan, 1988). The second subsystem depends on a comparison of the rates at which quanta are caught by the long- and middle-wave cones (Plate 2); and this subsystem appears to have arisen through the duplication of a gene that coded for the photopigment of an ancestral middle-wave cone. (By 'middle-wave' is meant here only that the peak sensitivity lay in the range 510 to 570 nm, a range that includes the present 'long-wave' cone.) Since the time of Donders and Ladd-Franklin, it has been suspected that the long- and middle-wave photopigments were differentiated only recently, in evolutionary terms. The traditional evidence for this idea has been the distribution of trichromacy among the mammals (Jacobs, 1981, 1990) and the fact that hereditary disorders of colour vision chiefly affect the long- and middle-wave receptors (Pokorny *et al.*, 1979). But particularly convincing evidence has now come from the molecular genetic results of Nathans and his collaborators (1986b), who have shown that the inferred amino-acid sequences of the middle- and long-wave human pigments are 96% identical. Moreover, the two genes remain juxtaposed in a tandem array on the q-arm of the X-chromosome (Vollrath *et al.*, 1988). The juxtaposition and the extreme homology of these genes render them vulnerable to misalignment when the X-chromosomes come together at meiosis; and Nathans and his collaborators suppose that the high incidence of human colour deficiency arises from the unequal crossing-over that can follow such misalignments (Nathans *et al.*, 1986a).

Coevolution of Fruit and of Primate Trichromacy?

Interspecific variations occur in the photopigments of salmonid fish that have been isolated in land-locked glacial lakes for only a short evolutionary period (Bridges and Yoshikami, 1970; Bridges and Delisle, 1974), and it has become customary to suppose that opsins (the protein moieties of the photopigments) can evolve very rapidly. Moreover, Nathans and his collaborators have suggested how, in man, hybrid genes can readily be formed that code for pigments with different spectral sensitivities (see above). The very consistency of the photopigments of catarrhine monkeys thus becomes a matter for remark. The monkeys so far examined – macaques (Bowmaker et al., 1978; Hárosi, 1982; Schnapf et al., 1987), baboons (Bowmaker et al., 1983), patas monkeys, talapoins and guenons (Bowmaker et al., 1991) – vary widely in their habitat, their size and their bodily colourings; yet they all exhibit a middle-wave pigment with peak sensitivity close to 535 nm and a long-wave pigment with a peak sensitivity close to 565 nm. What this set of monkeys do have in common is that a substantial part of their diet consists of fruit; the proportion varies for different species, but is as high as 85% in the case of the moustached guenon, *Cercopithecus cephus* (Sourd and Gautier-Hion, 1986).

Polyak (1957) explicitly proposed that the trichromatic colour sense of primates coevolved with coloured fruits, such as mangos, bananas, papaya, and the fruits of the citrus family. Recent ecological studies lend fresh credence to this view. Gautier-Hion and her collaborators (1985) have examined the relationships between frugivors and fruits in an African tropical rain forest, and it becomes clear that it is not enough to describe an animal as fruit-eating. There exists a category of fruit that is disproportionately taken by monkeys: such fruits are orange or yellow coloured, weigh between 5 and 50 g, and are either dehiscent with arillate seeds or are succulent and fleshy. In contrast, the fruits predominantly taken by birds are red or purple, and are smaller, while the fruits taken by ruminants, squirrels and rodents are dull-coloured (green or brown) and have a dry fibrous flesh.

Thus there exist fruits that appear to be specialized for attracting catarrhine monkeys. The fruiting tree serves the monkey by providing both food and water, and the monkey serves the tree by dispersing the tree's seed in one of two ways. When the seed is large and the soft flesh is free from the seed, the latter is often spat out; this usually happens at some distance from the parent tree, because the monkeys fill their cheek pouches and move to another place to eat the contents. When the seeds are small, they are usually swallowed and are excreted intact. Thus the monkey is a disperser for the tree, rather than a predator (Hallé, 1974; Gautier-Hion et al., 1985). And to find orange and yellow fruits amongst foliage, the monkey needs trichromatic colour vision; without it, the monkey would be at the very disadvantage emphasized in the early accounts of human colour blindness (see pp 306–310). It would be instructive to know whether the leaf-eating catarrhine monkeys differ from the frugivors, but nothing is yet known about their colour vision.

We can well summarize this subsection by quoting from an early admirer of Darwin:

> What insects are to bright-hued flowers, birds and mammals are to bright-hued fruits. And we might almost say, though with more reservation, what flowers are to the colour-sense in insects, fruits are to the colour-sense in birds and mammals.
> (Grant Allen, 1892)

The Second Subsystem and the Parvocellular Pathway

At the early stages of the visual system, there does not appear to be a channel devoted exclusively to carrying the second type of chromatic information, the information obtained by comparing the quantum catches of long- and middle-wave cones; rather the second subsystem is parasitic upon an existing channel that carries information about spatial detail. The substrate of this channel is the Pβ cell of the primate retina (Perry and Cowey, 1981) and the predominant type of unit in the parvocellular layers of the lateral geniculate nucleus (represented in red in Plate 3). The receptive fields of such cells are divided into antagonistic centre and surround regions, and thus the cells are very sensitive to spatial contrast. In the case of the Old World primates, the centre input to such cells is drawn from one class of cone (often perhaps, from one individual cone), whereas the surround is drawn either from a different class of cone, or perhaps indiscriminately from all middle- and long-wave cones (Shapley and Perry, 1986; Lennie et al., 1989). Thus, as Ingling and Martinez (1983) have emphasized, the response of any individual cell is ambiguous: the cell will be colour-specific at low spatial frequencies but will respond to all wavelengths at higher spatial frequencies.

Shapley and Perry (1986) have argued that the primary function of the parvocellular system is colour vision, and that the analysis of spatial contrast depends predominantly on the magnocellular pathway. Arguments against this position are given by Mollon and Jordan (1988).

Polymorphism of Colour Vision in Platyrrhine Primates

Until recently the platyrrhine (New World) monkeys were held to be protanopes, and thereby to represent an earlier stage of our own colour vision. We now know that platyrrhine colour vision is much more complicated and is characterized by a striking within-species variability (Jacobs, 1984). In a double-blind study that combined behavioural testing and microspectrophotometry (Mollon et al., 1984; Bowmaker et al., 1987), six phenotypes were identified in

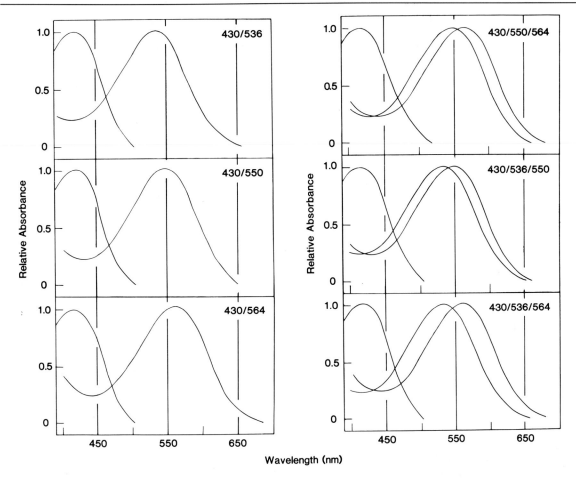

Fig. 15.3 *Six types of squirrel monkey (*Saimiri sciureus*). Each panel shows the absorbance spectra of the photopigments found microspectrophotometrically in a given phenotype. To the left are shown the three possible combinations of pigments that occur in dichromat phenotypes, that is, all males and some females. To the right are shown the combinations that occur in the three possible trichromatic phenotypes, which are always female.*

the squirrel monkey (*Saimiri sciureus*): all males and some females were dichromats, combining a short-wave pigment with one of three possible pigments in the red–green spectral region (Fig. 15.3), whereas other females were trichromatic, combining a short-wave pigment with two pigments in the red–green region. These findings have been explained by a genetic model (Mollon *et al.*, 1984) that makes the following assumptions:

1. The squirrel monkey has only a single genetic locus for a pigment in the red–green spectral region.
2. There are at least three alleles that can occur at this locus, the three alleles corresponding to three slightly different versions of the opsin of the photopigment.
3. The locus is on the X-chromosome.
4. In those females that are heterozygous at this locus, only one of the two alleles is expressed in any given cone cell, owing to the phenomenon of Lyonization or X-chromosome inactivation.

By this account, at least three photopigments are potentially available to the squirrel monkey in the red–green region. Male monkeys, it is supposed, can draw only one pigment from the set, since they have only one X-chromosome; so males are obligatory dichromats. Females may draw either one or two pigments from the set; if they inherit the same allele from both parents, then they will be dichromatic, but if they inherit two different alleles then they will be trichromatic. We must make the explicit assumption that the visual system of the heterozygous female is sufficiently plastic to exploit the presence of distinct subsets of signals from the two types of cone with peak sensitivities in the red–green region.

The theory draws support from the electroretinographic study of Jacobs and Neitz (1987), who examined 25 members of 9 *Saimiri* families and found that none of the pedigrees contradicted the pattern of inheritance required. For example, sons inherited only pigments that were present in the mother; and a dichromatic daughter inherited the pigment common to both her parents.

Such patterns of colour vision may be prove to be common in both the families of New World primates, the *Cebidae* and the *Callitrichidae*. For example, in the case of the callitricid species, *Callithrix jacchus* (the common marmoset), microspectrophotometric measurements of over 40 animals have revealed at least three distinct long-wave pigments (Travis et al., 1988; Tovée, 1990). As in the case of the cebid squirrel monkey, male marmosets are always dichromatic whereas some females are trichromats. The spectral positions of the three long-wave pigments (approximately 543, 556 and 565 nm) are different from those of the squirrel monkey but similar to those shown by electroretinography in another polymorphic callitricid, the saddle-backed tamarin (Jacobs et al., 1987). In general, pedigree studies of marmoset colour vision support a genetic theory analogous to that proposed by Mollon et al. (1984) for the squirrel monkey, but in both the sample studied by Travis et al. (1988) and that studied by Tovée (1990), we found a discrepancy – a dichromatic offspring with a peak sensitivity lying near 560 nm and shifted by 3–4 nm from the possible values predicted from the pedigree. A final genetic model of marmoset colour vision may prove to be complicated.

The Evolution of a Balanced Polymorphism

How might polymorphisms arise and be maintained in platyrrhine monkeys? Heterozygous advantage offers one plausible mechanism (Mollon, 1987). The most celebrated example of a trait maintained by heterozygous advantage is sickle cell anaemia, where the heterozygous carrier is more resistant to the malarial parasite than is the normal homozygote (Allison, 1964). There is no doubt that the heterozygous squirrel monkey is at an advantage in laboratory tests, where she is behaviourally able to make red–green discriminations that defeat her homozygous and hemizygous conspecifics (Jacobs, 1984); and the arguments of preceding sections suggest that she will also be at an advantage in finding ripe fruit in the dappled environment of the forest.

Consider now a homogeneous dichromatic population and suppose that a rare mutant gene arises at the X-chromosome locus that specifies the single ancestral pigment in the red–green range. Just one nucleotide substitution, when it occurs at a critical position in the sequence of the gene, may be enough to produce a pigment with a different spectral sensitivity. As long as the new allele is rare, it will almost invariably find itself paired with the wild-type allele whenever it is inherited by a female monkey. If we allow, as argued above, that the heterozygous female enjoys an advantage, then the frequency of the new allele should rise. But the higher the frequency of the second allele, the more likely it is to be paired with itself and so to be carried by a homozygous monkey which will not be at an advantage. *Ceteris paribus*, the second allele should reach an equilibrium when its frequency is 0.5: if its frequency were to rise above this value, the second gene would be at a disadvantage in comparison with the original gene, since it would be then more often paired with itself (i.e. carried by a dichromat) than would be the original gene. Suppose now that there arises in the gene pool a further mutant gene, which specifies a viable photopigment with a spectral sensitivity distinct from the preceding two. As long as this allele is rare, it will be almost always carried by heterozygotes and its frequency too should rise. A new equilibrium should be reached when each of the three alleles has a frequency of 0.33. If any of the alleles rises above this frequency, it will be disproportionately paired with itself in females and thus will place itself at a relative disadvantage.

Accounts of the evolution of an organ or system traditionally face the problem that several components must evolve before the phenotype is at an advantage. In the present case, the reader may ask how the female heterozygote is at an advantage before she has developed the neural machinery to exploit the new pigment. One possibility is that she has no need to develop new machinery, and that her new colour system (as suggested above for catarrhines) is parasitic on an existing parvocellular channel. X-chromosome inactivation ensures that the new pigment is segregated in a distinct subset of cones. And the presence in the *Saimiri* retina of midget bipolar cell comparable to those seen in Old World monkeys (B. Boycott, personal communication) suggests that there might be a subset of ganglion cells and parvocellular LGN units that draw their centre inputs from single cones. If the surround input is simply drawn promiscuously from all cones within a local area, such units will necessarily become chromatically opponent, as has been postulated in the case of Old World monkeys (Shapley and Perry, 1986; Lennie et al., 1989). If chromatically opponent units arise automatically in this way, we need only assume that the developing cortex has a general ability to identify correlated inputs; and then we are close to explaining how the heterozygote exploits her extra pigment.

In the above account we have assumed that the polymorphism is maintained by the heterozygous advantage of the trichromat. However, there are several other ways in which a polymorphism of colour vision could arise (Mollon et al., 1984) and one that especially should not be ruled out here is kinship selection. It is possible that species that forage in coordinated groups, as do tamarins, are most successful if the group contains a mixture of dichromats and trichromats. Menzel and Juno (1985), describing the foraging behaviour of saddle-backed tamarins in a semi-natural environment, report: 'On entering the test room the animals fanned out from one another and scanned different sections of the room, as if operating as a team and actively searching for the food... On first detecting the food, the animals very often made a distinctive 'food call', which brought others towards the object'. Since dichromats may penetrate camouflage that defeats trichromats (Anon, 1940; Morgan et al., 1989) and since trichromats can penetrate camouflage that defeats dichromats (see pp 306–309 above), the most successful foraging troop may be the one that includes a variety of visual systems. We should note here too that the visual tasks facing the New World monkeys may be different from those facing the Old World primate: the one available study of the colours of fruits taken by a platyrrhine monkey – *Callicebus torquatus* – showed that many were cryptic rather than conspicuous (Snodderly, 1979).

Do the Platyrrhini Represent a Stage in the Evolution of Catarrhine Trichromacy?

What then is the evolutionary relationship between the polymorphism seen in New World monkeys and the full trichromacy that we observe in Old World primates and man? Three possibilities can be distinguished. It may be worth setting them out here explicitly, since every genome carries with it a fossil record of its evolution, and in that redundant baggage of abandoned DNA we may soon find an empirical answer to our question.

1. The platyrrhine polymorphisms may represent an earlier stage in the evolution of our own trichromacy, a suggestion made by Bowmaker et al. (1987) and by Jacobs (1990). First, multiple alleles would be established at a single locus by a selective process such as described in the previous section, to give the situation now obtaining in New World monkeys. In the case of the Old World primates, we must imagine that an unequal crossing-over subsequently established two different alleles on one chromosome: for this to occur, two chromosomes carrying different alleles would have to be misaligned at meiosis in such a way that a subsequent break-point between the mispaired genes left one chromosome with two genes and one with none. We also need to postulate that a mechanism exists, or later evolves, to do the job that Lyonization does so prettily in the heterozygous platyrrhine – segregate the two gene products into different cone cells.

2. The second possibility is that platyrrhine polymorphism is *sui generis* and that Old World primates have gained their trichromacy by the route that is assumed in most discussions of the evolution of multi-gene families. By this more conventional assumption, duplication of one gene first establishes two identical copies on one chromosome. Later, one of the copies is changed by random mutations, and selective pressures then shape this gene for its new rôle. The essential difference from scenario (1) is that the duplication precedes the divergence of the genes, rather than *vice versa*. As in the case of (1), we require a mechanism that will ensure that the second gene is expressed in a distinct class of cone cells.

3. Logically there is a third possibility, that platyrrhine colour vision represents a degenerate form of the trichromacy seen in catarrhine monkeys. An evolution of this kind might have occurred if, say, there is an advantage to having a mixed troop of trichromats and dichromats (see pp 316–317). Trichromacy is not necessarily such a good thing that no species has ever given it up.

Acknowledgements

This chapter is a revised version of an essay originally published in the *Journal of Experimental Biology* (1989), 146, 21–38 and is reproduced with the kind permission of the Company of Biologists.

I am grateful to Gabriele Jordan, Professor M. J. Morgan, Mr E. Potma, Mr J. Reffin and Dr M. A. Webster for discussion. Plate 3 was designed by Gabriele Jordan.

Notes

1. An interesting self-report by a young physician of Dresden, Hans Haenel, who deliberately did not wear sunglasses during a two-day ski tour of the Riesengebirge in 1907. Although his acuity was afterwards only a little reduced, and his fundi were normal in appearance, he suffered a red-green colour blindness that took several weeks to clear. In his professional practice he had particular difficulty in judging complexions: from their waxen jaundiced faces and blue lips, all his patients looked as if they had severe heart failure (Best and Haenel, 1907).
2. Rodieck (1988) has drawn attention to a discrepancy, however: the axons of Mariani's 'blue-cone' bipolars end in the proximal part of the inner plexiform layer, whereas the dendrites of De Monasterio's short-wave ganglion cells make contacts in the distal part of the inner plexiform layer.

References

Allen, G. (1892). *The Colour-Sense: Its Origin and Development. An Essay in Comparative Psychology*, London: Kegan Paul.
Allison, A. C. (1964). Polymorphism and natural selection in human populations. *Cold Spring Harbor Symp. quant. Biol.*, 29, 137–149.
Anon (1940). Colour-blindness and camouflage. *Nature Lond.*, 146, 226.
Barlow, H. B. (1981). Critical limiting factors in the design of the eye and visual cortex. *Proc. R. Soc. Lond. Ser. B*, 212, 1–34.
Barlow, H. B. (1982). What causes trichromacy? A theoretical analysis using comb-filtered spectra. *Vision Res.*, 22, 635–643.
Best, F. and Haenel, H. (1907). Rotgrünblindheit durch Schneeblendung. *Klin. Monatsbl. Augenheilkd.*, 45, Beilageh. pp. 88–105.

Bowmaker, J. K., Astell, S., Hunt, D. M. and Mollon, J. D. (1991). Photosensitive and photostable pigments in the retinae of Old World monkeys. *J. Exp. Biol.*, in press.

Bowmaker, J. K., Dartnall, H. J. A., Lythgoe, J. N. and Mollon, J. D. (1978). The visual pigments of rods and cones in the rhesus monkey, *Macaca mulatta*. *J. Physiol.*, **272**, 329–348.

Bowmaker, J. K., Jacobs, G. H. and Mollon, J. D. (1987) Polymorphism of photopigments in the squirrel monkey: a sixth phenotype. *Proc. R. Soc. Lond. B*, **231**, 383–390.

Boyle, R. (1688). *Some Uncommon Observations About Vitiated Sight.* London: J. Taylor.

Boynton, R. M. (1979). *Human Color Vision*. New York: Holt, Rinehart.

Boynton, R. M. and Kambe, N. (1980). Chromatic difference steps of moderate size measured along theoretically critical axes. *Color Res. Applic.*, **5**, 12–23.

Bridges, C. D. B. and Delisle, C. E. (1974). Evolution of visual pigments. *Exper. Eye Res.*, **18**, 323–332.

Bridges, C. D. B. and Yoshikami, S. (1970). Distribution and evolution of visual pigment in salmonid fishes. *Vision Res.*, **10**, 609–626.

Brindley, G. S. (1964). The summation areas of human colour-receptive mechanisms at increment threshold. *J. Physiol.*, **124**, 400–408.

Chisholm, J. J. (1869). Colour blindness, an effect of neuritis. *Ophthalmic Hosp. Rep.*, April, 214–215.

Colquhoun, H. (1829). An account of two cases of insensibility of the eye to certain of the rays of colour. *Glasgow Med. J.*, **2**, 12–21.

Combe, G. (1824). Case of deficiency in the power of perceiving and distinguishing colours, accompanied with a small development of the organ, in Mr James Milne, Brassfounder in Edinburgh. *Trans. Phrenological Soc.* pp. 222–234.

Dartnall, H. J. A., Bowmaker, J. K. and Mollon, J. D. (1983). Human visual pigments: microspectrophotometric results from the eyes of seven persons. *Proc. R. Soc. Lond. B*, **220**, 115–130.

De Monasterio, F. M. (1979). Asymmetry of on- and off-pathways of blue-sensitive cones of the retina of macaques. *Brain Res.*, **166**, 39–84.

Derrington, A. M., Krauskopf, J. and Lennie, P. (1984). Chromatic mechanisms in lateral geniculate nucleus of macaque. *J. Physiol.*, **357**, 241–265.

Donders, F. C. (1883). *Nog eens: De Kleurstelsels, naar aanleiding van Hering's Kritiek*. Utrecht: Metzelaar.

Edwards, E. A. and Duntley, S. Q. (1939). The pigments and color of living human skin. *Am. J. Anat.*, **65**, 1–33.

Gouras, P. (1974). Opponent-color cells in different layers of foveal striate cortex. *J. Physiol.*, **238**, 582–602.

Gouras, P. (1990). Chromatic and achromatic contrast mechanisms in visual cortex. In *Colour Vision Deficiences*. ed. Ohta, Y. pp. 87–98. Amsterdam: Kugler & Ghedini.

Gouras, P. and Eggers, H. (1983). Responses of primate retinal ganglion cells to moving spectral contrast. *Vision Res.*, **23**, 1175–1182.

Gautier-Hion, A. *et al.* (1985). Fruit characteristics as a basis of fruit choice and seed dispersal in a tropical forest vertebrate community. *Oecologia*, **65**, 324–337.

Hallé, N. (1974). Attractivité visuelle des fruits pour les animaux. *J. Psychol. Norm. Path.*, **71**, 390–405.

Hárosi, F. I. (1982). Recent results from single-cell microspectrophotometry: cone pigment in frog, fish and monkey. *Colour Res. Applic.*, **7**, 135–141.

Huddart, J. (1777). An account of persons who could not distinguish colours. *Phil. Trans. R. Soc.*, **67**, 260–265.

Ingling, C. R. and Martinez, E. (1983). The spatiochromatic signal of the r-g channel. In *Colour Vision: Physiology and Psychophysics*. eds Mollon, J. D. and Sharpe, L. T. pp. 433–444. London: Academic.

Jacobs, G. H. (1981). *Comparative Color Vision*. New York: Academic.

Jacobs, G. H. (1984). Within-species variations in visual capacity among squirrel monkeys (*Saimiri sciureus*): colour vision. *Vision Res.*, **24**, 1267–1277.

Jacobs, G. H. (1990). Evolution of mechanisms for color vision. *Soc. Photo-Opt. Instrum. Engrs Proc.*, **1250**, 287–292.

Jacobs, G. H. and Neitz, J. (1986). Spectral mechanisms and color vision in the tree shrew (*Tupaia belangeri*). *Vision Res.*, **26**, 291–298.

Jacobs, G. H. and Neitz, J. (1987). Inheritance of color vision in a New World monkey (*Saimiri sciureus*). *Proc. Natl Acad. Sci. USA*, **84**, 2545–2549.

Jacobs, G. H., Neitz, J. and Crognale, M. (1987). Color vision polymorphism and its photopigment basis in a callitricid monkey (*Saguinus fuscicollis*). *Vision Res.*, **27**, 2089–2100.

Kalmus, H. (1955). The familial distribution of congenital tritanopia, with some remarks on some similar conditions. *Annals Hum. Genet.*, **20**, 39–56.

Katz, D. (1935). *The World of Colour*, London: Kegan Paul.

Kingdon, J. S. (1980). The role of visual signals and face patterns in African forest monkeys (guenons) of the genus *Cercopithecus*. *Trans. Zool. Soc. Lond.*, **35**, 425–475.

Kingdon, J. S. (1988). What are face patterns and do they contribute to reproductive isolation in guenons? In *A Primate Radiation: Evolutionary Biology of the African Guenons*. ed. Gautier-Hion, A., Bourliere, F., Gauter, J-P. and Kingdon, J. pp. 227–245. Cambridge: Cambridge University Press.

Koffka, K. (1935). *Principles of Gestalt Psychology*. London: Kegan Paul.

Koffka, K. and Harrower, M. R. (1932). Colour and organization. *Psychol. Forschung.*, **15**, 177–303.

Land, E. (1983). Recent advances in retinex theory and some implications for cortical computations: Color vision and the natural image. *Proc. Natl. Acad. Sci. USA*, **80**, 5163–5169.

Ladd-Franklin, C. (1982). A new theory of light sensation. In *International Congress of Psychology, 2nd Session*, Williams & Norgate, London, 1892, pp. 103–108.

Lennie, P., Haake, P. W. and Williams, D. R. (1989). Chromatic opponency through random connections to cones. *Invest. Ophthal. Vis. Sci.*, **30** (Suppl.), 322.

Liebmann, S. (1927). Über das Verhalten farbiger Formen bei Helligkeitsgleichheit von Figur und Grund. *Psychol. Forsch.*, **9**, 300–353.

Livingstone, M. and Hubel, D. (1984). Anatomy and physiology of a color system in the primate visual cortex. *J. Neurosci.*, **4**, 309–356.

Livingstone, M. and Hubel, D. (1988). Segregation of form, color, movement, and depth: anatomy, physiology, and perception. *Science Wash. DC*, **240**, 740–749.

Mariani, A. P. (1984). Bipolar cells in monkey retina selective for the cones likely to be blue-sensitive. *Nature Lond.*, **308**, 184–186.

Marshak, D. W., Aldrich, L. B., Del Valle, J. and Yamada, T. (1990). Localization of immunoreactive cholecystokinin precursor to amacrine cells and bipolar cells of the macaque monkey retina. *J. Neurosci.*, **10**, 3045–3055.

McDougall, W. (1901). Some new observations in support of Thomas Young's theory of light and colour vision (II). *Mind*, **10**, 210–245.

Meadows, J. C. (1974). Disturbed perception of colours associated with localized cerebral lesions. *Brain*, **97**, 615–632.

Menzel, E. W. and Juno, C. (1985). Social foraging in marmoset monkeys and the question of intelligence. *Phil. Trans. R. Soc. Lond. B*, **308**, 145–158.

Michael, C. R. (1988). Retinal afferent arborization patterns, dendritic field orientations, and the segregation of function in the lateral geniculate nucleus of the monkey. *Proc. Natl Acad. Sci. USA*, **85**, 4914–4918.

Mollon, J. D. (1980). Post-receptoral processes in colour vision. *Nature*, **283**, 623–624.

Mollon, J. D. (1982). A taxonomy of tritanopias. In *Colour Vision Deficiencies VI*. ed. Verriest G. (*Doc. Ophthal. Proc. Ser. No. 33*), pp 87–101. The Hague: Dr W. Junk.

Mollon, J. D. (1987). On the origins of polymorphisms. In *Frontiers of Visual Science*. ed. Committee on Vision, National Research Council, pp. 160–168. Washington, DC: National Academy Press.

Mollon, J. D. (1987). On the nature of models of colour vision. *Die Farbe*, **34**, 29–46.

Mollon, J. D., Bowmaker, J. K. and Jacobs, G. H. (1984). Variations of colour vision in a New World primate can be explained by polymorphism of retinal photopigments. *Proc. R. Soc. Lond. B*, **222**, 373–399.

Mollon, J. D., Estévez, O. and Cavonius, C. R. (1990). The two subsystems of colour vision and their roles in wavelength discrimination. In *Vision: Coding and Efficiency*. ed. Blakemore, C. B. pp. 119–131. Cambridge: Cambridge University Press.

Mollon, J. D. and Jordan, G. (1988). Eine evolutionäire Interpretation des menschlichen Farbensehens. *Die Farbe*, **35**, 139–170.

Mollon, J. D., Newcombe, F., Polden, P. G. and Ratcliff, G. (1980). On the presence of three cone mechanisms in a case of total achromatopsia. In *Colour Vision Deficiencies V*. ed. Verriest, G. pp. 130–135. Bristol: Hilger.

Monge, G. (1791). Mémoire sur quelques phénomènes de la vision. *Ann. Chim.*, **3**, 131–147.

Morgan, M. J., Mollon, J. D. and Adam, A. (1989). Dichromats break colour-camouflage of textural boundaries. *Invest. Ophthalmol. Vis. Sci.*, **30**, (Suppl.), 220.

Nathans, J., Piantanida, T. P., Eddy, R. L., Shows, T. B. and Hogness, D. S. (1986a). Molecular genetics of inherited variation in human color vision. *Science Wash. DC*, **232**, 203–210.

Nathans, J., Thomas, D. and Hogness, D. S. (1986b). Molecular genetics of human colour vision: the genes encoding blue, green and red pigments. *Science Wash. DC*, **232**, 193–202.

Nicholl, W. (1818). Account of a case of defective power to distinguish colours. *Med.-Chirurg. Trans.*, **9**, 359–363.

Norden, K. (1948). *Shadow and Diffusion in Illuminating Engineering*. London: Pitman.

Perry, V. H. and Cowey, A. (1981). The morphological correlates of X- and Y-like retinal ganglion cells in the retina of monkeys. *Exp. Brain Res.*, **43**, 226–228.

Pokorny, J., Smith, V. C., Verriest, G. and Pinckers, A. J. L. G. eds. (1979). *Congenital and Acquired Color Vision Defects*. New York: Grune & Stratton.

Polyak, S. (1957). *The Vertebrate Visual System*. Chicago: Chicago University Press.

Rodieck, R. W. (1988). The primate retina. In *Comparative Primate Biology*. ed. Stecklis, H. D. and Erwin, J. pp. 203–278. New York: Liss.

Rushton, W. A. H. (1972). Pigments and cones in colour vision. *J. Physiol.*, **220**, 1–31P.

Schiller, P. H. and Malpelli, J. G. (1978). Functional specificity of lateral geniculate nucleus laminae of the rhesus monkey. *J. Neurophysiol.*, **41**, 788–797.

Schnapf, J. L., Kraft, T. W. and Baylor, D. A. (1987). Spectral sensitivity of human cone receptors. *Nature Lond.*, **325**, 439–441.

Shapley, R. and Perry, V. H. (1986). Cat and monkey retinal ganglion cells and their visual functional roles. *Trends Neurosci*, **9**, 229–235 (May).

Smith, V. C. and Pokorny, J. (1975). Spectral sensitivity of the foveal cone photopigments between 400 and 500 nm. *Vision Res.*, **15**, 161–171.

Snodderly, D. M. (1979). Visual discriminations encountered in food foraging by a neotropical primate: implications for the evolution of color vision. In *Behavioral Significance of Color*. ed. Burtt, E. H. Jr. pp. 237–279. New York: Garland.

Sourd, C. and Gautier-Hion, A. (1986). Fruit selection by a forest guenon. *J. Animal Ecol.*, **55**, 235–244.

Steward, J. M. and Cole, L. (1989). What do color vision defectives say about everyday tasks? *Optom. Vis. Sci.*, **66**, 288–295.

Stiles, W. S. (1949). Increment thresholds and the mechanisms of colour vision. *Doc. Ophthal.*, **3**, 138–163.

Stiles, W. S. (1952). Colour vision: a retrospect. *Endeavour*, **11**, 33–40.

Stilling, J. (1877). *Die Prüfung des Farbensinnes beim Eisenbahn- und Marine-Personal*. Kassel: Fischer.

Stilling, J. (1883). *Pseudo-isochromatische Tafeln für die Prüfung des Farbensinnes*. Kassel & Berlin: Fischer.

Stromeyer, C. S., Kronauer, R. E. Madsen, J. C. and Cohen, M. A. (1980). Spatial adaptation of short-wavelength pathways in humans. *Science Wash. DC*, **207**, 555–557.

Stromeyer, C. S. and Lee, J. (1987). Motion and short-wave cone signals: influx to luminance mechanisms. *Invest. Ophthal. Vis. Sci.*, **28** (Suppl.), 232.

Tansley, B. W. and Boynton, R. M. (1976). A line, not a space, represents visual distinctness of borders formed by different colors. *Science Wash. DC*, **191**, 954–957.

Thoma, W. and Scheibner, H. (1980). Die spektrale tritanopische Sättigungsfunktion beschreibt die spektrale distinktibilität. *Farbe Des.*, **17**, 49–52.

Tovée, M. J. (1990). A polymorphism of the middle- to long-wave cone photopigments in the common marmoset (*Callithrix jacchus jacchus*): a behavioural and microspectrophotometric study. Ph.D. thesis, University of Cambridge.

Travis, D. S., Bowmaker, J. K. and Mollon, J. D. (1988). Polymorphism of visual pigments in a callitricid monkey. *Vision Res.*, **28**, 481–490.

Ts'o, Y. and Gilbert, C. D. (1988). The organization of chromatic and spatial interactions in the primate striate cortex. *J. Neurosci.*, **8**, 1712–1727.

Vallée, L. L. (1821). *Traité de la science du dessin, contenant la théorie générale des ombres, la perspective linéaire, la théorie générale des images d'optique, et al perspective aérienne appliquée au lavis*, Mme Ve. Courcier, Paris.

Vollrath, D., Nathans, J. and Davies, R. W. (1988). Tandem array of human visual pigment genes at Xq28. *Science Wash. DC*, **240**, 1669–1672.

Webster, M. A., De Valois, K. K., and Switkes, E. (1990). Orientation and spatial-frequency discrimination for luminance and chromatic gratings. *J. Opt. Soc. Am. A*, **7**, 1034–1049.

Wertheimer, M. (1923). Untersuchungen zur Lehre von der Gestalt. *Psychol. Forsch.*, **4**, 301–350.

Zeki, S. M. (1978). Uniformity and diversity of structure and function in rhesus monkey prestriate visual cortex. *J. Physiol.*, **277**, 273–290.

PART III

PHYLOGENETIC EVOLUTION OF THE EYE AND VISUAL SYSTEM

15 Photosensory Systems in Eukaryotic Algae

John D. Dodge

Introduction

It has long been known that many flagellated, and therefore motile, algae can perform movements in relation to the direction of light. In the nineteenth century the famous botanist Strasburger introduced the term phototaxis to describe this phenomenon. From an early time it was assumed that there was a part of the cell which acted as the photosensory receptor able to pass information to the locomotory apparatus. In many of these organisms there is a very obvious body located in a particular part of the cell and brightly coloured, red or orange. This was assumed to be the photoreceptor and was given the name eyespot or stigma. In a small number of organisms which are classified in the group Dinophyta there is a much larger body called an ocellus with both a pigmented area and what appears to be a type of lens. Ultrastructural studies over the past 30 years have shown that there are several distinct types of eyespot apparatus and other structural features are often associated with the pigmented stigma. These will be surveyed in the present chapter and have been the subject of a number of earlier reviews (Dodge, 1969; Piccinni and Omodeo, 1975; Foster and Smythe, 1980; Greuet, 1982). Eyespots have certain characteristics in common. They all contain pigmented lipid droplets which may be bound together by one or more membranes. The pigments consist of one or more carotenoids, related to those which may be found in chloroplasts. Finally, the eyespot is located in a fixed position in the cell which is characteristic for the organism.

The fact that the eyespot is so conspicuous has tended to obscure the fact that the actual photoreceptor is almost invariably some other part of the cell. Indeed, some phototactic organisms, such as the dinoflagellate *Gyrodinium dorsum*, which has a very precise photic response (Hand and Schmidt, 1975), show no trace of an eyespot even when subjected to detailed ultrastructural study (Dodge, unpublished). Structural studies and analysis of responses have shown that what has been called the 'light antenna', or the combination of eyespot and other structures, function in various ways (Foster and Smythe, 1980). This then raises the question of the evolution of the diverse mechanisms. The suggestion has been made by Kivic and Walne (1983) that there has been a degree of parallel evolution among these photosynthetic protists which are more generally classed as algae.

In the present chapter we will only consider photosensory responses in flagellate algae and flagellates which possess an obvious eyespot structure. For convenience an outline classification scheme of the algal groups covered in this chapter is given in Table 15.1.

Description of the Eyespots and Associated Structures

Here the eyespots will be classified in a system slightly modified from that of Dodge (1969), which is based on the position of the pigmented structure and its relationship with the flagella (Fig. 15.1).

Type A: Eyespot Part of a Chloroplast but not Obviously Associated with the Flagella

This type has been found in three of the algal groups: in the Chlorophyceae and Prasinophyceae, which are both 'green algae' (Fig. 15.2) and in the Cryptophyceae (Fig. 15.3). In the green algae the eyespot apparatus has been the subject of a very detailed review covering structure, function and evolution (Melkonian and Robenek, 1984), and the reader is referred to this for more details than can be given here.

Chlorophyceae
In this class the eyespot is situated on one side of the cell (Fig. 15.2(a)), where it may cause a slight bulge to the cell

Table 15.1 *A classification scheme for the eukaryotic algal groups with their types of eyespot apparatus listed. Some common names used in this chapter are shown on the right.*

Phylogenetic line	Group	Eyespot types A	B	C	Other	Common name
	Rhodophyta				none	red algae
	Cryptophyta	+				cryptophytes
	Dinophyta				various	dinoflagellates
	Prymnesiophyta		+			haptophytes
	Eustigmatophyta			+		—
	Raphidophyta				none	—
Chromophytes	Xanthophyta		+			—
	Chrysophyta		+			chrysophytes
	Bacillariophyta				none	diatoms
	Phaeophyta		+			brown algae
	Glaucophyta				none	—
	Euglenophyta			+		euglenas
Chlorophytes	Chlorophyta	+				green algae
	Prasinophyta	+				prasinophytes
	Charophyta				none	charophytes

covering. It forms a small part of the single cup-shaped chloroplast which is the norm in these organisms. The most common type consists of a single layer of closely packed globules, containing carotenoid pigment, lying between the outermost thylakoid (a pigment-containing membrane sac) and the two-layered chloroplast envelope. Often the eyespot area protrudes slightly from the surface of the chloroplast. In many organisms there are one or more additional layers of globules underneath the first layer (Fig. 15.2(c)) but each is normally separated by a single swollen thylakoid. The colonial flagellates of the order Volvocales, where some cell specialization is found, may have up to nine layers in the eyespots on one side of the colony. In addition to the variability in the number of layers and thus of pigmented globules, there is also much variety in the size of the globules. They range from 80 to 160 nm in diameter and the number of globules from 30 to approximately 2000. The surface area of the eyespot varies from 0.3 to 9.0 μm^2. There is some evidence that the number of globules is not precisely fixed for, in *Dunaliella*, it appears to increase by a factor of two between the logarithmic and stationary phases of culture (Hoshaw and Maluf, 1981).

Freeze-fracture studies (Melkonian and Robenek, 1979, 1984) have shown that the outer membrane of the chloroplast envelope over the eyespot has a much higher number of 6–8-nm intramembranous particles than other parts of the envelope. It is suggested that this membrane may be the actual photoreceptor involved in primary sensory transduction (Melkonian and Robenek, 1980a; 1980b). In some organisms one of the microtubular roots of the flagellar system has been found to run close to the eyespot. For example, in *Spermatozopsis* a four- or five-microtubule root lies between the chloroplast envelope

and plasma membrane, where it covers the eyespot (Preisig and Melkonian, 1984). In *Haematococcus* the microtubules pass close to the edge of the eyespot (Ristori and Rosati, 1983). In this organism experimental studies have suggested that an electrically active portion of the cell membrane lies over the eyespot (Ristori *et al.*, 1981).

Prasinophyceae

The majority of members of this class are flagellates which have scaly flagella and the few benthic representatives have a flagellate stage. Probably all species possess an eyespot and this is generally situated in a lateral position, within the chloroplast, as described for the Chlorophyceae (Fig. 15.1(a); Fig 15.2(d)). A number of taxa have been investigated. For example, in *Mantoniella* the single-layer eyespot contains about 70 close-packed globules which, in this asymmetric cell, are situated almost opposite the flagellar insertion (Barlow and Cattolico, 1980; Marchant *et al.*, 1989). In *Pyramimonas pseudoparkeae* the three- to four-layered eyespot is concave and the adjacent cell surface is depressed (Pienaar and Aken, 1985). By way of contrast, in *P. nanseni* the two-layered eyespot is fairly straight and there is a distinct bulge where it is pressed tightly against the plasma membrane (Thomsen, 1988). In this group the main flagellar roots run into the centre of the cell and there are no reports of any root structures adjoining the eyespot.

In summary, in the Prasinophyceae the size of the eyespot ranges from 0.3 to 3.8 μm^2, the globules are 80–190 nm in diameter and 30 to 350 in number and they are normally arranged in only one or two layers (Fig. 15.2(d)).

In both of the above groups the cells rotate as they swim forwards. Thus the eyespot, which is on one side of the cell

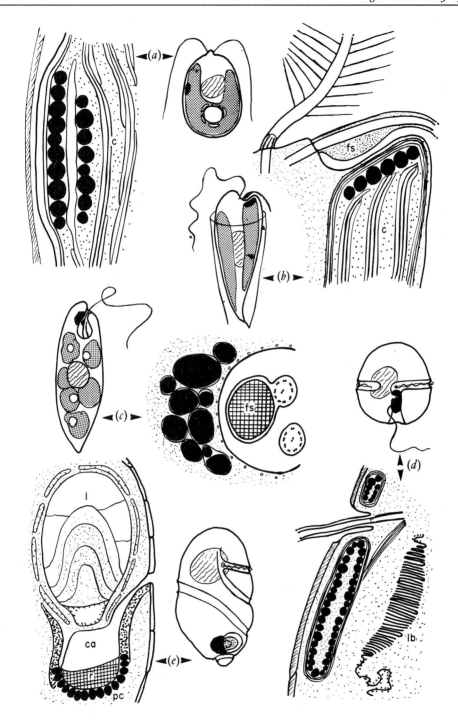

Fig. 15.1 *Drawings to illustrate the usual location and structure of the three main types of eyespot apparatus or photosensory system found in the algae (types a–c) and two of the variants found in the dinoflagellates: (d) that of* Glenodinium*; (e) the ocellus of* Nematodinium. *c, chloroplast; ca, chamber; fs, flagellar swelling; l, lens; lb, lamellar body; pc, pigment cup; r, retinoid. (After Dodge, 1969; not to scale.)*

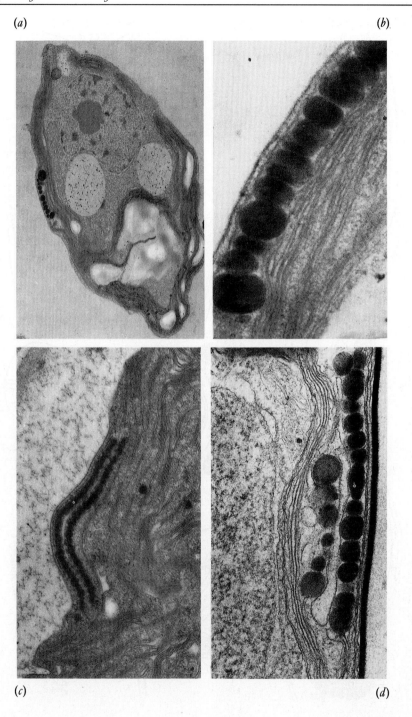

Fig. 15.2 *Electron micrographs of eyespots in the green algae. (a,b)* Dunaliella *(Chlorophyceae): (a) shows a longitudinal section of a cell with the eyespot situated at one side of the cup-shaped chloroplast (×6000); (b) is a higher magnification picture to show the single layer of pigmented droplets situated within the chloroplast (×43 000) (c) The two-layered eyespot in a species of* Chlamydomonas *(Chlorophyta). (×16 800). (d) A higher-magnification picture of the two-layered eyespot in* Platymonas *(Prasinophyta) (×40 000) ((a) and (b) are courtesy of K. Berubé).*

and consequently edge-on to the light, will be in a position of minimal light reception while the cell swims towards the light. Light falling on the eyespot will increase immediately the cell deviates from this course.

Cryptophyta, Class Cryptophyceae

In this distinctive group of flagellated organisms only a very small number of species are known to have eyespots, and these would all seem to belong to the genus *Chroomonas*. Here, the eyespot (designated type A2) is rather unusually situated near the centre of the cell and it consists of a single layer of about 35 closely packed globules placed at the end of a spur of the chloroplast which extends beyond the single pyrenoid (Fig. 15.3) (Dodge, 1969; Lucas, 1970; Meyer and Pienaar, 1984; Santore, 1987). The eyespot is bounded by the four membranes, two of the chloroplast envelope and two of the endoplasmic reticulum, which surround the plastidial compartment of the cell. Often there is a rather empty-looking vesicle lying over the face of the eyespot (Fig. 15.3, main picture) which could perhaps act as a simple lens. As yet there have been no reports of any flagellar root or other connections to the eyespot in these organisms. It has been suggested (Foster and Smyth, 1980) that the short thylakoids situated between the eyespot and the pyrenoid (Fig. 15.3, main picture) could be the photoreceptor, but not all the species with eyespots have this structure. It was also noted that the eyespot will absorb most strongly when the light is at right angles to the plane of movement (or the long axis of the cell), but it is difficult to see how this would help a cell swim towards a light source. However, other studies have shown that the red pigment, phycoerythrin, is the receptor for phototaxis (Watanabe and Furuya, 1974) and this is definitely present in the chloroplast thylakoids in these organisms (Gantt *et al.*, 1971; Ludwig and Gibbs, 1989).

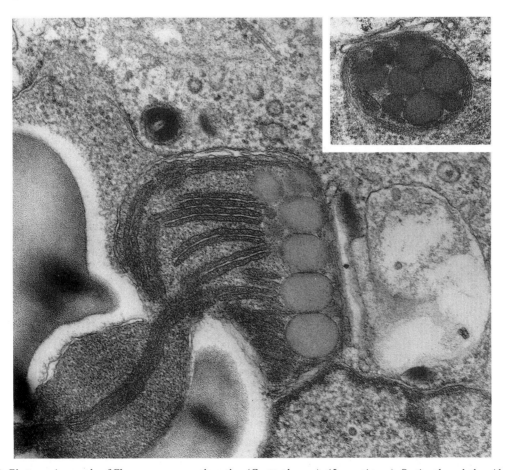

Fig. 15.3 *Electron micrographs of* Chroomonas mesostigmatica *(Cryptophyceae). (Large picture). Section through the mid-region of the cell showing the type A eyespot subtended by a short spur of chloroplast (with dense thylakoids) which projects beyond the pyrenoid. At the right a large vacuole lies in front of the eyespot. ($\times$ 56 000). (Inset) Section in the plane of the eyespot showing the arrangement of the approximately 9 carotenoid-containing droplets ($\times$ 48 000).*

Type B: Eyespot Part of a Chloroplast and Closely Associated with the Flagella

This type of eyespot (Fig. 15.1(b)) is found in four of the algal groups which make up the division of the algae often referred to as the Chromophyta: Chrysophyta, Xanthophyta, Phaeophyta and Prymnesiophyta. A large proportion of the flagellated species of the Chrysophyta, all of the zoospores (motile reproductive stages) of the Xanthophyta and male gametes of most Phaeophyta possess an eyespot. The basic arrangement is always the same in that the eyespot is part of a chloroplast situated in close proximity to the flagellar swelling, but there are a number of differences in location and in the condition of the associated flagellum. It should be noted that most of these algae have what are known as heterokont flagella, where one, usually bearing stiff hairs, points in a forward direction and the other is smooth and is bent to point towards the posterior. Often this flagellum is much reduced in length.

Chrysophyta and Xanthophyta

Here, typically, a single layer of about 40 lipid droplets is situated at one side of the anterior end of the two chloroplasts (Figs 15.1(b) and 15.4(a,b)). The eyespot is therefore bounded by the four membranes which constitute the chloroplast envelope and its associated endoplasmic reticulum. Within the chloroplast it seems to be contained by the outermost lamella of the photosynthetic system. Adjacent to the eyespot there is a groove formed by the cell-covering membrane and through this depression passes the posteriorly directed flagellum, which in this region is swollen on one side. This swelling contains amorphous or lamellate material and is situated directly above the eyespot (Chrysophyta: Dodge, 1969, 1973; Wujak, 1969; Hibberd, 1971, 1976; Anderson, 1987. Xanthophyta: Massalski and Leedale, 1969; Hibberd and Leedale, 1971).

In the chrysophyte *Dinobryon* a short structural connection has been found between the base of the posteriorly directed flagellum and the eyespot (Kristiansen and Walne, 1976, 1977). This takes the form of a fibrous, banded structure and it clearly could be responsible for the transfer of excitation from the eyespot to the cell motor apparatus. However, this would go against the normal view that the flagellar swelling is the photoreceptor and the pigmented eyespot merely a shading device. In *Chromulina placentula* and *Phaeaster pascheri* (Belcher and Swale, 1967, 1971) the eyespot is situated in a finger-like extension of the chloroplast envelope. Here the posterior flagellum is reduced to a short, swollen stump and so the combination of this and the eyespot situated away from the other cell organelles does make this structure appear like a specialized functional light antenna. A further development is found in *Anthophysa* where a red eyespot and a flagellar swelling are present, but the chloroplast is reduced to a colourless leucoplast (Belcher and Swale, 1972b). When this organism was illuminated with polarized light the authors noted a strong reflection from a colourless layer which backs the pigmented droplets. This would seem to be a modification which has followed the loss of photosynthetic capability by the chloroplasts.

There are many other variations in this group. Recent studies (Coleman, 1988; Kawai, 1988) have shown that the flagellar swelling and the remainder of the short flagellum of some other chrysophytes exhibits strong blue–green autofluoresence with maximal emission of 515–520 nm under excitation by light at 440 nm. This fluorescence has also been found in one flagellum of *Prymnesium* (Prymnesiophyceae—see below) and in the flagellar swelling of various phaeophytan male gametes. The material responsible for the fluorescence and possibly also the active photoreceptor is thought to be a flavin (Kawai, 1988).

A number of members of the Chrysophyta have no eyespots, but nevertheless flagellar swellings may be present (Belcher and Swale, 1972a; Hibberd, 1971, 1976; Anderson, 1985; 1987). In *Poterioochromonas* there is no eyespot, but instead a vesicular structure has developed from the periplastidial membranes and this lies beneath the plasma membrane (Schnepf *et al.*, 1977).

According to Foster and Smythe (1980) the basic light antenna design in the organisms described above represents a light screen (pigmented eyespot) acting in association with a photosensitive body (flagellar swelling). The antenna points more or less normal to the direction of movement.

Phaeophyta

In this group the only motile stages consist of biflagellated zoospores and male gametes. In general these have a lateral insertion of the heteromorphic flagella. Very few have been studied by electron microscopy, but in those which have there is a prominent eyespot (Figs. 15.4(c,d)) consisting of a much-reduced chloroplast which is concave in shape and contains a single layer of about 60 pigmented globules (Clayton, 1989; Dodge, 1969; Manton, 1964). The eyespot is situated in the posterior part of the cell behind a depression of the cell surface through which runs the posterior flagellum, which here is swollen. In some cases, such as *Scytosiphon*, the chloroplast is not so reduced and normal thylakoids are present in addition to the eyespot globules.

It is interesting that during the development of the spermatozoids, before they are released from the antheridium, the flagellar swelling is pressed tightly into the groove over the eyespot, and this seems to contact the flagellar base at one end (Moestrup, 1982). Little structural information is available concerning the flagellar swelling, but

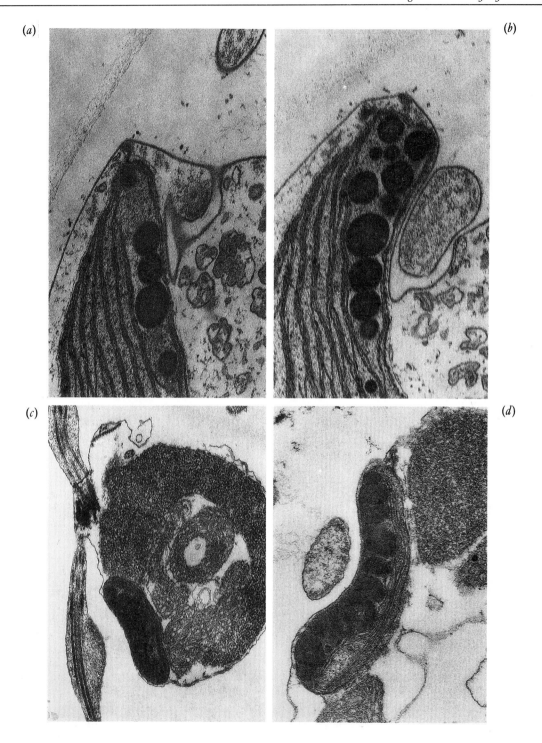

Fig. 15.4 *Electron micrographs showing eyespots of type B. (a,b). Two of a series of sections through the anterior end of a cell of the chrysophyte* Dinobryon. *The eyespot is situated in one of the chloroplasts and the posterior flagellum lies in a groove adjacent to this. The anterior flagellum can be seen at the top of (a) (×40 000). (c,d) Longitudinal and transverse sections through the male gamete of* Fucus *showing the position of the eyespot and the swelling on the posterior flagellum (c), (c, ×24 000; d, ×32 000).*

recent studies have shown that it has a strong autofluorescence (see above: Coleman, 1988; Kawai, 1988; Müller et al., 1987).

Prymnesiophyta

Eyespots are only found in a few species belonging to the order Pavlovales. Here, a single layer of carotenoid-containing droplets is situated at the anterior end of one of the chloroplasts. This is just beneath the posteriorly directed flagellum where it emerges from the cell (Green and Manton, 1970; Green, 1973, 1975; Van der Veer, 1976). There is a distinctive curved pit formed by an invagination of the plasma membrane which penetrates from near the base of the forward flagellum and takes a curved path to end above the eyespot. Minor variations in eyespot structure have been recorded in some species of the order Pavlovales (Green, 1980).

In the genus *Diacronema* the posterior flagellum is short and bears an elongated swelling of unique construction, where it passes over the eyespot, composed of a concave layer of approximately 40 droplets (Green and Hibberd, 1977). The swelling consists of a series of teeth-like projections which extend from one pair of doublets of the axoneme and which are surrounded by amorphus material. This structure is quite different from the amorphous, layered, or paracrystalline flagellar swelling found in other organisms. It is of interest that *Diacronema* is said to show no phototactic response in spite of having the eyespot and flagellar swelling (quoted in Foster and Smythe, 1980). No other information is available on the functioning of the eyespot apparatus in this group.

Type C: Eyespot Near the Flagella but not Part of a Chloroplast

This is the characteristic eyespot type of the Euglenophyta and a variant is also found in the chromophyte group Eustigmatophyta.

Euglenophyta

The eyespot here consists of a somewhat loose collection of carotenoid-containing lipid droplets situated on the

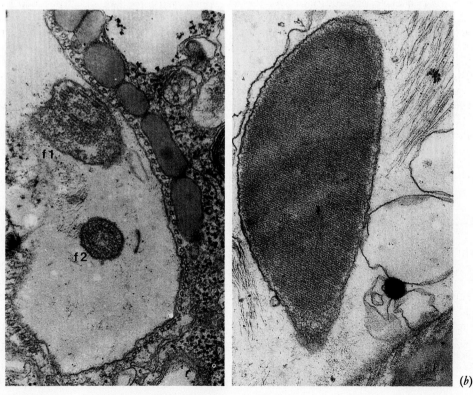

Fig. 15.5 *Electron micrographs of the type C eyespot in* Trachelomonas *(Euglenophyta). (a) Transverse section through the anterior invagination of the cell showing a single layer of large eyespot droplets on the right with an adjacent row of microtubules which surround the invagination. The two flagella seen in section are the emergent flagellum (f1) which bears the swelling and the reduced flagellum (f2) (× 40 000). (b) A longitudinal section which cuts through the photoreceptor swelling on the emergent flagellum and shows the paracrystalline contents. Some fine flagellar hairs are seen at top right (× 48 000).*

dorsal side of the anterior invagination (or reservoir) which is characteristic of these organisms (Figs. 15.1(c) and 15.5(a)). The droplets, which form a structure 3–7 μm in diameter, lie in a single layer or may be bunched together. Sometimes individual droplets appear to be membrane-bound, but there is never any membrane or envelope surrounding the eyespot and there is no association with any chloroplast. A single layer of microtubules which is part of the cell's cytoskeleton system lies between the eyespot and the plasma membrane where it lines the anterior invagination (Fig. 15.12) (Dodge, 1975; Leedale, 1982; Walne and Arnott, 1967; Walne, 1971; 1980). The carotenoid pigments of the eyespot have been extensively studied and numerous pigments including flavin have been identified (Pagni et al., 1976; see Walne, 1980, for a summary).

Associated with the eyespot is a paraflagellar swelling situated on the long or emergent flagellum (Fig. 15.5(b)). This body, which probably moves in front of the eyespot when the flagellum is beating, contains a paracrystalline material which is joined to the paraflagellar rod (West et al., 1980). The flagellar swelling is a dichroic structure which is said to contain flavin, a fact confirmed by the action spectrum for phototaxis (Beneditti and Checcucci, 1975; Benedetti and Lenci, 1977). It consists of a monoclinic or slightly distorted hexagonal unit cell with dimensions: $a = 8.9$ nm, $b = 7.7$ nm, $c = 8.3$ nm, $\beta = 110°$ (Piccinni and Mammi, 1978).

The euglenoid eyespot apparatus is said to be able to detect direction, intensity and quality of light (Walne, 1980). The traditional theory was that the eyespot globules provided a shading device to the paraflagellar swelling, which was the effective photoreceptor. However, Creutz and Diehn (1976), using polarized light, found that Euglena cells oriented themselves perpendicular to the beam, thus giving maximum light absorption. High-speed cine studies showed that in a high-intensity beam of light the cells reacted by turning towards the dorsal side of the cell (Diehn et al., 1975). There is still much to be resolved here.

The environment can affect the structure of the euglenoid eyespot, which diminishes in size when the cell is subject to heterotrophic growth in the dark. Presumably this is because the chloroplasts are unable to synthesize carotenoids. The restoration of light rapidly restores the eyespot to its normal state (Kivic and Vesk, 1972). It has recently been shown that in dark-grown cells of Euglena the eyespot becomes reduced to a patch of amorphous material. Bringing such cells into the light for 72–96 hours of continuous illumination results in the formation of a normal eyespot (Osafune et al., 1985).

Eustigmatophyta

In this small and rather obscure group of heterokont yellow–green algae the vegetative phase is non-motile, so it is the reproductive zoospores which have flagella. Here, an eyespot is present at the anterior end of the cell adjacent to the flagellar insertion. It consists of a somewhat irregular collection of pigment-containing lipid droplets situated in a slight bulge of the zoospore but not enclosed by a membrane (Hibberd and Leedale, 1972; Hibberd, 1980; Preisig and Wilhelm, 1989). The anterior hairy flagellum has a pronounced swelling, called a paraflagellar button by Foster and Smythe (1980), which fits alongside the eyespot. The flagellar swelling has one or two layers of electron dense material situated in the expanded part adjacent to the eyespot. It is interesting that in these organisms there is a rather unusual form of movement brought about by tip-generated sine wave movement of the anterior flagellum, which hauls the zoospore along in a fairly straight line with very little rotation (Hibberd, 1980).

Type D: Various Eyespots and Ocelli in the Dinophyta

In this group of mainly unicellular organisms only a small number of the several thousand motile species are known to have some form of eyespot apparatus. Rather surprisingly, they fall into several different categories. The most simple (Fig. 15.6(a)) are rather like type B described above and are found in two species of *Woloszynskia*, in *Gymnodinium* sp. and in *Peridinium cinctum* (Crawford et al., 1970; Bibby and Dodge, 1972; Schnepf and Deichgräber, 1972; Messer and Ben-Shaul, 1969). Here, the eyespot consists of a layer of carotenoid-containing droplets under the chloroplast envelope and situated behind the longitudinal groove or sulcus. In some cases there are microtubules between the eyespot and the cell covering and in *W. sanguineum* (= *Glenodinium sanguineum*) above the eyespot there is a vesicle which appears to contain guanine crystals (Dodge et al., 1987). In none of these cases is there any obvious modification to the chloroplast, and no flagellar swellings have been reported.

The second type has some similarities with type C, as described above, and this has so far only been found in *Woloszynskia coronata* (Crawford and Dodge, 1971; Dodge, 1984). Here, an irregular cluster of lipid droplets lies beneath the sulcus and immediately adjacent to the subthecal microtubules (Fig. 15.6(b)). Often droplets appear to have become fused together. There is no connection with a chloroplast, neither are there any membranes surrounding the eyespot. A striated root which runs from the base of the longitudinal flagellum seems to make contact with the top edge of the eyespot. A careful study of longitudinal sections showed that although the flagellum passes closely in front of the eyespot there seems to be no swelling or other special structure as part of the flagellum in this region. However, it should be noted that this flagellum has a complex paraflagellar rod.

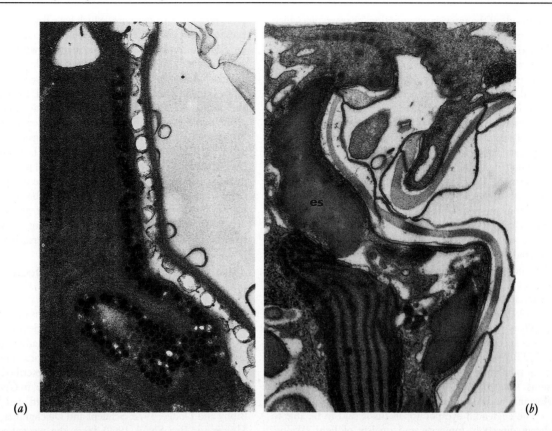

Fig. 15.6 *Sections of two of the types of eyespot found in the Dinophyta. (a) Eyespot, consisting of numerous globules, within the chloroplast in* Woloszynskia tenuissimum *(× 25 600). (b) Eyespot independent of the chloroplast in* W. coronata. *Here the globules appear to have partly fused together. Part of the sulcal groove and flagellar roots can also be seen at the top of the picture (× 24 000).*

The third type of dinoflagellate eyespot has so far been reported in two closely related organisms, *Glenodinium foliacaeum* and *Peridinium balticum* (Dodge and Crawford, 1969; Tomas and Cox, 1973). In these organisms the eyespot is a roughly triangular body situated behind the sulcus (longitudinal groove) with a gap towards the anterior end through which the longitudinal flagellum emerges (Fig. 15.7(a)). This eyespot is an independent structure bounded by a three-membrane envelope. Basically, there are two layers of pigmented globules, generally separated by a vesicle or granular material (Fig. 15.7(b)). In some cases the eyespot is folded back upon itself, thus making four or more layers. Just beneath or adjacent to the eyespot is a lamellate body (Fig. 15.7(b,d)), which appears to be connected to the endoplasmic reticulum system of the cell. It has been speculated that this lamellar body is part of the photosensory apparatus of the cell, but there is as yet no experimental evidence to support this idea. A layer of microtubules lies between the eyespot and the cell covering and fibrillar roots from both flagellar bases appear to run through or close to the eyespot (Fig. 15.7(c)). There is no evidence of a flagellar swelling in this species.

The fourth type of dinoflagellate eyespot is the much larger and more complex ocellus found in some members of the order Warnowiales. Since this has been described in detail by Greuet (1968; 1978; 1987) and Mornin and Francis (1967), only a brief description will be given here. The ocellus (Fig. 15.1(e)) is situated towards the left side of the ventral surface of the cell, in much the same position as other dinoflagellate eyespots. There are several distinct parts (Figs. 15.8 and 15.9). Most conspicuous is a large refractile structure, termed the hyalosome, thought to be capable of acting as a focusing lens. The hyalosome is constructed by the superimposition of a number of endoplasmic vesicles, which sit upon a basal plate. Beneath the hyalosome is a chamber which represents an invagination of the cell covering or theca and therefore allows the entry of external medium into the ocelloid. The third part of the organelle is the domed pigmented part or melanosome. This consists of two sections, a retinoid and a pigment cup. The retinoid is an extremely complex membranous construction made up of numerous regularly arranged layers giving an almost paracrystalline appearance. The outer boundary of the melanosome is formed from a

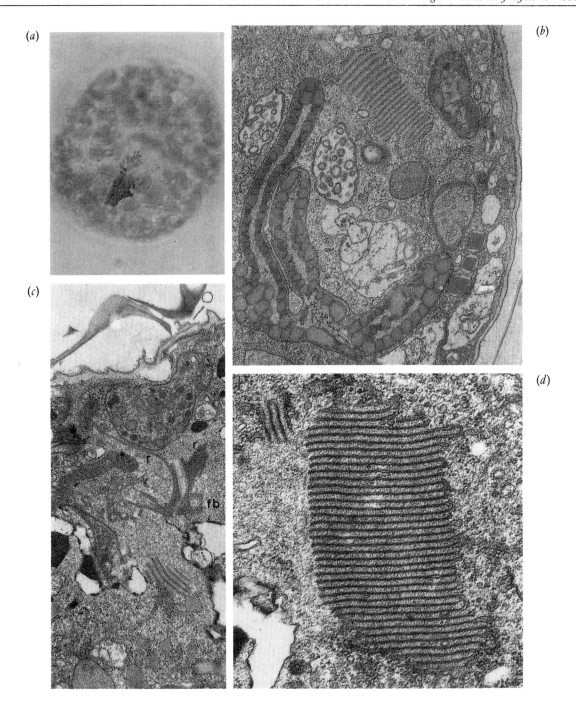

Fig. 15.7 *Micrographs of the complex eyespot apparatus in the dinoflagellate* Glenodinium foliaceum. *(a) Light micrograph to show the position and shape of the eyespot situated behind the sulcus. The faint grey bodies throughout the cell are the chloroplasts (×1600). (b) TEM section through the eyespot region showing the two layers of eyespot droplets within a triple envelope, the lamellar body, and various cell components (16 800). (c) The flagellar bases and their roots which run in the direction of the eyespot (×16 800). (d) Enlarged view of the lamellar body showing the stack of endoplasmic reticulum cisternae of which it is made up (×40 000).*

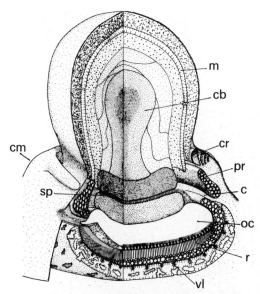

Fig. 15.8 *Diagram to illustrate the structure of the ocellus of* Erythropsidinium *as seen in longitudinal section: c, canal; cb, crystalline body; cm, cell membrane; cr, constriction ring; m, mitochondria; oc, ocelloid chamber; pr, pigmented ring; r, retinoid; sp, scalariform plate; vl, vesicular layer (After Greuet, 1968).*

vesicular layer made up of pigment-containing droplets. Small droplets are said to contain carotenoid pigments and larger droplets contain melanoid pigments. Studies of the reconstruction of the ocellus preceding cell division show what appear to be chloroplast thylakoids in pairs, which become reorganized to form the melanosome (Greuet, 1987). This suggests that, in part at least, the ocellus is in fact a modified chloroplast, as it clearly must be in the *Glenodinium* type described above. Experimental studies (Francis, 1967) have shown that the lens can make parallel rays of light converge so that an image can be focused on the retinoid.

Function of the Photosensory Apparatus

Because of the small size of algal cells, and the consequent difficulty of carrying out experiments on them, many questions regarding the functioning of the photosensory systems remain unanswered. The first concerns the ubiquitous carotenoid pigments of the eyespot. Are they purely for the absorption of light, as is implied by the shading model, or is the pigment affected by light such that chemical or electrical messages are produced? At present both theories are current, for the eyespot is regarded both as having a shading function in those organisms with what appears to be a separate photoreceptor and as a transducer of light energy in organisms where the eyespot is situated away from the flagella. However, one might have expected a fairly similar structure containing what appear to be similar pigments to have a unified function throughout the organisms which contain it.

Regarding the photoreceptor, we have examples in the Euglenophyta of a flagellar swelling (Fig. 15.5(b)), which is said to function as a dichoic crystal detector (Foster and Smythe, 1980). This apparently contains the pigment flavin, which has recently also been detected in the proximal portions of the posterior flagellum of various chromophyte algae (Coleman, 1988; Kawai, 1988). In a number of algae such as *Chlamydomonas*, with no obvious photoreceptor, the visual pigment rhodopsin has been detected (Foster *et al.*, 1984). The photoconversion of this pigment could be causing changes in the permeability of the adjacent plasma membrane which become converted into signals to the flagella, for it is clear that the action spectrum for phototaxis in *Chlamydomonas* is similar to the absorption band for rhodopsin. However, the relationship between rhodopsin and the carotenoids of the eyespots is not clear, and recent experiments utilizing a carotenoid-free blind mutant of *Chlamydomonas*, into which analogues of its retinal chromophore were inserted, showed that normal photobehaviour was restored (Foster *et al.*, 1984). By way of contrast, bleaching of rhodopsin by exposure of normal cells to hydroxylamine resulted in the loss of photoactivity until the bleaching agent was removed (Hegeman *et al.*, 1988). From the data on action spectra for various algae collected by Foster *et al.* (1984) it would seem that, as other chlorophyceans (*Volvox, Haematococcus*) and a prasinophycean (*Platymonas*) all have absorption peaks within the narrow band of 472–493 nm, they all have rhodopsin as the photoreceptor. This may also be the case in the dinoflagellate *Gymnodinium splendens*, where the absorption peak is a little lower at 457 nm.

As mentioned earlier, there is some evidence that in the Cryptophyta the red pigment phycoerythrin is linked with phototaxis. Experiments with a marine member of this group, *Cryptomonas maculata* (Haeder *et al.*, 1987), have shown that in low light of below $15 \, W \, m^{-2}$ this shows a weak positive phototaxis, whereas at higher fluence rates a more pronounced negative phototaxis is exhibited. This is in contrast to the response to freshwater members of the group, where only positive phototaxis has been found.

Various experiments on *Chlamydomonas* (e.g. Dolle *et al.*, 1986, 1987) have suggested that phototaxis is very dependent on calcium levels and, in *Gyrodinium dorsum*, low calcium levels below 10^{-3} M or the use of calcium

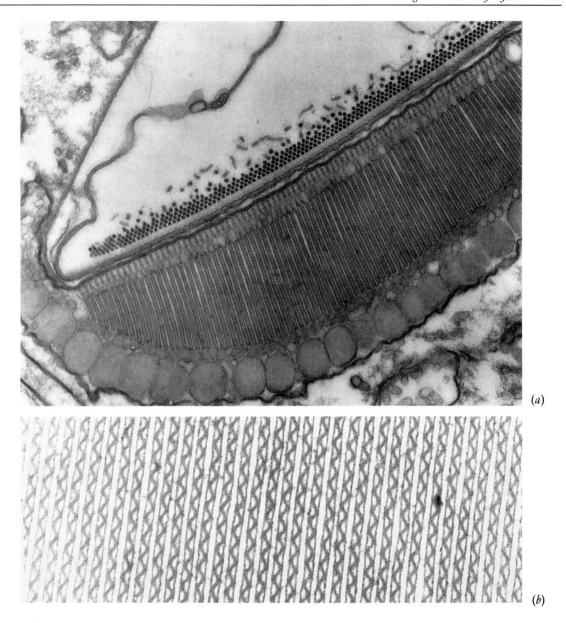

Fig. 15.9 *Ultrastructure of the ocellus. (a) Longitudinal section through part of the ocellus of* Erythropsidinium *showing part of the ocelloid chamber, the retinoid, and the vesicular layer which consists of droplets like those in other eyespots (× 32 000). (b) Transverse section of the retinoid showing the parallel rows of thylakoid-like structures of which it is composed with the thicker central sinusoidal membranes (× 72 000) (Micrographs courtesy of C. Greuet, from Greuet, 1987.)*

channel blockers resulted in a reduced photophobic response (Ekelund, 1989). The two flagella of *Chlamydomonas* are known to have different sensitivities to calcium ions, so it may be that voltage-sensitive calcium channels might be the means by which the light absorbed is eventually converted into a phototactic response. In the Prasinophyceae it has been shown that the concentration of calcium ions can exert a strong effect on the lengthening or contraction of the broad, fibrous, flagellar roots (Salisbury and Floyd, 1978). Clearly, this reaction could affect the angle of the flagellar beat with regard to the cell and thus alter the direction in which the cell swims. Such a hypothesis was put forward to explain the functioning of the eyespot in the dinoflagellate *Woloszynskia coronata* (Dodge, 1984). Here there is a microtubular root of the longitudinal flagellum, which probably makes contact with the eyespot and, by changing its length, could alter the angle of the flagellum. In most dinoflagellates the transverse flagellum provides most of the forward motion and rotates the cell, and the longitudinal flagellum appears to be responsible for phototactic orientation (Hand, 1970) and perhaps also for bursts of extra speed (Dodge, unpublished). Thus it could be that the microtubular root simply passes chemical and/or physical messages to the longitudinal flagellum, rather than adjusting the flagellar angle by the means suggested above.

In some organisms there is evidence that transduction between the photoreceptor and the motor apparatus takes place by changes in the electric potential. Experiments on the green alga *Haematococcus* (Litvin et al., 1978) have shown that there is a considerable change in electric potential difference over the cell surface when the cell is exposed to a flash of light.

Evolution of the Photosensory Apparatus

Since there is no possibility of a fossil record for the eyespot or associated structures, the only way an evolutionary scenario can be developed is by a comparative study of the present-day examples. Obviously, such an approach is open to error, owing to the possible complete lack of ancestral forms or missing links and to evolutionary lines being drawn in the wrong direction.

As a starting point we should remember that for a photosensory device to be effective it must function in conjunction with some means of movement, and since the main form of movement in these organisms is that brought about by flagella, then it would seem likely that eyespot evolution is closely linked to that of the flagellar systems. Secondly, since the function of the eyespot systems are generally to enable cells to swim towards light, presumably in order that the chloroplasts may receive adequate light energy, then it would seem reasonable to assume that there may be some link between the evolution of chloroplasts and that of eyespots. In the case of gametes it is difficult to see why they should need an eyespot, since they have only a rudimentary chloroplast, but perhaps it is advantageous for them to be able to swim towards or away from light.

Kivic and Walne (1983) have put forward the proposition that there have been as many parallel evolutions of algal phototactic mechanisms as there are types of plastid and types of light antenna (e.g. as elaborated by Foster and Smythe, 1980). They conclude that the minimum number of independent parallel evolutions is 10 or more! They further suggest that this can help to explain the apparent diversity of photoreceptor pigments, of which at least 6 are known. However, it must be said that this scenario is a rather extreme view, and it is much more likely that evolutionary developments have in large measure given rise to the variety of light-sensing systems we find today.

Much has been written about the phylogeny and evolution of flagella and chloroplasts. For flagella this has been most recently summarized by Moestrup (1982), who concludes that there are two main lines of development, with the dinoflagellates occupying a third, independent line owing to their numerous unique features. Of the two main lines, one includes all the heterokont or chromophyte groups (from Cryptophyta to Eustigmatophyta in Table 15.1). and the other all the green algae including the Euglenophyta. These groupings immediately raise problems with regard to the eyespot types. The first group contains all the representatives of eyespot type B, but in addition there are the eustigmatophytes with type C and the few cryptophytes with eyespots which fall most closely into type A. The second main flagellar evolutionary line contains the chlorophytan algae, all with eyespots which fall into type A, but the Euglenophyta with the quite distinct type C eyespots are a part of this group. From the above we might conclude that two eyespot types, A and B, are so closely associated with their respective evolutionary groups for flagella that it would seem reasonable to assume that they have evolved together. The remaining types of eyespot probably have other evolutionary affinities.

Many schemes have been proposed to explain the evolution of algal chloroplasts (e.g. Gibbs, 1978; Dodge, 1979; Whatley and Whatley, 1981; Whatley 1989) and it is generally accepted that symbiotic processes had much to do with the variety of types which are found. Two clear lines emerge (Fig. 15.10), which bear a strong similarity to those described for flagella. Once again, one line includes all the heterokont (or chromophyte) algae but with the starting point being represented by the present-day Rhodophyta where there are no flagella or eyespots. The second main line includes all the green algae including the euglenas, since it is thought that their chloroplasts have been obtained from other green algae by endosymbiosis. Once again the dinoflagellates occupy an uncertain position because it is clear that their chloroplasts have come from various sources (Dodge, 1989).

The net result of this study of flagella and chloroplast evolutionary relationships is the idea that for two of the eyespot types, A and B, there has probably been parallel evolution in conjunction with the evolution of both the flagella and the chloroplasts (Fig. 15.10). Variations, such as the numerous differences in flagellar root systems or in number of eyespot layers in the green algae, can be accounted for by the reasonable supposition that a certain amount of evolution has occurred within the group, as has been discussed by Melkonian (1984). Similarly, with the

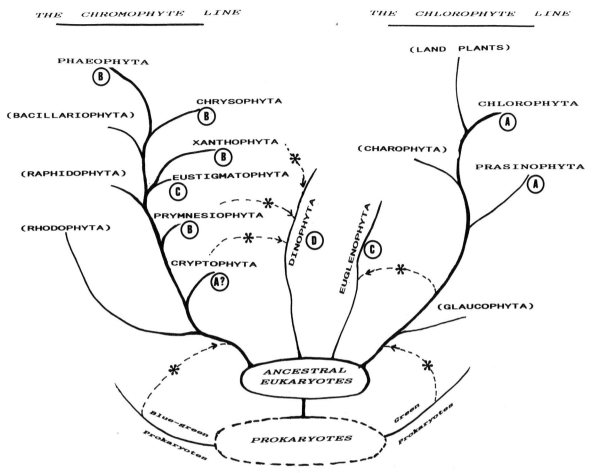

Fig. 15.10 *Diagram to illustrate the possible phylogenetic lines in the algae and their eyespot types. The groups in brackets do not have eyespots, but are shown here to give a more complete picture of algal phylogeny. The various eyespot types are indicated by letters but for the Dinophyta the 'D' includes all four types which have been found there. The assumed acquisitions of chloroplasts from other groups, by endosymbiosis, are shown by dotted lines and asterisks.*

chromophyte line or tree there is one major eyespot type with a range of variations on the theme.

Thus, we now have only to explain the origin of the eyespot in the cryptophytes, euglenophytes, eustigmatophytes and dinophytes. In the first of these groups the eyespot almost certainly has an independent origin, since the group is not closely related to any other, the structure is quite distinct and, if Foster and Smythe's (1980) analysis is correct, it is unique in its photosensory mechanism. At present this eyespot appears to involve only modifications of the chloroplast and no structural adaptations of the flagella have been noted. In the eustigmatophytes, where the eyespot is outside the chloroplast and the anterior flagellum is modified, the situation is radically different from that of other chromophytes and could well represent a separate evolutionary development. It is probably only analogy rather than homology that has produced a substantially similar eyespot in the euglenophytes, for otherwise this group appears to have no phylogenetic connection with the eustigmatophytes. Since the Euglenophyta are basically non-photosynthetic we have to address the question of what came first, the eyespot or the chloroplasts. There is a general assumption that chloroplasts are necessary for the synthesis of the carotenoid pigments found in eyespots, and although Lembi and Lang (1965) have made the reasonable suggestion that eyespot globules originate within the chloroplast, colourless euglenoids are known which have eyespots (Leedale, 1967). Perhaps leucoplasts or reduced plastidial structures were necessary before eyespots could have developed, although it is possible for certain higher organisms to obtain carotenoids from their food, as happens with some crustacea which obtain pigment from the phytoplankton which they consume. So far as the euglenoid flagellar swelling is concerned, this could well have developed from the already complex paraflagellar structure found in these organisms.

Finally we come back to the most problematical group, the Dinophyta. Since here the chloroplasts have been acquired from various other algae (see Dodge, 1989, for a summary) it might be reasonable to assume that in some cases the eyespot was acquired at the same time. For example, in *Woloszynskia tylota* and other species with eyespots within the chloroplasts this may have been the case, although it is still necessary to explain how the carotenoid droplets came to be concentrated in just one of the chloroplasts which is situated in a particular position behind the sulcus. The second dinoflagellate type, as found in *W. coronata*, would seem to have developed independently unless it is more easy for eyespot droplets to escape through the triple-membrane chloroplast envelope than would appear to be the case. In neither of these types, nor in that of *Glenodinium foliaceum* and the other binucleate dinoflagellates, has any modifications of the flagella been discovered. In the latter it is clear that the eyespot is derived from a reduced chloroplast, since it is bounded by a three-membrane envelope and the cell has a quite separate chloroplast system which is provided by the presence of a chrysophytic endosymbiont within the cell. The eyespot therefore probably developed from an earlier plastid system present in the cell before it obtained its endosymbiont. Thus it could have evolved from the first dinoflagellate type mentioned above. It would be very interesting to know what part if any is played by the lamellar body in the photoresponse of the *Glenodinium* eyespot type.

The ocellus is the most complex eyespot to be found in any unicellular organism. Greuet (1987) suggests that it could have evolved from the type of eyespot found in *Glenodinium* by the realignment and development of the lamellar body to give rise to the retinoid which then fills the pigment cup. The lens structure would presumably have developed from swollen vesicles, although vesicles do not normally seem to be a feature of the eyespot region in *Glenodinium*. There is an interesting problem with the ocellus, for none of the organisms which have it are photosynthetic. So, the function of the organelle has evolved from being involved in the sensing of light intensity and direction to some other purpose, which might be that of seeking prey or of avoiding predators. This then could explain the considerable development of the lens-like structure, which may be a focusing device. Thus the final question must be, did structures like these ocelli give rise to the primitive photosensory systems of lower invertebrates, or is that yet anther case of parallel evolution?

The net result of these discussions would seem to be the possibility that there may have been at least five independent or parallel developments of eyespot systems in the algae without including the dinoflagellates. These phylogenetic lines are summarized in Fig. 15.10 in relation to the general phylogenetic system for the algae. What now remains to be investigated are the potential links between the photosensory systems found in the unicellular algae and those present in multicellular higher organisms.

References

Anderson, R. A. (1985). The flagellar apparatus of the golden alga *Synura uvella*: four absolute orientations. *Protoplasma*, 128, 94–106.

Anderson, R. A. (1987). Synurophyceae classis nov., a new class of algae. *Am. J. Bot.*, 74, 337–353.

Barlow, S. B. and Cattolico, R. A. (1980). Fine structure of the scale-covered green flagellate *Mantoniella squamata* (Manton et Parke) Desikachary. *Br. Phycol. J.*, 15, 321–333.

Belcher, J. H. and Swale, E. M. F. (1971). The microanatomy of *Phaeaster pascheri* Scherffel (Chrysophyceae). *Br. Phycol. J.*, 6, 157–169.

Belcher, J. H. and Swale, E. M. F. (1967). *Cromulina placentula* sp. nov. (Chrysophyceae), a freshwater nannoplankton flagellate. *Br. Phycol. Bull.*, 3, 257–267.

Belcher, J. H. and Swale, E. M. F. (1972a). Some features of the microanatomy of *Chrysoccus cordiformis* Naumann. *Br. Phycol. J.* 7, 53–59.

Belcher, J. H. and Swale, E. M. F. (1972b). The morphology and fine structure of the colourless colonial flagellate *Anthophysa vegetans* (O. F. Muller) Stein. *Br. Phycol. J.*, 7, 335–346.

Benedetti, P. A. and Checcucci, A. (1975). Paraflagellar body pigments studied by fluorescence microscopy in *Euglena gracilis*. *Plant. Sci. Lett.*, 4, 47–51.

Benedetti, P. A. and Lenci, F. (1977). In vivo microspectrofluorometry of photoreceptor pigments in *Euglena gracilis*. *Photochem. Photobiol.*, 26, 315–318.

Bibby, B. T. and Dodge, J. D. (1972). The encystment of a freshwater dinoflagellate: a light and electron-microscopical study. *Br. Phycol. J.*, 7, 85–100.

Clayton, M. N. (1989). Brown algae and chromophyte phylogeny. In: *The Chromophyte Algae: Problems and Perspectives*. Eds, Green, J. C. Leadbetter, B. S. C. and Diver, W. L. pp. 229–253. Oxford: Clarendon.

Coleman, A. W. (1988). The autofluorescent flagellum: a new phylogenetic enigma. *J. Phycol.*, 24, 118–120.

Crawford, R. M. and Dodge, J. D. (1971). The dinoflagellate genus *Woloszynskia*. 2. The fine structure of *W. coronata*. *Nova Hedw.* 22, 699–719.

Crawford, R. M., Dodge, J. D. and Happey, C. M. (1970). The dinoflagellate genus *Woloszynskia*. 1. Fine Structure and ecology of *W. tenuissima* from Abbot's Pool, Somerset. *Nova Hedw.* 19, 825–840.

Creutz, C. and Diehn, B. (1976). Motor responses to polarised light and gravity sensing in *Euglena gracilis*. *J. Protozool.*, 23, 552–556.

Diehn, B., Fonseca, J. R. and Jahn, T. L. (1975). High speed cinemicrography of the direct photophobic responses of *Euglena* and the mechanism of negative phototaxis. *J. Protozool.*, 22, 492–494.

Dodge, J. D. (1969). A review of the fine structure of algal eyespots. *Br. Phycol., J.*, 4, 199–210.

Dodge, J. D. (1973). *The Fine Structure of Algal Cells*. London, New York: Academic.

Dodge, J. D. (1975). The fine structure of *Trachelomonas* (Euglenophyceae). *Arch. Protist.*, 117, 65–77.

Dodge, J. D. (1979). *The Phytoflagellates: Fine Structure and Phylogeny*. In: *Biochemistry and Physiology of Protozoa*. 2nd ed vol.1. Eds Levandowsky, M. and Hutner, S. H. pp. 7–57. New York: Academic.

Dodge, J. D. (1984). The functional and phylogenetic significance of dinoflagellate eyespots. *BioSystems*, 16, 259–267.

Dodge, J. D. (1989). Phylogenetic relationships of dinoflagellates and their plastids. *The Chromophyte Algae: Problems and Perspectives*. Eds Green, J. C., Leadbeater, B. S. C. and Diver, W. L. pp. 207–227. Oxford: Clarendon.

Dodge, J. D. and Crawford, R. M. (1969). Observations on the fine structure of the eyespot and associated organelles in the dinoflagellate *Glenodinium foliaceum*. *J. Cell Sci.*, 5, 479–493.

Dodge, J. D., Mariani, P., Paganelli, A. and Trevisan, R. (1987). Fine structure of the red-bloom dinoflagellate *Glenodinium sanguineum*, from Lake Tovel (N. Italy). *Arch. Hydrobiol., supp.*, 78, 125–138.

Dolle, R. and Nuptsch, W. (1987). Effects of calcium ions and of calcium channel blockers on galvanotaxis of *Chlamydomonas reinhardtii*. *Bot. Acta*, 101, 11–16.

Dolle, R., Pfau, J. and Nultsch, W. (1986). Role of calcium ions in motility and phototaxis of *Chlamydomonas reinhardtii*. *J. Plant Physiol.*, 126, 467–473.

Ekelund, N. G. A. (1989). Effects of calcium channel blockers and DCMU on motility and photophobic response of *Gyrodinium dorsum*. *Arch. Microbiol.*, 151, 187–190.

Foster, K. W. and Smythe, R. D. (1980). Light antennas in phototactic algae. *Microbiol. Rev.*, 44, 572–630.

Foster, K. W., Saranak, J., Patel, N., Zarilli, G., Okabe, M., Kline, T. and Nakanishi, K. (1984). A rhodopsin is the functional photoreceptor for phototaxis in the unicellular eukaryote *Chlamydomonas*. *Nature Lond.*, 311, 756–759.

Francis, D. (1967). On the eyespot of the dinoflagellate *Nematodinium*. *J. Exp. Biol.*, 47, 495–501.

Gantt, E., Edwards, M. R. and Provasoli, L. (1971). Chloroplast structure of the Cryptophyceae. Evidence for phycobiliproteins within intrathylakoidal spaces. *J. Cell Biol.*, 48, 280–290.

Gibbs, S. P. (1978). The chloroplasts of *Euglena* may have evolved from symbiotic green algae. *Can. J. Bot.*, 56, 2883–2889.

Green, J. C. (1973). Studies in the fine structure and taxonomy of flagellates in the genus *Pavlova*. II. A freshwater representative, *Pavlova granifera* (Mack) comb nov. *Br. Phycol. J.*, 8, 1–12.

Green, J. C. (1975). The fine structure and taxonomy of the haptophycean flagellate *Pavlova lutheri* (Droop) comb. nov. (= *Monochrysis lutheri* Droop). *J. Mar. Biol. Ass. U.K.*, 55, 785–793.

Green, J. C. (1980). The fine structure of *Pavlova pinguis* Green and a preliminary survey of the order Pavlovales (Prymnesiophyceae). *Br. Phycol. J.*, 15, 151–191.

Green, J. C. and Hibberd, D. J. (1977). The ultrastructure and taxonomy of *Diacronema vlkianum* (Prymnesiophyceae) with special reference to the haptonema and flagellar apparatus. *J. Mar. Biol. Ass. UK*, 57, 1125–1136.

Green, J. C. and Manton, I. (1970). Studies in the fine structure and taxonmy of flagellates in the genus *Pavlova*. I. A revision of *Pavlova gyrans*, the type species. *J. Mar. Biol. Asso. U.K.*, 50, 1113–1130.

Greuet, C. (1968). Organisation ultrastructurale de l'ocelle de deux peridiniens Warnowiidae, *Erythropsis pavillardi* Kofoid et Swezy et *Warnowia pulchra* Schiller. *Protistologica*, 4, 209–230.

Greuet, C. (1978). Organisation ultrastructurale de l'ocelloide de *Nematodinium*. Aspect phylogenetique de l'évolution du photorecepteur des Peridiniens Warnowiidae Lindemann. *Cytobiologie*, 17, 114–136.

Greuet, C. (1982). Photorécepteurs et photaxie des flagellés et des stades unicellulaires d'organismes inférieurs. *Annales de Biologie*, 21, 97–141.

Greuet, C. (1987). Complex organelles. In: *The Biology of Dinoflagellates*. Ed. Taylor, F. J. R. pp. 119–142. Oxford: Blackwell.

Haeder, D. P., Rhiel, E. and Wehrmeyer, W. (1987). Phototaxis in the marine flagellate *Cryptomonas maculata*. *J. Photochem. Photobiol., B.* 1, 115–122.

Hand, W. G. (1970). Phototactic orientation by the marine dinoflagellate *Gyrodinium dorsum* Kofoid. I. A mechanism model. *J. Exp. Zool.* 174, 33–38.

Hand, W. G. and Schmidt, J. A. (1975). Phototactic orientation by the marine dinoflagellate *Gyrodinium dorsum* Kofoid. II. Flagellar activity and overall response mechanism. *J. Protozool.*, 22, 494–498.

Hegemann, P., Hegemann, U. and Foster, K. W. (1988). Reversible bleaching of *Chlamydomonas reinhardtii* rhodopsin in vivo. *Photochem. Photobiol.*, 48, 123–128.

Hibberd, D. J. (1971). Observations on the cytology and ultrastructure of *Chrysamoeba radians* Klebs (Chrysophyceae). *Br. Phycol. J.*, 6, 207–223.

Hibberd, D. J. (1976). The ultrastructure and taxonomy of the Chrysophyceae and Prymnesiophyceae (Haptophyceae): a survey with some new observations on the ultrastructure of the Chrysophyceae. *Bot. J. Linnean Soc.*, 72, 55–80.

Hibberd, D. J. (1980). Eustigmatophytes. In: *Phytoflagellates*. Ed. Cox, E. R. pp. 319–334. New York: Elsevier/North Holland.

Hibberd, D. J. and Leedale, G. F. (1971). Cytology and ultrastructure of the Xanthophyceae. II. The zoospore and vegetative cell of coccoid forms, with special reference to *Ophiocytium majus* Naegeli. *Br. Phycol. J.*, 6, 1–23.

Hibberd, D. J. and Leedale, G. F. (1972). Observations on the cytology and ultrastructure of the new algal class, Eustigmatophyceae. *Ann. Botany*, 36, 49–71.

Hoshaw, R. W. and Maluf, L. Y. (1981). Ultrastructure of the green flagellate *Dunaliella tertiolecta* (Chlorophyceae, Volvocales), with comparative notes on three other species. *Phycologia*, 20, 199–206.

Kawai, H. (1988). A flavin-like autofluorescent substance in the posterior flagellum of golden and brown algae. *J. Phycol.*, 24, 114–117.

Kivic, P. A. and Vesk, M. (1972). Structure and function in the euglenoid eyespot apparatus: the fine structure, and response to environmental changes. *Planta Berlin*, 105, 1–14.

Kivic, P. A. and Walne, P. L. (1983). Algal photosensory apparatus probably represent multiple parallel evolutions. *BioSystems*, 16, 31–38.

Kristiansen, J. and Walne, P. L. (1976). Structural connections between flagellar base and stigma in *Dinobryon*. *Protoplasma*, 89, 371–374.

Kristiansen, J. and Walne, P. L. (1977). Fine structure of photokinetic systems in *Dinobryn cylindricum* var. *alpinum* (Chrysophyceae). *Br. Phycol. J.*, 12, 329–341.

Leedale, G. F. (1967). *Euglenoid Flagellates*. Englewood Cliffs, NJ: Prentice-Hall.

Leedale, G. F. (1982). Ultrastructure. In: *The Biology of Euglena*. Ed. Buetow, D. E. pp. 1–27. New York: Academic.

Lembi, C. A. and Lang, N. J. (1965). Electron microscopy of *Carteria* and *Chlamydomonas*. *Am. J. Bot.*, 52, 464–477.

Litvin, F. F., Sineshchekov, O. A. and Sineshchekov, V. A. (1978). Photoreceptor electric potential in the phototaxis of the alga *Haematoccus pluvialis*. *Nature Lond.*, 271, 476–478.

Lucas, I. A. N. (1970). Observations on the fine structure of the Cryptophyceae. I. The genus *Cryptomonas*. *J. Phycol.*, 6, 30–38.

Ludwig, M. and Gibbs, S. P. (1989). Localization of phycoerythrin at the lumenal surface of the thylakoid membrane in *Rhodomonas lens*. *J. Cell Biol.*, 108, 875–884.

Manton, I. (1964). A contribution towards understanding of 'the primitive fucoid'. *New Phytologist*, 63, 244–254.

Marchant, H. J., Buck, K. R. and Garrison, D. L. (1989). *Mantoniella* in antarctic waters including the description of *M. antarctica* sp. nov. (Prasinophyceae). *J. Phycol.*, 25, 167–174.

Massalski, A. and Leedale, G. F. (1969). Cytology and ultrastructure of the Xanthophyceae. I. Comparative morphology of the zoospores of *Bumilleria sicula* Borzi and *Tribonema vulgare* Pascher. *Br. Phycol. J.*, 4, 159–180.

Melkonian, M. (1984). Flagellar apparatus ultrastructure in relation to green algal classifications. In: *Systematics of the Green Algae.* Eds Irvine, D. E. G. and John, D. M. pp. 73–120. London: Academic.

Melkonian, M. and Robenek, H. (1979). The eyespot of the flagellate *Tetraselmis cordiformis* Stein (Chlorophyceae): structural specialization of the outer chloroplast membrane and its possible significance in phototaxis of green algae. *Protoplasma*, 100, 183–197.

Melkonian, M. and Robenek, H. (1980a). Eyespot membranes of *Chlamydomonas reinhardii*: a freeze-fracture study. *J. Ultrastruct. Res.*, 72, 90–102.

Melkonian, M. and Robenek, H. (1980b). Eyespot membranes in newly released zoospores of the green alga *Chlorosarcinopsis gelatinosa* (Chlorosarcinales) and their fate during zoospore settlement. *Protoplasma*, 104, 129–140.

Melkonian, M. and Robenek, H. (1984). The eyespot apparatus of flagellated green algae: a critical review. *Prog. Physiol. Res.*, 3, 193–268.

Messer, G. and Ben-Shaul, Y. (1969). Fine structure of *Peridinium westii* Lemm., a freshwater dinoflagellate. *J. Protozool.*, 16, 272–280.

Meyer, S. R. and Pienaar, R.N. (1984). The microanatomy of *Chroomonas africana* sp. nov. (Cryptophyceae). *S. Afr. J. Bot.*, 3, 306–319.

Moestrup, Ø. (1982). Flagellar structure in algae: a review, with new observations particularly on the Chrysophyceae, Phaeophyceae (Fucophyceae), Euglenophyceae and *Reckerita*. *Phycologia*, 21, 427–528.

Mornin, L. and Francis, D. (1967). The fine structure of *Nematodinium armatum*, a naked dinoflagellate. *J. Microsc.* 6, 759–772.

Müller, D. G., Maier, I. and Müller, H. (1987). Flagellum autofluorescence and photoaccumulation in heterokont algae. *Photochem. Photobiol.* 46, 1003–1008.

Osafune, T., Alhadeff, M. and Schiff, J. A. (1985). Light-triggered organization of the stigma in dark-grown nondividing cells of *Euglena gracilis*. *J. Ultrastruct. Res.*, 93, 27–32.

Pagni, P. G. S., Walne, P. L. and Wehry, E. (1976). Fluorometric evidence for flavins in isolated eyespots of *Euglena gracilis* var. *bacillaris*. *Photochem. Photobiol.*, 24, 373–375.

Piccinii, E. and Mammi, M. (1978). Motor apparatus of *Euglena gracilis*: ultrastructure of the basal portion of the flagellum and the paraflagellar body. *Boll. Zool.*, 45, 405–414.

Piccinii, E. and Omodeo, P. (1975). Photoreceptors and photatic programs in Protista. *Boll. Zool.*, 42, 57–79.

Pienaar, R. and Aken, M. E. (1985). The ultrastructure of *Pyramimonas pseudoparkaea* sp. nov. (Prasinophyceae) from South Africa. *J. Phycol.*, 21, 428–447.

Preisig, H. R.and Wilhelm, C. (1989). Ultrastructure, pigments and taxonomy of *Botryochloropsis similis* gen. et sp. nov. (Eustigmatophyceae). *Phycologia*, 28, 61–69.

Preisig, H. R., and Melkonian, M. (1984). A light and electron microscopical study of the green flagellate *Spermatozopsis similsi* spec. nova. *Plant Syst. Evol.*, 146, 57–74.

Ristori, T. and Rosati, G. (1983). The eyespot membranes of *Haematococcus pluvialis* Flotow (Chlorophyceae): their ultrastructure and possible significance in phototaxis. *Monit. Zool. Ital.*, 17, 401–408.

Ristori, T., Ascoli, C., Banchetti, R., Parrini, P. and Petracchi, D. (1981). Localization of photoreceptor and active membrane in the green alga *Haematoccus pluvialis*. *Abstracts IVth International Congress on Protozoology*, Warsaw, 314.

Salisbury, J. L. and Floyd, G. L. (1978). Calcium-induced contraction of the rhizoplast of a quadriflagellate green algae. *Science Wash. DC*, 202, 975–977.

Santore, U. J. (1987). A cytological survey of the genus *Chroomonas* — with comments on the taxonomy of this natural group of the Cryptophyceae. *Arch. Protist.*, 134, 83–114.

Schnepf, E. and Deichgräber, G. (1972). Uber den feinbau von theka, pusule und golgi-apparat bei dem dinoflagellaten *Gymnodinium* spec. *Protoplasma*, 74, 411–425.

Schnepf, E., Deichgräber, G., Röderer, G. and Herth, W. (1977). The flagellar root apparatus, the microtubular system and associated organelles in the chrysophycean flagellate, *Poterioochromonas malhamensis* Peterfi. *Protoplasma*, 92, 87–107.

Thomsen, H. A. (1988). Fine structure of *Pyramimonas nansenii* (Prasinophyceae) from Danish coastal waters. *Nord. J. Bot.*, 8, 305–318.

Tomas, R. N. and Cox, E. R. (1973). Observations on the symbiosis of *Peridinium balticum* and its intracellular alga. I. Ultrastructure. *J. Phycol.*, 9, 304–323.

Van der Veer, J. (1976). *Pavlova calceolata* (Haptophycae), a new species from the Tamar estuary, Cornwall, England. *J. Mar. Biol. Ass. UK*, 56, 21–30.

Walne, P. L. (1971). Comparative ultrastructure of eyespots in selected euglenoid flagellates. In: *Contributions in Phycology*. Eds Parker, B. C. and Brown, R. M. pp. 107–120, Lawrence Allen Press: Kansas.

Walne, P. L. (1980). Euglenoid flagellates. In: *Phytoflagellates*. Ed. Cox, E. R. pp. 165–212. New York: Elsevier/North Holland.

Walne, P. L. and Arnott, H. J. (1967). The comparative ultrastructure and possible function of eyespot: *Euglena granulata* and *Chlamydomonas eugametos*. *Planta Berlin*, 77, 325–353.

Watanabe, M. and Furuya, M. (1974). Action spectrum of phototaxis in a cryptomonad alga, *Cryptomonas* sp. *Plant Cell Physiol.*, 15, 413–420.

West, L. K., Walne, P. L. and Rosowski, J. R. (1980). *Trachelomonas hispida* var *coronata* (Euglenophyceae). I. Ultrastructure of cytoskeletal and flagellar systems. *J. Phycol.*, 16, 489–497.

Whatley, J. M. (1989). Chromophyte chloroplasts—a polyphyletic origin? In: *The Chromophyte Algae: Problems and Perspectives*. Eds Green, J. C., Leadbeater, B. S. C. and Diver, W. L. pp. 125–144. Oxford: Clarendon.

Whatley, J. M. and Whatley, F. R. (1981). Chloroplast evolution. *New Phytol.* 87, 233–247.

Wujak, D. E. (1969). Ultrastructure of flagellated chrysophytes. I. *Dinobryon*. *Cytologia*, 34, 71–79.

16 Evolution of the Cellular Organization of the Arthropod Compound Eye and Optic Lobe

I. A. Meinertzhagen

Introduction

The diversity of eye types and of associated visual systems is a self-indulgence of the arthropod groups, confounding not only resolution of relationships but also facile treatment of their evolutionary origins, and providing little that may be said with any certainty. This treatment will therefore deal with the topic restrictively, skating on areas where the ice appears to the author least thin.

Uncertainties over the evolution of visual systems, especially of the compound eyes, are the subject of longstanding discussions, still unresolved, that have picked their way through the battlefield of debates waged over the evolutionary relationships of the major arthropod groups themselves. For the non-specialist, and as an introduction to this introduction, Willmer (1990) provides a balanced treatment of invertebrate relationships in general, and of arthropod relationships in particular. Neglecting for the moment Arachnata (chelicerates and trilobites), two major views have been espoused for the origin of mandibulate groups, one that Crustacea and the myriapod–insect stemline arose independently, and the other that they are monophyletic. The former view, promulgated especially by Manton (Tiegs and Manton, 1958; Manton, 1973, 1977) and endorsed by the embryological evidence (Anderson, 1973), has held sway at least in the English-speaking world and until relatively recently, whereas the latter view (monophyly) has been reached by various other authorities (Snodgrass, 1935; Remane, 1959; Siewing, 1960; Sharov, 1966; Boudreaux, 1979). To these traditional views must now be added the younger voices from recent molecular phylogenies, but far from clarifying the issues these merely provoke more questions, so that no real credibility can be attached to any one view alone. The first comprehensive analysis, based on 18S RNA sequences (Field *et al.*, 1988), radically concluded not only that arthropods are polyphyletic but that Metazoa are too. Using improved methods of phylogenetic tree climbing, recent reanalyses of the same data (Lake, 1990) provide a conclusion more readily reconciled with conventional views based on comparative anatomy, in the form of five phylogenetic trees. Only one of these, reproduced here as Fig. 16.1 but without implied endorsement, reconstructs the protostome–deuterostome dichotomy to be found in every textbook of zoology. However, all five indicate that the arthropods are a paraphyletic assemblage (Lake, 1990), one in other words that contains some but not all of the descendents of a common ancestor, combining only those descendents of that ancestor that lack the feature(s) of the more recently derived sister groups. Thus if compound eyes arose once only in the common ancestor of all arthropod groups, they should be found in the non-arthropod sister groups too. Examples of compound eyes do in fact exist in both annelids and molluscs, in sabellid polychaetes (Krasne and Lawrence, 1966) and in the mantle eyes of arc clams (Nowikoff, 1926), but these are totally isolated convergences and not remnant homologous structures.

Thus, if the molecular phylogenies can be trusted, compound eyes are not homologous but arose separately in the different groups of a paraphyletic arthropod assemblage. Yet this flies in the face of the very detailed similarities to be found among the structural designs of the compound eyes and their optic lobes in those groups, similarities that are greatest in two of the groups presumably separated by the longest period of evolutionary divergence, insects and malacostracan crustaceans. These similarities have already been acknowledged as 'one of the most disconcerting problems of arthropod phylogeny' (Tiegs and Manton, 1958); others of course stress that these same similarities

The Compound Eye

Early Radiation

Compound eyes arose early in evolution. For at least 300 Myr ancestors of modern insect and crustacean groups had clearly developed compound eyes. The relationship of these to the eyes of early trilobites that existed 250 Myr ago back to the Cambrian 600 MYr ago (Levi-Setti, 1915) is uncertain. The holochroal type of trilobite eye had a close packing of ommatidia and a single cornea very much like the compound eyes in extant arthropods. It is believed to be ancestral to the schizochroal trilobite eye found in Phacopina, which differed from true compound eyes in having separately encased lenselets each with a separate cornea, and which is thought to have arisen by paedomorphosis (Clarkson, 1975).

Elsewhere, ancestral myriapods had well-developed compound eyes, and these persist in the modern chilopod, *Scutigera*, which needs closer examination. In Chelicerates, compound eyes existed in fossil eurypterids as well as in the distantly related Merostomata, which survive as the ancient genus *Limulus* and its relatives. The compound eyes of *Limulus* still leave no discerned hint of their ancestry, despite receiving intense scrutiny by vision physiologists, who compare it with the compound eyes of other arthropods without an evolutionary blush.

On functional grounds, Nilsson (1989) has proposed – as others have too – that compound eyes (or possibly *the* compound eye, if this arose but once) most likely derived from the fusion of a number of non-resolving eyespots. Not surprisingly, no post-fusion stage of compound eye organization representative of that which might have existed in ancestral forms is clearly preserved in any current form, although in terrestrial (oniscoidean) isopods a case has been made for the derivation of the tri-ommatidial compound eyes found in certain species from simple eyes represented in related species (Wolsky and Wolsky, 1985). Even if we shall never reconstruct the actual events with any certainty, the first major issue to acknowledge is whether there was only one ancestral eye, or many.

A Single Origin?

The clear development of similar compound eyes in both insects and crustaceans highlights the question of the origin of the two groups, whether monophyletic or independent. The issue of the evolution of the arthropod groups lies beyond the scope of this review, in which I propose to argue that the case for homology of compound eyes in different groups has not been given the airing it deserves, outside that in the very extensive treatment given by Paulus (1979). Furthermore, the strong morphogenetic dependence of at least the distal optic neuropiles upon development in the overlying retina

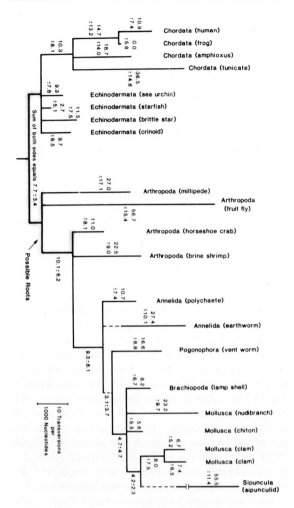

Fig. 16.1 *Phylogenetic tree of metazoan 18S RNA, derived from the molecular data of Field* et al. *(1988), purporting to show that arthropods are a paraphyletic assemblage arising from a protostomatous ancestor which also produced, as sister groups to Crustacea, the annelids and molluscs (From Lake, 1990).*

argue persuasively for the homology of the visual system (Paulus, 1979), and so deny the poly- or paraphyly of arthropods. Another reviewer valiantly faces the same conundrum (Osorio, 1991).

One major difference in the eyes of insects and Crustacea is that in many crustacean groups the eyes are carried on articulated stalks. Eyestalks are not lost in eyeless forms, or in those showing eye reduction, which has led Bowman (1984) to argue that the sessile condition is not readily derived from the stalked one and is thus ancestral, so resembling the eyes of all insect groups. The opposite view has also been reached, but primarily from embryological considerations (Paulus, 1979).

(Pflugfelder, 1958; Meinertzhagen, 1973) means that the evolution of the eye itself must also have been a profound influence upon the underlying optic lobe, which also bears on the status of polyphyly in the visual system. If the case for the monophyletic origin of the compound eye were to be upheld then the uncomfortable consequences of that conclusion for the origin of insects and crustaceans would have to be sorted out.

From the outset it is clear that photoreceptors have arisen many times in the course of evolution (Salvini-Plawen and Mayr, 1977) and that there is no reason in principle why this could not be as true amongst arthropods as in other groups. In considering this question in arthropods, different authors have in the past come to polarly opposing views. A carefully reasoned cladistic treatment based on cellular organization comes out in favour of monophyly (Paulus, 1979); another, based primarily on physiological optics (Land, 1981a, 1989) favours more than one independent origin; a third simply assumes the independence of origins to explain functional divergences (Horridge, 1977). The fact that these views diverge is no accident of personality; but it may be a consequence of the range of evolutionary arguments brought to bear. For example, Nilsson (1989) argues that the derivation of a compound eye is a simple matter, consisting of the fusion of non-resolving eyespots, and that this could have occurred on multiple occasions during the evolution of the different arthropod groups, requiring us then only to identify which of the different eyes are true homologues. This line of reasoning, that the similarity of compound eyes is only superficial (Nilsson, 1989), and easily derived during the course of evolution, is implicit in many other views based on evidence from optical mechanisms, but it is not supported on the other hand by genetic evidence. The ontogenetic assembly of an ommatidium, for example, is definitely not a simple matter, but instead involves many genes, which have been identified in *Drosophila* from the number of eye mutants isolated (Tomlinson, 1988).

Thus the divergence of opinion arises from which aspect of the compound eye's phenotype is examined. The cellular organization of the retina is by and large unified, built around a quasi-standard ommatidial cluster, which argues either for homology or for convergence of a most remarkable degree; while the optical mechanisms provided by the subcellular structure of these cells offer functional extremes that are not easily bridged by intermediate designs of any optical utility, and so appear independent in origin. The diversity of crustacean eyes illustrates this most clearly (Land, 1989; Cronin, 1986). An argument not yet explored in detail is that the eye has arisen by parallel evolution in different arthropod groups, from an ancestral structure, perhaps an eyespot or a cluster of eyespots, each with a limited number of cells, to form a proto-ommatidial eye.

Evolutionary Pressure on the Compound Eye

The adaptive significance of vision is of such magnitude that vision has almost certainly arisen independently many times (Land, 1989). In conjunction with an elaborate neural system for abstraction in the optic lobe, compound eyes confer arthropods with moderately high spatial resolution (Kirschfeld, 1976; Land, 1981a) and many other visual attributes. Why have these arisen particularly in arthropods? Land (1989) gives three more or less comprehensive options: prey capture, predator evasion and detection of self-motion, and cites the ubiquity of optomotor reflexes in all invertebrates with good vision as the evidence that a requirement for motion detection impelled evolution of their eyes. The result in arthropods was the compound eye; the parallel development of an associated optic lobe satisfied the requirement for a powerful system to detect motion, rendered all the more necessary by the later adoption of the aerial habit in insects. In a related idea, Edwards (1989) proposes that the rapid flight that he argues arose from stabilization of escape jump behaviour, led to the refinement of the compound eye and descending pathways, many encoding visual information to control flight. These ideas remind us that while considering the evolutionary origins of the sensory apparatus for vision we need also to examine its motor targets.

Different Compound Eyes: Conserved and Labile Features

The essential organizational similarity in all compound eyes, insect and crustacean, is that the architectural principle of both is overtly modular. The perfect geometrical precision of the retina is dictated by the optical principles in sampling visual space, and could – even if it arose convergently – impose many morphological similarities upon the cells within the component units. Although the modularity arises in the retina, with the ommatidial lattice, it is projected centrally through the patterns of photoreceptor innervation to the underlying first neuropile, the lamina (Meinertzhagen, 1976); the lamina depends on the overlying retina for the organization of that pattern (Meyerowitz and Kankel, 1978). Most morphological diversity in the eye and its neuropiles between different species is then the product of variation in the number, size, geometrical array and curvature of these cell clusters arranged in parallel. This is a problem in relative growth (Bernard, 1937; Bodenstein, 1953), which could have been solved independently in different arthropod groups once the iterative developmental programme to construct an ommatidium, for example, had been perfected. However, if the iterative organization of compound eyes did arise independently in different arthropods then it is much harder to imagine why the modules in the various groups adopted the same numerical composition within their cell clusters. This numerical constancy is easiest to see in the ommatidium, which is widely built around a cluster of eight photoreceptors, but also probably exists in the lamina cartridge, which appears to be built around five monopolar cells in many groups (Shaw and Meinertzhagen, 1986).

The best examples of conservation and diversity in eye design come from detailed consideration of groups with a clear phylogeny. Two examples from a single taxon will be treated here; others are considered in other chapters in this volume. It is well known that the forces acting during natural selection must have had a special interest in the outcome of Coleoptera in creating so many species (22% of known animal species: Imms, 1957). Eye structure in beetles has also received special evolutionary attention, in two contemporary studies. The structure of the ommatidium (lens system, corneal pigment, presence or absence of a clear zone and rhabdome organization) has been examined in Polyphaga (Caveney, 1986) and confirms most important phylogenetic divergences established for other characters. This analysis is significant because three major lens types (eucone, exocone and acone) and the two types of retinula (open- and fused-rhabdome) found in insect ommatidia are also found in the polyphagous groups. In a phylogenetic tree based upon eye characters (Fig. 16.2), ancestral groups have eucone ommatidia with small cones and narrow clear zones, while acone and exocone eye types are both derived, possibly having arisen several times. The clear zones of exocone and eucone eyes differ structurally and arose separately, and are thus not homologous. Within the Chrysomyeloidea, the largest polyphagous group and second largest group of Coleoptera, the open-rhabdomere ommatidium is arranged in one of two configurations, *insula* (in which the rhabdomeres of 2 central cells and those of the surrounding ring of 6 are separate) or *pontificus* (in which they contact each other), each with different subpatterns. Within the Chrysomyeloidea, the open-rhabdomere arrangement is homologous in all forms but the two basic rhabdome patterns and several of the insula-type subpatterns must have arisen convergently (Schmitt et al., 1982). Taken as a whole, these findings illustrate two points: that the cellular arrangement of ommatidia can change relatively rapidly within the evolution of a group, whereas the cellular composition, the number of receptors etc., does not. This theme will recur in later sections.

Evolution of the Ommatidium

The ommatidium is the key to the compound eye; to know one is to understand the other. Likewise, the evolution of the compound eye is really the evolution of this simple cell cluster, and the number and pattern of its array.

Photoreceptors

Two major lines of photoreceptors have emerged in all visual systems, so-called ciliary and rhabdomeric. The dichotomy between the two has provoked lively discussion over 30 years (Eakin, 1982; Vanfleteren, 1982; Salvini-Plawen, 1982), but little resolution of its significance. Since only rhabdomeric receptors are found in arthropods, that discussion is irrelevant to this review. Arthropod receptors are also uniform in their responses to light: all depolarize. Burr (1984) proposed that a master gene selects the different components of a photoreceptor, membrane domain, rhodopsin, and transduction mechanism, and that these are incorporated into the organelles of the receptor. Evolution performs a useful control experiment for this theory, in the photoreceptors of the evolutionarily

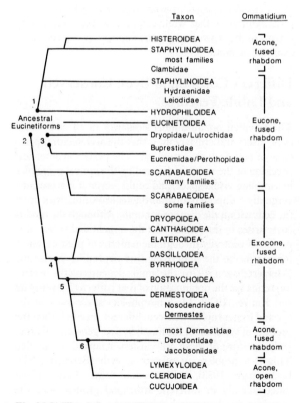

Fig. 16.2 *The phylogeny of Polyphaga (Coleoptera) derived from the structure of their ommatidia. The ancestral group, to which the modern eucinetoids are closest, probably had eucone eyes with fused rhabdomes (From Caveney, 1986).*

convergent compound eyes in the sabellid polychaete *Branchiomma*. These have rhabdomeric receptors (Krasne and Lawrence, 1966). They also depolarize to light (Leutscher-Hazelhoff, 1984), just like their arthropod counterparts, and have thus selected the same combination of membrane and transduction mechanism as the latter, although derived quite independently.

Within arthropods, a century of careful comparative studies has not really come to grips with the evolutionary origin of the ommatidial unit structure of the compound eye. The only two clear cases of ommatidial organization outside mandibulates offer little information about the evolution of the compound eye. The chilopod *Scutigera* has an ommatidium only superficially similar to mandibulate ommatidia, and probably independently derived (Paulus, 1979) (Fig. 16.3). There are neither corneagenous cells, as in Crustacea, nor is there a cone of 4 parts, as in both insects and crustaceans; there are 4 proximal receptor cells, beneath a distal retinula of at least 7 and up to 23 receptor cells, the number varying both within ommatidia and between ommatidia of different eye regions (Bähr, 1974; Paulus, 1979). Paulus (1979) thinks that general similarities to mandibulate ommatidia are a product of modification from an ancestral eye that was homologous to those in insects and crustaceans, but the scheme proposed for its derivation is mostly imagination. This connection is critical because myriapods are considered a sibling group of the insects by almost all authorities, and the inability to forge a strong link of homology between insect and myriapod ommatidia threatens any claim that insect and crustacean ommatidia can be evolutionarily close, despite their sometimes detailed similarities.

The ommatidium in *Limulus* is, like that in *Scutigera*, numerically indeterminate, containing between 4 and 20 receptors (Fahrenbach, 1969). The homologies of these are wholly obscure. In particular the eccentric cells, usually 2 per ommatidium, have totally eluded identification. They could be basal receptor cells, homologues of R8 in the insect's ommatidium (see below), but seem more closely to be neurones (Fahrenbach, 1975); they could be interneurones, homologues of L1/L2 in the lamina of insects (see below, also), but in that case differ in depolarizing to light and in supporting spiking potentials. In fact, such major differences exist as to make it most likely that they are one of the special features of an eye that arose quite independently.

Mandibulate Ommatidia

With these two precedents, the mandibulate ommatidium is distinguished in one feature: it almost invariably has a fixed complement of cells, and this number is generally the

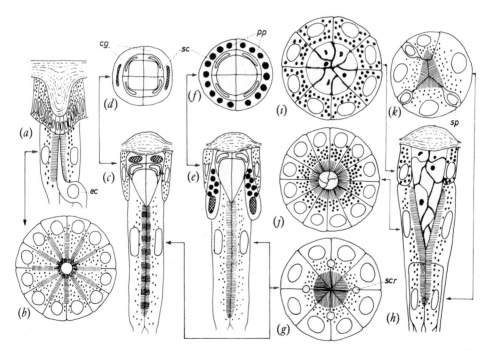

Fig. 16.3 *Different types of ommatidia that have arisen in different arthropod groups. (a, b) Limulus, in longitudinal (a) and transverse (b) section planes. (c, d) Crustacea. (e, f) Insecta. (g) Transverse section of c and d. (h–k) Scutigera coleoptera (Chilopoda). cg, corneagenous cell; ec, eccentric cell; pp, primary pigment cell; sc, Semper cell; scr, cone cell process; sp, supporting cell (From Paulus, 1979.)*

same (eight) as is found in other species. Given that the ommatidium's receptor cell complement is not fixed in either *Scutigera* or *Limulus*, cell number per cluster may not have been fixed in one or more ancestral ommatidia of mandibulates, either. Generating a highly regular ommatidial array apparently required the evolutionary acquisition of special genetic programmes for the developmental assembly of the ommatidium, and these have now been partially uncovered in insects (Tomlinson, 1988; Zipursky, 1989). Possibly these programmes could have been modified from those used to assemble sensilla in the insect's integument; the latter are small, receptor-bearing clusters, which unlike ommatidia occur singly and are individually generated from a mother cell, by a series of differentiative mitoses according to a fixed lineage (Peters, 1965; Lawrence, 1966). On the other hand, a clonal relationship does not exist for cells in sensilla of the anterior wing margin in *Drosophila* (Hartenstein, personal communication) and this is true in the ommatidium too. In the ommatidia of *Drosophila*, cells are assembled into a cluster from a pool of proliferated cells, without reference to lineage but with the subsequent determination of cell fate occurring through cell interactions within the cluster (Ready et al., 1976; Lawrence and Green, 1979; Tomlinson, 1988; Ready, 1989). The sequence of recruitment occurs in pairs, each cluster passing successively through 1-, 3-, 5-, 7- and 8-cell stages (Tomlinson and Ready, 1987a). These pairwise steps define four symmetry conditions, which Ready (1989) suggests are important in determining the cell interactions required to confer cell fate uniquely. In that case, acquiring the geometrically constrained structure of the cluster may have been a developmental constraint, perhaps an essential step in the establishment of unique cell fates within the ommatidium, for which the fixity of cell complement would have then been a prerequisite.

Studies on mandibulate ommatidia are legion, but most are directed to the vision of the arthropod bearer, so that many are innocent of (or, worse still, report erroneously) important details concerning the number of cells and their arrangement. As in *Drosophila*, 8 receptor cells are standard in many groups, particularly most insects, with an additional basal cell in some others, such as Hymenoptera and Lepidoptera. Less than 8 cells are recorded for various crustaceans: 7 in the mysid *Neomysis* (Strusfeld and Nässel, 1981), 6 in some branchiopods and 5 in most amphipods (Hallberg et al., 1980). Occasionally more than these numbers exist, or indeterminate numbers are found, such as: amongst insects, 9–12 in *Ephestia* (Fischer and Horstmann, 1971); in crustacean ommatidia, up to 17 in isopods (Strausfeld and Nässel, 1981), 12 in many mysids, or 16 in *Oniscus* (Isopoda) (Debaisieux, 1944) (Fig. 16.4). Receptor cells are numbered uniquely, either within each species or sometimes in taxa, if these have obvious simi-

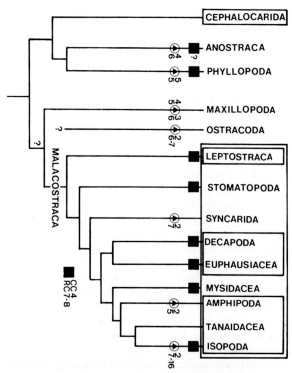

Fig. 16.4 *Dendrogram of crustacean groups showing the distribution of ommatidia having cellular organizations different from the typical form containing 8 receptors and 4 cone cells (square symbols). Departures from this are shown by circles within which the upper number denotes cone-cell complement and the lower one the receptor-cell complement (From Paulus, 1979.)*

larities between member species. Mistaken observation, especially missing small receptors or receptors at unusual depth levels in the retina, is a considerable possibility, and even likely in the old literature. Detailed citation of the many studies will not be given here, and interested parties should consult the published bibliographies in several reviews (Snodgrass, 1935; Horridge, 1965; Paulus, 1979; Shaw and Stowe, 1982; Mouze, 1984; Hallberg and Elofsson, 1989). However, the most remarkable conclusion to be made from this detailed if boring anatomical inventory is not the theme of its variations, but that no serious attempt has been made from it to number cells or discuss their comparative arrangement in different species on the grounds of evolutionary homology; however, for this a clever insight by Ready (1989) now provides a basis.

The Dipteran Ommatidium and Its Homologies

In insects, much is known about ommatidia in Diptera (Dietrich, 1909), the structure of which will be reviewed briefly here, both for immediate explanatory purposes and as an introduction to later sections. The photoreceptive

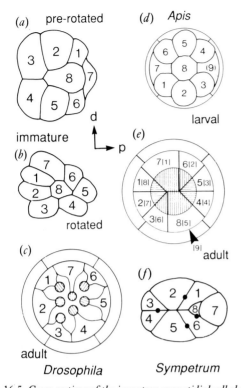

Fig. 16.5 Cross-sections of the immature ommatidial cell cluster found in the larva and its mature form in the adult ommatidium of insects. All the ommatidia are shown from the left ventral retina, as if from the outside looking in. Comparison between Drosophila (a–c) and the bee, Apis (d,e), shows the numerical and pattern basis for claimed receptor homologies between these two species (redrawn from Ready, 1989). The dipteran ommatidial cluster (a) rotates 90° (b) before differentiating into the adult form (c) (Tomlinson and Ready, 1987a). (d,e) In the bee, the cells are numbered here for their dipteran counterparts as given in Table 16.1 to which R9 is added (cf. a,d). (e) The adult bee ommatidial plan is found at the distal tip of the Y-type of rhabdome, which twists going proximally; cell numbers in brackets are those of the numbering scheme in Hymenoptera (Menzel and Blakers, 1976; Wehner, 1976). These in turn are compared with the adult ommatidium of the dragonfly Sympetrum (f) (redrawn from Meinertzhagen and Shaw, 1989) and also renumbered according to the dipteran scheme: Table 16.1). The ommatidial plan in Odonata, when different levels are collapsed upon a single section, is identical amongst Anisoptera (Armett-Kibel and Meinertzhagen, 1983) and Zygoptera (Ninomiya et al., 1969), although only clearly compared in the ventral eye; the dorsal eye differs in having three rudimentary cells (Eguchi, 1971). All cells are uniquely identifiable from the positions of four cone threads (black dots); R6 alone sits between two cone threads, so providing an asymmetry with respect to which R1, R7 and R8 occupy the position between R6 and the adjacent cone thread, and R2–R5 form two bilateral pairs separated by the remaining cone thread. Comparison of this pattern with the pre-rotated dipteran ommatidial cluster (a) reveals that the originating central cell in *Diptera, R8, lies in a position corresponding to the cell numbered R6 in the numbering scheme of Odonata, with its partner (R7 in both numbering schemata) separating R1 and R6 (R5 and R8 in the dragonfly) and readily suggesting homologies between the remaining cells, based upon evidence of the conserved number of eight receptor cells, and their organization into four pairs, one unmatched.*

rhabdomeres are separate from each other, in the so-called open-rhabdomere arrangement, having an asymmetrical pattern in the cross-section of the ommatidium, which permits all cells to be uniquely numbered as one of R1–R8. R1–R6 are uniform in their properties (Hardie, 1985) and have short axons that terminate within the cartridges of the first optic neuropile, or lamina (Braitenberg, 1967), while two central cells, R7 and R8, are unequal and have long visual fibres terminating in the second neuropile, or medulla (Melamed and Trujillo-Cenóz, 1968).

In the fused-rhabdome ommatidia of other groups, receptor cells are frequently arranged in cell pairs, lending a bilateral symmetry to the ommatidium in cross-section. Even here, the organization of the cells may also conceal an asymmetry (Meinertzhagen, 1976), enabling the unique labelling of all cells. A good example is the ommatidium in dragonflies (Armett-Kibel and Meinertzhagen, 1983; Fig. 16.5), a comparison chosen because of the long evolutionary lineage of dragonflies back to the Carboniferous, and because of the relatively recent derivation of the special open-rhabdomere ommatidium of Diptera. If the ommatidia in these two orders could be homologized (Meinertzhagen and Shaw, 1989), a basis would have been initiated to establish cell homologies between other insect orders. Although the special open-rhabdomere arrangement of the fly's ommatidium offers no ready comparison with other non-dipterous ommatidia in the adult form, ommatidia arise from preclusters which have a pattern known in Drosophila to be built up sequentially in pairs (see above). Although this cluster gradually rotates 90° during its assembly, dorsal and ventral clusters in opposite senses with an equator (Dietrich, 1909) forming between them, in its prerotated pattern it resembles the immature form of the precluster in the bee larva (Ready, 1989); it also compares with the adult ommatidium in dragonflies (Fig. 16.5). Receptor cell numbers have previously been homologized between odonate species (Armett-Kibel et al., 1977), but with no expressed evolutionary motive. The homologies proposed between ommatidia in the fly and the dragonfly are, by contrast, explicitly evolutionary, and based on the evidence of the conserved number of 8 receptors, and their organization into 4 pairs, with 1 pair unmatched (Table 16.1). The proposal implies that the receptors of the fly have an evolutionary continuum through the lineage of the pterygote insects, during more than 300 Myr. Confirming evidence will come only when

Table 16.1 *Proposed homologies of receptor cells in different insect ommatidia[a].*

Diptera[b]	Hymenoptera[c]	Odonata[d]
R8	R5	R6
R2	R7	R1
R5	R3	R4
R3	R6	R2
R4	R4	R3
R1	R8	R8
R6	R2	R5
R7	R1 (R9)	R7

[a]See Fig. 16.5.
[b]Numbering convention of Dietrich (1909), arranged in the sequence of incorporation of cells into the assembling ommatidium in *Drosophila* (Tomlinson and Ready, 1987a).
[c]Numbering convention as resolved in Menzel and Blakers (1976) and Wehner (1976), see Fig. 16.5.
[d]Numbering convention as reported in Armett-Kibel and Meinertzhagen (1983).

the sequence of ommatidial assembly is known in dragonflies, to compare with that already known in the fruitfly, and so delineate the phylogeny of ommatidial ontogenies.

A divergence from the pattern of ommatidial assembly in the fly includes the presence of a ninth basal cell, found in Hymenoptera, members of which have congruent ommatidial cell patterns (Menzel and Blakers, 1976; Wehner, 1976). R9 appears in a position paired with that of R7 (Ready, 1989; Fig. 16.5), confusing the asymmetry within the central cell pair (R7 and R8) by which all other receptors are numbered. R7 itself, which is added last to the fly's assembling ommatidium (Tomlinson and Ready, 1987a), has been suggested to have been promoted from a cone cell during the course of evolution (Tomlinson and Ready, 1987b); in *sevenless*, R7 is lacking, they suggest, because the mutant ommatidium lacks the *sev* geene product which is required for reception of a cue directing the prospective R7 cell to its fate. This line of reasoning predicts that ancestral insect forms may have had ommatidia comprising 7 cells. It has also been suggested (Ready, personal communication) that intermediate steps in attaining the fixity of ommatidial construction through pairwise recruitments have been retained in ancestral crustacean groups (e.g. 5 in amphipods, 7 in some mysids, etc.), suggesting their early divergence from this sequence. To play this sort of numbers game we have to presume that the phylogenetic sequence of ommatidial acquisition is preserved in the ontogenetic sequence of ommatidial assembly. The ontogeny is the proper subject of interspecies comparison, but one we have no direct knowledge of in Crustacea. However, one piece of evidence does fit the idea of an early divergence of insect and crustacean ommatidia. The ommatidium in decapod crustaceans has only a single long visual fibre (Strausfeld and Nässel, 1981); could a possible homologue of R7 have been added to the phylogenetically prototypical ommatidium only after the divergence of insect and crustacean ommatidia, and programmed in decapods to terminate in the lamina, instead of the medulla as for its partner R8? This speculation is of course highly derived. It is also circumscribed by the lack of evidence on long visual fibres in other crustacean groups and attenuated by the precedent, in male flies, for the redeployment of R7 as a lamina terminal, augmenting the cartridge input of R1–R6 in frontal cartridges (Hardie, 1983).

The preceding arguments imply the homology of each of an insect ommatidium's individual photoreceptors with a counterpart in the ommatidia of other species, at least of other insect species (Table 16.1). However, this does not mean that other ommatidial cells need be homologous (see below), nor that the ultrastructural components of the cells have been conserved. The organization of the ommatidium has in fact changed relatively rapidly in different groups, around the core of receptor cells (this section) and cone cells (below), for example in coleopterans (see above) and in amphipods (Hallberg *et al.*, 1980). This is particularly true for the elaboration of rhabdomeric microvilli; at least two rhabdome designs, open-rhabdomere and layered rhabdomes, have been convergently adopted in insects and crustaceans, or have arisen through parallel evolution (if there were a single ommatidial ancestor for the two groups), in both cases presumably derived from an ancestral fused-rhabdome (Elofsson, 1976). Thus we should be quite clear what it is we mean when we claim that the ommatidium is homologous in different compound eyes, for we certainly cannot mean that the subcellular components of the ommatidial cells are necessarily homologous (Elofsson, 1976; Vanfleteren, 1982). Rather the ontogenetic programme to assemble a fixed-complement receptor cluster is the proposed homology within different ommatidia, the cells being variably transformed to different functional designs, once assembled. This conclusion anticipates a similar one reached independently for the cells of the lamina (Shaw and Meinertzhagen, 1986), where cell number likewise seems to be fixed but homologous cells in different groups may be remodelled to generate different functions (Shaw and Moore, 1989).

Photopigments

The evolutionary history of the novel photopigment, 3-hydroxyretinal, has been tracked in the compound eyes of many different insect orders (Gleadall *et al.*, 1989; Vogt,

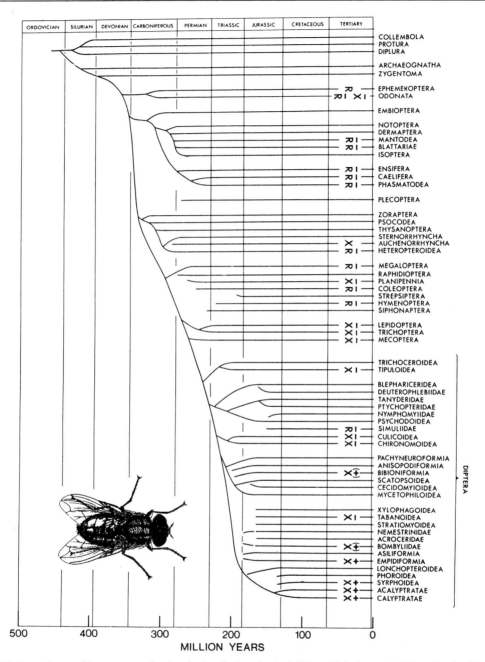

Fig. 16.6 *Phylogenetic tree of insect groups, showing the distribution of retinal (R) and 3-hydroxyretinal (xanthopsin: X) photopigments as a function of their implied geological antiquity (From Vogt, 1989).*

1989; Smith and Goldsmith, 1990) (Fig. 16.6). This pigment, a xanthopsin (Vogt, 1987), co-exists in recent Diptera with a second xanthopsin antenna pigment (3-hydroxyretinol: Vogt and Kirschfeld, 1984) the presence of which in these insects correlates with receptor sensitization in the UV (Vogt, 1984). Vogt has suggested that the transition to xanthopsins occurred somewhere among the holometabolous orders, so as to confer this chromophore upon extant members of the Diptera, Mecoptera, Lepidoptera, Trichoptera and Neuroptera (Fig. 16.7), thereby corroborating the monophyly of the mecopteroid orders (Hennig, 1973). If, according to this

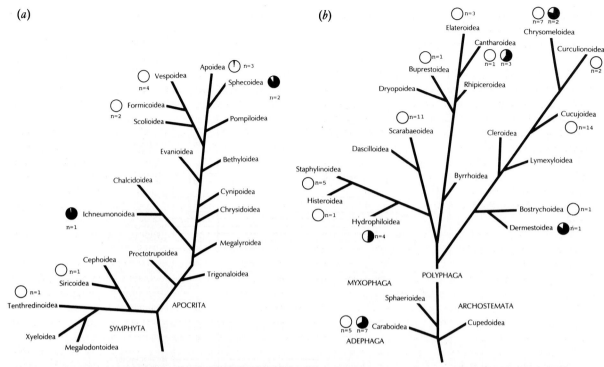

Fig. 16.7 *Variation in the distribution of xanthopsin in the families of two holometabolous orders: (a) Hymenoptera. (b) Coleoptera. Open circles are species having retinal alone; partially filled circles are those having both xanthopsin and retinal, with the unfilled sector indicating the relative amount of retinal;* n *is the number of species (From Smith and Goldsmith, 1990).*

reconstruction, the 3-hydroxy chromophore and its attendant protein changes were acquired only once and were not subsequently lost, then the evolutionary time scale for the transition may be derived from the fossil record and is close to 300 Myr ago, in the Upper Carboniferous (Fig. 16.6).

Four exceptions to this single-step progression were originally detected (Vogt, 1989). Odonata, which possess both retinal and xanthopsin pigments in the same eye and which are too distant from holometabolous ancestry to have been close to the appearance of xanthopsin in these groups, must therefore have come by their own xanthopsin in an independent step. Cicadas also possess the 3-hydroxy chromophore, despite falling within the retinal region of insect orders, which was originally interpreted as an independent acquisition imposed by a carotenoid-poor diet (Vogt, 1989) but is now known to be the case among other Hemiptera (Smith and Goldsmith, 1990). A strong case for the loss of xanthopsin can also be made. For example, in blackflies the chromophore is retinal (Vogt, 1989; Smith and Goldsmith, 1990), whereas all other Diptera possess xanthopsin and this was therefore presumably also true for the blackfly's ancestors. The same loss seems to have occurred among Neuroptera, with the regressive acquisition of retinal in *Ascalaphus*.

A yet more extensive inventory reveals that the xanthopsin distribution in insect eyes does not cleanly support the attractively simple idea of a single-step acquisition of the 3-OH chromophore (Gleadall *et al.*, 1989; Smith and Goldsmith, 1990). Thus, not only do many eyes contain both pigments but many species of holometabolous insects have retinal, not the predicted 3-OH retinal. This is most clearly illustrated in Hymenoptera (Fig. 16.7(a)) and Coleoptera (Fig. 16.7(b)), in which the distribution indicates that if there had been a switch in the ancestral group of Holometabola then it was neither unique nor exclusive (eyes can have both pigments) nor incapable of frequent reversal. For these cases the influence of both habit and habitat is invoked (Gleadall *et al.*, 1989; Smith and Goldsmith, 1990).

Other Cells

If the receptor cells could have arisen in mandibulate ommatidia from a common 8-cell cluster or its antecedents, then are the other cells of that cluster also fixed, so as to contribute homologues in the ommatidia of all groups? There are 2 corneagenous cells in Crustacea, which have no direct counterpart in insects, but which resemble the primary pigment cells of insect ommatidia

(Nilsson, 1989). Cone cells and screening pigment cells are also numerically conserved, like their receptors. Four cells form the crystalline cone in insects and Crustacea, but with exceptions in the latter group (Fig. 16.4); various numbers of cells form the cones in the few compound eyes outside these two classes (summarized in Nilsson, 1989). The cone cells produce a cone of one of various types, each with particular dioptric properties. This diversity in the optical properties of the cone contrasts with its immunoreactive constancy; a component of the crystalline cone epitope is recognized by an antibody (3G6) in both insect and crustacean eyes, which has been interpreted to imply the homology of the antigen, and thence the monophyly of the ommatidium in the two groups (Edwards and Meyer, 1990).

Screening pigment cells have such variable numbers, positions and types in the different groups that the search for homologies is frustrated. In insects each ommatidium has two primary pigment cells located distally by the cone, while in malacostracan Crustacea the distal pigment cells are shared among neighbouring ommatidia, 6 per set of 3 neighbours (Nilsson, 1989). A recent valuable review of the pigment screen elsewhere in Crustacea, and elsewhere in the crustacean ommatidium, indicates the existence of up to 6 different pigment cell types, based on the position each occupies in the depth of the retina (Hallberg and Elofsson, 1990). Crustacean ommatidia have been assigned to one of three grades: the first, in which only the photoreceptors are pigmented; the second, in which one class of pigment cell is added either axially or coaxially; and the third, in which various elaborations of these patterns involve more pigment cell types. Some orders, such as isopods, display an entire range of pigmentary complexity; clear examples of convergent organization can be found in groups that are related only distantly; while closely related species in yet other groups have diverged. Hallberg and Elofsson (1990) suggest that because it is found in ancient groups, pigmentation in the receptor cells is the ancestral condition in Crustacea, from which there have been many paths to the more advanced pigment screens of other eyes. They conclude that the pigment screen has been readily modified during the various lines of evolution in the crustacean ommatidium, so as to offer no clear insight to the evolution of the visual system as a whole.

Optical Mechanisms

The diversity of optical mechanisms in the compound eye is overwhelming, especially in Crustacea. Many different optical principles have been exploited, and the current list has recently been classified (Nilsson, 1989) as follows: apposition eyes, of several different types – simple, open-rhabdomere, afocal apposition (Nilsson *et al.*, 1984) and the transparent apposition; and superposition eyes – refracting, reflecting (Vogt, 1975; Land, 1976) and parabolic superposition (Nilsson, 1988) (Figs 16.8 and 16.9). In Crustacea, compound eye optics are suggested as features of such evolutionary stability as to base taxonomic judgements on them; for example, the common occurrence of refracting superposition optics in mysids and euphausiids suggests the close status of these two groups and favours the readoption of an early taxonomy, now abandoned, in which they were considered sister groups (Fincham, 1980; Land, 1981b). Taxonomic questions are beyond the scope of this review, but the idea that eye design is more stable than other characters has undoubtedly been influenced by the difficulty in seeing how eye types have evolved from one to the other, or more likely from an unknown intermediate, and this difficulty has been diminished in recent years.

Attempts to explain how the different types of optics may have evolved (e.g. Land, 1981b; Nilsson, 1989) have generally concluded that the simple apposition eye was ancestral in all groups. To derive refracting superposition optics, and especially reflection optics, from the presumed ancestral apposition optics is the essential trick, and it was not at first sight easy to see how during evolution this could ever have been accomplished (Land, 1981b), since in their final forms the different types have no reconcilable functional intermediates. However, the recent discovery of two types of intermediate optics, and consideration of how optical mechanisms transform during the life-cycle, now offer three clear paths to derive superposition optics (Nilsson, 1989; Figs 16.8 and 16.9). These are:

1. Via afocal apposition optics in butterflies (Nilsson *et al.*, 1984); if a butterfly with afocal apposition optics became nocturnal, or adopted a scotopic habit, then its optics could transform to the superposition type (Fig. 16.8), suggesting that modern butterfly (apposition) and moth (superposition) eyes arose in this way from a common ancestor with afocal apposition optics.

2. Via parabolic superposition optics in short-bodied decapods (crabs) (Nilsson, 1988; Fig. 16.8).

3. Via larval intermediates in pelagic species of decapods, euphausiids and mysids that have transparent larvae. As camouflage against predation the eyes of these larvae lack pigment, and instead have developed optical means of isolating their ommatidia to prevent cross-talk. The means employed in decapod larvae is to have the light totally internally reflected within the cone and this results in gratuitous superposition rays, which are of no use in the apposition eye of the larva, but which the adults later frequently use to transform their eyes into reflecting superposition optics. In euphausiids on the other hand the light is refracted in the cone and the eye later transformed into refracting superposition optics in the adult (Nilsson, 1983; Fig. 16.8). In both cases the transformation is accom-

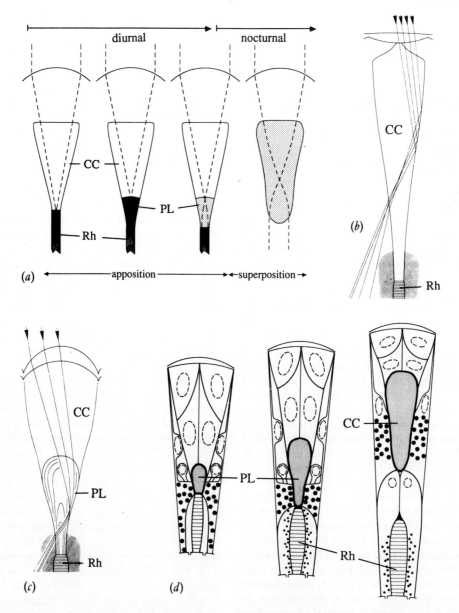

Fig. 16.8 *Transformations of optical design.* (a) The transformation between simple apposition to refracting superposition via afocal apposition optics. (b) Superposition rays reflected within the transparent apposition eye of a decapod larva. (c) Similar superposition rays lost by refraction in the transparent eye of euphausiid larvae. (d) Ontogeny of refracting superposition optics from transparent apposition optics in euphausiids and mysids. CC, crystalline cone; RH, rhabdome; PL, proximal lens (From Nilsson, 1989).

plished by the creation of a clear-zone between cone and receptor layers and the development of square facet lenses in the case of eyes with reflecting optics. Thus, with a few exceptions, the eyes in these two groups are preadapted as larvae to the reflecting (decapod) and refracting (euphausiid) superposition optics of their adults, suggesting the route by which apposition optics (in the larval stages of existing species) could have developed superposition optics during evolution (Nilsson, 1983, 1989). The eyes of mysids which, like euphausiids, also have refracting superposition optics, lack access to a larval stage in the life cycle. Nevertheless they pass through a transparent apposition stage during their ontogeny (Nilsson *et al.*, 1986), which however is no longer used for vision.

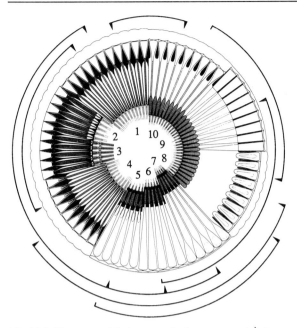

Fig. 16.9 *Summary of the known optical eye types and their possible evolutionary transitions. The ten types are: 1, simple apposition; 2, open-rhabdomere apposition; 3, neural superposition; 4, afocal apposition; 5–7, three types of transparent apposition optics; 8, refracting superposition; and 9, reflecting superposition optics (From Nilsson, 1989).*

Yet one must still be wary, even of plausible tales. In ampeliscid amphipods, for example, an array of ommatidia lies beneath a single vitreous and cornea. This has been interpreted as a compound eye that was able to substitute single-lens optics after ancestral forms had passed through a period of scotopic life and lost their compound eye optics, leaving a layer of photoreceptors that only later on, following resumption of a photic habit, became served by a single lens (Nilsson, 1989). This is an ingenious reconstruction of how the eye type first arose, but it cannot account for the topology of the layers in the eye, in particular the separation between the corneagenous cell layer and the rest of the retina, which Mayrat (1981) claims is evidence that the eye is in fact not a compound eye at all, but an ocellus. Since we must assume that ommatidia of the compound eye, ampeliscid or otherwise, develop from a single epithelium (Meinertzhagen, 1973), a clear ontogenetic account of how the additional two layers arise is called for.

The Optic Lobe

Relatively less attention has been accorded to the cellular organization of the optic lobes than has been given to that of the overlying compound eye, and treatment of the evolution of the optic lobe has been likewise overshadowed by consideration of how its photoreceptor inputs and their optics might have arisen. This is unfortunate, because the optic lobe, which is known in far less functional detail than is the retina, can by no means as easily be recruited to the case for the independent origins of insects and Crustacea. The particularly striking test case is the comparison between the optic lobes of insects and malacostracan Crustacea (see Osorio, 1991). In these two examples, the numbers of optic neuropiles, their laminations and type of fibre projections, and the cells which constitute them and their synaptic organization, all present such strikingly detailed similarities as to spell out a case for homology that, when taken alone, none but the unreasonable could deny. In evaluating these similarities it is important to bear in mind that to a considerable extent the structural cellular similarities in the optic lobe originate with its modularity. The subdivision of the neuropiles into cartridges imposes major geometrical constraints upon the component cells which could conceivably obscure major differences in lineage.

The Number of Neuropiles

There are three main columnar neuropiles in insects and decapod crustaceans: lamina, medulla and lobula; in Crustacea these were historically called the lamina ganglionaris, medulla externa and medulla interna. The most recent comparison of the two groups (Strausfeld and Nässel, 1981) homologizes the terminology for these neuropiles without addressing the deeper issue of their evolutionary homology. A fourth neuropile in decapod and other Crustacea, the medulla terminalis, is now dubbed as in insects the lateral protocerebrum, or one of its optic foci (Strausfeld and Nässel, 1981). The lobula in certain holometabolous insect orders (Lepidoptera, Diptera, Coleoptera) is a double structure comprising lobula and lobula plate neuropiles (Strausfeld and Nässel, 1981). It is not clear whether the division of the lobula neuropile was a single event in the evolution of Holometabola, for which the arrangement in Neuroptera, Trichoptera and Megaloptera needs description. The lobula in bees is a bipartite structure which possibly represents an intermediate condition; a bipartite structure is also visible in hemimetabolous forms, albeit less clearly, and is possibly mirrored in the organization of the medulla into distal and proximal divisions in all insects (Strausfeld and Nässel, 1981). There thus seems to be an evolutionary trend toward the division of neuropile layers which is clearest in the case of the holometabolous lobula complex. The extent to which old cells conserved from ancestral forms lacking a divided lobula were able to contribute to these rearrangements during the evolution of the holometabolous optic lobe is unclear; possibly new cells arose. Cajal and Sánchez (1915) suggest similarities between narrow-field trans-lobula monopolar cells in the bee and the dragonfly (both of which have only a single lobula) and T-cells of the fly which connect lobula and lobula plate of the divided lobula complex, but otherwise a carefully reasoned basis for homologies between the cells of the lobula complex and the single lobula neuropile in the earliest insect orders, or in Holometabola such as Hymenoptera, has not been systematically attempted. The status of satellite neuropiles such as flank the single lobula dorsally in Orthoptera and in species of Odonata and Ephemeroptera (Strausfeld, 1976) is unclear; they lack direct projections from the medulla and thus cannot be homologues of the holometabolous lobula plate.

If there is an evolutionary trend for the optic neuropiles to have arisen by splitting from an ancestral neuropile, what evidence is there for this process having occurred in the generation of the three neuropiles found in all extant insect groups – lamina, medulla and lobula? None that is direct. Strausfeld and Nässel (1981) consider that these are actually five: lamina (1), outer (2) and inner (3) medulla, outer (4) and inner (5) lobula (the latter two having split to yield lobula and lobula plate). Could it be that (1) and (2) split early in the evolution of the insect optic lobe from an ancestral outer neuropile, creating a chiasma between themselves, and that (3) and some part of the lobula (4 and 5) have also arisen by division, likewise creating the inner chiasma between these two? One reason to entertain this idea is that the long visual fibre of R8, which is the earliest cell of the developing ommatidial cluster (Tomlinson and Ready, 1987a), possibly retains its ancestral condition in projecting to layer (2), while other cells of the ommatidium project to the lamina (1). R7, which is possibly an evolutionary afterthought of the ommatidial cluster (Tomlinson and Ready, 1987b), mimics R8. A second reason is developmental. The cortices of the optic neuropiles are generated by the progeny of neuroblasts that are in two primordia, the optic anlagen (1) and (2) arise from the outer optic anlage, while (3)–(5) arise from the inner optic anlage (Meinertzhagen, 1973; Hofbauer and Campos-Ortega, 1990). This is the situation in the fly as far as it is understood, but it is too incompletely described in other groups to compare their ontogenies.

How well do these ideas fit the organization found in Crustacea? Decapod crustaceans resemble insects closely, as we have seen, except in lacking a second long visual fibre (Strausfeld and Nässel, 1981). Elsewhere in Crustacea, branchiopods reveal a trend toward midline fusion of the eyes (Hanström, 1927), from lateral-eyed forms such as *Artemia* to *Daphnia*, in which both eyes are fused to form a cyclopean eye beneath which sits a single fused lamina and a partially fused pair of medullae (Sims and Macagno, 1985) (Fig. 16.10). Beneath the eyes, the optic lobe there-

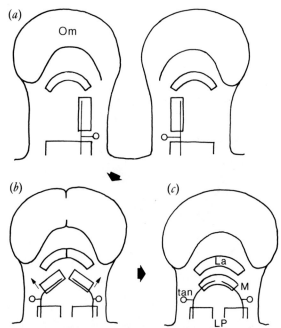

Fig. 16.10 *Progressive union of the compound eyes and optic neuropiles during the evolution of branchiopods, in the process of which the positions of the medullae (M) rotate, as shown by arrows in (b), and the ommatidia (Om), laminae (La) and medullae coalesce. The union of the latter is revealed in (c) by the positions of the retained tangential neurones (tan) projecting to the lateral protcerebrumo (Relabelled from Elofsson and Dahl, 1970.)*

fore has only a lamina and medulla, and lacks all chiasmata and central optic neuropiles. Possibly these two neuropiles actually represent the prefission stage of a primitive outer and inner neuropile, so that the so-called 'lamina' would actually correspond to the united lamina and distal medulla, and the 'medulla' to the proximal medulla and lobula. The absence of long visual fibres projecting to the medulla would then be explained as the mistaken identity of that neuropile in *Daphnia*. The predicted photoreceptor of origin would contribute the first axon to grow within each ommatidial bundle, the so-called lead fibre; this derives from R1 (Flaster *et al.*, 1982), which terminates only half-way down the depth of the lamina. Comparison of the cell types in *Daphnia* (Sims and Macagno, 1985) with those in decapods (Strausfeld and Nässel, 1981) does not reveal morphological similarities between the two, once past the LMc monopolar cells of the lamina (L1–L5); it thus does not clearly uphold this view of the branchiopod optic lobe.

Fibre Projections Between the Optic Neuropiles

Typically a direct projection maps the retinal visual field upon the lamina through the receptor axons, while the

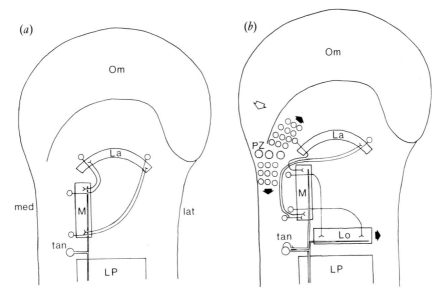

Fig. 16.11 *Polarity of axon growth in the embryonic malacostracan optic lobe (b), compared with the adult optic lobe of non-malacostracan forms, such as branchiopods (a). The production of optic lobe cells apparently occurs in a similar postero–anterior sequence in both, with cells of the posterior eye and their corresponding interneurones in the optic lobe generated first, but the patterns of fibre growth differ. In malacostracan optic lobes (b), the columnar cells of the lamina and medulla grow toward their target neuropiles, and enter them by way of the corresponding cortex before penetrating the neuropile proper. In so doing they generate a chiasmal projection, unlike the non-malacostracan, in which axons apparently grow directly towards the neuropile and generate a non-inverted projection (a) (From Elofsson and Dahl, 1970).*

lamina maps in horizontally inverted sequence upon the medulla, through the external chiasma, and the internal chiasma reinverts this sequence again, between medulla and lobula. The cardinal difference between direct and chiasmal projections is possibly developmental in origin and probably functionally trivial, having to do with the polarity of axon growth between neuropiles and the direction from which innervation approaches the neuropile face (Elofsson and Dahl, 1970) (Fig. 16.11). The acquisition of the divided lobula complex in three orders of holometabolous insects (Lepidoptera, Diptera, Coleoptera) required the establishment of new topographical relations between the cell populations and the rerouting of fibre pathways to establish new chiasmata between these. Rerouted fibre tracts are sometimes encountered either in mutant optic lobes (Fischbach et al., 1989; Boschert et al., 1990) or in flies in which the optic lobes were lesioned during development (Nässel and Geiger, 1983), suggesting that the transition to different projection patterns may be readily accomplished secondarily to more profound alterations in cell fate or topography. In branchiopod Crustacea chiasmata are lacking (Elofsson and Dahl, 1970; Sims and Macagno, 1985). Details in other crustacean groups are listed in Horridge (1965), chiefly from the older and not invariably reliable studies of Hanström.

The Neural Superposition Projection

The evolutionary acquisition in Diptera of the system of neural superposition (Kirschfeld, 1967) required the apparently fateful coincidence of two steps: first, a subcellular rearrangement of the apposition rhabdomeres of receptors and, second, a rearrangement of the receptors' axonal wiring patterns. These patterns arose from those occurring in the dipteran ancestor, resembling modern Mecoptera, which were presumably typical of forms with fused-rhabdome ommatidia elsewhere, in which receptor axons project to the cartridge underlying their ommatidium of origin (Meinertzhagen, 1976). To bring about the necessary alteration required that the trajectories of receptor growth cones be redirected across the face of the developing lamina, instead of associating with the lamina monopolar cells which they initially contact beneath their ommatidium (Trujillo-Cenóz and Melamed, 1973), so as to guide the receptor axons belonging to the appropriate pattern of neighbouring ommatidia (Braitenberg, 1967) to terminate in a single cartridge.

The groups of visually most studied Diptera are members of the recent Brachycera. The open-rhabdomere pattern of all recent brachyceran groups has the same asymmetrical pattern (Shaw, 1989a), in which all but the rhabdomere of R3 align optically along one of the three major axes of the facet array (Shaw, 1989b). Among Nematocera (the other major group of Diptera and sister group to the Brachycera), anisopodids are closest to the brachyceran ancestral stemline. Compared with the brachyceran pattern, their rhabdomere pattern is more symmetrical: the rhabdomeres align along the minor axes of the facet array at 30° to the major axes and the angular divergences are close to those required for the superposition of images from next-but-one neighbouring facets, not neighbouring ones as in more recently derived brachycerans. The receptor axons terminate in the corresponding pattern of underlying cartridges, so producing neural superposition, but they also synapse at non-superposition sites *en route* to their cartridge (Shaw, 1990b). Shaw (1990b) speculates that this is an adaptation to vision under dim light which could also have served in a similarly endowed ancestral form as a preadaptation to the advanced condition, enabling the advanced muscomorph projection (Braitenberg, 1967) to be generated by selectively withdrawing some of the synapses at non-superposition cartridges.

Some insight into how the open-rhabdomere pattern may have arisen in the first place can be gained by considering the open-rhabdomere arrangement that has arisen independently among apposition ommatidia elsewhere. Nilsson (1989) points out that in the absence of a system of axonal interweaving at the distal lamina, the open-rhabdomere arrangement of R1–R6 provides an additional receptor system having increased sensitivity but diminished resolution, and complements the high-resolution–low-sensitivity system of the two central cells, so providing a scotopic (R1–R6) system separate from the photopic system of R7/R8. In some species of Nematocera, which are as a group closest to the ancestral form of all Diptera, a structural transition occurs during dark adaptation which suggests the mode of origin of the open-rhabdomere ommatidium (Seifert and Wunderer, 1989). In the dark-adapted state the rhabdomeres join laterally to form what is essentially a fused-rhabdome ommatidium, with all rhabdomeres drawn closely together. After light adaptation the rhabdomere diameter reduces, diminishing the degree of this lateral fusion, in some cases to form an open-rhabdomere pattern with R1–R6 drawn away both from themselves and from R7/R8 (Seifert and Wunderer, 1989). In some species (such as the mycetophilid: Meyer-Rochow and Waldvogel, 1979; and tipulid: Williams, 1980) this separation is maintained proximally in the retina, even in the dark, so that the rhabdomeres resemble the roots of a molar tooth. Thus, the ommatidium exists in one of two states, and for only one of these is the ancestral projection of receptor axons (one ommatidium projecting exclusively to one cartridge: Meinertzhagen, 1976) appropriate.

Natural selection thus appears to have acted on a dual system which arose from a single one (the one equivalent to the present dark-adapted state), to produce an open-rhabdomere condition from the light-adapted state. The exact point at which the brachyceran rhabdomere pattern diverged from the patterns still found in extant Nematocera is unclear; presumably the open-rhabdomere organization evolved in parallel, and somewhat differently, in the two groups.

How did the corresponding projection pattern arise to utilize the new ommatidial optics? The steps in that evolutionary progression are not yet understood. In the nematoceran group, Bibionidae, there is a redistribution of receptor terminals from single ommatidia to different cartridges, but these lie in rows that do not precisely match the alignment of divergent optical axes in the overlying retina, so providing an independent but less than optimal solution to the mechanism of neural superposition (Zeil, 1983). In anisopodids, a substrate for neural superposition occurs too, but it seems to involve the reconstitution of images from next-but-one facets, as detailed above. Turning to Brachycera, the rhagionids and stratiomyids which diverged > 200 Myr ago as well as the dolichopodids, a third brachyceran of intermediate antiquity, all have projection patterns characteristic of more recent families (Shaw, 1990a), including forms such as muscids, drosophilids and calliphorids extensively studied in vision. This implies that the projection pattern and its elaborate ontogeny had already arisen early in the origin of brachyceran flies, at or close to the time at which the asymmetrical open-rhabdomeric pattern first appeared (Seifert and Wunderer, 1989; Shaw, 1989a).

Thus the evolution of these two innovations, open rhabdomeres and divergent axons, was probably accomplished during a switch from a scotopic to a photopic habitat, and possibly involved a series of smaller intermediate steps. Although our reconstruction of these events in the early evolution of Diptera is at best incomplete, it provides further examples of evolution acting through duplication (in this case via structural adaptations to light and dark adaptation) and the subsequent divergence of the duplicated structures.

Cellular Organization of Neuropiles

Various pieces of evidence collectively imply that many cells of the arthropod optic lobe may have been conserved for long periods of evolutionary history in the different groups. The evidence has been presented formally for the dipteran lamina cartridge (Shaw and Meinertzhagen, 1986; Meinertzhagen and Shaw, 1989; Shaw and Moore, 1989), and is recapitulated below. It consists of three parts: first, that neuronal isomorphs exist in related forms; second, that these occupy homotopic locations within the cartridge cross-section; third, that the total number of all cartridge elements has been conserved. From these observations it is argued that each columnar element of the dipteran cartridge is homologous with an isomorphic counterpart in other species, possibly with isomorphs in other orders. The existence of isomorphic cells in the deeper neuropiles of related species is compatible with the idea that many optic lobe neurones may have been conserved in much the same way as is reasoned for those in the lamina.

Cell Types

The realization that the same cell types may be repeatedly recognized in different species using morphological criteria was clear long ago, during the first Golgi studies (Cajal and Sánchez, 1915). It has persisted as the implicit theme underlying the adoption of unifying taxonomies of optic lobe neurones (Strausfeld, 1976; Strausfeld and Nässel, 1981), but an evolutionary basis in homology has never been explicitly claimed. As cell shapes progressively diverge between taxa, claims of cell homology are highly dependent upon the particular taxonomy of cells adopted. An important basis for comparison is the classification of cells into three major classes, columnar, tangential and amacrine, depending on the inferred developmental relationships of the growing cell's axon with respect to the face of the neuropile it innervates (Fischbach and Dittrich, 1989). Thereafter the columnar cells may be grouped by the position of their terminals with respect to their soma and by the number and strata of neuropiles in which their axon emits processes. All of these represent developmental programmes for which genomic instructions must have evolved. The most obvious case is the distinction between the short receptor terminals of R1–R6 and their homologues (see above), which terminate in the lamina, and the long visual fibres of the central cells, which terminate in the medulla of the insect's optic lobe.

Using features of cell morphology so defined, it is relatively easy to locate isomorphs in close species. This is the first piece of evidence that such cells may be evolutionary homologues, although taken alone it is no proof of an evolutionary relationship. Various classifications, notably those given by Strausfeld (1970, 1971, 1976, etc.), adopt a uniform scheme of cell nomenclature, but largely as a taxonomic tool. The second piece of evidence that cell isomorphs are actual homologues comes from cell position. Since the cross-section of cartridges in the lamina is filled with the profiles of uniquely identified cells, it is possible to demonstrate that the positions of cell isomorphs are homotopic in different species, as has been validated for flies (Meinertzhagen and Shaw, 1989; Shaw and Moore, 1989; Meinertzhagen and O'Neil, 1991). The final piece of evidence is that the total cell number is constant in different species, so that if one element had been either lost or gained at an evolutionary divergence then another element would improbably have to have been either acquired or lost at the same time, to offset the change (Shaw and Meinertzhagen, 1986). For example, there are 5 monopolar cells, L1–L5, in the lamina cartridge in flies. Each is a uniquely identified interneurone, with a distinct shape and position in the lamina, and projects through the chiasma to a unique terminal in the medulla (Strausfeld, 1971). One each of these per cartridge is found in flies and other Diptera (Shaw and Meinertzhagen, 1986; Shaw and Moore, 1989). The same number has also been found in the dragonfly (Meinertzhagen and Armett-Kibel, 1982), in the crayfish (Strausfeld and Nässel, 1981) and in the branchiopod *Daphnia* (Sims and Macagno, 1985), leading us to suspect that this number may be uncovered beneath all compound eyes.

By extension, the same three criteria (isomorphy, homotopy, and numerical constancy) are anticipated to be upheld in the medulla too, where indeed cell isomorphs are already well known, but the remaining evidence is far from clear. How much further can we go? The most striking case of cell isomorphy is recorded by Fischbach and Dittrich (1989) in the case of a medulla cell, T1. This cell has identical shapes, albeit different sizes, in *Drosophila* and decapods (Fig. 16.12). It is far too early to conclude that representatives of two of the most recently diverged groups of their respective classes have retained the same shape from a hypothetical common ancestor, possibly since the Cambrian, but clearly this possibility is raised by the evidence, and the greater implication is that there may have been a common ancestor with T1, and presumably a compound eye and optic lobe to carry it in. One can of course always argue that the presence of identical T1s is no more remarkable than the similar arrangement of the optic

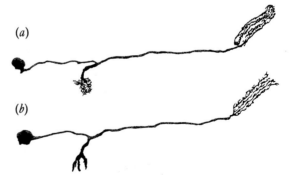

Fig. 16.12 *The neurone T1, with a cell body in the medulla cortex and a terminal in the neuropile, has a characteristic basket cell ending in the lamina neuropile, where it is predominantly postsynaptic (Meinertzhagen and O'Neil, 1991). Despite differences in absolute dimensions, the cell is strikingly isomorphic in (a) the fruitfly; and (b) a shrimp (From Fischbach and Dittrich, 1989).*

neuropiles in insects and decapods, and that the cartridge organization of both so constrains cell morphology as to ensure the similar shapes of cells we call T1.

Synaptic Organization

If the preceding evidence can be taken to establish the homology of cells in the insect optic lobe, at least in Diptera, then the synaptic contacts between identical combinations of these homologous cells should also be homologous. It is clear that these have been modified during the course of mandibulate evolution. For example, the geometry and postsynaptic composition of afferent receptor synapses has changed considerably, as is shown in the following examples.

Triads

In the dragonfly *Sympetrum*, for example, receptor synapses are triads incorporating MII (the homologue of the fly's L1) as the median element and MI (L2) as the two lateral elements of a postsynaptic triad (Armett-Kibel *et al.*, 1977; Meinertzhagen and Armett-Kibel, 1982). MI also forms feedback connections upon receptor terminals, just as does L2 in the fly, and these are concentrated in the proximal lamina (Armett-Kibel *et al.*, 1977). Triad receptor synapses are also found in the crayfish *Procambarus* (Hafner, 1974) and *Pacifastacus* (Nässel and Waterman, 1977). In the later, the median element is from one monopolar cell (M_2) while in the distal lamina a lateral element comes one each from two other monopolar cells (M_3 and M_1); in the proximal lamina they come instead from two other monopolar cells (M_4 and M_{1s}) (Nässel and Waterman, 1977; Strausfeld and Nässel, 1981).

Dyads

Receptor synapses have postsynaptic dyads in the cockroach *Periplaneta* (Ribi, 1977), in the honeybee *Apis* (Ribi, 1981) and in the branchiopod *Artemia* (Nässel *et al.*, 1978).

Monads

The lamina of *Daphnia* only has monad synapses (Macagno *et al.*, 1973), and this in particular underscores that no necessary evolutionary trend exists between the composition of receptor synapses in related species (*Artemia*, for example, having dyads but *Daphnia* monads). Indeed, the comparison highlights the dangers of making spot comparisons between too few examples, when the synaptic organization in *Daphnia* (apparently unique among arthropod optic neuropiles in having only monads) is possibly secondarily simplified.

The Tetrad and Other Synapses of the Fly's Lamina

Even so, in selected cases the synaptic sites have undergone evolutionary transitions, most clearly revealed in the case of the photoreceptor input synapse to the dipteran lamina (Shaw and Meinertzhagen, 1986) (Fig. 16.13). Despite the strikingly uniform composition of this synapse in recent Diptera in which it is a tetrad, this site is a dyad in forms closer to the ancestral dipteran stock; the tetrad is argued to have arisen from the dyad at a single transition about 120 Myr ago during which an amacrine cell incorporated itself into the original dyad of processes from two monopolar cells, L1 and L2 (Shaw and Meinertzhagen, 1986) (Fig. 16.14).

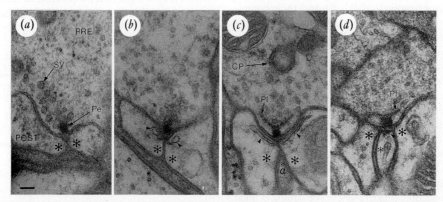

Fig. 16.13 *The photoreceptor synapses in representative species of four dipteran groups. They are presented in the sequence, left to right, of the divergence of their group from the ancestral stock (Myr before the present time, given in brackets: see Fig. 16.14), to reveal the evolutionary transition in ultrastructure and composition. (a) Anisopodid (205 Myr); (b) stratiomyid (170 Myr); (c) dolichopodid (120 Myr); (d) hippoboscid (60 Myr). The progressive elaboration in the components of the presynaptic ribbon (Pe, pedestal; Pl, platform) and the postsynaptic cisternae (arrowheads) contrasts with the single-step addition of two polar postsynaptic elements from amacrine cell (α) processes to derive a tetrad (in (c) and (d)) from the ancestral dyad (in (a) and (b)) of L1 and L2 (*). CP, capitate projection. Scale bar 0.1 μm. (From Shaw and Meinertzhagen, 1986.)*

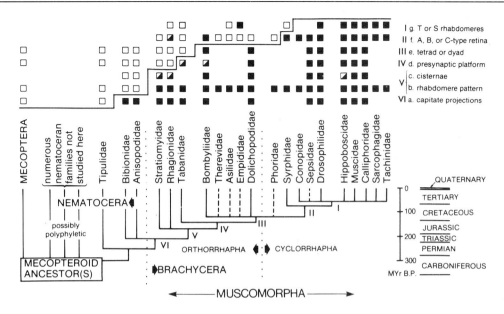

Fig. 16.14 *The features of the eye arranged in a checklist for different families of Diptera, in which characters e–c are from the photoreceptor synapse illustrated in Fig. 16.13. Other characters are: a, the presence of capitate projections (illustrated in Fig. 16.13); b, the asymmetrical trapezoidal rhabdomere pattern (Shaw, 1989a); f, A-, B- and C-type retinae (Wada, 1975) defined by the position of R8 in the ommatidial cross section; and g, tiered or segmented arrangement of the central rhabdome in dorsal ommatidia (Wada, 1974). Open symbols represent the ancestral condition, closed ones the derived condition and half-symbols the existence of an intermediate structure. The characters are arranged in the sequence (I–VI) of the implied antiquity of the transition between these states, so as to construct the dendrogram below. This dendrogram endorses existing ones, arbitrating on detailed differences, and is also compared with evidence from the fossil record to derive the geological timescale. (From Shaw and Meinertzhagen, 1986.)*

Discussion

This example of an evolutionary gain in synaptic complexity is matched by other cases of losses, for example, in *Musca* of the feedback synapse from amacrine cell processes upon receptor terminals (Shaw and Meinertzhagen, 1986). Possibly the main input synapses have been relatively protected from rapid change because of their critical input position to the synaptic networks of the lamina. Elsewhere, the connections of a third monopolar cell, L3, have changed much more radically. In the stratiomyid *Hermetia*, they bear little resemblance to their counterparts in muscids, for example receiving input from one of the long visual fibres (belonging to R8) as well as from L4 (Shaw and Moore, 1989). These examples provide a basis to the claim that synaptic change has been a major avenue for evolutionary modification in the optic lobe. In order for such changes to have occurred, natural selection must have acted not upon identical synaptic contacts assembled according to a perfect template but upon variants of these, since variations provide the fuel for evolution. It is thus interesting that the synaptic populations of one insect species (the dragonfly *Sympetrum*: Armett-Kibel *et al.*, 1976) show far more variations than those of another (the fly *Musca*: Nicol and Meinertzhagen, 1982). Although variation should be less in the system more exposed to natural selection (Wright, 1978), both parameters are in practice hard to quantify.

Each of the evolutionary changes of the sort seen in Diptera is an increment; the overall sum of such changes becomes major because of the large number of cell combinations at which synaptic contacts occur and which are liable to modification during evolutionary progression. Other structurally identifiable changes include modifications to the ultrastructure of synaptic organelles (Fig. 16.13), especially the acquisition of the platform of the presynaptic ribbon and the elaboration of postsynaptic cisternae in each of L1 and L2 (Shaw and Meinertzhagen, 1986); both of which seem to have been acquired in stages (Fig. 16.14). These seem to be fly synaptic specialities, however, supplementing the less differentiated ultrastructural features of ancestral forms such as are retained in the anisopodid shown in Fig. 16.13, and such as compare with the synapses in other groups like Odonata (Armett-Kibel *et al.*, 1977).

Although these are changes that have been positively identified, their functional significance is unclear and may be overshadowed by other evolutionary avenues for functional change which lack an anatomical basis and for which no evidence currently exists. The latter include the following: altered postsynaptic gain or polarity at a structurally unmodified synaptic site; altered transmitter phenotype; altered role of possible neuromodulators; dynamic changes in the configuration of neural networks. These possibilities have recently been considered in greater detail elsewhere (Arbas *et al.*, 1991).

Transmitter Phenotypes

Transmitter identity offers one further criterion upon which to judge the homology of isomorphic or homotopic cells in related groups. Alternatively, if evidence exists for the homology of a particular cell in one species with that in a related species, then comparisons between the transmitter phenotypes of the two cells reveal how closely the transmitter machinery has been conserved during evolution. The interpretation is seriously clouded by the quality of evidence for the transmitter in question, most readily but not always reliably indicated by immunocytochemical methods, by the assignment of that transmitter to a particular synapse in question (since many optic lobe cells form several classes of presynaptic contact), and by the question of colocalization of different transmitters in the same cell.

Overlooking these concerns, a simple and apparently clear case may be made for the transmitter at the afferent synapses of photoreceptor terminals, which appears to be histamine in the compound eyes of various arthropods (fly: Hardie, 1987; Nässel *et al.*, 1988; Sarthy, 1989; cockroach: Pirvola *et al.*, 1988; *Limulus*: Battelle *et al.*, 1989). Histamine is also the apparent transmitter at photoreceptor synapses of other eyes, however (insect ocelli: Nässel *et al.*, 1988; Simmons and Hardie, 1988; Schlemmermeyer *et al.*, 1989; barnacle Nauplius eye: Callaway and Stuart, 1989), with which they are not homologous. This suggests that strong functional constraints exist upon the use of this transmitter or its hyperpolarizing postsynaptic receptor sites at photoreceptor synapses in general. The most effective histamine antagonists are those which also have a strong action at cholinergic synapses; the pharmacology of the histamine receptor thus indicates similarities to the cholinoceptor (Hardie, 1989), from which it may therefore have arisen. This is an interesting coincidence, if the evolutionary history of ommatidia and sensilla were related as earlier proposed, because acetylcholine is a major sensory transmitter in insect sensilla (Sattelle, 1985).

Conclusions

What, if anything, may we conclude? Uncertainty dogs every step in any reconstruction of the evolutionary path of arthropod vision; this must be our first conclusion, as it is of all previous reviews. The homology of the compound eye seems less unlikely to this reviewer than it has to others, when considered in the light of recent discoveries concerning the range of differences in its optics and when we bear in mind the structural details of the optic lobe. The extent of parallel evolution, the independent derivation of the organization of the eye from a homologous precursor in two or more groups, has probably still to be seriously assessed; identification of the correct features upon which to base a cladistic comparison may be thwarted by the extreme conservatism of the cellular organization, on the one hand, and the total fluidity of the subcellular optical properties of the eye, on the other. A serious cladistic treatment of all visual systems, optic lobe as well as compound eye architectures, one that takes account of many different features, from cell numbers to synaptic sites, is badly needed, but lies beyond the scope of this review.

Several mechanisms may be identified at work. The conservation of individual cells can be validated in detail in certain cases (lamina cells, Diptera; receptor cells, Diptera and Hymenoptera); elsewhere they are surmised from suggestive evidence to have occurred much more widely. By contrast, synaptic connections have certainly changed. Changes are both qualitative and numerical, but the functional consequences of either are still unclear. These changes and others in the visual system probably occurred incrementally. Duplication of structures is a common path for their subsequent divergence from a common origin. This mechanism has been clearly identified at the molecular level (Ohno, 1970), but it is only now becoming recognized at the cellular level in nervous systems (Arbas *et al.*, 1991). Possible examples seen here in the visual systems include the splitting of optic neuropiles, the divergence of scotopic and photopic pathways in the genesis of the neural superposition principle, and the duplication of cell types in the lamina. The reverse process of fusion is often seen as an evolutionary trend in groups, as part of the relocation of eyes to different head regions.

Acknowledgements

The author's research on invertebrate visual systems is supported by NSERC grant A 0065. Thanks are due to the following for commenting on portions of the manuscript: D.-E. Nilsson; D. Osorio; D. R. Ready; K. Vogt.

References

Anderson, D. T. (1973). *Embryology and Phylogeny in Annelids and Arthropods*. Oxford: Pergamon.

Arbas, E. A., Meinertzhagen, I. A. and Shaw, S. R. (1991). Evolution

in nervous systems. *Ann. Rev. Neurosci.*, **14**, 9–38.
Armett-Kibel, C. and Meinertzhagen, I. A. (1983). Structural organization of the ommatidium in the ventral compound eye of the dragonfly *Sympetrum*. *J. Comp. Physiol.*, **151**, 285–294.
Armett-Kibel, C., Meinertzhagen, I. A. and Dowling, J. E. (1976). Cellular and synaptic organization in the lamina of the dragon-fly *Sympetrum rubicundulum*. *Proc. R. Soc. Lond. B*, **196**, 385–413.
Bähr, R. R. (1974). Contribution to the morphology of chilopod eyes. *Symp. Zool. Soc. Lond.*, **32**, 383–404.
Battelle, B.-A., Calman, B. G., Grieco, F. D., Mleziva, M. B., Callaway, J. C. and Stuart, A. E. (1989). Histamine: a putative afferent neurotransmitter in *Limulus* eyes. *Invest. Ophthalm. Vis. Sci.*, Suppl., **30**, 290.
Bernard, F. (1937). Recherches sur la morphogénèse des yeux composés d'arthropodes. Developpement, croissance, réduction. *Bull. Biol. France Belg.*, Suppl., **23**, 1–162.
Bodenstein, D. (1953). Postembryonic development. In: *Insect Physiology*. ed. Roeder, K. D. pp. 822–865. New York: Wiley.
Boschert, U., Ramos, R. G. P., Tix, S., Technau, G. M. and Fischbach, K.-F. (1990). Genetic and developmental analysis of *irreC*, a genetic function required for optic chiasm formation in *Drosophila*. *J. Neurogenet.*, **6**, 153–171.
Boudreaux, H. B. (1979). *Arthropod Phylogeny with Special Reference to Insects*. New York: Wiley.
Bowman, T. E. (1984). Stalking the wild crustacean: the significance of sessile and stalked eyes in phylogeny. *J. Crust. Biol.*, **4**, 7–11.
Braitenberg, V. (1967). Patterns of projection in the visual system of the fly. I. Retina–lamina projections. *Exp. Brain Res.*, **3**, 271–298.
Burr, A. H. (1984). Evolution of eyes and photoreceptor organelles in the lower phyla. In: *Photoreception and Vision in Invertebrates* (NATO ASI Ser. A, vol. 74). ed. Ali, M. A. pp. 131–178. New York: Plenum.
Callaway, J. C. and Stuart, A. E. (1989). Biochemical and physiological evidence that histamine is the transmitter of barnacle photoreceptors. *Vis. Neurosci.*, **3**, 311–325.
Cajal, S. Ramón y and Sánchez, D. (1915). Contribución al conocimiento de los centros nerviosos de los insectos. *Trab. Lab. Invest. Biol. Univ. Madrid*, **13**, 1–164.
Caveney, S. (1986). The phylogenetic significance of ommatidium structure in the compound eyes of polyphagan beetles. *Can. J. Zool.*, **64**, 1787–1819.
Clarkson, E. N. K. (1975). The evolution of the eye in trilobites. *Foss. Strata*, **4**, 7–32.
Cronin, T. W. (1986). Optical design and evolutionary adaptation in crustacean compound eyes. *J. Crust. Biol.*, **6**, 1–23.
Debaisieux, P. (1944). Les yeux de crustacés; structure développement, réactins à l'éclairement. *Cellule*, **50**, 9–122.
Dietrich, W. (1909) Die Facettenaugen der Dipteren. *Z. Wissen. Zool.*, **92**, 465–539.
Eakin, R. M. (1982). Continuity and diversity in photoreceptors. In: *Visual Cells in Evolution*. ed. Westfall, J. A. pp. 91–105. New York: Raven.
Edwards, J. S. (1989). Predator evasion systems and the origin of flight in insects. In: *Neural Mechanisms of Behavior*. eds. Erber, J., Menzel, R., Pflüger, H.-J. and Todt, D. pp. 120–121. Stuttgart: Thieme.
Edwards, J. S. and Meyer, M. R. (1990). Conservation of antigen 3G6, a crystalline cone constituent in the compound eye of arthropods. *J. Neurobiol.*, **21**, 441–452.
Eguchi, E. (1971). Fine structure and spectral sensitivity of retinular cells in the dorsal sector of compound eyes in the dragonfly *Aeschna*. *Z. vergl. Physiol.*, **71**, 201–218.
Elofsson, R. (1976). Rhabdom adaptation and its phylogenetic significance. *Zool. Scripta*, **5**, 97–101.
Elofsson, R. and Dahl, E. (1970). The optic neuropiles and chiasmata of Crustacea. *Z. Zellforsch.*, **107**, 343–360.

Fahrenbach, W. H. (1969). The morphology of the eyes of *Limulus*. II. Ommatidia of the compound eye. *Z. Zellforsch. Mikrosk. Anat.*, **93**, 451–483.
Fahrenbach, W. H. (1975). The visual system of the horseshoe crab *Limulus polyphemus*. *Int. Rev. Cytol.*, **41**, 285–349.
Field, K. G., Olsen, G. J., Lane, D. J., Giovannoni, S. J., Ghiselin, M. T., Raff, E. C., Pace, N. R. and Raff, R. A. (1988). Molecular phylogeny of the animal kingdom. *Science Wash. DC*, **239**, 748–753.
Fincham, A. A. (1980). Eyes and classification of malacostracan crustaceans. *Nature Lond.*, **287**, 729–731.
Fischbach, K.-F. and Dittrich, A. P. M. (1989). The optic lobe of *Drosophila melanogaster*. I. A Golgi analysis of wild-type structure. *Cell Tiss. Res.*, **258**, 441–475.
Fischer, A. and Horstmann, G. (1971). Der Feinbau des Auges der Mehlmotte, *Ephestia kuehniella* Zeller (Lepidoptera, Pyralididae). *Z. Zellforsch.*, **116**, 275–304.
Flaster, M. S., Macagno, E. R. and Schehr, R. S. (1982). Mechanisms for the formation of synaptic connections in the isogenic nervous system of *Daphnia magna*. In: *Neuronal Development*. ed. Spitzer, N. C. pp. 267–296. New York: Plenum.
Gleadall, I. G., Hariyama, T. and Tsukahara, Y. (1989). The visual pigment chromophores in the retina of insect compound eyes, with special reference to the Coleoptera. *J. Insect Physiol.*, **35**, 787–795.
Hafner, G. S. (1974). The ultrastructure of retinula cell endings in the compound eye of the crayfish. *J. Neurocytol.*, **3**, 295–311.
Halberg, E. and Elofsson, R. (1990). Construction of the pigment shield of the crustacean compound eye: a review. *J. Crust. Biol.*, **9**, 359–372.
Halberg, E. and Elofsson, R. (1990). Construction of the pigment shield of the crustacean compound eye: a review. *J. Crust. Biol.*, **9**, 359–372.
Hallberg, E., Nilsson, H. L. and Elofsson, R. (1980). Classification of amphipod compound eyes – the fine structure of the ommatidial units (Crustacea, Amphipoda). *Zoomorphol.*, **94**, 279–306.
Hanström, B. (1927). Neue Beobachtungen über Augen und Sehzentren von Entomostracen, Schizopoden und Pantopoden. *Zool. Anz.*, **70**, 236–251.
Hardie, R. C. (1983). Projection and connectivity of sex-specific photoreceptors in the compound eye of the male housefly (*Musca domestica*). *Cell Tiss. Res.*, **233**, 1–21.
Hardie, R. C. (1985). Functional organization of the fly retina. In: *Progress in Sensory Physiology*, vol. 5. ed. Ottoson, D. pp. 1–79. Berlin, Heidelberg: Springer.
Hardie, R. C. (1987). Is histamine a neurotransmitter in insect photoreceptors? *J. Comp. Physiol. A*, **161**, 201–213.
Hardie, R. C. (1989). Neurotransmitters in compound eyes. In: *Facets of Vision*. eds. Stavenga, D. G. and Hardie, R. C. pp. 235–256. Berlin, Heidelberg: Springer.
Hartenstein, V. and Posakony, J. W. (1989). Development of adult sensilla on the wing and notum of *Drosophilia melanogaster*. *Development*, **107**, 389–405.
Hennig, W. (1973). *Kükenthal's Handbuch der Zoologie: Diptera*, 2nd edn, vol. IV2. ed. Beier, M. 2.31. Berlin: de Gruyter.
Hofbauer, A. and Campos-Ortega, J. A. (1990). Proliferation pattern and early differentiation of the optic lobes in *Drosophila melanogaster*. *Roub's Arch. Dev. Biol.*, **198**, 264–274.
Horridge, G. A. (1965). Arthropoda: receptors for light, and optic lobe. In *Structure and Function in the Nervous Systems of Invertebrates*. ed. Bullock, T. H. and Horridge, G. A. San Francisco. W. H. Freeman and Co., pp. 1063–1113.
Horridge, G. A. (1977). Organs of Sight. In: *The Arthropoda: Habits, Functional Morphology and Evolution*. ed. Manton, S. M. pp. 274–278. Oxford: Oxford University Press.
Imms, A. D. (1957). *A General Textbook of Entomology*, 9th edn. Revised Richards, O. W. and Davies, R. G. London: Chapman & Hall.

Kirschfeld, K. (1967). Die Projektion der optischen Umwelt auf des Raster der Ommatidien und dem Raster der Rhabdomere im Komplexauge von *Musca. Exp. Brain Res.*, 3, 248–270.

Kirschfeld, K. (1976). The resolution of lens and compound eyes. In *Neural Principles in Vision*, eds. Zettler, F. and Weiler, R. pp. 354–370. Berlin, Heidelberg: Springer-Verlag.

Krasne, F. B. and Lawrence, P. A. (1966). Structure of the photoreceptors in the compound eyespots of *Branchiomma vesiculosum*. *J. Cell Sci.*, 1, 239–248.

• Lake, J. A. (1990). Origin of the Metazoa. *Proc. Natl Acad. Sci. USA*, 87, 763–766.

Land, M. F. (1976). Superposition images are formed by reflection in the eyes of some oceanic decapod crustacea. *Nature Lond.*, 263, 764–765.

Land, M. F. (1981a). Optics and vision in invertebrates. In: *Handbook of Sensory Physiology*, vol. VII/6B: *Comparative Physiology and Evolution of Vision in Invertebrates*. ed. Autrum, H. pp. 471–592. Berlin, Heidelberg: Springer.

Land, M. F. (1981b). Optical mechanisms in the higher Crustacea with a comment on their evolutionary origins. In: *Sense Organs*. eds. Laverack, M. S. and Cosens, D. J. pp. 31–48. Glasgow: Blackie.

Land, M. F. (1989). Evolution of invertebrate visual systems. In: *Neural Mechanisms of Behavior*. eds. Erber, J., Menzel, R., Pflüger, H.-J. and Todt, D. pp. 143–144. Stuttgart: G. Thieme.

Lawrence, P. A. (1966). Development and determination of hairs and bristles in the milkweed bug, *Oncopeltus fasciatus* (Lygaeidae, Hemiptera). *J. Cell Sci.*, 1, 475–498.

Lawrence, P. A. and Green, S. M. (1979). Cell lineage in the developing retina of *Drosophila. Dev. Biol.*, 71, 142–152.

Leutscher-Hazelhoff, J. T. (1984). Ciliary cells evolved for vision hyperpolarize – Why? *Naturwissen.*, 71, 213.

Levi-Setti, R. (1975). *Trilobites*. Chicago: University of Chicago Press.

Macagno, E. R., LoPresti, V. and Levinthal, C. (1973). Structure and development of neuronal connections in isogenic organisms: variations and similarities in the optic system of *Daphnia magna. Proc. Natl Acad. Sci. USA*, 70, 57–61.

Manton, S. M. (1973). Arthropod phylogeny – A modern synthesis. *J. Zool. Lond.*, 171, 111–130.

Manton, S. M. (1977). *The Arthropoda: Habits, Functional Morphology and Evolution*. Oxford: Oxford University Press.

Mayrat, A. (1981). Nouvelle définition des yeux simples et composés chez les Arthropodes. Le cas des Amphipodes et des Cumacés. *Arch. Zool. Exp. Gén.*, 122, 225–236.

Meinertzhagen, I. A. (1973). Development of the compound eye and optic lobe of insects. In: *Developmental Neurobiology of Arthropods*. ed. Young, D. pp. 51–104. Cambridge: Cambridge University Press.

Meinertzhagen, I. A. (1976). The organisation of perpendicular fibre pathways in the insect optic lobe. *Phil. Trans. R. Soc. Lond. B*, 274, 555–594.

Meinertzhagen, I. A. and Armett-Kibel, C. (1982). The lamina monopolar cells in the optic lobe of the dragonfly *Sympetrum. Phil. Trans. R. Soc. Lond. B*, 297, 27–49.

Meinertzhagen, I. A. and O'Neil, S. D. (1991). Synaptic organization of columnar elements in the lamina of the wild type in *Drosophila melanogaster. J. Comp. Neurol.*, 305, 232–263.

Meinertzhagen, I. A. and Shaw, S. R. (1989). Evolution of synaptic connections between homologous neurons in insects: new cells for old in the optic lobe. In: *Neural Mechanisms of Behaviour*. eds. Erber, J., Menzel, R., Pflüger, H.-J. and Todt, D. pp. 124–126. Stuttgart: Thieme.

Melamed, J. and Trujillo-Cenóz, O. (1968). The fine structure of the central cells in the ommatidium of dipterans. *J. Ultrastruct. Res.*, 21, 313–334.

Menzel, R. and Blakers, M. (1976). Colour receptors in the bee eye – morphology and spectral sensitivity. *J. Comp. Physiol.*, 108, 11–33.

Meyerowitz, E. M. and Kankel, D. R. (1978). A genetic analysis of visual system development in *Drosophila melanogaster. Dev. Biol.*, 62, 112–142.

Meyer-Rochow, V. B. and Waldvogel, H. (1979). Visual behaviour and the structure of dark- and light-adapted larval and adult eyes of the New Zealand glownorm *Arachnocampa luminosa* (Mycetophilidae: Diptera). *J. Insect Physiol.*, 25, 601–613.

Mouze, M. (1984). Morphologie et développement des yeux simples et composés des insectes. In: *Photoreception and Vision in Invertebrates* (NATO ASI Ser. A, vol. 74). ed. Ali, M. A. pp. 661–698. New York: Plenum.

Nässel, D. R. and Geiger, G. (1983). Neuronal organization in fly optic lobes altered by laser ablations early in development or by mutations of the eye. *J. Comp. Neurol.*, 217, 86–102.

Nässel, D. R. and Waterman, T. H. (1977). Golgi EM evidence for visual information channelling in the crayfish lamina ganglionaris. *Brain Res.*, 130, 556–563.

Nässel, D. R., Elofsson, R. and Odselius, R. (1978). Neuronal connectivity patterns in the compound eyes of *Artemia salina* and *Daphnia magna* (Crustacea: Branchiopoda). *Cell Tiss. Res.*, 190, 435–457.

Nässel, D. R., Holmqvist, M. H., Hardie, R. C., Håkanson, R. and Sundler, F. (1988). Histamine-like immunoreactivity in photoreceptors of the compound eyes and ocelli of the flies *Calliphora eythrocephala* and *Musca domestica. Cell Tiss. Res.*, 253, 639–646.

Nicol, D., and Meinertzhagen, I. A. (1982). An analysis of the number and composition of the synaptic populations formed by photoreceptors of the fly. *J. Comp. Neurol.*, 207, 29–44.

Nilsson, D.-E. (1983). Evolutionary links between apposition and superposition optics in crustacean eys. *Nature Lond.*, 302, 818–821.

Nilson, D.-E. (1988). A new type of imaging optics in compound eyes. *Nature Lond.*, 332, 76–78.

Nilsson, D.-E. (1989). Optics and evolution of the compound eye. In: *Facets of Vision*. eds. Stavenga, D. G. and Hardie, R. C. pp. 30–73. Berlin, Heidelberg: Springer.

Nilsson, D.-E., Hallberg, E. and Elofsson, R. (1986). The ontogenetic development of refracting superposition eyes in crustaceans: transformation of optical design. *Tissue Cell* 18, 509–519.

Nilsson, D.-E., Land, M. F. and Howard, J. (1984). Afocal apposition optics in butterfly eyes. *Nature Lond.*, 312, 561–563.

Ninomiya, N., Tominaga, Y. and Kuwabara, M. (1969). The fine structure of the compound eye of a damsel-fly. *Z. Zellforsch. mikrosk, Anat.*, 98, 17–32.

Nowikoff, M. (1926). Über die Komplexaugen der Gattung *Arca. Zool. Anz.*, 67, 277–289.

Ohno, S. (1970). *Evolution by Gene Duplication*. New York, Heidelberg, Berlin: Springer.

Osorio, D. (1991). Compound eyes, the optic lobe and arthropod evolution. In: *Evolution of the Eye and Visual System*. ed. Cronly-Dillon, J. R. *Vision and Visual Dysfunction*. Vol. 2. London: Macmillan.

Paulus, H. F. (1979). Eye structure and the monophyly of the arthropoda. In: *Arthropod Phylogeny*. ed. Gupta, A. P. pp. 299–383. New York: Van Nostrand Reinhold.

Peters, W. (1965). Die Sinnesorgane an den Labellen von *Calliphora erythrocephala* Mg. (Diptera). *Z. Morph. Ökol. Tiere*, 55, 259–320.

Pflugfelder, O. (1958). *Entwicklungsphysiologie der Insekten*, 2nd edn. Leipzig: Akademische.

Pirvola, U., Tuomisto, L., Yamatodani, A. and Panula, P. (1988). Distribution of histamine in the cockroach brain and visual system: an immunocytochemical and biochemical study. *J. Comp. Neurol.*, 276, 514–526.

Ready, D. F., Hanson, T. E. and Beanzer, S. (1976). Development of the *Drosophila* retina, a neurocrystalline lattice. *Dev. Biol.*, 53, 217–240.

Ready, D. F. (1989). A multifacetted approach to neural development. *Trends Neurosci.*, 12, 102–110.

Remane, A. (1959). Die Geschichte der Tiere. In: *Die Evolution der Organismen.* ed. Heberer, G. pp. 340–422. Stuttgart: Fischer.

Ribi, W. A. (1977). Fine structure of the first optic ganglion (lamina) of the cockroach, *Periplaneta americana. Tissue Cell,* 9, 57–72.

Ribi, W. A. (1981). The first optic ganglion of the bee. IV. Synaptic fine structure and connectivity patterns of receptor cell axons and first order interneurones. *Cell Tiss. Res.,* 215, 443–464.

Salvini-Plawen, L. von (1982). On the polyphyletic origin of photoreceptors. In: *Visual Cells in Evolution.* ed. Westfall, J. A. pp. 137–154. New York: Raven.

Salvini-Plawen, L. von and Mayr, E. (1977). On the evolution of photoreceptors and eyes. In: *Evolutionary Biology,* vol. 10. pp. 207–263. New York: Plenum.

Sarthy, P. V. (1989). Histamine: a neurotransmitter candidate for photoreceptors in *Drosophila melanogaster. Invest. Ophthalm. Vis. Sci. (Suppl.),* 30, 290.

Sattelle, D. B. (1985). Acetylcholine receptors. In: *Comprehensive Insect Physiology Biochemistry and Pharmacology,* vol. 11: *Pharmacology.* pp. 395–434. Oxford: Pergamon.

Schlemermeyer, E., Schütte, M. and Ammermüller, J. (1989). Immunohistochemical and electrophysiological evidence that histamine is a photoreceptor transmitter in locust ocellar retina. *Invest. Ophthalm. Vis. Sci., Suppl.,* 30, 290.

Schmitt, M., Mischke, U. and Wachmann, E. (1982). Phylogenetic and functional implications of the rhabdom patterns in the eyes of Chrysomeloidea (Coleoptera). *Zool. Scripta,* 11, 31–44.

Seifert, P. and Wunderer, H. (1989). Hell- und Dunkel-Adaptation bei Dipteren – eine funktionelle Vorbedigung zur Rhabdomentwicklung? *Mitt. Deuts. Ges. Allg. Angew. Ent.,* in press.

Sharov, A. G. (1966). *Basic Arthropodan Stock.* Oxford: Pergamon.

Shaw, S. R. (1989a). The retina–lamina pathway in insects, particularly Diptera, viewed from an evolutionary perspective. In: *Faces of Vision.* eds. Stavenga, D. G. and Hardie, R. C. pp. 186–212. Berlin, Heidelberg: Springer.

Shaw, S. R. (1989b). Visual optics predict neural connections in a primitive dipteran eye. *Soc. Neurosci. Abstr.,* 15, 1291.

Shaw, S. R. (1990a). The photoreceptor axon projection and its evolution in the neural superposition eyes of some primitive brachyceran Diptera. *Brain Behav. Evol.,* 35, 107–125.

Shaw, S. R. (1990b). Cellular evolution in a compound eye: coordinated optical and synaptic changes have functionally remodelled the post-Triassic dipteran visual system. *Proc. Int. Soc. Eye Res.,* vol. VI, p. 174. (Abstr. 9th International Congress of Eye Research.)

Shaw, S. R. and Meinertzhagen, I. A. (1986). Evolutionary progression at synaptic connections made by identified homologous neurones. *Proc. Natl Acad. Sci. USA,* 83, 7961–7965.

Shaw, S. R. and Moore, D. (1989). Evolutionary remodelling in a visual system through extensive changes in the synaptic connectivity of homologous neurons. *Vis. Neurosci.,* 3, 405–410.

Shaw, S. R. and Stowe, S. (1982). Photoreception. In: *The Biology of Crustacea.* vol. 3. eds. Atwood, H. L. and Sandeman, D. C. pp. 291–367. London: Academic.

Siewing, R. (1960). Zum Problem der Polyphylie der Arthropoda. *Z. wissen. Zool.,* 164, 238–270.

Simmons, P. J. and Hardie, R. C. (1988). Evidence that histamine is a neurotransmitter of photoreceptors in the locust ocellus. *J. Exp. Biol.,* 138, 205–219.

Sims, S. J. and Macagno, E. R. (1985). Computer reconstruction of all the neurons in the optic ganglion of *Daphnia magna. J. Comp. Neurol.,* 233, 12–29.

Smith, W. C. and Goldsmith, T. H. (1990). Phyletic aspects of the distribution of 3-hydroxyretinal in the class Insecta. *J. Mol. Evol.,* 30, 72–84.

Snodgrass, R. E. (1935). *Principles of Insect Morphology.* New York: McGraw-Hill.

Strausfeld, N. J. (1970). Golgi studies on insects. Part II. The optic lobes of Diptera. *Phil. Trans. Roy. Soc. Lond. B,* 258, 135–223.

Strausfeld, N. J. (1971). The organization of the insect visual system (light microscopy). I. Projections and arrangements of neurons in the lamina ganglionaris of Diptera. *Z. Zellforsch.,* 121, 377–441.

Strausfeld, N. J. (1976). *Atlas of an Insect Brain.* Berlin, Heidelberg: Springer.

Strausfeld, N. J. and Nässel, D. R. (1981). Neuroarchitectures serving compound eyes of Crustacea and insects. In: *Handbook of Sensory Physiology,* vol. VII/6B: *Comparative Physiology and Evolution of Vision in Invertebrates.* ed. Autrum, H. pp. 1–132. Berlin, Heidelberg: Springer.

Tiegs, O. W. and Manton, S. M. (1958). The evolution of the arthropoda. *Biol. Rev. Camb.,* 33, 255–337.

Tomlinson, A. (1988). Cellular interactions in the developing *Drosophila* eye. *Development,* 104, 183–193.

Tomlinson, A. and Ready, D. F. (1987). Neuronal differentiation in the *Drosophila* ommatidium. *Dev. Biol.,* 120, 366–376.

Tomlinson, A. and Ready, D. F. (1988). Cell fate in the *Drosophila* ommatidium. *Dev. Biol.,* 123, 264–275.

Trujillo-Cenóz, O. (1985). The eye: development, structure and neural connections. In: *Comprehensive Insect Physiology, Biochemistry and Pharmacology,* vol. 6: *Nervous System: Sensory.* eds. Kerkut, G. A. and Gilbert, L. I. pp. 171–223. Oxford: Pergamon Press.

Trujillo-Cenóz, O. and Melamed, J. (1973). The development of the retina-lamina complex in muscoid flies. *J. Ultrastruct. Res.,* 42, 554–581.

Vanfleteren, J. R. (1982). A monophyletic line of evolution? Ciliary induced photoreceptor membranes. In: *Visual Cells in Evolution.* ed. Westfall, J. A. pp. 107–136. New York: Raven.

Vogt, K. (1975). Zur Optik des Flusskrebsauges. *Z. Naturforsch.,* 30c, 691.

Vogt, K. (1984). The chromophore of the visual pigment in some insect orders. *Z. Naturforsch.,* 39c, 196–197.

Vogt, K. (1987). Chromophores of insect visual pigments. *Photobiochem. Photobiophys., Suppl.,* 273–296.

Vogt, K. (1989). Distribution of insect visual chromophores: functional and phylogenetic aspects. In: *Facets of Vision.* eds. Stavenga, D. G. and Hardie, R. C. pp. 134–151. Berlin, Heidelberg: Springer.

Vogt, K. and Kirschfeld, K. (1984). Chemical identity of the chromophores of fly visual pigment. *Naturwissen,* 71, 211–213.

Wada, S. (1974). Spezielle randzonale Ommatidien der Fliegen (Diptera: Brachycera): Architektur und Verteilung in den Komplexaugen. *Z. Morph. Tiere,* 77, 87–125.

Wada, S. (1975). Morphological duality of the retinal pattern in flies. *Experientia,* 31, 921–923.

Wehner, R. (1976). Structure and function of the peripheral visual pathway in hymenopterans. In: *Neural Principles in Vision.* eds. Zettler, F. and Weiler, R. pp. 280–333. Berlin, Heidelberg: Springer.

Williams, D. S. (1980). Organization of the compound eye of a tipulid fly during the day and night. *Zoomorphol.,* 95, 85–104.

Willmer, P. (1990). *Invertebrate Relationships.* Cambridge: Cambridge University Press.

Wolsky, A. and Wolsky, M. de I. (1985). Problems of the evolution and morphogenesis of the arthropodan eye. In: *Evolution and Morphogenesis.* eds. Mlíkovsky, J. and Novák, V. J. A. pp. 487–492. Praha: Academia.

Wright, S. (1978). *Evolution and the Genetics of Populations,* Vol. 4, *Variability Within and Among Natural Populations.* Chicago: Univ. Chicago Press.

Zeil, J. (1983). Sexual dimorphism in the visual system of flies: the compound eyes and neural superposition in Bibionidae (Diptera). *J. Comp. Physiol.,* 150, 379–393.

Zipursky, S. L. (1989). Molecular and genetic analysis of *Drosophila* eye development: *sevenless, bride of sevenless* and *rough. Trends Neurosci.,* 12, 183–189.

17 Photoreception and Vision in Molluscs

J. B. Messenger

Introduction

The molluscs present us with a staggering variety of eye types, the most varied of all the animal phyla. There are simple, cup-like eyes with or without a lens, compound eyes, fish-like eyes with lenses, a pinhole eye and even a scanning eye. Even more remarkable is the widespread use of extra-ocular photoreceptors. In every class of mollusc there are animals with some kind of photosensitivity that is not mediated by an eye: it may be a dermal light sense, or a light-sensitive neurone deep within the CNS, or a complex organ alongside a large, conventional eye.

Since variety is such a key feature of molluscan visual systems, and since this chapter of a book on the evolution of the eye may be read by biochemists, ophthalmologists, molecular biologists or others without a zoological training, it is necessary to begin with a reminder. Evolution should not be thought of as some simple progression, from protistans through molluscs and arthropods to humans. Just as fish should not be thought of as 'inferior' vertebrates that would have been mammals if they could have managed it, so invertebrate groups like molluscs should not be thought of as a step on the road to the vertebrates. Molluscs and vertebrates are completely unrelated groups that have been distinct for 400 Myr. There are therefore profound differences even between the complex eyes of cephalopods and those of the fish they so closely resemble.

The eyes of molluscs have evolved to provide their owners with the kind of information about the world that is relevant to their own particular life-style. Because molluscs have adopted a wide variety of life-styles they have many different kinds of eyes. So even *within* the group we should expect no simplistic evolutionary series of eyes, from chiton to octopus.

Classification

Authorities differ as to whether there are five or seven classes within the Mollusca (Fig. 17.1), but all agree that whether one judges on the numbers of species, on the biomass or on the amount of 'solar energy through-put' (Russell-Hunter, 1979) there are three major classes: the Gastropoda (marine, freshwater and land snails), the Bivalvia (clams) and the Cephalopoda (squids and octopuses). Unfortunately these groups have been distinct since the Devonian and their evolutionary relationships are unknown. Zoologists attempt to trace their descent from a hypothetical protomolluscan ancestor, with a foot, mantle, head, radula and ctenidium (or gill) and they tend to agree that such minor molluscan classes as the Monoplacophora, Polyplacophora (chitons) and Aplacophora are closer to that ancestor than are the modern snails, clams or squids. The relationships of the elephant-tusk shells (Scaphopoda) are obscure.

For a review of mollusc vision group by group the reader is referred to Messenger (1981), an extended treatment that lists many review papers. Since that date two or three brief reviews on aspects of molluscan vision have appeared, the most stimulating of which is that by Land (1984). This chapter considers the photoreceptors type by type, proceeding from those that process simple information to those handling the most complex.

Extra-ocular Photoreception

By this we mean a response to light that is not mediated by an organ that could conceivably be termed an eye. Photoreceptors of this category are widespread in molluscs and exhibit a surprising range of complexity. It is convenient to recognize four types.

Dermal Sensitivity

If a shadow falls on the stalked eyes of the garden snail, *Helix*, nothing happens. Yet if it crosses the mantle near the base of the shell the animal withdraws into the shell (Föh, 1932). This is an example of the 'shadow responses' that are so widespread in gastropods and bivalves. This

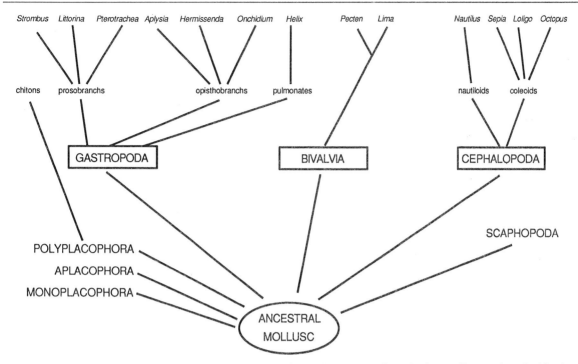

Fig. 17.1 *Classification of molluscs according to commonest usage, listing the genera cited most in the text. For a modern classification that includes fossil forms see Trueman and Clarke (1985).*

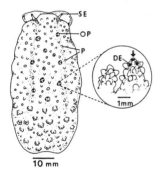

Fig. 17.2 Onchidium *showing the cephalic, stalked eyes (SE), the numerous dorsal eyes (DE) and the papillae (P) with ocelli (OP). Dermal photoreceptors not shown (From Katagiri, 1984).*

particular example is mediated by unidentified receptors in the skin. Another pulmonate, *Lymnaea stagnalis*, also exhibits a shadow response but this may be mediated by receptors (possibly ciliary: see below) in the tentacles, foot, mantle or lips. Certainly it persists after removal of the eyes (Stoll, 1972).

In *Nassarius*, a prosobranch gastropod, Crisp (1972) found two shadow responses that survived removal of the eyes. One of these, contraction of the siphon, almost certainly involves sensory endings in the external siphonal epithelium that have both microvilli and cilia.

Among opisthobranch gastropods, *Aplysia* is known to have photoreceptors in the mantle skin (Lukowiak and Jacklet, 1972), as well as in the oral tentacles (Block, 1975, cited in Lickey *et al.*, 1976). In *Onchidium* Katagiri (1984) has described 'numerous' dermal photoreceptor cells in the tubercles of each ocellus papilla (Fig. 17.2) apparently with microvilli. They appear to mediate light-evoked contraction of the dorsal papillae and to be independent of the CNS (Shimatani *et al.*, 1986). In *Elysia timida*, Rahat and Monselise (1979) showed that the parapodia are spread or retracted to expose or conceal the symbiotic chloroplasts to sunlight and that this involves no less than three sets of photoreceptors, two of which are extra-ocular.

In some chitons (Polyplacophora) the girdle and ventral surface are light-sensitive despite there being many hundreds of ocelli (see pp 368–369), and masking the ocelli does not prevent the animal from orienting away from a light source (Boyle, 1972), although it does slow down the rate of movement.

Scaphopods are a group of molluscs about which little is known. They live half-buried in the substrate and lack eyes. Yet J. Levitt (personal communication) reports that *Cadulus aberrans* appears negatively phototactic, in response to sunlight and artificial light. Moreover, they have distinct burrowing cycles that are apparently related to the day–night cycle and also exhibit a shadow response, burrowing into the substrate. Further experiments are clearly needed, but it is interesting that two recent ultrastructural studies describe ciliated receptors of unknown function in the mantle edge in several genera of scaphopods (Reynolds, 1988; Steiner, 1990).

Among bivalves we may note dermal receptors in the file shell, *Lima* (Mpitsos, 1973), which give 'off' responses and diffuse, unspecialized cells in the siphon of the clam, *Mya*, which subserve shell closure or siphon withdrawal (Messenger, 1981).

Light-sensitive Neurones

These are a special feature of opisthobranchs and bivalves, although they are also known in *Helix*, where Pasic and her co-workers (1977) report three photosensitive neurones that depolarize with light and one that hyperpolarizes.

In *Aplysia*, photosensitive neurones occur in the cerebral ganglion (Block and Smith, 1973) and abdominal ganglion, where they have attracted considerable attention (Arvanitaki and Chalazonitis, 1961; Kandel, 1976; Krauhs et al., 1977; and above all A. M. Brown et al., 1973, 1977). Some of these neurones are known to respond to light by slow hyperpolarization, and the opening of K^+ channels; Ca^{2+} has also been implicated in the photoresponse (Andresen and Brown, 1979; Andresen et al., 1979). The function of these photosensitive neurones is not known, but Brown et al. (1977) suggest that they are accessible to the ambient light in their natural habitat. One suggestion is that they may be involved in maintaining the locomotor rhythm (Lickey et al., 1977).

Onchidium is another opisthobranch with photosensitive neurones in its CNS (Hisano et al., 1972a,b; Gotow, 1975). In the photosensitive neurone A-P-1 of *Onchidium* Gotow (1989) and Nishi and Gotow (1989) have recently found that light induced a depolarization as a result of closure of the K^+ channel and also that cyclic GMP is involved in the photoresponse. Fujimoto and his colleagues (1966) have shown the shadow response still persists when the cephalic and dorsal eyes have been removed. It is not known whether this is achieved by means of the photosensitive neurones.

In bivalves, well-documented examples of photosensitive neurones occur in the clams, *Spisula*, where it is thought that part of the axon is itself light-sensitive (Kennedy, 1960), and *Mercenaria*, where Wiederhold et al. (1973) report off-responses for single axons that can increase in size for up to 8 *minutes* to a constant-intensity stimulus.

Photosensitive Vesicles

The photosensitive vesicles (PSV) are rather more elaborate photoreceptor organs that are found only in the cephalopods. When first described they were taken to be some kind of neurosecretory gland (Young, 1936) and it was not until the advent of electron microscopy that ultra-

Fig. 17.3 *An octopod and a decapod showing the position of the photosensitive vesicles (PSV) in the mantle and in the head respectively.*

structural evidence was forthcoming to show that they must be photoreceptors (Nishioka et al., 1962). Subsequently biochemical and physiological evidence has emerged to confirm this.

In octopods the PSV occur in the mantle. They are just visible to the naked eye as an orange spot at the posterior margin of the stellate ganglion, and Young (1936) gave them the name 'epistellar body'. In decapods (Fig. 17.3) they are found in the head, where they lie close to the olfactory lobe on each side of the central brain, whence the name 'parolfactory vesicles' proposed by Boycott and Young (1956a) on the basis of light microscopy. A curious feature of these two sets of PSV is that they lie just medial to the large optic lobes, which, of course, process visual information gathered by the enormous eyes, each with at least 20 million receptor cells in the retina.

In 1977 Mauro, adopting the suggestion of J. Z. Young and R. E. Young, introduced the term 'photosensitive vesicles' for both sets of organs, for both have the following features in common:

1. No dioptric apparatus.
2. Rhabdomeric receptor cells with microvilli. These are not precisely organized into a two-plane array as in the eye (Fig. 17.21) but neverthelesss look strikingly like the retina (Nishioka et al., 1962).
3. The visual pigment rhodopsin, with absorption characteristics very close to that of the rhodopsin in the eye. For the squid *Loligo* the values are PSV λ_{max} 495 nm, eye λ_{max} 493 nm. For the octopus *Eledone* they are PSV λ_{max} 475 nm, eye λ_{max} 470 nm (Nishioka et al., 1966). These figures are in good agreement with electrophysiological estimates of spectral sensitivity (Mauro, 1977).
4. The accessory photopigment retinochrome (Hara and Hara, 1980; Ozaki et al., 1983).

5. A depolarizing response to light (Mauro and Baumann, 1968; Mauro and Sten-Knudsen, 1972).

We know almost nothing about the function of the PSV. R. E. Young has evidence that in some genera of squid the PSV may monitor the quality of the downwelling light to regulate light production by the photophores for counter-illumination (see for example, R. E. Young, 1977; R. E. Young and Roper, 1976; Young, Roper and Walters, 1979). But beyond that we have only speculation. Comparative studies (R. E. Young, 1972; J. Z. Young, 1977) show that the PSV tends to be more developed in squids that live in deeper waters. For example, in the cranchiid *Liocranchia* (Messenger, 1967) the vesicles are numerous and organized into a chain (Fig. 17.4), whereas in *Loligo* there are just a few scattered PSV vesicles (Messenger, 1979a). In the blind *Cirrothauma* they are especially large (Aldred *et al.*, 1978) yet migratory squids, such as *Illex* and *Todarodes*, also have well-developed PSV. Do these organs monitor light levels to regulate diel (vertical) migrations?

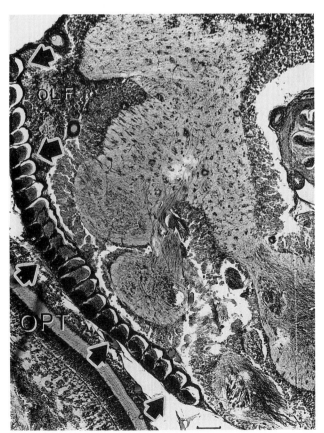

Fig. 17.4 *The mid-water squid,* Liocranchia, *showing the well-developed chain of PSV (arrows) lying between the large optic lobe (opt.) and the olfactory lobe (olf.)*

Or do these use light cues to drive longer-term physiological processes, such as gonad maturation?

In decapods the PSV nerves project to the 'cerebellum', the peduncle lobe that regulates posture and locomotion (Baumann *et al.*, 1970; Messenger, 1979a). However, nothing is known about the connectivity of the octopod PSV and the mystery remains: why should an animal with some 40 million photoreceptor cells in two large eyes need to have two small photoreceptors buried in the mantle?

As this account was nearing completion the very interesting paper of Sundermann (1990) appeared. Her electron micrographs reveal that there are structures with all the appearance of microvillar photoreceptors in the paired posterior vesicle organs, on the head of hatchling *Loligo* and *Sepia*. The function of these ectodermal organs is not known, nor is it clear whether they are lost in adults or become internalized as the PSV.

Summary

Some kind of extra-ocular photosensitivity is found in five of the molluscan classes, including the three major ones, and may be regarded as a key feature of the phylum. A general dermal sensitivity may mediate shadow responses, such as siphon or foot withdrawal. And some photosensitive neurones influence the locomotor rhythms, as in *Aplysia*, and perhaps help entrain circadian rhythms (see pp 372–373). In cephalopods, despite the well-developed eyes, there are PSV forming quite distinct structures, either in the head or in the mantle. In squids the PSV are closely associated with the brain and in some genera may be used to regulate light emission from photophores. In octopods the PSV lie on the stellate ganglion: their function remains enigmatic.

It is not always easy to know which is the structure responsible for a response: the limpet *Notoacmea*, for example, shows a negative phototropism that must be directed by some internal photoreceptor for if the translucent patches on the shell are covered the response is slowed significantly (Lindberg *et al.*, 1975). The whelk, *Bullia digitalis*, a burrowing form that lacks eyes, shows an interesting 'off' response. If kept in bright light, without a substrate to bury in, the animals respond to the cessation of light by a burst of increased locomotor activity lasting 5 minutes or more (Brown and Webb, 1985). The function of such a response is unknown. The pond snail, *Lymnaea*, exhibits a clear positive photo-orientation that survives bilateral eye removal (van Duivenboden, 1982). Again the receptor for this response is not known, though the author hazards that it might be a light-sensitive neurone.

Finally, to emphasize that biologists studying 'visual' behaviour in molluscs must always be cautious, consider the recent discovery (Jacklet, 1980) that in *Aplysia* the rhinophore usually thought of as a chemoreceptor also contains photoreceptors, despite there being a dermal light sense and paired cephalic eyes. The nature of the photoreceptive cells is not known, but Jacklet was able to show that the spectral sensitivity of the rhinophores was close to that of the eyes (λ_{max} 500 nm). The nature or function of these extra-ocular photoreceptors is not known: they may be involved in the circadian system (see pp 372–373), be modulators of ocular sensitivity or something more, because the rhinophores make orienting movements when the head is illuminated.

Non-cephalic Eyes

In the molluscs it is necessary to distinguish between true, 'cephalic' eyes borne on the head and 'non-cephalic' eyes that may occur all over the body. These non-cephalic eyes all contain large numbers of receptor cells and some kind of dioptric apparatus: to that extent they must be considered an advance over extra-ocular photoreceptors, although, as we shall see, they generally provide their owners with a rather limited amount of information. Following Land (1984), we can recognize five quite different kinds of non-cephalic eyes distributed among the chitons, gastropods and bivalves, a group that has been particularly inventive in producing novel eye designs.

Aesthetes and Ocelli

The chitons, or coat-of-mail shells, are slow-moving creatures, without cephalic eyes yet remarkable for having large numbers of tiny eyes embedded in their shell plates. In *Onithochiton*, Boyle (1969a) estimates that there are 1472 ocelli (Fig. 18.5) and in *Acanthopleura* there are apparently no fewer than 11 500 (Moseley, 1885). The lens is flattened, about 25–40 μm in diameter and the tulip-shaped retina contains about 100 receptors bearing microvilli, which is the 'typical' molluscan photoreceptor-type (cf. Eakin, 1972; see below). The microvilli lie parallel to the lens surface, rows of them forming a rhabdom. Boyle (1969b) also describes cells in the ocelli that are ciliary, but whether or not these are photoreceptors remains unclear. There is no physiological evidence about photoreception in chitons, though one might expect these microvillar receptors to depolarize at the onset of illumination.

Chitons are known to give shadow responses that are presumably mediated by the ocelli. However, they also give orienting, or phototactic, responses and since such responses are known to be mediated by ciliary receptors in *Pecten* and elsewhere (see p 371), these responses may be mediated by the putative ciliary photoreceptor cells. However, it should be noted that associated with the ocelli there are large numbers of even smaller structures,

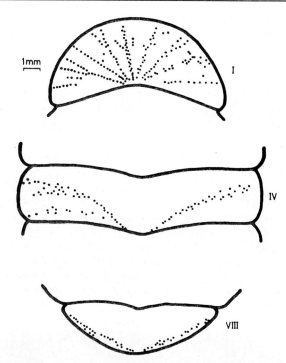

Fig. 17.5 *Numerous ocelli on the shell valves of a chiton*, Onithochiton. *About a quarter occur on the head valve (From Boyle, 1969a).*

termed aesthetes, that have long been held to be photoreceptors (Blumrich, 1891) though recent ultrastructural evidence puts this in doubt. However, until further ultrastructural and behavioural evidence appears, we can only conclude that chitons, whether via their ocelli, their aesthetes or some extra-ocular photoreceptor, can show off-responses, on-responses (Bauer, 1976) and orientation responses (for review see Boyle, 1977; Messenger, 1981).

Dorsal Eyes

These are restricted to one order of opisthobranch gastropods, the Onchidiacea. From the viewpoint of photoreception, *Onchidium verruculatum* is among the most fascinating of all molluscs in that it possesses photosensitive neurones, plus dermal photoreceptive cells, plus a pair of cephalic eyes, plus a set of between 18 and 46 *dorsal eyes* (Fig. 17.2).

The cephalic eyes are 'typically' molluscan in that the receptors are microvillar and show depolarizing 'on' responses to illumination (Fujimoto *et al.*, 1966); that is, they do not mediate shadow responses. However, the numerous dorsal eyes, which are borne on raised papillae that themselves contain photoreceptive cells, lack microvilli, and instead have cells with membranous whorls deriving from cilia (Yanase and Sakamoto, 1965). The eyes are 100–200 μm in diameter and are situated on papillae all over the animal's back. Fujimoto et al. (1966) found that these dorsal eyes mediate hyperpolarizing 'off' responses, which is consistent with the hypothesis that they mediate the shadow response, like the ciliary receptors in the distal retina of *Pecten* (p 271). The reason for the papillae having a light-mediated contraction response (p 365) is not clear.

Eye Spots

This rather heterogeneous collection of non-cephalic eyes includes examples from among the gastropods and bivalves.

Among opisthobranchs the pelagic pteropods, such as *Corolla*, have numerous eye spots around the edges of the 'wings' (Morton, 1967). Among the bivalves we can note the occurrence of eye spots in the file shells (*Lima*); in pholads, venerids, and *Mya* (Charles, 1966; Morton, 1967); and in *Tridacna*, the giant clam, which has been the subject of two very interesting papers recently (Wilkens, 1984, 1986). This animal has numerous eyes (up to several thousand in a large specimen) borne along the edge of the siphon lobes, which are spread out over the shell in bright sunlight. Intracellular recording has shown that each eye contains spiking and non-spiking photoreceptor cells, both of which are hyperpolarizing. The spiking cells are like the *Pecten* distal photoreceptors but the function of the non-spiking cells, not known in other mollusc, is unknown. Furthermore, Wilkens (1984) reported at least three classes of receptor, each with a different spectral sensitivity (Fig. 17.6) including one that is sensitive to ultraviolet (UV) light (λ_{max} 360). Sensitivity to UV has not been reported elsewhere in molluscs and it may be an adaptation related to the fact that *Tridacna* possess algal symbionts that utilize a broad spectrum of light (300–700 nm) for photosynthesis.

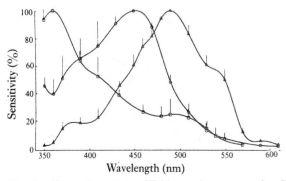

Fig. 17.6 *Spectral sensitivity of* Tridacna *photoreceptors, based on size of receptor potential at different wavelengths. Three peaks are evident, at 360, 450 and 490 nm. (From Wilkens, 1984.)*

The eye spot of the cockle, *Cardium*, has also been studied in some detail (Barber and Land, 1967; Barber and Wright, 1969). There are about 60 of them, each on a little tentacle: they are very small (50 μm in diameter), lack a proper lens, contain 12–20 ciliary photoreceptor cells, and mediate off-responses. The most interesting feature is that the eye is backed by a reflector, though it seems most unlikely that this can produce a usable image (Land, 1984).

Mirror Eyes

However, there is no doubt that in the scallop, *Pecten*, the image is formed by a reflector and this most bizarre of all mollusc eyes has attracted considerable attention from physiologists and zoologists for over 100 years (Patten, 1886; Hesse, 1900; Dakin, 1910) not least because of its curious double retina (Fig. 17.7).

The optics of the scallop eye have been beautifully elucidated by Land (1965, 1966a,b). He observed an inverted image of himself when he looked at a scallop eye through a dissecting microscope and realized that this image must be formed by reflection, from the hemispherical argentea (or tapetum) that lies behind the eye (Fig. 17.7). The reflected image lies about 140 μm in front of the reflector, so that it must lie on the distal retina and experiments showed that this retina responds to movements of stripes that do not cause any dimming. Land's physiological data were in agreement with the behavioural threshold to movement found by Buddenbrock and Moller-Racke (1953), which showed that scallops can respond to 1–2° movements of a dark stripe. There is a lens in the scallop eye but its refractive index is sufficiently low for the image formed by it to lie about 1 mm behind the back of the eye (Fig. 17.8). Moreover, its curious shape suggests that it is designed not to focus light but to correct for the optical defects of a hemispherical mirror. This mirror is composed of alternating layers of cytoplasm and guanine crystals so organized that they form an 'ideal' quarter-wave-length multi-layer reflector (Fig. 17.8). If the distal retina utilizes the image produced by the mirror, what is the function of the proximal retina, which must receive an out-of-focus image from the mirror? Land confirmed Hartline's (1938) finding that it responds to an increase in the level of illumination and suggested that the scallop utilizes its proximal retina to guide it either towards, or away from, regions of brighter light.

Why scallops should have as many as 100 eyes arranged around the edge of the mantle, each eye containing half a million proximal receptor cells and half a million distal receptors is not clear, especially since the field of vision of each eye is about 90–100° Land (1984a).

One other feature of the scallop eye has attracted considerable attention: the double retina. Although this has

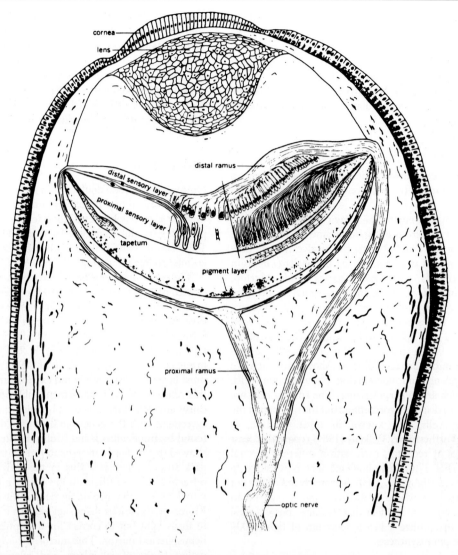

Fig. 17.7 *The eye of the scallop*, Pecten, *showing the double retina. (Modified from Dakin, 1910.)*

been reviewed elsewhere (Messenger, 1981; Land, 1984) it merits re-examination here because it emphasizes the dangers of drawing evolutionary conclusions too hastily from incomplete data. In the late 1960s it was becoming apparent that photoreceptors throughout the animal kingdom were of two basic types: in the annelids, arthropods and molluscs (cephalic type) they were rhabdomeric (microvillar), but in the cnidarians, echinoderms and chordates they were derived from cilia (Eakin, 1972). When the *Pecten* double retina, known in the light microscope for over 60 years, was first examined in the electron microscope (Barber *et al.*, 1967) the extraordinary fact emerged that the proximal retina contained microvillar receptor cells, while the distal cells contained ciliary photoreceptors (Fig. 17.9). The electrophysiological experiments of McReynolds and Gorman (1970a,b) showed that receptors in the proximal retina depolarized to light ('on' response), whereas the distal photoreceptors hyperpolarized ('off' response), like vertebrate photoreceptors (Toyoda *et al.*, 1969; Baylor and Fuortes, 1970).

It might be thought, then, that the scallop eye contains a mixture of 'vertebrate' and 'invertebrate' receptor types, if we follow Eakin's classification (1972). Further evidence shows this is simply not true. Gorman and McReynolds (1978) showed that the hyperpolarizing response in the *Pecten* distal retina derives from an *increase in the membrane permeability to* K^+, as it does also in the file shell, *Lima*. In the vertebrates (Baylor and Fuortes, 1970)

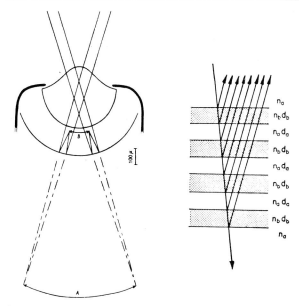

Fig. 17.8 Left: image formation in Pecten, by lens (A) and by lens/mirror (B). Right: the multilayer reflecting stack of guanine crystals and cytoplasm that forms the mirror. (From Land, 1966b.)

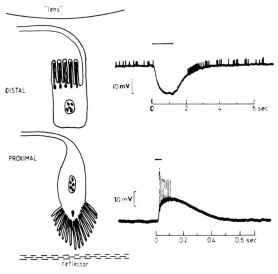

Fig. 17.9 The two cell types in Pecten retina: ciliary, hyperpolarizing cells in the distal retina, above, and microvillar, depolarizing cells in the proximal retina, below. (From Land, 1984.)

hyperpolarization is brought about by decreasing Na^+ permeability. In other words the bivalves have independently evolved their own way of hyperpolarizing their photoreceptors and their eyes can tell us nothing about the evolution of vertebrate photoreceptors (see Land, 1984).

The demonstration that in the Pecten ciliary photoreceptors it is the opening of the K^+ channels that is all important has been confirmed in two elegant papers by Cornwall and Gorman (1983a,b), to which all readers interested in the biophysical aspects of the visual process should refer.

The eyes of the scallop are fascinating at all levels, behavioural, optical, ultrastructural and biophysical, and will undoubtedly repay further study but, to repeat, they have little to tell us about the evolution of the molluscan (or any other) eye, being what Land (1984) terms a 'one-off invention'.

Compound Eyes

These are another example of independently evolved eyes. They are restricted to the bivalve family Arcidae (e.g. *Barbatia, Pectunculus*) and in *Arca* there are over 100 situated along each edge of the mantle (Patten, 1886). They are small, less than 200 μm, and each contains 10–80 facets (for figure see Land, 1984). Each facet represents a single photoreceptor cell with a peripheral transparent face (Patten, 1886; Hesse, 1900) and the eye must be able to detect moving objects that do not throw a direct shadow on the animal (Braun, 1954). There must be considerable overlap in the fields of view of each eye. There are apparently no ultrastructural data to confirm or deny the hypothesis that these eyes may contain ciliary photoreceptors (Land, 1984).

Summary

Several different kinds of non-cephalic photoreceptors, of sufficient complexity to be termed eyes, have evolved in the chitons, gastropods and bivalves. They often mediate a shadow response, a behaviour that is presumably primitive, and this may be handled by photoreceptors that are ciliary rather than microvillar. Although their ultrastructure appears superficially to be 'vertebrate' rather than molluscan, such a distinction is meaningless.

More complex non-cephalic eyes, such as the compound eye of the ark shell and the mirror eye of *Pecten* and its relatives, can yield a limited amount of information about movement and direction, without a shadow having to fall on the animal. Simple though such eyes are by vertebrate or cephalopod standards, they show the first signs of distance reception that so characterizes the sense organs of higher animals (Sherrington, 1906).

Cephalic Eyes

We turn finally to what could be termed true eyes, that is paired organs borne on the head, containing photoreceptor cells in a cup-shaped retina with pigment backing.

Nearly all these eyes have a lens and they range in size from less than 1 mm to 10 cm or more. It is convenient, if somewhat arbitrary, to recognize four categories.

Simple Eyes

We define these as eyes, either without a proper lens or so small that they could not form an image. They are characteristic of gastropods, apart from the tiny cephalic eyes of *Mytilus* that lie at the base of the first gill filament (Bullock, 1965; Rosen *et al.*, 1978).

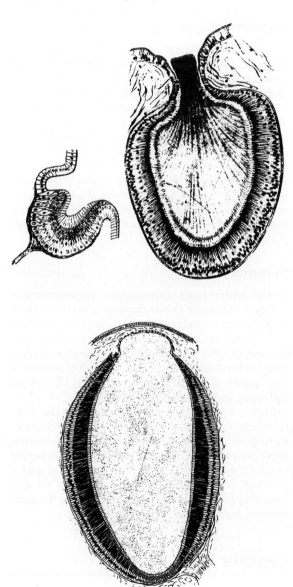

Fig. 17.10 *Prosobranch eyes. Top left:* Patella. *Top right:* Haliotis. *Bottom:* Nerita. *(Modified from Messenger, 1981.)*

Among prosobranchs we may note the eyes of the limpet, *Patella*, the abalone, *Haliotis*, and *Nerita* (Fig. 17.10), which appear increasingly complex although they should not be thought of as an evolutionary sequence. *Ilyanassa* is another prosobranch with a small eye much of which is occupied by the lens (Gibson, 1984). The photoreceptors of the abalone, *Nordotis*, bear strikingly regular microvilli (Kataoka and Yamamoto, 1981; Kataoka *et al.*, 1982) but the claim that their morphology changes dramatically during a 24-hour period has been challenged (see p 377).

The opisthobranchs' eyes tend to be very small (40 μm in *Doto* (Hughes, 1970) and *Glaucus* (Grassé, 1968)) and are often sunk below the skin. There is a lens, but the eye is so small that it is unlikely to be able to form an image. This is especially obvious in the nudibranchs, which are also characterized by having only a few photoreceptor cells. In *Rostanga*, for example, there are seven sensory cells only (Chia and Koss, 1983). In *Tritonia* (diameter 275 μm) and *Hermissenda* (40 μm) there are only *five* cells in each eye (three in the larva (Buchanan, 1986)), a fact that has led to detailed investigation of the *Hermissenda* eye as part of a model system for studying memory (see, for example, Alkon, 1987).

In *Aplysia* the paired eyes are slightly larger but the lens is large and abuts on the retina so that it almost certainly cannot form a sharp image. However, despite its small size the eye is surprisingly complex and in an important recent paper Herman and Strumwasser (1984) recognized for the first time that the retina comprises no fewer than five classes of receptors as well as two types of neurones, which is more than is found even in cephalopods (see pp 377–380). The most conspicuous photoreceptor has long microvilli, but the other four types bear cilia as well as microvilli. These four receptors are small, and so occupy only a fraction of the retina, but numerically they account for about half of all the receptors. Even odder, there is specialization within the retina. The microvillar receptor and one 'mixed' receptor are found throughout the eye; the other three receptor types are restricted to the ventral part of the retina. There are other regional specializations, too, such as length of rhabdom and cross-sectional area being larger dorsally and the pigmented layer being thicker there. One of the two types of neurone is also restricted to one part of the retina.

The functional significance of these findings is not clear, but they are a salutary reminder that in the molluscs even a 'simple' eye, which cannot give its owner a great deal of information because of its optics, can still be surprisingly complex. This is at least partly because this eye has another function as well as a visual one.

One of the remarkable features of *Aplysia* is that its eye contains a *neuronal circadian oscillator system* as well as a photoreceptor system. If an eye of *Aplysia* is excised and

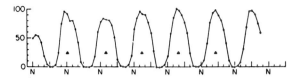

Fig. 17.11 Aplysis *eye; the circadian rhythm of compound action potentials in the optic nerve of an eye isolated in constant darkness at 15°C. Time scale marked at successive noons (N)* (*From Jacklet, 1980*).

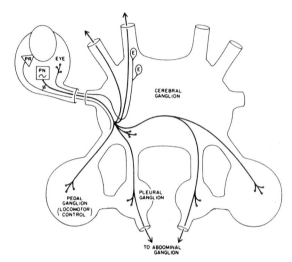

Fig. 17.12 Aplysis: *widespread projection of the circadian pacemaker (PN) in the CNS the optic nerve.* (*From Jacklet, 1984.*)

maintained in continuous darkness at a constant temperature fires spontaneously and the frequency of discharge shows a circadian rhythm (Fig. 17.11; Jacklet, 1969). This fact has been exploited heavily over the last few years and because the eye of *Aplysia* is such a good model for studying circadian organization in the CNS there is now a considerable literature on this aspect of the eye. There are complementary data for other opisthobranchs, notably *Bulla* (Block and Friesen, 1981; Block *et al.*, 1984), but also *Bursatella* (Block and Roberts, 1981) and *Navanax* (Eskin and Harcombe, 1977). All these genera seem to contain within the eye the pacemaker as well as the photoreceptors for the entrainment of the pacemaker. Jacklet (1984) has a useful review of the earlier literature; but there is a plethora of more recent papers implicating serotonin, neuropeptides (e.g. Khalsa and Block, 1990) and cGMP and calcium in the various stages of the photoreceptive process. And the debate continues about the relative importance of the brain and the endogenous pacemakers in each eye. Some workers (Jordan *et al.*, 1985) consider that there must be a third pacemaker outside the eyes.

In *Aplysia* it seems certain that pacemaker neurones in the eye project widely in the CNS (Fig. 17.12) to influence locomotor centres. However, there are significant differences in the different genera between the influence of the eyes and other photoreceptors (such as the rhinophore; Jacklet, 1980) on locomotion and no coherent picture has yet emerged of the precise role of the pacemakers in visual processing.

Most pulmonate gastropods have well-developed, 'simple' cephalic eyes, which either lie on the head or are borne at the tip of the posterior pair of tentacles. Typical microvillar photoreceptors (microvilli 10–12 μm long; 80–100 nm in diameter) have been described in the slugs

(a)

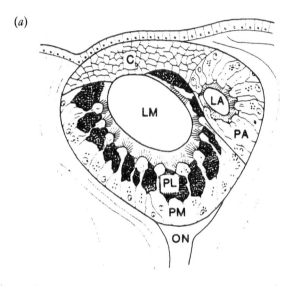

(b)

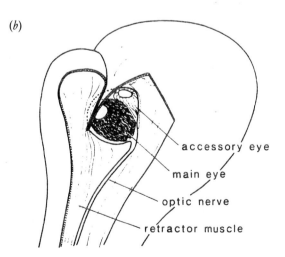

Fig. 17.13 *The eye of the pulmonate,* Achatina, *with its accessory lens (LA) and photoreceptors (PA), (a); in the partially retracted position, (b).* (*From Tamamski, 1989.*)

Agriolimax and *Limax*; and in the snails *Helix* and *Lymnaea* (Newell and Newell, 1968; Kataoka, 1975; Eakin and Brandenburger, 1975b; Stoll, 1973b). The retina also contains other cell types, including ganglion cells (Brandenburger, 1975; Eakin and Brandenburger, 1975b) and, in *Lymnaea*, secondary cells (Stoll, 1973b), which may or may not be the same thing. Recently, Tamamaki (1989a, b) has provided ultrastructural and electrophysiological evidence that the African giant land snail *Achatina* and the slug *Limax* have an *accessory eye*, within the main eye but separate from it, with a similar microvillar arrangement and identical spectral sensitivity (λ_{max} 480 nm and 460 nm respectively). The accessory eye has a lens that could not possibly form an image and its function may be to monitor light intensity while the tentacle is partly retracted (Fig. 17.13), when light is denied access to the main eye (see also Newell and Newell, 1968). This is another example of the almost profligate way in which molluscs seem able to 'invent' additional photoreceptors when they are needed.

The pulmonate eye probably provides little more than intensity and directional information (Eakin and Brandenburger, 1975a) that may subserve orienting responses. Hermann (1968) has reported that in the snail *Otala* negative phototaxis is abolished by bilateral eye removal; since unilateral removal is followed by increased head movements, we can assume this response is mediated by the eye.

Pinhole Eyes

These are found in one genus only, *Nautilus*, the relic shelled cephalopod from the Indo-Pacific Ocean and around Australia. For a variety of reasons it is quite distinct from all the other living coleoid cephalopods (the modern cuttlefish, squids and octopuses) and its eye is unique. There is no cornea and no lens, an open 'pinhole' (diameter 0.4 to 2.8 mm) leading into the interior (Fig. 17.14). The retina contains about 4×10^6 rhabdomeric cells (Messenger, 1981); this is fewer than the 20×10^6 in *Octopus* (Young, 1962a), but is two orders of magnitude more than the eye of *Strombus*, the largest gastropod lens eye. Moreover, the diameter of the receptor cells in the retina (5–10 μm) is the same as in *Octopus*. Hence the paradox of this eye: there is a large, fine-grained retina, but no lens.

Land (1984) calculates that with an eye like this the resolution must be very poor, around 2.3° for the minimum resolvable angle ($\Delta\phi$) (*Octopus*: 1.3') and the sensitivity worse: even with a wide-open pupil (at which stage the $\Delta\phi$ will be 16°!) the image in a *Nautilus* eye will be 13 times dimmer than in *Octopus*. Muntz and Raj (1984) actually tested *Nautilus* in an optomotor apparatus and obtained a value for $\Delta\phi$ of between 5.5° and 11.25°. They also reported that the position of the screening pigment is the same in light- and dark-adapted animals.

If these findings suggest that the eye is a poor device for forming an image, there are others that do not. Thus Muntz and Raj (1984) found signs of a dorso-ventral chiasma in the optic nerves leaving the back of the eye for the optic lobe (cf. *Octopus*; see below). And previously Hartline *et al.* (1979) had shown there are compensatory eye movements, driven by the ipsilateral statocyst. Although these are characteristic of coleoid cephalopods they are not known elsewhere in the molluscs. There were no eye movements to revolving stripes in a nystagmus drum, but there was an optomotor response, which again is unknown in gastropods (Dahmen, 1977). Moreover, Hurley *et al.* (1978) have described a pupillary response. Although this is extraordinarily slow by vertebrate or coleoid standards – 50 seconds to open fully and 90 seconds to constrict – there is no such thing in any gastropod. And finally, the eye is quite large: about 15×10 mm, with the long axis horizontal (Muntz and Raj, 1984).

Taken together these findings are difficult to reconcile, and indeed Land (1984) agonizes over the implausibility of such a 'bad' eye persisting for some 400 million years, when even the acquisition of a simple plug of jelly could have done so much to improve its optics.

Using a Y- or T-maze, Muntz (1986, 1987) has demonstrated positive phototaxis and postulated that perhaps the role of this is to take the animal towards areas of bioluminescence. Another hypothesis is that it could help mediate the diel vertical migration that *Nautilus* exhibits (Fig. 17.15) (Ward *et al.*, 1984). Zann (1984) observed a rhythmic activity in the behaviour of *Nautilus* kept in a constant dark–dark or light–light regime and noted that this was essentially crepuscular.

The evidence emerging from field studies emphasizes

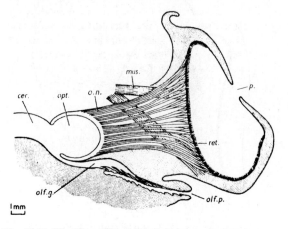

Fig. 17.14 Nautilus. *The pinhole (p) eye seen in transverse section. The optic nerves (o.n.) appear to enter the optic lobe (opt.) without chiasm, but see text (From Young, 1985).*

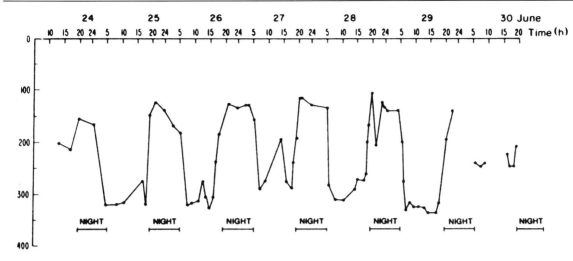

Fig. 17.15 *Vertical movements of an individual* Nautilus *measured by remote telemetry. Ordinate shows depths in metres (From Ward et al. 1984)*.

that *Nautilus* must spend much of its life in areas of low light intensity (Muntz, 1987). But if the eyes are only for phototactic purposes, why have as many as 8×10^6 receptors, much less compensatory eye movements?

Scanning Eyes

These are strictly a subset of the lens eye, but they merit separate treatment, if only to draw attention yet again to the great variety of mollusc eyes.

They are found in one group of prosobranch gastropods called heteropods. These are epipelagic, carnivorous creatures, such as *Pterotrachea*, and their eyes are unusually big for a gastropod with a large spherical lens (Fig. 17.16; Hess, 1900). What really distinguishes these animals is that the retina is ribbon-shaped, only 3–6 receptors wide but several hundred receptors long and Land (1982) has been able to show in one genus, *Oxygyrus*, that this ribbon functions like a 'line-scan' system. The small (1.2 mm) eyes are in almost constant motion, moving rapidly down ($250°\,s^{-1}$), then more slowly upwards ($80°\,s^{-1}$) quite smoothly through an arc of about 90°, such that they move from a horizontal resting position to scan through a visual field in the water below (Fig. 17.17). They are presumably seeking objects that glisten in the light from above.

The eyes of heteropods are also interesting from another point of view. The ultrastructural evidence of Dilly (1969) shows that the retina contains most unusual receptors not unlike vertebrate rods, with some 300 discs in each cell. These are apparently ciliary in origin; certainly they are not rhabdomeric. These wonderful eyes urgently need re-examination.

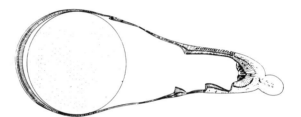

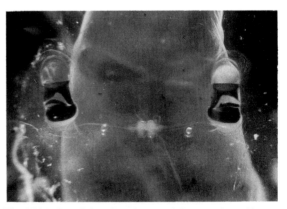

Fig. 17.16 *Eye of the pelagic heteropod*, Pterotrachea. *Above, in section (after Hesse, 1900); below, from life (From Land, 1984)*.

Lens Eyes

We are now referring to eyes with such dimensions that the lens must be able to produce an image on the retina so that some kind of form vision must obtain. Such eyes are found in the prosobranch and pulmonate gastropods and in the coleoid cephalopods.

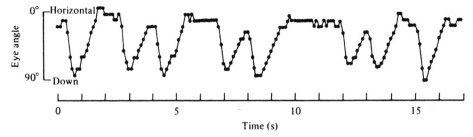

Fig. 17.17 *The scanning movements of another heteropod eye,* Oxygyrus. *(From Land, 1982.)*

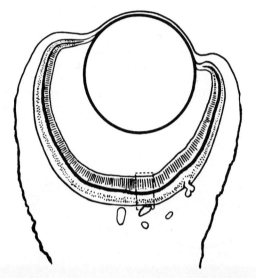

Fig. 17.18 Strombus *eye in cross-section (From Gillary, 1974).*

Among prosobranchs the eyes of the periwinkle, *Littorina* sp., a very successful and widespread form, have a spherical lens and there is good behavioural evidence (p 390) for rudimentary form vision, making it plain that an image is formed on the rhabdomeric retina, which agrees with the optics, especially with the fact that the lens is not homogeneous (see below) (Newell, 1965; Hamilton *et al.*, 1983).

However, if we exclude the heteropod eyes dealt with on p 275, then the prize for a prosobranch eye must go to the Pacific conch *Strombus luhuanus*, whose eyes, borne on stalks, reach 2 mm in diameter (Fig. 17.18). The lens is spherical, about 75 μm in diameter (Gillary and Gillary, 1979), and its focal length conforms to Matthiesen's ratio. The resolving power is good, with an inter-receptor angle of about 14′ (Land, 1984). Gillary and his colleagues have shown that the retina contains no fewer than four different cell types, and that there are about 50 000 photoreceptor cells. These are microvillar, but there are also some ciliary cells and ultrastructural studies have shown synaptic vesicles of different size and density in the neuropil.

That the retina might be complex is borne out by electrophysiological evidence. The ERG (Gillary, 1974) may have two distinct 'on' peaks and also some rhythmic oscillations when the illumination ceases: it depends on the state of light adaptation as well as the intensity and duration of the stimulus flash. Complementary evidence from intracellular recording confirms that a certain amount of interaction occurs in the retina between the different cell types, though its significance remains unclear. This is partly because there have been no behavioural studies on *Strombus*, which is a grazing herbivore. Indeed, we have no clues to the function of these remarkable eyes.

Among the pulmonates the pond snail, *Lymnaea*, uses its eyes for orientation (see Stoll, 1973a) and they are probably capable of forming images (Land, 1984).

We now turn to the well-known eyes of the coleoid cephalopods (hereafter termed simply 'cephalopods'). These are depicted in many elementary biology texts as an example of convergent evolution. The similarities between cephalopod and fish eyes are indeed striking (large size, large lens, iris), but it should be pointed out that there are important differences between them in the retina: in cephalopods the photoreceptors point towards the light, not away from it, and the photoreceptors are microvillar, like the cephalic eyes of gastropods as well as the eyes of crustaceans, insects and many other invertebrates (Eakin, 1972). Another important difference is that the retina of cephalopods is much simpler than the vertebrate retina and much of the early stages of visual processing are carried out in the brain in the outer regions of the optic lobe, which Cajal (1917) termed the 'retina profunda'.

Cephalopod eyes (Fig. 17.19) are generally very large and can be enormous: in small *Octopus vulgaris* the diameter is 10–15 mm but in the giant squid *Architeuthis* diameters of up to 40 cm (*sic*) have been reported (Akimushkin, 1965). In *Sepiola* (Bullock, 1965) the weight of each eye can be 25% of the body weight. In *Octopus*, the retina contains about 20×10^6 photoreceptors (Young, 1962b).

There is a retina, a choroid, a sclera and an external argentea, and an elaborate intrinsic musculature (Alexandrowicz, 1928). As well as these there are extrinsic eye

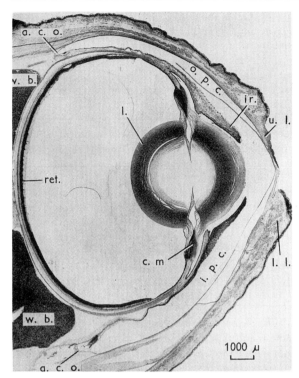

Fig. 17.19 Octopus *eye. c.m., ciliary muscle; ir., iris; l., lens; ret., retina (From Boycott and Young, 1956a).*

muscles which serve to move the eyes in the orbit as a result of either visual or mechanical information. Such movements, perhaps regarded as 'standard' by vertebrate biologists, are restricted to cephalopods among the molluscs and give advance notice, as it were, of a high-performance eye.

Eye Movements

Budelmann and Young (1984) have described the functional organization of the extra-ocular eye muscles in *Octopus* and stress the similarity of the 'vestibulo-ocular' reflex in cephalopods and vertebrates. However, McVean (1984) provides physiological evidence that there are important differences between cephalopod and mammalian eye muscles: in particular, the *Octopus* eye muscles contract very much more slowly. The mechanoreceptors providing stability for the eye, despite variation in the position of the head or body, lie on the macula of the statocyst (for reviews see Budelmann, 1977; Budelmann *et al.*, 1987). Collewijn (1970) has demonstrated their effects on eye movements very elegantly and has shown also that there are optokinetic influences (see also Messenger, 1970). The importance of gravity for keeping the eye correctly oriented was shown long ago by Wells (1960), who found that after bilateral statocyst ablation octopuses failed to make the otherwise simple vertical-horizontal discrimination.

Finally, it should be noted that *Sepia*, which fixates its prey binocularly, exhibits convergent eye movements. In this genus there is also evidence for distance perception (Messenger, 1968).

Lens

The cephalopod lens, like the fish lens, has a focal length approximately 2.5 times the lens radius (Matthiessen's ratio) (Pumphrey, 1961). This is achieved by having a gradient of refractive indices across the lens, high in the centre and progressively lower towards the periphery. Such an arrangement is essential for an aquatic animal needing to produce a high-quality image (Land, 1984), for in water the cornea can contribute nothing to refraction. We find a 'Matthiessen lens' only in the prosobranchs *Littorina* and *Strombus* and in the heteropods; in the pulmonate *Lymnaea*; and in the cephalopods. These may have evolved independently. The optics of the cephalopod lens have been studied recently by Sivak (1982) and by Sroczynski and Muntz (1985, 1987) and the well-marked pupillary response by Muntz (1977). Visual acuity has been examined in three species of *Octopus*: the best estimates show that stripe widths subtending 9.7' are discriminable (Muntz and Guyther, 1988).

Retina

The extensive literature on the cephalopod retina dates back to 1864 (see Messenger, 1981, for references). The modern description owes much to the important LM study of Young (1962a) coupled with the EM studies of his collaborators (Moody and Robertson, 1960; Moody and Parriss, 1961) on *Octopus*. There are now complementary ultrastructural data for the squid (*Loligo*) (Cohen, 1973a, b) and studies of several other genera (Yamada and Usukura, 1982; Yamamoto and Takasu, 1984; Yamamoto *et al.*, 1985; Saibil and Hewat, 1987; Tsukita *et al.*, 1988) using such modern techniques as freeze-fracture, X-ray diffraction and image analysis. Apart from an argument about the turnover of membranes these studies are in good agreement and there do not seem to be significant differences between the basic organization of the octopod and decapod retina. Yamamoto (1985) gives an important account of the developing retina in *Sepiella*.

Fig. 17.20 shows that there the retina comprises a limiting membrane, a long 'outer-segment' layer (directed towards the lens), a shorter 'inner-segment' layer and below this a plexiform layer. The retina comprises supporting cells, retinal glia (epithelial) cells and receptor (or retinula) cells (Young, 1962a). The retinula cell contains pigment granules that move distally in the light, and retract to the base of the outer segment in the dark. There is also screening pigment in the supporting cells (Cohen,

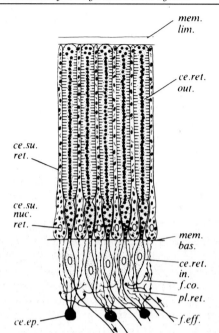

Fig. 17.20 Octopus *retina in vertical section, showing outer (ce. ret. out.) and inner (ce. ret. in.) segments and the retinal plexiform layer (pl. ret.), with its efferent fibres (f.eff.). Light comes from above. (From Young, 1962a.)*

1973a). Light- and dark-adaptation has been examined by Hagins and Liebman (1962), by Daw and Pearlman (1974) and by Young (1963), who emphasizes that the rate of pigment migration may vary in different regions of the retina. There are also changes in length of the receptor cell with illumination: they contract in the light (Young, 1963; Tasaki and Nakaye, 1984).

The receptor cells are of one type only. They are very long and thin, and bear two sets of microvilli at right angles to the main axis and opposite each other: there may be 200 000 to 700 000 microvilli in each set (Zonana, 1961). The *outer segment* is between 200 μm and 300 μm long: the microvilli are about 1 μm long but only 60 nm in diameter (see below). The receptors are therefore uncompromisingly microvillar or rhabdomeric (Fig. 17.21). Moreover, the retina constitutes a two-plane rhabdomere system (Eakin, 1972) because the retinular cells are oriented with the microvilli either *horizontal* or *vertical* as the eye is normally held in space, a feature thought to be associated with the undoubted polarized light sensitivity (PLS) exhibited by the cephalopod eye (see below).

The *inner segment* contains the nuclei of the cells and large numbers of mitochondria, together with sheets of thin membranes characteristically ordered and known as the somal, or myeloid bodies. Associated with these bodies is the photopigment retinochrome; the visual pigment,

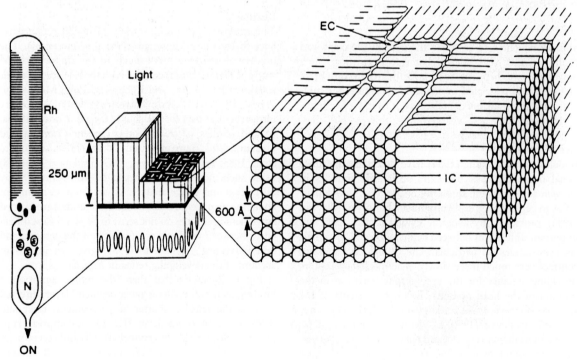

Fig. 17.21 *Diagram of the orthogonal arrangement of receptor cells (Rh) in the cephalopod retina (the so-called 'two-plane rhabdomere system'). Note the long, thin microvilli. EC, IC, extracellular, intracellular compartments (From Saibil and Hewat, 1987).*

rhodopsin, is localized in the outer segment (see below). The *plexiform layer* comprises glial cells, fine collaterals of the receptor cells and tangentially running *efferent* fibres from the optic lobe (Patterson and Silver, 1983). The important ultrastructural study of Yamamoto and Takasu (1984) has shown there are gap junctions between the visual cell inner segments and also between efferents and visual cell axons.

Axons of the retinula cells leave the back of the eye and run to the optic lobe after making a complex dorso-ventral chiasma (Fig. 32; Young, 1962b; Saidel, 1979). The significance of this chiasma is presumably that it 'corrects' the inverted retinal image produced by the lens and serves to project a visual map that is in register with a gravitational map (Young, 1962c).

Ultrastructure and Biophysics of the Outer Segment

The work of Moody and Parris (1961), Cohen (1973a,b,c) and, more recently, the elegant biophysical studies of Saibil and her collaborators (e.g. Saibil, 1982, 1990a,b; Saibil and Michel-Villaz, 1984; Saibil and Hewat, 1987) have established that the microvilli are packed into beautiful hexagonal arrays (Fig. 17.22) with an actomyosin skeleton and that the hydrophobic layer of the microvillar membrane has an extremely high protein content, more

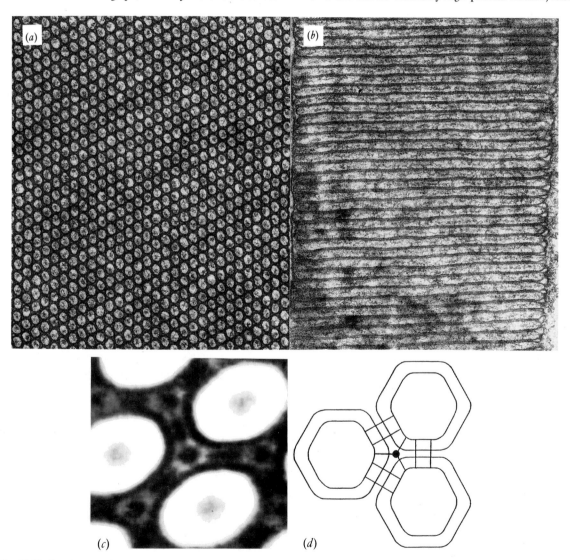

Fig. 17.22 *Squid microvilli in (a) transverse and (b) longitudinal section, showing the hexagonal packing, and the continuity of the microvilli with the cell cytoplasm. (c) High power obtained by image processing to show the substructure of the membrane and (d) a schematic representation of this (From Saibil, 1990a).*

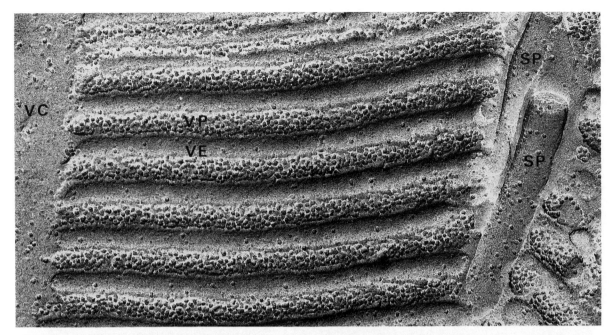

Fig. 17.23 *The microvilli of* Sepiella *after freeze-fracture, showing alternating P-faces (VP) and E-faces (VE) (× 108 000). Note the tiny vesicles thought to contain rhodopsin (From Yamamoto and Takasu, 1984).*

than half of which is probably rhodopsin, the visual pigment. This is localized entirely within the outer segment (Hara and Hara, 1976). Moody (1962) has shown that the outer segments are weakly birefringent, and the suggestion is that rhodopsin molecules are confined to the membrane and are axially aligned along it. Saibil's work emphasizes that the microvilli comprise an extensive and highly ordered array of membrane cytoskeleton connections. Tsukita *et al.* (1988) now have evidence, using rapid-freezing methods, that this actin skeleton breaks down when stimulated by light, on a time scale compatible with its being involved in phototransduction. The transmembrane links seem to be crucial for maintaining structural order in the microvillar array and may align the rhodopsin molecules to permit a high degree of PLS. Recent evidence implicates Ca^{2+}/calmodulin in the light-induced structural changes in the microvillar cytoskeleton (Asai *et al.*, 1989).

Freeze-fracture studies suggest that the rhodopsin molecules may be localized in small (10 nm) spheres on the P-face of the membrane (Fig. 17.23; Yamamoto and Takasu, 1984; Yamamoto, 1985). The bases of the microvilli are probably always open to the cytoplasm (Yamamoto and Yoshida, 1984; but see Kataoka and T. Y. Yamamoto, 1983) and according to Saibil (1990a) the basal loops of the microvilli are probably the site of the light-sensitive cation channels (see below).

The biochemical events underlying the visual process in cephalopods are beginning to emerge now, and squid photoreceptors are proving an important system for studying fundamental features of photoreception. This summary is based on Saibil (1990a; see Fig. 17.24). Photon capture by rhodopsin (R) located in the microvilli drives an enzyme cascade there, which may involve the intracellular messenger's calcium and possibly cyclic GMP. The G protein, which is related to the vertebrate transducin (Tsuda *et al.*, 1986), activates phospholipase C (PLC), which ultimately releases inositol (1,4,5) triphosphate (IP3) (Szuts *et al.*, 1986; Brown *et al.*, 1987; Wood *et al.*, 1989). This mobilizes internal calcium and the increase in calcium levels leads to depolarization. The IP3 is thought to diffuse to the central ends of the microvilli in order to do this, but the details of all this are still unclear.

There is an extensive literature on GTP binding proteins and phospholipase C in the squid retina (Vandenberg and Montal, 1984a,b; Saibil, 1990b).

Photopigments

The chemistry of cephalopod rhodopsin has been the subject of frequent examination ever since the important papers of Hubbard and St George (1958) and Kropf, Brown and Hubbard (1959) (see also Hara and Hara, 1972; Hagins, 1973). It is a conjugated protein, an opsin bound to retinaldehyde in the 11-*cis* configuration. Light will convert this to the all-*trans* configuration.

Until very recently it could be said that all cephalopods

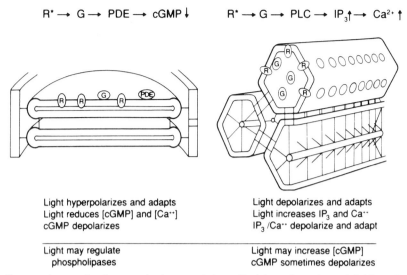

Fig. 17.24 *Current views on the phototransduction cascade in squid, right, and vertebrate, left (From Saibil, 1990a).*

had only one type of rhodopsin (although the maximum spectral sensitivity varies from 470 nm in the octopod *Eledone* to 500 nm in the cuttlefish *Sepiella*: see Table 17.1 and Messenger, 1981, for details). However, it is now known that in one species, a bioluminescent deep-sea squid, *Watasenia scintillans*, the retina contains three visual pigments, as follows:

1. A1 pigment λ_{max} 484 nm based on retinal.
2. A2 pigment λ_{max} 500 nm based on 3-dehydroretinal.
3. A4 pigment λ_{max} 470 nm based on 4-dehydroretinal.

(Matsui *et al.*, 1988; Seidou *et al.*, 1990). These authors looked at no fewer than 23 other species and found them all to contain only the A1 pigment. It is believed that in *Watasenia* these pigments have evolved so that intraspecific bioluminescence produced by the photophores can be distinguished from the downwelling light and perhaps enable the animals to produce signals invisible to other organisms. Unfortunately we know very little about the behaviour of these squids in the sea.

The other photopigment found in the cephalopod retina (and PSV) is *retinochrome*, the subject of a prolonged and elegant investigation by Hara and Hara, e.g. 1965; Hara *et al.* 1981; Ozaki *et al.* 1987; Terakita *et al.* 1989. This pigment is localized in the myeloid bodies of the inner segment but is transported, via a retinal-binding protein (RALBP), towards the rhabdomes where it interacts with meta-rhodopsin to help regenerate the rhodopsin. Immunohistochemical studies of both *Octopus* (Ozaki *et al.*, 1983) and *Todarodes* (Fukushima, 1985) demonstrate that retinochrome is present throughout the inner segments, but extends only into the basal region of the outer segments, where rhodopsin is located. The important work of Robles *et al.* (1984, 1987) and Breneman *et al.* (1986) supports and extends these findings about turnover within the receptor cell. The possession of a dual photopigment system is unusual by vertebrate standards, but it may be a fundamental molluscan feature because Ozaki *et al.* (1983; 1986) have found it in the gastropods *Limax* and *Strombus*, although there are significant differences in the proportion of rhodopsin and retinochrome between gastropods and cephalopods. Shimatani *et al.* (1988) also

Table 17.1 *Maximum absorbance (in nm) of cephalopod retinochromes and rhodopsins*[a].

	Sepia esculenta	*Sepiella japonica*	*Loligo pealeii*	*Todarodes pacificus*	*Octopus vulgaris*	*Octopus ocellatus*
Retinochrome	508	552	500[b]	495	490	490
Rhodopsin	486	500	493[c]	480	475	477

[a] Hara *et al.* (1967).
[b] Sperling and Hubbard (1975).
[c] Brown and Brown (1958).

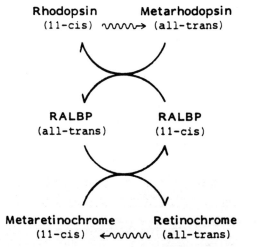

Fig. 17.25 *The relation between rhodopsin and retinochrome; scheme showing chromophore exchange through the intermediary RALBP.* (∿∿) *photoreaction;* (→) *dark reaction (From Terakita et al., 1989).*

report retinochrome in the stalk and dermal eyes of *Onchidium*. Fig. 17.25 summarizes current thinking about the interrelationship between these two photopigments.

Electrical Activity

Electrical activity associated with light stimulation has been known in the cephalopod eye for over 90 years (Beck, 1899). Fröhlich published his famous photograph of the electroretinogram (ERG) of *Eledone* in 1914 and 60 years later Weeks and Duncan obtained a similar response for the small sepioid *Sepiola* (Fig. 17.26). The ERG has often been used as a tool for investigating cephalopod visual function: dark-adaptation and spectral sensitivity (Fig. 17.26) as well as flicker fusion frequency and regional differentiation within the retina have all been investigated. In *Octopus*, Tasaki and his colleagues (Tasaki et al., 1963c; Norton et al., 1965) have some evidence for a 'physiological centre' in the retina, though the significance of this is not understood. (For review see Messenger, 1981.)

Perhaps the most significant finding to emerge using the ERG is that the isolated retina, deprived of lens, is sensitive to polarized light. Using *Octopus* and the squid *Todarodes*, Tasaki and Karita (1966a,b) showed that after the retina had light-adapted to polarized light of a fixed plane, the amplitude of the ERG varied with the plane of polarization of a stimulus light, being maximal when the e-vector was at right angles to that of the adapting light and minimal when the planes were coincident.

Sugawar et al. (1971) went further in that they recorded intracellularly from receptor cells in the isolated retina. Out of approximately 50 units they found that half

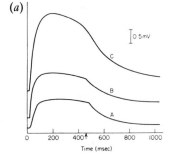

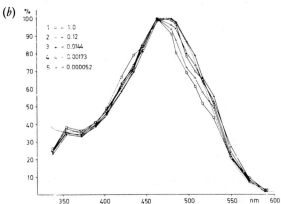

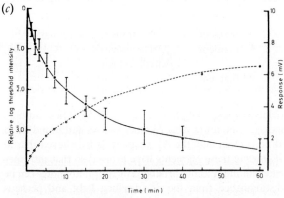

Fig. 17.26 *The cephalopod ERG and its application.* (a) *Sepiola ERG with long flashes (zero to arrow) of increasing intensity, A–C (From Weeks and Duncan, 1974);* (b) *Eledone, an octopod, showing the close fit between the responses at intensities indicated and the absorption curve for the pigment (dotted) (From Hamdorf et al., 1968);* (c) *Octopus dark adaptation curve (left ordinate) or increase in response amplitude (right), showing no evidence for a duplex retina (see text) (From Hamasaki, 1968).*

responded maximally when the plane of the e-vector was parallel to the horizontal axis as the eye is held in life, and half when it was vertical to it (Fig. 17.27). This was incontrovertible evidence that the two-plane retina, with micro-

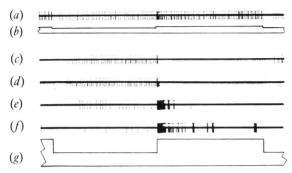

Fig. 17.27 Loligo: *complex optic nerve responses during adaptation. Square wave intensity modulation at low level (b) for trace (a), at high level (g) for traces (c)–(f). Tonic and phasic excitation in (a) become tonic off response in (c) and evolves to a transient* on *response in (f). (c) at time zero, (d) after 40 s, (e) after 2 min, (f) after 5 min) (From Hartline and Lange, 1974).*

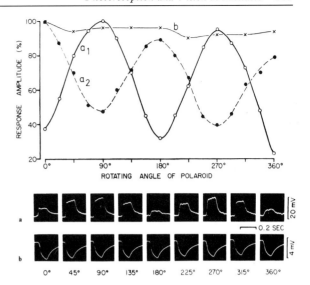

Fig. 17.28 *Polarized light sensitivity in isolated retina. Intracellular records from two cells (a_1, a_2) showing changes in response amplitude as the plane of polarization of stimulus light, measured from the horizontal, changes (From Sugawara et al., 1971).*

villar arrays orthogonal to one another, could subserve polarized light sensitivity, although this had been inferred previously on behavioural as well as structural grounds (Moody and Parriss, 1961; see p 388).

The electrochemical events accompanying absorption of light by the photoreceptors are also well documented (Hagins *et al.*, 1962; Hagins, 1965; Duncan and Weeks, 1973; Pinto and Brown, 1977; Duncan and Pynsent, 1979;for review see Messenger, 1981). The retinula cell depolarizes in the light, the membrane current being largely carried by Na^+, and in this the microvillar receptors can be thought of as typically invertebrate 'on' receptors. It is generally held that the waveform of the photoresponse reflects the time course of conductance changes in the microvillar membrane (but see Hartline and Lange, 1977, 1984; Lange *et al.*, 1979).

Finally we must consider the electrophysiological evidence derived from recording from the optic nerves, although this is tantalizingly fragmentary by arthropod or vertebrate standards. Despite a handful of earlier papers (see Messenger, 1981), the most comprehensive study remains that of Hartline and Lange (1974), working with an isolated eye–optic lobe preparation in the squid, *Loligo opalescens*. They found 'slow' and 'fast' units in the optic nerves. Some slow units are centrifugal. The 'fast' units, whose spike frequency is intensity-dependent, derive from the receptors. There are 'on', 'off' and 'on-off' units, but they are not separate categories and a change in the state of adaptation can alter the response of a unit. In fact, the most notable feature of the squid retina is that its responses are surprisingly complex. Prolonged illumination at constant intensity can produce two bursts instead of only one; responses change sign during adaptation (Fig. 17.28); and increasing the levels of background illumination can enhance the response to an identical flash.

These findings suggest not only that there may be strong centrifugal influences on the retina (cf. Boycott *et al.*, 1965) but also that the retina itself may have unusual properties. Clearly the evidence that there may be gap junctions needs urgent re-examination (see also Hartline and Lange, 1984). More recently Saidel *et al.* (1983) demonstrated polarized light sensitivity electrophysiologically in the optic nerves of the squid, *Loligo*. This shows up, of course, as an apparent sensitivity to the intensity of the stimulus, and the authors confirmed that there were two kinds of receptor, one maximally sensitive to a vertical, the other to a horizontal, plane of polarization.

Visual Processing

In this section we consider what is known anatomically and physiologically about where and how information gathered by the photoreceptors is subsequently handled in the CNS. It can be said at once that nothing is known about this in chitons or scaphopods and relatively little in gastropods and bivalves. Among the gastropods only heteropods are known to have a well-developed 'optic ganglion' (Fig. 17.29) that would clearly repay detailed study using modern techniques. However, in the opisthobranch *Hermissenda* the detailed investigation of Alkon and his collaborators has already produced a surprisingly complete account of central processing of visual information in this simple system, which comprises 5 photorecep-

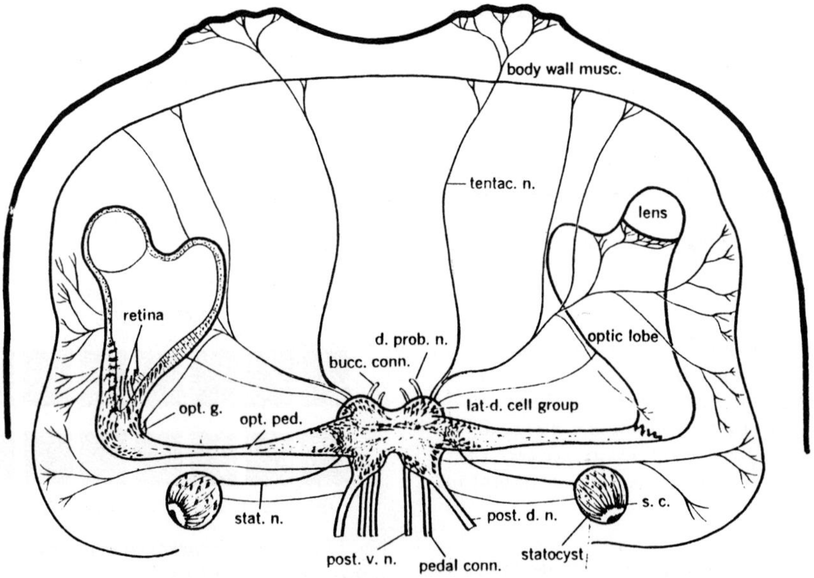

384 *Evolution of the Eye and Visual System*

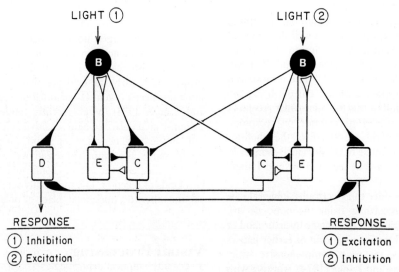

Fig. 17.29 *The brain of a heteropod, showing well defined optic ganglion (opt.g.) behind the large eye (From Bullock, 1965).*

Fig. 17.30 *The simple eyes of the opisthobranch,* Hermissenda *can nevertheless give complex responses because of the connectivity. Thus light 1 can cause inhibition, light 2 excitation of cell D in the left eye. Filled processes, inhibitory; open ones, excitatory (From Tabata and Alkon, 1982).*

tors in each eye associated with 14 optic ganglion cells and other second-order cells in the central ganglia (Fig. 17.30). Alkon has shown that the B-cells may mediate weak arousal behaviour but that the A-cell activity can lead to visually guided turning movements of the animal. In life *Hermissenda* orients and moves towards a light source until it reaches the most intense light within the gradient. It will then turn to remain at the edge of this bright light. The remarkable, and rather depressing, feature of the *Her-*

missenda visual system is that it is so simple yet can achieve quite a complex range of behaviours (Fig. 17.31). And this is an animal whose visual capabilities are, to say the least, banal. There can be no form vision, colour vision, PLS, nor motion detection. So much for trying to understand the eye of the cephalopods!

In the bivalve *Argopecten*, Spagnolia and Wilkens (1983) have used modern neuroanatomical methods to confirm Dakin's (1910) suggestion that the large lateral

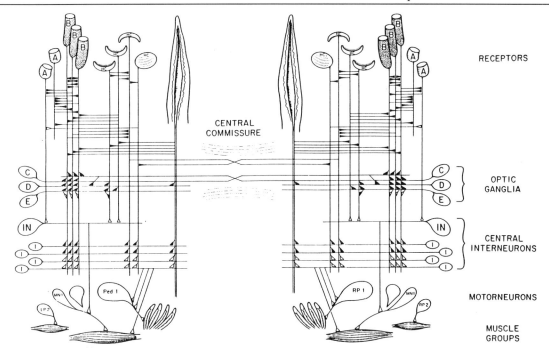

Fig. 17.31 Hermissenda *nervous system showing known connections of all 10 photoreceptor cells (A, B) with the motor system (From Alkon, 1987).*

lobes of the parieto-visceral ganglion function as the optic lobes of these very 'visual' animals. However, their organization is complex and it has not yet been possible to define precisely the sites where 'on' and 'off' receptors terminate (cf. Wilkens and Ache, 1977).

Processing in the cephalopod visual system has been studied extensively anatomically, but less so physiologically because the system is so refractory to microelectrode analysis (see Messenger, 1981). However, to offset this there has been an extensive series of detailed behavioural experiments with living *Octopus*, which has revealed a great deal about the capabilities of their eyes.

Mainly on the basis of Cajal silver preparations, Young has provided a thorough description of the optic lobe of *Octopus* (1960, 1962b) and of *Loligo* (1974). It is impossible here to give more than a summary of these very important papers at the light microscopic level, which have essentially been substantiated by the few electron microscopic studies available (Dilly *et al.*, 1963; Cohen, 1973c).

As Fig. 17.32 shows, the axons of the receptor cells terminate in the outer cortical region of the optic lobe (the 'deep retina' of Cajal). This comprises an *outer granule cell layer*, a *plexiform zone* (with radial and tangential layers) and an *inner granule cell layer*. In *Loligo* there is a *palisade layer* between this region and the centre of the optic lobe, the *medulla*.

The optic nerves terminate mostly in the plexiform zone, making contact with the *second-order visual cells*, whose cell bodies lie in the granule layers. The second-order cells in *Octopus* have oval-shaped dendritic fields in the plexiform layer and these fields are oriented predominantly vertically and horizontally as the head is normally held in space. They could thus serve as feature detectors coding for horizontal and vertical edges in the visual field and may provide the basis for the discrimination of vertical and horizontal (though not oblique) rectangles by octopuses (see below).

In *Loligo*, an epipelagic squid, there are at least four kinds of second-order visual cells, some with circular, others with elliptical dendritic fields. It is interesting that Daw and Pearlman (1969) succeeded in recording from (unidentified) cells in squid optic lobes and found cells with circular receptive fields of size ranging from 3° to 20°, with 'on' or 'off' centres. Unfortunately theirs is the only data of this kind available.

The plexiform zone contains also the processes of amacrine cells, whose cell bodies lie in the granule layers, as well as fibres that proceed deep into the medulla of the lobe. The latter is an area containing many tens of millions of cells, and the work of Boycott (1961) and Young (1962, 1974) has shown that it is one of the most complex areas of the brain, being part higher motor centre, part visual memory store, part 'association' area.

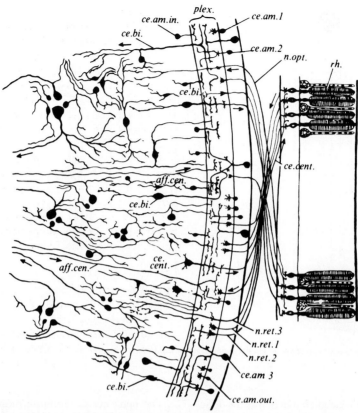

Fig. 17.32 Octopus *retina and 'deep retina' in the optic lobe. aff. cen., afferents to plexiform zone (plex.); and three types of retinal nerve fibres (n.ret.1,2,3) (From Young, 1962b).*

There are complementary findings about the neuropharmacology of the retina, deep retina and optic lobe medulla (see Tansey, 1980; Messenger, 1981). Acetylcholine is thought to be the transmitter of the receptor cells themselves, but there are undoubtedly biogenic amines in the retina and also in the deep retina. In the retina Silver et al. (1983) report dopamine (see also Lam et al., 1974), and Kito-Yamashita et al. (1990) have found 5-HT. There is also evidence of substance P (Osborne et al., 1986), FMRF-amide and somatostatin in the deep retina (Feldman, 1986). Hagighet et al. (1984) have evidence for three sorts of chemical synapse in the plexiform zone (lr) as well as gap junctions.

Because *Octopus vulgaris* is a hardy animal that survives well in aquaria and learns very quickly in a discrimination learning situation, a great deal is known, in this species at least, about the kinds of information the cephalopod eye makes available to the CNS as a basis for learning. Because of the paucity of electrophysiological data in these animals this work assumes great importance.

For the most part, the data derive from a discrimination training technique developed by Sutherland, and carried out by him and his colleagues (e.g. Sutherland, 1957, 1960, 1963, 1969). A complete summary of this work is given, in tabular form, in Messenger (1981); but see also Wells (1978). The account that follows is necessarily a summary one.

Octopuses can learn to attack (for a food reward) or not attack (because of a small electric shock) shapes that differ in terms of: brightness, size, orientation, form, and plane of polarization (Fig. 17.33). They *cannot* learn to discriminate shapes on the basis of hue.

Brightness
Not only black and white shapes, but also grey and white are easily discriminable (Messenger and Sanders, 1971). Moreover, in a nystagmus apparatus octopuses can discriminate between different intensity stripes of blue, cyan, green and red (Messenger et al., 1973).

Size
Different size squares (solid or outline) and circles are easily discriminable by *Octopus*. *Sepia* can be trained to discriminate (binocularly) between a pair of squares (4 × 4 cm and 2 × 2 cm) presented at 60 cm and is not subsequently confused by the smaller square being presented at 30 cm (Messenger, 1977a).

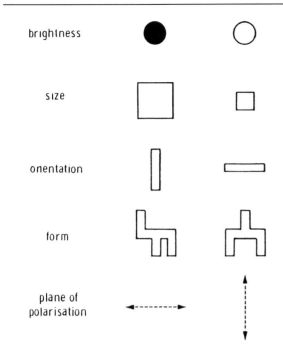

Fig. 17.33 *Five kinds of information that octopuses can extract from their visual world (From Messenger, 1979b).*

Orientation

The ability to discriminate between a vertical and horizontal rectangle has been shown repeatedly after Sutherland's (1957) original experiments. And it has also been confirmed that octopuses are quite unable to discriminate between the same sized rectangles presented as mirror-image obliques (Fig. 17.34). This finding is in agreement with the apparent absence of second-order visual neurones with oblique dendritic fields in the plexiform layer of the optic lobe (see above).

One curious feature that has emerged from these experiments is that octopuses have a strong preference for a rectangle that is moved along its axis. There are also preferences for large rather than small shapes (within limits) and for 'pointed' shapes (diamond or triangle) moving in the direction of the point. The significance of these preferences is not known.

Form

As Boycott and Young (1956b, 1957) first showed octopuses can easily discriminate form and not merely area or length or outline (Fig. 17.33). Subsequently Sutherland, in an attempt to establish a comprehensive theory of shape recognition (Sutherland, 1968), trained octopuses to discriminate between a whole variety of shapes. Some of

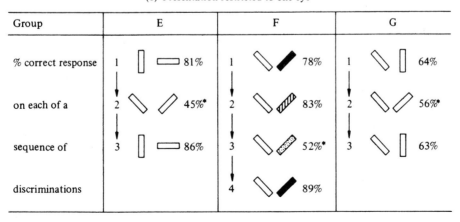

Fig. 17.34 *Inability to discrimination between obliquely oriented rectangles. a: all shapes 10 × 2 cm; b: data from Sutherland (1957, 1958); c: simultaneous presentation (Sutherland and Muntz, 1959); d: data from Messenger and Sanders (1971); *: chance score (From Messenger, 1981).*

these are shown in Fig. 17.35, which makes clear that their form vision is excellent and certainly comparable to that of lower vertebrates. There is no reason to suppose that *Octopus* is unusual in its visual capabilities. Recently, Allen et al. (1985) showed that a squid, *Lolliguncula*, can discriminate horizontal from vertical rectangles, black spheres from white and a white horizontal from a white sphere. There are no data from other species.

Plane of polarization
Moody and Parriss (1960, 1961) were the first to show that cephalopods are sensitive to the plane of polarized light (see above). They trained octopuses to attack (or not) a light from a torch covered with a Polaroid sheet. When the e-vectors to be discriminated were vertical *versus* horizontal the animals learned easily but when the e-vectors were obliquely oriented the discrimination was much poorer, though scores were claimed to be better than chance. There is also a claim that young *Euprymna* and *Sepioteuthis* can orient themselves in four directions relative to the e-vector: 0°, 45°, 90°, 135° (Jander et al., 1963).

Although these results seem at variance with the electrophysiological experiments of Tasaki and Karita (1966a, b), Sugawara et al. (1971) and Saidel et al. (1983), they could be explained on the assumption that not all the microvilli are perfectly regularly oriented within the retina.

Absence of colour vision
Visual discrimination training has also been employed to investigate colour vision in *Octopus* (Messenger et al., 1973; Messenger, 1977b): despite a variety of training methods, the results of such training were always negative (Table 17.2).

In the first series of experiments octopuses discriminated on the basis of brightness before and after they failed to discriminate on the basis of hue.

In the second series three groups of experimentally naive animals were trained on two-cue discriminations: brightness *and* hue relevant, brightness only relevant, hue only relevant. The first two groups discriminated easily but their scores were statistically indistinguishable, sug-

Table 17.2 *Octopuses fail to learn on the basis of wavelength.*

I: Successive presentation: one cue

Group	N	Stage 1 (brightness)	Result	Stage 2 (hue)	Result	Stage 3 (brightness)	Result
A	7	dark blue vs. light blue	yes	dark blue vs. dark red	no!	black vs. white	yes
B	9	light red vs. dark red	yes	light red vs. light green	no!	black vs. white	yes
C	15	medium green vs. light green	yes	medium green vs. medium blue	no!	black vs. white	yes

II: Successive presentation: two cues

Group	N	Positive shape		Negative Shape	Result
D	14	light blue	vs.	dark green	yes
E	14	light (blue or green)	vs.	dark (blue or green)	yes
F	14	blue (light or dark)	vs.	green (light or dark)	no!

III: Simultaneous presentation, alternating tasks
(group G; N = 6)

Stage 1	Result	Stage 2	Result	Stage 3	Result	Stage 4	Result
yellow vs. grey	no!	horizontal vs. vertical	yes	horizontal vs. vertical	yes	horizontal vs. vertical	yes
black vs. white	yes	black vs. white	yes	violet vs. grey	no!	black vs. white	yes
alternate tasks over 3 sessions		alternate tasks over 2 sessions		alternate tasks over 3 sessions		alternate tasks over 2 sessions	

This table summarizes the results of three experiments, using 79 experimentally naive *Octopus vulgaris*, in which no individual ever learned on the basis of wavelength. All hues were matched for brightness. Data from Messenger et al. (1973) and Messenger (1977).

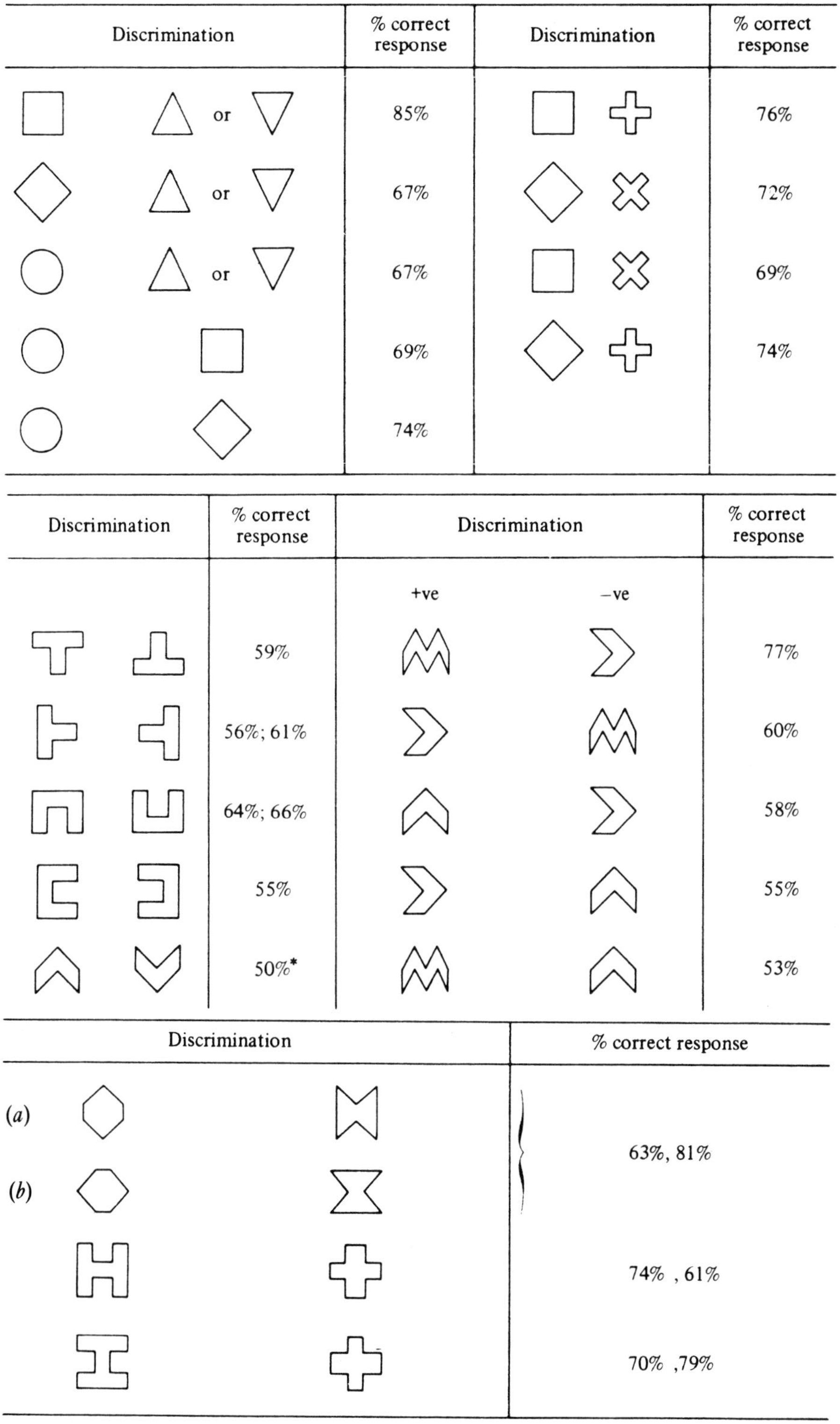

Fig. 17.35 *A selection of pairs of shapes that octopuses can discriminate with various degrees of ease,* *: *chance score (Based on data of Sutherland, modified from Messenger, 1981).*

gesting that in the discrimination based on brightness *and* hue the octopuses were attending only to brightness.

In the third experiment, naive octopuses were trained in a novel way. They were shown alternating simultaneous presentations of a pair of discriminanda. They learned to discriminate between black and white over the same period of time in which they did not learn to discriminate between yellow and grey; and they discriminated between horizontals and verticals over the same period in which they did not learn to discriminate violet from grey.

Further experiments that tested octopuses in a nystagmus apparatus with a sequence of different striped backgrounds support the idea that octopuses cannot resolve hue. Optomotor responses and eye movements were made to stripes that differed in intensity but not to stripes of the same intensity but different hue (Messenger *et al.*, 1973).

Thus *Octopus vulgaris* seems to be colour blind (as does *O. apollyon*, Roffe, 1975), which is in agreement with a whole variety of anatomical, physiological and biochemical evidence (reviewed in Messenger, 1979b, 1981) that there seems to be only one kind of photoreceptor with only one kind of photopigment.

Of course, whether all cephalopods are colour blind is not known. The recent evidence (see above) for three different visual pigments in the eye of the firefly squid *Watasenia scintillans* suggests that this species at least can resolve wavelength differences, specifically those that relate to light produced by its own light organs. This poses the question: if one deep-sea cephalopod has been under selective pressure to separate differences in wavelength, why have not the majority of shallow-water forms, which have to compete with other animals with good colour vision, notably teleosts?

Visually directed behaviour

Sense organs, including eyes, provide their owners with the information necessary to maintain life by helping lead them away from their enemies, towards their food or to each other.

Many of the more primitive molluscs use light cues to lead them to safety and we have seen that shadow responses are widespread among them. They may be mediated by extra-ocular photoreceptors (see pp 364–367) or by distinct ocelli, or non-cephalic eyes (see pp 368–371). More elaborate eyes, such as those of the ark-shells or scallops, can detect objects at short distances without a shadow falling on them so that the animal can close its shell valves for greater safety. Coleoid cephalopods also use their eyes to avoid their enemies but they are able to detect predators at much greater distances and take appropriate avoiding action earlier.

In at least one gastropod, the march periwinkle, *Littorina irrorata*, there is good evidence that the eyes are used to locate the plant stems up which the animals climb in order to avoid predators on the sea bed (Hamilton, 1977, 1978). This species and two other prosobranchs have also been shown to respond to simple shape stimuli and seem to prefer vertical to horizontal lines (Hamilton and Winter, 1984). Data like these teach us that vision may be more important in the lives of gastropods than is usually thought (Audesirk and Audesirk, 1985). For example, in the giant clam *Tridacna*, Wilkens (1986) gives a fascinating account of surprisingly complex visual behaviour. Shadows without abrupt changes in intensity or movement elicit local, slow retraction responses from the siphonal lobes projecting beyond the shell valves. A strong shadow, or a distant movement, elicits a synchronous contraction of the entire brightly coloured siphon and rapid adduction of the valves, forcing out a jet of water startling to a human observer. There is also a 'phototropic response', where the siphon is extended as a result of increase in light intensity. Wilkens also noted responses to passing clouds, and concludes that somehow the *Tridacna* nervous system must be complex enough to discriminate both the spatial and temporal features of visual stimuli received by its thousands of eyes.

The eyes of many molluscs are used to move the animal into lighter or dark regions of the environment, perhaps primarily for safety, perhaps for food. Scallops, for example, and *Nautilus*, make these kinds of simple 'phototactic' responses. But there is a certain amount of plasticity in these taxes. In the opisthobranchs, for example, Miller and Hadfield (1986) found that the positive phototaxis shown by *Phestilla* larvae decreases with age and in *Hermissenda* the detailed experiments of Crow (1985a,b) have shown that although increasing light intensity produces increased positive phototaxis this can be altered by conditioning, probably as a result of membrane changes in the B-photoreceptors (Fig. 17.31). More complex orientation movements are shown by the sea-hare *Aplysia brasiliana*, which can, in daylight, maintain a straight swimming path in the sea at least partly by using celestial cues (Hamilton and Russell, 1982a,b). The function of such movements could be to move the animal offshore, and hence away from danger.

Although coleoid cephalopods are well known to use their eyes to detect and capture prey, sometimes after binocular fixation (Messenger, 1968), this seems not to be a general rule for molluscs and most chitons and gastropods appear to rely on other kinds of information to find their food. There is the notable exception of the Heteropoda, with superb eyes and a curious scanning retina (p 275).

The kinds of visual cues that cephalopods use to detect their prey include movement, form (with perhaps some kind of inbuilt visual preferences) and brightness. Here we should recall the fact that their polarized-light sensitivity could enable them to 'break' the camouflage provided by fish scales (Denton, 1970) and make the fish more, not less, contrasting to the background (see Land, 1984). Incidentally, the use to which polarized light sensitivity is put is far from clear in these and other molluscs. For example, *Littorina* is known to be able to detect the plane of polarization (Charles, 1961a,b) and it is possible that these and other littoral gastropods could use this ability to complement sun compass reactions for orientation (Charles, 1966).

In some animals vision is also very important for finding a mate and for maintaining social organization. However, there is no evidence that any of the 'lower' molluscs utilize visual cues for such an end. Only in the cephalopods do we find animals that recognize their mates visually and even indulge in sexual displays. The cuttlefish, *Sepia*, is the best known example (L. Tinbergen, 1939; Hanlon and Messenger, 1988). Sexually mature males are black-and-white striped along their sides and along the lateral, fourth arm. Mature females lack the striping on the fourth arm and, as such, are recognized by courting males. There are other examples of body patterns communicating sex and species identity, especially in squids that school (Hanlon, 1982). The very fact that cephalopods have evolved elaborate chromatophore displays for intraspecific signalling reveals the importance of visual information in their life.

Finally it should be noted that many mid-water cephalopods exhibit diel rhythms that seem to be driven by light and that some cephalopods undertake long migrations prior to mating or spawning. Like sexual maturation itself, these may also be light-driven. And the navigation of such migratory schools of squids may also be brought about by visual cues, perhaps involving polarized light sensitivity. If this seems too speculative a note on which to end, it is worth remembering that there is a great deal that we do not know about the vision of cephalopods and, indeed, of the other molluscs with their curious and often underrated eyes.

owners. As Land puts it: 'If limpets needed better eyes they have had a very long time to acquire them.'

What we can learn from studying molluscan eyes are a few general principles, first among which is the zoologists' *cri de coeur*: that there are more ways than one of killing a cat. The emphasis throughout this account has been on the variety of devices that have evolved in the molluscs to collect photons. We have seen that many of them do not involve a lens. There are also many eyes that have a lens which, because of its nature and size, or the geometry of the eye, cannot form an image. However, wherever it is necessary to form an image, we find inhomogeneous lenses. And where good resolution is important we find large numbers of comparatively small receptors. In the retina of such eyes there is usually provision for lateral interaction. Complex visual processing requires additional hardware centrally, and the optic lobes of cephalopods are the largest of any invertebrate. Cephalopods also move their eyes in relation to the environment in a way very like a vertebrate, suggesting the importance of maintaining visual stability in any situation where form vision is important. Finally, the close parallels between the molecular biology of photoreception in squids and vertebrates (Cooper *et al.*, 1986) must also alert us to the constraints operating in any biological system that has only a limited number of ways of achieving similar ends (Pantin, 1951).

Acknowledgements

I am especially grateful to Lucia Misasi of the Stazione Zoologica di Napoli and to Sue Boxall of the Library of the Marine Biological Association, Plymouth for their invaluable help in conducting literature searches for me, a service unavailable to me at the Sheffield University Library because of its cost. I am also deeply indebted to Josie Healey and Angela Sellars for their help in producing the typescript. Finally it is a pleasure to record that this article was completed in the beautiful Harold A Miller library at the Hopkins Marine Station, Pacific Grove, California.

Summary

As we emphasized at the outset there is little point in scouring these pages for clues to the evolution of vertebrate eyes. Nor can we expect a neat evolutionary story within the mollusca, for we know little about the relationships of the various molluscan classes. Land (1984) has drawn attention to the fact that the three major groups of molluscs have been distinct and separate for 540 Myr. All three of these groups evidently found that the possession of photoreceptors could usefully improve their chances of survival but the degree of sophistication required for such organs was dependent on the habits and behaviour of their

References

Akimushkin, I. I. (1965). *Cephalopods of the Seas of the USSR*. Israel program for scientific translations. Jerusalem.

Aldred, R. G., Nixon, M. and Young, J. Z. (1978). The blind octopus, *Cirrothauma*. *Nature Lond.*, **275**, 547–549.

Alexandrowicz, J. S. (1928). Sur la fonction des muscles intrinsiques de l'oeil de céphalopodes. *C.R. Séanc. Soc. Biol.*, **99**, 1161–1163.

Alkon, D. L. (1987). *Memory Traces in the Brain*. Cambridge: University Press.

Allen, A., Michels, U. and Young, J. Z. (1985). Memory and visual discriminaton by squids. *Mar. Behav. Physiol.*, **11**, 271–282.

Andresen, M. C., Brown, A. M. (1979). Photoresponses of a sensitive extraretinal photoreceptor in *Aplysia*. *J. Physiol. Lond.*, **287**, 267–282.

Andresen, M. C., Brown, A. M. and Yasui, S. (1979). The role of diffusion in the photoresponse of an extraretinal photoreceptor of *Aplysia. J. Physiol. Lond.*, 287, 283–301.

Arvanitaki, A. and Chalazonitis, N. (1961). Excitatory and inhibitory processes initiated by light and infra-red radiations in single identifiable nerve cells (giant ganglion cells in *Aplysia*). In *Nervous Inhibition.* Ed. Florey, E. New York: Pergamon.

Asai, H., Arai, T., Fujii, T. and Matsumoto, G. (1989). Purification and characterisation of Ca^{2+}/calmodulin-dependent actin-binding proteins from squid retina. *FEBS Lett.*, 247, 377–380.

Audesirk, T. and Audesirk, G. (1985). Behaviour of gastropod molluscs. In *The Mollusca* Vol. 8. Ed. Willows, A. O. D. pp. 1–94. London: Academic Press.

Barber, V. C., Evans, E. M. and Land, M. F. (1967). The fine structure of the eye of the mollusc *Pecten maximus. Z. Zellforsch.*, 76, 295–312.

Barber, V. C. and Land, M. F. (1967). Eye of the cockle, *Cardium edule*: anatomical and physiological investigations. *Experientia*, 23, 677.

Barber, V. C. and Wright, D. E. (1969). The fine structure of the eye and optic tentacle of the mollusc *Cardium edule. J. Ultrastruct. Res.*, 26, 515–528.

Bauer, K. (1976). A behavioural light response in the chiton *Stenoplax heathiana. Veliger*, 18, suppl., 74–78.

Baumann, F., Mauro, A., Millecchia, R., Nightingale, S. and Young, J. Z. (1970). The extraocular light receptors of the squids *Todarodes* and *Illex. Brain Res.*, 21, 275–279.

Baylor, D. A. and Fuortes, M. G. F. (1970). Electrical responses of single cones in the retina of the turtle. *J. Physiol. Lond.*, 207, 77–92.

Beck, A. (1899). Über die bei Belichtung der Netzhaut von *Eledone moschata* entstehenden Aktionsströme. *Pflügers Arch.*, 78, 129–162.

Block, G. D. and Friesen, W. (1981). Electrophysiology of *Bulla* eyes: circadian rhythm and intracellular responses to illumination. *Neurosci. Abstr.*, 7, 45.

Block, G. D. and Roberts, M. H. (1981). Circadian pacemaker in the *Bursatella* eye: properties of the rhythm and its effect on locomotor behaviour. *J. Comp. Physiol.*, 142, 403–410.

Block, G. D. and Smith, J. T. (1973). Cerebral photoreceptors in *Aplysia. Comp. Biochem. Physiol.*, 46 A, 115–121.

Block, G. D., McMahon, D. G., Wallace, S. F. and Friesen, W. O. (1984). Cellular analysis of the *Bulla* ocular circadian pacemaker system. I A model for retinal organization. *J. Comp. Physiol. A*, 155, 365–378.

Blumrich, J. (1891). Das Integument der Chitonen. *Z. Wissen. Zool.*, 52, 404–476.

Boycott, B. B. (1961). The functional organisation of the brain of the cuttlefish, *Sepia officinalis. Proc. R. Soc. Lond. (Biol.)*, 153, 503–534.

Boycott, B. B. and Young, J. Z. (1956a). Subpedunculate body and nerve and other organs associated with the optic tract of cephalopods. In *Bertil Hanström: Zoological papers in honour of his sixty-fifth birthday.* Ed. Wingstrand, K. G. pp. 76–102. Lund: Zoological Institute.

Boycott, B. B. and Young, J. Z. (1956b). Reactions to shape in *Octopus vulgaris* Lamarck. *Proc. Zool. Soc. Lond.*, 126, 491–547.

Boycott, B. B. and Young, J. Z. (1957). Effects of interference with the vertical lobe on visual discriminations in *Octopus vulgaris* Lamarck. *Proc. R. Soc. Lond. (Biol.)*, 146, 439–459.

Boycott, B. B., Lettvin, J., Maturana, H. and Wall, P. D. (1965). *Octopus* optic responses. *Exp. Neurol.*, 12, 247–256.

Boyle, P. R. (1969a). Rhabdomeric ocellus in a chiton. *Nature Lond.*, 222, 895–896.

Boyle, P. R. (1969b). Fine structure of the eyes of *Onithochiton neglectus* (Mollusca: Polyplacophora). *Z. Zellforsch.*, 102, 313–332.

Boyle, P. R. (1972). The aesthetes of chitons. I. Role in the light response of whole animals. *Mar. Behav. Physio.*, 1, 171–184.

Boyle, P. R. (1977). Physiology and behaviour of chitons. *Oceanogr. Mar. Biol. Ann. Rev.*, 15, 461–509.

Brandenberger, J. L. (1975). Two new kinds of retinal cells in the eye of a snail, *Helix aspersa. J. Ultrastruct. Res.*, 50, 216–230.

Braun, R. (1954). Zum Lichtsinn facettenaugentragender Muscheln. *Zool. Jb., (Zool. Physiol.)*, 65, 91–125.

Breneman, J. W., Robles, L. V. and Bok, D. (1986). Light-activated retinoid transport in cephalopod photoreceptors. *Exp. Eye Res.*, 42, 645–658.

Brown, A. C. and Webb, S. C. (1985). The dark response of the whelk *Bullia digitalis* (Dillwyn). *J. mollusc. Stud.*, 51, 351–352.

Brown, A. M. and Brown, H. M. (1973). Light response of a giant *Aplysia* neuron. *J. Gen. Physiol.*, 62, 239–254.

Brown, A. M., Brodwick, M. S. and Eaton, D. C. (1977). Intracellular calcium and extra-retinal photoreception in *Aplysia* giant neurons. *J. Neurobiol.*, 8, 1–18.

Brown, J. E., Watkins, D. C. and Malbon, C. C. (1987). Light-induced changes in the content of inositol phosphates in squid (*Loligo pealei*) retina. *Biochem. J.*, 247, 293–297.

Brown, P. K. and Brown, P. S. (1958). Visual pigments of the octopus and cuttlefish. *Nature Lond.*, 182, 1288–1290.

Buchanan, J. A. (1986). Ultrastructure of the larval eyes of *Hermissenda crassicornis* (Mollusca: Nudibranchia). *J. Ultrastruct. Molec. Res.*, 94, 52–62.

Buddenbrock, W. and von Moller-Racke, I. (1953). Über den Lichtsinn von *Pecten. Pubbl. Staz. Zool. Napoli*, 24, 217–245.

Budelmann, B.-U. (1977). Structure and function of the angular acceleration receptor systems in the statocysts of cephalopods. *Symp. Zool. Soc. Lond.*, 38, 309–324.

Budelmann, B.-U. and Young, J. Z. (1984). The statocyst-oculomotor system of *Octopus vulgaris*: extraocular eye muscles, eye muscle nerves, statocyst nerves and the oculomotor centre in the central nervous system. *Phil. Trans. R. Soc. Lond. B*, 306, 159–189.

Budelmann, B.-U., Sachse, M. and Staudigl, M. (1987). The angular acceleration receptor system of the statocyst of *Octopus vulgaris*: morphometry, ultrastructure and neuronal and synaptic organisation. *Phil. Trans. R. Soc. Lond. B*, 315, 305–343.

Bullock, T. H. (1965). The mollusca. In *Structure and Function in the Nervous Systems of Invertebrates.* Eds Bullock, T. H. and Horridge, G. A. pp. 1273–1518. San Francisco: Freeman.

Cajal, S. Ramón y. (1917). Contribución al conocimiento de la rétina y centros ópticos de los cefalópodos. *Trab. Lab. Invest. Biol. Madr.*, 15, 1–82.

Charles, G. H. (1961a). The orientation of *Littorina* species to polarized light. *J. Exp. Biol.*, 38, 189–202.

Charles, G. H. (1961b). The mechanism of orientation of freely moving *Littorina littoralis* (L.) to polarized light. *J. Exp. Biol.*, 38, 203–212.

Charles, G. H. (1966). Sense organs (less cephalopods) In *Physiology of the Mollusca*, vol. 2. Eds Wilbur, K. M. and Yonge, C. M. pp. 455–521. London: Academic Press.

Chia, F.-S. and Koss, R. (1983). Fine structure of the larval eyes of *Rostanga pulchra* (Mollusca, Opisthobranchia, Nudibranchia). *Zoomorphology*, 102, 1–10.

Cohen, A. I. (1973a). An ultrastructural analysis of the photoreceptors of the squid and their synaptic connections. I. Photoreceptive and non-synaptic regions of the retina. *J. Comp. Neurol.*, 147, 351–378.

Cohn, A. I. (1973b). An ultrastructural analysis of the photoreceptors of the squid and their synaptic connections. II. Intraretinal synapses and plexus. *J. Comp. Neurol.*, 147, 379–398.

Cohen, A. I. (1973c). An ultrastructural analysis of the photoreceptors of the squid and their synaptic connections. III. Photoreceptor terminations in the optic lobes. *J. Comp. Neurol.*, 147, 399–426.

Collewijn, H. (1970). Oculomotor reactions in the cuttlefish, *Sepia officinalis. J. Exp. Biol.*, 52, 369–384.

Cooper, A., Dixon, S. F. and Tsuda, M. (1986). Photoenergetics of octopus rhodopsin. Convergent evolution of biological photon counters? *Eur. Biophys. J.*, 13, 195–201.

Cornwall, M. C. and Gorman, A. L. F. (1983a). The cation selectivity and voltage dependence of the light-activated potassium conductance in scallop distal photoreceptor. *J. Physiol. Lond.*, 340, 287–305.

Cornwall, M. C. and Gorman, A. L. F. (1983b). Colour dependence of the early receptor potential in scallop distal photoreceptor. *J. Physiol. Lond.*, 340, 307–334.

Crisp, M. (1972). Photoreceptive function of an epithelial receptor in *Nassarius reticulatus* (Gastropoda, Prosobranchia). *J. Mar. Biol. Ass. UK*, 52, 437–442.

Crow, T. (1985a). Conditioned modification of phototactic behaviour in *Hermissenda* I Analysis of light intensity *J. Neurosci.*, 5, 209–214.

Crow, T. (1985b). Conditioned modification of phototactic behaviour in *Hermissenda* II Differential adaptation of B-photoreceptors. *J. Neurosci.*, 5, 215–233.

Dahmen, H. J. (1977). The menotactic orientation of the prosobranch mollusc *Littorina littorea*. *Biol. Cybernetics*, 26, 17–23.

Dakin, W. J. (1910). The eye of *Pecten*. *Q. J. Microsc. Sci.*, 55, 49–112.

Daw, N. W. and Pearlman, A. L. (1969). Receptive field studies of single units in the optic lobe of the squid *Loligo pealeii*. *Biol. Bull.*, 137, 398.

Daw, N. W. and Pearlman, A. L. (1974). Pigment migration and adaptation in the eye of the squid. *J. Gen. Physiol.*, 63, 22–36.

Denton, E. J. (1970). On the organization of reflecting surfaces in some marine animals. *Phil. Trans. R. Soc. Lond. B*, 258, 285–313.

Dilly, P. N. (1969). The structure of a photoreceptor organelle in the eye of *Pterotrachea mutica*. *Z. Zellforsch.*, 99, 420–429.

Dilly, P. N., Gray, E. G. and Young, J. Z. (1963). Electron microscopy of optic nerves and optic lobes of *Octopus* and *Eledone*. *Proc. R. Soc. Lond. B.*, 158, 446–456.

Duncan, G., and Pynsent, P. B. (1979). An analysis of the wave forms of photoreceptor potentials in the retina of the cephalopod *Sepiola atlantica*. *J. Physiol. Lond.*, 288, 171–188.

Duncan, G. and Weeks, F. I. (1973). Photoreception by a cephalopod retina in vitro. *Exp. Eye Res.*, 17, 183–192.

Eakin, R. M. (1972). Structure of invertebrate photoreceptors. In *Photochemistry of Vision. Handbook of Sensory Physiology*, vol. VII/I. Ed. Dartnall, H. J. A. pp. 625–684. Berlin: Springer.

Eakin, R. M. and Brandenburger, J. L. (1975a). Understanding a snail's eye at a snail's pace. *Amer. Zool.*, 15, 851–863.

Eakin, R. M. and Brandenburger, J. L. (1975). Retinal differences between light-tolerant and light-avoiding slugs (Mollusca: Pulmonata). *J. Ultrastruct. Res.*, 53, 382–394.

Eskin, A. and Harcombe, E. (1977). Eye of *Navanax*: optic activity, circadian rhythm, and morphology. *Comp. Biochem. Physiol.*, 57A, 443–449.

Feldman, S. C. (1986). Distribution of immunoreactive somatostatin (ISRIF) in the nervous system of the squid, *Loligo pealei*. *J. Comp. Neurol.*, 245, 238–257.

Föh, H. (1932). Der Schattenreflex bei *Helix pomatia*. *Zool. Jb. (Zool. v. Physiol.)*, 52, 1–78.

Fröhlich, F. W. (1914). Beiträge zur allgemeinen Physiologie der Sinnesorgane. *Z. Sinnesphysiol.*, 48, 28–164.

Fujimoto, K., Yanase, T., Okuno, Y. and Iwata, K. (1966). Electrical response in the *Onchidium* eyes. *Memoirs of Osaka Kyoiku Univ.*, (3), B15, 98–108.

Fukushima, H. (1985). Immunocytochemical localization of retinochrome in the visual cells of the squid. *Zool. Sci.*, 2, 333–340.

Gibson, B. L. (1984). Cellular and ultrastructural features of the adult and the embryonic eye in the marine gastropod, *Ilyanassa obsoleta*. *J. Morph.*, 181, 205–220.

Gillary, H. L. (1974). Light-evoked and electrical potentials from the eye and optic nerve of *Strombus*: response waveform and spectral sensitivity. *J. Exp. Biol.*, 60, 383–396.

Gillary, H. L. and Gillary, E. W. (1979). Ultrastructural features of the retina and optic nerve of *Strombus luhuanus*, a marine gastropod. *J. Morph.*, 159, 89–116.

Gorman, A. L. F. and McReynolds, J. S. (1978). Ionic effects on the membrane potential of hyperpolarizing photoreceptors in scallop retina. *J. Physiol. Lond.*, 275, 345–355.

Gotow, T. (1975). Morphology and function of the photo-excitable neurons in the central ganglia of *Onchidium verruculatum*. *J. Comp. Physiol.*, 99, 139–152.

Gotow, T. (1989). Photoresponses of an extraocular photoreceptor associated with a decrease in membrane conductance in an opisthobranch mollusc. *Brain Res.*, 479, 120–129.

Grassé, P. P. (1968). *Traité de Zoologie*, vol. 5 (3). Paris: Masson.

Haghighet, N., Cohen, R. S. and Pappas, G. D. (1984). Fine structure of squid (*Loligo pealei*) optic lobe synapses. *Neurosci.*, 13, 527–546.

Hagins, F. M. (1973). Purification and partial characterization of the protein component of squid rhodopsin. *J. Biol. Chem.*, 248, 3298–3304.

Hagins, W. A. (1965). Electrical signs of information flow in photoreceptors. *Cold Spring Harbor Symp. Quant. Biol.*, 30, 403–418.

Hagins, W. A., Liebman, P. A. (1962). Light-induced pigment migration in the squid retina. *Biol. Bull.*, 123, 498.

Hagins, W. A., Zonana, H. V. and Adams R. G. (1962). Local membrane current in the outer segment of squid photoreceptors. *Nature Lond.*, 194, 844–847.

Hall, M. D., Hoon, M. A., Ryba, N. J. P., Pottinger, J. D. D., Keen, J. N., Saibil, H. R. and Findlay, J. B. C. (1991). Squid rhodopsin. Molecular cloning and primary structure of a phospholipase-C directed G-protein linked receptor. *Biochem. J.*, 273, (in press).

Hamasaki, D. I. (1968). The electroretinogram of the intact anesthetized octopus. *Vision Res.*, 8, 247–258.

Hamdorf, K., Schwemer, J. and Taüber, U. (1968). Der Sehfarbstoff, die Absorption der Rezeptoren und die spektrale Empfindlichkeit der Retina von *Eledone moschata*. *Z. Vergl. Physiol.*, 60, 375–415.

Hamilton, P. V. (1977). Daily movements and visual location of plant stems by *Littorina irrorata* (Mollusca: Gastropoda). *Mar. Behav. Physiol.*, 4, 293–304.

Hamilton, P. V. (1978). Adaptive visually-mediated movements of *Littorina irrorata* (Mollusca: Gastropoda) when displaced from their natural habitat. *Mar. Behav. Physiol.*, 5, 255–271.

Hamilton, P. V., Ardizzoni, S. C. and Penn, J. S. (1983). Eye structure and optics in the intertidal snail, *Littorina irrorata*. *J. Comp. Physiol.*, 152, 435–445.

Hamilton, P. V. and Russell, B. J. (1982a). Field-experiments on the sense organs and directional cues involved in off-shore-oriented swimming by *Aplysia brasiliana* Rang (Mollusc: Gastropoda) *J. Exp. Mar. Biol. Ecol.*, 56, 123–143.

Hamilton, P. V. and Russell, B. J. (1982b). Celestial orientation by surface-swimming *Aplysia brasiliana* Rang (Mollusca: Gastropoda). *J. Exp. Mar. Biol. Ecol.*, 56, 145–152.

Hamilton, P. V. and Winter, M. A. (1984). Behavioural responses to visual stimuli by the snails *Tectarius muricatus*, *Turbo castanea* and *Helix aspersa*. *Anim. Behav.*, 32, 51–57.

Hanlon, R. T. (1982). The functional organization of chromatophores and iridescent cells in the body patterning of *Loligo plei* (Cephalopoda, Myopsida). *Malacologia*, 23, 89–119.

Hanlon, R. T. and Messenger, J. B. (1988). Adaptive coloration in young cuttlefish (*Sepia officinalis* L): the morphology and development of body patterns and their relation to behaviour. *Phil. Trans. R. Soc. Lond. B*, 320, 437–487.

Hara, R., Hara, T., Tokunaga, F. and Yoshizawa, T. (1981). Photochemistry of retinochrome. *Photochem. Photobiol.*, 33, 883–891.

Hara, T. and Hara, R. (1965). New photosensitive pigment found in the retina of the squid *Ommastrephes*. *Nature Lond.*, 206, 1331–1334.

Hara, T. and Hara, R. (1972). Cephalopod retinochrome. In *Handbook of Sensory Physiology*, vol. VII/I. Ed. Dartnall, H. J. A. pp. 720–746. Berlin: Springer.

Hara, T. and Hara, R. (1976). Distribution of rhodopsin and retinochrome in the squid retina. *J. Gen. Physiol.*, 67, 791–805.

Hara, T. and Hara, R. (1980). Retinochrome and rhodopsin in the extraocular photoreceptor of the squid, *Todarodes*. *J. Gen. Physiol.*, 75, 1–19.

Hara, T., Hara, R. and Takeuchi, J. (1967). Rhodopsin and retinochrome in the octopus retina. *Nature Lond.*, 214, 572–573.

Hara-Nishimura, I., Matsumoto, T., Mori, H., Nishimura, N., Hara, R. and Hara, T. (1990). Cloning and nucleotide sequence of cDNA for retinochrome, retinal photoisomerase from the squid retina. *FEBS Letters*, 271, 106–110.

Hartline, H. K. (1938). The discharge of impulses in the optic nerve of *Pecten* in response to illumination of the eye. *J. Cell. Comp. Physiol.*, 11, 465–478.

Hartline, P. H. and Lange, G. D. (1974). Optic nerve responses to visual stimuli in squid. *J. Comp. Physiol.*, 93, 37–54.

Hartline, P. H. and Lange, G. D. (1977). Sinusoidal analysis of electroretinogram of squid and octopus. *J. Neurophysiol.*, 40, 174–187.

Hartline, P. H., Hurley, A. C. and Lange, G. D. (1979). Eye stabilization by statocyst mediated oculomotor reflex in *Nautilus*. *J. Comp. Physiol.*, 132, 117–126.

Hartline, P. H. and Lange, G. D. (1984). Visual systems of cephalopods. In *Comparative Physiology of Sensory Systems*. Eds Bolis, L., Keynes, R. D. and Maddrell, S. H. P. pp. 335–355. Cambridge: Cambridge University Press.

Herman, K. G. and Strumwasser, F. (1984). Regional specializations in the eye of *Aplysia*, a neuronal circadian oscillator. *J. Comp. Neurol.*, 230, 593–613.

Hermann, H. T. (1968). Optic guidance of locomotor behavior in the land snail, *Otala lactea*. *Vision Res.*, 8, 601–612.

Hesse, R. (1900). Untersuchungen über die Organe der Lichtempfindung bei niederen Thieren. VI: Die Augen einiger Mollusken. *Z. Wiss. Zool.*, 68, 379–477.

Hisano, N., Tateda, H. and Kuwabara, M. (1972a). Photosensitive neurones in the marine pulmonate snail *Onchidium verruculatum*. *J. Exp. Biol.*, 57, 651–660.

Hisano, N., Tateda, H. and Kuwabara, M. (1972b). An electrophysiological study of the photo-excitative neurones of *Onchidium verruculatum*. *J. Exp. Biol.*, 57, 661–671.

Hubbard, R. and St George, R. C. C. (1958). The rhodopsin system of the squid. *J. Gen. Physiol.*, 41, 501–528.

Hughes, H. P. I. (1970). A light and electron microscope study of some opisthobranch eyes. *Z. Zellforsch.*, 106, 79–98.

Hurley, A. C., Langle, G. D. and Hartline, P. H. (1978). The adjustable 'pinhole camera' eye of *Nautilus*. *J. Exp. Zool.*, 205, 37–44.

Jacklet, J. W. (1969). Circadian rhythm of optic nerve impulses recorded in darkness from isolated eye of *Aplysia*. *Science Wash. DC*, 164, 562–563.

Jacklet, J. W. (1980). Light sensitivity of the rhinophores and eyes of *Aplysia*. *J. Comp. Physiol.*, 136, 257–262.

Jacklet, J. W. (1984). Neuronal organization and cellular mechanisms of circadian pacemakers. *Int. Rev. Cytol.*, 89, 251–294.

Jander, R., Daumer, K. and Waterman, T. H. (1963). Polarized light orientation in two Hawaiian decapod cephalopods. *Z. Vergl. Physiol.*, 46, 383–394.

Jordan, W. P., Lickey, M. E. and Hiaasen, S. D. (1985). Circadian organization in *Aplysia*: internal desynchronization and amplitude of locomotor rhythm. *J. Comp. Physiol. A*, 156, 293–303.

Kandel, E. R. (1976). *Cellular Basis of Behavior*. San Francisco: Freeman.

Katagiri, N. (1984). Cytoplasmic characteristics of three different photoreceptor cells in a marine gastropod, *Onchidium verruculation*. *J. Electron Microsc.*, 33, 142–150.

Kataoka, S. (1975). Fine structure of the retina of a slug, *Limax flavus* L. *Vision Res.*, 15, 681–686.

Kataoka, S., Yamamoto, T. Y., Tonosaki, A. and Washioka, H. (1982). Ultrastructural study of photoreceptor membrane turnover in an invertebrate retina. In *The Structure of the Eye*. Ed. Hollyfield, J. G. pp. 35–44. New York: Elsevier.

Kataoka, S. and Yamamoto, T. Y. (1981). Diurnal changes in the fine structure of photoreceptors in an abalone, *Nordotis discus*. *Cell Tiss. Res.*, 218, 181–189.

Kataoka, S. and Yamamoto, T. Y. (1983). Fine structure and formation of the photoreceptor in *Octopus ocellatus*. *Biol. Cell.*, 49, 45–54.

Kennedy, D. (1960). Neural photoreception in a lamellibranch mollusc. *J. Gen. Physiol.*, 44, 277–299.

Khalsa, S. B. S. and Block, G. D. (1990). Calcium in phase control of the *Bulla* circadian pacemaker. *Brain Res.*, 506, 40–45.

Kito, Y., Seki, T., Suzuki, T. and Uchiyama, J. (1986). 3-Dehydroretinal in the eye of a bioluminescent squid, *Watasenia scintillans*. *Vision Res.*, 26, 275–279.

Kito-Yamashita, T., Haga, G., Hirai, K., Uemura, T., Kondo, H. and Kosaka, K. (1990). Localization of serotonin immunoreactivity in cephalopod visual system. *Brain Res.*, 521, 81–88.

Krauhs, J. M., Sordahl, L. A. and Brown, A. M. (1977). Isolation of pigmented granules involved in extra-retinal photoreception in *Aplysia californica* neurons. *Biochim. Biophys. Acta*, 471, 25–31.

Kropf, A., Brown, P. K. and Hubbard, R. (1959). Lumi- and metarhodopsins of squid and octopus. *Nature Lond.*, 183, 446–450.

Lam, D. M. K., Wiesel, T. N. and Kaneko, A. (1974). Neurotransmitter synthesis in cephalopod retina. *Brain Res.*, 82, 365–368.

Land, M. F. (1965). Image formation by a concave reflector in the eye of the scallop, *Pecten maximus*. *J. Physiol. Lond.*, 179, 138–153.

Land, M. F. (1966a). Activity in the optic nerve of *Pecten maximus* in response to changes in light intensity, and to pattern and movement in the optical environment. *J. Exp. Biol.*, 45, 83–99.

Land, M. F. (1966b). A multilayer interference reflector in the eye of the scallop, *Pecten maximus*. *J. Exp. Biol.*, 45, 433–447.

Land, M. F. (1982). Scanning eye movements in a heteropod mollusc. *J. Exp. Biol.*, 96, 427–430.

Land, M. F. (1984). Molluscs. In *Photoreception and vision in invertebrates*. Ed. Ali, M. A. pp. 699–725. New York: Plenum.

Lickey, M. E., Block, G. D., Hudson, D. J. and Smith, J. T. (1976). Circadian oscillators and photoreceptors in the gastropod, *Aplysia*. *Photochem. Photobiol.*, 23, 253–273.

Lickey, M. E., Wozniak, J. A., Block, G. D., Hudson, D. J. and Augter, G. K. (1977). The consequences of eye removal for the circadian rhythm of behavioural activity in *Aplysia*. *J. Comp. Physiol.*, 118, 121–143.

Lindberg, D. R., Kellog, M. G. and Hughes, W. E. (1975). Evidence of light reception through the shell of *Notoacmea persona* (Rathke, 1833). *Veliger*, 17, 383–386.

Lukowiak, K. and Jacklet, J. W. (1972). Habituation: a peripheral and central nervous system process in *Aplysia*. *Fed. Proc.*, 31, 405 (abstract).

McReynolds, J. S. and Gorman, A. L. F. (1970a). Photoreceptor potentials of opposite polarity in the eye of the scallop, *Pecten irradians*. *J. Gen. Physiol.*, 56, 376–391.

McReynolds, J. S. and Gorman, A. L. F. (1970b). Membrane conductances and spectral sensitivities of *Pecten* photoreceptors. *J. Gen. Physiol.*, 56, 392–406.

McVean, A. (1984). *Octopus* extraocular muscle. *Comp. Biochem. Physiol.*, 78A, 711–718.

Matsui, S., Seidou, M., Horiuchi, S., Uchiyama, I. and Kito, Y. (1988). Adaptation of a deep-sea cephalopod to the photic environment. *J. Gen. Physiol.*, 92, 55–66.

Mauro, A. (1977). Extra-ocular photoreceptors in cephalopods. *Symp. Zool. Soc. Lond.*, 38, 287–308.

Mauro, A. and Baumann, F. (1968). Electrophysiological evidence of photoreceptors in the epistellar body of *Eledone moschata*. *Nature Lond.*, **220**, 1332–1334.

Mauro, A. and Sten-Knudsen, O. (1972). Light-evoked impulses from extra-ocular photoreceptors in the squid *Todarodes*. *Nature Lond.*, **237**, 342–343.

Messenger, J. B. (1967). Parolfactory vesicles as photoreceptors in a deep-sea squid. *Nature Lond.*, **213**, 836–838.

Messenger, J. B. (1968). The visual attack of the cuttlefish, *Sepia officinalis*. *Anim. Behav.*, **16**, 342–357.

Messenger, J. B. (1970). Optomotor responses and nystagmus in intact, blinded, and statocystless cuttlefish (*Sepia officinalis* L.) *J. Exp. Biol.*, **53**, 789–796.

Messenger, J. B. (1977a). Prey-capture and learning in the cuttlefish. *Sepia*. *Symp. Zool. Soc. Lond.*, **38**, 347–376.

Messenger, J. B. (1977b). Evidence that *Octopus* is colour-blind. *J. Exp. Biol.*, **70**, 49–55.

Messenger, J. B. (1979a). The nervous system of *Loligo* IV. The peduncle and olfactory lobes. *Phil. Trans. R. Soc. Lond. (Biol.)*, **285**, 275–309.

Messenger, J. B. (1979b). The eyes and skin of *Octopus*: compensating for sensory deficiencies. *Endeavour*, **3**, 92–98.

Messenger, J. B. (1981). Comparative physiology of vision in molluscs. In *Handbook of Sensory Physiology*. Vol. VII/6C. Ed. Autrum, H. pp. 93–200. Berlin: Springer.

Messenger, J. B. and Sanders, G. D. (1971). The inability of *Octopus vulgaris* to discriminate monocularly between oblique rectangles. *J. Neurosci.*, **1**, 171–173.

Messenger, J. B., Wilson, A. P. and Hedge, A. (1973). Some evidence for colour-blindness in *Octopus*. *J. Exp. Biol.*, **59**, 77–94.

Miller, S. E. and Hadfield, M. G. (1986). Ontogeny of phototaxis and metamorphic competence in larvae of the nudibranch *Phestilla sibogae* Bergh (Gastropoda: Opisthobranchia). *J. Exp. Mar. Biol. Ecol.*, **97**, 95–112.

Moody, M. F. (1962). Evidence for the intraocular discrimination of vertically and horizontally polarized light by *Octopus*. *J. Exp. Biol.*, **39**, 21–30.

Moody, M. F. and Parriss, J. R. (1960). The discrimination of polarized light by *Octopus*. *Nature Lond.*, **186**, 839–840.

Moody, M. F. and Parriss, J. R. (1961). The discrimination of polarized light by *Octopus*: a behavioural and morphological study, *Z. Vergl. Physiol.*, **44**, 268–291.

Moody, M. F. and Robertson, J. D. (1960). The fine structure of some retinal photoreceptors.. *J. Biophys. Biochem. Cytol.*, **7**, 87–92.

Morton, J. E. (1967). *Molluscs*, 4th edn. London: Hutchinson.

Moseley, H. N. (1885). On the presence of eyes in the shells of certain Chitonidae and on the structure of these organs. *Q. J. Microsc. Sci.*, **25**, 37–60.

Mpitsos, G. J. (1973). Physiology of vision in the mollusk *Lima scabra*. *J. Neurophysiol.*, **36**, 371–383.

Muntz, W. R. A. (1977). Pupillary response of cephalopods. *Symp. Zool. Soc. Lond.*, **38**, 277–285.

Muntz, W. R. A. (1986). The spectral sensitivity of *Nautilus pompilius*. *J. Exp. Biol.*, **126**, 513–517.

Muntz, W. R. A. (1987). Visual behaviour and visual sensitivity of *Nautilus pompilius*. In *Nautilus. The Biology and Paleobiology of a Living Fossil*. Eds Saunder, W. B. and Landman, N. H. pp. 231–244. New York: Plenum.

Muntz, W. R. A. and Raj, U. (1984). On the visual system of *Nautilus pompilius*. *J. Exp. Biol.*, **109**, 253–263.

Muntz, W. R. A. and Guyther, J. (1988). Visual acuity in *Octopus pallidus* and *Octopus australis*. J. Exp. Biol., **134**, 119–129.

Newell, G. E. (1965). The eye of *Littorina littorea*. *Proc. Zool. Soc. Lond.*, **144**, 75–86.

Newell, P. F. and Newell, G. E. (1968). The eye of the slug, *Agriolimax reticulatus* (Müll). *Symp. Zool. Soc. Lond.*, **23**, 97–111.

Nishi, T. and Gotow, T. (1989). A light-induced decrease of cyclic GMP in the photoresponse of a molluscan extraocular photoreceptor. *Brain Res.*, **485**, 185–188.

Nishioka, R., Hagadorn, I. R. and Bern, H. A. (1962). Ultrastructure of the epistellar body of the octopus. *Z. Zellforsch.*, **57**, 406–421.

Nishioka, R., Yasumasu, I., Packard, A., Bern, H. A. and Young, J. Z. (1966). Nature of vesicles associated with the nervous system of cephalopods. *Z. Zellforsch.*, **75**, 301–316.

Norton, A. C., Fukada, Y., Motokawa, K. and Tasaki, K. (1965). An investigation of the lateral spread of potentials in the octopus retina. *Vision Res.*, **5**, 253–267.

Osborne, N. N., Beaton, D. W., Boyd, P. J. and Walker, R. J. (1986). Substance P – like immunoreactivity in the retina and optic lobe of the squid. *Neurosci. Lett.*, **70**, 65–68.

Ovchinnikov, Yu. A., Abdulaev, N. G., Zolotarev, A. S., Artamanov, I. D., Bespalov, I. A., Dergachev, A. E. and Tsuda, A. (1988). Octopus rhodopsin. Amino acid sequence deduced from cDNA. *FEBS Letters*, **232**, 69–72.

Ozaki, K., Hara, R. and Hara, T. (1983). Histochemical localization of retinochrome and rhodopsin studied by fluorescence microscopy. *Cell Tiss. Res.*, **233**, 335–345.

Ozaki, K., Terakita, A., Hara, R. and Hara, T. (1986). Rhodopsin and retinochrome in the retina of a marine gastropod, *Conomulex ruhuanus*. *Vision Res.*, **26**, 691–705.

Ozaki, K., Terakita, A., Hara, R. and Hara, T. (1987). Isolation and characterization of a retinal-binding protein from the squid retina. *Vision Res.*, **27**, 1057–1070.

Pantin, C. F. A. (1951). Organic design. *Adv. Sci.*, **8**, 138–150.

Pasic, M., Ristanovic, D., Zecevic, D. and Kartelija, G. (1977). Effects of light on identified *Helix pomatia* neurons. *Comp. Biochem. Physiol.*, **58A**, 81–85.

Patten, W. (1886). Eyes of molluscs and arthropods. *Mitt. Zool. Staz. Neapel.*, **6**, 542–756.

Patterson, J. A. and Silver, S. C. (1983). Afferent and efferent components of *Octopus* retina. *J. Comp. Physiol.*, **151**, 381–387.

Pinto, L. H. and Brown, J. E. (1977). Intracellular recordings from photoreceptors of the squid (*Loligo pealeii*). *J. Comp. Physiol.*, **122**, 241–250.

Pumphrey, R. J. (1961). Concerning vision. In *The Cell and The Organism*. eds Ramsay, J. A. and Wigglesworth, V. B. Cambridge: Cambridge University Press.

Rahat, M. and Monselise, E. B. (1979). Photobiology of the chloroplast hosting mollusc *Elysia timida* (Opisthobranchia). *J. Exp. Biol.*, **79**, 225–233.

Reynolds, P. D. (1988). The structure and distribution of ciliated sensory receptors in the scaphopoda (Mollusca). *Amer. Zool.*, **28**, 140a.

Robles, L. J., Cabebe, C. S., Aguilo, J. A., Anyakora, P. A. and Bok, D. (1984). Autoradiographic and biochemical analysis of photoreceptor membrane renewal in *Octopus* retina. *J. Neurocytol.*, **13**, 145–164.

Robles, L. J., Watanabe, A., Kremer, N. E., Wong, F. and Bok, D. (1987). Immunocytochemical localization of photopigments in cephalopod retinae. *J. Neurocytol.*, **16**, 403–415.

Roffe, T. (1975). Spectral perception in *Octopus*: a behavioural study. *Vision Res.*, **15**, 353–356.

Rosen, M. D., Stasek, C. R. and Hermans, C. O. (1978). The ultrastructure and evolutionary significance of the cerebral ocelli of *Mytilus edulis*, the bay mussel. *Veliger*, **21**, 10–18.

Russell-Hunter, W. D. (1979). *The Life of Invertebrates*. New York: Macmillan.

Saibil, H. (1982). An ordered membrane cytoskeleton network in squid photoreceptor microvilli. *J. Molec. Biol.*, **158**, 435–456.

Saibil, H. (1990a). Cell and molecular biology of photoreceptors. *Semin. Neurosci.*, **2**, 15–23.

Saibil, H. (1990b). Structure and function of the squid eye. In *Squid as Experimental Animals*. Eds. Gilbert, D. L., Adelman, W. J. and

Arnold, J. M. New York: Plenum.
Saibil, H. and Hewat, E. (1987). Ordered transmembrane and extracellular structure in squid photoreceptor microvilli. *J. Cell Biol.*, 105, 19–28.
Saibil, H. and Michel-Villaz, M. (1984). Squid rhodopsin and GTP-binding protein crossreact with vertebrate photoreceptor enzymes. *Proc. Natl. Acad. Sci. USA*, 81, 5111–5115.
Saidel, W. M. (1979). Relationship between photoreceptor terminations and centrifugal neurons in the optic lobe of *Octopus*. *Cell Tissue Res.*, 204, 463–472.
Saidel, W. M., Lettvin, J. Y. and MacNichol, E. G. (1987). Processing of polarised light by squid photoreceptors. *Nature Lond.*, 304, 534–536.
Seidou, M., Sugahara, M., Uchiyama, H., Hiraki, K., Hamanaka, T., Michinomae, M., Yoshihara, K. and Kito, Y. (1990). On the three visual pigments in the retina of the firefly squid, *Watasenia scintillans*. *J. Comp. Physiol. A*, 166, 769–773.
Sherrington, C. S. (1906). *The Integrative Action of the Nervous System*. New Haven: Yale University Press.
Shimatani, Y., Katagiri, Y. and Katagiri, N. (1986). Light-evoked peripheral reflex of isolated mantle papilla of *Onchidium verruculatum* (Gastropoda, Mollusca). *Zool. Sci.*, 3, PH 126, p. 993.
Shimatani, Y., Katagiri, N., Katagiri, Y. and Suzuki, T. (1988). Histochemical localization and chromophore of pigments in multiple photoreceptive system in *Onchidium verriculatum*. *Zool. Sci.*, 5, PH 36, p. 1200.
Silver, S. C., Patterson, J. A. and Mobbs, P. G. (1983). Biogenic amines in cephalopod retina. *Brain Res.*, 273, 366–368.
Sivak, J. G. (1982). Optical properties of a cephalopod eye (the short-finned squid, *Illex illecebrosus*). *J. Comp. Physiol.*, 147, 323–327.
Spagnolia, T. and Wilkens, L. A. (1983). Neurobiology of the scallop II. Structure of the parietovisceral ganglion lateral lobes in relation to afferent projections from the mantle eyes. *Mar. Behav. Physiol.*, 10, 23–55.
Sperling, L. and Hubbard, R. (1975). Squid retinochrome. *J. Gen. Physiol.*, 65, 235–251
Sroczynski, S. and Muntz, W. R. A. (1985). Image structure in *Eledone cirrhosa*, an octopus. *Zool. Jb. Physiol.*, 89, 157–168.
Sroczynski, S. and Muntz, W. R. A. (1987). The optics of oblique beams in the eye of *Eledone cirrhosa*, an octopus. *Zool. Jb. Physiol.*, 91, 419–446.
Steiner, G. (1990). Observations on the anatomy of the scaphopod mantle. *Western Society of Malacologists Annual Report*, 22, 3.
Stoll, C. J. (1972). Sensory systems involved in the shadow response of *Lymnaea stagnalis* (L.) as studied with the use of habituation phenomena. *Proc. K. Ned. Akad. Wet. C*, 75, 342–351.
Stoll, C. J. (1973a). On the role of eyes and non-ocular light receptors in orientational behaviour of *Lymnaea stagnalis* (L.) *Proc. K. Ned. Akad. C*, 76, 203–214.
Stoll, C. J. (1973b). Observations on the ultrastructure of the eye of the basommatophoran snail, *Lymnaea stagnalis* (L.) *Proc. K. Ned. Akad. Wet. C*, 76, 414–424.
Sugawara, K., Katagiri, Y. and Tomita, T. (1971). Polarized light responses from octopus single retinular cells. *J. Fac. Sci. Hokkaido. Univ. Ser. VI Zool.*, 17, 581–586.
Sundermann, G. (1990). Development and hatching state of ectodemal vesicle-organs in the head of *Sepia officinalis*, *Loligo vulgaris* and *Loligo forbesi* (Cephalopoda, Decabrachia). *Zoomorph.*, 109, 343–352.
Sutherland, N. S. (1957). Visual discrimination of orientation and shape by the octopus. *Nature Lond.*, 179, 11–13.
Sutherland, N. S. (1958). Visual discrimination of the orientation of rectangles by *Octopus vulgaris* Lamarck. *J. Comp. Physiol. Psychol.*, 52, 135–141.
Sutherland, N. S. (1960). Theories of shape discrimination in *Octopus*. *Nature Lond.*, 186, 840–844.
Sutherland, N. S. (1963). Shape discrimination and receptive fields. *Nature Lond.*, 197, 118–122.
Sutherland, N. S. (1968). Outlines of a theory of visual pattern recognition in animals and man. *Proc. R. Soc. Lond. (Biol.)*, 171, 297–317.
Sutherland, N. S. (1969). Shape discrimination in rat, octopus, and goldfish: a comparative study. *J. Comp. Physiol. Psychol.*, 67, 160–176.
Sutherland, N. S. and Muntz, W. R. A. (1959). Simultaneous discrimination training and preferred directions of motion in visual discrimination of shape in *Octopus vulgaris* Lamarck. *Pubbl. Staz. Zool. Napoli.*, 31, 109–126.
Szuts, E. Z., Wood, S. F., Reid, M. S. and Fein, A. (1986). Light stimulates the rapid formation of inositol triphosphate in squid retinas. *Biochem. J.*, 240, 929–932.
Tamamaki, N. (1989a). Visible light reception of accessory eye in the giant snail, *Achatina fulica*, as revealed by an electrophysiological study. *Zool. Sci.*, 6, 867–875.
Tamamaki, N. (1989b). The accessory photosensory organ of the terrestrial slug, *Limax flavus* L. (Gastropoda, Pulmonata): morphological and electrophysiological study. *Zool. Sci.*, 6, 877–883.
Tansey, E. M. (1980). Aminergic fluorescence in the cephalopod brain. *Phil. Trans. R. Soc. B*, 291, 127–145.
Tasaki, I. and Nakaye, T. (1984). Rapid mechanical responses of the dark-adapted squid retina to light pulses *Science Wash. DC*, 223, 411–413.
Tasaki, K. and Karita, K. (1966a). Intraretinal discrimination of horizontal and vertical planes of polarized light by *Octopus*. *Nature*, 209, 934–935.
Tasaki, K. and Karita, K. (1966b). Discrimination of horizontal and vertical planes of polarized light by the cephalopod retina. *Jap. J. Physiol.*, 16, 205–216.
Tasaki, K., Oikawa, T. and Norton, A. C. (1963c). The dual nature of the octopus electroretinogram. *Vision Res.*, 3, 61–73.
Terakita, A., Hara, R. and Hara, T. (1989). Retinal-binding protein as a shuttle for retinal in the rhodopsin-retinochrome system of the squid visual cells. *Vision Res.*, 29, 639–652.
Tinbergen, L. (1939). Zur Fortpflanzungsethologie von *Sepia officinalis* L. *Archs Neérl. Zool.*, 3, 323–364.
Toyoda, J., Nosaki, H. and Tomita, T. (1969). Light induced resistance change in single photoreceptors of *Necturus* and *Gecko*. *Vision Res.*, 9, 453–463.
Trueman, E. R. and Clarke, M. R. (1985). *The Mollusca*, vol. 10, *Evolution*. pp. xiii–xiv. Orlando: Academic Press.
Tsuda, M., Tsuda, T., Terayama, Y., Fukuda, Y., Akino, T., Yamanaka, G., Stryer, L., Katada, T., Ui, M. and Ebrey, T. (1986). Kinship of cephalopod photoreceptor G-protein with vertebrate transducin. *FEBS Lett.*, 198, 5–10.
Tsukita, S., Tsukita, S. and Matsumoto, G. (1988). Light-induced structural changes of cytoskeleton in squid photoreceptor microvilli detected by rapid-freeze method. *J. Cell Biol.*, 106, 1151–1160.
Van Duivenboden, Y. A. (1982). Non-ocular photoreceptors and photo-orientation in the pond snail *Lymnaea stagnalis* (L.). *J. Comp. Physiol.*, 149, 363–368.
Vandenberg, C. and Montal, M. (1984a). Light-regulated biochemical events in invertebrate photoreceptors. 1. Light-activated guanosinetriphosphatase, guanine nucleotide binding, and cholera toxin catalyzed labelling of squid photoreceptor membranes *Biochemistry*, 23, 2339–2347.
Vandenberg, C. and Montal, M. (1984b). Light-regulated biochemical events in invertebrate photoreceptors. 2. Light regulated phosphorylation of rhodopsin and phosphoinositides in squid photoreceptor membranes *Biochemistry*, 23, 2347–2352.
Ward, P., Carlson, B., Weekly, M. and Brumbaugh, B. (1984). Remote telemetry of daily vertical and horizontal movement of *Nautilus* in Palau. *Nature Lond.*, 309, 248–250.

Weeks, F. I. and Duncan, G. (1974). Photoreception by a cephalopod retina: response dynamics. *Exp. Eye Res.*, **19**, 493–509.

Wells, M. J. (1960). Proprioception and visual discrimination of orientation in Octopus. *J. Exp. Biol.*, **37**, 489–499.

Wells, M. J. (1978). *Octopus: Physiology and Behaviour of an Advanced Invertebrate.* London: Chapman & Hall.

Wiederhold, M. L., MacNichol, E. F. and Bell, A. L. (1973). Photoreceptor spike responses in the hard shell clam, *Mercenaria mercenaria*. *J. Gen. Physiol.*, **61**, 24–55.

Wilkens, L. A. (1984). Ultraviolet sensitivity in hyperpolarizing photoreceptors of the giant clam *Tridacna*. *Nature Lond.*, **309**, 446–448.

Wilkens, L. A. (1986). The visual system of the giant clam, *Tridacna*: behavioural adaptations. *Biol. Bull.*, **170**, 393–408.

Wilkens, L. A. and Ache, B. W. (1977). Visual responses in the central nervous system of the scallop *Pecten ziczac*. *Experientia*, **33**, 1338–1340.

Wood, S. F., Szuts, E. Z. and Fein, A. (1989). Inositol triphosphate production in squid photoreceptors. *J. Biol. Chem.*, **264**, 12970–12976.

Yamada, E. and Usukura, J. (1982). The plasmalemma specialization of retinal cells in the cuttlefish as revealed by the freeze-fracture method. In *The Structure of the Eye*. Ed. Hollyfield, J. G. New York: Elsevier.

Yamamoto, M. (1985). Ontogeny of the visual system in the cuttlefish, *Sepiella japonica*. I. Morphological differentiation of the visual cell. *J. Comp. Neurol.*, **232**, 347–361.

Yamamoto, M. and Takasu, N. (1984). Membrane particles and gap junctions in the retina of two species of cephalopods, *Octopus ocellatus* and *Sepiella japonica*. *Cell Tissue Res.*, **237**, 209–218.

Yamamoto, M., Takasu, N. and Uragami, I. (1985). Ontogeny of the visual system in the cuttlefish, *Sepiella japonica* II. Intramembrane particles, histofluorescence, and electrical responses in the developing retina. *J. Comp. Neurol.*, **232**, 362–371.

Yamamoto, M. and Yoshida, M. (1984). Reexamination of rhabdomeric microvilli in the *Octopus* photoreceptor. *Biol. cell.*, **52**, 83–86.

Yanase, T. and Sakamoto, S. (1965). Fine structure of the visual cells of the dorsal eye in molluscan *Onchidium verriculatum*. *Zool. Mag. Tokyo*, **74**, 238–242.

Young, J. Z. (1936). The giant nerve fibres and epistellar body of cephalopods. *Q. J. Microsc. Sci.*, **78**, 367–386.

Young, J.Z. (1960). Regularities in the retina and optic lobes of *Octopus* in relation to form discrimination. *Nature Lond.*, **186**, 836–839.

Young, J. Z. (1962a). The retina of cephalopods and its degeneration after optic nerve section. *Phil. Trans. R. Soc. Lond. (Biol.)*, **245**, 1–18.

Young, J. Z. (1962b). The optic lobes of *Octopus vulgaris*. *Phil. Trans. R. Soc. Lond. (Biol.)*, **245**, 19–58.

Young, J. Z. (1962c). Why do we have two brains? In *Interhemispheric Relations and Cerebral Dominance*. Ed. Mountcastle, V. B. pp. 7–24. Baltimore: Johns Hopkins.

Young, J. Z. (1963). Light- and dark-adaptation in the eyes of some cephalopods. *Proc. Zool. Soc. Lond.*, **140**, 255–272.

Young, J. Z. (1965). The central nervous system of *Nautilus*. *Phil. Trans. R. Soc. Lond. (Biol.)*, **249**, 1–25.

Young, J. Z. (1974). The central nervous system of *Loligo*. I. The optic lobe. *Phil. Trans. R. Soc. Lond. (Biol.)*, **267**, 263–302.

Young, J. Z. (1977). Brain, behaviour and evolution of cephalopods. *Symp. Zool. Soc. Lond.*, **38**, 377–434.

Young, R. E. (1972). Function of extra-ocular photoreceptors in bathypelagic cephalopods. *Deep-Sea Res.*, **19**, 651–660.

Young, R. E. (1977). Ventral bioluminscent countershading in midwater cephalopods. *Symp. Zool. Soc. Lond.*, **38**, 161–190.

Young, R. E. and Roper, C. F. E. (1976). Bioluminescent countershading in midwater animals: evidence from living squid. *Science Wash. DC*, **191**, 1046–1048.

Young, R. E., Roper, C. F. E. and Walters, J. F. (1979). Eyes and extraocular photoreceptors in midwater cephalopods and fishes: their roles in detecting downwelling light for counterillumination. *Mar. Biol.*, **51**, 371–380.

Zann, L. P. (1984). The rhythmic activity of *Nautilus pompilius* with notes on its ecology and behaviour in Fiji. *Veliger*, **27**, 19–28.

Zonana, H. V. (1961). Fine structure of the squid retina. *Bull. Johns Hopkins Hosp.*, **109**, 185–205.

Note added in proof

Rhodopsin from an octopus (Ovchinniikov *et al.*, 1988) and a squid (Hall *et al.*, 1991) has now been sequenced, as has retinochrome (Hara-Nishimura *et al.*, 1990).

18 Evolution of Vision in Fishes

Stuart M. Bunt

Introduction

When we examine the evolution of the fish eye and brain we are looking back at the origins of our own visual system. Have our early ancestors left us a recognizable legacy? Are there restrictions built into our type of eye, evolved aeons ago to suit conditions now long gone? Can we see, in the ancient fish, signs of preadaptation, structures to be used in innovative ways as the first vertebrates ventured into the new terrestrial visual environment?

I do not intend to use this review of the adaptive radiation of the fish visual system to analyse the relationships between the various taxa and by implication their evolutionary history. It is doubtful how illuminating such a study of the visual system of present-day fish can be, as most modern groups of fish have been evolving since the Devonian and even the Agnatha appear to have arisen more than 280 million years ago (Bardack and Zangerl, 1981). There can be no justification for selecting a particular recent fish (say the lamprey) and assuming it is some primitive teleost from which another has evolved. There is even a difficulty in selecting a particular fossil group as ancestral. Who is to say whether the ostracoderms ever gave rise to anything other than other ostracoderms (Nelson, 1969)?

Teleosts are arguably the most successful vertebrates, with nearly as many species as all other living vertebrates put together. They inhabit everything from waters below freezing in the Antarctic to hot springs at more than 40°C. They can survive in sunlit mountain streams so cold and torrential that man dare not enter and waters so deep and dark that they have yet to be fully explored. This long evolutionary history, the vast range of species, and varied underwater visual environments provide a living record of the diversity that the vertebrate visual system can attain (for recent reviews see Douglas and Djamgoz, 1990, and Atema *et al.*, 1988). We can compare this diversity with existing 'relict' species and extinct fossil forms. This may be as interesting for the limitations it reveals as for the range of adaptations displayed by the more than 20 000 teleost species. It may show how the constraints built into the developmental mechanisms and the preadaptations available in the teleost visual system may have limited the evolutionary pathways available to the ichthyostegalians as they dragged themselves out of the Devonian mud.

Constraints on Evolution

The constraints may be physical. There are (with the possible exception of the rotating bacterial flagellum) no wheels and axles in the biological world; presumably because of the impossibility of providing nutrients and innervation to a rotating part. Similarly, carbon-based life forms obviously cannot produce the temperatures needed to make glass lenses. However, the enormous advances in camera technology over the last decade have suggested a number of new ways of improving an image. It has been suggested that the function of the heart was only understood when water pumps were perfected. Now that we have CCD cameras, light guides, line scanners and photomultipliers can we find equivalents among the fish eyes?

Alternatively, the constraints may be historical; selection can act only on pre-existing structures. If development is committed to a certain direction by past or present selection pressures, any improvement that requires a radical reworking of these constraints may be impossible. The typewriter keyboard is a good analogy here. The layout of the keys was originally designed to prevent the mechanical levers of manual typewriters from becoming entangled. To this end, letters that were most often struck one after another were placed as far apart on the keyboard as possible. This made evident sense in the days of mechanical keyboards but makes no sense at all today, when one may use a wordprocessor that can process information at frequencies measured in megahertz!

In spite of this it is unlikely that the keyboard layout will ever change because all training (read 'development') is based on this particular keyboard layout. The millions of people trained only to use this sort of keyboard will continue to demand it. Thus there will never be a range of keyboards available for 'natural' selection (i.e. people) to select from. A secondary pressure has come to maintain the trait of an awkward keyboard even though the original selection pressure (mechanical typewriters) has all but disappeared.

Thus the absence of a particular adaptation may be as revealing as the presence of another. It may tell us a great deal about past or present selection pressures, but it may also reveal something about the genetic or developmental constraints placed on the animal. If the animal lacks the mechanism for producing a structure then there will be nothing for natural selection to select.

The Original Visual System?

One of the first steps is to look at what selection has had to act on in the past: what were the first visual systems like? From what little we know about the ancient environment, which includes sympatric animals, what can we deduce about the selective pressures that moulded the nature of these first eyes? We should be careful to note that this is not the same as saying our eyes evolved from these 'primitive' eyes. The fossil fish may tell us about the range of developments available to the vertebrate visual system, but they do not necessarily show us the ancestral condition.

Deuterostomes

If we go back beyond the first obvious vertebrates, the agnathans, the phylogenetic relationships become uncertain. At various times insects, annelids, arachnids, nemerteans and various deuterostomes have been proposed as ancestors of the vertebrates. The consensus of opinion now largely favours the deuterostome line. The solitary tunicate (Urochordata) larva possesses a dorsal nerve cord above a notochord, gill slits, and somite-like muscle blocks alongside the 'tail'. Embryological and biochemical evidence also favours the tunicate theory of vertebrate origin (see Romer and Parsons, 1977, for a more extensive account).

The larval tunicate *Ciona intestinalis* has a simple ocellus in the posterior wall of the cerebral vesicle (Dilly, 1961; Eakin and Kuda, 1971). The ocellus has a three-celled 'lens' and 15–20 'retinal' cells embedded in a large pigment-containing cell. These photosensitive cells have a number of similarities to the vertebrate receptor: a pile of membranous discs resting on an inner segment containing the basal body of the modified cilium connecting it to the cell body; and, in contrast to the graded depolarization of the invertebrate receptor, a hyperpolarization associated with a decrease in membrane conductance when exposed to light (Gorman *et al.*, 1971). However, Dilly was unable to identify any synapses between the cell and neighbouring neurones.

Among the deuterostomes the cephalochordates such as amphioxus (*Branchiostoma*) are clearly the most 'fish-like'. These translucent animals are found in the mud of shallow marine waters and can be found in large concentrations in the English channel. They are filter feeders, trapping small food particles in slime secreted inside the pharynx. They spend their time buried in the sea bed with just the anterior end projecting but are capable of swimming, albeit inefficiently, as they lack well-developed fins. Despite the piscine appearance these organisms are clearly more basic than any fish. They have very poor cephalization, lacking any anterior sense organs and perhaps as a result show only a slight anterior thickening of the neural tube. The notocord persists in the adult and extends right to the anterior tip of the animal. There are no vertebrae, ribs or skull. The segmental arrangement of the gonads and excretory apparatus is also clearly primitive. They even lack any blood cells, pigments or a recognizable heart (Romer and Parsons, 1977).

Ciliated light-sensitive cells surrounded by a pigment cup are found in its spinal cord, but not in the anterior portion. These 'eyes' can presumably register direction but, lacking any sort of accessory organs such as the lens, iris or cornea, and being buried in the neural tube, cannot form an image. However, they can provide information on intensity, temporal variations or flicker, and, if there is more than one pigment involved, some information on the wavelength of the light. A similar general sensitivity of the nervous system may be present in more advanced fish. Even blind cave fish have been reported to be light-sensitive (De la Motte, 1964).

The photoreceptor-like cells of amphioxus are distributed along the *inside* of the neural tube. Their position *inside* the neural tube might explain why the vertebrate receptors lie deep in the retina, rather than on the surface as is the case in most invertebrate eyes. The vertebrate receptors develop by evagination of the neural tube and are clearly derived from ciliated cells, such as are found in the neural tube of many simple vertebrate species including amphioxus (Gans, 1989). Such an evagination would result in processes, previously projecting into the centre of the neural tube, lying deep in the forming retina.

To look for the next steps in vertebrate evolution we can look at the fossil record as, unlike amphioxus, many of the earliest true vertebrates were heavily armoured and left well-preserved fossils. First seen in rocks from the Upper Cambrian (Repetski, 1978), larger numbers of archaic jawless fish (the ostracoderms) existed in the following Silurian but it was really with the evolution of jawed fish in the Devonian period that the evolution of fish exploded.

400 Evolution of the Eye and Visual System

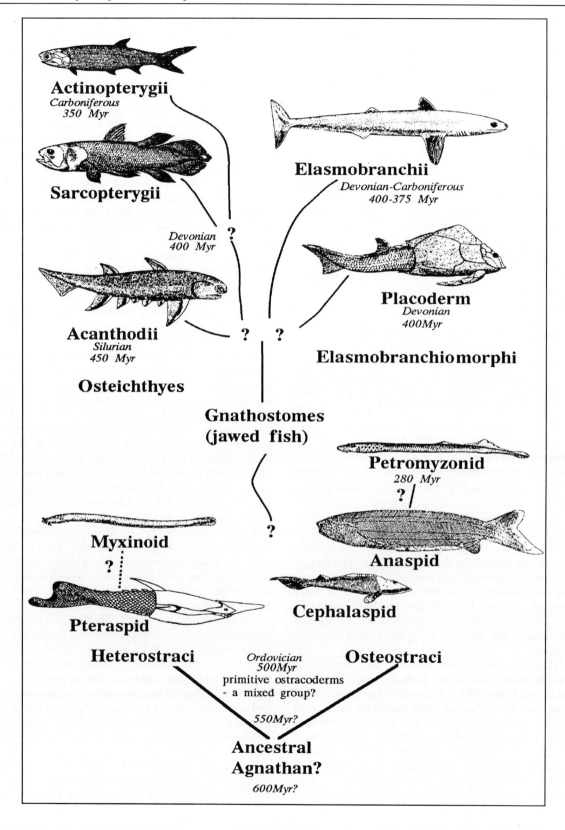

Fossils of fish are so abundant in sedimentary rocks of this period that it is sometimes referred to as the 'Age of Fishes'. Thus the first fish probably arose around 500 million years ago and by 400 million years ago most of the classes of fish we see today had evolved (Robertson, 1957, 1959; Schaeffer, 1969) (Fig. 18.1).

Agnatha

The first fish from the Upper Cambrian were jawless. The first jawed fish, the acanthodians, have not been found in deposits older than the Silurian. It is tempting to suggest an evolutionary sequence from the cyclostomes to the gnathostomes including, eventually, terrestrial vertebrates, as this is the sequence they appear in in the fossil record. However, most now reject this view (not least because of the difficulties of deriving the visceral arches of the gnathostomes from the fundamentally different structures of the agnathans) preferring instead to regard the agnathans and gnathostomes as sister groups of equal antiquity, but sharing a hypothetical and more remote ancestor (Hardisty, 1979).

Given the many unique traits and specializations of the present-day agnathans, it is highly unlikely that they were our ancestors. Present-day agnathans may well have preserved more of the characteristics of the original jawless fish which gave rise to the gnathostome than the gnathostomes themselves, presumably because they inhabit an ecological niche closer to that originally inhabited by the ancestral agnathans.

It is not the aim of this chapter to discuss the details of cladistics or the phylogeny of the vertebrates. These primitive fish are of interest because they may show something about the ancestral condition. What do we know of their primitive visual systems? What was their environment like? What selection pressures led to the development of the modern teleost visual system?

The fossil agnathans are divisible into two broad groupings, the Osteostraci and the Pteraspidomorphi. The cephalaspid and anaspid osteostracans share many of the structural features of the modern-day lampreys (petromyzonids), which is recognized by grouping them in one class, the Cephalaspidomorphi (Stensio, quoted in Romer and Parsons, 1977). The remaining fossil vertebrates, the Heterostraci (Pteraspida) and the Thelodonti (of which we have little more than scales to go on) are left grouped uncomfortably in a second class, the Pteraspidomorphi, and it has been suggested by some that the hagfishes (the myxinoids) have more in common with the Heterostraci than any of the Cephalaspidomorphi (Jarvik, 1968) (Fig. 18.1).

Pteraspidomorphs

Heterostracans

Fragments of these heavily armoured fish have been found from Upper Cambrian deposits in North America (Repetski, 1978). The head and trunk were covered in bony plates made of a characteristic material, aspidin. The gill pouches, like those of the hagfish *Myxine*, open by a common duct. Janvier (1974, 1975) compared the pteraspid eyeball to that of the hagfish *Eptatretus burgeri*, which still retains a rudimentary eye. He concluded that the pteraspid had an elongated conical eyeball like the hagfish. This shape would preclude any rotation of the eyeball, so Janvier inferred that, like the present-day hagfishes, the heterostracans would also have limited or no extraocular eye muscles. Tarlo and Whiting (1965) had previously suggested that the second and third somites in the heterostracans were complete and probably had not given rise to extraocular eye muscles. If the eyeball was movable at all then the extrinsic eye muscles would have to come

Fig. 18.1 *The possible relationships of the archaic fishes. The diagram indicates the periods when fossils of each group were most common, not the period when they were first identified. The question marks highlight the gaps in our knowledge of early fish evolution. It is not known what group of the varied Agnatha dominant in the early Silurian seas gave rise to the first gnathostomes. It appears that the early radiation of the jawed fish probably occurred in fresh water, but most fossils from that period come from marine rocks. In the early Devonian the seas were dominated by the varied and grotesque forms of the placoderms, which appear quite unlike any fish living today. There is some reason to believe that the placoderms may be related to the elasmobranchs and that the loss of bone in this latter group may be secondary.*

The origins of the Osteichthyes is no clearer. Fossils of the Acanthodii have been found in deposits older than the superficially more ostracoderm-like placoderms. These small fish have many osteichthyan characters but it is unlikely that these odd fish with their numerous spiked fins are closely related to any of the modern groups of gnathostomes. By the middle of the Devonian fresh waters were dominated by more modern-looking Osteichthyans. From the beginning they were divided into the lobe-finned Sarcopterygii and the ray-finned Actinopterygii. The Sarcopterygians dominated the Devonian fresh waters, particularly the streamlined Crossopterygians. They thus appeared slightly before the elasmobranchs became common in marine waters in the latter half of the Devonian. It was not until the latter part of the Palaeozoic and early Mesozoic era that the osteichthyans took over the seas. The Actinopterygii appeared rather late on the scene but became the dominant fish in all the seas from the Carboniferous onwards.

The modern agnathans (the cyclostomes) are probably two distinct groups, with the lampreys related to the Osteostraci and the myxinoids to the Heterostraci.

from the first (premandibular) segment and not from the first three preotic segments as in all other vertebrates.

The eyes were lateral and dorsal and a depression between them on the inner surface of the dorsal head plate has been identified as the position of the pineal. The covering of bone over the pineal may have prevented the pineal from detecting any more than gross changes in light level. In the absence of an ossified endocranium we know little about the internal structure of the head and therefore about the brain.

Myxinoids

The hagfish or myxinoids show less diversity than the lampreys, perhaps because of the restricted environments that they occupy. The hags are blind scavengers rather than active predators living on the sea bed in comparatively deep water. They can produce copious amounts of slime as a defensive measure. They differ from the lampreys in many quite fundamental ways quite apart from their life styles. The hags lack a cerebellum, have an aneural heart, have no choroid plexus, no external eye muscles and no lateral line system. On the other hand they do possess a gall bladder and bile duct, accessory hearts, and united spinal roots all of which are missing from the lamprey (for a complete list see Hardisty, 1979, pp. 10–11).

The hagfish has only rudimentary paired and lateral eyes covered with opaque skin. In spite of their degenerate eyes, hagfish, like ammocoetes, will congregate in the shaded part of an aquarium. Again, like the larval lamprey they possess skin photoreceptors concentrated mainly in the head region and around the cloaca, although these differ from those of the ammocoete in being innervated by spinal nerves rather than by a lateral line nerve. In *Myxine* these dermal receptors show their greatest spectral sensitivity to light of 500–520 nm, corresponding to the wavelengths of the light that would penetrate into their deep-water habitat (Steven, 1955). Their responses to light are characterized by very long reaction times. These vary from 10 seconds at the maximum intensities employed to between 2 and 5 minutes at threshold intensities, similar to those that the animal would experience in its natural environment (Newth and Ross, 1955).

Perhaps because of their rudimentary eyes, the tectum is poorly developed in *Myxine*. The cell groups of the tectum are ill defined and not clearly separated from the semicircular torus. Rostrally, the cells of the tectum cannot be separated from those of the posterior thalamus. The cranial nerve nuclei for eye muscles are absent (as are the muscles!). The degeneration of the eyes and the almost complete lack of skin pigmentation in the hagfish may be adaptations to the deep-sea life. In shallower living species of the genus *Eptatretus* a vitreous body is present in the eye and the retina is more differentiated (Hardisty, 1979).

Osteostraci

Cephalaspidomorphs

These heavily armoured fish had a flattened bony head shield followed by a trunk covered in thick, articulated scales. Although the soft parts have not been preserved the magnificent reconstructions of the head shield by Stensio (1968) have shown that the typical cephalaspid possessed a pair of large eyes on the dorsal surface of the head armour, with a single nasohypophyseal opening rostrally and an opening for the pineal caudally. Ventral casts of the head shield show large indentations behind the first branchial arch, corresponding with the orbital openings above. Endocasts show the possible possession of a well-developed optic tectum in these ancient armoured fish (Stensio, in Romer and Parsons, 1977).

Anaspids

These smaller fish (< 15 cm long) are shaped much more like a modern cyclostome but were covered in bone-like scales. They appear to have forsaken the heavy armour of the bottom-dwelling cephalaspids for speed and motility. The large, paired eyes are placed laterally, surrounded by a protective ring of sclerotic plates and there is a pineal foramen between them. The anaspid *Jamoytius* (Ritchie, 1968) has many characteristics in common with the modern-day lamprey.

Petromyzonids

The only extant agnathans are the modern-day lampreys or petromyzonids and the hagfishes or myxinoids. The lampreys are widely distributed in both marine and freshwater environments in both hemispheres but appear to avoid tropical and subtropical zones. Temperature seems to be the restricting factor, the only two species found in equatorial zones being the two freshwater Mexican forms restricted to high altitudes and therefore cool waters. 29–31°C is lethal for their larval stages (Potter and Beamish, 1975). An interesting quirk of the lampreys is the number of paired species. It appears that many of the dwarf, non-parasitic, freshwater lampreys have evolved from a coexisting parasitic form.

The oldest fossil lamprey, *Mayomyzon*, was found in Upper Carboniferous deposits (Bardack and Zangerl, 1971). In common with modern-day lampreys, in the macrophthalmic or adult stages, it possessed two large, paired lateral eyes. *Mayomyzon* seems remarkably similar to modern-day lampreys, suggesting that little change has occurred over the last 280 million years or more. The modern-day lampreys may therefore give us considerable insight into characteristics of the sort shown by our Palaeozoic ancestors.

The lamprey visual system

With the notable exception of the degenerate eyes of the hagfish, what is most noticeable is the advanced nature of the agnathan eye. In all fossil agnathans, where it can be identified, and in the living and fossil lampreys, there is a well-developed pair of laterally placed eyes. As these are soft-tissue parts we can tell little from the fossil forms other than that they possessed large sockets in the correct place, and endocasts of the braincase seems to suggest the presence of well-developed tecta.

We must rely on the living agnathan petromyzonids for details of the agnathan eye. It has been suggested that some of the simplicity of the lampreys stems from their 'degeneracy' due to their adoption of a parasitic way of life. However, there is much reason to believe that many of their unique features are a result of their maintaining a number of archaic features. To quote Hardisty, 'The striking parallels between the lampreys and the cephalaspidomorphs leave us in little doubt that they have been an extremely conservative group and that, in spite of specialized features related to their particular mode of life, they have nevertheless retained a basic plan of organization throughout the 600 million years which spans the entire history of vertebrate evolution.' (Hardisty, 1979) Thus, although some of the features of their visual system may result from their way of life, they may also show visual arrangements similar to the original Agnatha, albeit parasitic or mud-dwelling Agnatha.

The larval lamprey or ammocoete behaves and even looks superficially somewhat like amphioxus. It burrows into the mud in the daytime, leaving only its anterior portion exposed and feeds by filtering organisms from the water which are then trapped in a mucus strand which is swept by cilia towards the pharynx. It has a translucent spot on the dorsal surface of the head marking the position of the underlying pineal, but the rudimentary paired lateral eyes are buried beneath the skin and are not visible externally.

However, the skin of the ammocoete and even the spinal cord itself is sensitive to light. Specific photoreceptors have not been identified, but the tail region is particularly sensitive and the response is mediated by the lateral line nerves, as it is abolished when they are cut. The response is greatest in the blue–green region of the spectrum at 530 μm, close to the absorption maximum of porphyropsin (520–530 μm). The sensitivity is low and the reaction times very long.

In *Lampetra fluviatilis*, by the time the larva is 7 mm long, the eye vesicle has differentiated into an outer pigmented layer and a well-differentiated inner layer with a few visual cells near the optic nerve head. This strange situation becomes more obvious as the filter-feeding ammocoete increases in size. By the time it is 95 mm long the eye has increased by addition of cells to its margins to

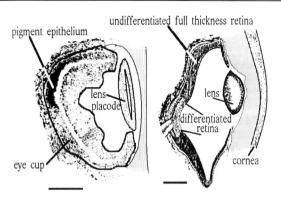

Fig. 18.2 *Diagrams of the eye of a 7-mm ammocoete larva (left) and a 95-mm ammocoete to show the suspended differentiation of a large part of the retina in the premetamorphic ammocoete. At the 7-mm stage the retina is thickened in the central part over a small patch of pigment epithelium. The retina then grows enormously around the periphery but remains undifferentiated. This condition may persist for 6 years and has no counterpart in other vertebrates (After Dickson and Graves, 1981). Scale bar 100 μm.*

about 500 μm in diameter but the area of differentiated retina extends only for approximately 40 μm each side of the optic nerve head. The rest of the 'retina' consists of a single layer of undifferentiated neuroblastic cells (Fig. 18.2) (Hardisty and Potter, 1981).

It seems possible that this small central area of differentiated retina is functional, as synapses are formed between the elements (Dickson and Collard, 1979) and there is a distinct retino-tectal projection (Kennedy and Rubison, 1977). The evagination of the optic vesicle from the anterior cerebral vesicle induces the formation of a lens, but this only enters the optic vesicle at a later stage of development (30 mm). It looks as if this underdeveloped eye could detect light but would have little chance of forming any sort of focused image on the poorly developed retina.

The full differentiation of the more peripheral retina does not take place until metamorphosis, which in *P. marinus* may be as long as six years after hatching (Beamish and Potter, 1975). At metamorphosis the eyes enlarge; this process is extraordinarily rapid, with the depression marking the future position of the eye developing into a large, fully developed eye in less than four or five weeks (Hardisty and Potter, 1971). The change is most marked in parasitic species where the eye is larger in proportion to the body than paired species with a non-parasitic life cycle (Hardisty *et al.*, 1970). This transition is so marked that it is often referred to as the macrophthalmic stage and is completed before many of the internal changes (such as the sealing off of the foregut) have been completed.

The delayed development of the peripheral retina is

unique to the lamprey; in all other vertebrate eyes the differentiation of the retina may have a centre-to-periphery gradient or continue to add to the peripheral retina throughout life, but there is never a lengthy stage where one part of the retina is fully differentiated and the rest is not (Johns and Fernald, 1981; Bunt et al., 1983). It appears as if the embryonic development of the eye has been interrupted during phylogeny by the interpolation of an extended larval phase (Hardisty, 1979). The only parallel among the vertebrates is that in the amphibia a binocular projection does not form until after metamorphosis (Keating, 1977). This is clearly related to the postmetamorphic change in the position of the eyes. There is no analogous delay in differentiation of the retina, even in such obviously neotenous species as the axolotl.

All the visual areas are undeveloped in the young ammocoete and increase enormously in size at metamorphosis as the paired eyes develop (Kennedy and Rubison, 1977). The lamination of the tectum becomes much more distinct at metamorphosis, when the tectum may increase four times in size (De Miguel et al., 1990).

The three preotic segments give rise to the extraocular eye muscles, innervated by cranial nerves II, IV and VI, although Bjerring (1977) has suggested that the first two preotic segments give rise to the two separate nerve bundles sometimes seen in the lamprey oculomotor nerve.

The cornea in the adult lamprey is unique in having two distinct layers. The outer, dermal, cornea consists of a transparent layer of dermal epithelium continuous with the skin. Below this is a mesodermal layer which is a transparent continuation of the sclera of the eyeball itself. Between these two layers is a capillary space containing mucus. This allows the eyeball to rotate freely beneath the flatter dermal cornea. Although some gnathostome embryos have a similar bilayered cornea, the two layers are fused in the adult. The iris forms as a continuation of the posterior epithelium of the cornea and in some species may be covered by an argentea layer containing reflective guanine crystals.

The large, almost spherical lens has a number of features which have been lost in more advanced species. The cell nuclei are distributed throughout the lens, rather than being limited to a specific zone as in most vertebrates. There are yellow pigment granules in the body of the lens and, while the centre consists of polygonal or round fibres, on the periphery they form an irregular array. The refractive power is highest in the centre and decreases towards the periphery (Hess, 1912, quoted in Kleerekoper, 1972).

In the lamprey the lens is unique among vertebrates in not being fixed to the walls of the eyeball. Only pressure from the 'vitreous humour' behind it and the cornea in front holds it in place (Romer and Parsons, 1977). The method of accommodation employed by the cyclostomes is also unique among living vertebrates. In the lamprey the lens is accommodated for near vision at rest (about 12 cm). The hard and inflexible lens is pushed back by a flattening of the cornea, which it directly underlies (Franz, 1932). The flattening is accomplished by a unique corneal muscle derived partly from dorsal somatic muscle and partly from visceral muscle. Such a muscle does not seem to have been present in the cephalaspids; Janvier (1979) believes that such a muscle may have been present in the anaspids. The other extraocular eye muscles almost completely surround the equator of the globe and are connected by a ligament. Contraction of these muscles may lengthen the globe and result in an increase in distance between the lens and retina. This would make the cyclostome the only vertebrate in which accommodation occurs in two directions: positive through contraction of the corneal muscle, negative through contraction of the superior and external rectii (Kleerekoper, 1972).

The visual cells are of two different lengths and their distribution and ratios vary between species. However, it is difficult to assign them to the category of rods or cones on any morphological or biochemical grounds. This suggests that the cyclostomes may show a primitive stage where the differentiation into the rods and cones of higher vertebrates is not complete. The concentration of visual pigment in the retina of *Petromyzon marinus* and *Lampetra fluviatilis* is very low. In newly metamorphosed animals the pigment is rhodopsin but in the sexually mature animal migrating upstream to spawn the visual pigment is replaced with porphyropsin (Wald, 1957).

Although the receptors are poorly differentiated the retina itself possesses the normal three-layered vertebrate pattern. Synaptic ribbons have been identified on the inner segments, but there appear to be few bipolar cells in the lamprey retina. Unlike the teleost retina, there do not appear to be any photomechanical migrations in the cyclostome retina.

The pineal is also light-sensitive in the lamprey. The pineal vesicle may even have a rudimentary lens (e.g. in *Geotria*) and a more ventral retina consisting of an inner pigmented epithelial layer containing a few cone-like sensory cells and an outer layer of ganglion cells and nerve fibres, the very opposite of the arrangements in the retina of the larger paired lateral 'eyes' (Meiniel, 1971). The smaller parapineal has an even more poorly differentiated pellucida and retina and may contain only a few poorly developed sensory cells. The pineal seems to be adapted for functioning at low light levels when any light signal causes an inhibition of nervous discharges.

The pineal is also a secretory organ producing, among other things, hydroxyindole-*O*-methyltransferase (HIOMT), which converts serotonin to melatonin. Pinealectomy prevents the contraction of the melanophores and diurnal changes in skin colour of the ammocoete, which can be reversed by administering extracts of pineal or melatonin (Eddy, 1972). The long time span over which some of the resultant pigment changes last, in spite of only shortlived bursts of HIOMT, has led to the suggestion, since confirmed in many vertebrates, that the role of the pineal may be to help set the biological clock accurately using the diurnal light changes it detects. Pinealectomy has been shown to inhibit metamorphosis in ammocoetes (Eddy, 1969) and to stimulate swimming activity in *Xenopus* tadpoles (Roberts, 1978).

In *Lampetra* the pineal stalk contains about 600 nerve fibres connecting directly to the right habenular ganglion through an adjoining ganglionic mass, some of whose neural elements may have been derived by rostral migration from the habenular ganglion itself (Meiniel and Collins, 1971). There is no parapineal in the Southern Hemisphere genus *Mordacia*, just a small vestigial structure known as the pineal appendix, although its lack of connections with the left habenular ganglion suggests it may not be a relic of a previously present parapineal in this species.

Because of the symmetrical right and left projections of the pineal and parapineal respectively, it has been suggested that the pineal and parapineal may have once been symmetrical, paired, lateral organs like the still-paired main eyes. In early developmental stages the two habenular ganglia are more equal in size. Whiting and Tarlo (1965) thought that the heterostracan brain had a symmetrical epithalamus with equally sized habenulae. However, any such condition must have predated the fossil agnathans, as none of them appear to possess paired epiphyseal foramina.

Behind the pineal the dorsal diencephalon receives important optic input (Kennedy and Rubison, 1977) to a dorso-medial group of bipolar neurones which has sometimes been called the equivalent of the lateral geniculate of mammals (see Northcutt and Wullimann, 1988, for a discussion). The lamprey mesencephalon is unique in that the two optic tecta are separated along the midline by a large choroid plexus, such as may have existed in the Heterostraci. Laterally and caudally the tecta unite behind the choroid plexus and project backwards over the cerebellum.

Transition from the Agnathan to the Gnathostome

As Dawkins has pointed out so expressively, 'A simple, rudimentary, half-cocked eye . . . is better than none at all. Without an eye you are totally blind. With half an eye you may at least be able to detect the general direction of a predator's movement, even if you can't focus a clear image. And this may make all the difference between life and death.' (Dawkins, 1986.)

However, this is *not* to say that we should expect to find many such 'eyes' in the fossil record. While it is true to say that a rudimentary eye is better than none at all, an efficient eye is still better than a rudimentary one. Once the evolution of an eye has started the basic eye should soon be replaced with a more efficient eye. Only when a structure has reached the best compromise (given the possible constraints on the way it can develop) between its own energy 'cost' and the survival advantage that it confers will the pressure for further evolution abate and evolution slow down.

This may be why, ignoring the degenerate hagfish, the agnathan eye looks so complete, and appears to possess most of the features of even the most advanced vertebrate eye. Only in extreme environments, where eye development has taken a specialized and possibly self-limiting excursion, can we see much variation in basic eye structure. Such examples are to be found in the tubular eyes of deep-sea fish and the odd double cornea of the 'four-eyed' *Anableps* (Schwassman and Kruger, 1965; Sivak, 1976).

It appears as if the early vertebrate history involved a change from a small, free-swimming, neotenous, tunicate-like larva to larger, filter-feeding, burrowing animals. It may be that at this stage simple detection of light or a shadow would be enough for the animal to withdraw into its burrow. This is the specialized niche still occupied by the ammocoete and, to a lesser extent, the hagfish living in the benthic darkness. There being no need to hunt for food or to recognize hunter from prey, a complex eye would not be required. In fact, it might be positively advantageous to have light detection spread out along the body.

Amphioxus has light-sensitive pits along the spinal cord. Ammocoetes appear to have light sensitivity along their body. The living lampreys possess a well-developed median eye. The presence of a prominent median socket in many of the ostracoderms indicates that they too possessed a conspicuous median eye, perhaps a pair of them. This central opening in the head plate was found in the placoderms but not in the cartilaginous fish. In the Devonian, primitive representatives of all three major groups of bone fish possessed a pineal 'eye' and it was especially common in the crossopterygians (Romer and Parsons, 1977). The pineal may have enabled the agnathans to avoid predators before lateral eyes developed (Roberts, 1978). However, a median light-sensitive area has been evolved independently several times and can be found in animals with well-developed lateral eyes. For example, although the Penaeidae (a group of prawn-like crustaceans) have well-developed lateral compound eyes, they also have a light-sensitive organ under the rostrum, complete with lens and photoreceptor layer (Dall et al., 1990).

The pineal organ or epiphysis cerebri develops as an evagination of the diencephalon in the same way as the retinas of the lateral eyes. In some fish the skin above the pineal is lacking in melanophores and forms a translucent 'pineal window'. For a review of the adaptive radiation of the pineal system, including the morphological, physiological and functional aspects see Hamaski and Eder (1977). This organ is certainly photosensitive in many species and in several species the photosensitivity has been accurately measured. The output from the pineal was totally inhibited at light levels down to 10^{-4} lx; the receptors seem most sensitive to blue–green light of 495–525 nm wavelength (Tamura and Hanyu, 1980). This strong inhibition by bright light would make the pineal an ideal 'shadow' detector. Its gradual increase in firing at lower light levels would also make it a good detector of changes in ambient light at dusk and dawn or with depth.

Many early terrestrial animals, including the precursors of the mammals, possessed such a median 'eye'. However, by the Triassic, few animals, not even the teleosts, possessed a distinct medial pineal eye. Today it is only found in the lampreys and some lizards. In *Sphenodon* it even has a lens-like cover and a basic 'retina'. Curiously, it is the pineal that is largest in most lampreys and the parapineal in the lizards such as *Sphenodon*.

While the pineal may have decreased in importance as a visual organ or taken on new roles, the transition from small burrowing filter feeders to larger, free-swimming scavengers required new and better light detectors. The amount of armour borne by the early ostracoderms bears witness to the dangerous environment in which they lived. Their fossils are often found with the remains of eurypterids, giant aquatic invertebrates (some up to 3 m long) looking like monstrous lobsters, with claws to match. Bony armour would be a good defence against such predators, as they lacked efficient jaws or masticatory apparatus to crush and separate the bone from flesh.

Lacking jaws, the ostracoderms could not become efficient predators themselves, and probably filled the niches taken by the modern day cyclostomes as scavengers and opportunist feeders. They would need good image-forming eyesight to tell food from predator, even to find the food on the muddy bottom. Many of the ostracoderms have their eyes (and median eyes) placed high on the head, similarly the shape of the head shield suggests a bottom-living habit where detection of predators above would be most important. The Anaspida look more like the free-living adult lamprey and had large, laterally placed eyes. With lighter body armour they may have relied on speed for escape, while the heavier and larger eurypterids were confined to the bottom.

Acanthodii

Like the earliest protomammals, the first jawed fish were small and apparently inconsequential. The first fragments of these earliest gnathostomes, the curious Acanthodii, were found in early Silurian deposits. Still armoured with bony plates and scales, and with a bony skeleton, they predate even the elasmobranchiomorphs (the placoderms and elasmobranchs) which gave rise to present-day sharks. The small, stickleback-sized acanthodians possessed numerous paired fins, each supported by a vicious spike. They must have made a most uncomfortable meal for any predator (Fig. 18.1).

Soon to be overshadowed by the more abundant and spectacularly grotesque placoderms, the elasmobranchs and other osteichthyans, these small fish mark the vanguard of the revolution to come. With jaws the fish could bite back, with better methods of feeding they could get larger, then in turn take larger prey, and soon would displace forever the large invertebrates so common in the early Silurian seas.

Elasmobranchiomorphii

Placoderms

Towards the end of the Silurian and in the following Devonian the placoderms became dominant. With true jaws, many variations of fins and well-developed eyes, often surrounded by supporting bony rings, the placoderms spread rapidly. Able to take more varied food they show a great variation in size and morphology. Many still possessed heavy armour and show many of the characteristics of bottom-dwelling species such as a flattened profile. There was a general tendency for reduction in the armour and an increase in size. *Dinichthys*, a predaceous Devonian placoderm, was up to 8 m long.

Elasmobranchii

At the end of the Devonian the wild excesses of the placoderms were replaced by fish which looked far more like present-day elasmobranchs. These were presumably the most successful variations on the placoderm theme. With mobility more important than armour, these later fish were more streamlined and the bony armour reduced or absent. If the chondrichthyans are descended from the placoderms they show that there was an advantage in losing even the internal ossification.

Osteichthyes

Probably even before the chondrichthyan explosion, in the early Devonian, the fresh waters were full of recognizable osteichthyans. Lungs appear to have been present in all these primitive bony fishes. In the drought-prone Devonian they may have been a necessity. With the transition to the sea they lost their importance and many were transformed into swim bladders. They gradually displaced the placoderms and chondrichthyans in the seas, so that by the Triassic they were the dominant form in both fresh and marine water. The inter-relationships and common ancestor of the gnathostomes continue to be one of the largest gaps in the fossil record. They probably arose in the early Silurian, perhaps in fresh water, but the Silurian deposits that we know of are almost entirely marine (Romer and Parsons, 1977).

At the very beginning of their history (leaving aside the odd acanthodians) the osteichthyans were already divided into two major groups, the subclasses Sarcopterygii (lobe finned) and Actinopterygii (ray finned).

Actinopterygii

Many ray-finned fish lines appeared in the Devonian and some modern representatives such as *Polypterus* and *Amia* survive today. Their taxonomic position is uncertain. The Chondrostei, represented by the sturgeons and spoonbills, were successful for a time. Late in the Permian period the Holostei appeared; they were so successful that they became the predominant fauna of the Triassic and Jurassic periods. Their numbers decreased through the Cretaceous and they were replaced by their descendants, the teleosts, that remain the dominant fish today. Only a few Holostei remain today, the bowfins and garfish.

Sarcopterygii

The Sarcopterygii contain the Crossopterygii, from which land vertebrates appear to have descended. In the Devonian they were the most common bony fish but were rare by the Carboniferous, and typical crossopterygians, termed rhipidistians, were extinct before the end of the Palaeozoic. The only living members are the relict coelacanth *Latimeria* and the Dipnoi or lungfishes (Crescitelli, 1972).

The Aquatic Visual Environment

In the Devonian our crossopterygian ancestors were aquatic predators; if *Latimeria* is any guide, many possessed the well-developed eyes needed for hunting mobile prey. The evolution of the crossopterygian visual system would have been dominated by the nature of the underwater visual environment. This environment differs in many ways from the gaseous terrestrial environment and may have burdened us with many adaptations or restrictions which are not so suited to our brighter, more transparent but less challenging visual environment (for a review see Lythgoe, 1980, 1988; Loew and McFarland, 1990; Muntz, 1990).

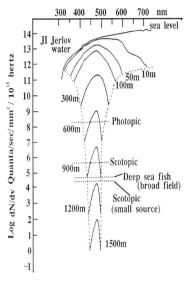

Fig. 18.3 *Spectral irradiance changes with depth in Jerlov type I water (the clearest open ocean waters). Note the cutoff of longer wavelengths by even the shallowest water and the rapid further narrowing of the spectra with depth. The absolute irradiance also falls dramatically, even in this very clear water. In more typically coloured and particulate filled water, scotopic vision may become impossible in water as shallow as 30 m (Adapted from Lythgoe, 1988).*

Air is, at the densities and distances normally encountered, colourless. In contrast, even pure water acts as a strong selective filter, removing reds and yellows even at shallow depths (see Fig. 18.3). The deep blue of the open ocean, or the nutrient-poor inshore waters of the Mediterranean, is due to the optical properties of the water itself.

In air, particulates are sometimes seen as smoke or fog, but they are normally short-term phenomena. On the other hand, water may frequently be turbid to the point of opacity. These suspended particles both scatter (reflect) light and absorb it. Rayleigh scatter from the water molecules and particles less than a wavelength of light in diameter contribute to the 'blueness' of water as shortwave blue light is scattered more than longer wavelengths. The scattered light adds to the general background 'noise' that the visual system must work with to identify a potential mate, food or foe (Lythgoe, 1972). When scattering is severe (as in dense fog) it may be very difficult to tell from which direction the illumination is coming (Fig. 18.4).

Because of the scattering and absorption of light even in 'clear' water, vision is likely to be of little use for seeing much more than general shapes beyond a few tens of metres in even the most advantageous conditions of high contrast and pattern (Nicol, 1975). There will be so much interference from the backscattered light coming from

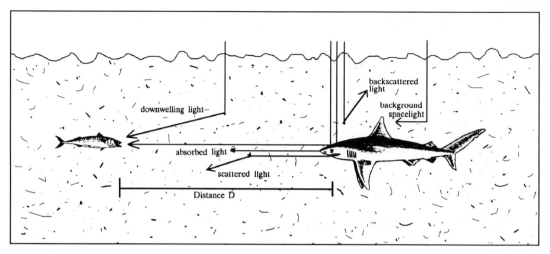

Fig. 18.4 *The main factors limiting the visibility over distance D in shallow waters where light intensity is not limiting. Downwelling light is deflected towards the eye by particles in the water creating underwater spacelight and veiling light between the mackerel and the shark, both of which will lower contrast. Light from the shark is attenuated by absorption and reflection by particles between the two fish. Deflection of light, not originally heading directly towards the fish's eye, into the fish's eye may further obscure the image (Modified from Lythgoe, 1988).*

particles between the observer and the object that contrast will be severely degraded even when the absorption is not enough to prevent light from the object reaching the eye.

Because of the aquatic environment, fish vision appears to have become more specialized for increasing sensitivity and contrast than for resolution or acuity. Behavioural and physiological studies indicate that the dark-adapted sensitivity (Northmore, 1977) and contrast thresholds (Northmore and Dvorak, 1979) of most fish exceeds that of man (Douglas and Hawryshyn, 1990). On the other hand, as underwater objects will be nearer to the fish before they are seen, very high resolving power, such as that possessed by many terrestrial vertebrates, will be of little use under water. The maximal contrast sensitivity may therefore be obtained at lower spatial frequencies than in man (Northmore and Dvorak, 1979). Acuity has been measured in many species and values from a minimum resolvable angle of 20' down to 3.7' have been recorded (listed in Douglas and Hawryshyn, 1990). This is much less than the 1' resolution recorded for vernier acuity in man.

The diffraction at the air–water interface limits vision of the aerial world to a circle with a radius of 48.5°: Snell's window. Beyond this angle there is total internal reflection from the underside of the water surface. A fish looking upwards will therefore see a clear view of the air directly above with increasing distortion and colour fringes caused by the refraction of the water out to 48.5°; beyond that the fish will see the reflection of the deeper water or sea bottom.

In calm water the line defining the edge of Snell's window is sharp, with a very obvious transition from the bright scene overhead to the darker reflection of the light backscattered from deeper water or the ocean floor. Transparent animals which make up the food of many pelagic fish are most easily spotted as they cross this line, as their different refractive index will cause them to appear lighter or darker than their background. Some plankton-feeding fish may possess a specialized area of retina which receives the edge of Snell's window (Munk, 1970; quoted in Lythgoe, 1988).

Many fish have evolved upwardly directed eyes to enable them to detect prey against the bright circle of light. Some prey species counteract this by having photophores along their ventral surface and even by matching their light output to that from above (Warner et al., 1979). Muntz (1976) has suggested the yellow lenses of some mesopelagic fish may be designed to pass most light in the wavelengths produced by such photophores and thereby make the prey appear bright against the broader spectrum of the background light (Fig. 18.5).

The scattered light from the bright image falling on the ventral retina may degrade the contrast of the darker image of the deep water below falling on the dorsal retina. Some fish have a septum derived from the falciform process which runs horizontally across the retina which may isolate optically the dorsal and ventral hemiretinae (Saidel and Braford, quoted in Lythgoe, 1988).

Water currents, changes in temperature, and therefore density, cause diffraction effects many times worse than the mirages encountered in areas of terrestrial temperature gradients. To further complicate matters, many fish have the ability, apparently missing from our particular line of

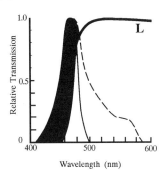

Fig. 18.5 *This diagram indicates how selective colour filters in the lens or cornea can increase contrast between the prey and background spacelight (compare with Fig. 18.4). By acting as a low band pass filter (indicated by the thick line marked L) the lens of* Scopelarchus analis *blocks out more of the downwelling light (indicated by the thin solid line, based on data from JIB water) than the bioluminescence of* Myctophum punctatum *(indicated by the dashed line) (Adapted from Muntz, 1990).*

evolution, of generating their own light, like aquatic fireflies.

Adaptations to the Visual Spectrum of Aquatic Environments

The addition of coloured particulates may further restrict the spectrum of the light that penetrates to any depth. The colour varies enormously and depends not only on inorganic matter such as dissolved soil particles but also on the products of organic decay and the autumnal colours of phytoplankton blooms. Bodies of water can even be classified on the basis of the amount of dissolved organic matter and chlorophyll-containing plankton (Baker and Smith, 1982). The colours may range from the deep blue of clear water to the black or brown of a Scottish loch where the water has been filtered through layers of peat and may transmit most strongly in the near-infrared. This may explain why the fish, more than any other group, have produced a vast array of visual pigments and further specializations such as tinted lenses or corneal chromatophores (Muntz, 1976) to add contrast to a scene (Muntz and Mouat, 1984). However, water blocks several wavelengths (see Fig. 18.3) to which air is transparent. Thus although some surface fish may have pigments sensitive to ultraviolet (Hawryshyn and Beauchamp, 1985; Neumeyer and Arnold, 1989) it is unlikely that many infrared sensors will be found in fish. The polar distribution of radiance also changes with depth. Below about 100 m the polarity of the remaining light becomes effectively vertical as more oblique rays are selectively absorbed along their longer routes to the depths.

Porphyropsins are common in freshwater animals and are found only rarely in marine and terrestrial species (Cohen *et al.*, 1990). They have not been found in mammals or birds. The proportion of the various pigments may vary with the seasons, day length, temperature, the developmental stage and whether the fish is in a fresh or sea water environment. The proportion may even vary among a population of fish (Bridges and Delisle, 1974). It has been suggested that, among schooling fish, it could be an advantage to have members of the population with different spectral sensitivities (Archer and Lythgoe, 1990). The changes in pigment type may occur in migrating fish before they enter a different aquatic environment, which suggests that in this case the change is not triggered directly by the visual environment (Beatty, 1984). However, by fitting opaque contact lenses to one eye of the rudd, Bridges and Yoshikami (1970) have shown that retinal input is important for shifting the rhodopsin/porphyropsin ratio in this species. Pinealectomy does not affect the ratio in goldfish (Beatty, 1984).

Rod Pigments

The correlation between maximum sensitivity of rod pigments and the wavelengths likely to penetrate the water is good only in deep-water fish where the pigments have maximum sensitivity at the deep blues seen in deep, clear water. Fish living in shallower water have pigments maximally sensitive at shorter wavelengths than would be expected, given the transparency of such water to more reddish hues, particularly in brackish or peat-coloured waters (Muntz and Mouat, 1984).

It may be that maximum contrast is available at shorter wavelengths than the wavelength giving maximum image irradiance (Lythgoe, 1988). Contrast is reduced by light scatter and backlight irradiance between the object and the viewer. Maximal contrast between an object and its background will be obtained at those wavelengths where the water is most transparent. Therefore, one might assume that the fish will experience less loss of contrast if its photoreceptors are sensitive to those wavelengths where the water is most transparent.

However, the situation may be different for pale objects in shallow, turbid water. Most of the light illuminating the object will have travelled directly from above to be reflected off the object towards the fish. On the other hand, the backscattered light against which the object is viewed will have taken a longer and more tortuous path through the water. The result is that the spectral radiance of the object has a wider frequency spectrum than the scattered light which has had to travel through more water. This in turn means that maximum contrast will be obtained at wavelengths *other* than those at which the water is most transparent (Fig. 18.4).

For objects that are dark or distant, maximum visibility will be obtained at the wavelength at which the water is most transparent, as the light coming from the object will have a spectrum closely matching the backscattered light. This may explain why deep-dwelling fish have pigments more closely matching the frequency of the practically monochromatic light which can penetrate to these depths, while shallow-water fish have visual pigments maximally sensitive to offset wavelengths (Lythgoe, 1988).

Another way of increasing this visual contrast is to use barrier filters to block off light of those wavelengths contributing most to the background 'noise'. Coloured oil droplets are common in bird retinas and colour filters, apparently made of a mass of mitochondria, absorbing at wavelengths similar to cytochrome C, are present in some fish cones (MacNichol et al., 1978). Yellow carotenoid filters in the lens or cornea are common in teleosts living in bright environments (Douglas and McGuigan, 1989). The yellow pigment is often concentrated in the dorsal area (Douglas, 1989) and in some species may become colourless at night to improve sensitivity (Muntz, 1990). This may cut out scattered downwelling light, and, like the patrolman's 'yellow sunglasses', increase the contrast. By further limiting the already reduced spectrum of light admitted by the water, the filters may also have the effect of reducing the chromatic aberration that large spherical lenses are prone to at short wavelengths. A similar contrast-enhancing effect may be obtained by the complex iridescent reflectors in the cornea of many fish (Lythgoe, 1974).

Cone Pigments

This matching between pigment sensitivity and the environment may be even more complicated among cone pigments. These function only in relatively bright light and the reduction in sensitivity that ensues from having some pigments that do not match the transparency of the water may be less of a disadvantage. It may be an advantage to possess both matched and offset pigments to cope with near, brightly lit objects and darker, more distant objects. The spectral range covered by cone pigments is certainly larger than that of the rods and appears to cover the range of frequencies a fish will meet in its environment, e.g. the shorter wavelengths in the deep-blue waters of a coral reef, the longer wavelengths in peaty freshwater lochs, or the narrow spectrum of blue light in the deep sea (Loew and Lythgoe, 1978).

In shallower water the aquatic environment can be quite colourful. Even though the light at depth may be practically monochromatic, bioluminescence may still provide some colour. Thus colour vision can be advantageous even in the relatively monochromatic aquatic environment. It is difficult to prove the existence of colour vision because of the difficulty of matching the brightness of objects of different colour. Muntz and Cronly-Dillon (1966) were the first to show that the goldfish could distinguish green from blue and red at equal photon fluxes. The brilliant colours displayed by tropical fish in shallow reef environments may testify to the ubiquity of colour vision in fish.

To provide colour vision at least two receptors with different spectral sensitivities are required. Each receptor on its own can provide only a measure of intensity; a green-sensitive photoreceptor will give the same signal when viewing a bright orange object as a dimly illuminated green one. Only by comparing the responses of a number of different photoreceptors can the colour spectrum of the object's irradiance be deduced. Multichromatic photoreceptor systems have even been identified in holosteans, suggesting that colour vision was an early adaptation in vertebrate evolution (Burkhardt et al., 1983).

Elasmobranchs may have only one type of cone receptor, precluding colour vision unless by comparing responses from the rods with the single type of cone (Cohen, 1990). Some teleosts have only one or two types of cones, but most have three and some four or more (Harosi and Hashimoto, 1983). In many freshwater fish one of the receptors in the paired cones is sensitive to light at about 600 nm. Such red-sensitive receptors are absent from most marine and terrestrial animals (Lythgoe, 1988). A particular feature of the fish retina is the presence of paired or even quadruple groups of cones, often with different spectral sensitivities. It has been suggested that this is a result of the short focal length of the typical fish lens, which results in the retina lying close to the lens. A high receptor density is therefore needed for decent resolving power (Fernald, 1990).

The maximal sensitivities of the cones matches to a great degree the environment of the fish. In fresh water (which is often yellow to brown water), diurnal fish living in shallow or mid-water levels have cones maximally sensitive at short wavelengths (less than 450 nm). Fish feeding at twilight lack these shortwave-sensitive cones and the photoreceptors in paired cones may both be sensitive to the same relatively longer wavelengths (approximately 530 nm or 630 nm). Deeper-living fish lack the paired cones and are maximally sensitive to the narrow spectrum of between 575 and 625 nm which penetrates through the reddish water. Marine fish (where the water has a more green or blue hue) lack the red-sensitive cones and most of their paired cones are sensitive at about 550 nm (Lythgoe, 1988).

Optical Adaptations for Low Light Levels

While deep-sea fish may encounter little 'shade', the intensity of the light will vary enormously as they dive from the surface to the depths. The normal diurnal pattern of light and dark is therefore further complicated in those many fish which have several diving and surfacing cycles in each 24 hours. Even in the clearest oceanic water, radiance falls below the critical $0.05\,\mu\text{W/cm}^2$ at which light-dependent behaviour ceases at 288 m. Even the theoretical maximum sensitivity of deep-sea fish of $10^{-9}\,\mu\text{W/cm}^2$ is breached at 700 m. Beyond this depth the only things the fish are going to see must be illuminated by bioluminescence (Loew and McFarland, 1970). In turbid coastal waters these figures may fall as low as 18–40 m. Thus fish may often be struggling to get enough light to detect the object.

For point sources of light the size of the pupil is crucial, but for extended sources the f-number of the eye determines the number of photons reaching the retina. The f-number is used in photography to denote the light-gathering power of the lens and is equal to the focal length divided by the aperture. The telephoto lens focuses light onto the film from a smaller area of the visual world than does a wide-angle lens. Given a uniform illumination of the scene, the telephoto lens must have a larger aperture than the wide-angle to deliver the same number of photons to the film. The f-number takes this into account and the photographer knows that if his light meter tells him that 'f 8' is required it does not matter if he uses a wide-angle at 'f 8' or a telephoto lens at 'f 8', since both will give correct exposure.

Living as they do in environments where levels of illumination may be low, many fish eyes appear to have evolved the best light-gathering methods they can. Most fish eyes have f-numbers of about 1.1–1.3 (Fernald, 1988), as good as the best lenses obtainable for single-lens reflex cameras (a 50 mm f 1.0 lens for a single-lens reflex camera can cost more than £1500; what price a good fish eye?). This attainment is all the more remarkable when one considers that, unlike the terrestrial eye, where the air–cornea interface is an important refractive element of the eye (in the human the lens only provides 13 D of refractive power compared with 43 D from the cornea, Westheimer, 1968) the fish eye has a single focusing element, the lens. The cornea, being immersed in water, can act as a lens only when its thickness varies. In a few cases the cornea *is* slightly thinner in the centre than at the edges, which may contribute, albeit in a small way, to the optics of the eye.

The Lens
The refractive index of most fish lenses is about 1.65 and the focal length is a curiously constant 2.55 times the radius of the lens (the Matthiessen's ratio). Because of this the f-number, and therefore the light-gathering power, is roughly constant, no matter what the size of the eye. However, a larger eye can increase resolution (if all other relevant factors such as lens quality and accommodation are equal). Resolving power equals focal length divided by double the receptor separation (Land, 1981). Receptor size and separation seem to remain fairly constant during eye growth (Johns and Fernald, 1981), leading to theoretically better resolution in the larger eyes. As a result, older fish will be able to detect prey at greater distances, and therefore there may be considerable selective advantage for continual retinal growth and increased eye size in fish living in clear water.

In man it has been calculated that the receptor acuity closely matches the resolution of the lens. In contrast, Fernald (1988) measured the optical characteristics of a number of fish lenses and found that their angular resolution was about ten times that of the retina; only in the very smallest fry did the resolution of the lens match that of the retina (Powers et al., 1988). In fish it appears that all but the smallest eyes have receptor densities greater than are required by the resolution of the optics. However, this should be taken with caution, as measurements of receptor density may not give an accurate view of maximum acuity. Human vernier acuity is an order of magnitude greater than the ability to resolve a fine line which more closely matches the theoretical receptor resolution. Convergence of receptors on single neurones and central processing can confuse interpretations of the acuity of the receptor layer. It seems that receptor packing becomes a real problem only in very small eyes. In the tiny eyes of the jumping spider all four pairs of eyes grossly under-sample the image (Land, 1990). The superior optics of the fish eye have probably been evolved to maximize contrast and sensitivity rather than acuity (Douglas and Hawryshyn, 1990).

Another way of increasing local acuity when the lens is not limiting is to pack more cones into an area centralis or fovea. Given the apparent over-sampling of the fish retina it is surprising that a number of species show a great range of such retinal specializations. For example, the rippled blenny *Choerodon edentulus* has a horizontal streak plus a nasal and temporal area centralis. This would seem to give the fish a better acuity along the plane of the surface along with a frontal and caudal viewpoint (Collin and Pettigrew, 1988; Collin, 1989).

Why should the ratio of focal length to lens radius be so constant, varying by less than 30% in all species examined (Fernald, 1988)? It may be because most fish attempt to gather as much light as possible. The retinal illuminance is given by the following equation:

$$E = \pi r^2 / f^2 \times L \qquad (18.1)$$

where f is the focal length of the eye, L is the luminance of the extended object and r is the radius of the pupil.

Thus the fish needs to maximize the pupil radius while minimizing the focal length in order to maximize retinal illuminance.

Focal length is determined by the refractive index (μ) of the lens (or, more strictly, the difference between the refractive index of the lens and the medium) and the radius of curvature of the lens (r_l). These are related by the following formula:

$$f = r_l/2(\mu - 1) \qquad (18.2)$$

Thus to minimize focal length the fish needs to increase the refractive index of the lens or reduce the radius of curvature. The minimum radius of curvature is obtained with a spherical lens, while the maximum refractive index is determined by the composition of the lens. Given a spherical lens and no significant iris, the radius of the fish pupil is the same as its radius of curvature. If the lens were anything less than spherical then the radius of curvature would be greater than the radius of the pupil. Focal length would increase for a given pupil radius, which would result in less retinal radiance; if $r_l = r_p$ then

$$f = r_p/2(\mu - 1) \qquad (18.3)$$

The constancy of the Matthiessen's ratio (f/r_p) indicates little more than that the μ of fish lenses is fairly constant; presumably fish have reached the maximum μ that can be produced with organic materials or perhaps the maximum at which spherical aberration can be contained.

However, the chromatic aberration is proportional to focal length, and this may limit the ultimate size of the eyes. Powerful spherical lenses are particularly prone to such spherical or chromatic aberration (the curvature of the retina removes the further difficulty of focusing light on a plane). The chromatic error of teleost lenses has been measured at from 2 to 5% of the focal length in various species (listed in Fernald, 1988). These low figures result from a gradient of refractive index across the lens which is maintained even as the eye grows (Fernald and Wright, 1983).

It has been suggested that the varying lengths of cones sensitive to different wavelengths could compensate for the chromatic aberration. Scholes (1975) in *Scardinius erythrophthalamus* and Fernald and Liebman (1980) in *Haplochromis burtoni* have shown that the variation in length of the cones, although in the appropriate direction, are not enough to compensate for the different focal planes of the various wavelengths of light that they are each maximally sensitive to. Such a mechanism may not be needed, as the coloured fringes around the image resulting from the chromatic aberration may be below the resolution of the retinal receptors anyway (Fernald and Wright, 1985).

As the fish lens protrudes through the pupil, each eye may have a field greater than 180°. This large field, combined with the narrowing of the head anteriorly, often results in a small binocular overlap (28° in the goldfish; Fernald, 1988), even though there is no known brain site that receives a binocular retinal projection.

Only a few species have other than spherical lenses. The skates and rays have proportionately larger eyes than their pupil size would indicate. This may suggest they have traded off light-gathering power for great resolution and have aspherical lenses. A few oddities such as the 'four-eyed' minnow *Anableps* (Sivak, 1976) and the mud skipper *Dialommus fuscus* (Stevens and Parsons, 1980) that live at the air–water interface also have aspherical lenses. In *Anableps* the lens is egg-shaped, light from the water passing through the lower, almost spherical part, while light from above the water, partially refracted by the cornea, passes through the less curved upper part of the lens.

Accommodation and Eye Shape

Given the high power of the spherical fish lens, most fish eyes are short and roughly hemispherical. As most fish focus by moving the lens, the globe may be elongated in the axis of accommodation to provide enough space. In most species this is along the naso-temporal axis (Fernald, 1988). This also implies that, at any one time, different areas of the fish retina may be in different states of focus. For example, in the trouts, the lens is slightly ellipsoid with the longest lens diameter parallel to the long axis of the fish. The lens thus has two focal lengths. Distant objects lateral to the fish are focused on central retina while nearer objects are focused on temporal retina. By moving the lens in a plane parallel to the long axis of the fish rather than inwards, the lateral objects remain in focus while the fish adjusts for near vision anteriorly (Lagler *et al.*, 1977). Some Elasmobranchii accommodate by small changes in lens convexity while others use a muscular papilla on the ciliary body to move the lens. The Actinopterygii have varying degrees of accommodation, depending on the development of the retractor lentis muscle. Focusing by moving the lens rather than altering its shape also enables the fish to have a very efficient crystalline lens.

In many deep-water fish the eye is tubular. It has been suggested that this is a way of increasing light capture without having a very large eye (Fernald, 1988). However, as eye size is independent of light-gathering power when the lens is spherical it would seem more likely this is a way of increasing focal length (and hence acuity) while maintaining high retinal illuminance and a relatively small eye. A similar tubular eye is found in the frontal eye pairs of hunting spiders (Land, 1990).

Why tubular eyes should be found in deep-water fish is not clear. Perhaps the relatively clear water at depth permits location of bioluminescent prey at greater distances than in more turbid surface waters. The decreased visual field caused by the tubular eye is mitigated in some species by the presence of a form of light guide that leads light from beyond the periphery of the normal field of view onto an accessory retina in the wall of the tubular eye (Locket, 1977). The hunting spider overcomes the problem of the narrow field of view by being able to scan the 'retina' across the base of its 'tube' eye. This enables the spider to remain still while visually 'hunting' for prey (Land, 1990). It would be interesting to see if any teleosts show a similar adaptation.

Many of these fish with tubular eyes have considerable overlap between the visual fields. This may provide stereoscopic vision; it would be very hard to judge the distance of the light spots on bioluminescent prey against the black featureless background found at depth. This also has the effect of increasing light sensitivity, catching twice the number of photons from the target. This arrangement may also help the signal-to-noise ratio and enhance contrast, as some noise can be filtered out if the inputs from the two eyes can be compared centrally. The axis of the tube is usually directed upwards. This may be because at depth the downwelling light is always brighter than ambient light, so that by looking up the fish can more easily spot its prey against the background light. Other fish compensate for this by having a row of photophores along their ventral surface. This may provide a selection pressure for living at depth, as it will always confer an advantage on the lower fish.

Retinal Specializations for Maximum Light Gathering

The eyes of deep-water fish possess several specializations for maximum light gathering. Some fish have retinae with multiple banks of cone-like receptors sometimes located in a fovea-like pit at the temporal pole of the retina. This retinal specialization is taken to an extreme in some fish, where light is reflected by crystals in a well-developed argenteum lateral to the lens into retinal diverticula containing rods with particularly long outer segments. This may serve two purposes, (a) to increase the visual field and (b) to provide a specialized, particularly light-sensitive area (though perhaps with poor resolution) (Locket, 1977).

Many eyes have a reflective layer behind the retina (the tapetum) which reflects any photons not absorbed by the retina back onto the rods and cones. There must be some loss of acuity because the tapetum is not a perfect mirror, but given the number of aquatic (and terrestrial) animals with such reflective layers the improvement in light detection must more than compensate for the image degradation (Muntz, 1972).

A similar trade-off of acuity for detection can be made by linking receptors together. This varies with the environment of the fish. Fish that rely on good light and vision for hunting such as the pikes (Esocidae) have few receptors linked to one neurone, while deep-sea fish may have many rods converging on one ganglion cell. Some deep-sea fish even go as far as to place a number of receptors in a small reflective cup of tapetum, so they must act as one large receptor. Neural summation can act similarly. In the human as many as 300 rods may synapse on a single ganglion cell; this may have the advantage over a 'physical connection' that these connections may be turned on or off at will by other interneurones.

Integration time is also a way of obtaining more photons from a given area with a possible loss in contrast and resolution. This is revealed by the fish's ability to distinguish separate light flashes at 10 Hz or less in dim light, but up to 60 Hz in bright light (Douglas and Hawryshyn, 1990). A number of deep-sea fish such as the anglers capture their prey by ambush. By remaining still the predator can reduce the rate of movement of the image of the prey over its retina (Lythgoe, 1979).

Physiological noise can also limit resolution. At very low light levels a single rod may receive only one photon every 40 minutes (Land, 1981). Each rod contains 10^9–10^{10} rhodopsin molecules, so these molecules must be very stable or else random conversions will swamp those triggered by photons. Such a phenomenon has been recognized in human manufactured ultra-low-light CCD cameras which may be cooled with liquid nitrogen to reduce background noise. The cold-blooded fish may have less thermal background noise than mammals, which may also be a reason why there are few retinal pigments with maximum sensitivity at long wavelengths. In the fish retina this random background (equivalent to the 'dark current' of the CCD) may limit the thickness of the rhodopsin layer that can be used (Lythgoe, 1988).

Retinomotor movements are a feature of many teleost retinas, particularly those with duplex retinae. In the light the rods elongate towards the pigment epithelium. The pigment granules of the retinal pigment epithelium spread towards the rod outer segments, completely immersing them in the light-absorbing pigment. At the same time the cones contract so that they move away from the pigment, probably in response to a signal from the rods (Kirsch et al., 1989). These movements may take from 20 to 70 minutes. In the dark the process reverses. The fish thus has effectively two different retinae, one for the dark and one for light (Ali and Wagner, 1975).

The retinomotor movements ensure that the rod outer segment is protected from bleaching in the bright light; the dispersed pigment may also absorb scattered light from the cones, raising photopic acuity. The movements of the receptors may also enable closer packing of the cones for better acuity in bright light and for less loss of photons into the cones in dim light. The closer cone packing may be important only in the smallest fish, as the diffraction acuity of the lens in larger eyes is normally limiting, not the resolution of the retina (Fernald, 1988).

In Elasmobranchii lacking a well-developed pigment epithelium the photoreceptor layer may be shielded by melanin which moves in and out in the choroid layer. This may be less effective than the retinomotor movements of the teleosts and may explain why the skates and sharks have developed effective contractile irises.

The Iris

The Elasmobranchii possess dilator and sphincter muscles in the iris and can reduce the pupil to a small slit. In the skates (Batoidei) the pupil may assume an intricate shape. In the winter skate, *Raja ocellata*, a fold from the dorsal iris hangs down over the pupil. The size and shape of this piece depends on the illumination. This operculum is comb-like, its teeth forming a series of about 15 opaque bars across the pupil. These specialized pupils are found in those species living in shallow water or with nocturnal habits which have to cope with wide variations in light intensity. The slit and bar-shaped pupils allow more complete (or even complete) closure of the pupil than our circular iris, where the maximum contraction is limited by the thickness of the muscle fibres in the sphincter pupillae.

Perhaps because of the generally dim environment they inhabit, few teleosts have an equivalent of the elasmobranch or mammalian's sphincteric iris. In most teleosts the pupil size and shape is fixed. The iris is usually circular and the pupil round or slightly oval. When light enters the eye obliquely the size of the effective pupil aperture, and hence the light-gathering power of the eye in that direction, is partially reduced by the margin of the pupil. In some predatory teleosts such as the sailfish *Istiophorus* a notch at the nasal edge of the pupil allows a wider field of view rostrally. This area of the pupil not occupied by the lens is known as the aphakic space. In certain deep-water fish the lens is completely surrounded by an aphakic ring. In a bright shallow-water environment this would lead to higher background 'noise', as the unfocussed light admitted by the aphakic ring spreads around the retina. However, in the deep-water environment where sources of light are mainly bioluminescence coming from discrete point sources, the aphakic space may permit identification of light flashes even at the edge of the field of view. The fish can then turn to view the object so that the light falls mainly on the central, lens-covered, area. A different approach may be taken by deep-sea fish with tubular eyes. Here light coming from oblique angles hits the sides of the 'tube' far from the image-forming retina at the base.

The Cornea and Eye Protection

The eye with all its specializations needs protection. Elasmobranchs have lid-like skin folds and most have a prominent nictitating membrane, particularly obvious in the Carchariidae. The rays do not have such specializations, but withdraw the eye into the head so that a ventral skin fold can cover the opening (Harder, 1975). Most teleosts do not have any such eye-closing structures; instead, some flounders (*Solea*) have retained a distinctly bilayered cornea into the adult to protect the eye of such a bottom-dwelling fish from abrasion by the silt. The mudskipper (*Periophthalmus*) also has a two-layered cornea and also a fold of skin below the eye which holds a pool of water to moisten the eye during the fish's lengthy periods out of the water.

The spherical form of the eyeball provides some strength (*not*, as often erroneously suggested, resistance to pressure; fish tissues are all at ambient pressure, whatever the depth). The internal pressure in small eyes provides some support but in larger eyes most protection is provided by the sclera, which consists of connective tissue with variable amounts of bone or cartilage. In the elasmobranchs there may be bars of cartilage embedded in the sclera. This is carried further in some teleosts, where there may be bony lamellae in the sclera. In the sturgeons (Acipenseriformes) there are two half-moon-shaped bony lamellae between the sclera and the cornea; similar platelets are found in a number of teleosts. These plates are independent of the scleral ossification but are not thought to be related to the pericorneal bony ring found in many fossil species. In agnathans a blood sinus behind the eye may provide some support (Harder, 1975).

Vascular Supply

A number of teleost families possess a unique orbital specialization, the choroideal body. This is a bipolar rete mirabile; it is not found in the Polypterriformes or Lepisosteiformes, but is found in *Amia*. It is not found in small eyes or fish that live in turbid water. It is also missing from animals without a pseudobranch. These factors may indicate that it has some nutritive or excretory function, perhaps providing extra oxygen to the retina (which in many teleosts lacks the surface blood vessels of amniotes). It may be coincidental, but fish with well-developed choroid rete such as the remoras have intraocular oxygen tensions (450–780 mmHg) which are several times that of their arterial blood. In contrast, fishes with small or no rete such as the eels and elasmobranchii have intraocular oxygen tensions lower than arterial blood (Lagler *et al.*, 1977). In some fast-swimming species the choroid body protrudes into the eye itself to form a falciform process which resembles the pecten found in many bird eyes (Collin, 1989). It has been calculated that in some retinas a photon may have a 40–50% chance of hitting a capillary before reaching a receptor (Snodderly and Weinhaus, 1990). By limiting the number of surface blood vessels fish may limit the optical filtering effect of retinal blood vessels which will (being red) tend to cut out the predominantly blue light of the aquatic environment.

The pseudobranch may also be a teleost specialization for improving the oxygenation of the eye. This gill-like organ is found on the inner surface of the gill cover; in some species it is reduced or absent (e.g. in *Amia*), while in others such as the Chondrostei and many teleosts it forms a true-gill-like structure beneath the operculum. It is known as a 'pseudo' branch because it receives arterial, oxygenated blood from the heart rather than the venous deoxygenated flow received by the true gill arches. This would appear to provide a second chance for oxygenation of the blood that has already passed through the main gills. From the capillary bed of the pseudobranch the blood enters the efferent pseudobranch artery, then runs rostrally along the parasphenoid and forms an anastomosis with the contralateral efferent pseudobranch artery. The artery then runs under the optic nerve into the eye (see Harder, 1975, for a review).

The Optic Nerve

Among the Actinopterygii a number of taxa have ribbon-shaped optic nerves, often pleated from side to side (Scholes, 1979; Bunt 1983), while others have more or less fasciculated optic nerves (Bunt, 1982; see Northcutt and Wulliman, 1988, for a comprehensive list). From the observed distribution of these characters among the various actinopterygian classes Northcutt and Wulliman conclude that a pleated optic nerve may be the original or 'plesiomorphic' condition among the teleosts. However, both pleated and fasciculated optic nerves have been found in most classes, including the archaic cladistans and Chondrostei, in roughly equal proportions. With only one example each of the Ginglymodi and Halecomorpha to go on, the fact that both have pleated nerves cannot be used as evidence for it being a primitive character.

It may well be that the degree of 'pleatedness' as opposed to fasciculation may be more related to eye and body size than to phylogenetic relationships. Both the pleated and fasciculated nerves allow more movement of a large eye on a short, thick optic nerve than would a large, solid nerve (Fernald, 1980). The ribbon nerve is often correlated with an extended 'slot'-shaped optic nerve head. This would also have the advantage of taking up less of the crucial central retinal field with a 'blind spot', than would the normal round optic nerve head. The ribbon nerve might be a consequence of the shape of the optic nerve head and have little to do with eye movement (Easter, quoted in Northcutt and Wullimann, 1988). Against this hypothesis, the chick has a slot-shaped optic nerve head but a very solid optic nerve with a circular cross-section (Bunt and Horder, 1983). When examined closely, even fasciculated nerves show an underlying 'pleat' that may be obscured by secondary splitting up of the nerve into fascicles, so the whole issue may be a *non-sequitur* (Bunt, 1982).

The Optic Chiasm

In all fish examined the optic nerves are almost entirely crossed. This crossing of nerves from one eye to the contralateral brainstem is common to all vertebrates. We can only conjecture as to its origins (see for example Sarnat and Netsky, 1974; Gregory, Chapter 3). In some fish such as the goldfish the nerves pass beside each other (Bunt, 1982). In many fish the optic fibres interdigitate at the chiasm, sometimes, as in the salmon, with one nerve penetrating the other as a single block (Stroer, 1940), in others, such as the catfish, in numerous large fascicles (Prasada Rao and Sharma, 1982). We do not have enough evidence on the distribution of such interdigitations to detect any phylogenetic trends. Large interdigitating fascicles appear to have evolved many times among the vertebrates, being found in birds and reptiles while in mammals the fibres are finely mixed at the chiasm. The significance of the number and size of fascicles is unknown. Interdigitation would seem to save space at the chiasm but *Cichlastoma*, with a large pleated nerve in a small head, has separate, non-interdigitating optic nerves (Bunt and Horder, 1983).

Many terrestrial vertebrates have developed a large ipsilateral projection as well. This is most well developed in animals with a large binocular field such as man, but is quite extensive even in animals with only a small visual field overlap such as the rat (Bunt *et al.*, 1983). Although a small ipsilateral projection can be identified in the primitive lamprey (De Miguel *et al.*, 1990), it has never become as well developed in fish. Even advanced teleosts have only a very small ipsilateral projection, and this does not go to the optic tectum (Springer and Gaffney, 1981; Springer and Mednick, 1985). This may seem surprising as many predatory fish, such as the pike, have forward-looking eyes with considerable binocular fields. Correlation between the two visual inputs must occur after termination on the contralateral optic tecta rather than at the level of the thalamus or superior colliculus as in mammals. It may be as a consequence of this that fish appear to judge distance solely by the visual angle subtended by an object, which requires no correlation of the input of the two eyes (Douglas and Hawryshyn, 1990).

Central Visual Projections

Remarkable as they are, the adaptations of the fish eye are only part of the story of evolution of fish vision. The evolution of the brain has been far more important than the development of the eye in the evolution of improvements in the terrestrial visual system. It is probable that the central nervous system has undergone as many changes and adaptations as the external organs. As the evolution of the brain must to some extent have accompanied the development of larger and more sophisticated eyes, it may also be true that, like the eye, most of the evolution occurred early in fish evolutionary history. When we examine existing fish, even the lowly cyclostomes, we may only be looking at the variations in an already well-refined system.

Unfortunately, the soft parts of the brain are not conserved in the fossil record; all we can examine are endocasts of the skulls of the more robust species. This may give some clues as to the gross size of externally observable structures such as the optic tectum, but will tell us nothing about the more important details of the number and type of nuclei subserving vision in the Devonian fish. It is rather like trying to decide the power of a computer by looking at the shape of its case. In fact it's probably even worse than that; it is like trying to see how the software works when all you have is an imperfectly preserved empty computer case with the contents long since destroyed!

It may be possible to detect some evolutionary trends by examining the nuclei and connections directly or indirectly involved in the analysis of vision across a number of taxa. Valiant attempts have been made in the past (see for example Northcutt and Wullimann, 1988). Such studies are beset by difficulties. We rarely have evidence of the functional significance of the various retinal recipient nuclei or their projections. Most studies consist of injections of neuroanatomical tracers such as tritiated amino acids (Repérant et al., 1982), cobaltous lysine (Springer and Mednick, 1985) or HRP (Prasada Rao and Sharma, 1982) into the eye. Such studies show only the direct retinal projections; they tell us nothing about the second-order neurones which may be involved in the processing of the visual input. Only a few species in a few restricted classes have been thoroughly investigated, with further tracing of the efferents and afferents of the nuclei receiving primary retinal input. So far most such studies have been restricted to the Cypriniformes and the Percomorpha (see, for example, Grover and Sharma, 1981; Northcutt and Braford, 1984; Northcutt and Wullimann, 1988) or to fish with particular unique features which make them interesting to study but make cross-taxa comparisons difficult (Meyer and Ebbesson, 1981; Wullimann and Northcutt, 1990).

Without electrophysiological or behavioural evidence we cannot tell the function of the various cell groups identified (Schellart, 1990). It is more than likely that the same cell groups may have been used for different functions at various points in evolution, but from gross appearance or even histological examination such changes may be less than obvious. Unless an adaptation involves radical changes in connectivity, pathway tracing will not pick it up.

Given these caveats, can any trends in the gross appearance of the retinal recipient nuclei be detected? The cyclostomes appear to have maintained more of the characteristics of the original agnathans than any other group (Nieuwenhuys, 1972, 1977). Degeneration studies (Northcutt and Przbylski, 1973), autoradiography (Kennedy and Rubison, 1977) and HRP studies (De Miguel et al., 1990a, b) of the adult lamprey have shown optic projections to the posterior third of the dorsal thalamus contralaterally, the pretectum bilaterally, the tectum contralaterally. At metamorphosis a unique projection appears to the ipsilateral ventrolateral margins of the tectum and pretectum. This may correspond to the accessory nucleus found in other species. In the optic tectum the layered pattern of optic terminals is similar to those in other vertebrates, although it differs from them in having a superficial layer of ependyma (Kennedy and Rubison, 1977).

In contrast to other vertebrates, there appear to be a number of visual fibres that bypass the optic tectum. Responses recorded from the medulla and spinal cord in response to stimulation of the optic nerve were not abolished by removal of the tectum (Veselkin, 1966). They were abolished by incisions made between the midbrain and medulla, suggesting that the pathway may be via the tegmental motor nucleus of the midbrain and the system of Müller neurones.

It has been suggested that the vertebrates show a gradual evolution of more specific pathways from an originally diffuse, almost nerve-net-like ancestral arrangement (Ebbeson, 1980). The lampreys may show a stage in the evolution of nervous system organization based on integration of sensory input in the mesencephalon with outputs to the bulbar nuclei and the descending reticulospinal system (Karamyan, 1975; Karamyan et al., 1975). It is only in later evolved species such as the teleosts that a well-developed cerebellum plays an important role in such sensorimotor integration.

The apparently diffuse nature of the nervous system in the more ancient vertebrate groups has confounded the attempts of many anatomists who have tried to analyse the connections of the brain by looking in detail at what they thought would be 'simpler' patterns of organization (see particularly Herrick, whose heroic but ultimately doomed attempts have filled many an issue of his own journal; Herrick, 1941). As early as 1930, Jansen commented on the number of diffuse connections in the Myxine brain, with numerous axons with long dendritic branches giving off many collaterals along the length of the axon, rather like the projections from the locus coeruleus in mammals.

Whether such a progression is evident in the visual system is very debatable. The most basic vertebrate, the lamprey, has a primary optic projection no more diffuse or 'primitive' than a rat (compare the remarkable similarity between the retinal projections in Kennedy and Rubison, 1977, and Bunt et al., 1983). Although improved techniques may be responsible in part, many more retinal recipient nuclei can be detected in the more 'advanced' teleosts (see Northcutt and Wullimann, 1988, for a review).

Comparative descriptions of retinofugal projections in the teleosts are confused by the difficulty of comparing results obtained with different methodology. For example, compare the studies of the optic projection in Carassius *auratus* by degeneration (Roth, 1969); cobaltous lysine (Springer and Gaffney, 1981); autoradiography (Braford and Northcutt, 1983); and HRP (Fraley and Sharma, 1984), where different projections are identified by each method. Personal experience of using all four methods on the goldfish suggests that the cobaltous lysine, when it works, is the most sensitive method, and individual fibres can be traced with ease so that any leakage or artifactual spread could easily be identified. Nor is there any standard nomenclature, so that some may consider nuclei described in two fish as equivalent while others would see a real difference. Reviews have therefore reported from 7 (Vanegas and Ito, 1983) to 14 (Springer and Gaffney, 1981) projection areas in the teleosts.

For a glimpse of the phylogenetic insights which a detailed comparison of the various details of the optic projections in various representatives of the different teleost taxa may provide, the reader is encouraged to consult two fascinating if contradictory reviews (Vanegas and Ito, 1983; Northcutt and Wullimann, 1988).

In spite of differences in the naming and grouping of nuclei, there is still a distinct pattern of visual input which seems constant over all vertebrate classes from lamprey to man. In some species, only some of the projections have been detected; whether this is due to technical difficulties or is a true absence is often not clear. A visual input to an area may seem to project to an inordinate number of little cell groups, but the general areas of projection are constant. To avoid getting bogged down in competing terminologies, we may say, in the grossest terms, that there is usually a 'suprachiasmatic' projection, inputs to a variable number of cell groups in the thalamus from which a distinct projection to several pretectal nuclei can be distinguished. A distinct fascicle leaves the optic tract to innervate the accessory optic nuclei while the main projection continues to the optic tectum. It has been suggested that the accessory optic nuclei control posture and motion and monitor the body orientation in relation to the environment, while the optic tectum is more concerned with the motion of observed objects such as prey (Frost *et al.*, 1990).

The consistency of this pattern throughout the vertebrates is somewhat surprising, given the evidence that even in normal development some optic fibres seem to overshoot their destinations and form temporary aberrant projections, for example to the contralateral retina (Bunt and Lund, 1981) or even to the auditory colliculus or the spinal cord (P. W. Land, personal communication). This would seem to provide the basis for the evolution of a number of wild projections if they conferred any selective advantage on the animal.

Conclusions

The fish have to cope with a far more difficult visual environment than do terrestrial vertebrates. They have evolved many and varied specializations to cope with low light levels, turbid water and the colour-filtering effects of different water types. Fish appear to have evolved methods of increasing contrast and sensitivity rather than acuity.

In spite of the wide range of adaptations, the basic layout of the retina and lens remains similar from cyclostome to teleost. A close examination of the agnathans, both living and fossil, shows an eye not unlike our own. The invertebrates show many more 'visual experiments'. There are no mirror lenses among the fish to compare with those of *Pecten* (Land, 1965) or other invertebrates (Land, 1978; Cronin, 1988), no multi-lensed or compound eyes to increase light-gathering power or counteract chromatic aberration. The visual pigments, though enormously varied, do not extend over as wide a frequency spectrum as those found in terrestrial insects.

It appears that the method of growth of the earliest eyes as an outpocketing of the early neural plate followed by the induction of a lens from the overlying ectoderm has locked the vertebrate eye into a pattern from which it can deviate only so far. The writer knows of no vertebrate with more than one lens, though, as in *Anableps*, the lens may be shaped to act as two. There is thus no equivalent of the zoom lens, although the tubular eye could be analogous to a fixed telephoto lens. Some fish have evolved optical globes of odd shapes, with tubular eyes and eyes with lateral pockets, but all possess only two well-developed lateral eyes. The median eyes, paired or single, appear never to have become well developed and may have lost most of their use as shadow detectors when the ostracoderms left their muddy habitat. Was, or is, there any restriction on the way these organs connect to the brain, which prevented them ever becoming useful as image-forming devices?

The terrestrial vertebrates have relied on increasing sophistication of central processing rather than improvements of the eye and retina to improve vision. The increasing complexity of the brain is more striking in the terrestrial vertebrates than the teleosts. A large and complex brain requires a high metabolic rate. One can conjecture that the even aquatic temperatures have led to the limited homiothermy shown by fish and this in turn has restricted the maximum development of their central nervous system. Once the eye and its accompanying central optic centres were evolved by the first agnathans, only limited changes could be made to the basic design. Thus, for all the wonders we can see with it, our eye remains little changed from that of the lowly cyclostome, which may never see more than the slime-encrusted sides of its muddy burrow.

References

Ali, M. A. and Wagner, H. H. (1975). Distribution and development of retinomotor responses. In *Vision in Fishes*. ed. Ali, M. A. pp. 369–396. New York: Plenum.

Archer, S. N. and Lythgoe, J. N. (1990). The visual pigment basis for cone polymorphism in the guppy *Poecilia reticulata*. *Vision Res.*, 30, 225–233.

Atema, J., Fay, R. R., Popper, A. N. and Tavolga, W. N. (eds) (1988). *Sensory Biology of Aquatic Animals*. New York: Springer.

Bardack, D. and Zangerl, R. (1971). Lampreys in the fossil record. In *The Biology of Lampreys*, Vol. 1. eds. Hardisty, M. W. and Potter, I. C. pp. 67–84. London: Academic.

Beamish, F. W. H. and Potter, I. C (1975). The biology of the anadromous sea lamprey (*Petromyzon marinus*) in New Brunswick. *J. Zool. Lond.*, 177, 57–72.

Beatty, D. D. (1984). Visual pigments and the labile scotopic visual system of fish. *Vision Res.*, 24, 1563–1573.

Bjerring, H. (1977). A contribution to structural analysis of the head of craniate animals. *Zool. Scripta*, 6, 127–183.

Braford, M. R. Jr and Northcutt, R. G. (1983). Organization of the diencephalon and pretectum of the ray-finned fishes. In *Fish Neurobiology*, Vol. 2. eds. Davis, R. E. and Northcutt, R. G. pp. 117–163. Ann Arbor: University of Michigan Press.

Bridges, C. D. B. and Delisle, C. E. (1974). Postglacial evolution of the visual pigments of the smelt, *Osmerus eperlanus mordax*. *Vision Res.*, 14, 345–356.

Bridges, C. D. B. and Yoshikami, S. (1970). The rhodopsin–porphyropsin system in freshwater fishes. 2. Turnover and interconversion of visual pigment prosthetic groups in light and darkness: role of the pigment epithelium. *Vision Res.*, 10, 1333–1345.

Bunt, S. M. (1982). Retinotopic and temporal organization of the optic nerve and tracts in the adult goldfish, *J. Comp. Neurol.*, 206, 209–226.

Bunt, S. M. and Lund, R. D. (1981). Development of a transient retino-retinal pathway in albino and hooded rats. *Brain Res.*, 211, 399–404.

Bunt, S. M. and Horder, T. J. (1983). Evidence for an orderly arrangement of optic axons within the optic nerves of the major non-mammalian vertebrate classes. *J. Comp. Neurol.*, 213, 94–114.

Bunt, S. M., Lund, R. D. and Land, P. W. (1983). Prenatal development of the optic projection in albino and hooded rats. *Dev. Brain Res.*, 6, 149–168.

Burkhardt, D. A., Gottesman, J., Levine, J. S. and MacNichol, E. F. Jr (1983). Cellular mechanisms for color-coding in holostean retinas and the evolution of colour vision. *Vision Res.*, 23, 1031–1041.

Cohen, J. L. (1990). Vision in elasmobranchs. In *The Visual System of Fish*. eds. Douglas, R. H. and Djamgoz, M. B. A. pp. 465–490. London: Chapman & Hall.

Cohen, J. L., Hueter, R. E. and Organisciak, D. T. (1990). The presence of a porphyropsin-based visual pigment in the juvenile lemon shark (*Negaprion brevirostris*). *Vision Res.*, 30, 1949–1953.

Collin, S. P. (1989). Topographic organization of the ganglion cell layer and intraocular vascularization in the retinae of the reef teleosts. *Vision Res.*, 29, 765–775.

Collin, S. P. and Pettigrew, J. D. (1988). Retina and ganglion cell topography in teleosts: A comparison between Nissl-stained material and retrograde labelling from the orbit. *J. Comp. Neurol.*, 276, 412–422.

Crescitelli, F. (1972). The visual cells and visual pigments of the vertebrate eye. In *Photochemistry of Vision*. ed. Dartnall, H. J. A. pp. 245–363. Berlin: Springer.

Cronly-Dillon, J. R. and Muntz, W. R. A. (1965). The spectral sensitivity of the goldfish and the clawed toad tadpole under photopic conditions. *J. Exp. Biol.*, 42, 481–493.

Cronin, T. W. (1988). Vision in marine invertebrates. In *Sensory Biology of Aquatic Animals*. eds. Atema, J., Fay, R. R., Popper, A. N. and Tavolga, W. N. pp. 403–418. New York: Springer.

Dall, W., Hill, B. J., Rothlisberg, P. C. and Sharples, D. J. (1990). *The Biology of the Penaeidae*. Vol. 27. *Advances in Marine Biology*.

Dawkins, R. (1986). *The Blind Watchmaker*. London: Penguin.

De la Motte, I. (1964). Untersuchungen zur vergleichenden Physiologie der Lichtempfindlichkeit geblendeter Fische. *Z. Vergl. Physiol.*, 49, 58–90.

De Miguel, E., Rodicio, M. C. and Anadon, R. (1990a). Organization of the visual system in larval lampreys: an HRP study. *J. Comp. Neurol.*, 302, 529–542.

De Miguel, E., Rodicio, M. C. and Anadon, R. (1990b). Growth cones and retinopetal fibres of the larval lamprey retina. An HRP ultrastructural study. *Neurosci. Lett.*, 106, 1–6.

Dickson, D. H. and Collard, T. R. (1979). Retinal development in the lamprey (*Petromyzon marinus* L.): premetamorphic ammocoete eye. *Am. J. Anat.*, 154, 321–336.

Dickson, D. H. and Graves, D. A. (1981). The ultrastructure and development of the eye. In *The Biology of Lampreys*. eds. Hardisty, M. W. and Potter, I. C. pp. 43–94. London: Academic.

Dilly, N. (1961). Electron microscopic observations of the receptors in the sensory vesicle of the ascidian tadpole. *Nature Lond.*, 191, 786–787.

Douglas, R. H. (1989). The spectral transmission of the lens and cornea of the brown trout (*Salmo trutta*) and goldfish (*Carassius auratus*) — effect of age and implication for UV vision. *Vision Res.*, 29, 861–869.

Douglas, R. H. and McGuigan, C. M. (1989). The spectral transmission of freshwater teleost ocular media: an interspecific comparison and a guide to potential ultraviolet sensitivity. *Vision Res.*, 29, 871–879.

Douglas, R. H. and Djamgoz, M. B. A. (eds) (1990). *The Visual System of Fish*. London: Chapman & Hall.

Douglas, R. H. and Hawryshyn, C. W. (1990). In *The Visual System of Fish*. eds. Douglas, R. H. and Djamgoz, M. B. A. pp. 373–418. London: Chapman & Hall.

Eakin, R. M. and Kuda, A. (1971). Ultrastructure of sensory receptors in ascidian tadpoles. *Z. Zellforsch.*, 112, 287–312.

Ebbeson, S. O. E. (1980). The parcellation theory and its relation to interspecific variability in brain organization, evolutionary and ontogenetic development and neuronal plasticity. *Cell Tiss. Res.*, 213, 505–508.

Eddy, J. M. P. (1969). Metamorphosis and the pineal complex in the brook lamprey, *Lampetra planeri*. *J. Endocrinol.*, 44, 415–452.

Eddy, J. M. P. (1972). The pineal complex. In *The Biology of Lampreys*, Vol. 2. eds. Hardisty, M. W. and Potter, I. C. pp. 91–103. London: Academic.

Fernald, R. D. (1980). Optic nerve distension in a cichlid fish. *Vision Res.*, 20, 1015–1019.

Fernald, R. D. (1988). Aquatic adaptations in fish eyes. In *Sensory Biology of Aquatic Animals*. eds. Atema, J., Fay, R. R., Popper, A. N. and Tavolga, W. N. pp. 434–466. New York: Springer.

Fernald, R. D. (1990). *Haplochromis burtoni*: a case study. In *The Visual System of Fish*. eds. Douglas, R. H. and Djamgoz, M. B. A. pp. 443–464. London: Chapman & Hall.

Fernald, R. D. and Liebman, P. (1980). Visual receptor pigments in the African cichlid fish *Haplochromis burtoni*. *Vision Res.*, 20, 857–864.

Fernald, R. D. and Wright, S. (1983). Maintenance of optical quality during crystalline lens growth. *Nature Lond.*, **301**, 618–620.

Fernald, R. D. and Wright, S. (1985). Growth of the visual system of the African cichlid fish *Haplochromis burtoni*: optics. *Vision Res.*, **25**, 155–161.

Fraley, S. M. and Sharma, S. C. (1984). Topography of retinal axons in the diencephalon of goldfish. *Cell Tissue Res.*, **238**, 529–538.

Franz, V. (1932). Auge und Akkommodation von *Petromyzon (Lampetra) fluviatilis L. Zool. J.*, **52**, 118–178.

Frost, B. J., Wylie, D. R. and Wong, Y.-C. (1990). The processing of object and self motion in the tectofugal and accessory optic pathways of birds. *Vision Res.*, **43**, 1677–1688.

Gans, C. (1989). Stages in the origin of vertebrates: analysis by means of scenarios. *Biol. Rev.*, **64**, 221–268.

Gorman, A. L. F., McReynolds, J. S. and Barnes, S. N. (1971). Photoreceptors in primitive chordates: fine structure, hyperpolarizing receptor potentials and evolution. *Science Wash. DC*, **172**, 1052–1054.

Grover, B. G. and Sharma, S. C. (1981). Organization of extrinsic tectal connections in goldfish (*Carassius auratus*). *J. Comp. Neurol.*, **196**, 471–488.

Hamaski, D. I. and Eder, D. J. (1977). Adaptive radiation of the pineal. In *Handbook of Sensory Physiology*, Vol. VII/5. ed. Crescitelli, F. pp. 497–548. Berlin: Springer.

Hawryshyn, C. W. and Beauchamp, R. D. (1985). Ultraviolet photosensitivity in goldfish: an independent u.v. retinal mechanism. *Vision Res.*, **25**, 11–20.

Harder, W. (1975). *Anatomy of Fishes*. Stuttgart: Schweizerbart'sche.

Hardisty, M. W. (1979). *Biology of the Cyclostomes*. London: Chapman & Hall.

Hardisty, M. W. and Potter, I. C. (1971). *The Biology of Lampreys*, Vol. 1. London: Academic.

Hardisty, M. W. and Potter, I. C. (1981). *The Biology of Lampreys*, Vol. 3. London: Academic.

Hardisty, M. W., Potter, I. C. and Sturge, R. A. (1970). Comparison of metamorphosing and macrophthalmia stages of the lampreys *Lampetra fluviatilis* and *L. planeri. J. Zool. Lond.*, **162**, 383–400.

Harosi, F. I. and Hashimoto, Y. (1983). Ultraviolet visual pigment in a vertebrate: a tetrachromatic cone system in the dace. *Science Wash. DC*, **222**, 1021–1023.

Herrick, C. J. (1941). The eyes and optic paths of the catfish, *Ameiurus. J. Comp. Neurol.*, **75**, 487–544.

Jansen, J. K. S. (1930). The brain of *Myxine glutinosa. J. Comp. Neurol.*, **22**, 359–507.

Janvier, P. (1974). The structure of the naso-hypophysial complex and the mouth in fossil and extant cyclostomes with remarks on amphiaspiformes. *Zool. Scripta*, **3**, 193–200.

Janvier, P. (1975). Les yeux des Cyclostomes fossiles et le problème de l'origine des Myxinoides. *Acta Zool. Stockh.*, **56**, 1–9.

Janvier, P. (1978). Les nageoires paires des Ostéostraces et la position systématique des Céphalaspidomorphes. *Ann. Palaeontol.*, **64**, 113–142.

Jarvik, E. (1968). In *Current Problems in Lower Vertebrate Phylogeny*, Nobel Symposium, Vol. 4. ed. Orvig, T. pp. 497–527. New York: Interscience.

Johns, P. R. and Fernald, R. D. (1981). Genesis of rods in teleost fish retina. *Nature Lond.*, **293**, 141–142.

Karamyan, A. I., Zagorul'ko, T. M., Belekhova, M. G., Veselkin, N. P. and Kosavera, A. A. (1975). Corticalisation of two divisions of the visual system during vertebrate evolution. *J. Evol. Biochem. Physiol.*, **7**, 12–20.

Karamyan, A. I. (1975). The views of A. A. Ukntomskii on the hierarchic organisation of the central nervous system and the modern achievements of evolutionary neurophysiology. *J. Evol. Biochem. Physiol.*, **11**, 218–224.

Keating, M. J. (1977). Evidence for a plasticity of intertectal connections in adult *Xenopus. Phil. Trans. R. Soc. Lond. B.*, **278**, 277–294.

Kennedy, M. C. and Rubison, K. (1977). Retinal projections in larval, transforming and adult sea lamprey, *Petromyzon marinus. J. Comp. Neurol.*, **171**, 465–479.

Kirsch, M., Wagner, H-J. and Douglas, R. M. (1989). Rods trigger light adaptive retinomotor movements in all spectral cone types in a teleost fish. *Vision Res.*, **29**, 389–396.

Kleerekoper, H. (1972). In *The Biology of Lampreys*. eds. Hardisty, M. W. and Potter, I. C. pp. 373–404. London: Academic.

Lagler, K. F., Bardach, J. E., Miller, R. R. and May Passino, D. R. (1977). *Ichthyology*, 2nd edn. New York: Wiley.

Land, M. F. (1965). Image formation by a concave reflector in the eye of the scallop, *Pecten maximus. J. Physiol. Lond.*, **179**, 138–153.

Land, M. F. (1978). Animal eyes with mirror optics. *Sci. Am.*, **239**, 126–134.

Land, M. F. (1981). Optics and vision in invertebrates. In *Handbook of Sensory Physiology*, Vol. VII/6B. ed. Autrum, H. J. pp. 472–592. Berlin: Springer.

Land, M. F. (1990). Direct observation of receptors and images in single and compound eyes. *Vision Res.*, **30**, 1721–1734.

Locket, N. A. (1977). Adapations to the deep-sea environment. In *Handbook of Sensory Physiology*, Vol. VII/5. ed. Crescitelli, F. pp. 67–192. Berlin: Springer.

Loew, E. R. and Lythgoe, J. N. (1978). The ecology of cone pigments in teleost fishes. *Vision Res.*, **18**, 715–722.

Loew, E. R. and McFarland, W. N. (1990). The underwater visual environment. In *The Visual System of Fish*. eds. Douglas, R. H. and Djamgoz, M. B. A. pp. 1–43. London: Chapman & Hall.

Lythgoe, J. N. (1972). The adaptation of visual pigments to the photic environment. In *Photochemistry of Vision*, Handbook of Sensory Physiology, Vol. VII/1. ed. Dartnall, H. T. A. pp. 529–565. Berlin: Springer.

Lythgoe, J. N. (1974). The structure and physiology of iridescent corneas in fishes. In *Vision in Fishes*. ed. Ali, M. A. pp. 253–262. New York: Plenum.

Lythgoe, J. N. (1979). *The Ecology of Vision*. Oxford: Clarendon Press.

Lythgoe, J. N. (1980). Vision in fishes: ecological adaptations. In *Environmental Physiology of Fishes*. ed. Ali, M. A. pp. 431–445. New York: Plenum.

Lythgoe, J. N. (1988). Light and vision in the aquatic environment. In *Sensory Biology of Aquatic Animals*. eds. Atema, J., Fay, R. R., Popper, A. N. and Tavolga, W. N. pp. 57–82. New York: Springer.

MacNichol, E. F. Jr, Kunz, Y. W., Levine, J. S., Harosi, F. I. and Collins, B. A. (1978). Ellipsosomes: organelles containing a cytochrome-like pigment in the retinal cones of certain fishes. *Science Wash. DC*, **200**, 549–552.

Meiniel, A. (1971). Etude cytophysiologique de l'organe parapinéal de *Lampetra planeri. J. Neurovisc. Rel.*, **32**, 157–199.

Meiniel, A. and Collins, J. P. (1971). Le complex pinéal de l'ammocoète (*Lampetra planeri*). *Z. Zellforsch.*, **117**, 354–380.

Meyer, D. L. and Ebbesson, S. O. E. (1981). Retinofugal and retinopetal connections in the upside-down catfish (*Synodontis nigriventris*). *Cell Tiss. Res.*, **218**, 389–401.

Muntz, W. R. A. (1972). Inert absorbing and reflecting pigments. In *Photochemistry of Vision*, Handbook of Sensory Physiology, Vol. VII/1. ed. Dartnall, H. J. A. pp. 529–565. Berlin: Springer.

Muntz, W. R. A. (1976). On yellow lenses in mesopelagic animals. *J. Mar. Biol. Ass. UK*, **56**, 963–976.

Muntz, W. R. A. (1990). Stimulus, environment and vision in fishes. In *The Visual System of Fish*. eds. Douglas, R. H. and Djamgoz, M. B. A. pp. 491–511. London: Chapman & Hall.

Muntz, W. R. A. and Mouat, G. S. V. (1984). Annual variation in the visual pigments of brown trout inhabiting lochs providing different light environments. *Vision Res.*, **24**, 1575–1580.

Nelson, G. J. (1969). Origin and diversification of teleostean fishes. In *Comparative and Evolutionary Aspects of the Vertebrate Central Ner-

vous System. eds. Petras J. M. and Noback, C. R. *Ann. NY Acad. Sci.*, **167**, 18–30.

Neumeyer, C. and Arnold, K. (1989). Tetrachromatic color vision in the goldfish becomes trichromatic under white adaptation light of moderate intensity. *Vision Res.*, **29**, 1719–1727.

Newth, D. R. and Ross, D. M. (1955). On the reaction to light of *Myxine glutinosa*. L. *J. Exp. Biol.*, **32**, 4–21.

Nicol, J. A. C. (1975). Studies on the eyes of fishes: structure and ultrastructure. In *Vision in Fishes*. ed. Ali, M. A. pp. 579–607. New York: Plenum.

Nieuwenhuys, R. (1972). Topographical analysis of the brainstem of the lamprey, *Lampetra fluviatilis*. *J. Comp. Neurol.*, **145**, 165–178.

Nieuwenhuys, R. (1977). The brain of the lamprey in a comparative perspective. *Ann. NY Acad. Sci.*, **299**, 97–145.

Northcutt, R. G. and Braford, M. R. (1984). Some efferent connections of the superficial pretectum in the goldfish. *Brain Res.*, **296**, 181–184.

Northcutt, R. G. and Przbylski, R. J. (1973). Retinal projections in the lamprey, *Petromyzon marinus* L. *Anat. Rec.*, **175**, 400.

Northcutt, R. G. and Wullimann, M. F. (1988). The visual system in teleost fishes: morphological patterns and trends. In *Sensory Biology of Aquatic Animals*. eds Atema, J., Fay, R. R., Popper, A. N. and Tavolga, W. N. pp. 515–552. New York: Springer.

Northmore, D. P. M. (1977). Spatial summation and light adaptation in the goldfish visual system. *Nature Lond.*, **268**, 450–451.

Northmore, D. P. M. and Dvorak, C. A. (1979). Contrast sensitivity and acuity of the goldfish. *Vision Res.*, **19**, 255–261.

Potter, I. C. and Beamish, F. W. H. (1975). Lethal temperatures in ammocoetes of four species of lampreys. *Acta Zool. Stockh.*, **56**, 88–91.

Powers, M. K., Bani, C. J., Rose, L. A. and Raymond, P. A. (1988). Visual detection by the rod system in goldfish of different sizes. *Vision Res.*, **28**, 211–221.

Prasada Rao, P. D. and Sharma, S. C. (1982). Retinofugal pathways in juvenile and adult channel catfish, *Ictalurus (Ameiurus) punctatus*: an HRP and autoradiographic study. *J. Comp. Neurol.*, **210**, 37–48.

Repérant, J., Vesselkin, N. P., Ermakova, T. V., Rustamov, E. K., Rio, J. P., Palatnikov, G. K., Perichoux, J. and Kasimov, R. V. (1982). The retinofugal pathways in a primitive actinopterygian, the chondrostean *Acipenser guldenstadti*: an experimental study using degeneration, radioautographic and HRP methods. *Brain Res.*, **251**, 1–23.

Repetski, J. E. (1978). A fish from the Upper Cambrian of North America. *Science Wash. DC*, **200**, 529–531.

Ritchie, A. (1968). New evidence on *Jamoytius kerwoodi* White, an important Ostracodern from the Silurian of Lanarkshire, Scotland. *Palaeontology*, **11**, 21–39.

Roberts, A. (1978). Pineal eye and behaviour in *Xenopus* tadpoles. *Nature Lond.*, **273**, 774–775.

Robertson, J. D. (1957). The habitat of the early vertebrates. *Biol. Rev.*, **32**, 156–187.

Robertson, J. D. (1959). The origin of the vertebrates — marine or freshwater. *Adv. Sci.*, **61**, 516–520.

Romer, A. S. and Parsons, T. S. (1977). *The Vertebrate Body*, 5th Edn. London: Saunders.

Roth, R. L. (1969). Optic tract projections in representatives of two fresh-water teleost families. *Anat. Rec.*, **163**, 253–254.

Sarnat, H. B. and Netsky, M. G. (1974). *Evolution of the Nervous System*. New York: Oxford University Press.

Schaeffer, B. (1969). Adaptive radiation of the fishes and the fish-amphibian transition. In *Comparative and evolutionary aspects of the vertebrate central nervous system*. eds. Petras, J. M. and Noback, C. R. *Ann. NY Acad. Sci.*, **167**, 5–17.

Schellart, N. A. M. (1990). The visual pathways and central non-tectal processing. In *The Visual System of Fish*. eds. Douglas, R. H. and Djamgoz, M. B. A. pp. 345–372. London: Chapman & Hall.

Scholes, J. H. (1975). Colour receptors and the synaptic connexions in the retina of a cyprinid fish. *Phil. Trans. R. Soc. Lond. B*, **270**, 61–118.

Scholes, J. H. (1979). Nerve fibre topography in the retinal projection to the tectum, *Nature Lond.*, **278**, 620–624.

Schwassmann, H. O. and Kruger, L. (1965). Experimental analysis of the visual system of the four-eyed fish (*Anableps*). *Vision Res.*, **14**, 209–213.

Sivak, J. G. (1976). Optics of the 'four-eyed fish' (*Anableps*). *Vision Res.*, **16**, 531–534.

Snodderly, D. M. and Weinhaus, R. S. (1990). Retinal vasculature of the fovea of the squirrel monkey *Saimiri sciureus*. Three dimensional architecture, visual screening, and relationships to the neuronal layers. *J. Comp. Neurol.*, **297**, 145–163.

Springer, A. D. and Gaffney, J. S. (1981). Retinal projections in the goldfish: a study using cobaltous-lysine. *J. Comp. Neurol.*, **203**, 401–424.

Springer, A. D. and Mednick, A. S. (1985). Topography of the retinal projections to the superficial pretectal parvicellular nucleus of goldfish: a cobaltous-lysine study. *J. Comp. Neurol.*, **237**, 239–250.

Stensio, E. A. (1968). The cyclostomes with special reference to the diphyletic origin of the Petromyzontida and the Myxinoidea. In *Current Problems in Lower Vertebrate Physiology*. ed. Orvig, T. pp. 13–71. Stockholm: Almquist & Wiksell.

Steven, D. M. (1955). Experiments on the light sense of the hag, *Myxine glutinosa* L. *J. Exp. Biol.*, **32**, 22–38.

Stevens, J. K. and Parsons, K. E. (1980). A fish with double vision. *Nat. Hist.*, **89**, 62–67.

Stroer, W. F. H. (1940). Zur vergleichenden Anatomie des primaren optischen Systems bei Wirbeltieren. *Z. Anat. Entwick.*, **101**, 301–321.

Tamura, T. and Hanyu, I. (1980). Pineal photosensitivity in fishes. In *Environmental Physiology of Fishes*. ed. Ali, M. A. pp. 477–496. New York: Plenum.

Tarlo, L. B. H. and Whiting, H. P. (1965). A new interpretation of the internal anatomy of the Heterostraci (Agnatha). *Nature Lond.*, **206**, 148–150.

Vanegas, H. and Ito, H. (1983). Morphological aspects of the teleostean visual system: a review. *Brain Res. Rev.*, **6**, 117–137.

Veselkin, N. P. (1966). Electrical reactions in midbrain, medulla and spinal cord of the lamprey to visual stimulation. *Fedn Proc. Fedn Am. Socs Exp. Biol.*, **25**, 957–960.

Wald, G. (1957). The metamorphosis of visual system in the sea lamprey. *J. Gen. Physiol.*, **40**, 901–904.

Warner, J. A., Latz, M. I. and Case, J. F. (1979). Cryptic biolumine-scence in a midwater shrimp. *Science Wash. DC*, **203**, 1109–1110.

Westheimer, G. (1968). In *The Eye, Medical Physiology*, 12th Edn. ed. Mountcastle, V. B. pp. 1532–1553. St Louis: Mosby.

Whiting, H. P. and Tarlo, L. B. H. (1965). The brain of the Heterostraci (agnatha). *Nature Lond.*, **207**, 829–831.

Wullimann, M. F. and Northcutt, R. G. (1990). Visual and electrosensory circuits of the diencephalon in Mormyrids: an evolutionary perspective. *J. Comp. Neurol.*, **297**, 537–552.

19 Central Visual Pathways in Reptiles and Birds: Evolution of the Visual System

Toru Shimizu and Harvey J. Karten

Introduction

Studies of the visual system have proved to be among the most important and fruitful topics for comparative analysis of the nervous system in all vertebrates. The field of modern comparative neurobiology is dominated by studies of the visual system and has provided the richest basis for analysis of the central nervous system. In large measure, this is consequent to the prominence of vision in most vertebrates and the accessibility of the retina and optic tectum to anatomical, physiological, biochemical, molecular biological and developmental investigations.

The greatest effort and resultant progress in this regard has unquestionably been in studies of the retina (e.g. Walls, 1942; Karten et al., 1990). Far less was known of the central organization of the visual system in non-mammalian vertebrates. For many decades, the optic tectum was almost the sole topic of interest in studies of the non-mammalian visual system and was, by default, considered to be the major structure involved in visual discrimination, orientation and general visuomotor coordination. This narrow and erroneous attitude led to the mistaken belief that the non-mammalian optic tectum was the 'highest visual centre' whose functions were 'taken over by the striate cortex' in mammals. This notion seemed particularly attractive in view of the elegant and distinctive laminar differentiation of the avian optic tectum.

However, research on the central visual pathways in birds and reptiles within the past two decades, though still modest in scope compared with the vast literature on the mammalian visual system, has provided a tantalizing view of the organization of the visual system in non-mammalian vertebrates. These recent studies have led to dramatically altered concepts of the probable evolutionary origins of 'higher' visual centres in mammals. The purpose of this chapter is to provide an overview of the organization of central visual pathways in reptiles and birds, with hopeful consequences that readers unfamiliar with non-mammalian visual systems will recognize the prevalence of patterns of organization common to all amniotes (i.e. reptiles, birds and mammals), and indeed perhaps all vertebrates.

Phylogeny: Reptiles, Birds and Mammals

The field of vertebrate taxonomy and phylogeny remains an area of active research and scientific advances. The main concepts in the following section are based largely on the works of Romer (1974), Macphail (1982) and Carroll (1987) and reflect the generally accepted views of the organization and relationships of amniotic vertebrates. It is intended as a guide for those readers with less familiarity with the taxonomic and morphological features of the amniotes.

Among vertebrates, reptiles, birds and mammals have an amnion and are collectively classified as amniotes. All the amniotes are considered to have evolved from the primitive tetrapods known as 'stem reptiles,' or cotylosaurs about 300 Myr ago (Romer, 1974; Carroll, 1987). However, the lines of descent from cotylosaurs to modern reptiles, birds and mammals diverged at a very early stage in the evolution of reptiles (see Fig. 19.1). The relationships among living amniotes are, therefore, quite distant. Thus, similarities between the brain and the visual system of the ancestors of present-day mammals and those of the present-day reptiles and birds may reflect their common origins. Yet there may also be many differences consequent to their evolutionary diversities. The brain and its

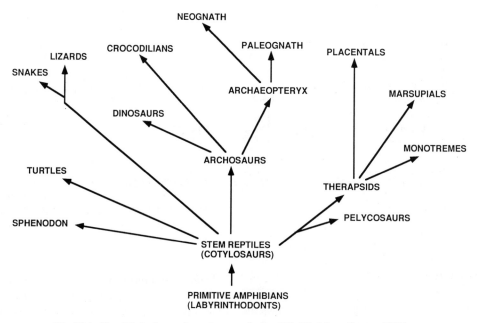

Fig. 19.1 *Simplified scheme of vertebrate evolution (Modified from Romer, 1974).*

functions (including the visual system) of existing reptiles and birds are not primitive versions of mammals (Hodos and Campbell, 1969; Macphail, 1982).

Reptiles

The first reptiles (cotylosaurs) appear to have emerged during the Carboniferous period (345–281 Myr ago), and greatly proliferated during the Mesozoic era (225–66 Myr ago), commonly referred to as the Age of Reptiles. During the Mesozoic era, the large dinosaurs evolved, became dominant on land, sea and in the air, and then became extinct rather abruptly by the end of the Cretaceous period (the last period of the Mesozoic era). Only four orders of reptiles (Rhynchocephalia, Squamata, Chelonia and Crocodilia) have survived.

The only surviving genus of the order Rhynchocephalia is the tuatara (*Sphenodon punctatus*) found in New Zealand. The *Sphenodon*, which is probably the most primitive living reptile, is similar to a lizard and has a rudimentary third eye. The order Squamata consists of three suborders: Sauria (Lacertilia, or lizards, about 3000 species); Amphisbaenia (Annulata, or worm-lizards, about 130 species); and Serpentes (Ophidia, or snakes, about 3000 species). Both Rhynchocephalia and Squamata belong to the same subclass lepidosaur, although they evolved independently (280–226 Myr ago). On the basis of the development of the dorsal ventricular ridge in the telencephalon (see pp 425–426), lizards have been subdivided into two groups: type I and type II (Northcutt and Butler,

1974b). Type I group includes the majority of lizards, such as *Gekkonoidae*, *Xantusidae* and *Lacertida*. Type II group includes *Iguanidae*, *Teiidae* and *Varanidae*.

The order Chelonia (turtles and tortoises) includes 220 species. They evolved directly from the stem reptiles cotylosaurs. They diverged from other reptilian groups at a very early stage in the reptilian radiation at least by the late Triassic period (225–196 Myr ago). They evolved various specialized features (modification for aquatic environment, development of shell and loss of dentition), and through their evolutionary history they have showed very little advancement in many respects since then. Thus, the visual system of turtles and tortoises may retain the features found in their ancestors, but in concert with their other specialized adaptations, their brain may represent a marked divergence from the main pattern of amniote visual systems.

The order Crocodilia (crocodiles and alligators) has 21 living species. They are large amphibious animals with coarse bony plates in their skin. The Crocodilia are the sole extant descendants of the ruling reptiles (archosaurs) from which dinosaurs, the saurischians and the ornithischians, evolved. Crocodilians have many neural characteristics – including the visual system – similar to those of birds, possibly because birds also evolved from archosaurs.

It was during the early Permian period (about 280–226 Myr ago) that the ancestors of lepidosaurs, chelonians and archosaurs evolved from cotylosaurs. Therefore, there has been no common ancestry among surviving reptilian orders for at least 225 Myr.

Birds

The most conclusive evidence that the evolution of birds derived from reptiles is the earliest fossils of *Archaeopteryx*, which has a reptilian body and a long tail yet undeniably birdlike wings and feathers (e.g. Wellnhofer, 1988). *Archaeopteryx*, a bird about the size of a pigeon, lived in the late Jurassic period (about 150 Myr ago).

Marked radiations of birds occurred during the Cretacean period (135–66 Myr ago) and the first, Tertiary, period of the Cenozoic era (65–1.9 Myr ago). Ancestors of all non-passerine and most passerine (perching birds) families had appeared by the end of the Miocene epoch (5.5 Myr ago). However, none of the actual species that existed during that epoch survives today. All modern species appear to have evolved within the last 1 Myr.

Birds are warm-blooded and the only vertebrates with feathers. Despite the diversities in adaptations (from aerial to flightless), their general body form has remained unchanged. They have a horny beak and their anterior limbs are specialized into wings. Modern birds are categorized into at least 8600 species and 28 orders. The two superorders of birds are the Palaeognathae (or Ratitae) and the Neognathae. The ostrich, emu, cassowary, rhea, kiwi and tinamous of South America belong to the former. Most of them have lost the power of flight. All carinate birds belong to the latter (22 orders), which were classified into three groups: aquatic birds, birds of prey and terrestrial birds (Carroll, 1987). Among them, a great number of studies of the central nervous system have focused on only a few species, including the following orders: (1) Strigiformes (owls), (2) Columbiformes (pigeons), (3) Galliformes (quail, chickens) and (4) Passeriformes (swallows, sparrows). The Passeriformes comprise about half of the present bird population.

Mammals

The ancestors of mammals, a group of reptiles known as the pelycosaurs, first emerged in the late carboniferous period (about 290 Myr ago). They became abundant during the early Permian period. The therapsids, a reptilian group derived from the pelycosaurs, succeeded them during the late Permian and early Triassic periods. In the late Triassic (about 200 Myr ago), mammals evolved from the therapsids order. Thus, ancestors of the present mammals, including monotremes, marsupials and placentals, were derived from other reptilian groups (ancestors of modern reptiles and birds) at least 280 Myr ago, and evolved independently from these groups. We do not know when the mammalian types of brain first appeared, but we shall assume that the brains of pelycosaurs and early therapsids more closely resembled that of the primitive reptilian type of brain. This interpretation is largely speculative as fossil evidence cannot directly reveal those features of microhistological organization of the brain that distinguish the varying patterns of brain organization of reptiles *vis-à-vis* mammals. Endocast data, being sparse, cannot directly clarify this issue.

Basic Sauropsida Plan (Reptiles and Birds)

In this section, we will provide a brief overview of the major visual structures and pathways common to reptiles and birds. A more detailed description of the organization in each major order of sauropsids will be discussed in the next section.

Retinal Targets

As in other sighted vertebrates, the optic tract of sauropsids terminates in six main nuclear masses within the brainstem (Cowan *et al.*, 1961). These are the optic tectum, the dorsal division of the lateral geniculate nucleus (LGNd), the ventral division of the lateral geniculate nucleus (LGNv), the hypothalamus, the pretectal area, and the accessory optic nuclei. Among them, two visual areas have been extensively studied: LGNd and the optic tectum. Each of them is an important visual centre in the two major visual pathways from the retina to the telencephalon.

Two Visual Pathways to the Telencephalon

All the amniotes have two major pathways for visual information from the retina to the telencephalon. One of these pathways, the retino-thalamo-telencephalic pathway (the *thalamofugal pathway*) runs from the retina to the group of nuclei of the dorsal thalamus to the telencephalon. This pathway is comparable to the retino-geniculo-striate (RGS) pathway in mammals. The area in the telencephalon which receives thalamofugal visual input is designated as the visual cortex in reptiles and mammals, and is the visual wulst in birds. In contrast to mammals, the thalamofugal (RGS) pathway is not necessarily dominant in visual information processing in many nonmammals. Rather the *tectofugal pathway* is *generally* the major system. The tectofugal pathway in reptiles and birds travels from the retina to the telencephalon via the optic tectum and then to a major nucleus of the thalamus (the nucleus rotundus). The area in the telencephalon that receives tectofugal visual input lies within a large, major subdivision of the telencephalon known as the dorsal ventricular ridge (DVR) in reptiles and birds. The DVR is a complex region with many functional subdivisions comparable to the diversity of the neocortex. In mammals, a similar pathway runs from the retina to the extrastriate visual cortex, via the superior colliculus and the lateral posterior nucleus and/or inferior pulvinar (see Harting and Huerta, 1984).

Optic Tectum

The optic tectum of reptiles and birds is generally the target of the majority of retinal projections. Although the precise number and percentage of retinal axons projecting to each of the six targets is unknown, probably more than 90% of retinal axons enter and terminate within the tectal layers. As in all other vertebrates, the reptilian and avian optic tectum receives not only visual input but also other sensory inputs, including auditory and somatosensory information. The tectum of many reptiles and birds shows a prominent pattern of lamination. There is considerable variation in the lamination of different reptilian and avian taxa (Huber and Crosby, 1933). There are many laminar classifications, but none provides a satisfactory universal nomenclature. In reptiles, Huber and Crosby (1933) subdivided the tectum into six major divisions, based on their data using alligators, lizards, snakes and turtles. These six strata comprise, *from the surface inward*: (a) the stratum opticum, a superficial layer of the optic fibres; (b) the stratum fibrosum et griseum superficiale; (c) the stratum griseum centrale; (d) the stratum album centrale, the main efferent layer of the tectum; (e) the stratum griseum periventriculare; and (f) the stratum fibrosum periventriculare. An earlier terminology provided a more detailed description of layers. Pedro Ramón (1891) recognized 14 layers, and labelled the laminae of the reptilian tectum *from the ependymal surface outward*. Other researchers have also reported these 14 layers in various reptiles (see Northcutt, 1984, for a review).

In birds, Cajal (1911) subdivided the avian tectum into 15 laminae, and numbered the laminae from 1 to 15 *starting from the most external laminae*. Thus, the stratum opticum of Huber and Crosby correspond to lamina 1 of Cajal, and to lamina 14 of Ramón. Cajal included the periventricular grey layer as lamina 15. Cowan *et al*. (1961) subdivided the avian tectum into five major groups, expanding on the system of Huber and Crosby, *from the surface inward*: (a) the stratum opticum; (b) the stratum griseum et fibrosum superficiale; (c) the stratum griseum centrale; (d) the stratum album centrale, the main efferent layer of the tectum; and (e) the stratum griseum periventriculare. Cowan *et al*. (1961) further subdivided the stratum griseum et fibrosum superficiale into 10 divisions, designated as a–j. These sublayers correspond closely to Cajal's layers 2–12. In this essay, we adopt the numerical system of Ramón for reptiles and that of Cajal for birds, since these systems are more commonly accepted than the classification of Huber and Crosby or that of Cowan *et al*. (see Northcutt, 1984, and Hunt and Brecha, 1984, for more details of the laminar classifications of the reptilian and avian tectum).

Both reptiles and birds have ascending tectal efferents projecting to the ipsilateral and contralateral nuclei in pretectum and thalamus. In contrast to mammals, which often possess only ipsilateral projections, sauropsids have well-developed contralateral ascending projections. Among the tectal-recipient nuclei in pretectum and thalamus, the nucleus rotundus is generally the most prominent visual nucleus. Other nuclei include several pretectal nuclei and LGNv (e.g. Hunt and Künzle, 1976; Northcutt, 1984).

Descending tectal efferents project massively to the hindbrain reticular formation. These projections can be regarded as a motor output of the tectum (e.g. Hunt and Künzle, 1976; Hunt and Brecha, 1984; Northcutt, 1984). In many reptiles and birds, several isthmic nuclei receive an ipsilateral tectal input. These isthmic nuclei, in turn, send a projection to the ipsilateral and contralateral tectum.

Lateral Geniculate Nucleus, pars dorsalis

Cell groups in the dorsal thalamus receive a direct visual input from the retina. These cell groups are known as the lateral geniculate nucleus pars dorsalis in mammals and many reptiles and as the principal optic nucleus in birds. They have common characteristics among vertebrates in their topological location in the dorsal thalamus and their reciprocal connections with the telencephalon. For many years, LGNd has been considered merely as a relay centre between the retina and the telencephalon. Recently, however, a rather active role for LGNd has been suggested in terms of the interactions with the telencephalon (e.g. Crick, 1984).

Lateral Geniculate Nucleus, pars ventralis

The LGNv is a prominent retinorecipient area in the diencephalon of reptiles and birds. In mammals, the LGNv has connections with a variety of subcortical sites, including superior colliculus, pretectal nuclei, nuclei of the accessory optic system, vestibular complex, and cerebellum (see Jones, 1985, for a review). These connections, taken together with unit recording in LGNv, suggest that LGNv may be related to several functions, such as pupillary light reflex and vestibular and oculomotor systems (see Jones, 1985, for a review).

Hypothalamus

In mammals, the suprachiasmatic nuclei (SCN) of the hypothalamus receive direct retinal projections and are considered to play an important role in circadian rhythms (see Swanson, 1987, for a review). Similar projections to areas in the hypothalamus have been reported in reptiles and birds, although the functions of these connections are unknown and the terminology is rather inconsistent among researchers and species.

Pretectum and Accessory Optic System

Several cell groups located at the mesodiencephalic junction receive optic fibres from the contralateral eye. Among them, at least three or four pretectal nuclei have been identified in reptiles and birds. The nucleus of the basal optic root (nBOR), the central component of the accessory optic system, also receives a direct visual input. The nBOR has been referred to as the nucleus ectomammilaris in birds, the nucleus opticus tegmenti in reptiles, and the medial terminal nucleus in mammals. In many vertebrates, some portions of the pretectum and the accessory optic system have connections with the vestibulocerebellum directly or indirectly via the inferior olive. Thus, they are considered to play an important role in visuomotor behaviours (see Simpson, 1984; Fite, 1985; McKenna and Wallman, 1985, for reviews.)

Centrifugal Pathway

Centrifugal systems from the brain to the retina have been found in many vertebrates (see Uchiyama, 1989, for a review). Such circuits have been found in reptiles and are well developed in birds. Ground-feeding birds (e.g. chicks, pigeons) exhibit a particularly well-developed centrifugal system (e.g. Weidner et al. 1987). These connections from the brain to the retina have two common characteristics: (a) that cell groups originating the centrifugal pathways are located in the isthmic region at the caudal margin of the mesencephalon; and (b) that these cell groups receive tectal projections (Uchiyama, 1989). Functions of the centrifugal pathway are not known, although a role of mediating visuomotor behaviour has been suggested for the system (e.g. Uchiyama, 1989).

Dorsal Ventricular Ridge

The dorsal ventricular ridge (DVR), a telencephalic structure of reptiles and birds, receives afferents from the brainstem, dorsal thalamus and other telencephalic areas. As shown in Fig. 19.2, the DVR is located in the lateral wall of the telencephalon as an intraventricular protrusion of cells within the core of the cerebral hemispheres. For a long time, the DVR has been misidentified as a hyper-

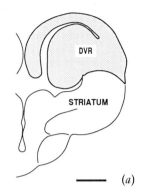

(a)

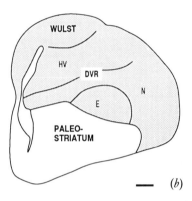

(b)

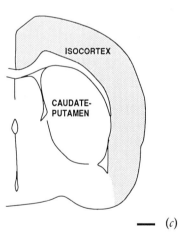

(c)

Fig. 19.2 *Organization of the telencephalon of a reptile, a bird, and a mammal. Shaded areas indicate the region of the telencephalon equivalent to the mammalian cortex. (a) Transverse section of the brain of an adult tegu lizard. (b) Transverse section of the brain of an adult pigeon. (c) Transverse section of the brain of an adult rat. Each bar indicates 1 mm (From Shimizu and Karten, 1991, a and b).*

trophic expanded form of the basal ganglia because of its topographical and topological resemblance to the caudate-putamen of mammals (Ariëns Kappers et al., 1936). However, on the basis of studies using tract-tracing methods and bio- and histochemistry, the region corresponding to the caudate-putamen and globus pallidus (basal ganglia or 'striatum') of mammals proved to be only a small portion of the sauropsid telencephalon, the region designated as the striatum in reptiles and the paleostriatal complex in birds (Reiner et al., 1984). The DVR is functionally best understood as equivalent to the pallium of mammals. Despite the lack of obvious lamination and pyramidal cells with apical dendrites, the DVR has several features in common with the neocortex in mammals (e.g. Karten and Shimizu, 1989; Shimizu and Karten, 1991a,b). Anatomically discrete regions of the DVR of sauropsids receive sensory projections from thalamic nuclei of different modalities, including visual, auditory and somatosensory inputs (see Ulinski, 1983, for a review). The DVR contains these modality specific regions, each of which has connections with specific sensory nuclei in the thalamus. The visual centre which receives input via the thalamic nuclei is located laterally in the DVR (Karten, 1979). No obvious retinotopic organization has been found in the visual centres in the DVR or the nucleus rotundus. Neurones in the visual centres in the DVR have very large receptive fields that are sensitive to moving stimuli and show distinctive directional selectivity. The basic connectivity of the sauropsid tectofugal visual pathway is that the retinal ganglion cells project to the optic tectum, which projects to the nucleus rotundus, which projects to the DVR. This pattern is similar to the connection of the mammalian retino-collicular pathway to the extrastriate visual cortex. In mammals, retinal input terminates in the superior colliculus, which projects to the inferior pulvinar and/or lateral posterior nucleus of the thalamus, which in turn sends efferents mainly to layer IV of the extrastriatal cortices of the temporal area.

The anterior portion of the DVR is known as the anterior dorsal ventricular ridge (ADVR), which is differentiated from the basal dorsal ventricular ridge (BDVR) (Ulinski, 1983). The ADVR includes areas receiving sensory input from various thalamus nuclei while BDVR is the major source area of efferents to the hypothalamus (Ulinski, 1983). The ADVR of reptiles is divided into four regions with distinct boundaries: three periventricular areas (dorsal area, medial area and ventral area) and a central area. The periventricular areas are characterized with differential elaboration of a common zonal pattern of organization, and the central area is mainly occupied with fibre tracts. Among these four areas, the dorsal area appears to be associated with visual information processing. In birds, ADVR corresponds to at least five distinct regions: the hyperstriatum ventrale, the neostriatum, the ectostriatum, the nucleus basalis and the temporo-parieto-occipital area. (The suffix '-striatum' is based on old nomenclature based on the erroneous identification of the DVR as the striatum. These '-striatum' structures do not refer to the structures with the same names in mammals.) Among them, the ectostriatum receives the tectofugal visual input from the nucleus rotundus. The BDVR in birds is called the archistriatum.

Dorsal Cortex

The final target of the thalamofugal pathway is the dorsal portion of the telencephalon. In mammals, this area is the visual cortex or the striate cortex. Equivalent areas are called the dorsal cortex in reptiles and the visual wulst in birds. All these areas are retinotopically organized.

Reptilian Plan

Chelonia (Turtles and Tortoises)

Thalamofugal Pathway

In turtles, the optic tract terminates in several structures in the diencephalon and mesencephalon (*Emys*: Kosareva, 1967; *Chelydra*: Knapp and Kang, 1968a; *Podocnemis*: Knapp and Kang, 1968b; *Pseudemys*: Hall and Ebner, 1970a,b), and the number of retinal targets in the thalamus varies among studies. The major target in the thalamus, however, is the contralateral dorsal portion of the lateral geniculate nucleus (LGNd) (*Chrysemys*: Bass and Northcutt, 1981a; *Caretta*: Bass and Northcutt, 1981b; Künzle and Schnyder, 1984; *Pseudemys* and *Chrysemys*: Rainey and Ulinski, 1986; Ulinski and Nautiyal, 1988). The LGNd includes several subnuclei (Bass and Northcutt, 1981a; Rainey and Ulinski, 1986; Ulinski and Nautiyal, 1988) which exhibit different sizes and densities of neurones (Rainey and Ulinski, 1986) and which receive information from different regions of the retina (Ulinski and Nautiyal, 1988). The LGNd is retinotopically organized: the nasotemporal axis of the retina projects along the rostrocaudal axis of the nucleus and the dorsoventral axis of the retina projects along the dorsoventral axis of the nucleus (Ulinski and Nautiyal, 1988). The turtle LGNd has a form of laminar organization – cell plate and neuropil layers (Ulinski and Nautiyal, 1988) – which is a different type of lamination found in the mammalian LGNd. Neurones of this nuclear complex have 30°, or smaller, excitatory receptive fields and respond to moving stimuli (Boiko, 1980).

The LGNd in turn gives rise to an ipsilateral projection to a structure in the telencephalon via the dorsal peduncle of the lateral forebrain bundle (Hall and Ebner, 1979a,b; Ebner and Colonnier, 1975; Hall et al., 1977; Belekhova et *al.*, 1979; Smith *et al.*, 1980; Ouimet *et al.*, 1985; Heller and Ulinski, 1987; Mulligan and Ulinski, 1990). This area of the dorsal telencephalon responds to visual stimuli or stimulation of the optic nerve (Mazurskaya, 1972; Bass *et al.*, 1983; Kriegstein, 1987), and is referred to as the visual cortex or area D2 by Desan (1988). There is an orderly representation of the rostrocaudal axis of the LGNd along the rostrocaudal axis of the visual cortex (Mulligan and Ulinski, 1990). The turtle visual cortex is organized as a series of approximately transverse slabs (lamellae), each of which contains a representation of a particular vertical meridian (azimuth) of visual space (Mulligan and Ulinski, 1990). The neurones of such isoazimuth lamellae possess axons running along and between lamellae (Cosans and Ulinski, 1990).

Tectofugal Pathway
The retinal ganglion cells in turtles also give rise to a bilateral, although mainly contralateral, projection to the superficial layers of the optic tectum (*Emys*: Kosareva, 1967; *Chrysemys*: Bass and Northcutt, 1981a). Bass and Northcutt (1981a) reported that retinal fibres terminate within the superficial zone – laminae 9, 12, and 14 – of the optic tectum, and that ipsilateral fibres terminate in laminae 9 and 12. The optic tectum is retinotopically organized: the nasotemporal visual axis is oriented along the rostrocaudal axis, whereas the dorsoventral visual axis is oriented along the mediolateral tectal axis (*Emys*: Gusel'nikov *et al.*, 1970). The tectum projects massively and bilaterally to the nucleus rotundus of the thalamus (Hall and Ebner, 1970b; Rainey, 1979; Rainey and Ulinski, 1982a,b). This nucleus is well developed in turtles, and contains morphologically homogeneous populations of neurones (Rainey and Ulinski, 1982a). There is no apparent retinotopic organization in the nucleus rotundus.

The neurones of the nucleus rotundus are the source of a projection to a restricted area in ADVR via the lateral forebrain bundle (Hall and Ebner, 1970b; Kosareva, 1974; Parent, 1976; Balban and Ulinski, 1981; Belekhova *et al.*, 1979). This target region is referred to as zone 4 of the dorsal area (Balban and Ulinski, 1981).

Hypothalamus
In two species, snapping turtles (Knapp and Kang, 1968a) and sidenecked turtles (Knapp and Kang, 1968b), a retinal projection to the lateral hypothalamus has been reported. No retinal projection to the medial hypothalamus has been found.

Pretectum and Accessory Optic System
At least four pretectal nuclei – the nucleus geniculatus pretectalis (GP), the nucleus lentiformis mesencephali (nLM), the nucleus posterodorsalis (Pd), and the external pretectal nucleus – receive a direct retinal input in two species, *Chrysemys picta* and *Caretta caretta* (Bass and Northcutt, 1981a,b).

The nBOR receives an input from giant ganglion cells and displaced ganglion cells (DGCs) in the retina (Reiner, 1981). Reiner and Karten (1978) showed a bisynaptic lemniscal route from the eye to the cerebellum via the accessory optic nuclei (nBOR and its associated nuclei).

Centrifugal Pathway
Schnyder and Künzle (1983) showed that cells located in a mesencephalic reticular area are the source of centrifugal projection to the retina in *Pseudemys*. However, the number of these cells is only in the range of tens.

Lepidosauria (Lizards and Snakes)
Thalamofugal Pathway
Sphenodon
Only one study has been conducted regarding the retinal projections in *Sphenodon* (Northcutt *et al.*, 1974). No ipsilateral projection was found. The LGNd in *Sphenodon* receives a massive retinal projection as well as other nuclei in the brainstem, including the optic tectum, LGNv, the ventral part of the nucleus ventrolateralis, and an area caudodorsal to the nucleus dorsolateralis anterior.

Snakes
The LGNd of snakes receives an almost entirely crossed projection from the contralateral eye (*Thamnophis*: Halpern and Frumin, 1973; Allen and Crews, 1989; *Natrix*: Northcutt and Butler, 1974a). Only an extremely small ipsilateral projection has been observed in a confined area in the LGNd. The LGNd projects via the ipsilateral forebrain bundle to an area located in ADVR (Wang and Halpern, 1977).

Lizards
The retinal projection to the LGNd is not completely crossed in lizards, unlike snakes (*Tupinambis*: Cruce and Cruce, 1978; *Cnemidophorus*: Allen and Crews, 1989). The LGNd is also known to send a projection to the telencephalon in these animals (*Tupinambis*: Lohman and van Woerdern-Verkley, 1978).

Tectofugal Pathway
Sphenodon
The projection from the retina to the optic tectum is completely crossed in *Sphenodon* (Northcutt *et al.*, 1974). Layers 8–14 are heavily labelled after a tritiated proline injection into the contralateral eye.

Snakes
The retinal projection to the optic tectum in snakes is also entirely crossed (Halpern and Frumin, 1973; Northcutt

and Butler, 1974a). However, one snake – *Vipera* – has been reported to possess ipsilateral retino-tectal projections (Repérant and Rio, 1976). The superficial layers of the optic tectum (layers 8–13 of Ramón) receive a contralateral retinal projection (Northcutt and Butler, 1974a). The nucleus rotundus of snakes is not as well developed as in other reptiles. Indeed, the existence of the nucleus itself had been in question. Only an 'extremely small' population of cells was found in *Natrix* (Northcutt and Butler, 1974a). However, Ulinski (*Natrix*; 1977) and Schroeder (*Crotalus*; 1981) could not decidedly identify the nucleus rotundus. Later, Dacey and Ulinski (1983) documented the existence of the nucleus in garter snakes (*Thamnophis sirtalis*), and more recently Berson and Hartline (1988) identified one in *Crotalus viridis* (rattlesnakes). The nucleus rotundus of snakes also innervates the rostral ADVR. Rattlesnakes have a vision-like sense that responds to infrared radiation. The nucleus rotundus of these snakes conveys infrared radiation signals as well as visual signals and/or vibrational stimuli from the tectum to the ADVR (Berson and Hartline, 1988).

Lizards
In addition to contralateral retinotectal projections, several lizards have been reported to possess projections from the ipsilateral eye (*Gekko*: Northcutt and Butler, 1974b; *Tupinambis*: Cruce and Cruce, 1975; Ebbesson and Karten, 1981; *Xantusia*: Butler, 1974; *Tarentola*: Repérant et al., 1978). All studies showed that contralateral fibres enter through laminae 12 and 14 and terminate in laminae 9 through 13. However, terminals in lamina 8 have also been suggested by several studies (Northcutt and Butler, 1974b; Butler, 1974; Cruce and Cruce, 1975). Ipsilateral retinal fibres were found in laminae 8, 11 and 13 (*Gekko*: Northcutt and Butler, 1974b), laminae 8–14 (*Xantusia*: Butler, 1974), and lamina 9 (*Tupinambis*: Cruce and Cruce, 1975). As in other reptiles, lizards had retinotopic representation in the optic tectum (*Iguana*: Gaither and Stein, 1979; Stein and Gaither, 1981). In *Iguana* (Stein and Gaither, 1981), the central 10 degrees of the visual field has an expanded representation in the optic tectum (about 20% of the tectal surface). Thus, *Iguana* has a magnified foveal or *area centralis* representation. The tecto-rotundal projections are bilateral in *Iguana* (Butler and Northcutt, 1971a) and *Gekko* (Butler, 1976). Most lizards possess a well-developed nucleus rotundus. The nucleus projects ipsilaterally to ADVR in iguanas (Bruce and Butler, 1979), tegus (Lohman and van Woerden-Verkley, 1978), monitors (Distel and Ebbesson, 1975) and *Gekko* (Butler, 1976).

In snakes (*Natrix*), the thalamofugal pathway is massive while the tectofugal pathway is rather marginal. In contrast, both visual pathways to the dorsal thalamus are well developed – more fibres in the pathways and larger retinal recipient areas – in lizards. Such differences in development of the two visual pathways between snakes and lizards might be related to differences in their evolutionary history. Many modern lizards (suborder Lacertilia) are descended from non-fossorial ancestors whereas ancestors of modern snakes (suborder Ophidia) were fossorial (Senn and Northcutt, 1973). Ancestral snakes possibly underwent marked changes in the visual system through adaptation to a fossorial niche (Walls, 1942). Subsequently, modern snakes have invaded a surface niche, and have redeveloped their visual system. However, in the course of reinvasion of the surface by ancestral fossorial snakes, the thalamofugal pathway, in contrast to the tectofugal pathway, may have been somehow increased (Northcutt and Butler, 1974b). The functions of the thalamofugal pathway of snakes may be related to such adaptive pressure.

Hypothalamus

Sphenodon
No study has been reported on the retinal projection to the hypothalamus.

Snakes
In *Vipera*, two nuclei in the hypothalamus – a nucleus in the medial hypothalamus (referred to as the suprachiasmatic nucleus) and a nucleus in the lateral hypothalamus (referred to as the supraoptic nucleus) have been found to receive direct retinal projections (Repérant and Rio, 1976). Similarly, in *Natrix*, Northcutt and Butler (1974a) used the Nauta method (silver impregnation) and found some argentophilic debris in these nuclei.

Lizards
Butler (1974) found some signs of a retinal projection to the medial hypothalamus in *Xantusia*. An autoradiographic study by Repérant and co-workers (1978) also showed a retino-mediohypothalamic projection in five species: *Acanthodactylus*, *Scincus*, *Tarentola*, *Uromastix* and *Zonosaurus*. In tegu lizards, this projection to the medial hypothalamus was found to be relatively large (Ebbesson and Karten, 1981). Furthermore, the tegu lizard had a retinal projection to the area immediately rostral to the supraoptic nucleus in the lateral hypothalamus (Ebbesson and Karten, 1981).

Pretectum and Accessory Optic System

Sphenodon
The GP, the nLM and a poorly developed Pd have been found to receive a retinal input (Northcutt et al., 1974).

Snakes
Three of four pretectal nuclei have been identified in snakes. These include GP, nLM, Pd (*Natrix*: Northcutt

and Butler, 1974a; Ulinski, 1977) and the nucleus pretectalis (*Vipera*: Repérant and Rio, 1976). A direct connection of nBOR with the vestibulocerebellum has not been found in snakes (Bangma and ten Donkelaar, 1982). However, among vertebrates only nBOR of snakes has been reported to receive direct tectal efferents (*Natrix*: Ulinski, 1977).

Lizards

Pretectal nuclei are developed and distinct in *Iguana* (Butler and Northcutt, 1971a) and some of them receive tectal efferents bilaterally (Butler and Northcutt, 1971b). The nBOR sends a projection to the cerebellum in *Varanus* (Bangma and ten Donkelaar, 1982).

Centrifugal Pathway

Snakes

Halpern and co-workers (1976) showed that a cell group (the nucleus ventral supraoptic decussation located longitudinally from the diencephalon to the mesencephalon) sends centrifugal fibres to the retina in garter snakes *Thamnophis*. In *Python*, Hoogland and Welker (1981) found that a cell group in the basal telencephalon sends a projection to the retina.

Lizards

Halpern and co-workers (1976) found retinopetal cells in the diencephalon (girdled lizard *Cordylus*) and in the mesencephalon (alligator lizard *Gerrhonotus*).

Crocodilia (Caiman)

Thalamofugal Pathway

Completely crossed retinal projections to the LGNd were found in *Caiman* (Braford, 1973). No report on the ascending efferents of the LGNd in crocodilians has been published.

Tectofugal Pathway

Burns and Goodman (1967) reported ipsilateral retinotectal projections in *Caiman*. However, Braford (1973) could not find such projections, although he noted small bilateral projections to the preoptic area. Laminae 8–14 receive the contralateral projections from the retina (Braford, 1973). Retinotopic projections in *Alligator* have been reported in an electrophysiological study (Heric and Kruger, 1965). The tectum projects bilaterally to the nucleus rotundus (*Caiman*: Braford, 1972). The rotundus is well developed in crocodilians (Huber and Crosby, 1926). This nucleus projects, via the lateral portion of the lateral forebrain bundle, to ADVR in *Caiman* (Pritz, 1975). The latter region is described as a lateral part of the rostral dorsolateral area.

Pretectum and Accessory Optic System

At least three pretectal nuclei – GP, nLM and Pd – receive retinal input from the contralateral retina (Braford, 1973) and a bilateral input from the tectum (Braford, 1972). A direct retinal projection to nBOR was found in *Caiman* (Braford, 1973).

Centrifugal Pathway

Ferguson and co-workers (1978) found that isthmic field in *Caiman* gives rise to a bilateral projection to the retina.

Avian Plan

Birds with Lateral Eyes and Birds with Frontal Eyes

There are significant differences in the development of the two visual pathways and related structures among different avian species. In birds with frontal eyes and a large binocular field (such as owls), the thalamofugal pathway is quite well developed and extremely well differentiated (Pettigrew, 1979). In contrast, birds with lateral eyes and a smaller binocular field (such as quail, zebra finches and pigeons) possess structures in the tectofugal pathway that are more developed than those in the thalamofugal pathway. Numerous studies have examined effects of lesions in the two visual pathways, but only in birds with lateral eyes. Birds with lateral eyes tend to have more severe visual deficits after damage to the tectofugal pathway than to the thalamofugal pathway (e.g. Macko and Hodos, 1984; Hodos et al., 1984; Bessette and Hodos, 1989). This is indirectly supported by the fact that lesions in the wulst do not have severe effects on visual information processing, unlike the visual cortical lesions in mammals, but rather more indirect effects (Shimizu and Hodos, 1989). Functions of the two visual pathways, therefore, may not be identical among different avian species.

Central Visual Pathways in Birds

Thalamofugal Pathway

The principal optic nucleus of the thalamus

Retinal projections to the brainstem in birds are completely crossed (Meier et al., 1974; but see Repérant, 1973). A group of nuclei in the dorsal thalamus receives direct retinal input and in turn sends a projection to the telencephalon. This nuclear group, including the nuclei lateralis anterior (LA), dorsolateralis anterior pars lateralis (DLL), and dorsolateralis anterior pars magnocellularis (DLAmc), is designated as the principal optic nucleus of the thalamus (OPT), and has been compared to the LGNd of reptiles and mammals. In chicks, circumscribed lesions in the retina did not reveal a precise retinotopic representation within the OPT (Ehrlich and Mark, 1984). In pigeons, Burkhalter et al. (1979) injected HRP into the OPT and found a retinotopic map: the rostrocaudal axis of the OPT corresponds to the temporal–nasal retinal axis. In owls, the OPT also seems to be retinotopically organized. Electrophysiologically, Pettigrew (1979) showed that the temporal–nasal axis of the foveal region in the retina corresponds to the rostral-caudal axis of the thalamus.

Neurones of the OPT give rise to bilateral projections to an area of the dorsal telencephalon called the 'wulst' (Karten et al., 1973; Mihailovic et al., 1974; Miceli et al., 1979; Streit et al., 1980a,b; Bagnoli and Burkhalter, 1983; Porciatti et al., 1990; Bagnoli et al., 1990). In birds with laterally placed eyes, the ventral portion of the OPT tends to project to the ipsilateral wulst while the dorsal portion projects to the contralateral wulst (Hunt and Webster, 1972; Miceli et al., 1975, 1979; Bagnoli and Burkhalter, 1983). On the other hand, birds with frontal eyes (*Athene*) have a more orderly pattern of the OPT–wulst projection: the rostral thalamus projects mostly to the contralateral wulst while the caudal thalamus projects to the ipsilateral one (Bagnoli et al., 1990). A more massive projection is found in the ipsilateral wulst in pigeons, whereas comparable sizes of projections are found in both hemispheres in owls (Bagnoli et al., 1990).

The wulst

The wulst ('bulge') is a parasagittal elevation located in the dorsomedial region of the avian hemisphere. The wulst consists of at least two distinct regions (see Fig. 19.3): (a) a medial portion purportedly similar to the hippocampus and associated areas (wulst regio hippocampalis, Wrh), and (b) a lateral portion, including regions comparable to some sensory areas of the mammalian neocortex (wulst regio hyperstriatica, Whs). The Whs is similar to the neocortex in that it has a laminar configuration (Karten et al., 1973; Pettigrew, 1979; Reiner and Karten, 1983), equivalent afferent and efferent connections (e.g. Karten et al., 1973; Miceli et al., 1987; Streit et al., 1980a,b; Nixdorf and Bischof, 1982; Bagnoli and Burkhalter, 1983), similar single-unit responses (e.g. Pettigrew and Konishi, 1976; Miceli et al., 1979) and extensive monoaminergic innervation (Bagnoli and Casini, 1985; Yamada and Sano, 1985; Shimizu and Karten, 1990). Although there are many similarities between the two structures, significant differences in cell morphology (Watanabe et al., 1983), hodology (Reiner and Karten, 1983), and chemistry (Shimizu and Karten, 1990) also exist. The anterior portion of Whs is considered to be the avian equivalent of the somatosensory–motor cortex of mammals (Karten, 1971; Delius and Bennetto, 1972; Karten et al., 1978; Wild, 1987; Funke, 1989a,b), the posterior portion to the striate cortex (Karten et al., 1973; Karten, 1979; Bagnoli et al., 1982), and the medial portion of Whs may be comparable to the limbic cortex (Karten et al., 1973; Berk and Hawkin, 1985). The topological relationships of these areas in the wulst are schematically shown in Fig. 19.3.

Cytoarchitectonically, the Whs is a laminated structure with at least four constituents. These include, from the dorsal surface inward, the hyperstriatum accessorium (HA), the intercalated nucleus of the hyperstriatum accessorium (IHA), the hyperstriatum intercalatus superior (HIS), and the hyperstriatum dorsale (HD) (Karten et al., 1973). The IHA, a thin granular layer, contains small cells. The IHA is the major recipient of input from the avian equivalent of the dorsal division of the lateral geniculate nucleus (Karten et al., 1973; Streit et al., 1980a, b; Watanabe et al., 1983). *The lateral portion of HD* also receives a projection from the visual nuclei of the dorsal thalamus (Karten et al., 1973; Watanabe et al., 1983). The HA is the major source of direct extratelencephalic projections to various subcortical nuclei (Karten et al., 1973; Bagnoli et al., 1980; Bravo and Pettigrew, 1981; Reiner and Karten, 1983). Axons of these cells in HA descend along the medial wall of the telencephalon as the tractus septomesencephalicus. This tract terminates in various visual nuclei in the brainstem, including OPT, the optic tectum, nBOR and LGNv (Karten et al., 1973; Miceli et al., 1987).

Neurones sensitive to binocular visual inputs have been found in the wulst of owls (Pettigrew and Konishi, 1976; Pettigrew, 1979). In contrast, in pigeons and chicks (Perisic et al., 1971; Wilson, 1980; Denton, 1981), only a small area in the wulst responds weakly to ipsilateral visual stimuli. The wulst is also retinotopically organized. In owls, the nasal–temporal axis of the contralateral retina is mapped along the lateral–medial axis of the wulst, while the superior–inferior axis of the retina corresponds to the caudal–rostral wulst (Pettigrew and Konishi, 1976). In pigeons and chicks, the superior–inferior axis of the visual field is also mapped on the caudal–rostral wulst (Perisic et al., 1971; Wilson, 1980; Denton, 1981).

Tectofugal Pathway

The optic tectum

The retinal projection terminates in laminae 2 through 7 of the contralateral optic tectum (Cajal, 1911; Cowan et al., 1961; Gray and Hamlyn, 1962; Repérant, 1973; Hayes and Webster, 1975; Hunt and Webster, 1975; Angaut and Repérant, 1976; Repérant and Angaut, 1977). Recently, birds have become a convenient model for studies of plasticity of the visual system. One of the major advantages is the complete decussation of the retinal input to the optic tectum. The entirely crossed projection of the retina to the tectum provides an effective experimental condition in which a treatment to one retinal pathway can be compared

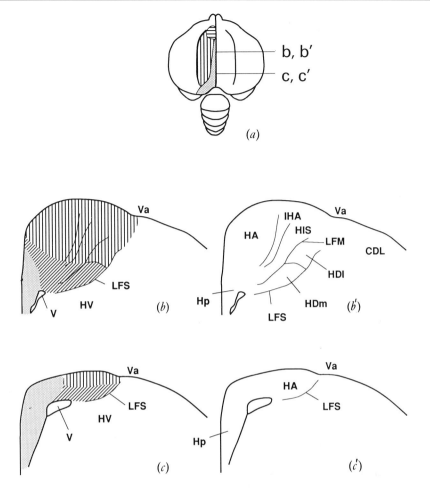

Fig. 19.3 Schematic drawings of a dorsal view of the pigeon brain (a), and transverse sections of the rostral (b, b') and caudal (c, c') wulst. (b, c) Topological relationships of Wrh, the medial Whs, and the posterior Whs (the boundaries of which are not clear; see text). (b', c') Cytoarchitectonically defined subdivisions of the wulst. Different hatchings indicate subregions of the wulst: area with dots = Wrh; area with horizontal lines = anterior Whs; area with vertical lines = posterolateral Whs; area with diagonal lines = posteromedial Whs (From Shimizu and Karten, 1990).

with the other pathway as a control (e.g. Bagnoli et al., 1989; Nixdorf, 1990). The retinal projection is topographically organized in the superficial layers of the tectum (Hamdi and Whitteridge, 1954; McGill et al., 1966; Clarke and Whitteridge, 1976; Acheson et al., 1980; Jassik-Gerschenfeld and Hardy, 1984). The superior half of the retina is mapped in the postero-ventral part of the tectum while the anterior quadrants of the retina are mapped in the posterior and dorsal parts (see Jassik-Gerschenfeld and Hardy, 1984).

Units in the optic tectum respond to moving stimuli and are selective for direction, size and orientation of a stimulus. These tectal cells have different properties of the receptive fields depending on the depth from the surface layer (see Jassik-Gerschenfeld and Hardy, 1984, for a review). The receptive fields of units in the superficial layers (laminae 1 to 12 of Cajal) are rather small, less than 4° of visual arc, and sensitive to motion with directional selectivity. The deeper the units are located, the larger and less selective to stimulus qualities do the receptive fields become. Cells in deeper layers (laminae 13 and 14) of the tectum tend to have larger receptive fields (larger than 10°), but often lack directional selectivity (Revzin, 1970).

Benowitz and Karten (1976) showed that neurones of laminae 13 throughout the tectum send a projection to the nucleus rotundus, without the obvious retinotopic organization. These authors also found that there are cytoarchitectonic subdivisions in the rotundus, and each subdivision receives projections from different sublayers of lamina 13.

Descending tectal efferents project to ipsilateral isthmic nuclei – the nucleus isthmi pars parvocellularis, the nucleus semilunaris, and the nucleus isthmo-opticus (see Hunt and Brecha, 1984, for a review). These projections are topographically organized (Hunt and Künzle, 1976). Among these nuclei, neurones in the nucleus isthmo-opticus are the sources of the centrifugal pathway to the contralateral retina (see p 433). The other nuclei send projections back to the tectum.

The nucleus rotundus

The rotundus has been subdivided into at least five subdivisions based on cytoarchitectonics (Benowitz and Karten, 1976; Nixdorf and Bischof, 1982) and the distribution of different intensities of acetylcholinesterase activity (Martínez-de-la-Torre et al., 1990). The nucleus rotundus sends a projection to a region in DVR known as the ectostriatum core (Karten and Hodos, 1970; Benowitz and Karten, 1976; Nixdorf and Bischof, 1982; Watanabe et al., 1985). Each subdivision appears to originate input to different regions within the ipsilateral ectostriatum core (Benowitz and Karten, 1976).

Many of the rotundal neurones (about 80%) tend to have extremely large receptive fields ranging from 100° to 175°, and are sensitive to motion (Revzin, 1979). Rotundal cells can be categorized into directionally selective groups and non-selective groups. The non-selective groups tend to cluster in the posterior third of the nucleus (Revzin, 1979). Wang and Frost (1990) also found units responding to Z-axis movement (expansion or contraction of a visual stimulus). Furthermore, there are cells selective to a particular colour (i.e. between 500 and 600 nm) (Maxwell and Granda, 1979; Wang and Frost, 1990).

The ectostriatum

The ectostriatum core receives a projection from the ipsilateral rotundus. Ritchie and Cohen (1979) and Shimizu et al. (1989) showed that the ectostriatum core projects to its surrounding region (including the periectostriatal belt region and the neostriatal intermedium laterale) in the DVR, and this region in turn projects to another restricted area in the telencephalon, the archistriatum. This area in the archistriatum then projects on laminae 11, 12 and 13 of the ipsilateral optic tectum (Brecha et al., 1976). Such a pattern of sequential projection through the various entities in the DVR may resemble the interlaminar connections of the neocortex; i.e., from layer IV (ectostriatum core) to layers II and III (the periectostriatal belt region and/or the neostriatum intermedium laterale), then to layers V and VI (the archistriatum), then to the optic tectum (Karten, 1969; Nauta and Karten, 1970; Karten and Shimizu, 1989). Figure 19.4 illustrates sche-

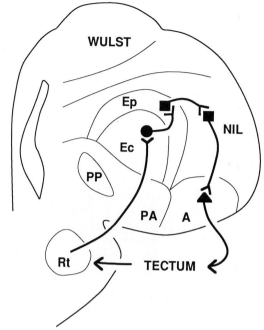

Fig. 19.4 *Visual-information circuit of the tectofugal pathway in the avian brain. Circles indicate thalamic recipient neurones; squares indicate intrinsic neurones; triangles indicate neurones projecting to the optic tectum. A, archistriatum; Ec, ectostriatum core; Ep, periectostriatal belt; HV, hyperstriatum ventrale; NIL, neostriatum intermedium laterale; PA, paleostriatum augmentatum (avian equivalent of caudate-putamen); PP, paleostriatum primitivum (equivalent of globus pallidus) (From Karten and Shimizu, 1989).*

matically this circuitry. Properties of the ectostriatal neurones have been reported as similar to those of the rotundal neurones (Revzin, 1970). Thus, the majority of these cells have large receptive fields and respond to moving stimuli with directional selectivity.
moving stimuli with directional selectivity.

In addition, another tectofugal pathway has been reported. The thalamic nucleus dorsolateralis posterior (DLP) receives tectal afferents and, in turn, sends a projection to an area in the neostriatum intermediale, immediately medial to the ectostriatum (Gamlin and Cohen, 1986). Since this small thalamic nucleus also receives somatosensory and auditory input, DLP is compared to the posterior complex of nuclei of mammals (Korzeniewska and Güntürkün, 1990).

Relatively little is known of the detailed sequence of development of the visual pathways. Tsai et al. (1981), using tritiated thymidine autoradiography to determine birthdates of various nuclei of the chick telencephalon, indicated that the neurones of the ectostriatum are born well prior to those of the wulst. Using ^{14}C-2-

deoxyglucose, Roger and Bell (1989) showed that the tectofugal pathway and its related nuclei appear to be fully operational at birth in chicks. In contrast, the thalamofugal pathway and its related nuclei are not established until approximately three weeks after hatching (Rogers and Bell, 1989).

Interactions of the Two Visual Pathways

The existence of interactions between the thalamofugal pathway and the tectofugal pathway has been suggested in physiological studies (e.g. Engelage and Bischof, 1988) and behavioural studies (e.g. Riley et al., 1988). Anatomical studies of the interconnections of the visual areas (Karten et al., 1973; Ritchie, 1979; Shimizu et al., 1989, 1990) suggested that several regions in the telencephalon receive visual information from the primary visual targets – the wulst and the ectostriatum core. For instance, Fig. 19.5 shows the distribution of inputs from the wulst to the DVR after a deposit of an anterograde tracer (*Phaseourus vulgaris* leucoagglutinin) in the wulst. However, most of the circuitry of the visual telencephalon is still unknown.

Lateral Geniculate Nucleus, pars ventralis

At least some of the retinal projections to the LGNv are collaterals of axons going to the optic tectum (e.g. Britto et al., 1989). The LGNv itself has reciprocal connections with the optic tectum (Crossland and Uchwat, 1979; Bravo and Pettigrew, 1981; Reiner and Karten, 1982), sends efferents to the pretectal area (Reiner and Karten, 1982; Gamlin et al., 1984), and receives telencephalic projections (Karten et al., 1973). Yet no ascending projections of LGNv to the telencephalon have been found. The roles of the avian LGNv in visual processing are still unknown, although some units in LGNv respond to movements of a visual stimulus (Pateromikelakis, 1979) and some units are sensitive to particular wavelengths of light (Maturana and Varela, 1982).

Hypothalamus

Direct retinal projections to the hypothalamus – the medial hypothalamus and/or the lateral hypothalamus – have been demonstrated in many species. However, several names have been assigned to these hypothalamic areas, depending on researchers and species (see Nogren and Silver, 1989, for a review). As a result, our understanding of the retinohypothalamic connections in birds has been rather ambiguous. In order to organize this confusing terminology of the retinorecipient hypothalamic areas, Norgren and Silver (1989) suggested use of the terms medial hypothalamic nucleus (MHN) and lateral hypothalamic retinorecipient nucleus (LHRN). Norgren and Silver (1989) placed horseradish peroxidase into the vitreous or the cut end of the optic nerve in five species: ringdoves, budgerigars, quail, starlings and song sparrows. They found consistent labelling in the LHRN, but none in MHN. The LHRN is similar to the mammalian SCN in that these nuclei receive the densest retinal projection in the hypothalamus. However, functions of LHRN in birds are still unknown (e.g. Simpson and Follet, 1981).

Pretectum and Accessory Optic System

In pigeons, three pretectal nuclei (i.e. nLM, the tectal grey – GT, and the area pretectalis – AP) receive heavy retinal input and a fourth pretectal nucleus (the pretectalis diffusus – PD) receives a slight input (Gamlin and Cohen, 1988a). The same number of pretectal areas has also been found in chicks (Ehrlich and Mark, 1984). Some of these nuclei have connections with the caudal cerebellum (Gamlin and Cohen, 1988b). Several lines of evidence suggest that the nLM is related to the optokinetic nystagmus for temporal-to-nasal movements in the contralateral eye (see McKenna and Wallman, 1985, for a review). The AP is associated with pupillary light reflex mediated by a connection of AP with the ciliary ganglion via the Edinger-Westphal nucleus (Gamlin et al., 1984).

The major afferent source of the avian nBOR is displaced ganglion cells (DGCs) of the contralateral retina (Karten et al., 1977; Reiner et al., 1979; Fite et al., 1981). Other afferent sources include pretectal nuclei, LGNv, wulst, preoculomotor nuclei, and contralateral nBOR (see McKenna and Wallman, 1985, for a review). The nBOR in turn has connections with various visuomotor areas including the oculomotor complex and vestibulocerebellum (e.g. Brecha et al., 1980). On the basis of such connections and physiological data, the avian nBOR, as well as the nLM, is considered to be related to optomotor reflexes (e.g. McKenna and Wallman, 1985; Britto et al., 1990).

Centrifugal Pathway

Among vertebrates, birds have a highly developed centrifugal pathway (see Uchiyama, 1989, for a review). Neurones of the nucleus isthmo-opticus (ION, located at the medial edge of the optic tectum) and the neighbouring ectopic cells project to the retina and make contact with amacrine cells and DGCs (e.g. Hayes and Holden, 1983). The ION also receives a retinotopic projection from the tectum (e.g. Hunt and Kunzle, 1976; Woodson et al., 1991). Although the functions of ION in birds are not known, a possible suggestion has been the role of discriminating and/or orienting behaviour mainly for food selection (e.g. Uchiyama, 1989). Indeed, the number of cells in ION is about 10 000 in ground-feeding birds (e.g. pigeons, chicks and quail) whereas non-ground-feeding birds, like raptors (e.g. falcon, hawks, and owls), have only 900–2000 cells in ION (Shortess and Klose, 1975; Weidner et al., 1987).

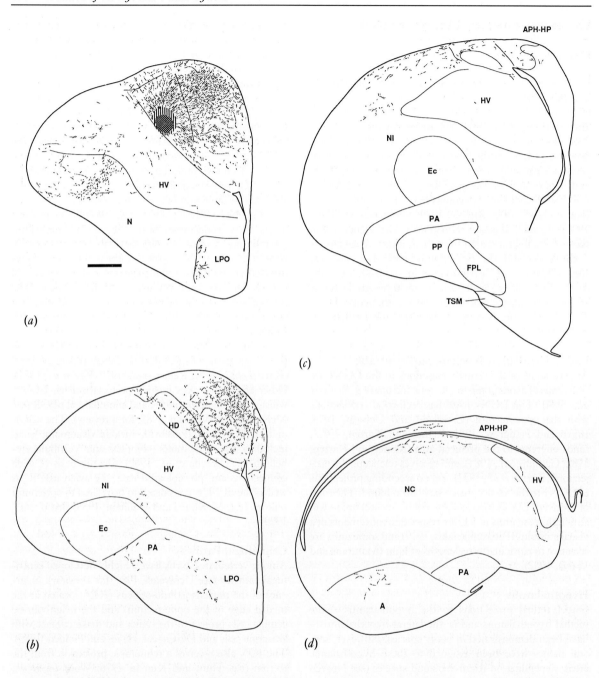

Fig. 19.5 *A series of line drawings through the rostrocaudal (a–d) extent of the pigeon telencephalon, showing anterogradely labelled fibres following an injection (hatched area) of* Phaseolus vulgaris *leucoagglutinin in the wulst. Note that fibres are found not only within the wulst, but also in the DVR (e.g. HV, N, and A) and in the basal ganglia (e.g. LPO and PA). A, archistriatum; APH-HP, area parahippocamalis and hippocampus; Ec, ectostriatum core; FPL, fasciculus prosenphali lateralis; HD, hyperstriatum dorsale; HV, hyperstriatum ventrale; LPO, lobus parolfactorius; N, neostriatum; NI, neostriatum intermedium; NC, neostriatum caudale; PA, paleostriatum augmentatum (avian equivalent of caudate-putamen); PP, paleostriatum primitivum (equivalent of globus pallidus); TSM, tractus septomesencephalicus. Bar indicates 1 mm (From Shimizu* et al., *1990).*

Evolution of the Visual System

Encephalization

In general, the mammalian visual system possesses a well-developed thalamofugal pathway in comparison with the non-mammalian visual system. In non-mammals the optic tectum is often the major visual structure in which most retinal fibres terminate. In birds (with laterally placed eyes), lesions in the tectofugal pathway result in severe deficits, and lesions in the thalamofugal pathway have no major effects on visual behaviours. In contrast, mammalian visual systems are often characterized by the great development of LGNd and the striate cortex. Damages to the thalamofugal pathway cause rather serious deficits in mammals, and the deficits are more severe in monkeys and humans than in non-primates (e.g. cats). Such observations may suggest that ancestral reptiles also possessed a visual system of the major tectofugal-type. Since the tissues of the brain decay rapidly after the death of an animal, there is no direct method to investigate the nervous system of ancestral reptiles. The study of palaeoneurobiology thus uses the remaining fossilized hard parts and the size and shape of the endocranial cavity of the skull of extinct vertebrates (Hopson, 1979). According to Hopson (1979), the outline of the optic tectum is often obscured in these materials, since the tectum is overlain by thick dura and a venous sinus. Nevertheless, the optic tectum seems well developed in some archosaurs (ruling reptiles) including pterosaurs (flying reptiles in the late Triassic to late Cretaceous periods) and some dinosaurs (*Stenonychosaurus* and *Tyrannosaurus*) (Hopson, 1979). These observations from comparative neurobiology and palaeoneurobiology may support the traditional concept of 'encephalization', namely that functions of the optic tectum were gradually taken over by the striate cortex during evolution. This concept regarding the visual structures may have some validity, but may not be entirely true. For instance, the tectofugal pathway and its related structures, such as the optic tectum (the superior colliculus) and the pulvinar, are still distinct in many mammals, including humans. The functions of these structures in mammals are not fully known, although various functions (e.g. visual attention, ocular movements) have been suggested (Chalupa, 1991). Another argument against the 'encephalization' concept is that non-mammals often possess well-developed *telencephalic* structures of the tectofugal pathway, as well as a large optic tectum. Most non-mammals have the tectothalamic connections and the subsequent thalamotelencephalic connections. For instance, in pigeons, lesions of the ectostriatum (i.e. the telencephalic target of the tectofugal pathway) produce severe effects on various visual abilities. The telencephalic structures of vision thus may play a significant role in some non-mammals, and therefore 'encephalization' of vision might have occurred at least before mammals evolved from the ancestral reptiles.

Bottleneck Theory

Nevertheless, the non-mammalian tectofugal pathway is *generally* the massive visual system, while the thalamofugal pathway is the major system in mammals. One attempt to account for such a great deal of difference between mammalian and non-mammalian visual systems is presented by the so-called *bottleneck* theory (Masterton and Glendenning, 1978). When ancestral mammals evolved, these animals possessed the tectofugal pathway-type visual system which they may have shared with other reptilians. However, since predatory and carnivorous reptiles dominated the land, the ancestral mammals lived in nocturnal or subterranean habitats and consequently their visual systems severely reduced. When most of the predatory reptiles became extinct and the 'bottleneck' was passed, the ancestral mammals started to occupy diurnal niches. Under much reduced predatory pressure, the visual system redeveloped. However, since the neocortex of mammals had already developed in response to more complex adaptations, the visual system expanded into the neocortex and developed a system of the thalamofugal type rather than redeveloping the system of the tectofugal type.

Birds evolved after the ancestral mammals appeared. However, according to the bottleneck theory (Masterton and Glendenning, 1978), modern birds have a prominent tectofugal pathway instead of the redeveloped thalamofugal pathway. Since birds have the ability to fly, they did not have to experience a nocturnal life with predatory pressure. Rather than reducing and redeveloping the visual system, birds simply developed the tectofugal type of system they had inherited from their reptilian forebears.

The bottleneck theory is one of explanations for the differences between mammals and non-mammals in the degree of development of the visual pathways. Nevertheless, many questions remain regarding the functions of the two pathways and their evolutionary origins (Masterton and Glendenning, 1978). Did the newly developed thalamofugal type of system in ancestral mammals take over the functions of the tectofugal type of system in the ancestral reptiles? Or did the new thalamofugal type of system acquire completely different functions from those in the premammalian tectofugal type?

Binocular Vision and the Thalamofugal Pathway

At least in birds, there are some positive correlations between the size of the thalamofugal pathway and the development of binocular vision. Birds with frontally placed eyes, thus with a large binocular field (e.g. owls), have a more extensively developed OPT and visual wulst (Karten et al., 1973). On the other hand, birds with laterally placed eyes and a rather small binocular field (e.g. pigeons) have a smaller OPT and visual wulst. Indeed, electrophysiological data indicated that extensive binocular processing is performed in the wulst of owls (e.g. Pettigrew, 1979). However, in pigeons Güntürkün et al. (1989) reported that the OPT of pigeons receives retinal input mostly from the region for lateral viewing rather than the frontal and binocular region. Nevertheless, some relationships have been found between descending projections from the pigeon wulst and binocular processing. Thus, descending projections from the wulst via the tractus septomesencephalicus terminate with various subcortical visual centres, such as OPT, the optic tectum and nBOR. In OPT, the highest density of these terminations is located in an area which originates bilateral projections to the wulst (Miceli et al., 1979). Similarly, in the optic tectum and nBOR, areas related to the binocular visual field receive the input from the wulst (Hamdi and Whitteridge, 1954; McGill et al., 1966; L. R. G. Britto, personal communication).

Such processing has been compared with that found in the mammalian striate cortex, and their surprising similarities (e.g. retinotopy, receptive field characteristics, binocularity, disparity detection, and ocular dominance) have been pointed out (Pettigrew, 1979). Since birds and mammals have a number of structural differences (e.g. a completely crossed retinofugal projection to the brain in birds), the mechanism of visual processing in birds must be fundamentally different from that of the mammalian striate cortex. Nevertheless, such striking similarities in functional organization between the owl's wulst and mammalian striate cortex may suggest some correlations of the thalamofugal pathway and binocular processing, not only in birds but also in mammals. For instance, cats and monkeys have extensive binocular overlap with stereopsis and well-developed striate cortices.

Conclusion

We are confronted by two substantial problems in understanding the organization and evolution of the visual system in amniotes. The first problem is the structure of the forebrain in non-mammalian amniotes, with its seeming lack of an area corresponding to the obvious mammalian neocortex. Our interpretation of this issue (i.e. the origins of the neocortex) has been dealt with briefly on p 432 and in detail in recent reviews (Karten and Shimizu, 1989; Shimizu and Karten, 1991a and b).

The second problem is the seemingly unusual visual pathways in non-mammals, with a predominantly tectally derived ascending visual pathway. The predominant emphasis in the study of mammalian telencephalic pathways has been on the anatomy, physiology and development of the retino-geniculo-striate pathway. One only familiar with the geniculo-striate system in mammals might view the reptilian/avian telencephalic visual pathways as having only vaguely similar organizational features when compared with the 'mammalian' visual system. However, there is a substantial variation in the degree of development, relative size, and location of various components of the visual system in different mammalian species. For instance, the nocturnal burrowing rat, with its impoverished visual system, cannot be taken as representing rodents in general. A simple comparison with several species of ground squirrels will reveal rodents with extensive colour vision, massive superior colliculi, and striate and extrastriate cortices. Thus, one familiar with the extrageniculo-telencephalic pathways in mammals might find that many features of the major visual pathways in reptiles and birds are surprisingly similar to those of even 'higher' primates. There is no singular pattern of the 'mammalian' visual systems that adequately encompasses the diverse organizational schemata found in the variety of ascending visual pathways to the telencephalon.

Dedication

This chapter is dedicated to Professor Walle J. H. Nauta, in honour of his many contributions to the field of neuroscience. His interests in evolution and behaviour have shaped the thoughts of a generation of young neuroscientists.

Acknowledgements

Supported by ONR Contract N00014-88-K-0504 and NINDS Grant NS24560. Drs Giovanni Casini, Luiz R. G. Britto, R. Glenn Northcutt and Anton Reiner provided valuable comments on the manuscript. The authors are grateful to Mr Kevin Cox for histological assistance, Ms Ellie B. Watelet for secretarial assistance and Ms Cynthia K. Oldmen for editorial assistance.

References

Acheson, D. W. K., Kempley, S. K. and Webster, K. E. (1980). Quantitative analysis of optic terminal profile distribution within the pigeon optic tectum. *Neuroscience*, 5, 1067–1084.

Allen, E. E. and Crews, D. (1989). 2-Deoxyglucose uptake following visual stimulation in squamate reptiles. *Brain, Behav. Evol.*, 34, 294–300.

Angaut, P. and Repérant, J. (1976). Fine structure of the optic fiber termination layers in the pigeon optic tectum: A Golgi and electron microscope study. *Neuroscience*, 1, 93–105.

Ariëns Kappers, C. U., Huber, G. C. and Crosby, E. C. (1936). *The Comparative Anatomy of the Nervous System of Vertebrates, Including Man*. Republished 1960. New York: Hafner.

Bagnoli, P. and Burkhalter, A. (1983). Organization of the afferent projections to the Wulst in the pigeon. *J. Comp. Neurol.*, 214, 103–113.

Bagnoli, P. and Casini, G. (1985). Regional distribution of catecholaminergic terminals in the pigeon visual system. *Brain Res.*, 247, 277–286.

Bagnoli, P., Fontanesi, G., Casini, G. and Porciatti, V. (1990). Binocularity in the little owl, *Athene noctua*. I. Anatomical investigation of the thalamo-wulst pathway. *Brain Behav. Evol.*, 35, 31–39.

Bagnoli, P., Fontanesi, G., Streit, P., Domenici, L. and Alesci, R. (1989). Changing distribution of GABA-like immunoreactivity in pigeon visual areas during the early posthatching period and effects of retinal removal on tectal GABAergic systems. *Vis. Neurosci.*, 3, 491–508.

Bagnoli, P., Francesconi, W. and Magni, F. (1982). Visual wulst-optic tectum relationships in birds: a comparison with the mammalian corticotectal system. *Arch. Ital. Biol.*, 120, 212–235.

Bagnoli, P., Grassi, S.and Magni, F. (1980). A direct connection between visual wulst and tectum opticum in the pigeon (*Columba livia*) demonstrated by horseradish peroxidase. *Arch. Ital. Biol.*, 118, 72–88.

Balban, C. D. and Ulinski, P. S. (1981). Organization of thalamic afferents to anterior dorsal ventricular ridge in turtles. I. Projections of thalamic nuclei. *J. Comp. Neurol.*, 200, 95–130.

Bangma, G. C. and ten Donkelaar, H. J. (1982). Afferent connections of the cerebellum in various types of reptiles. *J. Comp. Neurol.*, 207, 255–273.

Bass, A. H. and Northcutt, R. G. (1981a). Retinal recipient nuclei in the painted turtle, *Chrysemys picta*: An autoradiographic and HRP study. *J. Comp. Neurol.*, 199, 97–112.

Bass, A. H. and Northcutt, R. G. (1981b). Primary retinal targets in the Atlantic loggerhead sea turtle *Caretta caretta*. *Cell. Tiss. Res.*, 218, 253–264.

Bass, A. H., Andry, M. L. and Northcutt, R. G. (1983). Visual activity in the telencephalon of the painted turtle, *Chrysemys picta*. *Brain Res.*, 263, 201–210.

Belekhova, M. V., Kosareva, A. A., Veselkin, N. P. and Ermakova, T. V. (1979). Telencephalic afferent connections in the turtle *Emys orbicularis*: A peroxidase study. *Zh. Evol. Biochim. Fiz.*, 15, 97–103.

Benowitz, L. I. and Karten, H. J. (1976). Organization of the tectofugal visual pathway in the pigeon: a retrograde transport study. *J. Comp. Neurol.*, 167, 503–520.

Berk, M. L. and Hawkin, R. F. (1985). Ascending projections of the mammillary region in the pigeon: emphasis on telencephalic connections. *J. Comp. Neurol.*, 239, 330–340.

Berson, D. M. and Hartline, P. H. (1988). A tecto-rotundo-telencephalic pathway in the rattlesnake: Evidence for a forebrain representation of the infrared sense. *J. Neurosci.*, 8, 1074–1088.

Bessette, B. B. and Hodos, W. (1989). Intensity, color, and pattern discrimination deficits after lesions of the core and belt regions of the ectostriatum. *Vis. Neurosci.*, 2, 27–34.

Boiko, V. P. (1980). Responses to visual stimuli in thalamic neurons of the turtle *Emys orbicularis*. *Neurosci. Behav. Physiol.*, 10, 183–188.

Braford, M. R. Jr. (1972). Ascending efferent tectal projections in the South American spectacled caiman. *Anat. Rec.*, 172, 275–276.

Braford, M. R. Jr. (1973). Retinal projections in *Caiman crocodilus*. *Am. Zool.*, 13, 1345.

Bravo, H. and Pettigrew, J. D. (1981). The distribution of neurons projecting from the retina and visual cortex to the thalamus and tectum opticum of the barn owl, *Tyto alba*, and the burrowing owl, *Speotyto cunicularia*. *J. Comp. Neurol.*, 199, 419–441.

Brecha, N., Hunt, S. P. and Karten, H. J. (1976). Relations between the optic tectum and basal ganglia in the pigeon. *Soc. Neurosci. Abstr.*, 1, 95.

Brecha, N., Karten, H. J. and Hunt, S. P. (1980). Projections of the nucleus of the basal optic root in the pigeon: an autoradiographic and horseradish peroxidase study. *J. Comp. Neurol.*, 189, 615–670.

Britto, L. R. G., Gasparotto, O. C. and Hamassaki, D. E. (1990). Visual telencephalon modulates the directional selectivity of accessory optic neurons in pigeons. *Vis. Neurosci.*, 4, 3–10.

Britto, L. R. G., Keyser, K. T., Hamassaki, D. E., Shimizu, T. and Karten, H. J. (1989). Chemically specific retinal ganglion cells collateralize to the pars ventralis of the lateral geniculate nucleus and optic tectum in the pigeon (*Columba livia*). *Vis. Neurosci.*, 3, 477–482.

Bruce, L. L. and Butler, A. B. (1979). Afferent projections to the anterior dorsal ventricular ridge in the lizard *Iguana iguana*. *Soc. Neurosci. Abstr.*, 5, 140.

Burkhalter, A., Wang, S. J. and Streit, P. (1979). Thalamic projection of retinal ganglion cells: Distribution and classification. *Neurosci. Lett. (Suppl.)*, 3, 285.

Burns, A. H. and Goodman, D. C. (1967). Retinal projections of *Caiman sclerops*. *Exp. Neurol.*, 18, 105–115.

Butler, A. B. (1976). Telencephalon of the lizard *Gekko gecko* (linnaeus): some connections of the cortex and dorsal ventricular ridge. *Brain, Behav. Evol.*, 13, 396–417.

Butler, A. B. (1974). Retinal projections in the night lizard, *Xantusia vigilis* Baird. *Brain Res.*, 80, 116–121.

Butler, A. B. and Northcutt, R. G. (1971a). Retinal projections in *Iguana iguana* and *Anolis carolinensis*. *Brain Res.*, 26, 1–13.

Butler, A. B. and Northcutt, R. G. (1971b). Ascending tectal efferent projections in the lizard *Iguana iguana*. *Brain Res.*, 35, 597–601.

Cajal, S. Ramón y. (1911). *Histologie du Système Nerveux de l'Homme et des Vertébrés*. Paris: Maloine.

Carroll, R. L. (1987). *Vertebrate Paleontology and Evolution*. New York: Freeman.

Chalupa, L. M. (1991). The visual function of the pulvinar. In *The Neural Basis of Visual Function*. ed. Leventhal, A. G. *Vision and Visual Dysfunction*. Vol. 4. pp. 140–159. London: Macmillan.

Clarke, P. G. H. and Whitteridge, D. (1976). The projection of the retina, including the 'red area' on the optic tectum of the pigeon. *Q. J. Exp. Physiol.*, 61, 351–358.

Cosans, C. E. and Ulinski, P. S. (1990). Spatial organization of axons in turtle visual cortex: Intralamellar and interlamellar projections. *J. Comp. Neurol.*, 296, 548–558.

Cowan, W. M., Adamson, L. and Powell, T. P. S. (1961). An experimental study of the avian visual system. *J. Anat.*, 95, 545–563.

Crick, F. H. C. (1984). Function of the thalamic reticular complex: The searchlight hypothesis. *Proc. Natl. Acad. Sci. USA*, 81, 4586–4590.

Crossland, W. J. and Uchwat, C. J. (1979). Topographic projections of the retina and optic tectum upon the ventral lateral geniculate nucleus in the chick. *J. Comp. Neurol.*, 185, 87–106.

Cruce, J. A. F. and Cruce, W. L. R. (1978). Analysis of the visual system in a lizard, *Tupinambis nigropunctatus*. In: *Behavior and Neurobiology of Lizards*. eds. Greenberg, N. and MacLean, P. D. pp. 79–90. Rockville, Maryland: National Institute of Mental Health.

Cruce, W. L. R. and Cruce, J. A. F. (1975). Projections from the retina to the lateral geniculate nucleus and mesencephalic tectum in a reptile (*Tupinambis nigropunctatus*): A comparison of anterograde transport and anterograde degeneration. *Brain Res.*, 85, 221–228.

Dacey, D. M. and Ulinski, P. S. (1983). Nucleus rotundus in a snake, *Thamnophis sirtalis*: An analysis of non-retinotopic projection. *J. Comp. Neurol.*, 216, 175–191.

Delius, J. D. and Bennetto, K. (1972). Cutaneous sensory projections to the avian forebrain. *Brain Res.*, 37, 205–221.

Denton, C. J. (1981). Topography of the hyperstriatal visual projection area in the young domestic chicken. *Exp. Neurol.*, 74, 482–498.

Desan, P. H. (1988). Organization of cerebral cortex in turtle. In *The Forebrain of Reptiles.* eds. Schwerdtfeger, W. K. and Smeets, W. J. A. J. pp. 1–11. Basel: Karger.

Distel, H. and Ebbesson, S. O. E. (1975). Connections of the thalamus in the monitor lizard. *Soc. Neurosci. Abstr.*, 1, 559.

Ebbesson, S. O. E. and Karten, H. J. (1981). Terminal distribution of retinal fibers in the tegu lizard (*Tupinambis nigropunctatus*). *Cell Tiss. Res.*, 215, 591–606.

Ebner, F. F. and Colonnier, M. (1975). Synaptic patterns in the visual cortex of turtles: An electron microscopic study. *J. Comp. Neurol.*, 160, 51–80.

Ehrlich, D. and Mark, R. (1984). Topography of primary visual centers in the brain of the chick *Gallus gallus*. *J. Comp. Neurol.*, 223, 611–625.

Engelage, J. and Bischof, H. J. (1989). Enucleation enhances ipsilateral flash-evoked responses in the ectostriatum of the zebra finch (*Taeniopygia castanotis* Gould). *Exp. Brain Res.*, 70, 79–89.

Ferguson, J. L., Mulvanny, P. J. and Brauth, S. E. (1978). Distribution of neurons projecting to the retina of *Caiman crocodilus*. *Brain, Behav. Evol.*, 15, 294–306.

Fite, K. V. (1985). Pretectal and accessory-optic visual nuclei of fish, amphibia and reptiles: theme and variations. *Brain, Behav. Evol.*, 26, 71–90.

Fite, K. V., Brecha, N., Karten, H. J. and Hunt, S. P. (1981). Displaced ganglion cells and the accessory optic system of pigeon. *J. Comp. Neurol.*, 195, 279–288.

Funke, K. (1989a). Somatosensory areas in the telencephalon of the pigeon. I. Response characteristics. *Exp. Brain Res.*, 76, 603–619.

Funke, K. (1989b). Somatosensory areas in the telencephalon of the pigeon. II. Spinal pathways and afferent connections. *Exp. Brain Res.*, 76, 620–638.

Gaither, N. S. and Stein, B. E. (1979). Reptiles and mammals use similar sensory organizations in the midbrain, *Science Wash. DC*, 205, 595–597.

Gamlin, P. D. R., Reiner, A., Erichsen, J. T., Karten, H. J. and Cohen, D. H. (1984). The neural substrate for the pupillary light reflex in the pigeon (*Columba livia*). *J. Comp. Neurol.*, 226, 523–543.

Gamlin, P. D. R. and Cohen, D. H. (1986). A second ascending visual pathway from the optic tectum to the telencephalon in the pigeon (*Columba livia*). *J. Comp. Neurol.*, 250, 296–310.

Gamlin, P. D. R. and Cohen, D. H. (1988a). Retinal projections to the pretectum in the pigeon (*Columba livia*). *J. Comp. Neurol.*, 269, 1–17.

Gamlin, P. D. R. and Cohen, D. H. (1988b). Projections of the retinorecipient pretectal nuclei in the pigeon (*Columba livia*). *J. Comp. Neurol.*, 269, 18–46.

Gray, E. G. and Hamlyn, L. H. (1962). Electron microscopy of experimental degeneration in the avian optic tectum, *J. Anat.*, 96, 309–316.

Gusel'nikov, V. I., Morenkov, E. D. and Pivavarov, A. S. (1970). On functional organization of the visual system of the tortoise (*Emys orbicularis*), *Fiziol. Zh. SSSR Sech.*, 56, 1377–1385.

Güntürkün, O., Emmerton, J. and Delius, J. D. (1989). Neural asymmetries and visual behavior in birds. In *Biological Signal Processing.* eds. Lüttgau, H. C. and Necker, R. pp. 122–145. New York: VCH.

Hall, J. A., Foster, R. E., Ebner, F. F. and Hall, W. C. (1977). Visual cortex in the reptile, turtle (*Pseudemys scripta* and *Chrysemys picta*). *Brain Res.*, 130, 197–216.

Hall, W. C. and Ebner, F. F. (1970a). Parallels in the visual afferent projections of the thalamus in the hedgehog (*Paraechinus hypomeals*) and the turtle (*Pseudemys scripta*). *Brain, Behav. Evol.*, 3, 135–154.

Hall, W. C. and Ebner, F. F. (1970b). Thalamotelencephalic projections in the turtle (*Pseudemys scripta*). *J. Comp. Neurol.*, 140, 101–122.

Halpern, M., Wang, R. T. and Colman, D. R. (1976). Centrifugal fibers to the eye in a nonavian vertebrate: Source revealed by horseradish peroxidase studies. *Science Wash. DC*, 194, 1185–1188.

Halpern, M. and Frumin, N. (1973). Retinal projections in a snake, *Thamnophis sirtalis*. *J. Morphol.*, 141, 359–382.

Hamdi, F. A. and Whitteridge, D. (1954). The representation of the retina on the optic tectum of the pigeon. *Q. J. Exp. Physiol.*, 39, 111–119.

Harting, J. K. and Huerta, M. F. (1984). The mammalian superior colliculus: Studies of its morphology and connections. In *Comparative Neurology of the Optic Tectum.* ed. Vanegas, H. pp. 687–773. New York: Plenum.

Hayes, B. P. and Holden, A. L. (1983). The distribution of centrifugal terminals in the pigeon retina. *Exp. Brain Res.*, 49, 189–197.

Hayes, B. P. and Webster, K. E. (1975). An electron microscope study of the retino-receptive layers of the pigeon optic tectum. *J. Comp. Neurol.*, 162, 447–466.

Heller, S. B. and Ulinksi, P. S. (1987). Morphology of geniculocortical axons in turtles of the genera *Pseudemys* and *Chrysemys*. *Anat. Embryol.*, 175, 505–515.

Heric, T. M. and Kruger, L. (1965). Organization of the visual projection upon the optic tectum of a reptile (*Alligator mississippiensis*). *J. Comp. Neurol.*, 124, 101–111.

Hodos, W., Macko, K. A., and Bessette, B. B. (1984). Near-field acuity changes after visual system lesions in pigeons. II. Telencephalon. *Behav. Brain Res.*, 13, 15–30.

Hodos, W. and Campbell, C. B. G. (1969). *Scala nature*: Why there is no theory in comparative psychology. *Psychol. Rev.*, 76, 337–350.

Hopson, J. A. (1979). Paleoneurology. In *Biology of the Reptilia*, vol. 9. eds. Gans, C. G., Northcutt, R. G. and Ulinski, P. S. pp. 39–146. London: Academic.

Hoogland, P. V. and Welker, E. (1981). Telencephalic projections to the eye in *Python reticulatus*. *Brain Res.*, 213, 173–176.

Huber, G. C. and Crosby, E. C. (1926). On thalamic and tectal nuclei and fiber paths in the brain of the American alligator. *J. Comp. Neurol.*, 40, 97–227.

Huber, G. C. and Crosby, E. C. (1933). The reptilian optic tectum. *J. Comp. Neurol.*, 57, 57–163.

Hunt, S. P. and Brecha, N. (1984). The avian optic tectum: a synthesis of morphology and biochemistry. In *Comparative Neurology of the Optic Tectum.* ed. Vanegas, H. pp. 619–648. New York: Plenum.

Hunt, S. P. and Künzle, H. (1976). Observations on projections and intrinsic organization of the pigeon optic tectum: an autoradiographic study based on anterograde and retrograde, axonal and dendritic flow. *J. Comp. Neurol.*, 170, 153–172.

Hunt, S. P. and Webster, K. E. (1972). Thalamo-hyperstriate interrelations in the pigeon. *Brain Res.*, 44, 647–651.

Hunt, S. P. and Webster, K. E. (1975). The projection of the retina upon the optic tectum of the pigeon. *J. Comp. Neurol.*, 162, 433–466.

Jassik-Gerschenfeld, D. and Hardy, O. (1984). The avian optic tectum: neurophysiology and behavioural correlations. In *Comparative Neurology of the Optic Tectum.* ed. Vanegas, H. pp. 649–686. New York: Plenum.

Jones, E. G. (1985). *The Thalamus.* New York: Plenum.

Karten, H. J. (1969). The organization of the avian telencephalon and some speculations on the phylogeny of the amniote telencephalon. In *Comparative and Evolutionary Aspects of the Vertebrate Central Nervous System.* eds. Noback, C. and Petras, J. *Ann. NY Acad. Sci.*, 167, 146–179.

Karten, H. J. (1971). Efferent projections of the Wulst of the owl. *Anat. Rec.*, 169, 353.

Karten, H. J. (1979). Visual lemniscal pathways in birds. In *Neural*

Mechanisms of Behavior in the Pigeon. eds. Granda, A. M. and Maxwell, J. H. pp. 409–430. NY: Plenum.

Karten, H. J. and Hodos, W. (1970). Telencephalic projection of the nucleus rotundus in the pigeon (*Columba livia*). *J. Comp. Neurol.*, 140, 35–52.

Karten, H. J. and Shimizu, T. (1989). The origins of neocortex: connections and lamination as distinct events in evolution. *J. Cogn. Neurosci.*, 1, 291–301.

Karten, H. J., Fite, K. V. and Brecha, N. C. (1977). Specific projection of displaced retinal ganglion cells upon the accessory optic system in the pigeon (*Columba livia*). *Proc. Natl. Acad. Sci. USA*, 74, 1753–1756.

Karten, H. J., Keyser, K. T. and Brecha, N. C. (1990). Biochemical and morphological heterogeneity of retinal ganglion cells. In *Vision and the Brain.* eds. Cohen, B. and Bodis-Wollner, I. pp. 19–33. New York: Raven.

Karten, H. J., Konishi, M. and Pettigrew, J. D. (1978). Somatosensory representation in the anterior Wulst of the owl (*Speotyto cunicularia*). *Soc. Neurosci. Abstr.*, 4, 554.

Karten, H. J., Hodos, W., Nauta, W. J. H. and Revzin, A. M. (1973). Neural connections of the 'visual wulst' of the avian telencephalon. Experimental studies in the pigeon (*Columba livia*) and owl (*Speotyto cunicularia*). *J. Comp. Neurol.*, 150, 253–278.

Knapp, H. and Kang, D. S. (1968a). The visual pathways of the snapping turtle (*Chelydra serpentia*). *Brain, Behav. Evol.*, 1, 19–42.

Knapp, H. and Kang, D. S. (1968b). The retinal projections of the sidenecked turtle (*Podocnemeis unifilis*) with some notes on the possible origin of the pars dorsalis of the lateral geniculate body. *Brain, Behav. Evol.*, 1, 369–404.

Korzeniewska, E. and Güntürkün, O. (1990). Sensory properties and afferents of the n. dorsolateralis posterior thalami of the pigeon. *J. Comp. Neurol.*, 292, 457–479.

Kosareva, A. A. (1967). Projection of the optic fibers to visual centers in a turtles (*Emys orbicularis*). *J. Comp. Neurol.*, 130, 263–276.

Kosareva, A. A. (1974). Afferent and efferent connections of the nucleus rotundus in the tortoise *Emys orbicularis*. *Zh. Evol. Biokhim. Fiziol.*, 10, 395–399.

Kriegstein, A. R. (1987). Synaptic responses of cortical pyramidal neurons to light stimulation in the isolated visual system. *J. Neurosci.*, 7, 2488–2492.

Künzle, H. and Schnyder, H. (1984). Do retinal and spinal projections overlap within the turtle thalamus? *Neurosci.*, 10, 161–168.

Lohman, A. H. M. and van Woerden-Verkley, I. (1978). Ascending connections to the forebrain in the tegu lizards. *J. Comp. Neurol.*, 182, 555–594.

Macko, K. A. and Hodos, W. (1984). Near-field acuity changes after visual system lesions in pigeons. I. Thalamus. *Behav. Brain Res.*, 13, 1–14.

Macphail, E. M. (1982). *Brain and Intelligence in Vertebrates.* New York: Oxford University Press.

Martínez-de-la Torre, M., Martínez, S., and Puelles, L. (1990). Acetylcholinesterase-histochemical differential staining of subdivisions within the nucleus rotundus in the chick. *Anat. Embryol.*, 181, 129–135.

Masterton, R. B. and Glendenning, K. K. (1978). Phylogeny of the vertebrate sensory systems. In *Handbook of Behavioral Neurobiology,* vol. 1, *Sensory Integration.* ed. Masterton, R. B. pp. 1–38. New York: Plenum.

Maturana, H. R. and Varela, F. (1982). Color-opponent responses in the avian lateral geniculate: A study in the quail (*Coturnix coturnix japonica*). *Brain Res.*, 247, 227–241.

Mazurskaya, P. Z. (1972). Study of projection of the retina to the forebrain of the tortoise *Emys orbicularis*. *Zh. Evol. Biokhim. Fiziol.*, 8, 550–555.

Maxwell, J. H. and Granda, A. M. (1979). Receptive fields of movement sensitive cells in the pigeon thalamus. In *Neural Mechanisms of Behavior in the Pigeon.* eds. Granda, A. M. and Maxwell, J. H. pp. 178–198. New York: Plenum.

McGill, J. I., Powell, T. P. S. and Cowan, W. M. (1966). The retinal representation upon the tectum and isthmo-optic nucleus in the pigeon. *J. Anat.*, 100, 5–33.

McKenna, O. C. and Wallman, J. (1985). Accessory optic system and pretectum of birds: Comparisons with those of other vertebrates. *Brain Behav. Evol.*, 26, 91–116.

Meier, R. E., Mihailović, J. and Cuénod, M. (1974). Thalamic organization of the retino-thalamo-hyperstriatal pathway in the pigeon (*Columba livia*). *Exp. Brain Res.*, 19, 351–364.

Miceli, D., Gioanni, H., Repérant, J. and Peyrichoux, J. (1979). The avian visual wulst. I. An anatomical study of afferent and efferent pathways. II. An electrophysiological study of the functional properties of single neurons. In *Neural Mechanisms of Behavior in the Pigeon.* eds. Granda, A. M. and Maxwell, J. H. pp. 223–254. New York: Plenum.

Miceli, D., Peyrichoux, J. and Repérant, J. (1975). The retino-thalamo-hyperstriatal pathway in the pigeon (*Columba livia*). *Brain Res.*, 100, 125–131.

Miceli, D. Repérant, J., Villalobos, J. and Dionne, L. (1987). Extratelencephalic projections of the avian visual wulst. A quantitative autoradiographic study in the pigeon (*Columba livia*). *J. Hirnforsch.*, 28, 45–57.

Mihailovic, J., Perisic, M., Bergonzi, R. and Meier, R. E. (1974). The dorsolateral thalamus as a relay in the retino-wulst pathway in pigeon (*Columba livia*): an electrophysiological study. *Exp. Brain Res.*, 21, 229–240.

Mulligan, K. A. and Ulinski, P. S. (1990). Organization of geniculocortical projections in turtles: Isoazimuth lamellae in the visual cortex. *J. Comp. Neurol.*, 296, 531–547.

Nauta, W. J. H. and Karten, H. J. (1970). A general profile of the vertebrate brain with sidelights on the ancestry of the cerebral cortex. In *The Neurosciences: Second Study Program.* ed. Schmitt, F. O. pp. 6–27. New York: Rockefeller.

Nixdorf, B. E. (1990). Monocular deprivation alters the development of synaptic structure in the ectostriatum of the zebra finch. *Synapse*, 5, 224–232.

Nixdorf, B. and Bischof, H. J. (1982). Afferent connections of the ectostriatum and visual wulst in the zebra finch (*Taeniopygia guttata castanotis* Gould) — an HRP study. *Brain Res.*, 248, 9–17.

Norgren, R. B. Jr. and Silver, R. (1989). Retinohypothalamic projections and the suprachiasmatic nucleus in birds. *Brain, Behav. Evol.*, 34, 73–83.

Northcutt, R. G. (1984). Anatomical organization of the optic tectum in reptiles. In *Comparative Neurology of the Optic Tectum.* ed. Vanegas, H. pp. 547–600. New York: Plenum.

Northcutt, R. G. and Butler, A. B. (1974a). Retinal projections in the northern water snake *Natrix sipedon sipedon* (L.). *J. Morphol.*, 142, 117–136.

Northcutt, R. G. and Butler, A. B. (1974b). Evolution of reptilian visual systems: retinal projections in a nocturnal lizard, *Gekko gecko* (Linnaeus). *J. Comp. Neurol.*, 157, 453–466.

Northcutt, R. G., Braford, M. R. and Landreth, G. E. (1974). Retinal projections in the tuatara, *Sphenodon punctatus*. An autoradiographic study. *Anat. Rec.*, 178, 428.

Ouimet, C. L., Patrick, R. L. and Ebner, F. F. (1985). The projection of three extra thalamic cell groups to the cerebral cortex of the turtle *Pseudemys*. *J. Comp. Neurol.*, 237, 77–84.

Parent, A. (1976). Striatal afferent connections in the turtles (*Chrysemys picta*) as revealed by retrograde axonal transport of horseradish peroxidase. *Brain Res.*, 108, 25–36.

Pateromikelakis, S. (1979). Response properties of the units in the lateral geniculate nucleus of the domestic chick (*Gallus domesticus*). *Brain Res.*, 167, 281–296.

Perisic, M., Mihailovic, J. and Cuénod, M. (1971). Electrophysiology of the contralateral and ipsilateral projections to the Wulst in pigeon (*Columba livia*). *Int. J. Neurosci.*, **2**, 7–14.

Pettigrew, J. D. (1979). Binocular visual processing in the owl's telencephalon. *Proc. R. Soc. Lond. Ser. B.*, **204**, 435–454.

Pettigrew, J. D. and Konishi, M. (1976). Neurons selective for orientation and binocular disparity in the visual Wulst of the barn owl (*Tyto alba*). *Science Wash. DC*, **193**, 675–678.

Porciatti, V., Fontanesi, G., Raffaelli, A. and Bagnoli, P. (1990). Binocularity in the little owl, *Athene noctua*. II. Properties of visually evoked potentials from the wulst in response to monocular and binocular stimulation with sine wave gratings. *Brain, Behav. Evol.*, **35**, 40–48.

Pritz, M. B. (1975). Anatomical identification of a telencephalic visual area in crocodilus: ascending connections of nucleus rotundus in *Caiman crocodilus*. *J. Comp. Neurol.*, **164**, 323–338.

Rainey, W. T. (1979). Organization of nucleus rotundus, a tectofugal thalamic nucleus in turtles. I. Nissl and Golgi analyses. *J. Morphol.*, **160**, 121–142.

Rainey, W. T. and Ulinski, P. S. (1982a). Organization of nucleus rotundus, a tectofugal thalamic nucleus in turtles. II. Ultrastructural analyses. *J. Comp. Neurol.*, **209**, 187–207.

Rainey, W. T. and Ulinski, P. S. (1982b). Organization of nucleus rotundus, a tectofugal thalamic nucleus in turtles. III. The tectorotundal projection. *J. Comp. Neurol.*, **209**, 208–223.

Rainey, W. T. and Ulinski, P. S. (1986). Morphology of neurons in the dorsal lateral geniculate complex in turtles of the genera *Pseudemys* and *Chrysemys*. *J. Comp. Neurol.*, **253**, 440–465.

Ramón, P. (1891). *El Encéfelo de los Reptiles*. Zaragoza: Cited in Cajal (1911).

Reiner, A. (1981). A projection of displaced ganglion cells and giant ganglion cells to the accessory optic nuclei in turtles. *Brain Res.*, **204**, 403–409.

Reiner, A. and Karten, H. J. (1978). A bisynaptic retinocerebellar pathway in the turtle. *Brain Res.*, **150**, 163–169.

Reiner, A. and Karten, H. J. (1982). The laminar distribution of the cells of origin of the descending tectofugal pathways in the pigeon (*Columba livia*). *J. Comp. Neurol.*, **204**, 165–187.

Reiner, A. and Karten, H. J. (1983). The laminar source of efferent projections from the avian Wulst. *Brain Res.*, **275**, 349–354.

Reiner, A., Brauth, S. E. and Karten, H. J. (1984). Evolution of the amniote basal ganglia. *Trends Neurosci.*, **7**, 320–325.

Reiner, A., Brecha, N. and Karten, H. J. (1979). A specific projection of retinal displaced ganglion cells to the nucleus of the basal optic root in the chicken. *Neuroscience.*, **4**, 1679–1688.

Repérant, J. (1973). Nouvelles données sur les projections visuelles chez le pigeon (Columba livia). *J. Hirnforsch.*, **14**, 151–188.

Repérant, J. and Angaut, P. (1977). The retinotectal projection in the pigeon. An experimental optical and electron microscope study. *Neuroscience.*, **2**, 119–140.

Repérant, J. and Rio, J.-P. (1976). Retinal projections in *Vipera aspis*. A reinvestigation using light radiographic and electron microscopic degeneration techniques. *Brain Res.*, **107**, 603–609.

Repérant, J., Rio, J.-P., Miceli, D. and Lemire, M. (1978). A radioautographic study of retinal projections in type I and type II lizards. *Brain Res.*, **142**, 401–411.

Revzin, A. M. (1970). Some characteristics of wide-field units in the brain of the pigeon. *Brain, Behav. Evol.*, **3**, 195–204.

Revzin, A. M. (1979). Functional localization in the nucleus rotundus. In *Neural Mechanisms of Behavior in the Pigeon*. eds. Granda, A. M. and Maxwell, J. H. pp. 165–176. New York: Plenum.

Riley, N. M., Hodos, W. and Pasternak, T. (1988). Effects of serial lesions of telencephalic components of the visual system in pigeons. *Visual Neurosci.*, **1**, 387–394.

Ritchie, T. L. C. (1979). Intratelencephalic visual connections and their relationship to the archistriatum in the pigeon (*Columba livia*). Unpublished Ph.D. dissertation. University of Virginia, USA.

Ritchie, T. L. C. and Cohen, D. H. (1979). The avian tectofugal visual pathway: projections of its telencephalon target ectostriatal complex. *Soc. Neurosci. Abstr.*, **2**, 119.

Rogers, L. J. and Bell, G. A. (1989). Differential rates of functional development in the two visual systems of the chicken revealed by [^{14}C]2-deoxyglucose. *Dev. Brain Res.*, **49**, 161–172.

Romer, A. S. (1974). *Vertebrate Paleontology*. Chicago: University of Chicago Press.

Schnyder, H. and Künzle, H. (1983). The retinopetal system in the turtle *Pseudemys scripta elegans*. *Cell Tiss. Res.*, **234**, 219–224.

Schroeder, D. M. (1981). Tectal projections of an infrared sensitive snake, *Crotalus viridid*. *J. Comp. Neurol.*, **195**, 477–500.

Senn, D. G. and Northcutt, R. G. (1973). The forebrain and midbrain of some squamates and their bearing on the origin of snakes. *J. Morphol.*, **140**, 135–152.

Shimizu, T. and Hodos, W. (1989). Reversal learning in pigeons: effects of selective lesions of the wulst. *Behav. Neurosci.*, **103**, 262–272.

Shimizu, T. and Karten, H. J. (1990). Immunohistochemical analysis of the visual wulst of the pigeon (*Columba livia*). *J. Comp. Neurol.*, **300**, 346–369.

Shimizu, T. and Karten, H. J. (1991a). Multiple origins of neocortex: contributions of the dorsal ventricular ridge. In *The Neocortex: Ontogeny and Phylogeny*. Finlay, B. L., Innocenti, G. and Scheich, H. NATO ASI series. New York: Plenum.

Shimizu, T. and Karten, H. J. (1991b). Computational significance of lamination of the telencephalon. In *Visual Structures and Integrated Functions: Research Notes in Neural Computing*. eds, Arbib, M. A. and Ewert, J.-P. New York: Springer.

Shimizu, T., Woodson, W., Karten, H. J. Schimke, J. B. (1989). Intratelencephalic connections of the visual areas in birds (*Columba livia*). *Soc. Neurosci. Abstr.*, **15**, 1398.

Shimizu, T., Karten, H. J. and Cox, K. (1990). Intratelencephalic projections of the visual wulst in birds (*Columba livia*): A *Phaseolus vulgaris* leucoagglutinin study. *Soc. Neurosci. Abstr.*, **16**, 246.

Shortess, G. K. and Klose, E. (1975). The area of the nucleus isthmoopticus in the American kestrel (*Falco sparverius*) and the red-tailed hawk (*Butero jamaicensis*). *Brain Res.*, **88**, 525–531.

Simpson, J. I. (1984). The accessory optic system. *Ann. Rev. Neurosci.*, **7**, 13–41.

Simpson, S. M. and Follett, B. K. (1981). Pineal and hypothalamic pacemakers: their role in regulating circadian rhythmicity in Japanese quail. *J. Comp. Neurol.*, **144**, 381–389.

Smith, L. M., Ebner, F. F. and Colonnier, M. (1980). The thalamocortical projection in *Pseudemys* turtles: A quantitative electron microscope study. *J. Comp. Neurol.*, **190**, 445–461.

Stein, B. E. and Gaither, N. S. (1981). Sensory representation in reptilian optic tectum: Some comparisons with mammals. *J. Comp Neurol.*, **202**, 69–87.

Streit, P., Burkhalter, A., Stella, M. and Cuénod, M. (1980a). Patterns of activity in pigeon brain's visual relays as revealed by the [^{14}C]2-deoxyglucose method. *Neurosci.*, **5**, 1053–1066.

Streit, P., Stella, M. and Cuénod, M. (1980b). Transneuronal labeling in the pigeon visual system. *Neurosci.*, **5**, 763–775.

Swanson, L. W. (1987). The hypothalamus. In *Handbook of Chemical Neuroanatomy*, vol. 5. *Integrated Systems of the CNS. Part 1. Hypothalamus, Hippocampus, Amygdala, Retina*, pp. 1–124. eds. Björklund, A. and Hökfelt, T. New York: Elsevier.

Tsai, H. M., Garber, B. B. and Larramendi, L. M. H. (1981). ^{3}H-thymidine autoradiographic analysis of telencephalic histogenesis in the chick embryo: I. Neuronal birthdates of telencephalic compartments *in situ*. *J. Comp. Neurol.*, **198**, 275–292.

Uchiyama, H. (1989). Centrifugal pathways to the retina: Influence of the optic tectum. *Vis. Neurosci.*, **3**, 183–206.

Ulinski, P. S. (1977). Tectal afferents in the banded water snake, *Natrix sipedon*. *J. Comp. Neurol.*, **173**, 251–274.

Ulinski, P. S. (1983). *Dorsal Ventricular Ridge: A Treatise on Forebrain Organization in Reptiles and Birds.* New York: Wiley.

Ulinski, P. S. and Nautiyal, J. (1988). Organization of retinogeniculate projections in turtles of the Genera *Pseudemys* and *Chrysemys.* *J. Comp. Neurol.,* 276, 92–112.

Walls, G. L. (1942). *The Vertebrate Eye and its Adaptive Radiation.* Republished 1963. New York: Hafner.

Wang, R. T and Halpern, M. (1977). Afferent and efferent connections of thalamic nuclei of the visual system of garter snakes. *Anat. Rec.,* 187, 741–742.

Wang, Y.-C. and Frost, B. J. (1990). Functional organizations in the nucleus rotundus of pigeon. *Soc. Neurosci. Abstr.,* 16, 1314.

Watanabe, M., Ito, H. and Masai, H. (1983). Cytoarchitecture and visual receptive neurons in the wulst of the Japanese Quail (*Coturnix coturnix japonica*). *J. Comp. Neurol.,* 213, 188–198.

Watanabe, M., Ito, H. and Ikushima, M. (1985). Cytoarchitecture and ultrastructure of the avian ectostriatum: Afferent terminals from the dorsal telencephalon and some nuclei in the thalamus. *J. Comp. Neurol.,* 236, 241–257.

Weidner, C., Repérant, J., Desroches, A.-M., Miceli, D., and Vesselkin, N. P. (1987). Nuclear origin of the centrifugal visual pathway in birds of prey. *Brain Res.,* 436, 153–160.

Wellnhofer, P. (1988). A new specimen of *Archaeopteryx*. *Science Wash. DC,* 240, 1790–1792.

Wild, J. M. (1987). The avian somatosensory system: connections of regions of body representation in the forebrain of the pigeon. *Brain Res.,* 412, 205–223.

Wilson, P. (1980). The organization of the visual hyperstriatum in the domestic chick. I. Topology and topography of the visual projection. *Brain Res.,* 188, 319–332.

Woodson, W., Reiner, A., Anderson, K. and Karten, H. J. (1991). Distribution laminar location, and morphology of tectal neurons projecting to the isthmo-optic nucleus and the nucleus isthmi, pars parvocellularis in the pigeon (*Columba livia*) and chick (*Gallus domesticus*): a retrograde labelling study. *J. Comp. Neurol.,* 305, 1–19.

Yamada, H. and Sano, Y. (1985). Immunohistochemical studies in the serotonin neuron system in the brain of the chicken (*Gallus domesticus*). II. The distribution of the nerve fibers. *Biogen. Amines,* 2, 21–36.

20 Evolution of Mammalian Visual Pathways

G. H. Henry and T. R. Vidyasagar

Introduction

Polyak (1957), in studying the embryological development of the eye in search of clues for phylogeny, was struck, as many must have been, with the precociousness and relative largeness of the visual primordia. Most of what is to be the brain belongs to the developing eye, and 'the huge brain, so characteristic of the advanced vertebrates, appears to receive its first and most potent impetus from visual stimuli'. In a lyrical turn, Polyak wondered if sight provided the 'magical formula by which the vertebrates climbed from the twilight of the ocean floor into the upper reaches of the water and finally out into the unlimited photic freedom of the air'.

To follow the evolution of the mammals in this upward climb it is necessary to name the major mammalian subclasses. The need for these subdivisions arose first with the identification of unique mammals in Australia and America. By 1816, de Blainville (cited in Tyndale-Biscoe, 1973) had appreciated that the mode of reproduction set these animals apart from other mammals and that they formed two distinctive orders. De Blainville adopted an earlier name of *Didelphus* for marsupials to indicate the presence of two uteri, one internal and one external, and in accord placental mammals were called *Monodelphia*. The other new order, which included the platypus and echidna, he called *Ornithodelphia* because their oviducts resembled those of birds. The alternatives to these names that have been introduced at various times since are presented in Table 20.1, which has been reproduced from Tyndale-Biscoe (1973).

Tyndale-Biscoe sees the Huxley nomenclature as one that has exerted 'a long and baneful influence on the understanding of marsupials and monotremes' since 'it encouraged people to think that by studying these mammals they could ride a sort of Wellsian Time Machine back to the origin of mammals'. It remained for 'Huxley's grandson Julian to redress the balance by emphasizing that all living animals must be viewed in the context of their adaptations to the present environment'.

In this review we have adopted the names in bold face type in Table 20.1, but this only scratches the surface of the nomenclature question. There are 17 orders of living placental mammals and a total of about 4000 species (Walker, 1975) and 250 species of marsupials in basically two orders (Kirsch and Calaby, 1977). There are two surviving species of monotremes, the platypus and the echidna. Physiological information, in particular, is available for only a few of these species and this will limit our endeavour to trace the course of evolution of mammalian visual pathways. Among the placentals, particular attention will be paid to the evolution of the primates, since it is currently a topic of much interest (Pettigrew *et al.*, 1989) and also because there is some justification, from the distinctiveness of their visual pathways, to regard the emergence of the primates as a major step in mammalian evolution.

Mammalian Characteristics in the Primary Visual Pathway

In most of the non-mammalian species, with the obvious exception of some birds such as the owl and some deep-sea fish (Walls, 1942), the eyes are laterally positioned and there is little binocular overlap of the visual fields of each eye. The development of frontal eyes among the mammals leads to larger binocular fields, which are most extensive amongst the primates (see Hughes, 1977, for review).

Table 20.1

de Blainville	Bonaparte	Huxley	Illiger, Owen	
Ornithodelphia	**Monotremata**	Prototheria		
Didelphia	Ditremata	Metatheria	**Marsupialia**	
Monodelphia		Eutheria	**Placentalia**	

The development of frontal vision has been accompanied by an increase in the density of packing of ganglion cells in an area of frontally directed retina, which has led to an increase in visual acuity. In eyes with laterally directed visual axes, these areas of higher acuity are found in the temporal retina but are more centrally positioned in species with greater binocular overlap. Hence the name, *area centralis*, for the region of higher density and *fovea centralis* for the pit that lies at its centre in the case of the primates. The area centralis is more centrally placed in primates than in other mammals and foveation reaches its highest development in diurnal primates.

The requirement to move the area centralis of both eyes to an object of interest in the binocular field means that the eye movements of the mammals are immediately set apart from those of other vertebrates (Walls, 1942). Thus, whenever eye movements occur in mammals the two eyes never move independently but always as a conjugated pair. Where a mammal, such as the rabbit, has laterally placed eyes the area centralis is replaced by a horizontal visual streak and there is little in the way of spontaneous eye movements. There is now a tendency, however, to regard the horizontal streak as a basic feature of most eyes (Hughes, 1977). The greatest eye mobility occurs in the foveated primates where the greatest binocular overlap is also present. Therefore, conjugated eye movements align the two eyes so that the fovea or the points of highest cell density in each retina are directed to a point in front of the animal, the point of fixation, which divides visual space into right and left hemifields.

In all vertebrates the optic nerves merge under the brain to pass through the optic chiasm and emerge as the optic tracts. In non-mammalian vertebrates the decussation at the chiasm is complete and partial decussation is an exclusive feature of the mammalian visual pathways. As a result of partial decussation, the lateral geniculate nucleus (LGN) receives its inputs from the two half-retinae projecting to the contralateral hemifield (e.g. R LGN from R temporal and L nasal retinae). There is, therefore, a vertical line of decussation which separates the retina into, what is in reality, two unequal 'halves'. The fibres from the nasal half of the retina are the only ones to decussate and generally they contribute the higher proportion of fibres to the optic nerve. The line of decussation divides the area centralis and the two 'half' retinae become more nearly equal as the binocular field grows in size. Therefore, in most primates about half the fibres in the optic nerve remain uncrossed as they pass through the chiasm to the LGN.

The optic tract fibres passing via the brachium of the superior colliculus contribute to another pathway that can justifiably be called primary. This projection passes to the accessory optic tract in the tegmentum of the midbrain. It is likely that this pathway has a role in the appreciation of self-motion and acts complementarily to the vestibular system (see Simpson, 1984, for review).

The emergence of the neocortex stands as the most significant evolutionary development in the vertebrate brain and with it the primary target for the optic tract begins to change from the optic tectum to the occipital cortex. However, in terms of numbers of orders the transition has been slight and the superior colliculus (mammalian optic tectum) is still the recipient of the majority of retinal ganglion cell fibres for all mammalian orders other than the primates. The change of prime destination from tectum to neocortex is also accompanied by a change in retinotopic representation, so that the pattern in the superior colliculus of the primate is distinct from that of other mammals and it comes to resemble more closely the pattern in the LGN (Pettigrew, 1986a; Allman, 1977). However, in the superior colliculus the ipsilateral and contralateral representations mingle with each other and do not reside in separate layers as in the LGN (Pettigrew *et al.*, 1989).

The presence of binocular vision in the mammals is accompanied by the occurrence of lamination in the LGN. Examples of conspicuous lamination are present in most orders, but it is difficult, as will become apparent below, to fit all examples into a generalized pattern. However, there is a common design feature in that each layer, acting as a relay in the path to the visual cortex, receives a topographical projection from a hemiretina which is in positional register with the hemiretinal representation of the adjacent layer. These neighbouring representations may arise from the same or from opposite eyes. A pairing, in which layers of similar-sized cells receive inputs from opposite eyes, usually indicates that the two layers are carrying information from a common functional stream. Therefore one purpose of lamination appears to be the morphological alignment of corresponding retinal points from the two eyes and one design feature, the gap in the ipsilateral layer representing the optic nerve head, indicates that this binocular ordering is quite strict (Kaas *et al.*, 1973).

The inputs from the two eyes come together in the visual cortex. In most of the primates but not all (the squirrel monkey and owl monkey are exceptions), the projections from different layers of the LGN go to separate dominance bands in layer IV (Hubel *et al.*, 1976). In the majority of primates equivalent binocular responses do not occur in single cortical cells until the pathway reaches the layers above or below lamina IV or the extra-striate cortex. The stage at which single cortical cells respond to binocular stimulation appears to be delayed in primates so that while there is a high incidence of binocular cells in cat striate cortex, even in lamina IV, similar high proportions are not found in the monkey until area V2 (Hubel and Wiesel, 1970).

In most placental mammals there are three transverse commissures that link the two hemispheres of the forebrain. These are the large anterior commissure, the small hippocampal commissure and the very large corpus callosum. The posterior or splenial segment of the corpus callosum carries a large bundle of fibres linking different areas of visual cortex with their counterparts in the other hemisphere. There is evidence that these are fast conducting fibres that link with cells receiving inputs from similar fast conducting axons in the optic radiation (McCourt et al., 1990). There is also evidence that the anterior commissure in the cat also carries visual fibres to link the ectosylvian visual areas in both hemispheres (Boyapati, McCourt and Henry, unpublished results). However, the corpus callosum is absent from the forebrains of marsupials (Loo, 1930) and of monotremes but its place as a conduit for visual fibres appears to be taken by another commissure, the fasciculus aberrans, which lies dorsal to the anterior commissure (Heath and Jones, 1971). An interhemispheric commissure linking the visual centres appears to have a survival advantage and alternative designs to make the link have evolved both in marsupials and placentals.

Comparative View of Features of the Visual Pathway

In developing this review we now propose to look at the functional and structural design of different features along the visual pathways and wherever possible to make comparisons in examples taken from the marsupials and the placentals. Unfortunately, information is sparse on the visual pathways of the monotremes and few examples will come from this group.

The Ganglion Cell Layer of the Retina

As mentioned earlier there is a growing acceptance that the horizontal streak in the contour lines joining points of iso-density in the distribution of retinal ganglion cells may form part of the basic design for the mammalian retina and that later specializations have evolved from this base pattern. If this is the case then there is some interest to see if there are systematic species variations in the nature of the streak. Pettigrew et al. (1989) sought a possible distinction in retinal location of the streak, whether it runs above or below the optic nerve head. In the majority of mammals the streak is superior to the optic nerve head, but this is not unique to the mammals and also the exceptions among the mammals, where the streak is inferior to the nerve head, are a rather mixed group that includes the elephant (Halasz and Stone, 1989); the elephant shrew, edentates and hyrax (Pettigrew et al., 1989); the rabbit (Hughes, 1977), some rodents and microchiropteran bats (Pettigrew et al., 1988). The inclusion of the microchiropteran bats in this group, as a point of distinction from the megachiropteran bats, is of significance to those interested in placing the two types of bats in different taxa, but generally the position of the streak gives little insight on evolutionary paths among the mammals.

Rather than looking at the distribution of the total population of retinal ganglion cells it may be possible to discover more about evolutionary trends from a study of functional and morphological types in the ganglion cell population. In a review of the functional differences in the retinal ganglion cells of cat and macaque, taken as examples of carnivore and primate, Shapley and Perry (1986) have classed cells on the basis of their visual responses and then compared the distribution of their axon terminals in the LGN. For the cat, it was concluded that X cells are driven by linear receptive field centre and surround mechanisms while Y cells receive additional signals from non-linear subunits (Hochstein and Shapley, 1976). The responses of both X and Y cells are highly sensitive to contrast changes. The remaining cells of the ganglion cell population in the cat come under the umbrella title of W cells, which project in the main to the superior colliculus. The X-cells project almost entirely to the LGN while the Y and W cells often branch to send one arm to the LGN and the other to the superior colliculus. In the LGN, cells receiving from X and Y cells are mixed in the A and A_1 layers while the W cells are found in the C layers (terminology of Hickey and Guillery, 1974).

In the monkey, retinal ganglion cells are classified according to their LGN destination: the P cells project to the parvocellular layers of the LGN, have small receptive fields, are wavelength-selective and are poorly sensitive to luminance contrast (Blakemore and Vital Durand, 1981; Hicks et al., 1983; Shapley et al., 1983; Purpura et al., 1988); the M cells on the other hand, which project to the magnocellular layers, are contrast-sensitive but not wavelength-selective. There are two groups of M cells, one – the M_X – where the response is linear and the other – the M_Y – where it is not. There are possibly also two groups of P cells (P_X and P_Y) but very few P_Y cells have ever been reported. The hypothetical 'K' cells (those projecting to the koniocellular group in the LGN and possibly equivalent to the W cells of the cat) are yet to be identified in the primate retina. The poor contrast sensitivity in P cells has prompted Shapley and Perry to propose that they are distinctive from any cell type found in the cat. If homologies are to be drawn, the X and Y cells of the cat compare to the M_X and M_Y cells of the monkey and the P cells are thought to be a new evolutionary development bringing colour vision to the primate. These views have not received universal acceptance and others (in particular

Kaas, 1986) have proposed that the parvocellular layers receive their input from X cells and the magnocellular from Y cells. However, the Shapley–Perry model is based on the most thorough quantitative analysis conducted to date and it has been accepted in developing this review. A detailed interpretation of the response properties of primate retinal ganglion cells and the prospective contribution to human psychophysics is contained in a review by Kaplan et al. (1990).

The morphological correlates of the functional cell types are more surely established for the cat than for the primate. Physiological recordings related to histologically reconstructed retinal wholemounts suggest that, in the cat, the Y cells (4% of the population) are correlated with the large alpha cells, the X cells with the middle-sized beta cells and the W cells with the small gamma cells. There is some uncertainty about the percentages of X and W cells in the cat retina and the manner in which they vary from streak to periphery, but there is no great error in deciding on equal percentages of 48% (Hughes, 1981). Direct functional and morphological relationships are still to be established in the primate, but the retrograde labelling of retinal ganglion cells after the injection of horseradish peroxidase into the parvo- and magnocellular layers of the LGN (Leventhal et al., 1981; Perry et al., 1984) showed that the cells, at a given eccentricity, projecting to the two LGN subdivisions had ranges of dendritic trees of distinctly different sizes. For these two groups, the cells projecting to the M layers (A cells of Leventhal et al., 1981; P_α cells of Perry et al., 1984) made up approximately 10% of the total population, while those going to the parvocellular layers (B cells of Leventhal et al., 1981; P_β cells of Perry et al., 1984) made up 80%. The remaining 10% (called C and E by Leventhal et al., 1981) project to the superior colliculus, although others (Pettigrew, 1990) suggest that the proportion of retinotectal cells may be as low as 2%. Largely from their morphology, these retinotectal cells are thought to resemble W cells more than M or P cells. In these percentages there appears to be a dramatic reduction in the number of ganglion cells projecting to the superior colliculus in the primate and the functional usefulness of this projection in humans has sometimes even been called into doubt (see Pettigrew, 1990). However, if actual fibre numbers are counted the cat has 85 000 (50% of 170 000 say) and the monkey 100 000 (10% of 1 000 000; if we use Leventhal et al.'s percentage) projecting to the superior colliculus. In terms of absolute numbers, the principal evolutionary change, associated with the appearance of the primates, is one that leads to the pre-eminence of the P cells. The equivalent of the P cell does not exist in the cat but cells with the X property are much more numerous in the primate than in the cat; 880 000 (88% ($P_X + M_X$) of 1 000 000) compared with 81 600 (48% of 170 000), which is much closer to the number of M_X cells in the primate –

80 000 (8% of 1 000 000). The cells with Y properties in the primate, the M_Y cells, which contribute something like 14 000 (1.4% of 1 000 000), are not greatly in excess of the numbers in the cat, where the count is 6800 (4% of 170 000). Irrespective of the functional model adopted for the LGN, it is the introduction of P cells with their linear response and wavelength selectivity that sets the primate apart from other mammals.

In summary, it appears that the retinal cell types that project to the superior colliculus, the Y and W cells, are phylogenetically robust and may have projected to the optic tectum exclusively prior to the appearance of the neocortex. At some time around the emergence of the neocortex these cells sent a branch into the new geniculostriate pathway and at the same time a new cell type, the X cell, appeared and projected only to the LGN. The X cell may have undergone a further adaptation to lose contrast sensitivity and gain wavelength selectivity and create the P cell. Or alternatively the P cell in the primate is a new form, arising at the threshold of primate evolution.

Another morphological aid in the cross-species correlation of ganglion cell types is in the recognition of a class of retinal ganglion cells that are revealed with a neurofibrillar stain. In the cat these cells have been equated with alpha or Y cells (Wassle et al., 1975) and similar populations are found in a wide variety of mammalian retinae (Peichl et al., 1987). The similarity of the neurofibrillar carrying cells in so many mammalian species would also suggest that the Y cell population has changed little and that it reached an optimum design level at an early point in phylogenetic development.

From cell size it has now become possible to identify three groups of cells in Nissl stained retinal wholemounts of the cat retina (Hughes, 1975). The cell densities of these different size-classes correspond closely to those registered for Y, X and W cells (or for alpha, beta and gamma cells) (Hughes, 1981) and there is some justification for believing that the differentiation seen in the cat, if replicated in Nissl stained material from other species, may provide strong hints to existence of similar functional classes. The distribution of ganglion cell sizes in Nissl stained material in a marsupial, the tammar wallaby, is similar to that in the cat (Wong et al., 1986). In the wallaby, the large cells, which also take up the neurofibrillar stain, form a distinct group, while the smaller cells, which in some samples at certain eccentricities display a bimodal distribution, have a sufficiently broad distribution at all retinal locations to accommodate two overlapping populations as expected if X and W cells are also present in this marsupial. However, there is one striking difference in the Nissl stained appearance of the ganglion cell layer in the cat and the wallaby; the ratio of microcells or displaced amacrine cells to ganglion cells is much lower in the tammar wallaby (1:4) that in the cat (2:1) (Wong et al.,

1986). The organizational consequences of this distinction remain a mystery but in this feature it seems the wallaby is closer to the rabbit than the cat and it would be interesting to see how the ratio varies across a greater range of mammalian retinae.

The Optic Chiasm

Partial decussation in the optic chiasm is restricted to mammals and the explanation for its existence has usually been related to the development of frontal eyes and binocular vision. The increase in uncrossed fibres as the visual axes become closer to being parallel has prompted the development of a number of theories based on the fusion of the neural images in the two hemispheres. This kind of modelling was a pastime for a bygone age, but the ideas on partial decussation are worth recalling if only because it is now possible to assess them in the light of more recent findings. Walls (1942) threads his way through the logical ramifications of fusion theories proposed by Ramón y Cajal (1894; cited by Walls, 1942) and Ovio (1927; cited by Walls, 1942) and finally concludes that the fusion of interhemispheric images would produce similar psychic images irrespective of whether there was either total decussation, as shown in Fig. 20.1(a), or partial discussation, as in Fig. 20.1(b).

The only distinction between the two schemes in Fig. 20.1 is the nature of the interhemispheric representation; in the case of partial decussation each hemisphere has a hemiretinal representation, while in complete decussation each hemisphere has a full retinal representation. In Walls' view the fusional requirement is then the same; that is, the fusion of the whole right-eyed view of the object with the whole left-eyed view. The fusibility of these images and consequent development of stereoscopic vision cannot, in this interpretation, depend upon the character of the optic chiasm and it is necessary to look elsewhere for an explanation of partial decussation and reject the 'firmly-rooted traditional one that without partial decussation there could be no fusion and no stereopsis' (Walls, 1942).

As pointed out above, the appearance of partial decussation in mammals is accompanied not only by the development of binocular perception but also by conjugate eye movements. Walls stresses the essential contribution made by conjugate eye movements to binocular vision and how partial decussation of the optic nerves 'accomplishes a desirable tying-up of both retinae to both the left-brain and right-brain centres of eye-muscle control'. Partial decussation, in providing a direct two-eyed link with eye movement centres, is more essential to the retinotectal projection than to the geniculostriate pathway; it is more involved with motor function than with conscious sensory perception. Such an interpretation, in Walls' view,

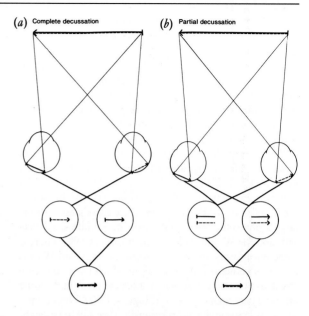

Fig. 20.1 *Schematic representation of the hemispheric images formed (a) in complete decussation and (b) in partial decussation. The lower circles represent the psychic act of interhemispheric fusion (Adapted from Walls, 1942, where it is argued that with either partial or complete decussation the resulting fusion image is of the same character).*

explains why partial decussation is absent in owls and frontal-eyed deep-sea fishes: 'not because they are not mammals, but because their eyes are motionless'.

The strength of this argument is influenced by information not available to Walls on the nature of binocular projection to the superior colliculus, which is considered in the next section. For example, there is more or less complete decussation to the superior colliculus of the cat, despite the presence of mobile eyes.

The Superior Colliculus

Pettigrew (Chapter 12) points out that binocular vision is present in all mammals (with the possible exception of the whales) and this ubiquity makes it more difficult to unravel the evolutionary history of binocular vision. However, there are some significant species differences in the ordering of the retinotectal pathway and these not only help in following the thread of mammalian binocular vision but also have a bearing on the Wallsian ideas developed above. These findings (Lane et al., 1971, 1973; Kaas et al., 1973) showed that tree shrews and squirrels differ from primates in the extent to which the contralateral retina is represented on each colliculus. It had previously been thought that the projection to the superior

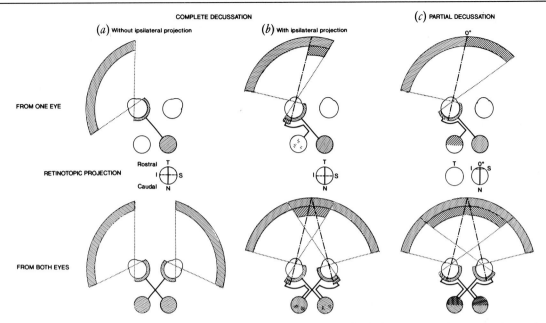

Fig. 20.2 *Schematic representation of projections from the two eyes to the superior colliculi showing, in (a) decussation of axons from the complete contralateral retina with no additional ipsilateral component and, in (b), complete decussation with an ipsilateral component. In (c), decussation is limited to the contralateral hemifield, the nasal retina, while ipsilateral hemifield, or temporal retina, is not undergoing decussation. The upper diagrams show the input from one eye only while the lower ones show the input from both eyes. In between the two sets of diagrams there is a plan view of the topographic representation of the retina on the surface of the superior colliculus. Stippled areas represent ipsilateral projections; hatched areas represent contralateral projections. N, T, I and S stand for nasal, temporal, inferior and superior retina, respectively, and are not to be confused with designations for visual fields.*

colliculus resembled that to the LGN (Walls even seemed to think it went via the LGN) and that each colliculus received a representation of the contralateral hemifield. Now, it has been shown that each superior colliculus, in a wide variety of non-primate mammals (which in addition to squirrels and tree shrews, includes cats, rabbits, tammar wallabies and opossums) contains a representation of the complete contralateral retina as represented in Fig. 20.2(b). In all these examples it is difficult, in physiological experiments, to find cells driven from the ipsilateral eye and the presence of an ipsilateral projection can usually be demonstrated only with an axonal tracer, which appears in scattered patches in the nucleus (see Fig. 20.2(b)). In this pattern of projection, now often termed the non-primate pattern, the front of the superior colliculus represents the far temporal edge of the retina.[1]

The weakness of the ipsilateral retinotectal projection in many non-primates (there is often doubt about its existence) does not seem to imply that it was a precursor of partial decussation. Nor does the common finding that the ipsilateral representation is more highly developed in the LGN than the superior colliculus stand well with Walls's belief that partial decussation is of greater aid to tectal than to cortical function.

The proportion of ipsilateral fibres to the retinotectal projection increases in the 'primate pattern'. Here, as shown in Fig. 20.2(c), the nasal hemiretina is represented on the contralateral superior colliculus and the temporal hemiretina on the ipsilateral nucleus. The front of the superior colliculus, instead of representing the temporal retina, now represents the line of decussation or the zero retinal (vertical) meridian and corresponding points from the two retinae mix together in each superior colliculus. The taxonomic significance of the 'primate pattern' has taken on a new significance as reports come forward that this is the pattern in flying foxes or megabats (Pettigrew, 1986b) and the gliding lemur (Pettigrew and Cooper, 1986), which have led Pettigrew and Cooper to suggest that 'an early branch of the primates gave rise to the dermopterans (gliding lemurs) some of which developed true flight to give rise to the megachiropterans (flying foxes)'.

From a comprehensive survey, Pettigrew et al. (1989) have pointed out that the size of the ipsilateral representation varies from zero in the microbats through increasing fractions in the rodents and elephant shrews until it becomes a significant proportion in the primates and megabats. In these latter groups the ipsilateral projection reaches the anterior pole of the superior colliculus and, in what appears to be the ultimate case, it covers the entire surface of the nucleus in the gibbon.

Whatever the taxonomic value of partial decussation and its subsequent influence on target nuclei, there is still a great deal of uncertainty about its value to the visual performance and survival of the animal. If it is accepted that partial decussation increases as the eyes become more frontal then there may be clues in looking at the survival advantages that go with frontal eyes (see Hughes, 1977, for review). Binocular overlap, the natural consequence of frontal eyes, is greatest in primates (120° to 150°), but it also reaches 90° in cats, and all these species have well-developed stereoscopic vision. These relationships led to an early proposal (Collins, 1921) that frontal vision and its attendant benefits were necessary for the arboreal, branch-jumping life of the early primates. A large binocular overlap is not essential for a climbing life, however, and squirrels lead a successful arboreal existence with only one-fifth of each visual field in binocular overlap (Hall *et al.*, 1971; Kaas *et al.*, 1972). In addition, stereoscopic vision has evolved in the absence of an arboreal life style. The tammar wallaby, a marsupial, an animal living among low bushes, has laterally directed eyes with only 60° of binocular overlap yet much of its visual cortex is devoted to the binocular segment, where many cells are tuned for binocular disparity (Vidyasagar *et al.*, 1990). Overall, it does not seem that the advantages that come with stereoscopic vision have led to the evolution of frontally directed eyes and the enlargement of partial decussation.

Alternative advantages, which come with frontal vision, act to increase the efficiency of predators (Cartmill, 1972, 1974; Allman, 1977). Frontal eyes may help in detecting prey and in particular stereopsis may assist in breaking the effect of camouflage (Julesz, 1971). Frontally directed eyes also have merit in that they bring the visual axis closer to the optical axis of the eye and so improve the quality of the retinal image (Jenkins and White, 1957; Leibowitz *et al.*, 1972). This optical improvement would assist the vision of the nocturnal predator by helping to counter the loss of retinal image quality caused by a dilated pupil.

In a review that finds shortcomings in the arboreal and predation hypotheses and finds that these life styles show little correlation with a wide binocular field, Hughes (1977) does find evidence that 'the width of the binocular field increases *pari passu* with that of the praxic field'. Praxic behaviour (Trevarthen, 1968) involves the use of visually controlled activity to manipulate the environment or, in other words, it occurs in the visually guided work space. Hughes produces a variety of examples to show that the binocular field is of similar dimension to the praxic space, but in the end seems to suggest that the implication of a *pari passu* relationship may go too far, and there are a number of out-of-step examples.

However, the advantages of frontal vision must be pitted against the loss of panoramic vision, which may well increase the likelihood that the predator becomes the prey.

One way of compensating for the loss of lateral vision is to improve the quality of sound localization. It is not known if sound localization is superior in animals with frontal eyes, nor if the superiority is more a feature of the space beyond the visual fields of the frontal eyes. It is known that the visual, auditory and somatosensory maps in the mammalian superior colliculus are in topographic register so that stimuli at the same locus in space, or in the equivalent somatotopic location, activate tectal cells in vertical alignment (see Vidyasagar, 1991). This alignment extends ventrally in the nucleus to a motor map for saccadic eye movements (Schiller and Stryker, 1972). The alignment of the visual map with the maps of other modalities is disturbed if the eyes move and there is not compensating head or body movement. The cat seems to avoid this outcome by moving the head to keep the eyes centred in the orbit (Harris *et al.*, 1980) but in the primate there appears to be a realignment of the auditory map (Jay and Sparks, 1984, 1987). Such a labile auditory map, monitoring the whole spatial environment, could prompt the eyes and/or the head to move to a new object of interest in a way that diminishes the loss of panoramic vision. It may be argued that partial decussation in the visual pathways facilitates the alignment with the contralateral auditory map, since both are representing a lateral half of the surrounding space.

In the evolutionary balance sheet there can be little doubt that stereoscopic vision is an asset and that frontal eyes make for better predators and for more efficient tree dwellers, and perhaps even help those that come from the trees onto the plains. At the quantitative level it may even be possible to relate its development with a greater use of visual work space. In reaping these benefits there may not be a loss in awareness if other sensory modalities, in close neural alignment with the visual process, can compensate for the loss of panoramic vision. Having made all these observations we are not much closer to solving the riddle of partial decussation and perhaps the best that can be said is that it has been retained as a design feature because it delivers many blessings: it assists in the creation of the stereoscopic percept, aids in cross-modal alignment of topographic maps, promotes conjugate eye movements and it does these things at the input to the processing nuclei rather than later through the various interhemispheric pathways.

The Accessory Optic Tract

The accessory optic tract in mammals (see Simpson, 1984, for review) consists of two sets of optic fibres of contralateral retinal origin. These form the superior and inferior fasciculi which project to three target nuclei, the dorsal, medial and lateral terminal nuclei (Hayhow, 1959, 1966; Hayhow *et al.*, 1960) in the midbrain tegmentum.

The medial terminal nucleus receives inputs from both the superior and the inferior fasciculi while the input to the dorsal and lateral terminal nuclei comes from the superior fasciculus. In non-mammals the accessory optic tract consists of one fasciculus, which leaves the optic nerve after the chiasm, terminates in a single ventral terminal nucleus, the nucleus of the basal optic root (nBOR; Brecha et al., 1980). The accessory optic tract, in one of these forms, has now been identified in fish, birds, reptiles and amphibians.

In many non-mammals, in particular the bird, the projection to the nBOR arises from displaced ganglion cells but in mammals such as rabbit and cat no displaced ganglion cells appear to contribute to the projection to the medial terminal nucleus. In the cat the accessory optic tract arises from approximately 2000 gamma (in the terminology of Boycott and Wassle, 1974) ganglion cells (Giolli, 1961) which are likely to have W-type receptive fields.

As mentioned earlier, the dorsal terminal nucleus lies adjacent to the nucleus of the optic tract in the pretectum and both are likely to send fibres to the dorsal cap of the inferior olive, which in turn projects to the flocculonodular lobe of the cerebellum (Alley et al., 1975).

Physiological experiments, conducted mainly in the rabbit, have revealed that the cells of the terminal nuclei prefer slow movements in a particular direction, upward (and posterior) in the medial nucleus, downward (and posterior) in the lateral nucleus and horizontally in the dorsal nucleus. Similar directional preferences are found in ON-direction selective retinal ganglion cells (Oyster and Barlow, 1967; Oyster et al., 1972; Oyster et al., 1980), suggesting that this cell type provides the input to the accessory optic tract. The preferred excitatory and inhibitory directions are not aligned in the medial and lateral terminal nuclei of the rabbit, and Simpson et al. (1979) have linked this non-collinearity with motion of the animal in relation to the axes of the semicircular canals in the vestibular apparatus. The outcome would then be one that leads to excitation when rotation occurs around the principal axis of one vertical canal and inhibition for rotation round the other vertical canal principal axis. Therefore, the accessory optic tract is believed to have a role in the detection of self-movement and in the consequent reflex stabilization of the eyes and head.

Hoffman and Distler (1989) found cells in the dorsal terminal nucleus of the primate responding in a similar manner to those in the nucleus of the optic tract. It will be of interest to see the contribution from the retina to these cells in the primate accessory optic pathway.

The Lateral Geniculate Nucleus

The fact that different subdivisions in the laminated LGN have been used in naming afferent retinal ganglion cells, which are grouped according to their response properties, is testimony to the thesis that lamination in the LGN has a functional basis. There is a school of thought that believes that lamination, revealed histologically from differences in cell size, provides a guide to the path taken by different functional streams as they pass through the LGN. A model developed by Kaas (1986) has been based on the

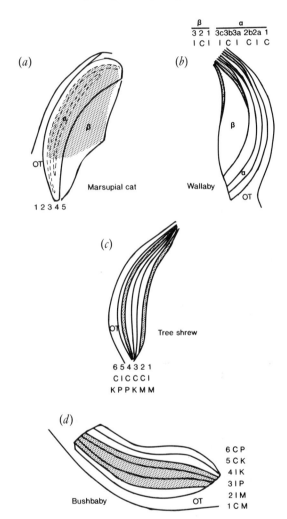

Fig. 20.3 *Patterns of LGN layering. (a) Prototype proposed by Sanderson (1986) where the ipsilateral projection is non-layered and overlies the contralateral projection, which displays rather ill-defined layering. (b) Marsupial pattern, where both ipsi- and contralateral projections form segregated layers. (c) Vertical pattern, a common basis to many non-primate mammals. (d) Horizontal pattern, a basic design for many primates. The stippled layers represent layers with an ipsilateral input. I and C stand for ipsilateral and contralateral respectively; M, P and K stand for magno-, parvo- and konio-cellular, respectively; OT, for the optic tract. The layer numbers have common but not universal usage.*

observation that layers or clusters of cells of three different sizes can be recognized in the LGN of a number of primates (Kaas et al., 1978). The most orderly arrangement of the three groupings occurs perhaps in the prosimian, *Galago senegalensis*, where large, small and dust-like cells are gathered together in three paired layers, with one member of each pair receiving its input from one or other eye. Kaas has proposed that the magnocellular layers receive their inputs from Y cells, the parvocellular theirs from X cells and the koniocellular theirs from W cells. In a general pattern, the magno-, parvo- and koniocellular subdivisions each have two layers, which receive inputs for the appropriate stream from each eye. Where four parvocellular layers are present in some primate LGNs there is usually evidence that the two primal layers have split and interdigitated. In the simians the koniocellular layers are dispersed to appear as cell clusters around and within the magnocellular layers.

At first blush, it seems that the Kaas model does not differ greatly from the Shapley–Perry interpretation. The difference turns mainly on whether the retinal ganglion cells projecting to the layers of small cells in the LGN should be called P or X cells. Shapley and Perry acknowledge the X-like properties of these cells and are even prepared to call them P_X cells, but the X cell appellation is rejected on the grounds that there are significant functional differences between P_X cells and the X cell of the cat, and to ignore them would open the way for a misleading homology. Further, the majority of M cells are X-like (Blakemore and Vital-Durand, 1981; Shapley et al., 1983). Even though their responses are transient they exhibit linearity of spatial summation, the defining characteristic of X cells.

Even with a reasonable scheme for lamination in the primate LGN, and the possibility of expanding it to include the cat LGN, it is still not a simple matter to extend the model to the wide diversity of LGN lamination patterns found amongst the non-primate mammals.

Sanderson (1986), following in the footsteps of a number of others (Brauer et al., 1978; Kaas et al., 1978) has attempted to arrange various mammalian orders into different groups with matching patterns of LGN lamination. In formulating a simple prototype from the LGN of living insectivores and opossums, Sanderson envisaged an LGN with no distinct cell layers and a single patch of ipsilateral retinal input, either superimposed on the contralateral representation, as in marsupials, or surrounded by it, as in placentals (see Fig. 20.3).

Two subsequent marsupial radiations have emerged; one, resembling the prototype, with marked overlap in the representations of each eye, is found in the polyprotodonts (carnivorous dasyurids and bandicoots and the opossum); the other pattern, where the inputs from the two eyes are segregated into separate layers and sub-layers, is found in the diprotodonts such as the kangaroos, wallabies, possums, wombats and koalas. In the second pattern the nucleus is also divided into alpha and beta segments, with the alpha containing the layers of more densely packed cells. The laminar patterns, which often amount to 11 layers, do not appear to be correlated with habitat or with a nocturnal or diurnal life style.

The placental radiation also has one pattern like the prototype, which is found in insectivores, rodents, lagomorphs, microchiropteran bats, edentates and cetaceans. The alternative pattern, one of a layered LGN with several layers for inputs from each eye, is found in primates, megachiropteran bats, squirrels, carnivores, seals, hyraxes and ungulates.

From a more distant perspective, it could be said that placental and marsupial radiations have mimicked one another and that each has a group close to the prototype and another where there is marked lamination. We have seen in the placentals that lamination is related to cellular function and may serve to segregate the inputs from each eye or from different classes of ganglion cells. Among the primates, segregation in the LGN is for M and P cells and perhaps for cells with different wavelength selectivity; among the carnivores it is for X, Y or W cells in the cat (for a review, see Rodieck, 1979) and also for cells with ON or OFF centre receptive fields in species such as the ferret (Stryker and Zahs, 1983) and the mink (LeVay and McConnell, 1982). Evidence is lacking on the type of functional segregation occurring in the marsupial LGN, although preliminary results from our laboratory (G. H. Henry and R. F. Mark, unpublished results) suggest that in the tammar wallaby the alpha segment contains cells with some of the characteristics of X and Y cells (brisk responses that are either sustained or transient), while the beta segment contains a high proportion of cells with the sluggish response characteristic of W cells. Also, from the appearance of Nissl stained material the beta segment contains a mix of large and very small cells. In all these features in the arrangement of the wallaby LGN there is much that is reminiscent of the cat LGN, with the alpha layers resembling A and A_1, and the beta layers the C complex. These findings seem to suggest that rather than looking for primal patterns in the marsupial LGN it may be more productive to seek evidence of parallel or convergent evolution. This would point to an impressive potential for design duplication, since the first marsupials and placentals appear in the fossil record in the Cretaceous period more than 65 million years ago. The geographical separation of the two groups almost certainly preceded the development of a laminated LGN which has, none the less, emerged in both with many design features in common.

Recent discussions of the possibility that the megabat is a primate have highlighted the need for a model to describe the character of lamination in the LGN. As if frustrated in the search for a universal model, Kaas and Preuss (1990) in a practical approach have adopted two basic placental patterns to go with the Sanderson prototypes. These two are from extensively studied species, and take one of two cytoarchitectural forms, either a vertical arrangement of the layers or a horizontal one. Dealing first with the horizontal model, Kaas and Preuss selected the owl monkey since they believe it provides the simplest pattern for the primate LGN. For the vertical conformation the tree shrew, *Tupaia*, was thought to be typical of many non-primates. This leaves out the laminated pattern of the marsupial LGN, and although it, like that of the tree shew, is vertically oriented there appear to be many dissimilar features and there may be a need to create a third model.

In the LGN of the owl monkey (Diamond *et al.*, 1985a) there are four main layers lying on top of one another and fed from below by the optic tract. The two dorsal layers are parvocellular and the two ventral are magnocellular. In this pattern the two external layers receive their inputs from the contralateral eye. The arrangement of the parvocellular layers is more complicated in catarrhine primates, where parts of the two layers interdigitate with one another. In most prosimians, as mentioned above for the *Galago*, the parvocellular layers are separated by two koniocellular layers. In a slight modification of the Kaas-Preuss base pattern we have adopted the *Galago* LGN as a more comprehensive model (see Fig. 20.3). There may be some prosimian deviations from this pattern and at the moment there is some discussion on whether the input to the two magnocellular layers is reversed in the tarsiers (Pettigrew *et al.*, 1989).

The LGN of the tree shrew, with its vertical orientation, has the layers arranged medial to lateral, and the optic tract is lateral to the nucleus (see Fig. 20.3). The layers, six in number, are thinner and less well differentiated by cell size than in the primate and it is more difficult to define their functional role. The most medial pair, 1 and 2 (numbering of Lund *et al.*, 1985), have slightly larger cells, suggesting a magnocellular designation, which would not parallel the primate pattern since these layers are distal rather than proximal to the optic tract. The functional roles of the other layers, 3 to 6, are not clear and X- and Y-like inputs appear to mix in most layers (Sherman *et al.*, 1975), although there is evidence of functional segregation for cells with ON centre and OFF centre receptive fields (Conway and Schiller, 1983); those with ON centres are in layers 1 and 2 and those with OFF centres in layers 4 and 5. The cells of layer 6 showed many of the characteristics of W cells of the cat; they have ON/OFF receptive fields and have the slowest conducting afferents (Conway and Schiller, 1983).

Kaas and Preuss (1990) conclude that the marked differences in the laminar pattern of tree shrews and primates suggest that these two groups, in emphasizing vision, have independently developed different laminar patterns in the LGN. Few would gainsay this conclusion, but it can be said that the primate pattern is more representative of the group and can be simply adjusted to accommodate most members. From a pragmatic point of view, there is value in being able to distinguish the primate pattern as being different from that of non-primates. Therefore, it is important, if the megachiropterans are to be accepted as primates, for them to have some derivative of the primate LGN pattern.

Therefore, a test for the application of the LGN model arises with the topical question of whether megabats are primates in distinction to microbats even though the two are traditionally placed in the same order. It transpires that some of the LGN qualities used to separate mega- and microbats (Pettigrew *et al.*, 1989) are microbat idiosyncrasies and would be just as effective in separating microbats from carnivores, for example, and may say more of the primitive adaptation in the microbat. They include the following:

1. The microbats lack laminar differentiation and sharp demarcation between the regions of ipsilateral and contralateral eye representation, which otherwise occurs only in the marsupial native cat (Sanderson and Pearson, 1977) and in the edentate tree sloth (M. L. Cooper, unpublished results – cited in Pettigrew *et al.*, 1989).
2. In megabats the magnocellular layers are adjacent to the optic tract and the relative position of the ipsi- and contralateral magnocellular layers is as found in the primate model.
3. The megabats have three paired layers in the LGN; one pair is magnocellular and the other two pairs, yet to be identified, could be parvo- and koniocellular, in keeping with the primate model.
4. A laminated parvocellular subdivision is present in megabats.

The question of how the megachiroptera fit into the plan for mammalian evolution is still being debated and in the view of Kaas and Preuss (1990) the laminar pattern of the megachiroptera LGN has yet to be resolved. We shall not enter the debate but simply compliment the participants for breathing life into old taxonomic and evolutionary questions.

The Optic Radiations

The use of modern tracing techniques in unravelling the composition of the optic radiations, particularly with regard to the detection of bifurcating axons, has resulted in a blossoming of information on the pathways of cats and primates, but few other species. The potential for bifurcating axons in the optic radiations is much less in the primate than in the carnivore, and generally the optic radiations of the primate have a single cortical destination, the striate cortex or V1 (Bullier and Kennedy, 1983). By contrast, projections from the LGN in the cat diverge to enter a number of cortical areas in a pattern that has all LGN layers (A, A1, the C layers and medial interlaminar nucleus–MIN) projecting to areas 17 and 18, while the C layers and MIN also project to area 19 and PMLS (Stone et al., 1979). There have been isolated reports of greater divergence but if these linkages are present they involve only a small number of cells.

The occurrence of bifurcation in the optic radiations of the cat can be treated as an extension of the divergence occurring in the optic tract. Just as the Y and W ganglion cells bifurcate to send axons to the tectum and the LGN, it also appears that the LGN cells, with an input from these two afferent streams, have bifurcating axons in the optic radiations. The numbers of bifurcating axons in the optic radiations of the cat make up about 10% of the total population of axons leaving the LGN (Geisert, 1980; Bullier et al., 1984; but see Birnbacher and Albus, 1987). Therefore, bifurcation appears to be a feature common to W and Y cells in the retina and their counterparts in the LGN. In contrast, the X cells of the retina, and their LGN recipients, have a very low level of axonal branching and provide a non-diverging path from retina to cortex. Traditionally bifurcation has been associated with the axons of large cells, but the existence of bifurcation in W cells and their LGN recipients, which generally are smaller than X cells, suggests that cell size is not a limiting factor. It is more likely that the amount of axonal space supported by a cell and hence the size of the axonal arborization (for the sizes of X, Y and W cell arbours, see Ferster and LeVay, 1978) may be better related to cell size. Perhaps phylogeny may have an influence on the pattern of bifurcation, since as mentioned above it is possible that the W and Y cells alone project to the superior colliculus and its antecedent, the optic tectum, and that the link with the developing neocortex was established when these cells sent an axonal branch to relay in the LGN. The capacity to carry on the axonal bifurcation to like cells of the next stage may be to assist the functional divergence of the first stage. Once the initial divergence has involved the functionally distinct areas then processing may be assisted by dispatching the information to disparate cortical areas with an interactive influence on the two primary functions. Since X cells are a late phylogenetic development, there has been no pressure for bifurcation at the first stage to start the process.

The phylogenetic need for bifurcation outlined above can be tested in the primates with mixed success. While the bifurcation may be a common property of all mammalian retinal Y cells, there is little available information on the homologue of the retinal W cell in the primate. For LGN cells, as suggested earlier, the possibilities for bifurcation are limited, since most fibres of the optic radiations terminate in V1. However, recent refinements in the HRP labelling technique have revealed a projection from the LGN to V2 (Wong-Riley, 1976; Benevento and Yoshida, 1981; Fries, 1981; Yukie and Iwai, 1981; Bullier and Kennedy, 1983). This is a minor projection, which arises from the intercalated cells of the LGN. These LGN cells in turn receive afferents from the superior colliculus, which could be bifurcating W cells, and are generally regarded as components of the koniocellular layers of the primate. Therefore, the ground exists for bifurcation to occur in retinal W cells and their koniocellular counterparts in the LGN. The carry over of Y cell bifurcation into the optic radiations is not obvious, although a few scattered larger cells were found in the LGN projection to the prestriate cortex (Benevento and Yoshida, 1981; Yukie and Iwai, 1981). However, in general terms the pre-eminence of P_X and M_X cells in the primate retina results in a small proportion of bifurcating axons in the optic tract and radiations and a corresponding low level of divergence to cortical areas outside the striate cortex. The koniocellular layers of the LGN are more apparent in the prosimian *Galago senegalensis* than in other primates, and the branching of the projection from these layers should shed light on the genesis of bifurcation.

The Striate Cortex

Afferent Projections

As described above, area 17 of the striate cortex is the primary receiving area in the cortex of most if not all mammals. In a wide range, which includes the tammar wallaby, the tree shrew and most of the primates, area 17 lives up to the name of striate cortex, which was first used by Grafton Elliott Smith (Polyak, 1957). However, the morphological basis of the striated pattern may not always be the same in all species, and there is a risk in regarding the striped cortex as the sign of a particular functional attribute. In the primate, and in the human cortex in particular, the stripe is often equated with the macroscopically visible stria of Gennari but in Nissl stained preparations there is evidence of additional stratification in the arrangement of cortical cells. Just inferior to the stria of Gennari, there is a layer of relatively low cell density which in Lund's (1973) terminology is lamina $4C_\alpha$ (4A in the terminology of Hassler, 1966). Lamina $4C_\beta$ (4B of Hassler), which is immediately adjacent, is composed of smaller and more densely packed cells and the difference in density between these two sublaminae forms an obvious stratification. There is a functional substrate to this layering in the primates since the upper sublayer marks the point of termination of the cells of the magnocellular layers of the LGN, and the lower the termination of cells of the parvocellular layers (Hubel and Wiesel, 1972). This means that cells with Y properties project only to the upper sublayer ($4C_\alpha$), whereas X input reaches both sublayers (M_X to $4C_\alpha$ and P_X to $4C_\beta$). This second arrangement is the same as that of the cat striate cortex, although the pattern of stratification is not nearly so obvious in the cat (Humphrey et al., 1985). The stratification in Nissl stained material, which is absent in the cat, is again more obvious in the tree shrew and the wallaby. Both these species have LGNs that are highly laminated and vertically oriented, but, in both, the X and Y cells are intermingled while the W-like cells appear to be confined to particular layers. In the striate cortex of the tree shrew and similarly in the wallaby there is a layer of densely packed cells that seems to encompass the full extent of lamina 4 and to date no sublayers have been identified as providing the ground for functional segregation.

In addition to the layers of LGN terminals occupying lamina 4 of the striate cortex of the primate there are also LGN terminals that form patches at the base of lamina 3. In examples taken from prosimians, old and new world monkeys, these patches receive their input from the S or superficial layers in the LGN, which in turn have been designated as koniocellular and linked with the W cell system (Fitzpatrick et al., 1983; Diamond et al., 1985a). Similar patches or blobs in the base of lamina 3 stain for cytochrome oxidase (Livingstone and Hubel, 1984) and this also seems to be a primate characteristic that is not found in non-primates such as the cat and the tree shrew.

In the primate, the cells in these cytochrome oxidase blobs have been reported to have non-oriented receptive fields and often a double colour opponency (Livingstone and Hubel, 1984), which distinguishes them from the other striate neurones that have colour-coded and non-oriented receptive fields, which reside in lamina $4C_\beta$ (4B of Hassler, 1966) and generally display single opponency. An inability to create a double opponent receptive field by summing two single opponent fields of $4C_\beta$ has led Livingstone and Hubel (1984) to speculate on the possibility that the direct inputs from the intercalated LGN layers act as the precursors to create the double opponent receptive fields of the cytochrome oxidase blobs. This is a doubtful possibility, which fails to account for the contribution made by the mass of cells with single opponent receptive fields found in the parvocellular layers of the LGN and in lamina $4C_\beta$. In addition, more recent investigations (Ts'o and Gilbert, 1988; Schein and Desimone, 1990) question the presence of double opponent cells in the striate cortex.

Cortical Magnification

We have spoken of the phenomenal development of the neocortex, particularly the visual cortex in mammalian evolution. In a lower mammal like the rat, the striate cortex of each hemisphere occupies about 7 mm^2 (Espinoza and Thomas, 1983), whereas in the cat it comprises 380 mm^2 (Tuse et al., 1978), in the monkey 1400 mm^2 (Daniel and Whitteridge, 1961) and in man 3000 mm^2 (Drasdo, 1977). In the marsupial tammar wallaby, the preliminary estimate of the striate area is 135 mm^2 (Vidyasagar et al., 1990). Certain factors must have exerted an enormous selection pressure in driving the striate cortex to enlarge. In every species studied there are large asymmetries in the area of cortex devoted to different parts of the visual field and an understanding of the factors responsible for this variation can give clues about the forces that were instrumental in leading to the evolutionary expansion of the visual cortex. This section will therefore deal with the issue of cortical magnification in some detail.

With the exception of movement perception, performance in most visual tasks falls off with eccentricity. Visual acuity, for example, drops by a factor of 40 between fovea and the periphery in man (Wertheim, 1894; Weymouth, 1958). The amount of cortical tissue coding information from a unit area of the visual field, the cortical magnification factor (CMF), expressed either as linear (in mm/deg) or as areal (in mm^2/deg^2) magnification factor, also falls off with eccentricity. This has been measured in monkey (Talbot and Marshall, 1941; Daniel and Whitteridge, 1961; Cowey, 1964; Rolls and Cowey, 1970; Hubel and Wiesel, 1974; Dow et al., 1981; Tootell et al., 1982, 1988; Van Essen et al., 1984) and cat (Talbot and Marshall, 1941; Bilge et al., 1967; Sanderson, 1971; Tusa et al., 1978) using electrophysiological or anatomical mapping techniques. For man CMF has been calculated (Cowey and Rolls, 1974; Dobelle et al., 1979) by using the data on the loci of phosphenes that a blind person could see on stimulation of electrodes permanently implanted on the visual cortex (Brindley and Lewin, 1968).

It has been difficult to obtain direct data from the monkey foveal cortex, and the linear CMF for this region differs by a factor of 7 among the various studies (for reviews, see Pointer, 1986; Tolhurst and Ling, 1988). The most realistic estimate may be around 13–20 mm/deg (Van Essen et al., 1984). Away from the fovea, there is agreement between the various authors. The linear CMF is around 4 mm/deg at 1° eccentricity and about 1 mm/deg at 10°.

In the extra-foveal range, the human magnification factors (linear CMF) were approximately 1.6 times higher than the monkey's (Cowey and Rolls, 1974). However, the mapping of the human striate cortex using positron-emission tomography has yielded magnification factors lower than those of the monkey (Fox et al., 1987). The foveal CMF values for the cat were between 1 and 2 mm/deg (Talbot and Marshall, 1941; Bilge et al., 1967;

Sanderson, 1971; Tusa et al., 1978). In terms of areal magnification, the primate devotes 50 to 100 times more cortex to unit area of visual field than the cat in central vision.

The various studies that quantified the change of cortical magnification with eccentricity have in general supported the idea that visual spatial resolution parallels CMF (Rolls and Cowey, 1970; Cowey and Rolls, 1974; Drasdo, 1977). This relationship has in turn been attributed to a constant relationship between cortical magnification and retinal ganglion cell density, and it has become crucial to know whether the CMF simply reflects the varying density of retinal ganglion cells across the retina. Alternatively, there could be an inhomogeneity in the amount of target area innervated by each retino-geniculate or geniculo-cortical afferent. The number of neurones innervated by a central afferent may be more than that by an afferent of peripheral vision.

As a result, many attempts have been made to correlate cortical magnification factor and ganglion cell density, especially in the primate (Weymouth, 1958; Rolls and Cowey, 1970; Drasdo, 1977; Schein and DeMonasterio, 1987; Schein, 1988; Wassle et al., 1990). These studies, as well as those in the cat (Wilson and Sherman, 1976; Tusa et al., 1978), led to the suggestion that areal cortical magnification (M^2) is directly proportional to ganglion cell density (D). Human cortical magnification factors, calculated from the density distribution of retinal ganglion cells, have been successfully applied to predict the psychophysical performance in contrast sensitivity and spatial resolution tasks at various eccentricities (Rovamo and Virsu, 1979).

However, performances in other visual tasks do not follow the same drop-off with eccentricity recorded for visual acuity. For example, at 10° eccentricity, thresholds for hyperacuity are 10 times higher than at the fovea, whereas visual resolution thresholds are only 4–5 times higher (Westheimer, 1982; Fahle and Schmidt, 1988). Performance in stereoacuity also drops off more steeply than visual resolution (Fendick and Westheimer, 1983). Levi et al. (1985) have argued that the human M calculated by Rovamo and Virsu (1979) reflects only the retinal ganglion cell density and may explain performance based on the retinal grain alone. The performance in other tasks that required further processing at a cortical level would more truly reflect the real cortical magnification factor.

A number of studies deny the existence of a strict correlation between cortical magnification and ganglion cell density (Malpeli and Baker, 1975; Myerson et al., 1977; Connolly and Van Essen, 1984; Van Essen et al., 1984; Perry and Cowey, 1985, 1988). The area of striate cortex representing a degree² of visual field varies nearly a thousandfold between 1° and 60° eccentricity (Hubel and Freeman, 1977; Tootell et al., 1982; Schein, 1988), whereas ganglion cell density varies only by a factor of 100 between these eccentricities (Perry et al., 1984; Wassle et al., 1990).

Since the major difficulty in arriving at the real relationship between M^2 and D in the monkey arises from the steep decline in ganglion cell density from the foveal region, and the problems in measuring central cell density in an animal with a foveal pit, it would appear worth while to study an animal that has a more gradual fall-off in ganglion cell density and no foveal pit. Many species show a pronounced visual streak (for review, see Hughes, 1977), with the ganglion cell density remaining high along most of the horizontal meridian. The marsupial tammar wallaby is one such species where data are now available on both cortical magnification and ganglion cell densities. The retina shows only a limited fall in ganglion cell density along the horizontal meridian (Wong et al., 1986), but the striate cortex exhibits a massive expansion of the central representation that has been measured both anatomically (Mark and Wye-Dvorak, personal communication) and electrophysiologically (Vidyasagar et al., 1990). The 30° of binocular segment of each visual hemifield occupies three-quarters of the striate cortical retinotopic map. Between the representations of the vertical meridian and of an azimuth of 80° along the horizontal meridian, the ganglion cell density drops only by a factor of about two, but the cortical magnification M^2 drops by nearly a factor of 50. This is clearly not the case in the retinotopic representation on the superior colliculus (Mark and Sheng, personal communication), where the peripheral streak is well represented. A similar disparity between cortical magnification (M^2) and D has been reported also for the South American rodent, the agouti (Oswaldo-Cruz et al., 1990).

The wallaby thus provides one example of a species where changes in cortical magnification with retinal location cannot be explained by variation in ganglion cell density. Single-unit recordings also revealed that the central representation contains a large number of binocular neurones that are tuned to interocular disparities (Vidyasagar et al., 1990). Such specifically cortical mechanisms may be the decisive factors in determining cortical magnification.

Our search for an interspecies relationship between ganglion cell numbers and visual cortical area, in an attempt to explain the expansion of the striate cortex in mammalian evolution, has produced a mixed result. The difference in overall striate cortical area between cat and monkey roughly matches the difference in total ganglion cell numbers. The monkey has about 6 times as many ganglion cells as the cat and its striate cortical area is about 4 times that of the cat. On the other hand, the rat has around 115 000 retinal ganglion cells (Fukuda, 1977), which is around two-thirds of the 170 000 in the cat (Wong and Hughes, 1987), even though the cat's striate area is over 50 times that of rat's. The tammar wallaby's retina has 360 000 ganglion cells (Wong et al., 1986), around double the number in the cat, but its striate area is only about 35% of the cat's. So there could be no simple relationship between ganglion cell numbers and striate cortical surface areas that is applicable across all mammals.

However, what sets the rat and the wallaby clearly apart from the cat and the primate is the degree of binocular overlap. The rat has largely laterally placed eyes. It also lacks an area centralis which is comparable to that in the cat or the primate. In contrast, the cat has reasonably well developed binocular vision with around 90° of binocular overlap and possesses an area of specialization for central vision and, in some ways, is comparable to primates. Thus if we take frontalization of eyes and the development of a significant region of binocular and high-acuity vision as the most likely reasons for the phylogenetic expansion of the striate area, it is possible to explain both the evolutionary changes in surface area of the striate cortex and the gradient of cortical magnification with eccentricity. The development of a central area of higher ganglion cell density may partially explain the higher CMFs seen in central vision. But, in addition, there are specific functions related to central vision, like stereopsis and hyperacuity, that are likely to be strongly correlated with the amount of striate cortical areas seen in any one species or for that matter across species that display different degrees of binocular and high acuity vision.

Using improved anatomical techniques, Wassle et al. (1990) have shown that over shorter ranges (see their Fig. 8, for example), there is excellent correlation between ganglion cell density and cortical mangification. However, taking the entire retina, there may be significant differences in the amount of cortical area innervated by a single retinal field, depending on the nature of the visual functions it supports.

Brain Stem Projections

In addition to the retinal projection to the tectum, the optic tract sends fibres to the nucleus of the optic tract in the pretectum and to the dorsal terminal nuclei of the accessory optic tract, which lie together as a scattered group of cells along the ventral margin of the brachium of the superior colliculus (Hoffmann et al., 1988). The output of the nucleus of the optic tract goes to the dorsal cap of the inferior olive, the nucleus prepositus hypoglossi and the nucleus reticulus tegmenti pontis (Precht et al., 1980; Lannou et al., 1984) and presumably links through the dorsal terminal nucleus to the accessory optic tract (Simpson, 1984). From experiments on the cat that measured the latency and threshold of retinal ganglion cell response to electral stimulation of the nucleus of the optic tract, Hoffmann and Stone (1985) argued that cells in the nucleus receive a direct contralateral input from W ganglion cells and an indirect binocular drive from Y ganglion cells, which travelled via lamina 5 of the visual cortex. The cells of the nucleus of the optic tract are thought to provide the input for horizontal optokinetic nystagmus. Support for this view has come from experiments showing that electrical stimulation of these cells produces eye movements similar to those of the slow phase of optokinetic nystagmus and also these cells fire only during this phase of naturally occurring optokinetic nystagmus.

The cells of the right nucleus of the optic tract respond only to stimuli moving to the right, and those of the left nucleus only to the left. Then, since the W cells only project to the contralateral nucleus they will provide an effective stimulus to nucleus of optic tract cells only for movement from temporal to nasal. In contrast, the Y pathway through the cortex, being binocular, provides an effective input as a result of movement in either direction. As a result a comparison of the gain for monocular optokinetic nystagmus in the two horizontal directions (the gain records how closely the eye movement follows that of the stimulus) gives an indication of the contribution coming from the W and Y pathways. Therefore, temporal to nasal movement is effective in activating the nucleus of the optic tract through the W and Y paths, while nasal to temporal movement is effective only through the Y path. In the cat there is an asymmetry in the responses to movement in the two directions and the temporonasal movements were more effective than the nasotemporal, and Hoffman (1983) has argued that this is due to the input from the monocular W pathway supplementing the binocular Y pathway.

From the comparative point of view these results in the cat are interesting because they are not replicated in primates where the responses of horizontal optokinetic nystagmus are symmetrical (Hoffman et al., 1988). In terms of the Hoffmann interpretation, this would indicate a decline in the effectiveness of the W path. In support of this conclusion, 100% of nucleus of optic tract cells appear to be binocularly driven in the monkey, and since these cells remain binocular after callosal section the balanced inputs from the ipsilateral and contralateral retinas presumably employ mostly ipsilateral geniculostriate pathways (Hoffmann et al., 1988). These findings seem to suggest that the W path is less active in the primate even though the number of axons projecting from the optic tract into the brachium of the superior colliculus is not greatly reduced in absolute terms.

The Corticotectal Pathway

The corticotectal pathway which arises from the complex cells of lamina V in the primary visual cortex is a homogeneous pathway which receives either direct or indirect inputs from Y cells contributing to the optic radiations. In keeping with the phylogenetic stability of the retinal Y cells, the corticotectal pathway displays consistent characteristics in a wide variety of mammalian species that extends from wallabies to rabbits, cats and monkeys. The same type of cortical cells appear to project both to the nucleus of the optic tract and to the superior colliculus (Schoppmann, 1981). It is not a simple experiment to discover if the retinotopic input from the cortex to the superior colliculus corresponds with that coming directly from the retina, but logically they would have to be the same or there would be a blurring of resolution in the topographic map. Sharp resolution is a feature of the superior colliculus map and, to cite two examples, it is similar in the wallaby and the cat. Therefore, there is good reason to believe that the retinotectal and the retinogeniculate pathways would benefit from a parallel organizational paradigm if their retinotopic maps are to come into register downstream from the point where they make their initial branching.

Extra-striate Areas

Thalamic Input to Extra-striate Areas

The way in which the structure and function of neocortical areas enable the higher level processing of the sensory input necessary for a correct and useful evaluation of the environment has triggered the imagination of neuroscientists for over a century. In the visual system alone, a large number of cortical areas have been identified, more than a dozen in most species studied; some of them have been targets of intensive investigation recently (for reviews, see Van Essen, 1985; Diamond et al., 1985b; Baylis et al., 1987; Maunsell and Newsome, 1987). It was Campbell (1905) who first attempted a clear categorization of cortical regions based upon their structural relationships and functional demands. His view separated the primary sensory areas that served simple sensations from association areas that were responsible for the complex function of learning. He thus referred to the striate cortex as 'visuosensory' and to the extra-striate areas as 'visuo-psychic', and he proposed that all visual information is first processed in the striate area before being dealt with by 'higher' areas.

There were many lines of work that provided the inspiration for Campbell to undertake an extensive histological definition of the primary sensory and association areas. One was Flechsig's finding (1901) that certain cortical areas were myelinated much earlier than others and this was interpreted by both Flechsig and Campbell as meaning that a newborn infant relied more on basic sensations than on their semantic contents. It was also reported by Flechsig (1901) that the association areas did not receive afferents from the thalamus. The idea that all visual information gets routed through the striate cortex was also supported by Schafer's finding (1888) that the ablation of the striate cortex in the monkey led to a severe blindness in the contralateral half of the visual field. The monkeys were very poor in discriminating pattern, colour, depth or brightness. Further primate studies have amply corroborated the perceptual blindness that follows striate lesions (Kluver 1942; Cowey and Weiskrantz, 1963; for review, see Mishkin, 1972).

These ideas have received only limited support from comparative studies. In non-primate mammals, like the cat, thalamic afferents project to a number of cortical areas, besides the striate cortex (see pp. 451–452). Lesions of areas 17 and 18 in the cat produce only deficits in acuity, but not in form perception, whereas lesions of area 19 and the suprasylvian cortex lead to deficits in form discrimination, leaving acuity normal (Berlucchi and Sprague, 1981; Sprague et al., 1981; Krueger et al., 1986, 1988; Kiefer et al., 1989). Such demonstrations of 'double dissociation' emphasize the parallel nature of the systems in the cat.

Ablation studies have revealed a similar pattern in another non-primate, namely the tree shrew. Only with removal of *both* striate and extrastriate areas did the *Tupaia* exhibit severe deficits in visual discrimination (Snyder and Diamond, 1968; Killackey et al., 1971, 1972). However, the extra-striate inputs in the *Tupaia* do not arise from the lateral geniculate, but largely from the pulvinar (Harting et al., 1973; Luppino et al., 1988).

Only in the monkey has the primary importance of striate cortex for visual functions, as stressed by Campbell and Flechsig, been vindicated (for review, see Diamond et al., 1985b). There have been extensive studies of the extrastriate visual areas in the monkey, which have documented the complex response properties of neurones in these areas or shown specific deficits following lesions (for review, see Van Essen, 1985). Much of the extra-striate function apparently relies on the integrity of the striate cortex. Mishkin (1966) did an elegant double dissociation experiment to show not only the importance of the striate cortex in the monkey, but also establish the existence of a pathway from the striate cortex to the inferotemporal cortex that is essential for the performance of visual functions. After inferotemporal lesion on one side, the monkey was normal. In a second stage, he removed the striate cortex on the other hemisphere. This again produced no behavioural effect, but at the next stage, when the corpus callosum was cut, there was a profound deficit in visual discrimination and recognition.

The effects of striate lesions in a prosimian, *Galago*, were intermediate between those in the primate and non-primate (Atencio et al., 1975; Caldwell and Ward, 1982). The deficits were less severe than those seen in the monkey. The extra-striate areas seem to depend less on the striate region than in the monkey and the *Galago* may thus represent an intermediate step in the evolution of primates.

The preservation of visual functions after striate lesions in non-primates must be largely due to the inputs to the extra-striate areas bypassing the geniculo-striate system. There is ample evidence for such extrageniculate projections to the cortex in a number of species (for review, see Tigges and Tigges, 1985). One important pathway is the one through the pulvinar. The LP-pulvinar complex has been shown to receive inputs from the visual cortical areas, the superior colliculus and the pretectal area in the cat (Kawamura et al., 1974, 1980; Updyke, 1977; Graybiel and Berson, 1980, 1981; Hughes, 1980; Mason and Gross, 1981; Berson and Graybiel, 1983; Benedek et al., 1983; Raczkowski and Rosenquist, 1983; for review, see Rosenquist, 1985), the hedgehog (Harting et al., 1972), the bushbaby (Glendenning et al., 1975; Symonds and Kaas, 1978; Raczkowski and Diamond, 1980, 1981), rat (Mason and Gross, 1981), the tree shrew (Harting et al., 1973; Luppino et al., 1988), the grey squirrel (Robson and Hall, 1977) and the monkey (Benevento and Rezak, 1976; Ogren and Hendrickson, 1976; Wong-Riley, 1977; Lin and Kaas, 1979, 1980; Benevento and Standage, 1983; Yeterian and Pandya, 1989). These investigations have revealed that the pulvinar, in all animals studied so far, has extensive afferent and efferent connections with many areas of the neocortex and also receives afferents from the tectum. The inputs from the striate cortex and the superior colliculus generally tend to remain separate in the pulvinar nucleus, even though there may be some overlap (for example, Benedek et al., 1983). Apart from being a possible site where geniculo-striate and tectal information can be integrated (Chalupa, 1977), the pulvinar can be an important alternative route for visual information to reach the neocortex.

The cortical targets of the tecto-recipient regions of the pulvinar have been well demarcated in many species. In the cat (for example, Raczkowski and Rosenquist, 1983), these are largely directed to the suprasylvian cortices, namely the anterior lateral lateral suprasylvian area (ALLS), posterior lateral lateral suprasylvian area (PLLS), dorsal lateral suprasylvian area (DLS), and the posterior suprasylvian area (PS). The input from the pretectum gets relayed onto areas 19, 5, 7, the splenial visual area and the cingulate gyrus. In the tree-shrew (*Tupaia*), it has recently been shown that there are two systems of tectal projections to the pulvinar, one diffuse and the other topographically specific (Luppino et al., 1988). The specific zone of the pulvinar nucleus projects to area 18 and to a broad band adjacent to area 18 called the dorsal temporal (Td) area. Most of the diffuse projection is to a region ventral to the dorsal temporal (Td) area and rostral to the temporal posterior (Tp) area. In the *Galago* the tecto-recipient zone has been shown to project to a region in the inferior temporal lobe ventral to the middle temporal (MT) area (Raczkowski and Diamond, 1980).

From the foregoing, it appears that the tecto-thalamo-cortical pathway could be mediating visual functions in non-primate mammals, and possibly prosimians, to an extent that striate lesions cause only minor deficits in visual tasks. Does the fact that striate lesions lead to severe blindness in the monkey mean that the alternative path through the pulvinar is not very well developed in primates? The crucial issue is whether the regions in the temporal lobe that are necessary for visual discrimination and recognition tasks receive a functionally important tectal input via the pulvinar.

As mentioned earlier, the monkey pulvinar does receive inputs from the superior colliculus in addition to that from the striate cortex. However, unlike most of the non-primate mammals there is considerable overlap of the tectal and striate inputs in the pulvinar of the monkey. Bender (1982) recorded the visual responses of neurones in the inferior pulvinar, which is the primary recipient of tectal inputs in the monkey (Benevento and Fallon, 1975; Partlow et al., 1977). He concluded that the striate cortex exerts a dominant influence on the neural responses and also found that even large collicular lesions failed to eliminate the visual responses of inferior pulvinar cells (Bender, 1981). It is possible that the increasing overlap of the tectal and striate inputs to the primate pulvinar may represent an evolutionary trend of a growing dominance of the geniculo-striate pathway. This is in line with the suggestion (Diamond and Hall, 1969) that the tecto-thalamo-cortical pathway, the major neocortical input for visual information in the reptiles and early mammals, has been largely taken over by the geniculo-striate and geniculo-striate-pulvino-cortical pathways in primates. In reptiles the afferents to the thalamus are almost entirely from the tectum and the neocortex gets little direct retinal input via the thalamus. In non-primate mammals, we see the coexistence of both pathways as important input channels to the neocortex, but with the phenomenal development of the geniculo-striate system in the primate, the tectal pathway to the cortex seems to have been relegated to the background.

If visual information via the tecto-pulvinar afferents is of minor significance to visual tasks of the neocortex, what indeed is the role of the collicular input to the pulvinar? Ungerleider and Christensen (1977, 1979) have in fact shown that pulvinar lesions produce an abnormal pattern of eye movements. Given the importance of the superior colliculus for eye movements in the monkey, the primary function of the tectal input to the pulvinar may be to integrate visual and oculomotor information rather than to provide a parallel visual input to the neocortex. This may be another unique evolutionary development in the primate lineage.

Multiplicity of Extra-striate Areas

The neocortical visual regions of the mammals have undergone a most dramatic expansion and diversification through 100 million years of evolution. Whereas primitive mammals have only 2 visual areas (Kaas, 1978), the cat has at least 12 (for review see Rosenquist, 1985) and the macaque nearly 20 (Van Essen, 1985). There are excellent reviews of the extra-striate areas, including their comparative aspects (Allman, 1977; Van Essen, 1985; Kaas, Volume 3; Sereno and Allman, Volume 4). The aim of this section is only to highlight certain conspicuous phylogenetic trends. However, this can draw upon data only from a limited number of mammalian species, since extra-striate areas have been described in detail only for the cat, the owl monkey and the macaque.

As noted in an earlier section, one major difference between primates and non-primates in the afferent innervation of the visual neocortex is the fact that the geniculate input in the monkey is almost entirely to the striate cortex. On the other hand, in the cat, areas 17, 18 and 19 all receive geniculate inputs. The thalamic input to the primate striate cortex is characterized also by a massive increase in the numbers brought about by a separate class of cells, namely the cells of the parvocellular pathway.

Processing of visual information beyond the thalamus takes on a unique character in the primates. The P and M pathways are kept separate to a significant extent in the striate cortex and this segregation seems to continue in the projection to extrastriate areas (for reviews, see De Yoe and Van Essen, 1988; Livingstone and Hubel, 1988; Zeki and Schipp, 1988). Zeki (1969, 1970, 1971, 1973) had earlier identified areas within the pre-striate region that were distinct functionally and in their intracortical and transcallosal connections. A host of such pre-striate areas have now been identified clearly in a number of primate species (reviewed in Van Essen, 1985; Krubitzer and Kaas, 1990; Kaas, Volume 3). Not only are the blobs of cytochrome-oxidase in area 17 unique to primates, but the segregation of the P and M streams in a bandlike fashion in the striate cortex (V1) is also characteristic of primates (Krubitzer and Kaas, 1990). These morphologically and functionally defined bands appear to project to distinct bands in the prestriate area, V2 (Schipp and Zeki, 1985; Livingstone and Hubel, 1987; De Yoe and Van Essen, 1988). The segregation is retained beyond V2 as well. Area V4 receives information largely from the P pathway and area V5, now known usually as middle temporal (MT) area, from the M pathway.

Despite wide quantitative differences among authors (Zeki, 1973; Van Essen and Zeki, 1978; Schein et al., 1982; Tanaka et al., 1986; Schein and Desimone, 1990), area V4 does seem to specialize in colour. Lesions restricted to area V4 also produce impairment of colour constancy (Wild et al., 1985; Butler et al., 1988). Neurones of area MT, on the other hand, are sensitive to direction of motion (Dubner and Zeki, 1971; Maunsell and Van Essen, 1983; Albright, 1984). Lesions of this area lead to a selective impairment of motion perception (Newsome and Pare, 1988).

From their cortical ablation studies, Mishkin and co-workers have also identified two streams of visual information processing in the primate neocortex (Ungerleider and Mishkin, 1982; Mishkin et al., 1983). They suggested that one pathway, coursing through V4, and directed towards the inferior temporal cortex, carries form information necessary for object identification. The other pathway, carrying information from the M stream, courses through MT towards the parietal cortex and is credited with mediating visual location of objects. The two streams appear to be concerned with the 'what' and 'where' of objects. Such a dichotomy had earlier been reported for lower mammals with regard to the cortical and collicular projections (Schneider, 1969).

There has been an impressive series of anatomical, physiological and psychophysical experiments that have sought to isolate and identify the nature of information that is processed by the two pathways at various levels of the visual system (for reviews, Shapley and Perry, 1986; Livingstone and Hubel, 1988; Kaplan et al., 1990; Schiller and Logothetis, 1990). While it is beyond the scope of this chapter to review the detailed literature on this which has shown profound differences between authors, it is our impression that the segregation is not very clear-cut and there are important regions of overlap and cross-talk between the two systems.

From the standpoint of this chapter, it would be premature to speculate upon the evolution of these parallel streams in the cerebral cortex, since little work has been done in other species. It would also be risky to draw homologies between the cortical areas of different mammalian species. It is only very recently that the homologies between the various extra-striate areas of different primates have been established with reasonable confidence (Krubitzer and Kaas, 1990). However, it is clear that the colour pathway as it exists in most new and old world monkeys is a unique primate feature. Similarly, area MT, by virtue of its afferent connectivity and myeloarchitectonics, has been said to be characteristic of primates (Allman, 1977; Krubitzer and Kaas, 1990). A cortical area with heavy myelination, possibly equivalent to MT, has also been identified in the megabats, which have been classified along with primates by Pettigrew et al. (1989). As early as the turn of the century Flechsig (1901) had described an early myelinating region at the junction of the occipital and temporal lobes of man. It is quite likely that this area is the homologue of MT in the human.

Even though the pathway directed towards the inferior temporal lobe and then on to the limbic and paralimbic structures cannot itself be phylogenetically new, the interposition of a host of areas around the superior temporal sulcus is a feature dominant in primates, just like the parvocellular pathway providing its major input. It is still a matter of intense debate (for example Shapley and Perry, 1986; Kaplan et al., 1990; Schiller and Logothetis, 1990) as to which perceptual processes are mediated by the P stream. It may turn out that, besides colour, the P stream is also responsible for texture perception, fine pattern vision and fine stereopsis. We may ultimately look upon this pathway as an evolutionary elaboration of object vision and the multiplicity of areas in the primate temporal neocortex (for review of these areas, see Bayliss et al., 1987) is probably one of the structural changes that has made this step feasible. Neurones in some of these areas show very sophisticated processing. One oft-cited example is the presence of neurones that respond selectively to faces in the fundus of the superior temporal sulcus (Perret et al., 1982). Many of the temporal neocortical areas are either polysensory or contiguous to areas receiving inputs from other modalities (see Vidyasagar, Volume 11). Thus there exists the infrastructure for the cross-modal integration essential for higher cognitive and semantic categorization of sensory information.

We have suggested that if the parvocellular pathway is a primate invention, the magnocellular (M) system might be seen as the basic mammalian visual system. This does not mean that the M pathway ceased to evolve. The overall performance of the M system is in many ways better than the performance of say, cat visual neurones. For example, the spatial resolution of the M cells in the monkey retina is remarkably good (Crook et al., 1988). Further, area MT, which is primarily a target for the M stream, is a primate feature. It is likely that the development of frontal vision with its attendant visual tracking of objects using conjugate eye movements needed a cortical area that emphasized motion perception in all its complexity.

Conclusions

The great advance in neural design to occur in the 100 million years of mammalian evolution came with the development of the neocortex and the subsequent growth of a multiregional visual cortex. Accompanying the move from tectum to visual cortex, the eyes became frontal, their movement was conjugated, the optic nerve underwent partial decussation and the topographic representation of the retina in the old tectum and the new lateral geniculate converted to that of a binocular hemifield, intermingled in the one and segregated into layers in the other. This binocular layering marked the beginning of a neural ordering that led finally to the creation of a three-dimensional percept, which aided survival and opened the way for visual performance to match the motor skills fostered by the growing brain. The ancient paths to the optic tectum and brain stem were almost certainly composed of W- and Y-type cells; cells that were concerned with the detection of external movement and self-motion, the stabilization of the visual image, the beginnings of pattern recognition and finally the initiation of eye, head and body movements.

Two paths arose to link with the neocortex; one extended from the tectum through the posterior thalamus to the visual cortex while the other formed a branch from the optic tract. The W and Y cells of original retinotectal path either bifurcated or contributed directly to this branch line, which passed through the lateral geniculate nucleus to the new cortex. Soon, we presume, the old cell types were joined by a new cell, the X cell of the placentals and the M_X cell of the primates. This step brought with it an improved pattern recognition, which was enhanced by the additional development of another cell type, the P_X cell, which increased the capacity for visual detection, improved the chances of survival and added the aesthetic bonus of bringing colour to the visual world. The success of the geniculate pathway rapidly seemed to diminish the reliance placed on the tectocortical link and its functional role is now obscure.

For some neural reason that is still not clear, there is an organizational advantage in keeping these cell systems separate as they make their way through the LGN to the cerebral cortex and on into the cortex itself, where the cells' axons branch and link with such complexity that segregation seems impossible. The elegance of this design is impressive and indeed, although we started with Polyak marvelling over the bulk of the visual primordia in ontological development, we end with a feeling of awe for the subtlety of the changes in cellular diversity and visual pathway design that have accompanied the phylogenetic development of the new cortex.

Note

[1] The 'salute and present arms' algorithm will assist in recalling the topography of the contralateral projection of the complete retina on the superior colliculus. If the open hand is positioned for a vertical salute then:

the fingertips represent the superior retina,
the thumb the medial retina,
the little finger the lateral and
the heel of the hand the inferior retina.

The representation on the contralateral superior colliculus is given by presenting arms with the hand, palm down, still retaining its retinal representation.

Then, the position of the hand links the following:
little finger: temporal retina—rostral sup. colliculus;
thumb: nasal retina—caudal superior colliculus;
finger tips: superior retina—lateral sup. colliculus;
and heel of hand: inferior retina—medial sup. colliculus.

References

Albright, T. D. (1984). Direction and orientation selectivity of neurons in visual area MT of the macaque. *J. Neurophysiol.*, 52, 1106–1130.

Alley, K., Baker, R. and Simpson, J. I. (1975). Afferents to the vestibulo-cerebellum and the origin of the visual climbing fibers in the rabbit. *Brain Res.*, 98, 582–589.

Allman, J. M. (1977). Evolution of the visual system in the early primates. In *Progress in Psychobiology and Physiological Psychology*, vol. 7. eds. Sprague, J. M. and Epstein, A. N. pp. 1–53. New York: Academic.

Atencio, F. W., Diamond, I. T. and Ward, J. P. (1975). Behavioral study of the visual cortex of *Galago senegalensis*. *J. Comp. Neurol.*, 89, 1109–1135.

Baylis, G. C., Rolls, E. T. and Leonard, C. M. (1987). Functional subdivisions of the temporal lobe neocortex. *J. Neurosci.*, 7, 330–342.

Bender, D. B. (1981). Effects of superior colliculus and striate cortex lesions on the visual response properties of inferior pulvinar neurons in the monkey. *Soc. Neurosci. Abstr.*, 7, 760.

Bender, D. B. (1982). Retinotopic organization of macaque pulvinar. *J. Neurophysiol.*, 46, 672–693.

Benedek, G., Norita, M. and Creutzfeldt, O. D. (1983). Electrophysiological and anatomical demonstration of an overlapping striate and tectal projection to the lateral posterior–pulvinar complex of the cat. *Exp. Brain Res.*, 52, 157–169.

Benevento, L. A. and Yoshida, K. (1981). The afferent and efferent organization of the lateral geniculo-prestriate pathways in the macaque monkey. *J. Comp. Neurol*, 203, 455–474.

Benevento, L. A. and Fallon, J. H. (1975). The ascending projections of the superior colliculus in the rhesus monkey (*Macaca mulatta*). *J. Comp. Neurol.*, 160, 339–361.

Benevento, L. A. and Rezak, M. (1976). The cortical projections of the inferior pulvinar and adjacent lateral pulvinar in the rhesus monkey (*Macaca mulatta*): an autoradiographic study. *Brain Res.*, 108, 1–24.

Benevento, L. A. and Standage, G. P. (1983). The organization of projections of the retinorecipient and nonretinorecipient nuclei of the pretectal complex and layers of the superior colliculus to the lateral pulvinar and medial pulvinar in the macaque monkey. *J. Comp. Neurol.*, 217, 307–336.

Berlucchi, G. and Sprague, J. M. (1981). The cerebral cortex in visual learning and memory, and in interhemispheric transfer in the cat. In *The Organization of the Cerebral Cortex*. eds. Schmitt, F. O., Worden, F. G., Adelman, G. and Dennis, J. G. Cambridge, MA: MIT Press.

Berson, D. M. and Graybiel, A. M. (1983). Organization of the striate-recipient zone of the cat's lateralis posterior-pulvinar complex and its relations with the geniculostriate system. *Neuroscience*, 9, 337–372.

Bilge, M., Bingle, A., Seneviratne, K. N. and Whitteridge, D. (1967). A map of the visual cortex in the cat. *J. Physiol. Lond.*, 191, 116P–118P.

Birnbacher, D. and Albus, K. (1987). Divergence of single axons in afferent projections to the cat's visual cortical areas 17, 18 and 19: a perimetric study. *J. Comp. Neurol.* 261, 543–561.

Blakemore, C. B. and Vital-Durand, F. (1981). Distribution of X and Y cells in the monkey's lateral geniculate nucleus. *J. Physiol.*, 320, 17P–18P.

Boycott, B. B. and Wassle, H. (1974). The morphological types of ganglion cells of the domestic cat's retina. *J. Physiol.*, 240, 397–419.

Brauer, K., Schober, W. and Winkelmann, E. (1978). Phylogenetic changes and functional specializations in the dorsal lateral geniculate nucleus (dLGN) of the mammals. *J. Hirnforsch.*, 19, 177–187.

Brecha, N., Karten, H. J. and Hunt, S. P. (1980). Projections of the nucleus of the basal optic root in the pigeon: an autoradiographic and horseradish peroxidase study. *J. Comp. Neurol.*, 189, 615–670.

Brindley, G. S. and Lewin, W. S. (1968). The sensations produced by electrical stimulation of the visual cortex. *J. Physiol. Lond.*, 196, 479–493.

Bullier, J. and Kennedy, H. (1983). Projection of the lateral geniculate nucleus onto cortical area V2 in the macaque monkey. *Exp. Brain Res.*, 53, 168–172.

Bullier, J., Kennedy, H. and Salinger, W. (1984). Bifurcation of cortical afferents to visual areas 17, 18 and 19 in the cat cortex. *J. Comp. Neurol.*, 228, 309–328.

Butler, S. R., Carden, C., Hilken, H. and Kulikowski, J. J. (1988). Deficit in colour constancy following V4 ablation in rhesus monkeys. *J. Physiol.*, 403, 72P.

Caldwell, R. B. and Ward, J. P. (1982). Central visual field representation in striate-prestriate cortex as the functional unit of pattern and form discrimination in bushbaby (*Galago senegalensis*). *Brain Behav. Evol.*, 21, 161–174.

Campbell, A. W. (1905). *Histological Studies on the Localization of Cerebral Function*. London: Cambridge University Press.

Cartmill, M. (1972). Arboreal adaptations and the origin of the order Primates. In *The Functional and Evolutionary Biology of Primates*. ed. Tuttle, R. H. pp. 97–122. Chicago: Aldine, Atherton.

Cartmill, M. (1974). Rethinking primate origins. *Science*, 184, 436–443.

Chalupa, L. M. (1977). A review of cat and monkey studies implicating the pulvinar in visual function. *Behav. Biol.*, 20, 149–167.

Collins, E. T. (1921). Changes in the visual organs correlated with the adoption of arboreal life with the assumption of the erect posture. *Trans. Ophthalmol. Soc. UK*, 41, 10–90.

Connolly, M. and Van Essen, D. C. (1984). The representation of the visual field in parvocellular and magnocellular layers of the lateral geniculate nucleus in the macaque monkey. *J. Comp. Neurol.*, 226, 544–564.

Conway, J. and Schiller, P. H. (1983). Laminar organization in the tree shrew lateral geniculate nucleus. *J. Neurophysiol.* 50, 1330–1342.

Cowey, A. (1964). The projection of the retina on to striate and prestriate cortex in the squirrel monkey *Saimiri sciureus*. *J. Neurophysiol.*, 27, 366–393.

Cowey, A. and Rolls, E. T. (1974). Human cortical magnification factor and its relation to visual acuity. *Exp. Brain Res.*, 21, 447–454.

Cowey, A. and Weiskrantz, L. (1963). A perimetric study of visual field defects in monkeys. *Q. J. Exp. Psychol.*, 15, 91–115.

Crook, J. M., Lange-Malecki, B., Lee, B. B. and Valberg, A. (1988). Visual resolution of macaque retinal ganglion cells. *J. Physiol.*, 396, 205–224.

Daniel, P. M. and Whitteridge, D. (1961). The representation of the visual field on the cerebral cortex in monkey. *J. Physiol. Lond.*, 159, 203–221.

De Yoe, E A. and Van Essen, D. C. (1988). Concurrent processing streams in monkey visual cortex. *Trends Neurosci.*, 11, 219–226.

Diamond, I. T., Conley, M., Itoch, K. and Fitzpatrick, D. (1985a). Laminar organization of geniculocortical projections in *Galago senegalensis* and *Aotus trivirgatus*. *J. Comp. Neurol.*, 242, 584–610.

Diamond, I. T., Fitzpatrick, D. and Sprague, J. M. (1985b). The extrastriate visual cortex. In *Cerebral Cortex*. vol. 4. eds Peters, A. and Jones, E. G. pp. 63–87. New York: Plenum.

Diamond, I. T. and Hall, W. C. (1969). Evolution of neocortex. *Science*, 164, 251–262.

Dobelle, W. H., Turkel, J., Henderson, D. C. and Evans, J. R. (1979). Mapping the representation of the visual field by electrical stimulation of human visual cortex. *Am. J. Ophthalmol.*, 88, 727–735.

Dow, B. M., Snyder, A. Z., Vautin, R. G. and Bauer, R. (1981). Magnification factor and receptive field size in foveal striate cortex of the monkey. *Exp. Brain Res.*, **44**, 213–228.

Drasdo, N. (1977). The neural representation of visual space. *Nature Lond.*, **266**, 554–556.

Dubner, R. and Zeki, S. M. (1971). Response properties and receptive fields of cells in an anatomically defined region of the superior temporal sulcus. *Brain Res.*, **35**, 528–532.

Espinoza, S. G. and Thomas, H. C. (1983). Retinotopic organization of striate and extrastriate visual cortex in the hooded rat. *Brain Res.*, **272**, 137–144.

Fahle, M. and Schmid, M. (1988). Naso-temporal asymmetry of visual perception and of the visual cortex. *Vision Res.*, **28**, 293–300.

Fendick, M. and Westheimer, G. (1983). Effects of practice and the separation of test targets on foveal and peripheral stereoacuity. *Vision Res.*, **23**, 145–150.

Ferster, D. and LeVay, S. (1978). The axonal arborizations of lateral geniculate neurons in the striate cortex of the cat. *J. Comp. Neurol.*, **182**, 923–944.

Fitzpatrick, D., Itoh, K. and Diamond, I. T. (1983). The laminar organization of the lateral geniculate body and the striate cortex in the squirrel monkey (*Saimiri sciureus*). *J. Neurosci.*, **3**, 673–702.

Flechsig, P. (1901). Development (myelogenetic) localisation of the cerebral cortex in the human subject. *Lancet*, ii, 1027–1029.

Fox, P. T., Miezin, F. M., Allman, J. M., Van Essen, D. C. and Raichle, M. E. (1987). Retinotopic organization of human visual cortex mapped with positron-emission tomography. *J. Neurosci.*, **7**, 913–922.

Fries, W. (1981). The projection from the lateral geniculate nucleus to the prestriate cortex of the macaque monkey. *Proc. R. Soc. Lond.*, **213**, 73–80.

Fukuda, Y. (1977). A three-group classification of rat retinal ganglion cells: histological and physiological studies. *Brain Res.*, **119**, 327–344.

Geisert, E. E. (1980). Cortical projections of the lateral geniculate nucleus in the cat. *J. Comp. Neurol.*, **190**, 793–812.

Giolli, R. A. (1961). An experimental study of the accessory optic tracts (transpeduncular tract and anterior accessory optic tracts) in the rabbit. *J. Comp. Neurol.*, **117**, 77–95.

Glendenning, K. K., Hall, J. A., Diamond, I. T. and Hall, W. C. (1975). The pulvinar nucleus of *Galago senegalensis*. *J. Comp. Neurol.*, **161**, 419–458.

Graybiel, A. M. (1972). Some extrageniculate visual pathways in the cat. *Invest. Ophthalmol.*, **11**, 322–332.

Graybiel, A. M. and Berson, D. M. (1980). Histochemical identification and afferent connections of subdivisions in the lateralis posterior–pulvinar complex and related thalamic nuclei in the cat. *Neuroscience*, **5**, 1175–1238.

Graybiel, A. M. and Berson, D. M. (1981). On the relation between transthalamic and transcortical pathways in the visual system. In *The Organisation of the Cerebral Cortex*. eds. Schmitt, F. O., Worden, F. G. and Dennis, F. pp. 286–319. Cambridge, MA: MIT Press.

Halasz, P. and Stone J. (1989). Retinal topography in the elephant. *Brain, Behav. Evol.*, **34**, 84–95.

Hall, W. C., Kaas, J. H., Killackey, H. and Diamond, I. T. (1971). Cortical visual areas in the grey squirrel (*Sciurus carolinensis*): a correlation between cortical evoked potential maps and architectonic subdivisions. *J. Neurophysiol.*, **34**, 437–452.

Harris, C. S., Blakemore, C. and Donaghy, M. (1980). Integration of visual and auditory space in the mammalian superior colliculus. *Nature*, **288**, 56–59.

Harting, J. K., Hall, W. C. and Diamond, I. T. (1972). Evolution of the pulvinar. *Brain, Behav. Evol.*, **6**, 424–452.

Harting, J. K., Hall, W. C., Diamond, I. T. and Martin, G. F. (1973). Anterograde degeneration study of the superior colliculus in *Tupaia glis*: evidence for a subdivision between superficial and deep layers. *J. Comp. Neurol.*, **148**, 361–386.

Hassler, R. (1966). Comparative anatomy of the central visual systems in day- and night-active primates. In *Evolution of the forebrain*. eds Hassler, R. and Stephen, H. pp. 419–434. Stuttgart: Thieme.

Hayhow, W. R. (1959). An experimental study of the accessory optic fiber system in the cat. *J. Comp. Neurol.*, **113**, 281–314.

Hayhow, W. R. (1966). The accessory optic system in the marsupial phalanger, *Trichosurus vulpecula*. An experimental degeneration study. *J. Comp. Neurol.*, **126**, 653–672.

Hayhow, W. R., Webb, C., Jervie, A. (1960). The accessory optic fiber system in the rat. *J. Comp. Neurol.*, **115**, 187–215.

Heath, C. J. and Jones, E. G. (1971). Interhemispheric paths in the absence of the corpus callosum: An experimental study of commissural connections in the marsupial phalanger. *J. Anat.*, **109**, 253–270.

Hickey, T. L. and Guillery, R. W. (1974). An autoradiographic study of retinogeniculate pathways in the cat and in the fox. *J. Comp. Neurol.*, **156**, 239–254.

Hicks, T. P., Lee, B. B. and Vidyasagar, T. R. (1983). The responses of cells in the macaque lateral geniculate nucleus to sinusoidal gratings. *J. Physiol.*, **337**, 183–200.

Hochstein, S. and Shapley, R. M. (1976). Linear and non-linear spatial subunits in Y cat retinal ganglion cells. *J. Physiol.*, **262**, 265–284.

Hoffmann, K.-P. (1983). Control of the optokinetic reflex by the nucleus of the optic tract in the cat. In *Spatially Oriented Behavior*. eds. Ennerod, M. and Hein, A. pp. 135–153. New York: Springer.

Hoffmann, K.-P. and Stone, J. (1985). Retinal input to the nucleus of the optic tract of the cat assessed by antidromic activation of ganglion cells. *Exp. Brain Res.*, **59**, 395–403.

Hoffmann, K.-P., Distler, C., Erickson, R. G. and Mader, W. (1988). Physiological and anatomical identification of the nucleus of the optic tract and dorsal terminal nucleus of the accessory optic tract in monkeys. *Exp. Brain Res.*, **69**, 635–644.

Hoffman, K.-P. and Distler, C. (1989). Quantitative analysis of visual receptive fields of neurons in nucleus of the optic tract and dorsal terminal nucleus of the accessory optic tract in macaque monkey. *J. Neurophysiol.*, **62**, 416–428.

Hubel, D. H. and Freeman, D. C. (1977). Projection into the visual field of ocular dominance columns in macaque monkey. *Brain Res.*, **122**, 336–343.

Hubel, D. H. and Wiesel, T. S. (1970). Stereoscopic vision in macaque monkey. *Nature Lond.*, **225**, 41–43.

Hubel, D. H. and Wiesel, T. N. (1972). Laminar and columnar distribution of geniculocortical fibers in the macaque monkey. *J. Comp. Neurol.*, **146**, 421–450.

Hubel, D. H. and Wiesel, T. N. (1974). Uniformity of monkey striate cortex: a parallel relationship between field size, scatter, and magnification factor. *J. Comp. Neurol.*, **158**, 295–305.

Hubel, D. H., Wiesel, T. S. and LeVay, S. (1976). Functional architecture in area 17 in normal and monocularly deprived macaque monkeys. *Cold Spring Harbor Symp. Quant. Biol.*, **40**, 581–589.

Hughes, A. (1975). A quantitative analysis of cat retinal ganglion cell topography. *J. Comp. Neurol.*, **163**, 107–128.

Hughes, A. (1977). The topography of vision in mammals of contrasting lifestyle. In *Handbook of Sensory Physiology*, vol. VII/5. ed. Crescitelli, F. pp. 613–756. Berlin: Springer.

Hughes, A. (1981). Population magnitudes and distribution of the major modal classes of cat retinal ganglion cells as estimated from HRP filling and a systematic survey of the soma diameter spectra for classical neurones. *J. Comp. Neurol.*, **197**, 303–339.

Hughes, H. C. (1980). Efferent organization of the cat pulvinar complex, with a note on bilateral claustrocortical and reticulocortical connections. *J. Comp. Neurol.*, **193**, 937–963.

Humphrey, A. L., Sur, M., Uhlrich, D. J. and Sherman, S. M. (1985). Projection of patterns of individual X- and Y-cells from the lateral geniculate nucleus to cortical area 17 in the cat. *J. Comp. Neurol.*, **233**, 159–189.

Jay, M. F. and Sparks, D. L. (1984). Auditory receptive fields in primate superior colliculus shift with changes in eye position.

Nature, **309**, 345–347.
Jay, M. F. and Sparks, D. L. (1987). Sensorimotor integration in the primate superior colliculus. II. Coordinates of auditory signals. *J. Neurophysiol.*, **57**, 35–55.
Jenkins, F. and White, M. (1957). *Fundamentals of Optics*. 3rd edn. New York: McGraw-Hill.
Julesz, B. (1971). *Foundations of Cyclopean Perception*. Chicago: University of Chicago Press.
Kaas, J. H. (1978). The organization of visual cortex in primates. *In Sensory Systems of Primates*. ed. Nobak, C. R. pp. 151–180. New York: Plenum Press.
Kaas, J. H. (1986). The structural basis for information processing in the primate visual system. In *Visual Neuroscience*. eds Pettigrew, J. D., Sanderson, K. J. and Levick, W. R. Cambridge: Cambridge University Press.
Kaas, J. H., Guillery, R. W. and Allman, J. M. (1973). Discontinuities in the dorsal lateral geniculate nucleus corresponding to the optic disc: a comparative study. *J. Comp. Neurol.*, **147**, 163–179.
Kaas, J. H., Hall, W. C. and Diamond, I. T. (1972). Visual cortex of the grey squirrel (*Sciurus carolinensis*): Architectonic subdivisions and connections from the visual thalamus. *J. Comp. Neurol.*, **145**, 273–306.
Kaas, J. H., Harting, J. K. and Guillery, R. W. (1973). Representation of the complete retina in the contralateral superior colliculus of some mammals. *Brain Res.*, **65**, 343–346.
Kaas, J. H., Huerta, M. F., Weber, J. T. and Harting, J. K. (1978). Patterns of retinal terminations and laminar organization of the lateral geniculate nucleus of primate. *J. Comp. Neurol.*, **182**, 517–554.
Kaas, J. H. and Preuss, T. M. (1990). Archontan affinities as reflected in the visual system. Submitted for publication.
Kaplan, E., Lee, B. B. and Shapley, R. M. (1990). New views of primate retinal function. *Prog. Ret. Res.*, **9**, 273–336.
Kawamura, S., Fukushima, N., Hattori, S. and Kudo, M. (1980). Laminar segregation of cells of origin of ascending projections from the superficial layers of the superior colliculus in the cat. *Brain Res.*, **184**, 486–490.
Kawamura, S., Sprague, J. M. and Niimi, K. (1974). Corticofugal projections from the visual cortices to the thalamus, pretectum and superior colliculus in the cat. *J. Comp. Neurol.*, **158**, 339–362.
Kiefer, W., Kruger, K., Strauss, G. and Berlucchi, G. (1989). Considerable deficits in the detection performance of the cat after lesion of the suprasylvian visual cortex. *Exp. Brain Res.*, **75**, 208–212.
Killackey, H., Synder, M. and Diamond, I. T. (1971). Function of the striate and temporal cortex in the tree shrew. *J. Comp. Physiol. Psychol.*, **74**, 1–29.
Killackey, H., Wilson, M. and Diamond, I. T. (1972). Further studies of the striate and extrastriate visual cortex in the tree shrew. *J. Comp. Physiol. Psychol.*, **81**, 45–63.
Kirsch, J. A. W. and Calaby, J. H. (1977). The species of living marsupials. In *The Biology of Marsupials*. eds Stonehouse, B. and Gilmore, D. pp. 9–26. London: Macmillan.
Kluver, H. (1942). Functional significance of the geniculostriate system. *Biol. Symp.*, **7**, 253–299.
Krubitzer, L. A. and Kaas, J. H. (1990). Cortical connections of MT in four species of primates: Areal, modular and retinotopic patterns. *Visual Neurosci.*, **5**, 165–204.
Kruger, K., Donicht, M., Muller Kusdian, G., Kiefer, W. and Berlucchi, G. (1988). Lesion of areas 17/18/19: effects on the cat's performance in a binary detection task. *Exp. Brain Res.*, **72**, 510–516.
Kruger, K., Heitlander Fansa, H., Dinse, H. and Berlucchi, G. (1986). Detection performance of normal cats and those lacking areas 17 and 18: a behavioral approach to analyse pattern recognition deficits. *Exp. Brain Res.*, **63**, 233–247.
Lane, R. H., Allman, J. M. and Kaas, J. H. (1971). Representation of the visual field in the superior colliculus of the grey squirrel (*Sciurus carolinensis*) and the tree shrew (*Tupaia glis*). *Brain Res.*, **26**, 277–292.
Lane, R. H., Allman, J. M., Kaas, J. H. and Miezin, F. M. (1973). The visuotopic organization of the superior colliculus of the owl monkey (*Aotus trivirgatus*) and the bush baby (*Galago senegalensis*). *Brain Res.*, **70**, 413–430.
Lannou, J., Cazin, L., Precht, W. and LeTaillanter, M. (1984). Responses of the prepositus hypoglossi neurons to optokinetic and vestibular stimulations in the rat. *Brain Res.*, **301**, 39–45.
Leibowitz, H., Johnson, C. and Isabelle, E. (1972). Peripheral motion detection and refractive error. *Science Wash. DC*, **177**, 1207–1208.
LeVay, S. and McConnell, S. K. (1982). ON and OFF layers in the lateral geniculate nucleus of the mink. *Nature Lond.*, **300**, 350–351.
Leventhal, A. G., Rodieck, R. W. and Dreher, B. (1981). Retinal ganglion cell classes in the old world monkey: morphology and central projections. *Science Wash. DC*, **213**, 1139–1142.
Levi, D. M., Klein, S. A. and Aitsebaomo, A. P. (1985). Vernier acuity, crowding and cortical magnification. *Vision Res.*, **25**, 963–977.
Lin, C. S. and Kaas, J. H. (1979). The inferior pulvinar complex in owl monkeys: architectonic subdivisions and patterns of input from the superior colliculus and subdivisions of visual cortex. *J. Comp. Neurol.*, **187**, 655–678.
Lin, C. S. and Kaas, J. H. (1980). Projections from the medial nucleus of the inferior pulvinar complex to the middle temporal area of the visual cortex. *Neuroscience*, **5**, 2219–2228.
Livingstone, M. S. and Hubel, D. H. (1984). Anatomy and physiology of a colour system in the primate visual cortex. *J. Neurosci.*, **4**, 309–356.
Livingstone, M. S. and Hubel, D. H. (1987). Connections between layer 4B of area 17 and thick cytochrome oxidase stripes of area 18 in the squirrel monkey. *J. Neurosci.*, **17**, 3371–3377.
Livingstone, M. S. and Hubel, D. H. (1988). Segregation of form, color, movement and depth: Anatomy, physiology and perception. *Science Wash. DC*, **240**, 740–749.
Loo, Y. T. (1930). The forebrain of the opossum, *Didelphis virginiana*. *J. Comp. Neurol.*, **51**, 13–64.
Lund, J. S. (1973). Organization of neurons in the visual cortex, area 17, of the monkey (*Macaca mulatta*). *J. Comp. Neurol.*, **147**, 455–496.
Lund, J. S., Fitzpatrick, D. and Humphrey, A. L. (1985). The striate visual cortex of the tree shrew. In *Cerebral Cortex*. eds Peters, A. and Jones, E. G. pp. 157–206. New York: Plenum.
Luppino, G., Matelli, M., Carey, R. G., Fitzpatrick, D. and Diamond, I. T. (1988). New view of the organization of the pulvinar nucleus in *Tupaia* as revealed by tectopulvinar and pulvinarcortical projections. *J. Comp. Neurol.*, **273**, 67–86.
Malpeli, J. G. and Baker, F. H. (1975). The representation of the visual field in the lateral geniculate nucleus of *Macaca mulatta*. *J. Comp. Neurol.*, **161**, 569–594.
Mason, R. and Gross, G. A. (1981). Cortico-recipient and tecto-recipient visual zones in the rat's lateral posterior (pulvinar) nucleus: an anatomical study. *Neurosci. Lett.*, **25**, 107–112.
Maunsell, J. H. R. and Newsome, W. T. (1987). Visual processing in monkey extrastriate cortex. *Ann. Rev. Neurosci.*, **10**, 363–401.
Maunsell, J. H. R. and Van Essen, D. C. (1983). Functional properties of neurons in the middle visual temporal area (MT) of the macaque monkey: I. Selectivity for stimulus direction, speed and orientation. *J. Neurophysiol.*, **49**, 1127–1147.
McCourt, M. E., Thalluri, J. and Henry, G. H. (1990). Properties of area 17/18 border neurons contributing to the visual transcallosal pathway in the cat. *Vis. Neurosci.*, **5**, 83–98.
Mishkin, M. (1966). Visual mechanisms beyond the striate cortex. In *Frontiers in Physiological Psychology*. pp. 187–208. New York: Academic.
Mishkin, M. (1972). Cortical visual areas and their interactions. In *Brain and Human Behavior*. eds Karczmar, A. G. and Eccles, J. C.

Berlin: Springer.
Mishkin, M., Ungerleider, L. G. and Macko, K. A. (1983). Object vision and spatial vision: two cortical pathways. *Trends Neurosci.*, **6**, 414–417.
Myerson, J., Manis, P. B., Miezin, F. M. and Allman, J. M. (1977). Magnification in striate cortex and retinal ganglion cell layer of owl monkey: a quantitative comparison. *Science Wash. DC*, **198**, 855–857.
Newsome, W. T. and Pare, E. B. (1988). A selective impairment of motion perception following lesions of the middle temporal visual area (MT). *J. Neurosci.*, **8**, 2201–2211.
Ogren, M. P. and Hendrickson, A. E. (1976). Pathways between striate cortex and subcortical regions in *Macaca mulatta* and *Saimiri sciureus:* Evidence for a reciprocal pulvinar connection. *Exp. Neurol.*, **53**, 780–800.
Oswaldo-Cruz, E., Picanco-Diniz, C. W., Pompeu, M. S. S. and Silveira, L. C. L. (1990). Asymmetries of the representation of the contralateral visual field on area 17 of the cortex of the anaesthetized agouti do not match the retinal ganglion cell regional densities. *J. Physiol.*, **422**, 8P.
Oyster, C. W. and Barlow, H. B. (1967). Direction-selective units in rabbit retina: distribution of preferred direction. *Science Wash. DC*, **155**, 841–842.
Oyster, C. W., Simpson, J. I., Takahashi, E. S. and Soodak, R. E. (1980). Retinal ganglion cells projecting to the rabbit accessory optic system. *J. Comp. Neurol.*, **190**, 49–61.
Oyster, C. W., Takahashi, E. and Collewijn, H. (1972). Direction-selective retinal ganglion cells and control of optokinetic nystagmus in the rabbit. *Vision Res.*, **12**, 183–193.
Partlow, G. D., Colonnier, M. and Szabo, J. (1977). Thalamic projections of the superior colliculus in the rhesus monkey. *Macaca mulatta*. A light and electron microscopic study. *J. Comp. Neurol.*, **72**, 285–318.
Peichl, L., Ott, H. and Boycott, B. B. (1987). Alpha ganglion cells in mammalian retinae. *Proc. Roy. Soc. Lond. Ser. B.*, **231**, 169–197.
Perret, D., Rolls, E. T. and Caan, W. (1982). Visual neurons responsive to faces in the monkey temporal cortex. *Exp. Brain Res.*, **47**, 329–342.
Perry, V. H. and Cowey, A. (1985). The ganglion cell and cone distributions in the monkey's retina: implications for central magnification factors. *Vision Res.*, **25**, 1795–1810.
Perry, V. H. and Cowey, A. (1988). The lengths of the fibres of Henle in the retina of macaque monkeys: implications for vision. *Neuroscience*, **25**, 225–236.
Perry, V. H., Oehler, R. and Cowey, A. (1984). Retinal ganglion cells that project to the dorsal lateral geniculate nucleus in the macaque monkey. *Neuroscience*, **12**, 1101–1123.
Pettigrew, J. D. (1986). Flying primates? Megabats have the advanced pathway from eye to midbrain. *Science Wash. DC*, **231**, 1304–1306.
Pettigrew, J. D. (1990). Wings or brain? Convergent evolution in the origins of bats. *System. Zool.*, in press.
Pettigrew, J. D., Jamieson, B. G. M., Robson, S. K., Hall, I. S., McNally, K. I. and Cooper, H. M. (1989). Phylogenetic relations between microbats, megabats and primates (Mammalia: Chiroptera and Primates). *Phil. Trans. R. Soc. Lond.*, **325**, 489–559.
Pettigrew, J. D., Cooper, H. M. (1986). Aerial primates: advanced visual pathways in megabats and flying lemurs. *Soc. Neurosci. Abstr.*, **12**, 1035.
Pettigrew, J. D., Dreher, B., Hopkins, C. S., McCall, M. J. and Brown, M. (1988). Peak density and distribution of ganglion cells in the retinae of microchiropteran bats: implications for visual acuity. *Brain Behav. Evol.*, **32**, 39–56.
Pointer, J. S. (1986). The cortical magnification factor and photopic vision. *Biol. Rev. Camb. Philos. Soc.*, **61**, 97–119.
Polyak, S. (1957) *The Vertebrate Visual System*. Chicago: University of Chicago Press.
Precht, W., Montarolo, P. G. and Strata, P. (1980). The role of the crossed and uncrossed retinal fibers in mediating the horizontal optokinetic nystagmus in the cat. *Neurosci. Lett.*, **17**, 39–42.
Purpura, K., Shapley, R. A. and Kaplan, E. (1988). Background light and the contrast gain of primate P and M retinal ganglion cells. *Proc. Natl Acad. Sci. USA*, **85**, 4534–4537.
Raczkowski, D. and Diamond, I. T. (1980). Cortical connections of the pulvinar nucleus in Galago. *J. Comp. Neurol.*, **193**, 1–40.
Raczkowski, D. and Diamond, I. T. (1981). Projections from the superior colliculus and the neocortex to the pulvinar nucleus in Galago. *J. Comp. Neurol.*, **200**, 231–254.
Raczkowski, D. and Rosenquist, A. C. (1983). Connections of the multiple visual cortical areas with the lateral posterior–pulvinar complex and adjacent thalamic nuclei in the cat. *J. Neurosci.*, **3**, 1912–1942.
Robson, J. A. and Hall, W. C. (1977). The organization of the pulvinar in the grey squirrel (*Sciurus carolinensis*). I. Cytoarchitecture and connections. *J. Comp. Neurol.*, **173**, 355–388.
Rodieck, R. W. (1979). Visual pathways. *Ann. Rev. Neurosci.*, **2**, 193–225.
Rolls, E. T. and Cowey, A. (1970). Topography of the retina and striate cortex and its relationship to visual acuity in rhesus monkeys and squirrel monkeys. *Exp. Brain Res.*, **10**, 298–310.
Rosenquist, A. C. (1985). Connections of visual cortical areas in the cat. In *Cerebral Cortex*, vol. 3. eds. Peters, A. and Jones, E. G. pp. 81–117. New York: Plenum.
Rovamo, J. and Virsu, V. (1979). An estimation and application of the human cortical magnification factor. *Exp. Brain Res.*, **37**, 495–510.
Sanderson, K. J. (1971). Visual field projection columns and magnification factors in the lateral geniculate nucleus of the cat. *Exp. Brain Res.*, **13**, 159–177.
Sanderson, K. J. (1986). Evolution of the lateral geniculate nucleus. In *Visual Neuroscience*. eds. Pettigrew, J. D., Sanderson, K. J. and Levick, W. R. pp. 183–195. Cambridge: Cambridge University Press.
Sanderson, K. H. and Pearson, L. J. (1977). Retinal projections in the native cat, *Dasyurus viverrinus*. *J. Comp. Neurol.*, **174**, 347–358.
Schafer, E. A. (1888). Experiments on special localization in the cortex cerebri of the monkey. *Brain*, **10**, 362–380.
Schein, S. J. (1988). Anatomy of macaque fovea and spatial densities of neurons in foveal representation. *J. Comp. Neurol.*, **269**, 479–505.
Schein, S. J. and DeMonasterio, F. M. (1987). Mapping of retinal and geniculate neurons onto striate cortex of macaque. *J. Neurosci.*, **7**, 996–1009.
Schein, S. J. and Desimone, R. (1990). Spectral properties of V4 neurones in the Macaque. *J. Neurosci.*, **10**, 3369–3389.
Schein, S. J., Marrocco, R. T. and DeMonasterio, F. M. (1982). Is there a high concentration of color-selective cells in area V4 of monkey visual cortex? *J. Neurophysiol.*, **47**, 193–213.
Schiller, P. H. and Logothetis, N. K. (1990). The color-opponent and broad-band channels of the primate visual system. *Trends Neurosci.*, **13**, 392–398.
Schiller, P. H. and Stryker, M. (1972). Single unit recording and stimulation in superior colliculus of the alert rhesus monkey. *J. Neurophysiol.*, **35**, 915–924.
Schneider, E. G. (1969). Two visual systems. Brain mechanisms for localization and discrimination are dissociated by tectal and cortical lesions. *Science Wash. DC*, **163**, 895–902.
Schoppmann, A. (1981). Projections from areas 17 and 18 of the visual cortex to the nucleus of the optic tract. *Brain Res.*, **223**, 1–17.
Shapley, R. and Perry, V. H. (1986). Cat and monkey retinal ganglion cells and their visual functional roles. *Trends Neurosci.*, **9**, 229–235.
Shapley, R., Kaplan, E. and Soodak, R. (1983). Spatial summation and contrast sensitivity of X and Y cells in the lateral geniculate nucleus of the macaque. *Nature Lond.*, **292**, 543–545.

Sherman, S. M., Norton, T. T. and Casagrande, V. A. (1975). X- and Y-cells in the dorsal lateral geniculate of the tree shrew (*Tupaia glis*). *Brain Res.*, **93**, 152–157.

Simpson, J. I. (1984). The accessory optic system. *Ann. Rev. Neurosci.*, **7**, 13–41.

Simpson, J. I., Soodak, R. E. and Hess, R. (1979). The accessory optic system and its relation to the vestibulocerebellum. *Prog. Brain Res.*, **50**, 715–724.

Snyder, M. and Diamond, I. T. (1968). The organization and function of the visual cortex in the tree shrew. *Brain Behav. Evol.*, **1**, 244–288.

Sprague, J. M., Hughes, C. H. and Berlucchi, G. (1981). Cortical mechanisms in pattern and form perception. In *Brain Mechanisms and Perceptual Awareness*. eds Pompeiano, O., Ajmone and Marsan, C. pp. 107–132. New York: Raven.

Stone, J., Dreher, B. and Leventhal, A. G. (1979). Hierarchical and parallel mechanisms in the organization of visual cortex. *Brain Res. Rev.*, **1**, 345–394.

Stryker, M. P. and Zahs, K. R. (1983). ON and OFF sublaminae in the lateral geniculate nucleus of the ferret. *J. Neurosci.*, **3**, 1943–1951.

Symonds, L. L. and Kaas, J. H. (1978). Connections of striate cortex in the prosimian, *Galago senegalensis*. *J. Comp. Neurol.*, **181**, 477–512.

Talbot, S. A. and Marshall, W. H. (1941). Physiological studies on neural mechanisms of visual localisation and discrimination. *Am. J. Ophthalmol.*, **24**, 1255–1264.

Tanaka, M., Weber, H. and Creutzfeldt, O. D. (1986). Visual properties and spatial distribution of neurones in the visual association area in the prelunate gyrus of the awake monkey. *Exp. Brain Res.*, **65**, 11–37.

Tigges, J. and Tigges, M. (1985). Subcortical sources of direct projections to visual cortex. In *Cerebral Cortex*, vol. 3. eds. Peters, A. and Jones, E. G. pp. 351–378. New York: Plenum.

Tolhurst, D. J. and Ling, L. (1988). Magnification factors and the organization of the human striate cortex. *Hum. Neurobiol.*, **6**, 247–254.

Tootell, R. B. H., Silverman, M. S., Switkes, E. and De Valois, R. L. (1982). Deoxyglucose analysis of retinotopic organization in primate striate cortex. *Proc. R. Soc. Lond. (Biol.)*, **213**, 435–450.

Tootell, R. B., Switkes, E., Silverman, M. S. and Hamilton, S. L. (1988). Functional anatomy of macaque striate cortex. II. Retinotopic organization. *J. Neurosci.*, **8**, 1531–1568.

Trevarthen, C. B. (1968). Two mechanisms of vision in primates. *Psychol. Forsch.* **31**, 299–337.

Ts'o, D. Y. and Gilbert, C. D. (1988). The organization of chromatic and spatial interaction in the primate striate cortex. *J. Neurosci.*, **8**, 1712–1727.

Tusa, R. J., Palmer, L. A. and Rosenquist, A. C. (1978). The retinotopic organization of area 17 (striate cortex) in the cat. *J. Comp. Neurol.*, **177**, 213–236.

Tyndale-Biscoe, H. (1973). *Life of Marsupials*. London: Arnold.

Ungerleider, L. G. and Christensen, C. A. (1977). Pulvinar lesions in monkeys produce abnormal eye movements during visual discrimination training. *Brain Res.*, **136**, 189–196.

Ungerleider, L. G. and Christensen, C. A. (1979). Pulvinar lesions in monkeys produce abnormal scanning of a complex visual array. *Neuropsychologia*, **17**, 493–501.

Ungerleider, L. G. and Mishkin, M. (1982). Two cortical visual systems. In *Advances in the Analysis of Visual Behaviour*. eds Ingle, D. J., Goodale, M. A. and Mansfield, R. J. W., pp. 549–586. Cambridge, MA: MIT Press.

Updyke, B. V. (1977). Topographic organization of the projections from cortical areas 17, 18 and 19 onto the thalamus, pretectum and superior colliculus in the cat. *J. Comp. Neurol.*, **173**, 81–122.

Van Essen, D. C. (1985). Functional organization of primate visual cortex. In *Cerebral Cortex*, vol. 3. eds. Peters, A. and Jones, E. G. pp. 259–329. New York: Plenum.

Van Essen, D. C., Newsome, W. T. and Maunsell, J. H. (1984). The visual field representation in striate cortex of the macaque monkey: asymmetries, anisotropies, and individual variability. *Vision Res.*, **24**, 429–448.

Van Essen, D. C. and Zeki, S. M. (1978). The topographic organization of rhesus monkey prestriate cortex. *J. Physiol. Lond.*, **277**, 193–226.

Vidyasagar, T. R. (1991). Interactions between the visual and other sensory modalities and their development. In *Development and Plasticity of the Visual System*. ed. Cronly-Dillon, J. R. *Vision and Visual Dysfunction*. vol. 11. London: Macmillan.

Vidyasagar, T. R., Mark, R. F., Henry, G. H. and Marotte, L. R. (1990). Retinotopic representation of the visual field on the cortex of the wallaby (*Macropus eugenii*) in relation to developmental studies. In *Brain–Perception–Cognition* (Proceedings of the 18th Goettingen Neurobiology Conference). eds. Elsner, N. and Roth, G. Stuttgart and New York: Thieme.

Walker, E. P. (1975). In *Mammals of the World*, 3rd edn. Baltimore: Johns Hopkins.

Walls, G. L. (1942). *The Vertebrate Eye and its Adaptive Radiation*. London: Hafner.

Wassle, H., Levick, W. R. and Cleland, B. G. (1975). The distribution of the alpha type of ganglion cells in the cat's retina. *J. Comp. Neurol.*, **59**, 419–438.

Wassle, H., Grunert, U., Rohrenbeck, J. and Boycott, B. (1990). Retinal ganglion cell density and cortical magnification factor in the primate. *Vision Res.*, **30**, 1897–1911.

Wertheim, T. (1894). Uber die indirekte Sehscharfe. *Zeitschrift fur Psychologie und Physiologie der Sinnes-organe.*, **7**, 172–187.

Westheimer, G. (1982). The spatial grain of the perifoveal visual field. *Vision Res.*, **22**, 157–162.

Weymouth, F. W. (1958). Visual sensory units and the minimal angle of resolution. *Am. J. Ophthalmol.*, **46**, 102–113.

Wild, H. M., Butler, S. R., Carden, D. and Kulikowski, J. J. (1985). Primate cortical area V4 important for colour consistency but not wavelength discrimination. *Nature Lond.*, **313**, 133–135.

Wilson, J. R. and Sherman, S. M. (1976). Receptive field characteristics of neurones in cat striate cortex: changes with visual field eccentricity. *J. Neurophysiol.*, **39**, 512–533.

Wong, R. O. L. and Hughes, A. (1987). The morphology, number and distribution of a large population of confirmed displaced amacrine cells in the adult cat retina. *J. Comp. Neurol.*, **255**, 159–177.

Wong, R. O. L., Wye-Dvorak, J. and Henry, G. H. (1986). Morphology and distribution of neurons in the retina of the Tammar wallaby (*Macropus eugenii*). *J. Comp. Neurol.*, **253**, 1–12.

Wong-Riley, M. T. (1976). Projections from the dorsal lateral geniculate nucleus to prestriate cortex in the squirrel monkey as demonstrated by retrograde transport of horseradish peroxidase. *Brain Res.*, **109**, 595–600.

Wong-Riley, M. T. (1977). Connections between the pulvinar nucleus and the prestriate cortex in the squirrel monkey as revealed by peroxidase histochemistry and autoradiography. *Brain Res.*, **134**, 249–267.

Yeterian, E. H. and Pandya, D. N. (1989). Thalamic connections of the cortex of the superior temporal sulcus in the rhesus monkey. *J. Comp. Neurol.*, **282**, 80–97.

Yukie and Iwai, E. (1981). Direct projection from the dorsal lateral geniculate nucleus to the prestriate cortex in macaque monkeys. *J. Comp. Neurol.*, **201**, 81–97.

Zeki, S. M. (1969). Representation of central visual fields in prestriate cortex of monkey. *Brain Res.*, **14**, 271–291.

Zeki, S. M. (1970). Interhemispheric connections of prestriate cortex in monkey. *Brain Res.*, **19**, 63–75.

Zeki, S. M. (1971). Cortical projections from two prestriate areas in the monkey. *Brain Res.*, **34**, 19–35.
Zeki, S. M. (1973). Colour coding in rhesus monkey prestriate cortex. *Brain Res.*, **53**, 422–427.
Zeki, S. M. and Schipp, S. (1988). The functional logic of cortical connections. *Nature Lond.*, **335**, 311–317.

Appendix: Mammals

Gordon L. Walls

This appendix is from Gordon Lynn Walls' The Vertebrate Eye and Its Adaptive Radiation, originally published in 1942, reprinted by Hafner Publishing Co. (New York) in 1963, pp. 663–691. It is included in this volume to allow today's reader ready access to this seminal work. References by relevant authors have been extracted from Walls' Bibliography.

Introduction

The Class Mammalia contains three major divisions which are not serially related, but represent three branches from a single stem. The lowest mammals, closest to the reptiles, are the monotremes. These egg-laying forms include only the duck-bill or platypus (*Ornithorhynchus*) and the echidnas or 'spiny ant-eaters' (*Tachyglossus* and *Zaglossus* [= *Echidna* and *Proechidna*]). Ranking higher in point of specialization and anatomical distinctness from the reptiles, but not derived directly from monotremes like those now living, are the marsupials. These likewise have yolky eggs, but hatch them inside the body and bear the young alive in an embryonic condition. The young complete their development on a milk diet, outside the mother but usually inside an abdominal pouch. In the common opossum, *Didelphis virginiana*, the 'embryology' of the eye continues for 30–40 days after birth. The highest (placental) mammals nourish their young inside the mother's body by means of a 'placenta'. They were not derived from marsupials, but with them, as one of two branches.

Monotremes and Marsupials

In these 'lower' mammals the eye alone would prove the reptilian origin of the whole mammalian class. Indeed, with the exception of exactly two features – one of them outside the eyeball (in the oculo-rotatory musculature) and the other one inside (in the ciliary body) – the monotreme eye is so completely reptilian that it affords no ammunition for use against those few mammalogists who claim separate reptilian origins for the monotremes and for all other mammals.

The marsupials originated as opossum-like animals, and only such forms (together with *Cænolestes*) have been able to survive in the American home of the group. In Australia however, where they became isolated from placental flesh-eaters, the marsupials differentiated into a number of types, many of them imitative of placental types. Thus, there are marsupial mice, rats, marmots, rabbits, flying-squirrels,[1] jerboas, bears, cats, wolves, anteaters and golden moles. There was once even a marsupial 'lion', though it was probably a mild-mannered vegetarian. The marsupials have avoided the water, so there are no marsupial seals or porpoises – the tropical American water opossum, *Chironectes*, is the only aquatic marsupial. Nor have the marsupials developed any hoofed types; but the larger kangaroos fill about the same ecological niche.

The lower marsupials are mostly carnivorous and the higher types (phalangers, kangaroos) herbivorous. Most marsupials, like the monotremes, are crepuscular or nocturnal to some degree; but the larger kangaroos are arrhythmic and a few are quite strongly diurnal. In keeping with the adaptive radiation of the marsupials, their eyes show great differences from form to form. In proportion to the number of species, they have had woefully little attention as compared with the placentals. The marsupials are really the central group of mammals, and deserve much more thorough exploration, from all biological viewpoints, than they have ever yet received.

The Monotreme Eye

The eye of *Ornithorhynchus* has been described only once, by Gunn (1884) from material preserved in whisky by a

Mr. Sinclair, who clearly took his science very seriously. The eyes of the two genera of echidnas have been described by Kolmer (1925), by Franz (1934) and by Gresser and Noback (1935). None of these accounts is entirely accurate – all incorporate particular serious errors in regard to the shape of the globe (which is 'avian' only when collapsed) and the presence of a ciliary muscle (which is wholly lacking, though two of these authors describe it as having the same three types of fibres – meridional, 'radial', and circular – as the ciliary muscle of man). The ensuing descriptions are based upon preparations of *Tachyglossus* and *Ornithorhynchus* made by Kevin O'Day (1938), and upon correspondence with him. Statements of earlier workers which happen not to be refuted by O'Day's splendid material are also incorporated.[2]

Ornithorhynchus has an excellent nictitating membrane. *Tachyglossus* has none; but both genera have retractor bulbi muscles. The lids are plump and small in both, and in *Tachyglossus* are closed by swinging rather than by sliding. Small Meibomian glands, still with relation to hair follicles, are present in *Ornithorhynchus*. These may be orimentary; but the same situation occurs in one placental, the hedgehog (*Erinaceus*). They are lacking in *Tachyglossus* and *Zaglossus*. Like most Sauropsida, *Tachyglossus* has a tarsus in the lower lid only, while *Ornithorhynchus* has one in each lid. Both genera are supposed to have both lacrimal and Harderian glands (but Kolmer found only serous glandular tissue in *Zaglossus*). The adnexa in *Ornithorhynchus* thus show no specialization for the amphibious life of the animal. In fact, those of the echidnas exhibit rather more reduction, which seems largely explained by the presence in those forms of a keratinization of the corneal epithelium, no doubt in adaptation to the ant-eating habit (as in armadillos and aardvarks).

In the arrangement of the superior oblique muscle, the monotremes are wholly 'mammalian'. In the echidnas there is a slip which runs from the old sub-mammalian origin (on the anterior nasal orbital wall) to an insertion on the globe; but merging in this same insertion is a second slip, muscular almost to the globe, which comes through a pulley from an origin only a few millimetres anterior to the deep point-of-origin of the four recti. The duck-bill has only this long portion, and moreover has it as in higher mammals, i.e., originating with the recti and becoming tendinous before reaching the pulley, with the latter chondroid rather than soft as in the echidnas. This seems too strong a similarity to the higher mammals to be dismissed as a coincidence by those who consider the monotremes to have originated from a separate reptilian stock. It is not certain what called forth the elongation of the mammalian superior oblique. Such an elongation may have occurred twice. In this connection, it would be nice to know whether the optic chiasmata of the monotremes are only partially decussated. Both types have wide binocular fields, that of the echidnas being projected forward and that of the duck-bill largely upward.

The eyeball is usually figured with a short axis and a pronounced circumcorneal scleral sulcus, both of which are collapse artefacts. Correspondingly, its shape has most often been called 'avian'.[3] Actually, the eyeball is everywhere convex and is spherical in all monotremes. This sphericity, so reminiscent of the snakes, has the same basis – a total disappearance of the ancestral scleral ossicles (Fig. 194a).

The eyeball of *Tachyglossus* is eight or nine millimetres in diameter, that of *Ornithorhynchus* about six. In all monotremes the sclera contains the cartilage cup with which we have become so familiar in preceding chapters. In *Tachyglossus* the cartilage is 27 µm thick in the region of the optic nerve, 14 µm thick near its sharp anterior lip. In *Zaglossus* (a larger animal) it averages 160 µm in thickness. In the duck-bill it is even thicker fundally (400 µm) but tapers to 25 µm near its knife-edge termination. The cartilage reaches to the posterior ends of the ciliary processes in *Ornithorhynchus*, but stops opposite the ora terminalis in *Zaglossus* and a little behind the ora in *Tachyglossus*. An outer layer of fibrous scleral tissue about equal in thickness to the cartilage (but only 96 µm in *Zaglossus*), continues forward (receiving an addition which replaces the cartilage) through a zone formerly occupied by the scleral ossicles, and blends with the substantia propria of the cornea. In *Tachyglossus* at least, an outer fraction of the substantia propria is easily seen to be continuous with the conjunctival corium or 'episcleral' connective tissue. A loose layer of episcleral blood vessels, from which capillaries are sent into the cornea for some distance, marks the boundary. Nowhere else above the teleosts is it so readily to be seen that an outer portion of the substantia propria is homologous with the dermis rather than with the dura (Fig. 194a). A Bowman's membrane has been claimed for *Ornithorhynchus*, but none can be made out in *Tachyglossus*. Both these genera have the usual Descemet's layers, but Kolmer (1925) could not make out the elastic membrane in *Zaglossus*.

In keeping with its aquatic habits the duck-bill has a relatively broader cornea than *Tachyglossus*, but it has a deeper anterior chamber (cornea 4.0 mm in diameter in a 6.0 mm eye, *vs* 3.4 mm in a 8.0 mm eye; chamber 1.25 mm deep *vs* 0.9 mm). The duck-bill's corneal substantia propria is only one-quarter as thick as the echidna's, but its epithelium is much thicker and nearly equals the propria – such thickening being highly characteristic of aquatic vertebrates in general. The duck-bill cornea is 100 µm thick peripherally, only 55 µm apically. *Zaglossus* reverses this relationship, with its whole cornea 320 µm thick centrally (with 264 µm of propria) and 540 µm peripherally (460 µm of propria). Comparable figures for *Tachyglossus* are 350–290, 330–210.

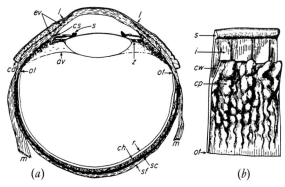

Fig. 194 *The monotreme eye. (a) Section of eye of* Tachyglossus *sp. × 8. Drawn from a preparation of O'Day. (b) Inner surface of segment of anterior uvea of* Tachyglossus. *After Franz (1934).* av – *anterior surface of vitreous;* cd – *conjunctival dermis;* ch – *chorioid;* cp – *ciliary process;* cs – *canal of Schlemm;* cw – *ciliary web;* ev – *episcleral vessels, marking boundary between dermis and fibrous tunic;* i – *iris;* l, l – *external limits of cornea;* m, m – *rectus muscles;* ot, ot – *ora terminalis retinæ;* r – *retina;* s – *sphincter;* sc – *scleral cartilage (black);* sf – *fibrous layer of sclera;* z – *zonule (main portion).*

The chorioid is only 50 μm thick in the duck-bill, a little more than twice this thick in *Tachyglossus*. Histologically, it is ambiguous – as turtle-like as it is 'mammalian'. The pigmented, laminated suprachorioidal layer or 'lamina fusca' is conspicuous, as is the choriocapillaris, whose elements are unusually large in lumen and are readily seen to be connected with the large veins.

In all three genera the iris is most simple, its web consisting of little more than the two heavily pigmented retinal layers and few small blood vessels attached loosely to the anterior face. There is no dilatator, but there is a massive sphincter around which the pigmented retinal layers are rolled so that their mutual edge lies on the anterior face of the iris. The root of the iris lies opposite the limbus in the duck-bill, but well back of this landmark in the echidnas. There is no pectinate ligament; but, as in reptiles which lack one, there is a thin anterior continuation, past the iris root, of ciliary-body connective tissue, which is adherent to the inner surface of the fibrous tunic and tapers to a knife-edge aligned with the peripheral margins of the Descemet's layers. The canal of Schlemm is embedded in this uveal meshwork tissue, as it is in sauropsidans in general. The iris is dark brown in life, the pupil always circular.

The anterior continuation of the chorioid forming the uveal portion of the ciliary body is thin, only lightly pigmented, and not sharply demarcated from the inner layers of scleral fibres except where it underlies the tallest portions of the ciliary processes. There is no trace of a ciliary muscle, and the writer is quite unable to imagine what it may be that others have mistaken for one. The ciliary processes are low, puffy, and tortuous, and number about 60 in *Tachyglossus*. Their anterior ends are interconnected by an annular shelf-like structure – like a miniature iris – the 'sims'. This German term has never been translated; perhaps it is high time that it was. Since the sims connects the ciliary processes, which give the ciliary body its name (cilia = hairs or threads), after the fashion of the webbing which connects the toes of a duck or a frog, it will be called here the 'ciliary web' (Fig. 194b, *cw*).

The ciliary web is a decidedly mammalian character, shared by many marsupials and placentals but by no sauropsidans. Every other feature of the monotreme eyeball – whether the feature is a structure, or the absence of a structure – occurs in some living reptilian group. The ciliary web alone[4] thus keeps the eye of the monotreme from being entirely reptilian, with its closest morphological resemblance to the eye of the likewise nocturnal crocodilian.

The lens is unexpectedly small, flat and anterior in position. The topography of the monotreme anterior segment, particularly in the echidnas, is in fact not at all sauropsidan but more like that of the sirenians and primates. *Tachyglossus* has the flattest of all lenses, with a flatness index (diameter divided by thickness) of 2.75.[5] This value is closely approximated elsewhere only in some of the higher primates, including man (ca. 2.7). At its equator, the lens epithelium is twice as tall as at the anterior pole, constituting perhaps a vestigial ringwulst. A similar situation obtains in the duck-bill, and also in some marsupials. The lens of the duck-bill, in keeping with the acquatic habit, is much less flat – 2.66/1.93 = 1.38 (Kahmann, 1930), 2.45/1.75 = 1.4 (Gunn, 1884), or 1.5 (from a photograph of O'Day's – scale not given). O'Day (1939) compares its form with that of the lens of the local Murray turtle, *Chelodina longicollis*.

No monotreme has any demonstrable accommodation, and there are no reports as to refractive conditions. It is not known whether *Ornithorhynchus* approaches emmetropia in either air or water, but the implications are that the eye is better adjusted to the latter medium. The echidna eye looks as though it must be extremely hypermetropic; but only a study of the living animal can settle the matter.

In both *Ornithorhynchus* and *Tachyglossus* the numerous zonule fibres arise from the coronal zone of the ciliary body and from the free portions of the ciliary web (including its very edge), and insert compactly on the extreme periphery of the lens, largely just in front of its equator.

The Monotreme Retina

The rather thin sensory retina extends farther forward temporally than nasally in *Ornithorhynchus* (but not in *Tachyglossus*?), suggesting an importance of the binocular

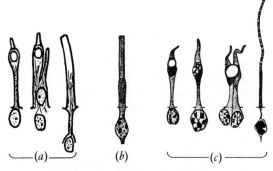

Fig. 195 *The visual cells of the lower mammals (× 1000). (a) Single cone, double cone, and rod of* Ornithorhynchus. *After O'Day (1938). (b) Element from pure-rod retina of* Tachyglossus. *Drawn from a preparation of O'Day. (c) Droplet-bearing and droplet-free single cones, double cone, and rod of an opossum.* Marmosa mexicana *(Australian marsupials have no droplet-free cones and have droplets in both members of their double cones).*

field. There are no blood vessels either in the retinal tissue (as in a few marsupials and many placentals) or lying on its surface like the hyaloid or vitreal systems of lower vertebrates. No monotreme has any trace of a conus papillaris. This complete nutritional dependence of the retina upon the chorioid is characteristic of light-shunning vertebrates. The disc is small, smooth, and unpigmented in both genera, circular in *Ornithorhynchus* and vertically oval in *Tachyglossus*. Kolmer (1925) describes a peculiar mass of connective tissue which is embedded in the bulbar portion of the optic nerve in *Zaglossus*.

Not only in its avascularity, but in its entire histology, the monotreme retina is sauropsidan and might easily be taken for that of a nocturnal reptile. In *Ornithorhynchus*, O'Day figures three rows of outer nuclei, four of inner, and a single row of ganglion cells, and says that the nerve-fibre layer is thin even near the disc. *Tachyglossus*, which is pure rod, has three layers of outer nuclei (*Zaglossus* has four), only two of inner (*Zaglossus* has three), and a decidedly scattered single row of ganglion cells. Some of the latter are ectopic and lie at various levels in the inner plexiform layer. The greater extent of summation in the echidnas, and the total absence of cones,[6] implies a stricter nocturnality than that of the duck-bill; but no great difference in habits seems to have been noted.

The types of visual cells are direct derivatives of those of the Sauropsida (Fig. 195a,b and see Plate I). In *Ornithorhynchus* the single and double cones have lost the paraboloid but have retained the oil-droplet, which was very recently found to be colourless. The rod and cone nuclei are not differentiated, but are both 'cone-like' as in all sauropsidans excepting nocturnal birds. In *Tachyglossus* the cones themselves have gone. The complete monotreme visual-cell pattern (of *Ornithorhynchus*) fits equally well the accepted idea that the monotremes are a lateral branch of the stock which culminated in the marsupials, and the minority notion that the monotremes evolved independently from reptiles. The simplification of the cones in the duck-bill, and their discard in the echidnas, are natural consequences of adaptation for dim-light activity.

The Marsupial Eye

Marsupials have a nictitating membrane, but it is never highly developed. Its gland (the Harderian) is present, along with the lacrimal. A retractor bulbi is present; but no details are on record concerning the extra-ocular muscles.

The eyeball is perfectly spherical in a very few species and is practically spherical in all others. The horizontal and vertical diameters are always equal, and usually exceed the axial length (by up to 10%). This relationship is reversed in some opossums. The topography of a sagittal section is always like that in nocturnal and arrhythmic placentals (Fig. 196a). The diameter of the cornea is always great in proportion to the diameter of the eyeball – 66–80% in kangaroos, 82% and 87% in opossums (*Didelphis virginiana* and *Marmosa mexicana* respectively), 91% in the cuscus (*Trichosurus vulpecula*). The cornea is horizontally ovoid only in large kangaroos, in simulation of their ungulate counterparts.

The sclera is fibrous, entirely devoid of cartilage (except for some questionable nodules in the marsupial 'golden mole', *Notoryctes*). It has thus taken the final step in the elimination of the cartilage-and-bone system of the reptilian eyeball wall, and the basically spherical form of the marsupial eyeball is the expression of this elimination (*cf.* snakes). Wherever in higher (i.e. placental) mammals the eye departs from this fundamental sphericity and gains the appearance of having a circumcorneal scleral sulcus, it is owing to the production of a cornea whose radius of curvature is substantially less than that of the sclera (e.g., man).

The cornea has a 4–5 layered epithelium, and no Bowman's membrane; but Descemet's membrane is ordinarily very thick (not, however, in *Marmosa*). The cornea is usually uniform in thickness throughout, but is thinned peripherally in opossums and perhaps in other small-eyed forms.

The chorioid is usually about as thick as the sclera – hence, thin in small eyes and thick in large ones. It is heavily pigmented, richly vascular, and ordinarily is built quite as in the placentals. In *Didelphis*, however, a choriocapillaris is present only in the pouch young. During growth to adulthood, *pari passu* with the maturation of the tapetum formed from the retinal pigment epithelium and the unusual invasion of the outer nuclear layer by retinal capillaries, the choriocapillaris is replaced by (or becomes) a plexus of plump, thin-walled veins which occupy the same position against the back of the glass membrane. These alterations bespeak a turning of the visual cells from the chorioid to the retinal circulation as their source of supplies, owing to the impermeability of the tapetum. In a very few other marsupials (*Dasyurus, Thylacinus*, possibly *Sarcophilus* and *Petaurus*) the chorioid is modified by the presence of a tapetum fibrosum. In *Dasyurus viverrinus* this nearly fills the chorioid – squeezing the large vessels and the few thin, pigmented lamellae out against the sclera – and runs practically from ora to ora, though permitted to reflect light back through the retina only in the superior half of the eyeground, where the retinal pigment epithelium is devoid of pigment. It is probably significant that it is only in *Dasyurus, Sarcophilus, Didelphis* and *Marmosa* (which may once have had a tapetum like its close relative *Didelphis*) that any retinal vessels are known to occur – necessitated, apparently, by the interference of the tapetum with the nourishment of the retina by the chorioid. A vestigial conus papillaris may also occur in marsupials (*Perameles, Hypsiprymnus,* and kangaroos generally). The supply of the retinal capillary bed, where present, is from paired veins and arteries which radiate over the retina from the disc. In *Dasyurus viverrinus* each of these veins (and their larger branches) is triangular in cross section and is embedded in the inner layers of the retina, with its round arterial companion lying on top of it in a low glial ridge which projects a trifle into the vitreous.

The iris contains an unstriated sphincter near the pupil margin (as in other mammals); but no marsupial is known to have a dilatator. Both retinal layers are therefore heavily pigmented. The stroma is likewise densely pigmented and is richly vascularized, often with many vessels partially extruded from its anterior surface.

In large eyes the ciliary body forms a broad zone with well-marked orbicular and coronal regions; but in small eyes, whose lenses are enormous, the ciliary body is reduced about as it is in snakes. In large eyes the ciliary processes are regular, tall and thin (Fig. 196b); and they are about as numerous as in comparable placentals (e.g., 120 in *Trichosurus*). Those of small eyes are low, tortuous, and not so readily counted. A ciliary web can usually be made out. Though no marsupial has yet been demonstrated to have any accommodation whatever, a small ciliary muscle is always present. This may present itself as a meridional Brücke's muscle with exactly the same relationship to the corneal margin as in reptiles (Fig. 196c, *cm*). More often, apparently, it contain both circular and meridional fibres. The circular ones occupy the anterior half of the muscle in *Dasyurus* and *Marmosa*, and in the former genus they lie toward the scleral side of the muscle. In *Didelphis* the meridional and circular fibres are intermingled in small bundles. The anterior tendon in these three

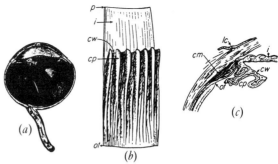

Fig. 196 *The marsupial eye. (a) Ventral half of left eyeball of a kangaroo,* Macropus giganteus *(× 1) After Soemmerring. (b) Inner surface of segment of anterior uvea of a kangaroo,* Macropus agilis. *After Franz. (c) Iris-angle region of cuscus.* Trichosurus vulpecula. *Redrawn from Franz.* cm – *ciliary muscle;* cp – *ciliary process;* cw – *ciliary web;* i – *iris;* lc – *limbus corneæ;* ot – *ora terminalis retinæ;* p – *pupil margin.*

genera, and probably in many others, is formed by a small mass of unpigmented uveal 'meshwork' tissue which lies against the anteriormost part of the sclera and extends forward beyond the iris root, where it tapers to meet the edge of the membrane of Descemet. Between this tissue and the sclera lies the endothelial canal of Schlemm. These relationships are essentially those of the human eye. Within the marsupial group, then, the ciliary muscle may be situated either as it is in reptiles, or as it is in placental mammals. The transition seems to be made simply by the creation of the meshwork 'tendon', dropping the anterior end of the muscle farther back from the lumbus corneæ.

The lens is always relatively large; and in the smaller, more strongly nocturnal types it may nearly fill the globe – as in many small-eyed nocturnal placentals, e.g., *Mus*. The flatter lenses occur, as would be expected, in the large-eyed arrhythmic forms (Fig. 196a). Even in the flattest lenses there are never more than traces of the reptilian ringwulst, and the lens is always quite out of contact with the ciliary processes. The flatness index of the lens may be little more than 1.0, or as high (in kangaroos) as 1.5. Some sample values are given in Table 1.

The Marsupial Retina

Through its loss of all the accessory structures involved in sauropsidan accommodation (except the ciliary processes – and these no longer bear upon the lens), the marsupial eye as a whole is thoroughly mammalian – i.e., placentalian. The retina, however, is as reptilian as that of the monotreme. The visual-cell types are the same ones as in *Ornithorhynchus* – single and double droplet-bearing cones, and filamentous rods (which always outnumber the cones very greatly, in contrast to *Ornithorhynchus*). Only minor modifications of the full monotreme pattern occur

Table 1. *Sample values of the flatness index of the lens.*

Species	Horiz. ϕ of eye (mm)	Lens ϕ D (mm)	Lens thickness T (mm)	Index D/T
Marmosa mexicana (a mouse opossum)	6.3	4.7	4.5	1.05
Didelphis virginiana (the common opossum)	11.0	7.3	6.0	1.22
Dendrolagus bennetti (a tree kangaroo)	15.0	8.3	6.7	1.24
Osphranter (robustus?) (a rock dwelling kangaroo)		13.0	10.0	1.30
Macropus giganteus (a ground kangaroo)	27.0	13.0	10.0	1.30

in marsupials. Thus in the American opossums some of the single cones lack oil-droplets (see Fig. 195c); and in all Australian marsupials so far examined by O'Day (1938), the double cones have oil-droplets in both their members – a curious situation which occurs in American marsupials (and in some birds) only as an occasional anomaly.[7].

The rod nuclei in marsupials contain only one or two chunks of chromatin – a differentiation, from the larger and open nuclei of the cones, which is characteristic of the placentals but not of the monotremes. It is not known whether the rod and cone foot pieces are also differentiated in marsupials. The cones of marsupials, like those of all other mammals, lack paraboloids. This seems a point of some value in defence of the monophyletic derivation of all the mammals from a single reptilian stock.

The single and double cones of marsupials and monotremes, from their oil-droplets (and despite their loss of the paraboloids), are clearly the 'same' elements as the corresponding ones of the reptiles. The monotreme-marsupial rod is left to be homologized with the droplet-free element of the sauropsidans (see Plate I). Its increased (nuclear) differentiation in the marsupials, over that in the monotremes, coupled with the persistence of the useless oil-droplets in both groups (these are gone entirely in the placentals!), suggests that the ancestral monotremes and the original marsupials were diurnal, and that the monotreme-marsupial line acquired its rods secondarily through transmutation and perfected them within the confines of the line (*cf.* reptiles, birds).

Placentals

The earliest placentals were 'insectivores' of the *Deltatheridium* type. In the Mesozoic, these primitive forms diversified and established several separate lines of ascent. The insectivore type itself persisted (giving off the still extant Lipotyphla and, later, forms ancestral to the whales) and culminated in the 'creodonts' – archaic carnivorous forms, of which the modern orders Carnivora and Pinnipedia (seals) are fairly direct descendants. From precreodonts, there diverged a line which produced the artiodactyl 'ungulates'[8] (peccaries, pigs, hippopotami, tylopods [camels etc.], deer, antelopes, cattle).

The Lipotyphla (hedgehogs, tenrecs, otter-shrews, shrews, moles, golden moles etc.) comprise the larger of the two groups of *living* insectivores. From this stock diverged the smaller branch called the Menotyphla, the living members of which comprise the tree-shrews and elephant-shrews. Along the way, the Menotyphla gave off a branch which bifurcated into the Dermoptera (taguans or 'flying lemurs' – *Galeopithecus* and *Galeopterus*) and the Chiroptera (bats). The order Primates also branched off from menotyphlous stock, close to the tree-shrews; and the latter, like the higher primates, have secondarily become diurnal – perhaps the most primitive placentals to have done so.

From a second of the groups of Mesozoic insectivore-like forms, the rodents and lagomorphs arose. In their highest specializations, the rodents have risen above some other groups whose origins were not as ancient.

A third assemblage of Mesozoic forms gave rise to the modern Xenarthra, comprising the sloths, armadillos and ant-bears. To these American forms the African pangolins or 'scaly ant-eaters', the Nomarthra may be closely related. The Xenarthra and Nomarthra, if they are thus related, form a natural order, the Edentata; otherwise the Nomarthra, deserve ordinal rank. To the 'edentates' in a former, looser sense, the Tubulidentata (now considered quite unrelated) were once assigned.

The Tubulidentata, represented today only by the aardvarks (*Orycteropus* spp.), the hyracoid-proboscidean-sirenian bouquet, and the perissodactyl ungulates (horses, zebras, tapirs, rhinoceroses) are all derivates of pro-ungulates which flourished in Cretaceous time and radiated from still a fourth branch of the Mesozoic radiation of insectivore-derivatives.

The Eye as a Whole

In so diversified a group of vertebrates – in contrast to the birds – the eye naturally exhibits a profuse adaptive radiation paralleling that of the group itself. The placental-mammalian eye has been carried along the ground – rapidly or slowly – and into trees, into the free air, into the fresh waters, and a mile below the surface of the ocean. It has been required to work in brightest sunlight and faintest starlight. It has been asked to inform its owner of an enemy miles away, and to analyse a tiny object held close before the face. The placental eye has been able to cope with all of these situations. Only in complete and permanent lightlessness has it given up, and shrivelled to a pinhead hidden beneath the skin. This sort of degeneration has occurred several times – in two distinct families of lipotyphlous insectivores, the true, talpid moles (*Talpa, Scalopus*, etc). and the golden moles (*Chrysochloris* spp.); in two families of rodent 'moles', the Spalacidæ and the Bathyergidæ; and in one additional genus of rodent (*Ellobius*) which belongs to the hamster branch of the mouse family. There remains a great deal which could be said about placentalian eyes, not much of which can be squeezed into the space allotted here. For detailed anatomical information the reader will have to turn to such compendia as that of Franz (1934), and to the works cited therein.

Functional, harmonious, placentalian eyes range in size from about a millimetre in the shrews and the smallest bats to that of the great blue whale, *Balænoptera musculus* ($145 \times 129 \times 107$ mm). Carnivores, diurnal primates, and ungulates have the largest eyes relative to body size. In the lowest orders (Insectivora, Chiroptera, Edentata, Rodentia) the eye is both relatively and absolutely small, in sympathy with the nocturnality of these animals and the unimportance of vision in their lives. *Orycteropus* however has a large eye (22×22 mm, 20.5 mm axis), which aligns this form with its ungulate relatives).

The basic shape of the eyeball is the sphere; but a horizontal ellipsoidality, at maximum about as great as it ever is in birds, occurs in some ungulates and in many large-eyed aquatic forms. The cornea may protrude from the sphere formed by the rest of the globe when it is small and sharply curved throughout (e.g., man), or its apex may be acutely curved even though the rest of the cornea blends with the curvature of the sclera (carnivores). The axis is somewhat shortened in many ungulates, in which the lens has been moved forward, and also in the more fish-like aquatic eyeballs. In *Galago* and *Tarsius*, and to a lesser extent in some other nocturnal prosimians (e.g., *Nycticebus*), the eye is 'tubular'. In large-eyed mammals, it is common for the lens and cornea to be shifted nasally as in birds, and for the ciliary body to be consequently narrower nasally than temporally, chiefly at the expense of the nasal orbiculus, which may be quite abolished.

The sclera never contains any traces of cartilage. It is usually thickest in the fundus and thinnest at the equator; but the cornea may be much thicker than any part of the sclera, or much thinner – the local differences of the thickness of the fibrous tunic are so various that they cannot be covered in a few words. A Bowman's membrane is seldom discriminable; but Descemet's layers are always present, and the elastic membrane may be enormously thick in large eyes. An exceptional cornea is that of the armadillo (*Dasypus novemcinctus*), in which the substantia propria contains many capillaries, even at the apex. These are perhaps required by the fact that the corneal epithelium, being strongly keratinized, can derive no sustenance from the tear fluid.

Except where a tapetum lucidum has been produced in it, the chorioid is usually built as it is in man, but is seldom so thick. It is exceptionally thin in the squirrel family; but the most unusual chorioid is that of the large bats (Megachiroptera). In these forms there are 20 000–30 000 conical, vascular papillæ which are protrusions of the chorioid, interdigitated with the retina and deforming the latter's visual-cell layer. Kolmer found this situation in all 16 of the species he studied, but not in any of an equal number of microchiropteran species. Five structural types of papillæ can be recognized; and more than one type may occur in one species, in different retinal regions.

The iris in large eyes (carnivores and seals, ungulates, whales, primates) has essentially the same constitutents as in man. All of these mammals have a dilatator, histologically and embryologically resembling that of man, but with a topographical arrangement which depends upon the shape of the contracted pupil. In well-adapted aquatic placentals (otters, seals, whales), and also in the pigs, the sphincter occupies the entire width of the iris; and the dilatator may send fibres into the ciliary body for firmer anchorage. In the smaller nocturnally-adapted eyes of all the lower orders of placentals, a dilatator is ordinarily lacking; but the sphincter is always in evidence and sometimes very large, though always compactly massed near the pupil margin. Towards the root of the iris, stromal strands may cross the filtration angle and join to the cornea, thus contributing to the 'pectinate ligament' (or sometimes forming the whole of it). At the other 'end' of the iris – the pupil margin – cystic protrusions of the pigmented retinal layers form the corpora nigra or 'grape-seed bodies' which are characteristic of the highest artiodactyls (tylopods, and ruminants except *Tragulus*) and also of the highest perissodactyls (horses). Where a dilatator is present, the anterior retinal layer is pigmented only slightly or not at all; otherwise, it is as dark as the posterior layer, as in lower vertebrates in which it has not partly differentiated into muscle. The colour of the iris is usually dark brown. Where it is not, the colour is generally optical, as in the 'blue' human eye; but lipophores and iridocytes may be present in the stroma, as in the cats and some prosimians.

The organization of the ciliary body in all placentals is basically the same as has been described earlier for man. A corona (bearing true, vascular, ciliary processes) and an orbiculus (smooth, or bearing only low meridional ridges) can usually be distinguished. In carnivores, however, the posterior ends of the processes are practically at the ora (Fig. 197); and in *Orycteropus* and ungulates, whose corneas are markedly ovoid horizontally, the obligation of the coronal zone to remain circular (to 'fit' the lens) results in an encroachment upon the iris, nasally and temporally, by the anterior ends of the processes – so that these portions of the iris serve as extensions of the base-plate of the ciliary body, and are rendered immobile as regards changes involved in the operation of the pupil. In ungulates and in many carnivores the orbiculus is practically eliminated nasally owing to the existence of marked nasad asymmetry.

Except in very small eyes, the main part of the ciliary body (apart from the processes, that is) gradually thickens toward its anterior end, as in man. This bulk of uveal tissue is not, however, solid muscle as in the primate eye. Muscle – sometimes considerable of it (*sic*), as in carnivores – is almost always present, but is in the form of slender fascicles interspersed with much connective tissue. Anteriorly, the ciliary muscle tends to have two anchorages: one, by means of the meshwork tissue which terminates at the margin of Descemet's membrane, and another attachment into the anteriormost portion of the base plate, practically in the root of the iris. Between these two anterior leaves of the muscle lies a nearly empty space, best visualized by imagining the human filtration angle to be eroded or extended backward deep into the ciliary body. This space, 'Fontana's space(s)', is traversed by delicate strands of uveal tissue which join the base-plate to the sclera. The anterior limit of Fontana's space – its boundary with the anterior chamber (with which it is of course actually continuous, between the strands) – is fixed by the strands or struts which make up the true pectinate ligament: These are heavy connective-tissue fibres, coated with mesothelium, which run from the limbal region of the fibrous tunic to the root of the iris, and support the latter against the tug of the part of the ciliary muscle attached thereto, and the pull of the sphincter during the partial closure of the pupil which ordinarily occurs during accommodation. The pectinate ligament gets its name from the word 'pecten', meaning 'a comb', and referring to the fact that its strands are like the teeth of a comb which has been bent into a circle with the teeth pointing inward. The strands are best developed in horses, artiodactyls, *Orycteropus*, carnivores, and especially in seals (where there may be not one 'tooth' but several in a given meridian, forming a fan, somewhat like the situation in reptiles. In the horse at least, they appear to be continuous with and identical with the material of Descemet's membrane; and the horse has very similar fibres, with a *circumferential* course, massed anteriorly in the meshwork of the iris angle.

In small, large-lensed eyes with very extensive corneas (in murids and similar rodents, armadillos, etc.) the whole ciliary body is reduced greatly and occupies a relatively narrow zone – sometimes, as in shrews, forming a simple roll without meridional folds or ridges, quite as in the snakes. The uveal meshwork tissue, covering the canal of Schlemm and tapering to meet Descemet's membrane, which so often serves as a tendon of the ciliary muscle, is still present in these eyes; but the ciliary muscle is usually wholly lacking. Fontana's space is either tiny, or else is confluent with the anterior chamber owing to the absence of a pectinate ligament (as also in some large eyes, e.g., the human). Such eyes have no accommodation; and for that matter none has ever been convincingly demonstrated for ungulates – domestic ones, at any rate – despite the presence of considerable tissue of supposedly contractile character. In these small eyes, the ciliary processes are so blobby and irregular that they can scarcely be counted. A very different situation exists in large placental eyes.

The ciliary processes in large eyes vary in number with the general size of the eye, as in birds – actually, with the size of the cornea, since it is this which the ciliary body must be thought of as surrounding. They number about 50–100 in carnivores, 60–100 in seals, 90–130 in ungulates and whales, and up to 135 in large-eyed rodents and lagomorphs (hares, beavers). A ciliary web is often present (see Fig. 194b); and, in a vestigial condition, it can be made out in man. The tips of the ciliary processes touch the lens in a number of mammals, but they are never fused with it and probably never exert any effective pressure on it in the few mammals which have useful accommodation (primates, squirrels, large carnivores). The mechanics of mammalian accommodation are entirely unlike those of the sauropsidan process, and the difference may be wholly ascribed to the fact that the primitive mammals allowed a 'circumlental space' to be opened up between the ciliary processes and the lens, when they threw away the ossicular ring of their reptilian forebears.

Two chief types of processes are distinguished. The more primitive type is puffy and rugose, like that in monotremes (Fig. 194b). This type occurs in all of the lower orders and also in some of the highest – the artiodactyls and perissodactyls, for example. A more specialized type, whose differences from the other have no known functional significance, is the thin, smooth-surfaced, 'knife-blade-like' process seen in most carnivores and pinnipeds and in some primates (Fig. 197a). This type has also been evolved by the higher marsupials (Fig. 196b).

The two kinds of ciliary processes are associated with fundamental differences in the organization of the zonule which are perhaps related to the extent of accommodation. In forms with thick processes, some of the zonule fibres arising from the inner surface of the base-plate run for a space along the floors of the valleys between the processes, and others run alongside the faces of the processes. As all these fibres curve out toward the lens, they are quite uniformly distributed both in the aspect of a sagittal section and in the view of the zonule obtained by removing the cornea and iris from a gross specimen. The attachments of the fibres to the periphery of the lens are uniformly distributed both circumferentially and meridionally (Fig. 197d). One cannot speak here of anterior and posterior 'leaves' of the zonule, for there is no canal of Hannover.

The greatest contrast with the situation in seen in the carnivores, as exemplified by the domestic cat, studied by Kahmann (1930). Here the ciliary processes are knife-like, and between every two major ones there is a low secondary process. Zonule fibres arising from the orbiculus segregate into paired bundles as they enter the ciliary valleys, and those in each bundle form a fan plastered against the face of a major process – one fan against each face. These fibres insert anterior to the lens equator (Fig. 197c). Other fibres arise from the ciliary epithelium alongside the roots of the major processes, and pass along their faces and across the circumlental space to insert posterior to the lens equator. Again there is no canal of Hannover; but the insertions of the fibres are not uniformly scattered around the lens, but are grouped at a number of discrete places – twice the number of the major processes. There is thus an even more free communication (between the bundles) from the anterior chamber to the posterior than in the case of ungulates, etc. The anterior surface of the vitreous is plicated where it bulges forward a bit into each ciliary valley, and its pressure against the posterior zonule fibres keeps them bowed; but the anterior fibres take straight courses. The periphery of the lens is scalloped by the discontinuous attachment of the zonule.

The minor processes also have sheets of zonule fibres against their flat surfaces. These arise perpendicularly from the anterior part of the ciliary body and pass to the posterior insertion zone on the lens. The insertions are in meridians intermediate between those of the major fans (Fig. 197b). All zonule fibres thus lie against ciliary-process surfaces. A frontal section through the ciliary body shows no fibre cross-sections on the floors of the valleys or in the open spaces of the valleys themselves – a great contrast with the ungulates and lower placentals.

According to Kahmann (1930), the primate zonule exhibits still another fundamental plan: fibres from the greater part of the orbiculus pass only to the anterior lens capsule (forming the anterior leaf), and others from near the posterior ends of the ciliary processes pass only to the posterior capsule, forming a posterior leaf. These two masses of fibres thus cross each other in the coronal zone (the writer is not at all convinced of this). A few of the fibres with orbicular origins insert at the lens equator, and thus travel across the otherwise 'empty' canal of Hannover. Still other fibres, originating far anteriorly, pass to the posterior capsule and thus compare with the 'perpendicular' fans of the cat; but in man there are no regular minor ciliary processes for them to cling to. From these descriptions, it will be seen that where accommodation is considerable (carnivores; and, to a much greater extent, primates[9]), the zonule fibres which will be most relaxed by the contraction of the ciliary muscle are delegated to the anterior surface of the lens, which exhibits the most elastic deformation. Other fibres, the perpendicular ones, seem to have been oriented favourably to serve as check ligaments, keeping minimal the change in curvature of the posterior surface of the lens.

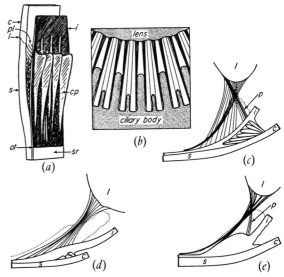

Fig. 197 *The ciliary region in placental mammals. (a) Portion of anterior segment of a carnivore,* Felis lybica. *Redrawn, modified, from Franz.* c – *cornea;* cp – *ciliary process;* i – *iris;* l – *limbus corneæ;* ot – *ora terminalis retinæ;* pl – *pectinate ligament (three fibres show);* s – *sclera;* sr – *sensory retina. (b) Diagrammatic thick frontal section in a carnivore (cat), showing paired bundles of zonule fibres against the faces of the major ciliary processes, and the smaller 'perpendicular' bundles associated with the minor processes. Based upon a photomicrograph of Kahmann (1930). (c) Carnivore (cat); (d) Ungulate (pig); (e) Primate (man). (c), (d) and (e) are combined from figures of Kahmann (1930). Dotted line shows profile of major ciliary process.* c – *cornea;* i – *iris;* l – *lens;* p – *perpendicular fans (see text);* s – *sclera.*

The lens is nowhere so flat as in man and other diurnal primates. It is perfectly spherical only in seals and in some toothed whales; but it is very nearly so in many murid rodents and in a few other small-eyed, nocturnal, lower placentals. In carnivores and ungulates it is variously intermediate in shape; and its relative size is always related to the habits of the animal with respect to light.

The Retina

The pigment epithelium usually contains relatively little pigment, which is never migratory. It may contain reflective material and serve as a tapetum lucidum in itself (opossum, Megachiroptera), or in aid of a chorioidal tapetum (dog). Not all fruit-bats have the reflective substance – it is lacking in most species of *Pteropus*, but is abundant in *Pteropus h. condorensis, Hypsignathus, Cynopterus* and *Epomophorus*. It is not apparent whether these differences relate to differences in the strictness of nocturnality of the various genera.

Usually the placentalian retina is described as being, typically, vascularized. Actually, retinal vessels, with capillary branches passing out ordinarily as far as the outer plexiform layer, are numerous only in primates, sciurids, carnivores and artiodactyls – all, characteristically diurnal or arrhythmic. In the primitive ruminant *Tragulus* (the mouse-deer or chevrotain), there are only superficial vessels, like the hyaloid or vitreal vessels of ichthyopsidans and snakes. In most perissodactyls there are no vessels, and in the horse they are restricted to a 6 mm circle concentric with the disc. There are but few vessels in lagomorphs, associated there with the horizontal band of medullated nerve fibres; and there are few or none in the various rodents outside the Sciuridæ. There are no vessels in the Xenarthra, or in the Chiroptera except for a few superficial capillaries in *Pipistrellus pipistrellus*. Retinal vessels thus seem to have arisen several times, independently, in those placental mammals with the most cones in their retinæ; and certain embryological differences appear to bear out this conclusion. In murids, for example, the few adult retinal vessels are formed directly by the embryonic vasa hyaloidea propria, whereas in primates these atrophy, and the definitive vessels bud out from the central retinal artery and vein in the optic-nerve head.

The lamination and the laminal purity of the placentalian retina are only ordinary, and quite well exemplified by the human retina. Only in the diurnal squirrels and particularly in the prairie-dogs (*Cynomys* spp.), and there only in the dorsal region, does the mammalian retina approach that of the birds in the segregation of inner-nuclear elements and in the stratification of the inner plexiform layer.

Most placentalian groups have duplex (rod-and-cone) retinæ, but in the lowest orders it seems to be the rule for only rods to be present. The cones are the simplest imaginable – all single, without paraboloids, oil-droplets, or myoid extensibility. There are no cones in the armadillo, possibly none in any edentate. All of the bats have only rods. Among the Lipotyphla, the hedgehogs are pure-rod according to most investigators, though Menner found a few cone-type nuclei. The tree-shrews (*Tupaia et al.*) should have many cones; but the shrews have few or none.[10] There are probably many more pure-rod rodents besides the guinea-pig, and no cones have been reported by modern investigators for any prosimian below the true lemurs,[11] or by any of the half-dozen investigators who have studied various species of *Aotus*.

The rodents characteristically have great numbers of excessively slender rods, like those of the rodent-like opossums. Slender rods are also the rule in nocturnal primates and carnivores, and in the fruit-bats. In all such forms the outer nuclear layer is naturally very thick, with up to 16–17 rows of nuclei. The inner nuclear layer in placentals rarely contains more than four or five rows, except in *Tupaia* and in diurnal squirrels, where it may be several times as thick as the outer. The ganglion cells usually form but one layer (in which they are often widely separated), except in an area centralis (where any) and in the neighbourhood of the primate fovea.

The more slender and numerous the rods, and the fewer the cones, the more likely it is that the rod and cone nuclei will be found markedly differentiated from each other in size, shape, and chromatin distribution.

The cones of most of the placentals which many of them are much like those of man as a rule. In flying-squirrels and ungulates, however, their 'myoid' regions are more or less elongated; and in diurnal squirrels (except prairie-dogs) there appear to be two types of single cones, one bulky proximally and slender distally, the other slender proximally and plumper distally. In prairie-dogs, however the cones are all alike, very slender, and not thus pseudostratified. Pure-cone retinæ are unknown in mammals outside of the Sciuridæ – none occur even in primates, though some of these (e.g., *Callithrix jacchus, Cercoebus torquatus*) do have many more cones than man. Favourable material of *Tupaia* has never been studied; and there are still other mammals outside the squirrel and monkey tribes which are reputedly strongly diurnal, and whose retinæ would bear investigation: *Ochotona, Dolichotis, Procavia, Cynictis, Suricata, et al.*

The Early History of the Placentalian Eye

The simplicity of the placentalian visual-cell pattern is striking, when one considers that in the lower mammals each of the standard reptilian-avian cell types is easily recognizable. No placental is known to have double cones, or oil-droplets in its single ones.[12] Obviously the whole sub-class must have been pulled through some sort of ancestral knot-hole: the 'original' placental mammal must have had a way of life which brought about these peculiarities and doomed all of its descendants to exhibit them.

The whole organization of the monotreme eye is, as we have seen, reptilian. If we think of it as a reptilian eye, its oddities seem logical consequences of a strong nocturnality of long standing. The reversion of the intra-ocular muscles from a striated to an unstriated condition shows that in the first mammals accommodation became unimportant, and it was never necessary for them to close the pupil quickly – presumably because they never exposed themselves to bright light. Accommodation is of no value to a strongly nocturnal eye – especially one which, though perhaps relatively large for the animal, is small in absolute dimensions. The discard of the scleral ossicles and the practical discard of the ringwulst of the lens allowed the monotreme eye to become rotund, took the ciliary body out of contact with the lens, and made forever impossible any return to the sauropsidan method of accommodation. Though the persistence of the retinal oil-droplets suggests that the early monotremes may have been sufficiently diurnal to have retained the reptilian eye quite unchanged, the nocturnality which eventually supervened accounts for the condition of the modern monotreme organ.

The marsupial eye, though secondarily arrhythmic in capacity in its highest expression (in ground kangaroos), bears the very same stigmata of a former universal nocturnality – perhaps even more complete than that of any monotremes, living or dead, since even the scleral cartilage has not been kept. In the opossums, which are the most archaic of living marsupials and hence, so to say, have had the most time in which to get rid of useless structures, some of the single cones have lost their oil-droplets.

The placental mammals must have gone farther in adaptation for dim-light activity, early in their history, than the marsupials have ever done. Their eyes are in fact best understood not by comparison with those of the lower mammals, but by comparison with those of the snakes. The early snakes so completely lost the reptilian assortment of special ocular structures that when the snake eye was rebuilt, upon the return of the snakes to the earth's surface, it ended up as a spherical organ with an entirely fibrous wall, with the lens and ciliary body out of contact (necessitating a new and special method of accommodation), with a wholly new set of visual cells, and (eventually) a yellow lens as a substitute for the ancestral diurnal-lacertilian yellow oil-droplets. To a degree, the placentalian eye incorporates equivalent changes and substitutions. The 'original' placentalian eye was of course not really degenerate like that of a mole or mole-rat, but it did take several steps down the same path which the eye of the incipient snake followed to its bitter end.

Whether the placentals evolved directly from nocturnal marsupials, or turned nocturnal after a derivation from diurnal common ancestors of the modern marsupials and the placentals, we cannot know; nor would the knowledge have much importance. We can be sure that at an early period in placentalian evolution, the only placentals on earth were so thoroughly nocturnal that their eyes had no stiffening structures to keep them from being spherical, had large pupils and large, simple lenses with no trace of a ringwulst and no contact with the ciliary body, had rudimented intra-ocular muscles which were unstriated and did not include a dilatator pupillæ, and had no accommodation whatever.

Now, what was the retina like in these strictly nocturnal, 'bottle-neck' insectivores? Apparently all of the lowest living orders of placental mammals have pure-rod retinæ. But the higher ones have both rods and cones. Do the cones of the higher placentals represent sauropsidan-monotreme-marsupial cones which squeezed through the primitive insectivoran knot-hole, or are they somehow new?

Placentalian cones are all alike in certain respects: they are all only single, without paraboloids, and without oil-droplets. These similarities are negative, and really mean that placental cones are cones reduced to their lowest structural terms. Naturally they would be alike, even if those of the tree-shrews, the higher primates, the duplex descendants of the pre-creodonts (i.e., carnivores, artiodactyls), and the duplex descendants of the Cretaceous pro-ungulates (i.e., hyracoids, proboscideans, perissodactyls) all represent independent productions of new cones in erstwhile pure-rod retinæ.

The absence of the paraboloid in placentals is no proof of an identity of placental cones with those of monotremes and marsupials. The latter groups have lost the paraboloid, to be sure; but the cones of placentals would not be expected to have evolved them even if those cones are 'new'. Paraboloids occur only in the cones of groups which have retinal photomechanical changes, and the paraboloid has been claimed to be a reserve of food which furnishes the energy for the activity of the cone myoid. The cones of lampreys and elasmobranchs naturally have never produced them, nor have the cones of snakes which are certainly 'new' cones.

If the placentalian cone represents the reptilian droplet-bearing single cone, then one can understand its lack of the oil-droplet (*cf.* opossums); but what has become of the reptilian double cone, so stubbornly persisting in even the most strongly nocturnal of the lower mammals except where all cones have been lost (*Tachyglossus*)? Elsewhere above the fishes,[13] double cones have never been either discarded, or transmuted into rods, without the matching single cones also undergoing discard or transmutation.

It seems highly significant that the placentalian cone

has no consequential capacity for colour vision except in the primates, where colour vision has evolved within the group. If the duplex placental mammals had had continuously duplex retinæ ever since the placentals originated, then all such mammals, and not the simians alone, should have as complete a colour-vision system as that which characterizes the Sauropsida; for, they should have retained that same system – having retained the same cones.

All in all, it seems most probable that at one time the only living placentals had no cones, but only the rods which we see in the lower mammals, and that subsequent placentals evolved duplex retinæ from pure-rod ones just as the Boidæ or their immediate ancestors had to do (see Plate I). The eye of man, with its pretty good accommodation, its fovea, its miscellaneous yellow filters, and its capacity for colour vision, possesses in substantial degree and physiological capacities of the standard sauropsidan eye as we see it in the lizard or the bird. But it has gained these powers through a length process of re-differentiation, which was carried out largely within the confines of the primate order itself.

Notes

[1] A fascinating coincidence is that the flying-squirrel type has been evolved more than once in the rodents – by the true flying-squirrels and in the Anomaluridæ – and more than once also in the marsupials: there are three kinds of flying phalangers, each a close relative of a different non-flying form.

[2] Dr O'Day has been trying for several years to find time to prepare a monograph on the eyes of the monotremes. When this does finally appear it will greatly extend, and no doubt partially contradict, the present treatment. In the meantime, because of the slowness and uncertainty of communication with Australia, the writer has made bold to discuss O'Day's findings without seeking his permission – they seem much too important to be left out of this book.

[3] And all the sauropsidoid internal features are likewise called avian by those who are familiar with their occurrence in birds but ignorant of their occurrence also in the reptiles. Attempts to derive the monotreme eye from the avian, and coy insinuations that the two eyes are identical through convergence (justifying the 'bill', webbed feet, spurs and egg-laying habit of the platypus), are naïve in the extreme; but they continue to be made.

The astute Franz indicates in several places that he suspects that the 'avian' form of the usual preserved echidna eye is a result of collapse. O'Day (1936, 1938) finds that this collapse occurs very readily in both *Tachyglossus* and *Ornithorhynchus*.

[4] And the unstriated condition of the sphincter pupillæ; but there is no reason to think that this is a new muscle. Iris muscles have been independently evolved several times of course; but the mammalian sphincter has, in all probability, been inherited directly from the reptiles. Not so the mammalian dilatator.

[5] Measured in O'Day's preparations (3.3 mm/1.2 mm); Franz gives $3.0/0.8 = 3.7$, but expresses doubt as to the validity of these figures. Kolmer gives $2.88/0.96 = 3.0$ for *Zaglossus*, but his material was preserved many hours post mortem.

[6] Certain in the case of *Tachyglossus*; probable in *Zaglossus*, but Kolmer's material was too badly histolized to make possible any study of the visual cells.

[7] According to Albarenque, *Didelphis marsupialis* and '*Azara*' ($=D.$ *azaræ*?) have only rods. This seems improbable in view of the extremely close relationship of these forms to *D. virginiana*.

[8] The mammalogical reader will have noticed that throughout this book the old term 'ungulate' has been employed. It embraces several orders which are of course widely separated in modern classification: the Artiodactyla (even-toed) and the Perissodactyla (odd-toed hoofed forms), the Hyracoidea (hyraxes) and their close relatives the Proboscidea (elephants). (The Sirenia, though never classed as 'ungulates', are connected with the base of the elephant branch.) The nowadays artificial term 'ungulate' has seemed here a convenient word-saver, for the orders embraced by it have eyes which are much alike. From comparative ophthalmological evidence, no one would be led to believe that the artiodactyl and perissodactyl lines of descent have actually been separate since almost the inception of the Placentalia.

[9] The zonules of the squirrels should receive as careful a study as Kahmann (1930) has given those of the other strongly-accommodating mammals.

[10] Verrier (1935) found all the cells alike (and, from her drawings, rods) in *Crocidura mimula*; but in *C. leucodon* and *C. aranea* there are more cones than in mice, according to Schwarz (1935).

[11] Kolmer (1925) claims a few for *Nycticebus tardigradus*, but Detwiler found none in this loris.

[12] Little shrinkage spaces at the distal ends of the cone inner segments have been all too often mistaken for oil-droplets – even by such careful workers as Kolmer. Examination of the retina in its fresh condition, and after fixation with osmic acid, will always demonstrate the presence or absence of real oil-droplets.

[13] The chondrosteans and *Neoceratodus* have apparently lost ancestral double cones – see Plate I.

References

Detwiler, S. R. (1939). Comparative studies upon the eyes of nocturnal lemuroids, monkeys and man. *Anat. Rec.*, 74, 129–145.

Detwiler, S. R. (1940). Comparative anatomical studies of the eye with especial reference to the photoreceptors. *J. Opt. Soc. Am.*, 30, 42–50.

Detwiler, S. R. (1941). The eye of the owl monkey (*Nyctipithecus*). *Anat. Rec.*, 80, 233–241.

Franz, V. (1913). Sehorgan. In: *Lehrbuch der vergleichenden mikroskopischen Anatomie der Wirbeltiere*, Teil 7, ed. Oppel (Jena: Gustav Fischer), pp. 417.

Franz, V. (1934). Vergleichende Anatomie des Wirbeltierauges. In: *Handbuch der vergleichenden Anatomie der Wirbeltiere*, Band 2, ed. Bolk et al. (Berlin: Urban & Schwarzenberg), pp. 989–1292.

Gresser, E. B. and Noback, C. V. (1935). The eye of the monotreme, *Echidna hystrix*. *J. Morphol.*, 58, 279–284.

Gunn, R. M. (1884). On the eye of *Ornithorhynchus paradoxus*. *J. Anat. Physiol.*, 18, 400–405.

Kahmann, H. (1930). Untersuchungen uber die Linse, die Zonula ciliaris, Refraktion und Akkommodation von Saugetieren. *Zool. Jahrb.*, 68, 509–588.

Kolmer, W. (1925) Zur Organologie und mikroskopischen Anatomie von *Proechidna (Zaglossus) bruynii*. 1. Mitteilung. *Z. Wiss. Zool.*, 125, 448–482.

Menner, E. (1929). Untersuchungen über die retina mit besonderer Berücksichtigung der ausseren Könerschicht. *Z. vergl. Physiol.*, 8, 761–826.

O'Day, K. J. (1938). The visual cells of the platypus (*Ornithorhynchus*). *Br. J. Ophthalmol.*, 22, 321–328.

O'Day, K. J. (1939). The visual cells of Australian reptiles and mammals. *Trans. Ophth. Soc. Aus.*, **1**, 12–20.

Schwartz, S. (1935). Über das Mausauge, seine Akkommodation, und über das Spitzen-mausauge. *Z. Naturwiss.*, **70**, 113–158.

Verrier, M. L. (1935). Les variations morphologiques de la rétine et leurs conséquences physiologiques. A propos de la rétine d'une musaraigne (*Crocidura mimula* Miller). *Ann. Sci. Nat. Zool.*, **18**, 205–216.

Walls, G. L. (1939). Origin of the vertebrate eye. *Arch. Ophthalmol.*, **22**, 452–486.

Plate I. Tentative schema of the evolution of the visual cells in vertebrates.

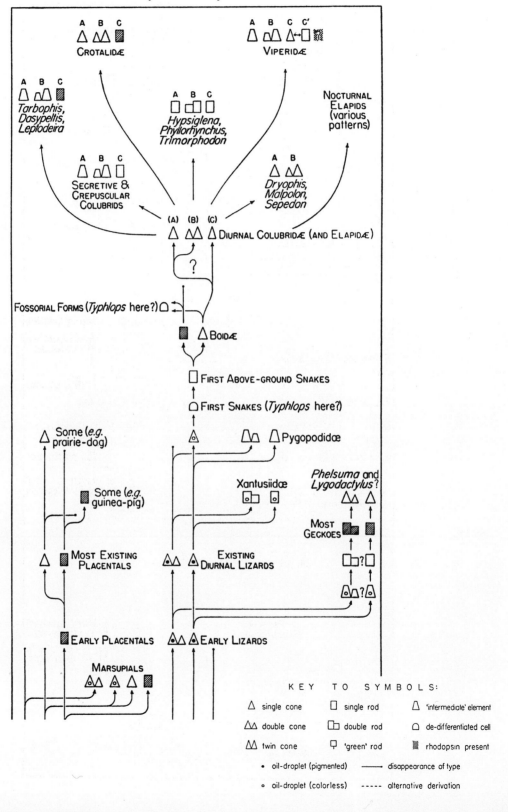

(Plate I continued)

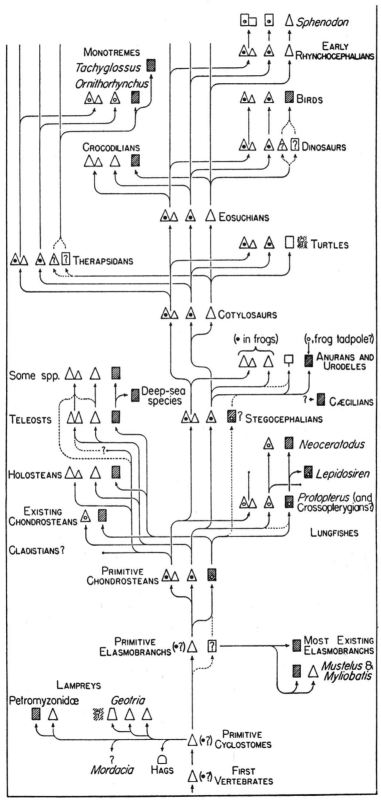

Index

Aberration 85
  chromatic 6, 94–5, 312
    in evolution of retina 139
    and evolution of vision 410, 412
    in eye evolution 417
    and eye optics 120
  in dioptrics 86, 89, 92–9, 105, 109
  lateral chromatic 94–5
  longitudinal chromatic 84, 94–5
  off-axis 94, 119–20
  spherical 84, 94–7, 119–21
Absorption
  light 407–8
  maxima 295
  selective 8
  spectra 229–30, 284, 290, 292, 294, 299, 301
  spectral 9
Acanthodii 406
Accommodation 104–12
  active 109–10, 112
  amplitude of 107–11
  in dioptrics 85, 100
  dynamic response 107
  in evolution of eye 21
  and evolution of vision 404, 411–2
  in mammals 469, 471, 474–5, 477–8
  mechanisms 107–12, 119
  over-accommodation 108
  and pretectum 157
  static response 107
  under-accommodation 108
Achromatopsia 306
Across-fibre processing 249, 251
Actinopterygii 400–1, 407, 412, 415
Acute zones 266
Adaptation
  dark 168–70, 374, 378, 382
  light 168, 170, 356, 374, 376, 378
  luminance 168, 171
Adaptationist programme 136, 144–7
Aesthetes 368
Afterimage 165–7, 169
Agnatha 398–405, 414, 416–7
Airy diffraction pattern 92
Algorithm, machine stereopsis 276–7
Algorithm, motion detection 252
Aliasing 102–3
Alignment, vertical 448
Alignment in binocular vision 272–3

Ametropia 82–3, 104–12
Ammocoetes 402–5
Amphibians 71–2
  colour vision 285–7, 290, 299
  dioptrics 89, 110
  feeding selectivity 157
  photoreceptor system 68–9
  and visual cortex 152
  visuomotor systems 153–7, 159–62
Amphioxus 15, 17–8, 24, 31, 48, 399, 403, 405
Anableps 405, 412, 417
Anaspids 402
Angle alpha 272–3
Anguilla (eel) 69
Angular sensitivity 232, 235
Anisotropy 276–7
Annelid worms 122
Anomalous pigment 78
Aperture, aphakic 97
Aperture, numerical 91–2, 96–7, 100, 102, 104, 110
Aphakic space 414
Aplysia 365–7, 372–3
Apogon brachygrammus (cardinal fish) 69
Arachnata 341
Archistriatum 426
Architeuthis (giant squid) 376
Area centralis
  and binocular vision 271–2, 278
  in evolution of retina 143–4
  in evolution of vision 411
  in mammals 476
  and visual pathways 428, 443, 455
Aristostomias scintillans (deep-sea fish) 88
Array, orthogonal 249, 254
Array, sampling 263, 266–7
Arthropodization 213
Arthropods
  compound eye and optic lobe 341–60
  history of vision 220–2
  optic lobe 203–24
  origins 213–4
Aspheric surface 96
Asphericity 109
Astigmatism 98, 111
  oblique 94, 97
Autocorrelation theory 251
Autoradiography, tritiated thymidine 432

Background 261
  and colour vision 300
  dappled 306–8, 316
  noise 410, 413–4
  in visual processing 239, 254–5, 257
Backlight irradiance 409
Backscattering 407–9
Bacteriorhodopsin 39, 44
Balfour's theory 31, 35
Bandwidth, spectral 6, 11
Barrier avoidance 153–6, 158–60, 162–3
Bats 145, 451, 458
Bee
  celestial compass 265
  colour vision 285, 287–8, 294, 297–9, 301–2
  eidetic image 262
  visual discrimination 261
  visual illusion 263
  visual processing 255, 261
Behaviour 260
  detour 158–60, 163
  visually directed 390–1
Binocular fusion theories 446
Binocular overlap 271, 275, 436, 442–3
  and visual pathways 448, 455
Binocular vision 435–6, 443, 446
  evolution of 271–81
  functional 271–5
  mammals 274–6
  organization 280–1
  and visual pathways 455
Biological clock 18, 33, 35
Biology, developmental 136–7
Biology, evolutionary 137
Bioluminescence 12, 68, 414
  and dioptrics 86–7
  in evolution of vision 409–10
  in photoreception 374, 381
Birds 72–3
  binocular vision 271–3, 276–7, 280
  colour vision 286–9, 291–2, 295–6, 298–9, 302
  dioptrics 86–7, 89, 94, 98, 110
  diurnal 77–8, 287, 292
  diving 85
  in evolution of retina 142–3, 146–7
  nocturnal 470
  visual pathways 421–36, 442
Blindness 456–7

483

Blobs, cytochrome oxidase 313, 453, 458
Blur 94–5, 166, 231
  circles 105, 232
Bottleneck theory 435
Bowman's membrane 470, 473
Boyle, Robert 306–8
Brain
  and eye 237
  and origins of eye 53–5
  and visual pathways 442–3
Brain stem
  in pursuit system 172–3, 176, 179–89, 191
  and visual pathways 423, 425, 429–30, 455, 459
Branchiopoda 222
Burst-tonic units 181–2
Bush baby 75, 91
Butterfly, colour vision 294–5, 298–9, 302
Butterfly, optics 132

Callicebus moloch (dusky titi) 76
Calliphora 207, 215
Callithricidae 76
Callithrix jacchus (common marmoset) 76–7, 316
Cambrian explosion 220
Camera 127
  and eye optics 118
  ultra low-light CCD 413
  and visual processing 229, 232, 236–7, 245
Canal of Schlemm 469, 471, 474
Canis familiaris (domestic dog) 74
Carassius auratus (goldfish) 69–70
Carotenoid 72
Cartridges 344, 348, 353, 355–8
Cat
  binocular vision 276–7, 279
  colour vision 293, 300
  dioptrics 85, 95, 98–100
  photoreceptor system 74
Cataract 88
Catarrhines 73, 76–7
Categories of vision 263, 268
Categorization 254, 268
  object 260
Cebidae 76
Cebus apella (tufted capuchin) 76
Celestial compass of bee 265
Celestial orientation 390
Cells
  alpha 278–9, 445
  amacrine 136, 140–1, 143
  beta 278, 445
  binocular 443
  bipolar 208, 313
    in evolution of retina 136, 140–1
    OFF-centre 140
    ON-centre 140
  broad-band 170
  C-type 140–1
  colour-opponent 170, 298
  cone 351
  corneagenous 350
  diffuse bipolar 140
  duplication of types 360
  fate 346, 355
  gamma 445
  horizontal 136, 140
  interplexiform 136, 141
  isomorphic 360
  isomorphy 357
  large monopolar (LMC) 204, 206–8, 210, 213, 218–9, 224
  light-sensitive 19–21, 23
  luminosity-type 140–1
  M 444, 450
  magnocellular 171, 444, 450
  midget bipolar 140
  non-concentric 170
  P 444–5, 450
  parvocellular 171, 444–5
  pigment 350
  primary pigment 351
  Purkinje 180–3, 188
  rarely-encountered 170, 172
  rod-cone bipolar 140, 147
  screening pigment 351
  W 278, 444, 450–3, 455, 459
  X 444–5, 450, 452–3, 459
    in pursuit system 170–2, 175, 191
  Y 278, 444–5, 450, 452–3, 455–6, 459
    in pursuit system 170–2, 175–6, 191
Centrifugal pathway 425, 427, 429, 432–3
Cephalaspidomorphs 402–3
Cephalic organ 15, 17–8, 21, 25, 30–1, 35, 37–40, 42, 48
  and origin of lens 44–5
Cephalopods 3–4, 13
  in evolution of eye 29–31, 39–40, 46–7
  eyes 21, 23
  origin of lens 45
  origins of eye 54
  photoreception 364–5, 367, 371–2, 374–8, 380–2, 384–5, 388, 390–1
Cercopithecus 77, 310
Cercopithecus cephus 314
Cerebellum
  in pursuit system 174, 179–89, 191
  and visual pathways 424, 427, 429, 449
Cerebral cortex in pursuit system 174–80, 183
Cetaceae 274
Charybdea 238
Chelicerata 213, 217
Chelonia 426–7
Chemoreception 249
Chiasm in eye evolution 415
Chiasma 354–5, 357
Chitons 364–5, 368, 371, 383, 390
Chlorophyll 8, 11
Choroideal body 414
Chromatic channels 312–3
Chromatin 472, 476
Chrysopa 234
Cilia
  in evolution of eye 26–9, 40–3
  photoreceptors 27–8, 365–6, 368–72, 375–6
Ciliary body 469, 471, 473, 475, 477
  muscle 469, 471, 475
  web 469, 471, 474
Circadian oscillator 372
Circadian rythm 367, 373, 425
Circadian system 368
Clammys caspica (turtle) 287
Classification, laminar 424
Cod dioptrics 83
Coelacanths 71–2
Coevolution 12–3, 302, 313–4
Coleoptera optics 130
Coloration 301–2
  body 302
  cryptic 302
Colour
  circle 301
  coding, opponent 298
  complementary 297, 301
  constancy 153, 267, 300–1, 310, 458
  contrast phenomena 301
  'derived' 297
  discrimination 230, 249, 254, 258, 300, 302, 456
  identification by 310
  opponency, double 453
  perception 297–8
  'pure' 297–8
  'unique' 297–8
Colour blindness 74, 242, 308, 314, 390
Colour vision 6–11
  absence of 388, 390
  advantages of 306–10
  arthropod 204, 223–4
  decachromatic 295
  dichromatic 73, 75–6, 78, 284, 289, 292–3, 296, 299–300
  dimensionality 299
  evolution 284–302
    dimensionality 289–95
    phylogeny 284, 295–7
    wavelength discrimination 285–8
  and evolution of eye 43
  and evolution of retina 142
  and evolution of vision 410
  human 297, 299–301
  insect 254, 257
  loss of 306
  in mammals 478
  mechanisms 13
  motion blind 259
  pentachromatic 71–3, 289, 291
  phylogenetic tree 295–6
  polychromatic 73
  primate 306–17, 444
  primordial subsystem 312–3
  second subsystem 313–4
  tetrachromatic 69, 289–92, 299
  theory 248
  trichromatic 72–3, 75–7, 284, 289, 292–4, 296, 314
Columba livia (pigeon) 70, 286, 288
Coma 94, 97
Combinatorial explosion 262
Commissures 444
Communication theory 73
Complexion 310

Cone fundamentals 289, 291
Cones
  in colour vision 296, 316–7
  double 66, 68, 70–2, 296, 470, 472, 477
  in evolution of eye 5, 8, 10, 12–6
    vertebrate and invertebrate 35, 37, 39–44
  in evolution of retina 136–40, 143
  in evolution of vision 404, 412–3
  in fish colour vision 290
  in mammals 476–8
  mosaics 139
  pigment 410
  in pursuit system 168–70
  short-wave 309, 311–3
  single 470, 472, 476–7
  in trichromacy 310–3
  ultraviolet-sensitive 290, 292
Constringence 95
Contour, articulation of 309
Contrast 168–9, 300
  reversed 276
  temporal 243, 245, 252
Contrast sensitivity 445, 454
Contrast sensitivity functions 106–7
Convergence 118
  in arthropods 203–4, 206, 212–4, 217, 219–21, 223
  and binocular vision 273–5, 277
  in evolution of retina 139, 143
  and origins of eye 57
Convolution 93
Copepods optics 121–2, 125
Copilia quadrata 55–9
Cornea
  aspheric 99
  curvature 109
  in dioptrics 82, 84–5, 88–9, 91, 95–6, 99, 103, 107–8, 110
  and evolution of vision 399, 404, 409–12
  in eye evolution 17, 414
  and eye optics 120–1
  iridescence 89
  in mammals 468, 470, 473, 475
Corner reflectors 130
Corpus callosum 444, 456
Correlation theory 252
Correlations, simultaneous 254
Cortex, parietal 173–4
Cortex, striate, in pursuit system 170, 172
Cortical magnification 453–5
  areal 454
Cortical magnification factor (CMF) 453–5
Corticonuclearmicrocomplexes (CNMC) 180–1
Corticotectal pathway 455–6
Counting statistics 89–90
Crocodilia 429
Crotalus viridis (rattlesnake) 428
Crustacea
  ancestral 348
  colour vision 295
  eye evolution 341–3, 345–6, 348, 350–1, 353–5
  eye optics 118–9
  macruran decapod 130

optics 121–2, 125–7, 129
  and origins of eye 54
  visual processing 203–4, 208, 213–4, 217–8, 220–4, 238, 241
Cubomedusa 238
Cuttlefish optics 122
Cyclostomes 66–7
  dioptrics 85, 109–10
  in evolution of retina 137, 140
Cyprinids 69–70, 290
Cytoarchitectonics 432, 451
Cytochrome C 66, 73
Cytochrome oxidase blobs 313, 453, 458

Darwin, Charles 52–3
Decussation
  double 280
  in evolution of eye 16, 46–8
  naso-temporal 279
  partial 273, 280, 443, 446–8, 459, 468
  patterns 278–80
  total 273, 446
  in visual pathways 430, 443
Defocus 104–6
Degeneracy 142–3
Dehydroretinal 9
3-dehydroretinal 64–5, 69
Dendritic fields 385, 387
Depolarization 28, 42–3
  in colour vision 298
  in photoreception 366–8, 370, 380, 383
  in visual processing 218, 230
Depth-of-focus *see* Eye
Dermal sensitivity in molluscs 364–6
Descemet's membrane 470–1, 473–4
Detour behaviour 158–60, 163
Deuteranopes 310
Deuteranopia 311
Deuterostomes 399–401
Dichromats 290, 293, 295, 315–7
Didelphis virginiana (common opossum) 467, 471
Diffraction 92–4, 231–2, 408
  X-ray 377
Diffraction limits 5
  in dioptrics 86, 103
  and eye optics 120, 127
  in visual processing 235–7
Dinopsis (net-casting spider) 120
Dioptrics
  basic optics 82–6
  depth-of-focus, ametropia and accommodation 104–12
  optical image quality limits 92–9
  physical constraints on design 86–92
  retinal aspects 100–4
  vertebrate 82–112
Diptera
  eye evolution in 346–9, 355–9
  visual processing in 204–12
Directional motion 210, 212–3
Directional selectivity 210–3, 219, 426, 431–2
Directional sensitivity 213
Disparity 275

binocular 448
  convergent 279
  detection 273–4, 436
  divergent 278
  horizontal 277
  interocular 454
  retinal 277–8
  selectivity 277–8
  vertical 276–7
Dissolved organic matter (DOM) 8
Dog colour vision 293
Dog photoreceptor system 74
Dolphin dioptrics 83
Donders, F.C. 311, 313
Dorsal cortex 426
Dorsal ventricular ridge (DVR) 423–6, 432–4
Double dissociation 456
Dragonfly
  eye evolution 354, 357–9
  ommatidium 347
  visual processing 233, 255
Duplicity theory of vision 137
Dyads 358

Eagle dioptrics 100, 102–3
Echidna 442, 467–70
Ectoderm 35
  surface 17, 25–6, 31, 44
Ectostriatum 432–3, 435
Edge detection 313
Edge polarity 252–3, 262
Eel 69
Elasmobranchs 67–8, 400, 406, 410, 412–4, 477
  in evolution of retina 137–9, 141
Electrical activity 382–3
Electromagnetic theory 89
Electrophysiology and evolution of eye 6
Electrophysiology and trichromacy 310
Electroretinogram (ERG) 73, 376, 382
Electroretinography 74–5
Ellipsosomes 66, 73, 78
Emergent properties 229
Emmetropia 84, 97–9, 105, 107, 111, 469
'Emmetropization' 84
Encephalization 146–7, 435–6
Ensemble processing 241, 251, 260
Equiluminance 306, 308–9, 312
Erythrocebus patas (patas monkey) 77
Evolution
  coevolution 12–3, 302, 313–4
  convergent 217, 273, 376, 450
  fish vision 398–417
  parallel 343, 348, 360, 450
  through duplication 356
Extrafovea 168
Eye
  accessory 374
  agnathan 402–3, 405
  apposition 351, 353
  aquatic lens 121–2
  and brain 237
  camera 4, 13, 119–21, 238
  camera-type 266

cephalic 368, 371–83
cephalopod 382, 384, 386
chambered 47
changes in length 109, 111
Chrysopa 234
compound 47
　apposition 4, 127, 132, 134
　arthropod 203–4, 213–4, 217–22, 341–60
　bee 265
　evolution 3–4, 13, 132, 222–4
　fish 417
　invertebrate 26, 28
　mollusc 364, 371
　ommatidia 57
　optics 118–20, 125–7, 130, 134
　origins 54–5, 59
　reflecting superposition 134
　refracting superposition 134
　superposition 4, 127, 132
　in visual processing 261
Cubomedusa 238
cyclopean 354
day 232
degenerate 142
depth-of-focus 85, 104–12
dermal 382
design 119–20, 134, 351
diffraction-limited 93, 99, 104–7
dorsal 368–9
dragonfly 233, 235
echidna 468
electronic 55–6
embryological development 442
embryology 467
embryonic development 403
emmetropic 83
evolution 3–5, 118, 230–7
frontal 442–3, 446, 448, 459
　bird 429–30, 436
frontalization 455
geometry 235, 237, 266
'gradations' 53
human 119–20
intracellular 118
invertebrate, chambered 21
invertebrate, origins 15–48
lateral 442–3, 448
　bird 429–30, 436
lens 3–4, 126, 375–83
'lobster' 130
mantis 231
marsupial 470–1, 477
medusae 237–8
mirror 369–71
monotreme 467–9, 477
multi-lensed 417
multicellular 21
muscles 187–9
nauplius 222–3
neural superposition 128
'neurocrystalline' 222
nocturnal 91, 232, 234, 477
non-cephalic 390
origins 52–9
oxygenation 414

parabolic reflecting 125
parabolic superposition 131
parameter 237
parietal 31–5, 138
physical constraints on design 86–92
pineal 31–6, 406
pinhole 3, 21, 23, 25, 54, 122
　mollusc 364, 374–5
placentalian 473, 476–8
prevertebrate 19
protection 414
proto compound 133
real 105–7
reduced 85, 91–2, 94–5, 105
reflecting superposition 129–30
reflector 4
refracting superposition 129–30
scanning 52, 55–9, 122, 237
　mollusc 364, 375–6
schematic 82–6, 91
sexual dimorphism 123
shape 412–3
simple 21–9, 39, 133
　aggregate 21–2, 28
　arthropod 218
　composite 21–2, 28
　compound 22, 24
　evolution 220, 222–3
　mirror 134
　mollusc 372–4
　origins 54–5, 59
single-lens 4, 54
spherical lens 134
stalked 222, 382
　mollusc 364–5, 376
stationary 239
superposition 351, 353
transmittance characteristics 86
tubular 91–2, 473
unicellular 21
vertebrate, cerebral origin 17–8
vertebrate, origins 15–48
without brain 237
worm 237–8
Eye cup 3, 22, 54, 58, 123, 133–4
Eye fields, supplementary 180, 187, 191
Eye movements
　binocular 186
　compensatory slow 186
　conjugate 443, 446, 448, 459
　conjugate saccadic 186
　control 187–8
　and dioptrics 92, 103
　disjunctive 279
　in eye evolution 415
　horizontal quick 186
　miniature 167
　mollusc 374–5, 377, 390
　and origins of eye 58–9
　postpursuit (PPEM) 188, 191
　rapid 167, 186
　saccadic 166, 172, 448
　sinusoidal passive 188
　smooth 179–80, 182
　smooth disconjugate 180
　spontaneous 185, 187, 443

　spontaneous saccadic 186
　stepwise 188
　vertical 187
　vertical rapid 186
　and visual pathways 443, 455, 457
　and visual processing 268
　visually induced 179
　visually induced saccadic 186
　see also Pursuit system
Eye muscles, external 402
Eye muscles, extra-ocular 401, 404
Eye spots 19–21, 24, 26, 35, 342–3, 369
Eye stalks 342
Eye-opening 145
Eyeball, spherical 473

F-number 232, 235–6, 411
Falciform process 408
Feature discrimination 259
Felis domesticus (domestic cat) 74
Fibre optics, retinal 100–2
Field of view 412, 414
Filter 409
　barrier 410
　spectral 87
　yellow 95
Fish
　colour vision 285–6, 290, 295–6, 298–300, 302
　deep-sea 68, 77, 88, 112
　　colour vision 290
　　dioptrics 91, 103
　　visual pathways 442, 446
　dioptrics 84–5, 87, 89, 94–7, 103, 109–10
　evolution of retina in 138–9, 141–2
　feeding selectivity 157
　four-eyed 110, 112
　optics 122
　photoreceptor system 67–71, 78
　shape recognition 162–3
　visual cortex 152
Fixation
　accuracy 172
　in arthropods 212
　point 443
　in pursuit system 165–8, 172, 175, 178–85, 188, 191
　in visual processing 254–60, 262, 268
Flocculus 180–4, 187–8
Flux, light 87
Flux, photon 89–92, 103, 410
Fontana's space 474
Forebrain 436, 444
Form discrimination 387–8, 456
Form perception 456
Fossil record 350, 450
Fossorial niche 428
Fourier transform 92–3
　inverse 94
Fovea 168–9, 272–3
　binocular 273
　in evolution of retina 139, 144
　in evolution of vision 411
　in eye evolution 413
　in mammals 476, 478
　Mantis 231

monocular 273
  in pursuit system 166–72, 177, 180, 183, 191
  and visual pathways 443, 453
  in visual processing 255–8, 266
Foveal pit 103–4, 139, 143, 443, 454
Freeze-fracture 377, 380
Frog 71
  dioptrics 83–4, 95, 99
  prey tracking 161–2
  shape recognition 163
  visuomotor system 153–9
Frontal eye fields 174, 176, 178–9, 188, 191
Fruit 313, 316–7
  coloured 314

G-protein 63–4, 380
Gadus morrhua (cod) 68
Gain in synaptic complexity 359
Galago 456–7, 473
Galago crassicaudatus (bush baby) 75, 91
Ganglion cell layer 137, 139, 145, 444–6
Ganglion cells 139–40
  density 454–5
  displaced (DGC) 427, 433, 449
  giant 427
  in mammals 470, 476
  retinal 141–7
    and binocular vision 274–5, 278
    classes 143, 145–7
    distribution 143–5, 147
    in evolution of retina 136–7
    numbers 142–3
    in pursuit system 170–3
    and visual pathways 443, 445, 450, 452, 454
Gap junctions 386
Gaussian thick-lens theory 82
Gecko 73, 111, 428
Gelbstoff 8
Genes
  coding visual pigments 78
  in colour vision 284, 316–7
  hybrid 314
  opsin 295
    in photoreceptor system 67, 69, 71, 77
  photopigment 293, 297, 311
  rare mutant 316
  in trichromacy 313–4
  visual pigment 26, 44
Gerbil photoreceptor system 74
Gigantocypris mulleri 125
Goldfish 69–70
  colour vision 286–7, 289–92, 298, 301
  dioptrics 85
Golgi stain 141–2
Guanine crystals 126
Guppy 12–3, 69–71, 73

Hagfish 31, 33, 401–2, 405, 416
Hallucination 268
Halobacterium habolium 39
Haplochromis burtoni (cichlid fish) 290
Hemiretina 443
Hermissenda 372, 383–5

Heterochrony 145, 147
Heteropods 375–7, 383–4, 390
Heterozygous advantage 316
Histamine 218, 360
Homology
  antigen 351
  apomorphic 136
  cell 357–8, 360
  of compound eye 342–3, 360
  dipteran ommatidium 346–8
  evolutionary 346
  ommatidia 345
  photoreceptors 348
  plesiomorphic 136, 138, 140–1, 145
  of visual system 342
Horse dioptrics 83
Hough transform 248, 256
Hughes' terrain hypothesis 143–4, 147
Hybrid offspring 70
Hybrid population 71
Hybridization of visual pigments 71
Hybrids 71
3-hydroxyretinal 348–50
Hyperacuity 239, 255, 454–5
Hypermetropia 111
  apparent 84
Hyperpolarization 27–8, 42–3
  in colour vision 298
  in evolution of retina 140–1
  in photoreception 366, 369–71
  in visual processing 206, 217
Hypothalamus 17, 31, 45
  and visual pathways 423, 425–8, 433

Illuminance, retinal 98–9
Illusion, visual 263, 267
Image
  analysis 377
  contrast 126
  degradation 6, 413
  eidetic 261–3
  geometry 127, 239
  off-axis 85–6
  one-dimensional 243, 245, 255–8
  quality 5
    in dioptrics 85–6, 89, 102
    limits on 92–5
    and eye optics 119–20
    and visual pathways 448
  retinal 5
  segmentation 209
  spatial 244
  spatio-temporal 246
  structure 259
  X-ray 130
Immunocytochemistry 67, 146
Inferotemporal cortex 456
Inferotemporal lobe 310
Inhibition, binocular 277–8
Insects
  colour vision 294, 299, 302
  eye evolution 341–3, 345–6, 348–51, 353–4, 358, 360
  pattern recognition 262
  stick 260

visual processing 235, 238–42, 246, 251, 253–4, 260
Interlaminar zones 177
Interneurones, intranuclear 187–8
Intraocular flare 89
Invertebrates, colour vision 288, 294–5, 297
Ion equilibrium 42–4
Ion translocation 40, 42–4
Ion transport 16, 28, 39
Ionochromic shift 73
Iris
  circular 414
  in evolution of vision 399, 404, 412
  in eye evolution 413–4
  in mammals 469, 471, 473–5
Irradiance, backlight 409
Isihara pseudo-isochromatic plates 306, 308–10
Isomerization, thermal 299
Isomorphs 357
Isomorphy, cell 357, 360

Kryptopterus bicirrhis (cat-fish) 290

Labidocera 121–3, 134
Ladd-Franklin, Mrs Christine 311, 313
Lamina
  arthropod 213, 216–9
  cartridges 344, 348, 353, 355–8
  cholinergic activity 218–9
  in eye evolution 343–5, 347–8, 353–4, 356–8, 360
  optic lobe 204–6, 208, 211–2, 220, 222–4
  synaptic connections in fly 217
Lamination 426, 443, 449–51, 476
Lamprey 66–7, 477
  dioptrics 110
  in evolution of retina 137–8, 140
  evolution of vision 398, 401–2, 404–6
  in eye evolution 415–7
  photoreceptor system 69
Landmarks 263
Lateral geniculate lamination 278–9
Lateral geniculate nucleus (LGN) 16
  and binocular vision 274–5, 277–8
  in colour vision 316
  in pursuit system 170–2, 174–5, 177, 188–9
  in trichromacy 312–4
  and visual pathways 423–30, 433, 435, 443–5, 449–53, 459
Lateral inhibition 224
  in visual processing 208, 206, 213, 245–7
Latimeria chalumnae 71
Lemur 77, 476
  black 76
  colour vision 293
  ring-tailed 75
Lemur catta (ring-tailed lemur) 75, 293
Lemur macaco (black lemur) 76
Lens
  as accommodative mechanism 119
  aquatic 121–2
  aspherical 412

camera 123
cellular 44
cephalopod 377
corneal 119–21, 132
cylinder 130–3
in dioptrics 82, 84–6, 88–9, 91, 95–7, 99, 103, 108–9
in evolution of eye 3–4, 6, 17–8, 21, 24, 26, 47, 417
in evolution of vision 399, 403, 405, 409–12
gradient index 96
in mammals 469, 471, 473–7
Matthiessen 122, 128, 130
microlens 89
mirror 417
movement 108–9, 111
multi-element 122–3
origin of vertebrate 44–6
and origins of eye 55, 57
in photoreception 369, 372, 374
quality 411
refracting 118
resolution 230–8
shape 109–11
simple 131
single 128
spectral transmittance 88
spherical 4, 97, 121, 134, 410, 412
telephoto 411, 417
triplet 123
vertebrate 38–9
wide-angle 411
yellowing 88
zoom 417
Lepidoptera optics 130
Lepomis macrochirus (bluegill fish) 69
Liebmann, S. 312
Liebmann effect 309, 312
Light
achromatic 293, 297–8
constancy 300
entopic stray 97
flux 87
guides 230–2, 235
long-wavelength 288
low 410–1, 417
monochromatic 289, 294, 297–8
nature of 3, 5–7, 13
polarized 298, 382
retinal stray 97
scattered 6, 8, 139, 265, 408, 410, 413
short-wavelength 6–7, 287–8
stray 97–8, 102
ultraviolet 6–7, 33, 35, 69–73
in colour vision 287–90, 292, 294–5, 298–9, 302
'white' 289–90, 292, 294–5, 297–8
Limbic cortex 430
Limulus 132, 217–8, 342, 345–6
Line-labelling 246, 248, 250, 267
Line-spread function 99
Linear systems theory 166
Lineus 30, 35, 37
Lizard, dioptrics 84–5, 103
Lizard, visual pathways 422, 427–9

Lobula
arthropod 216–7, 219
complex 354–5
in eye evolution 353–5
optic lobe 204–6, 210–1
Locust, optic lobe 222–4
Locust, visual processing 206, 208–12
Loligo opalescens (squid) 383, 385
Lolliguncula (squid) 388
Long visual fibres (LVF) 347–8, 354, 357, 359
photoreceptors 204–5, 208, 214, 217–8, 221
Looming 242
Low light 410–1, 417
Low-level vision 267–8
Luminance, equivalent veiling 97
Luminescence in dioptrics 88
Lynx dioptrics 91
Lyonization 315, 317

Macaca fasicularis 168
Macaca mulatta 168
Macaca nemestrina 168
Macaca speciosa 168
Macular pigment 89, 95
Magnification gradient 264
Magnocellular layer
in pursuit system 171, 175
and visual pathways 444–5, 450–2, 458–9
Malacostraca 204, 214, 216, 219
eye evolution 341, 351, 353, 355
visual processing 221–3
Mammals 467–78
binocular vision 271, 274–6
colour vision 288, 292–5, 299
monotremes and marsupials 467–72
photoreceptor system 73
placentals 472–8
visual pathways 424
Mandibulata 213, 217
Mantis range measurement 265–6
Mantis visual processing 231, 261
Map
auditory 448
motor 448
retinotopic 456
somatosensory 448
spatio-temporal 245–6, 248–9, 254, 256–8
topographic 240, 456
visual 448
Marmoset colour vision 316
Marsupials 442, 444–5, 448, 450–1, 453–6, 467–72
Matthiessen's ratio 97, 121–2, 376–7, 411–2
Medulla
arthropod 216–7, 219–20
in eye evolution 347–8, 353–5, 357, 416
optic lobe 204–6, 208–12, 222
Medusae 237–8
Megabats 451, 458
Melatonin 33
Membrane
Bowman's 470, 473
Descemet's 470–1, 473–4

nictitating 85, 470
Memory and visual processing 268
Meriones unguiculatus (gerbil) 74
Microbats 451
Microelectrode analysis 385
Microlens 89
Microspectrophotometry 6
and colour vision 290–3, 300, 314, 316
and photoreceptor system 67, 72, 75–7
and trichromacy 310, 312
Microvilli
in evolution of eye 22, 26–9, 35, 37, 39–43
photoreceptors 365–8, 370–3, 376, 378–80, 383
Midbrain 443
Minnow 285
Miosis, accommodative 108
Mirrors 128–31
biological 126–7
concave 123–6
and eye optics 119
interference 127
lens 417
reflecting 118
Misalignment 279, 313, 317
Models
autocorrelation visual processing 251–3
energy visual processing 252
Reichardt's motion perception 240
template visual processing 243–51, 266–7
Modulation 235–7
Modulation transfer function (MTF) 92–3, 99–100, 102, 104, 106
Molecular genetics 311, 313
Molecular markers 220
Molluscs
bivalve 123, 125
classification 364–5
optics 122
photoreception 364–91
Monads 358
Monkey
binocular vision 276–7
cebus 173
colour vision 287–8
macaque 76, 168, 173, 458
New World
colour vision 288, 293, 295–6, 314–7
photoreceptor system 76
visual pathways 453, 458
Old World
colour vision 288, 293, 295–6, 310, 314, 316–7
photoreceptor system 73, 76–8
visual pathways 453, 458
owl 169
visual pathways 443, 451, 458
rhesus 173
squirrel 173, 293, 443
visual pathways 456–7
Monochromats 290, 300
Monocular vision 276
panoramic 271
Monotremes 442, 444, 467–72
Morphology of pursuit system 169–70,

174–5, 178–82, 184, 187
Motion
  colour blind 258–9
  detection 212
    in visual processing 237, 239, 241, 243, 252
    detectors, insect 240
    direction 458
    directional 210, 212–3
    relative 161
    sensitivity 160–1
Motion parallax 160–1, 274, 281
Mouse dioptrics 91, 95
Muscles, ciliary 469, 471, 475
Muscles, extra-ocular 188–9, 470
Myopia 84, 98, 110–1
Myriapods 220, 222, 342, 345
Myxine (hagfish) 31, 33, 401–2, 405, 416
Myxinoids 401–2

Natural selection 356, 359
Natural vision copying 268
Nautilus 3–4, 54, 122, 374–5
Near triad 107
Negaprion brevirostris (lemon shark) 67
Nemertines 15, 17–8, 21, 25, 40–2
  and evolution of eye 34–5, 37
  and origin of lens 45–6, 48
  and vertebrates 29–32
Neocortex and visual pathways 424, 426, 430, 432, 435–6, 443, 445, 452–3, 457–9
Neoteny 222–3
Neural network 167, 183, 189–91
Neural tube 45
  in eye evolution 15–8, 24, 31, 35, 37–8, 42, 46
Neurobiology 137
  comparative 421, 435
Neuroblasts 354, 403
Neurones 42–3
  binocular 175, 273–5, 278–80, 454
    disparity-selective 276
    disparity-sensitive 277
  burst 185–6
  burst-tonic 184, 186–7
  colour-opponent 298, 301
  DCMD 242–3
  directional 241
  eye-velocity 185
  grating 241
  immediate-phase 185
  light-sensitive 178, 364, 366–7
  line-labelled 251
  magnocellular 171
  multi-modal 184
  non-oriented 175
  object-detector 261
  optic lobe 240, 246
  optomotor 241, 243, 252
  oriented 175
  parvocellular 171
  pursuit 185–6
  pursuit-related 179–80, 184, 189, 191
  real-motion 175
  saccade 178
  saccade-related 179–80
  visual fixation (VF) 178
  in visual pathways 427, 430–2, 456–9
  visual tracking (VT) 178–9
  wide-field 241, 243, 251
Neurophysiology of pursuit system 170–1, 175–7, 181–5, 187–8
Neuropile 343, 347, 353–5, 360
Neuropile layers 353
Neuroptera 234
Neutral points 293, 297
Nictitating membrane 85, 470
Nissl stain 141, 143
Noise 6
  environmental 7–8
  physiological 8, 413
  shot 7
Nucleus isthmus 158
Nucleus rotundus 423–4, 426–9, 431–2
Numerical aperture 91–2, 96–7, 100, 102, 104, 110
Nyquist limit 102
Nystagmus
  optokinetic (OKN) 166, 169, 172–4, 178–80, 182
    and visual pathways 433, 455
    in photoreception 374, 386, 390
    in pursuit system 180, 184
  Schau (look) 166
  Stier (stare) 166

Object detectors, insect 255
Object discrimination 262
Object recognition 255–8
Occipital cortex 174–6, 178, 443
Ocellus *see* Eye, simple
Octopus 122
  photoreception 366, 374, 376–8, 382, 385–90
Ocular dominance 436
Ocular media 94–5
Ocular transmittance 86–9
Oculomotor nuclear complex 172, 179, 185–7
Oculomotor nuclei 173–4, 183, 187–9
Oculomotor subsystems 165–6, 169, 174, 191
Oculomotor system 165, 169–70, 174, 187–9, 191, 424
Oil droplets 6
  in colour vision 287, 289, 291–2, 296, 299
  in dioptrics 89, 101
  in evolution of retina 138–9
  in evolution of vision 410
  in mammals 470, 472, 476–7
  in photoreceptor system 65–6, 71–3, 78
Oilbird 271, 280–1
Ommatidium 132
  arthropod 214, 217, 342–4
    evolution 344–53
    developmental assembly of 346, 348
  in eye evolution 4, 24, 353, 359–60
  fused-rhabdome 347–8, 355–6
  mandibulate 345–8, 350
  open-rhabdome 347–8, 355–6

optic lobe 224
  and origins of eye 55, 57
  in visual processing 238, 265
Onchidium 365–6, 368
One-dimensional vision 259–60, 262
Opossum 467, 471, 477
  dioptrics 84, 91
Opsins
  in colour vision 284, 295–6, 315
  in evolution of eye 9–10, 39–40, 43–4
  in photoreceptor system 67–9, 71–4, 76–8
  in trichromacy 314
  and visual pigments 63–5
  in visual processing 229
Optic chiasm 48, 415
  mammalian 468
  and visual pathways 443, 446, 449
Optic cup in evolution of eye 17–8, 42, 47
Optic lobe
  arthropod 203–24, 341–60
    evolution 212–24
    visual processing organization 204–12
  in eye evolution 353–60
  insect 240, 258
  pharmacology of 223
  in photoreception 376, 379, 385–6, 391
  retinotopic organization 204, 217, 222
Optic nerve 403
  in binocular vision 273
  in evolution of eye 16, 35, 46–8, 415
  in mammals 468
  and origins of eye 55, 57
  in photoreception 383
  regeneration 154
  and visual pathways 427, 433, 443–4, 446, 449, 459
Optic nuclei, accessory 416–7, 423–4
Optic radiations 451–2, 455
Optic stalk 17
Optic system, accessory 424–5, 427–9, 433
Optic tectum
  binocular neurones 274
  in evolution of eye 15–6, 47, 415–7
  in evolution of retina 145–6
  in evolution of vision 402, 405
  retinotopically organized 427
  and visual cortex 152
  and visual pathways 421–2, 424, 426–8, 430–3, 435–6, 443, 445, 452, 459
  and visuomotor systems 153–6
Optic tract
  accessory 172–3–3, 443, 448–9, 455
  in pursuit system 172
  and visual pathways 423, 426, 443, 451–2, 455–6, 459
Optic vesicle 44–5, 403
  in evolution of eye 15–8, 24, 31, 38, 42, 46
Optical axis 272–3, 448
Optical mechanisms 351–3
Optical modulation transfer 102–3
Optical system, diffraction-limited 92
Optical systems, duplicated 112
Optical technology 118–9
Optical transfer function (OTF) 92–4
Optics
  aberration-free 4

of animals 118–34
  apposition 127–8, 352
  compound eye 214
  inhomogeneous 128
  laws of 230–7
  reflecting superposition 351–3
  refracting superposition 351–3
  and retina 214
  single-lens 353
  superposition 128–33, 214
  of vertebrate eyes 82–6
Optokinetic nystagmus see Nystagmus
Optomotor reflexes 343
Optomotor responses 206, 210, 239, 241, 374, 390
Organelles, ultrastructure of synaptic 359
Organs of Hesse 24
Ornithorhynchus (platypus) 467–71
Osteichthyes 406
Ostracods 125
Ostrich dioptrics 83
Over-accommodation 108
Over-sampling 411
Owl 91
  binocular vision 271, 273, 276–7, 280–1
  visual pathways 429–30, 436, 442, 446
Oxygyrus (snail) 122

Paedomorphy 223
Palaeoneurobiology 435
Panoramic vision 273, 448
Papilio xuthus (swallowtail butterfly) 294
Papio papio (baboon) 77
Paraflocculus 180–2
Parafovea in pursuit system 166, 168–70
Parallax 239, 242, 254–5, 257–61
Parallel circuits 267
Parallel processing, distributed 229, 239, 244, 246, 249, 251, 253, 259
Parallelism 214
Parapineal 404–6
Paravermis 182
Parietal cortex 177–80, 174, 191, 458
Parvocellular layer
  in pursuit system 171–2, 175
  and visual pathways 444–5, 450–1, 453, 458–9
Pattern
  discrimination 263, 456
  disruption 261, 263
  recognition 459
Pecten (scallop) 40, 124–6, 417
  eye 4, 26–8
  photoreception 368–71
Pectinate ligament 474
Pentachromacy 295
Perca falvescens (perch) 69
Perca fluviatilis (perch) 290
Perceptual organization 308–9
Perigeniculate nucleus (PGN) 171–2, 174–6, 184
Peripheral vision 454
Pharmacology 218
Phase lead 181–2, 185–8
Phase transfer function (PTF) 93

Photo-oxidation 299
Photochemical changes 103
Photochemical damage 95
Photoisomerization 64, 66
Photon
  count 5–6, 8
  counters 242–3, 248
  in dioptrics 86–7, 103
  flux 89–92, 103, 410
  nature of light 3, 5–7, 13
  noise 237
Photophores 367, 381, 408, 413
Photopic system 168–70
Photopic vision 66–8, 74, 76, 78
Photopigments 39
  in colour vision 284, 287, 289–97, 299, 315–6
  in evolution of retina 138–9
  in eye evolution 348–50
  in photoreception 378, 380–3, 390
  in trichromacy 311–4
Photoreception 16, 18–9, 21, 26, 39
  cephalic eyes 371–83
  extra-ocular 364–8
  in molluscs 364–91
  non-cephalic eyes 368–71
  vertebrate 28
  visual processing 383–91
Photoreceptor organs 19–29
Photoreceptor systems 19, 21, 26–9
Photoreceptors
  absorption of light by 383
  ancestral 137–8
  ciliated 27–8, 365–6, 368–72, 375–6
  in colour vision 292, 294–7, 299–300
  distribution 139
  in evolution of eye 3–11
    arthropod 343–4, 351, 353–4, 358–60, 413
    vertebrate and invertebrate 16–21, 24–7, 35–44, 63–78
  in evolution of retina 136–40, 142–3
  in evolution of vision 405, 409–10
  extra-ocular 368, 390
  long visual fibres (LVF) 204–5, 208, 214, 217–8, 221
  in molluscs 364–7
  noise 299
  as photon counters 242–3, 248
  in pursuit system 170
  rhabdomeric 344, 366, 370, 375, 378
  short visual fibres (SVF) 204–7, 214, 217
  skin 402–3
  specializations 138–9
  and trichromacy 310–1
  types 204, 284, 287–9, 295, 297, 300
  in visual processing 210, 229–30, 232, 235, 237, 242, 245, 265
Photorefraction 107
  dynamic 112
Photosensitive vesicles (PSV) 366–7, 381
Photosensitivity 406
  extra-ocular 367
Phototaxis 212, 374–5, 390
Phototransduction 380
Phototropic response 390

Phototropism 367
Phoxinus laevis (minnow) 285
Phylogenetic tree 341–2, 344, 349
Phylogeny
  birds 421–3
  mammals 421–3
  molecular 341
  reptiles 421–3
Physiology, comparative sensory 285
Pig dioptrics 103
Pig photoreceptor system 74
Pigeon 70
  colour vision 286, 288, 291–2, 298–9
  dioptrics 86, 97–8
Pigment
  anomalous 78
  rod 409
  visual 6
Pigment epithelium 40–2
  mammalian 470–1, 476
Pineal 401–6
Pineal stalk 405
Pineal window 405
Pituitary complex 17, 31
Placentals 444, 450, 459, 467–9, 471–8
Placoderms 406
Platypus 442, 467–71
Plexiform layers 139–47
  inner 139–41, 145–6
  in mammals 470, 476
Poecilia reticulata (guppy) 69–70
Point-spread function (PSF) 92–3
Poissonian statistics 89
Polarity, edge 252–3, 262
Polarization 230
  in dioptrics 87, 89, 100
  plane 230, 265, 382–3, 388, 391
Polarized light sensitivity (PLS) 378, 383
Polarized-light sensitivity (PLS) 390–1
Pollachius pollachius (pollack fish) 69
Polymorphism 13, 70–1, 76
  colour vision 288, 314–7
  photopigments 293–4
Pontella 122–3
Pontine nuclei 175–83, 189
Porphyropsin
  in colour vision 291, 299
  in evolution of eye 6, 8–9
  in evolution of vision 403–4, 409
  in photoreceptor system 65–72
Portia fimbriata (jumping spider) 120
Precentral cortex 179–80, 189, 191
Predators 448
Pregeniculomesencephalic tract 172
Prepositus hypoglossi 184–7
Prestriate cortex 452
Pretectum 152–4
  and detour movements 158–60
  in eye evolution 416–7
  feeding selectivity by 157
  in pursuit system 173–4, 183–4
  and visual pathways 424–5, 427–9, 433, 449, 455, 457
Prey, bioluminescent 412
Prey selection 157–8
Prey-catching 154–5, 161

Primates 476
  New World 13
  visual pathways 442, 450–2, 458–9
Principle of univariance 284, 311
Projection, binocular 404
Prosimians 74–7, 451–2, 473–4, 476
  colour vision 293
  visual pathways 453, 456–7
Protanopes 314
Protochordates 15, 17, 24–5, 32
Pseudemys scripta elegans (turtle) 72, 286–7, 291–2, 299, 427
Pseudobranchs 414
Pseudoisochromatic plates 306–10, 312
Psychophysics 168–9, 310
Pteromyzonids 402
Pulvinar 173–5
  in pursuit system 176–8
  and visual pathways 424, 426, 435, 456–7
Pupil
  active 100
  aperture 3–6
  control 21
  in dioptrics 85, 92–3, 96–8, 100, 105, 108, 110
  entrance 85, 94, 98–9
  and evolution of vision 411–2
  exit 85, 100–2
  in eye evolution 413–4
  in mammals 469, 474, 477
  reflex 172
  slit 93, 111
  slit-shaped 99
  stenopaic 111
Pupillary light reflex 424, 433
Purkinje shift 73
Pursuit deficits 178–80, 184
Pursuit eye movements (PEM) *see* Pursuit system
Pursuit movements
  Sigma 166–7
  step-ramp 167, 176
Pursuit system 165–92
  cerebellum and brain stem 180–9
  smooth 167–74
    control 174–80
Python 429

Quantum catch 284, 296, 300
Quantum statistics 91
Quarter-wavelength multi-layers 126

Rabbit
  binocular vision 272–3, 275
  dioptrics 83–5, 88, 97
Radiation
  adaptive 136
  blackbody 86–7
  and dioptrics 88, 103
  electromagnetic 86, 289
  infrared 86, 428
  and ocular transmittance 86–9
  ultraviolet 18, 95, 299
Raja (skate) 67
Rana esculata (frog) 99
Rana pipiens (frog) 71

Range measurement 242, 254–5, 261, 265–7
Rat 74
  dioptrics 84–5, 95, 97
Rattus norvegicus (rat) 74
Rayleigh criteria 92, 104–5
Rayleigh scatter 8, 407
Receptive field centre 444
Receptive fields
  non-oriented 453
  OFF-centre 450–1
  ON-centre 450–1
  properties 275, 280
  and visual pathways 426
  in visual pathways 431–2
  and visual pathways 436, 449
Receptors, mosaic 102–3
Reflectance 126–7, 306, 309
  spectral 300–1, 310
Refraction
  in dioptrics 103, 107, 109
  in evolution of vision 408
  instruments 111
  in photoreception 377
Refractive index 118, 126
  in basic optics 82, 84
  and dioptrics 92, 94, 97, 100, 102–3, 105
  in evolution of eye 4
  and evolution of vision 408, 411–2
  graded 55
  and photoreception 369
  in visual processing 232
Refractive index gradient 120–3, 377
  and dioptrics 85–6, 95–7
  in evolution of vision 412
  and optics 129–31
Reichardt's autocorrelation theory 251
Reichardt's motion perception model 240
Reptiles 72–3, 78
  colour vision 287–9, 291–2, 296, 299, 302
  dioptrics 89
  visual pathways 421–36
Resolution
  least-motion 231
  lens 230–8
  spatial 231, 236–8, 240–3, 257, 259–60, 266, 454, 459
  spatial sampling 235, 237
  and visual pathways 456
  in visual processing 230
Response ratios 242
Responses
  ratios of template 260–3
  template 256–7, 267
Retina
  accessory 412
  bifoveate 272–3
  and binocular vision 277–9
  cephalopod 376
  of compound eye 204
  corrugated 111
  cup-shaped 371
  development 137, 144–5, 147
  double 369–70
  duplex 478
  in evolution of eye 16–8, 21–2, 24–6, 28, 40, 42–4, 47, 412–4, 417

evolution of vertebrate neural 136–47
  in evolution of vision 402–4, 406, 408, 410
illuminance 98–9
inverted 82
  in evolution of eye 23, 25, 35
  invertebrate 25–7
linear scanning 122
mammalian 471, 473, 476
mantis 231
marsupial 471–2
monofoveate 272
monotreme 469–70
origin of 18
  and origins of eye 57–8
  in photoreception 377–9
profunda 376, 385–6
in pursuit system 167–71, 189–91
'ramp' 111
ribbon 58, 375
scanning 390
specialization 274
squid 383
surface shape 98
temporal specialization 272
tiered 139
triggers 275
and visual pathways 421–2, 424–5, 427, 430–1, 443–7, 449, 452, 454–6, 459
Retinal 63–5, 69, 349–50, 381
Retinal-binding protein (RALBP) 381–2
Retinochrome 366, 378, 381–2
Retinogeniculate pathway 456
Retinomotor movements 413
Retinoscopy 83–4
Retinotectal pathway 446, 456
Rhabdom 344, 381
  fused 344, 347
  open 344, 347
  and optics 127–8, 132
  in photoreception 368, 372
  in visual processing 230–4, 238
Rhabdomeres 128, 378
  in evolution of eye 4, 39, 347, 355–6
  in visual processing 230, 234–5, 238
Rhodopsin
  in coevolution 12
  in colour vision 290, 292, 294, 299
  in evolution of eye 6–10, 39–40, 413
  in evolution of vision 404, 409
  and eye optics 118
  in photoreception 366, 379–82
  in photoreceptor system 64–72, 74, 77
  in trichromacy 311–2
  in visual processing 229
Roach 70–1
Rods
  in colour vision 296
  in evolution of eye 5–8, 12, 15–6, 37, 39–44, 413
  in evolution of retina 136–41, 143
  in evolution of vision 404
  in fish colour vision 290
  in mammals 472, 476–8
  pigment 409
  in pursuit system 168–70

Rudd 70–1
Rutilus rutilus (roach) 70–1

Saccades 59
  correctional 167
  divergent 271
  foveating 167, 172, 176
  in pursuit system 165–7, 172–3, 176, 179–82, 184–6
Saguinus fusicollis (saddle-backed tamarin) 76
Saimiri sciureus (squirrel monkey) 76–7, 287–8, 293, 315–6
Salamandra salamandra (tiger salamander) 285–7, 290, 295, 299
Salinity 21, 39–40, 42–5
  in evolution of eye 16–7, 31
Salmo trutta (brown trout) 69, 290
Salmonids 70–1
Salvelinus fontinalis (brook char) 71
Salvelinus namaycush (lake char) 71
Sampling array 240, 242
Sampling density 266
Sampling theory 73, 102–3, 237, 299
Sarcopterygii 400–1, 407
Sauropsida 423–4, 468–70
Saurospida 426
Scallop 124–7, 130, 368–71, 390
Scanning 122
  dynamic 103
Scardinius erythrophthalmus (rudd) 70–1
Scatter 87
  intra-retinal 103
  light 88–9, 97–8, 100, 408–9
  Rayleigh 407
Scheiner double-pinhole 111
Schiff's base, protonated 64–5
Schultze's duplicity theory of vision 137
Sciuridae 74
Sciuromorpha 73
Sciurus carolinensis (grey tree squirrel) 74
Sclera 414
  mammalian 468, 470–1, 473
Scotoma, central 168–9
Scotopic vision 66–8, 74, 78, 168–70, 407
Screening pigment 351, 374, 377
Scutigera 220, 342, 345–6
Segregation 308–9, 312
  in colour vision 316–7
  in mammals 476
  and visual pathways 450–1, 453, 458
Self-absorption 230
Self-inhibition 208
Self-motion 443, 459
Semivision 229, 253–60
  image recognition by 260–3
  insect 259
  mechanisms 248–9, 266
  templates 268
Sensory coding 18
Shadow 390
Shadow responses 364–9, 371, 390
Shape
  constancy 162
  discrimination 162, 389

from shading 308–9
recognition 162–3
of retinal surface 98
Short visual fibres (SVF) 204–7, 214, 217
Shrew, tree 74, 77, 293, 472, 476
  visual pathways 446–7, 451–3, 456–7
Sickle cell anaemia 316
Signal, delayed 252
'Silent substitution' 289
Size, anisotropy 276
Size, constancy 157, 263–5, 267
Size–distance constancy 156, 163
Slip velocity 167–8, 170, 180–4, 189, 191
Snail 122
Snake 477
  cat-eyed 103
  dioptrics 84–7, 100, 102, 110
  garter 102–3, 428–9
  visual pathways 422, 427–9
Snell's window 408
Solar spectrum 87
Somatosensory-motor cortex 430
Sound localization 448
'Spectacle' 84–5, 110
Spectral range 284–5, 288–9, 295, 298–9, 301
Spectral sensitivity functions 285, 287, 289–95, 299
Spectral 'window' 299
Spectrum, solar 87
Spermophilus (Citellus) lateralis (ground squirrel) 74
Sphenodon 427–8
Spider, jumping 120
Spider, net-casting 120
Spider optics 120
Squid optics 121–2
Squid photoreception 366–7, 374, 376–83, 385, 388, 390–1
Squirrel 74
  colour vision 286, 292
  diurnal 476
  tree 446–8
  visual pathways 450, 457
Steatornis caripensis (oilbird) 271, 280–1
Stereoacuity 454
Stereoblindness 281
Stereogram, Julesz random-dot 276
Stereogram, random-dot 271
Stereopsis 274, 436, 446, 448, 455, 459
  evolution of avian 280–1
  machine 273, 275–7
Stereoscopic vision *see* Stereopsis
Stiles-Crawford effect 100–1
Stilling pseudo-isochromatic plates 306–9
Strabismus 271
Striate cortex
  in pursuit system 174, 179
  and visual pathways 426, 430, 435–6, 452–5
Strombus 374, 376–7
Sun compass 391
Superior colliculus 145, 152, 415
  in pursuit system 171–4, 176–9, 184, 186–9
  and visual pathways 424, 426, 435–6,

443–8, 452, 454–7
Superior temporal sulcus (STS) 176–7, 459
Superposition, neural 355–6
Suprachiasmatic nuclei (SCN) 425
Sus scrofa (domestic pig) 74
Synapse organization 358–60
Synapses, receptor 358
Synapses, ribbon 140–1

Tachyglossus (spiny ant-eater) 467–70
Tapetum 26, 84, 92
  cat 4–5
  and dioptrics 97
  in eye evolution 413
  in mammals 470–1, 473
  and optics 127
Target movements, ramp 166
Target movements, step-ramp 166
Tectofugal pathway 427–33, 435
Tectum 402–4, 416
Telencephalon 423–7, 429–30, 432–4, 436
Teleosts 78, 281
  colour vision 290, 295, 300
  dioptrics 85, 103, 109–10
  in evolution of retina 138–9, 141–4, 146
  evolution of vision 398, 406–7, 410, 412
  in eye evolution 412–7
  feeding selectivity 157
  freshwater 290
  and photoreceptor system 65–71
  plesiomorphic 415
  and shape recognition 162–3
  and visual cortex 152
Telescope, Newtonian 125
Template 240–1, 254–7
  larger 260–3
  line-labelled 246, 248–50
  model of visual processing 243–51, 266–7
  motion-detection 235
  one-dimensional 245, 254
  properties 246
  responses 249–50
  semivision 258–60, 268
  theory 259, 262–3
  two-dimensional 253–4
  whole-eye 263–7
Temporal cortex 174–6
Terrain hypothesis 143–4, 147
Tetrachromacy 295
Tetrachromats 295
Tetrads 358
Thalamofugal pathway 426–9, 433, 435–6
Thalamus 402
  central 173–6, 178–9, 184
  in evolution of eye 16, 415–7
  in pursuit system 171, 189
  and visual pathways 423–30, 456–8
Thamnophis sirtalis (garter snake) 428–9
Toad dioptrics 100, 103, 110
Todarodes (squid) 382
Tomography, positron-emission 453
Transducin 63–4, 66, 380
Transients 246, 259, 267
Transillumination 86
Transmittance, ocular 86–9

Transmittance, water 87
Transmutation 73
Triads 358
Trichromacy 77, 294, 310–4
   anomalous 78
Trichromats 290, 293, 295, 300, 312, 315–7
Tridacna (giant clam) 369, 390
Trilobites 341–2
Tritanopes 309, 312
Tupaia see Shrew, tree
Turtle 68, 72–3
   colour vision 286–7, 291–2, 295, 298–9
   dioptrics 101
   visual pathways 422, 426–7
Two-dimensional vision 253–4, 260

Ultraviolet *see* Light
Ultraviolet vision 69
Under-accommodation 108
Under-sampling 102–3, 411

Velocity, angular 239, 242, 246
Vermis 182–3, 188–9
Vernier acuity 408, 411
Vestibular nuclear complex 174, 182–4, 187, 424
Vestibular system 424
Vestibular units 181
Vestibulo-ocular reflex (VOR) 165, 174, 181–2, 184–8
Vipera 428
Vision
   anthropocentric 229
   basis of 239
   evolution of 237–9
Visual acuity
   and binocular vision 272–3
   and evolution of vision 407–8, 411
   in eye evolution 413, 417
   and nature of light 5
   and photoreception 377
   and pursuit system 168
   and visual pathways 443, 453–4, 456
Visual areas
   area 17 173–6, 452, 456, 458
   area 18 173–4, 177, 456, 458
   area 19 173–4, 176–7, 458
   areas 7 and 8 173, 178–80, 184, 189–91
   ectosylvian 444

extrastriate 456–9
middle temporal 173, 175–7, 179, 191, 458–9
V1 171, 173–7, 189–91
V2 171, 173, 176–7, 179, 191, 443, 458
V4 177–8, 458
and visual pathways 423, 433
Visual axis 264, 272–3, 427, 448
Visual cortex 152–3, 159–60
   in binocular vision 274–5, 277, 279–81
   extrastriate 424, 426, 443
   mammalian 423
   primary 455
   in pursuit system 172–4, 188
   reptile 423
   and shape recognition 162–3
   and tracking 161–2
   and visual pathways 427, 444, 448, 453, 455, 459
Visual environment, aquatic 407–17
Visual field 246
   and colour vision 300
   in dioptrics 86, 91, 95, 97–100
   representation 279
Visual pathways, mammalian 442–59
Visual pathways, primary 442–4
Visual pigment 88, 120
   in colour vision 290–3, 299–300
   in evolution of eye 8–13, 26, 39–40, 44, 417
   evolution of vertebrate 63–78
   in evolution of vision 404, 409
   in photoreception 366, 378, 380–1, 390
   in visual processing 229–30, 234
Visual processing 433
   in arthropods 203
   autocorrelation model 251–3
   evolution of 229–68
      evolution of vision 237–9
      larger templates 260–3
      laws of optics 230–7
      semivision 254–60
      template model 243–51
      whole-eye templates 263–7
   low-level 212
   mechanisms 212–3, 238–9, 242, 252, 254, 261, 267, 436
   mollusc photoreception 383–91
   organization 204–12
   'top-down' effects 268
Visual scene, spatio-temporal 243, 247

Visual streak 86, 94, 97, 102, 445
   and evolution of retina 143–4
   horizontal 443–4
   and visual pathways 454
Visual system
   afferent 168–9, 171, 184, 191
   evolution in reptiles and birds 421–36
   lamprey 402–5
   mammalian 435
   non-mammalian 435
   plasticity 430
Visual wulst 423, 426, 429–34, 436
Visuomotor functions 152, 154
Visuomotor mechanisms 159–60
Visuomotor systems 153–7, 162
Vitamin A 16, 26, 39, 42
Vitreous 83–4

Wall's, Gordon 136–8, 142
Watasenia scintillans (firefly squid) 381, 390
Water transmittance 87
Waveguide modes 132
Waveguides 100, 102, 112
Wavelength
   colour vision 284
   discrimination 11, 285–9, 292–4, 298, 300–1
   information-carrying 7–8
   light 399, 402, 406, 408–10, 412, 433
   nature of light 5–6
   noise 407
   in photoreception 388, 390
   selectivity 445, 450
   short 8–9
Willmer, E.N. 15–8, 29–31, 33–5, 37–42, 44–5, 48
Worms 122, 237–8

X-chromosome 313, 315–6
X-chromosome inactivation 76
X-ray image 130
Xanthopsin 349–50

Yellow substance 8

Zaglossus (spiny ant-eater) 467–8, 470